Introduction to the
ALGAE

PRENTICE-HALL BIOLOGICAL SCIENCES SERIES

William D. McElroy and Carl P. Swanson, *Editors*

Introduction to the
ALGAE

STRUCTURE AND REPRODUCTION

Harold C. Bold

C. L. Lundell Professor of Systematic Botany
The University of Texas, Austin

Michael J.Wynne

Associate Professor of Botany
The University of Michigan, Ann Arbor

PRENTICE-HALL, INC., Englewood Cliffs, New Jersey   07632

Library of Congress Cataloging in Publication Data

BOLD, HAROLD CHARLES (date)
  Introduction to the algae.

  (Prentice-Hall biological sciences series)
  Bibliography: p.
  Includes index.
  1. Algology. 2. Algae—Anatomy. 3. Algae—
Reproduction. I. Wynne, Michael James, joint
author. II. Title.
QK566.B64   589′.3   77–11118
ISBN 0–13–477786–7

©1978 by PRENTICE-HALL, INC.
Englewood Cliffs, New Jersey 07632

Printed in the United States of America

10 9 8 7 6 5 4 3 2 1

PRENTICE-HALL INTERNATIONAL, INC., *London*
PRENTICE-HALL OF AUSTRALIA PTY. LIMITED, *Sydney*
PRENTICE-HALL OF CANADA, LTD., *Toronto*
PRENTICE-HALL OF INDIA PRIVATE LIMITED, *New Delhi*
PRENTICE-HALL OF JAPAN, INC., *Tokyo*
PRENTICE-HALL OF SOUTHEAST ASIA PTE. LTD., *Singapore*
WHITEHALL BOOKS LIMITED, *Wellington, New Zealand*

Dedicated to
Professors George F. Papenfuss
and
William Randolph Taylor,
who have
so greatly enriched our knowledge of the algae.

Contents

ix   Contents

8

Division Pyrrhophycophyta   417

9

Division Rhodophycophyta   451

Preface

It seems to the authors, from their own experience in offering a semester-long introductory course in general phycology, that there has been a need for a teachable textbook in the field. The great syntheses of Smith (1950) and Fritsch (1935, 1945) contain much useful information, but they are too encyclopedic in scope for an introductory course, and phycology has advanced greatly on a number of fronts since they were published.

The authors have emphasized the structure and reproduction of representative algae in the present volume and have omitted in-depth coverage of algal physiology and biochemistry because of their special competence in the areas emphasized. The excellent volume edited by Stewart (1974) has quite adequately summarized algal physiology and biochemistry and, furthermore, a volume on algal genetics (Lewin, 1976) has been published recently.

No effort has been spared in citing *relevant* literature regarding the organisms and phenomena included in this volume, but the authors have not attempted to be comprehensive in their coverage. In general, it has been their purpose to include references to significant phycological literature that appeared since Smith's and Fritsch's volumes were published.

A special effort has been made in choosing representative types of algae for discussion to include as many as possible that the reader might hopefully be able to observe in the living condition in the laboratory. The ease with which many species of algae can be grown and maintained in laboratory culture, collected in the field, or procured from biological supply houses or culture collections should shame those who provide introductory students only with specimens bleached in preservative. A brief summary of methods of cultivating algae is included in the Appendix. It is hoped that both students and instructors will engage in cultivating algae in the laboratory. A glossary has been included to assist the reader in mastering phycological terminology.

The authors acknowledge with gratitude assistance from Professors P. A. Archibald, E. R. Cox, T. Delevoryas, J. M. King, N. J. Lang, P. A. Lebednik, R. A. Lewin,

R. C. Starr, and C. Van Baalen, all of whom have read parts of the manuscript and offered critical suggestions. The careful and constructive review of the entire manuscript by Professors Robert W. Hoshaw, C. Carroll Kuehnert, and Richard E. Norris is appreciated. The authors are grateful to many phycological colleagues who generously provided illustrative materials. Acknowledgment of these is made, in each case, in the figure captions. The authors appreciate the effort and interest of Miss Felicia Bond who prepared the line drawings. Finally, the assistance of Mrs. Evelyn B. Edwards, Frank J. MacEntee, S.J., and of Mrs. Mary Douthit Bold in preparing the manuscript and index is gratefully acknowledged.

HAROLD C. BOLD

MICHAEL J. WYNNE

1

Introduction to the Algae[1]

Definition

The term *algae* (sing. *alga*) means different things to different people, and even the professional botanist and biologist find algae embarrassingly elusive of definition. Thus, laymen have given them such names as "pond scums," "frog spittle," "water mosses," and "seaweeds," while some professionals shrink from defining them. The reasons for this are that algae share their more obvious characteristics with other plants, while their really unique features are more subtle. There are a number of liverworts, mosses, ferns, and angiosperms that live in aquatic habitats with freshwater algae, and even marine environments contain angiosperms ("seagrasses," which are *not* members of the grass family!) as well as algae. There are also a great many terrestrial and subterranean algae, so that aquatic habitat is an untrustworthy criterion on which to base a distinction. A delightfully written and informal account of the algae has been published by Tiffany (1958).

How then does one distinguish algae from other chlorophyllous plants? The distinguishing characteristics reside in the phenomenon of sexual reproduction as it occurs in algae in which it differs from that in other green plants as follows: (1) In unicellular algae, the organisms themselves may function as gametes (Fig. 1.1a); (2) in some multicellular algae, the gametes may be produced in special *unicellular* containers or gametangia (Fig. 1.1b); or (3) in others, the gametangia are *multicellular* (Fig. 1.1c), every gametangial cell being fertile, that is, producing a gamete. None of these characteristics occurs in liverworts (Fig. 1d, e), mosses, ferns, or angiosperms. In their asexual reproduction, many algae produce flagellated spores and/or nonmotile spores in unicellular sporangia, or if the latter are multicellular, every cell is fertile.

[1]The blue-green algae are very similar to bacteria and by some biologists (e.g., Stanier, et al., 1971) said to be bacteria. See, however, p. 31.

1

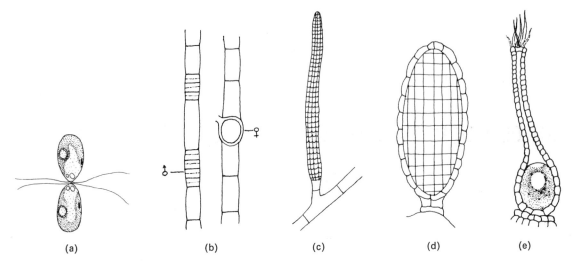

(a)   (b)   (c)   (d)   (e)

Fig. 1.1 Characteristics of the sexual reproduction of algae (*a*)–(*c*) and nonalgal plants (*d*), (*e*). (*a*) Uniting gametes of a unicellular alga, *Chlamydomonas*. (*b*) Unicellular gametangia of a filamentous alga, *Oedogonium*. (*c*) Multicellular gametangium of *Ectocarpus*. Note that every cell is gametogenous. (*d*) Archegonium and (*e*) Antheridium of a liverwort, a representative nonalgal plant. Sex organs multicellular and consisting of both gametic and sterile (vegetative) cells.

The Occurrence and Distribution of Algae[2]

When it is said that algae are ubiquitous, as in fact they are, such a statement might seem to tax credulity by one, for example, who was observing a desert scene or a permanent snowfield, but even in such diverse habitats algae are present.

Algae are aquatic or subaerial. By the latter is meant that they are exposed to the atmosphere rather than being submerged in water. Aquatic algae grow in waters of low salinity (as low as 10 ppm),[3] called *freshwater*, and in marine waters where the solutes are usually 33–40‰[4], although some algae occur in such locations as the Laguna Madre of Texas where the salinity may rise to 100‰ in dry seasons; this is in contrast to such an algal habitat as Mountain Lake, Virginia, in which the solutes total only 3.6 ppm. Some algae are remarkably tolerant to varying salinities such as those (e.g., *Enteromorpha*, p. 171) that live on ships that ply both freshwater and oceans. A number of algae live in brackish water; the latter is unpalatable for drinking but contains less salts than ocean water; for such algae the salinity optimum is less than that of seawater.

[2]No attempt has been made here to present a comprehensive account of algal ecology in view of the availability of such in-depth treatments of it as those of Dawson (1966), Boney (1966), Prescott (1968), and Round (1973).
[3]Parts per million.
[4]Parts per thousand.

With respect to their solutes, lakes are sometimes classified as **oligotrophic** or **eutrophic**. The former have been defined as those having less than 100 ppm of solutes, while the latter may have considerably higher concentrations. Oligotrophic lakes, as would be expected, support a sparser algal flora, with respect to numbers of organisms, than eutrophic lakes, but the number of species may be greater. Bodies of freshwater have also been classified as alkaline, hard-water lakes (with pH > 7) and as acid, soft-water lakes (with pH < 7), and their floras differ. Some species of algae can tolerate a broad range of pH, while others are more restricted. Bodies of water in which algae grow differ also in the concentration of dissolved oxygen and carbon dioxide and that of many other substances, as well as in temperature and turbidity (which affects depth of light penetration and hence photosynthesis and algal growth). All these factors have been examined carefully by many investigators as they affect algal growth, and these data have been summarized in part by Dawson (1966), Boney (1966), Prescott (1968), and Round (1973), among others.

Aquatic algae may be suspended (planktonic) or attached and living on the bottom (benthic). The **plankton** consists of a flora and fauna, together with bacteria and often fungi, of suspended organisms, while the **benthos** is composed of attached and bottom-dwelling organisms. A few algae are **neustonic**; that is, they live at the interface of water and the atmosphere.

Planktonic algae may be collected by drawing a plankton net through the water. Plankton nets are composed of silk with finely woven meshes, commonly 180 pores to the square inch. This serves as a strainer that filters out and concentrates many planktonic algae and other microorganisms. Planktonic algae may also be concentrated for microscopic study by centrifugation. Planktonic algae under certain combinations of nutrition favorable to them increase enormously in number and form **water blooms** (Fitzgerald, 1971). Diatoms, green algae, *Euglena* (p. 258), and blue-green algae (Fig. 2.1) are most frequently present in blooms.

Benthic algae grow attached to various substrates and may be classified as **epilithic** (living on stones), **epipelic** (attached to mud or sand), **epiphytic** (attached to plants), and **epizoic** (attached to animals). Examples of algae growing on all of these substrates are included in later pages.

In addition to the classification of algal habitats above, certain other categories have been erected to describe growth habits of marine algae. Some are **subaerial** or said to be **supralittoral**, since they grow above the water level and in the spray zone (e.g., *Gloeocapsa*, p. 45; *Calothrix*, p. 61; and *Prasiola*, p. 177). Others are **intertidal** in that they are exposed periodically in accordance with variations in water level due to tides. Still others are **sublittoral**; that is, they are constantly submerged and, depending on turbidity, may grow at depths as great as 100–200 m, the latter in clear tropical waters. Subaerial algae may be **edaphic** (growing in and on soil), epilithic, epiphytic, epizoic, and **corticolous** (growing on tree bark) and a few are parasitic (see p. 535).

The presence of algae on moist rocks, wood, living trees, and the surface of moist soil is readily observable, but the occurrence of algae beneath the surface of the soil is not so obvious. To educe evidence of their presence it is only necessary to moisten soil and to keep it under illumination or to introduce soil into sterile nutrient solutions

(p. 572) under illumination (Starr, 1973). Shtina (1974) recorded that 1410 species and forms of algae had been reported to be present in Russian soils.

Many of the blue-green algae, like certain bacteria, in soil fix gaseous nitrogen into a combined form and are thus of great importance in improving soil fertility (see p. 34). The role of other algae in the soil is less clear, although it is certain that they are involved in various types of relationships, either stimulatory or inhibitory, with other soil organisms (Parker and Bold, 1961). Bailey, et al. (1973) suggested, on the basis of their experiments, that algae are important in stabilizing and in improving the physical properties of soil by aggregating particles and by adding organic matter.

Some, but not all, subterranean algae have been proved to be facultatively hetero-trophic in darkness (Parker, 1961, 1971D; Parker, et al., 1961), but the nature of the nutrition of the others remains an enigma.

Algae have been found in desert soils when they were not always obvious macro-scopically (Chantanachat and Bold, 1962; Friedmann, et al., 1967; Friedmann, 1971), although they are important as primary producers (Friedmann and Ocampo, 1976). Friedmann and his associates found that although the macroenvironment may be hostile to algae, a surprising variety is present in the desert, their source of water probably being dew. These authors have classified desert soil algae as **endedaphic** (living in soil), **epidaphic** (living on the soil surface), **hypolithic** (on the lower surface of stones on soil)—and as rock algae, including **chasmolithic** algae (in rock fissures) and **endolithic** algae (rock penetrating). Trainor (1970) reported that algae survived in desiccated soils for more than 10 years in his laboratory. Booth (1941) reported that several filamentous blue-green algae are pioneers in plant succession on bare soil, where they form crusts that cut down on evaporation from the soil and also prevent erosion.

It is of interest that Brook (1968) reported the discoloration of roofs in the United States and Canada by certain blue-green and green algae, and the authors have observed discoloration of buildings by algae.

Corticolous or tree bark-inhabiting algae have been studied by Edwards (1968), Cox and Hightower (1972), and Wylie and Schlichting (1973), who reported a consider-able number of algae from such habitats. A number of algae, both blue-green and green, grow as members of lichen associations (Ahmadjian, 1967); of these, *Trebouxia* is one of the most frequently encountered.

Some algae live **endozoically** in various protozoa, coelenterates, molluscs, and worms. *Chlorella*-like algae are present within *Paramecium*, *Hydra*, molluscs (Cooke, 1975), and some freshwater sponges. Others, zooxanthellae (division Pyrrhophyco-phyta), live in intimate association with corals, where their photosynthetic activity is of primary importance to the reef community (Benson and Muscatine, 1974). Trench (1971) demonstrated that ^{14}C-labeled products of algae rapidly appear in the lipids and proteins of the host animals; these extracellular products range between 20 and 50% of the algal photosynthate. Pearse (1974) demonstrated that sea anemones con-taining algae were phototactic, while those lacking them did not move to or from light of varying intensity. A rather comprehensive review of algal associations with animals has been prepared by D. Smith, et al., (1969). Much of the literature on endosymbionts of *Hydra* has been summarized by Pardy (1974). Other reports on this subject are those of Droop (1963), Karakashian and Karakashian (1965), Oschman (1967), and Trench (1971). Of considerable interest is the association of the green alga *Platymonas* with the flatworm *Convoluta roscoffensis;* the latter seemingly is dependent for its development on the presence within it of the alga (Oschman, 1966; Provasoli, et al., 1968). Muscatine, et al. (1974) reported that the alga releases amino acids to the animal that is not holozoic and entirely dependent on the alga. Muscatine, et al. (1975) have reviewed the host-symbiont interfaces in various algal-invertebrate associations. The cell walls of the symbionts may be modified, unmodified, or lacking. Several algae grow as endophytes within other plants. Here may be mentioned the blue-green alga *Anabaena azollae*, which grows within the water fern *Azolla*, and the species of *Anabaena* or *Nostoc*, which live within the thalli of the hornwort *Anthoceros* and in the roots of cycads (Grilli, 1974) and those of *Gunnera*, an angiosperm. Lewin and Cheng (1975) have reported the consistent association of a marine *Synechocystis* (a blue-green alga) with ascidians and its occurrence on mangrove roots. The organism is remarkable in that, although it is prokaryotic in organization, it lacks phycobilin pigments (Lewin, 1975) and contains *both* chlorophylls *a* and *b*. D. Smith (1973) has summarized our knowledge of some cases of algal-animal symbiosis.

In this discussion of algal habitats, mention must be made of some of the effects of the algae on their environment. It has been demonstrated that algae secrete a number of substances from their cells (Hellebust, 1974). For example, Aaronson, et al. (1971) have reported that the unicellular *Ochromonas* (p. 363) contributes DNA, RNA, carbohydrates, vitamins, and proteins (including enzymes) to the surrounding medium. Other evidences of such extracellular secretion are presented in the reports of Lefevre (1964), Nalewajko (1966), Fogg (1971), Nalewajko and Lean (1972), Huntsman (1972), Belly, et al. (1973), W. Smith (1974), and Walsby (1974A, B). That transfer of these substances between algae and algae and between algae and other organisms occurs has been demonstrated by Lange (1970), Harlin (1973), and Bauld and Brock (1974). It has also been demonstrated in laboratory cultures that stimulatory and inhibitory effects occur between algae and algae and among algae, bacteria, and fungi (Parker and Bold, 1961; Fitzgerald, 1969; and Kroes, 1971, among others). It is probable that similar mutualistic effects occur among algae and other organisms in nature. Some of the secreted substances have antibiotic effects (Burkholder, et al. 1960; Berland, et al., 1972; and Khaleafa, et al., 1975). Of particular interest are recent observations (Deig, et al., 1974) that liquid extracts from a number of common red seaweeds were effective in halting replication of herpes virus types I and II, the agents responsible for such ailments as cold sores and sores in the genital area.

Finally, a number of algae or their products are toxic to various animals (Gorham, 1964; Harris and James, 1974). The most notable in this respect are the blue-green algae (see p. 47) and dinoflagellates (p. 417), although one member of the Prymnesiophyceae (*Prymnesium parvum*) is responsible for the death of fish.

The question also arises regarding the distribution of algae and the vectors that accomplish it. In aquatic habitats, vectors such as tides, currents, and agitation by wind are obvious factors, as are movements of animals and of ships. Milliger, et al.

(1971) have reported that beetles are active in dispersing algae, having recovered species of 101 different algal genera from 23 species of beetles. Proctor (1966) and Atkinson (1972) have demonstrated the role of aquatic birds as vectors in algal distribution. Schlichting's (1970, 1971) reports suggest that aquatic algae might be transported by bursting bubbles and air currents.

Edaphic algae and those present in drying ditches and pond margins are distributed by air currents, some of the literature on this topic and in algal survival having been summarized by Schlichting (1964, 1970, 1974A, B). McElhenney, et al. (1962) and McGovern, et al. (1966) reported that some of such airborne algae were allergenic. In this connection, Bernstein and Safferman (1970) found viable algae in house dusts. R. Brown (1971) studied dispersal of airborne algae in Hawaii. Schwimmer and Schwimmer (1964) and Schlichting and James (1972) have reviewed the relation of algae to medicine.

Finally, among algae of extreme habitats must be mentioned those that thrive on long-persistent snows (Stein and Brooke, 1964; Stein and Amundsen, 1967; Kol, 1968; Thomas, 1972; Gerrath and Nicholls, 1974; Hoham, 1973; 1974A, B; 1975A, B; 1976; Hoham and Mullet, 1977; and Fjeringstad, et al., 1974) and, by contrast, those thermophilic algae that inhabit hot springs. The latter, mostly blue-green algae, grow at temperatures between 50 and 73°C (Bauld and Brock, 1974). Castenholz (1969) has reviewed the occurrence of these algae and their environment.

Place of Algae in the Plant Kingdom

In surveys of the plant kingdom, the algae are usually studied first for several reasons. First, the fossil record indicates that the most ancient organisms that contained chlorophyll *a* were probably blue-green algae[5] with a fossil record extending back possibly 3 billion years into the Precambrian (Schopf, 1970). There are suggestions that these were followed in later Precambrian times by the several groups of eukaryotic algae (Schopf and Blacic, 1971). Thus, the antiquity of algae in the history of living organisms argues for their primacy in the plant kingdom. A second reason for this primacy is the relative simplicity of organization of most algal plant bodies, as compared with other groups of plants, especially the vascular plants, although in this connection, the kelps (p. 317) suggest caution. A third reason is that algae illustrate so elegantly and with great clarity many important biological phenomena (e.g., sexual reproduction) that in other plants are complicated by secondary characteristics. Most botanists, accordingly, view the algae, especially the green algae, as likely progenitors for the remaining members of the plant kingdom (other than algal groups) because they are similar in pigmentation (having chlorophyll *a* and *b*) and in the nature of their storage reserves (starch).

Form of the Algal Plant Body

The form of the plant body of algae (Fig. 1.2) varies from the *relative* simplicity of the single cell to the complexity exhibited by the giant kelps and the rockweeds.

[5]Cyanochloronta, Chapter 2, p. 31; see, however, Schopf (1976.)

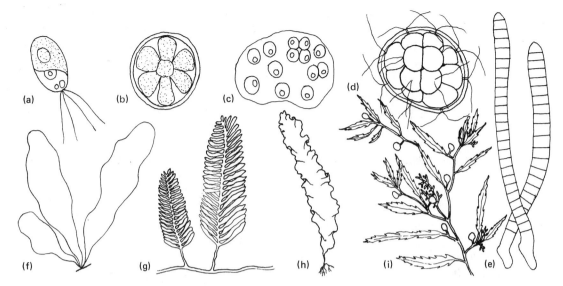

Fig. 1.2 Types of algal plant body, diagrammatic. (*a*) Unicellular, motile. (*b*) Unicellular, nonmotile. (*c*) Colonial, noncoenobic. (*d*) Colonial, coenobic. (*e*) Filamentous. (*f*) Membranous or foliar. (*g*) Tubular, coenocytic. (*h*) Blade-like, kelp. (*i*) Leafy axis.

While small algae like *Micromonas pusilla* (1 × 1.5 μm) and species of *Chlorella* (5–8 μm) are in the range of bacterial size, although eukaryotic, kelps, some of which are the largest of algae, may attain a length of 60 m.

Unicellular[6] (Fig. 1.2*a, b*), colonial (Fig. 1.2*c, d*), filamentous (Fig. 1.2*e*), membranous or foliose (Fig. 1.2*f*), and tubular (Fig. 1.2*g*) types of algal plant body occur, together with more highly differentiated blade-like types (Fig. 1.2*h*) and those that have rootlike organs, stems, and leaves (Fig. 1.2*i*), albeit these organs are lacking in vascular tissue, although phloem-like conducting cells occur in some (see p. 320). Two types of colonial algae are known. In the first (Fig. 1.2*c*), the aggregate is indefinite in cellular number, continues to grow by cell division of its components, and reproduces by fragmentation. The second type of colony, called a **coenobium** (Fig. 1.2*d*), has a fixed number of cells at its origin, and this number is not augmented during the individual's existence, even if some cells are accidentally lost or destroyed.

Growth of the various multicellular algae may be diffuse or generalized or it may be localized. In **generalized growth**, all of the cells may undergo division, so that the organism undergoes overall increase in size as in *Ulva*, the sea lettuce (Fig. 1.2*f*). In **localized growth**, cell multiplication is restricted to certain parts of the organism. Localized growth may be apical, basal, or intercalary. **Apical growth** is that which is restricted to the extremities of the organism or at its tips and occurs, for example, in *Cladophora* (Fig. 3.113), *Fucus* (Fig. 6.75), and *Dictyota* (Fig. 6.36), among other

---

[6]May be motile by means of flagella (see Fig. 3.2, p. 66) or nonmotile; the basal holdfast cell of some filamentous algae excluded.

algae. **Basal growth** is less common but may be observed in one pattern of ontogeny of *Bulbochaete* (Fig. 3.99). **Intercalary growth** is localized neither at the apex nor base but at one or several other loci. This is well illustrated in the green alga *Oedogonium* (Fig. 3.95) in which only certain cells of the filament undergo division, in the trichothallic growth of certain brown algae like *Desmarestia* (Fig. 6.27), and in the development of the blade of *Laminaria* (Fig. 6.50).

It has been suggested that the putative, ancestral unicellular flagellate algae evolved the diversity of algal organization just summarized. From such ancestors, it has been postulated, arose two major series, namely algae nonmotile in the vegetative stage and those that retained motility by flagella. The latter series is manifested by the colonial motile habit in several divisions of algae including the green algae (Chlorophycophyta) and golden algae (Chrysophyceae). The colonial motile type of organization has been called **volvocine**, an allusion to the genus *Volvox* (p. 94).

Among the nonmotile algae, certain types of organization occur in parallel fashion among the several divisions. Thus, gelatinous, nonmotile, palmelloid colonies occur in the Chlorophycophyta (*Tetraspora*, p. 111), Xanthophyceae (*Chlorosaccus*), and Chrysophyceae (*Phaeosphaera*) (Rhodes and Stofan, 1967). Such colonies illustrate the **tetrasporine** series. Nonmotile unicellular and colonial organisms occur in most algal divisions, the Phaeophycophyta being a conspicuous exception. These may have or lack the capacity to form flagellate motile cells.

Parallelism of other types of algal organization also is apparent when one considers the several algal divisions. The filamentous type occurs among the blue-green, green, yellow-green, brown, and red algae and in certain diatoms. Membranous or blade-like plant bodies are exemplified by certain genera of green, brown, golden, and red algae.

It should be emphasized that the organization of algae into filaments and membranes and into few-celled complexes (e.g., *Chlorosarcinopsis*, p. 128) originated when their capacity to undergo vegetative cell division or **desmoschisis** (see p. 127) evolved. The latter was a fundamental innovation which made possible development of complex, integrated, multicellular plant bodies.

It is noteworthy that the multinucleate, tubular (siphonous) type of organization (p. 190) apparently evolved in only the division Chlorophycophyta and in the class Xanthophyceae of the division Chrysophycophyta. The **rhizopodial**, or amoeboid, type of organization is manifested only in two classes of the latter division, namely, the Chrysophyceae and the Xanthophyceae.

Algal Reproduction

Both sexual and asexual reproduction are of widespread occurrence in algae; in some, however, sexual reproduction does not occur, either through its phylogenetic loss or because, seemingly, it has not developed. By sexual reproduction at the cellular level is meant the union of cells, **plasmogamy**; the union of their nuclei, **karyogamy**; the association of their chromosomes and genes; and **meiosis**. Sexual reproduction affords the opportunity for exchange and formation of new combinations of genetic materials. Sexual reproduction has apparently not yet been confirmed in the Eugleno-phycophyta but occurs in at least some members of all the other divisions of algae. Asexual reproduction is increase in progeny not involving cellular and nuclear union and association of parental genetic materials.

Asexual Reproduction

In some unicellular algae the organism reproduces by cell division. These divisions may be repeated in rapid succession, designated **repeated bipartition** (Fig. 1.3a–c), to form new individuals like the parental cell. This process is also sometimes called **binary fission**. In colonial and other multicellular types of algae, cell division and subsequent enlargement result in growth. In the division of some unicellular algae the division

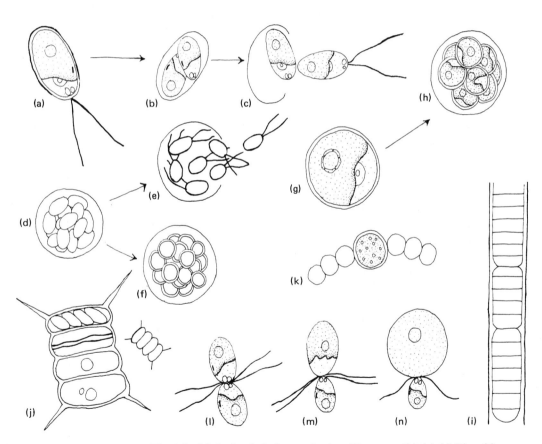

Fig. 1.3 Methods of algal reproduction (diagrammatic) (a)–(c) Bipartition or binary fission. (d), (e) Zoospore formation. (f) Aplanospore formation. (g), (h) Autospore formation. (i) Fragmentation or hormogonium formation. (j) Auto-colony formation. (k) Akinete formation. (l) Isogamy. (m) Anisogamy. (n) Oogamy.

products (aplanospores and autospores) remain associated for a period within the persistent and enlarging parental cell wall, a phenomenon that has been interpreted as incipient colony formation (Figs. 1.3*f*, 2.12).

Noncoenobic colonial, filamentous, and other types of multicellular algae reproduce by various types of **fragmentation**, the fragments having the capacity through continuing growth of developing into new individuals. Even fragments of such highly developed algae as *Enteromorpha* (p. 171) and *Polysiphonia* (p. 561) among others, when abcised from parent plants, can develop into new individuals. In the filamentous blue-green algae, the fragments, which exhibit gliding motility, are called **hormogonia** (Figs. 1.3*i*, 2.7). In some algae there are special buds or gemma-like fragments that are detached from the parent plant as agents of propagation. The **propagules** of *Sphacelaria* (Fig. 6.33) exemplify these. It should be noted that fragmentation is not a method of reproduction in coenobic algae. Instead, these undergo autocolony formation. An autocolony is a miniature colony produced by a parental colony that it resembles. In this process some or all of the component cells of a coenobium form miniature colonies by repeated cellular bipartition (Fig. 3.21*d*). Such colonies in other algae may be organized from zoospores as in *Hydrodictyon* and *Pediastrum* (Fig. 3.56).

In addition to fragments, algae produce a variety of (usually) unicellular agents of asexual reproduction called **spores**. Among these, **akinetes** are of widespread occurrence in the blue-green and green algae. An **akinete** (Fig. 2.8*a*) is essentially a vegetative (somatic) algal cell that has thickened its wall and thus has become able to withstand desiccation and other conditions hostile to vegetative development.

Many green, yellow-green, and brown algae produce flagellate agents of asexual reproduction known as **zoospores** (Fig. 1.3*d, e*), the name implying that their motility is an animal-like trait. In a number of instances the potential zoospores may omit their motile phase and begin their development within the parental cell wall. Such *ontogenetically* potential zoospores are then called **aplanospores** (Fig. 1.3*f*). Aplanospores that thicken their walls are known as **hypnospores**.

A number of other types of nonmotile spores are produced by various algae. Among these are autospores (Fig. 1.3*h*) of green and yellow-green algae. **Autospores** are superficially like aplanospores but differ in lacking the ontogenetic capacity for motility. Furthermore, they are in form miniatures of the parental cells that produced them. The monospores, tetraspores, paraspores, and carpospores of red algae and the statospores and auxospores of the golden algae are treated in the discussion of those organisms later in this book.

Spores may be produced within and by ordinary vegetative cells or within special cells or groups of cells in the algae. The specialized spore-producing structures are various types of **sporangia** (e.g., zoosporangium and aplanosporangium).

Sexual Reproduction and Life Cycles

Sexual reproduction is widespread in the algae as noted above. Certain algae have proved to be instructive in that whether their flagellate reproductive cells function as asexual zoospores or as gametes seemingly depends, at least in part, on environmental conditions. Of the latter, level of concentration of nitrogen in the surrounding medium seems to be of primary importance. Such lack of differentiation between asexual cells and gametes occurs, for example, in *Chlamydomonas* (p. 75) and in certain species of *Chlorococcum* (p. 115). Other algae, by contrast, produce zoospores that differ morphologically from gametes, although in some of these (e.g., *Ulva*) the differentiated gametes have the capacity to grow into new individuals without sexual union, i.e., **parthenogenetically**.

As noted on p. 1 dealing with the definition of algae, in certain unicellular algae, e.g., *Chlamydomonas*, the organisms themselves may function as gametes. The latter may be *morphologically*[7] indistinguishable or **isogamous** (Fig. 1.3*l*); or one member of the uniting pair may consistently be smaller than the other, the gametes, therefore, being designated as **anisogamous** (Fig. 1.3*m*); or the gametes may be extremely dimorphic, the larger, nonmotile and called an egg, and the smaller, motile by flagella, the sperm (Fig. 1.3*n*). This last type of sexual reproduction is called **oogamy**. **Heterogamy** is a more general term that includes anisogamy and oogamy. In the oogamous red algae the male gamete is nonflagellate but in some cases is amoeboid in movement. In the order Zygnematales (p. 225) of the green algae and in certain diatoms, the gametes also are nonflagellate and amoeboid. Isogamy, anisogamy, and oogamy are sometimes interpreted as a progressively more advanced evolutionary series. The evidence cited to support this interpretation is the indiscriminate pairing of gametes of different sizes in some cases of isogamy (occasioned by the different ages of members of the pair) and an intermediate condition between heterogamy, with the unequal gametes both flagellate, and oogamy (as in *Sphaeroplea*, p. 185) in which the female gametes may or may not be flagellate. Parker (1971C) has discussed the relative advantages to the organism of these three types of sexual reproduction. It is of interest that sexual reproduction in the single genus *Chlamydomonas* may be isogamous, anisogamous, or oogamous, depending on the species (Fig. 3.11).

The gametes may be *morphologically* identical with vegetative cells (e.g., *Chlamydomonas*) or, as in the case of many multicellular algae, they differ markedly from vegetative cells. They may arise from relatively unmodified vegetative cells that function as **gametangia** (e.g., *Sphaeroplea*, Fig. 3.116), or the gametangia may be morphologically specialized (*Oedogonium*, Fig. 1.1*b*; *Fucus*, Fig. 6.76; *Nemalion*, Fig. 9.28).

The female gametangium in oogamous algae is the unicellular **oogonium** (Fig. 1.1*b*). In the red algae and in several species of the green alga *Coleochaete*, the oogonium bears a protuberance or **trichogyne**, which is the receptive site for male gametes (Fig. 3.92*f*). The oogonia in red algae are called **carpogonia**.

Flagellate male gametes are produced in special gametangia called **antheridia**. The nonflagellate male gametes of red algae are called **spermatia** and are borne within

---

[7]"Morphologically" is italicized because physiological and biochemical differences prevail between the compatible strains of certain isogamous gametes. Furthermore, Triemer and Brown (1975B) have shown that the so-called isogametes of *Chlamydomonas reinhardtii* differ morphologically at the ultrastructural level.

minute gametangia called **spermatangia** (Fig. 9.28*d*). Further details regarding algal gametangia are deferred to the discussion of the algal genera themselves. Moestrup (1975) has reviewed a number of aspects of sexuality in algae.

The distribution or occurrence of compatible sexual potentialities in the algae varies among the species and genera. In those algae in which the gametes are differentiated into male and female (the anisogamous and oogamous algae), the male and female gametes may occur on the same individual of the species. The individual and species in this case are said to be **bisexual, monoecious,** or **hermaphroditic**. In other algae, the male and female gametes are produced always on different individuals; such individuals, accordingly, are unisexual, and the species to which they belong is **dioecious**. In algae, the compatible gametes involved in unions may arise from one individual or they may come from different ones. The authors would designate the first case as **monoecism** or **bisexuality** and the second as one of **dioecism** or **unisexuality** of individuals or populations of such individuals. Others (e.g., G. Smith, 1950) have designated the monoecism of isogamous algae as homothallism and their dioecism as heterothallism. These terms have been avoided here because they are used with somewhat different connotation in reference to the fungi.[8]

Finally, in relation to sexual reproduction, we must consider the nature of and variations in the algal life cycle and the site of meiosis (Heywood and Magee, 1976). The life cycle of algae in which sexual reproduction occurs may belong to one of three fundamentally different patterns illustrated diagrammatically in Fig. 1.4. In the first type (*a*), which up to the year 1921 was thought to characterize all green algae, the organism at maturity produces gametes which may unite to form zygotes which undergo a period of dormancy. The gamete-producing organism may be unisexual or bisexual, depending on the species. Upon germination of the zygote, its nucleus undergoes meiosis, so that the products of its germination (zoospores, aplanospores, or juvenile plants), like the adults, are haploid; the zygote alone, in this type of life cycle, is diploid and the organism may be said to undergo **zygotic meiosis**. Note in Fig. 1.4*a* that the organism may reproduce itself asexually. This type of life cycle was long ago (Svedelius, 1931) said to be **haplobiontic**, meaning that only a single type of free-living[9] individual is involved in the life cycle. Note again that in the Fig. 1.4*a* type of life cycle the organism is haploid in chromosome constitution. Accordingly, it will be useful for later discussion to designate the type *a* life cycle in abbreviation as "H, h," the first (uppercase) letter meaning haplobiontic and the second (lowercase) letter meaning haploid. H, h life cycles are widespread among green algae.

A second type of algal life cycle is illustrated in Fig. 1.4*b*. Here the sole free-living organism present in the life cycle is diploid, meiosis occurring (as in all animals) during gametogenesis; this is **gametic meiosis**. In this type of haplobiontic cycle only the gametes are haploid, the organisms that produce them being diploid. Accordingly, this type of life cycle may be designated as H, d, the lowercase letter referring to the

---

[8]For example, although both male and female gametes are produced on the mycelia of clonal cultures of *Neurospora sitophila*, it is said to be heterothallic because gametes from the same mycelium are self-incompatible.

[9]By *free-living* is meant not physically attached to another organism of the same species.

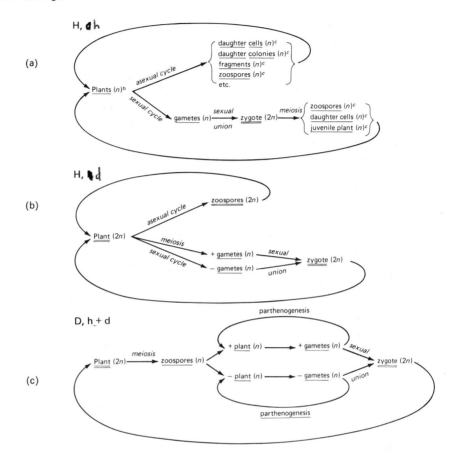

Fig. 1.4 (a)–(c) Types of algal life cycle[a]. (After Bold.)

[a]Haploid phases are underlined once; diploid phases twice. Both haploid and diploid plants may replicate themselves asexually by fragments and/or various types of spores.

[b]If the plants are unisexual, + and − or male and female plants occur.

[c]Alternate phases that characterize specific algae.

chromosome constitution of the free-living organisms in the cycle. Here again, the diploid phase may propagate itself by asexual agents. The H, d type of life cycle occurs in tubular (siphonous) green algae (p.190) and, according to one interpretation, in the rockweeds (Fucales, p. 343) among the brown algae, as well as in the diatoms.

Figure 1.4c illustrates a third type of life cycle that is rather widespread among green, brown, and red algae, namely, the **diplobiontic** type. As indicated by this designation, two free-living types of individuals occur in such a life cycle, namely, a haploid, gamete-producing plant (**gametophyte**) and a diploid, spore-producing one (**sporophyte**). The gametes (which develop from unisexual or bisexual individuals) unite to form zygotes, which, without undergoing a period of dormancy, grow into diploid sporophytes. As these mature and form spores, meiosis occurs during sporogenesis, so that **sporic meiosis** is said to occur. These spores develop into gametophytes. The sporophytes themselves are diploid and the gametophytes haploid. Note also in Fig. 1.4c that both the gametophytes and sporophytes may reproduce themselves by asexual agents. This type of life cycle may be designated in abbreviation D, h + d, because both haploid and diploid free-living plants occur. These plants may be morphologically similar, in which case the cycle is said to be **isomorphic**, or they may differ morphologically in a **heteromorphic** life cycle. To illustrate this in abbreviation, the life cycles may be written as D^i, h + d or D^h, h + d.

The Fig. 1.4c type of life cycle has one important variation that occurs in red algae and is included in the discussion of them (p. 460). It is similar to that of ferns but unlike that of mosses and liverworts in that the latter are not diplobiontic. The D, h + d life cycle, furthermore, illustrates the broad phenomenon of **alternation of generations**, which means that a haploid gametophyte is followed in the life cycle by a diploid sporophyte and the latter, in turn, by a haploid gametophyte.

A word of caution is indicated in this discussion of life cycles. It is sometimes implied or inferred that they are absolute and inexorable, but in the following pages it will be seen that deviations from them occur both in nature and in the laboratory. It is also sometimes inferred or even implied that differences in chromosome constitution (haploidy or diploidy) explain the differences between sporophytes and gametophytes and that they are the primary *cause* of alternation of generations. The reported occurrence, however, of haploid sporophytes in certain algae (e.g., *Ulva* and *Cladophora*, Føyn, 1934, 1958) and the occurrence of both haploid sporophytes and diploid gametophytes in ferns and in *Ectocarpus* (Müller, 1972) indicates caution in ascribing the cause of alternation to ploidy.

One further phenomenon is relevant to this discussion, namely, **parthenogenesis**, alluded to briefly earlier (p. 11). What becomes of gametes that fail to take part in sexual union? In some algae (e.g., *Scenedesmus*, Trainor and Burg, 1965A) they disintegrate. In others (e.g., *Ulva lactuca*, personal observation of author, H.C.B.) both the male and female gametes develop into new haploid gametophytes. The development of single gametes into new individuals without gametic union is known as *parthenogenesis*. Many aspects of algal reproduction have been summarized by Ettl, et al. (1967).

Cultivation of Algae in the Laboratory

When author Bold was an undergraduate student (1926–1929), other than *Chlorella* and *Scenedesmus*, which were then being used for investigations of mineral nutrition and photosynthesis, very few algae were being grown in laboratory culture in the United States. Instead, instruction in phycology depended almost exclusively on collections from the field that, although still very valuable, often fail to provide all the organisms desired because of their sporadic occurrence in nature. Thus, many laboratory periods were of necessity spent studying algae preserved in formaldehyde, a far from dynamic and inspiring exercise for all except the specialist. Some of the history, methods, and impact of growing algae in the laboratory and their significance for the advancement of phycology in particular and of biology in general have been summarized by Bold (1942, 1974). Briefly stated here, improvements in growing algae in the laboratory have made possible great advances in our knowledge of algal taxonomy, physiology, life cycles, ultrastructure, biochemistry, ecology, and genetics.

Bold (1942), Pringsheim (1946), and many others (see Stein, 1973) have discussed in detail methods of growing algae in the laboratory, which accordingly will not be repeated here. There are several kinds of cultures that serve a variety of purposes. Fresh collections of algae kept in the laboratory under illumination sometimes show a succession of different species as they age. These have been designated as **maintenance cultures**. Sometimes if one wishes to encourage the multiplication of one or more species, he may add special nutriments (e.g., a split pea, a rice grain, nitrate, or phosphate) to a maintenance culture. This would then be called an **enrichment culture**. By various techniques it is possible to separate algae from other algae into **unialgal culture** (in which protozoa, fungi, and/or bacteria may be present). It is also possible to separate an algal species from *all* other living organisms and to grow it as a **pure** or **axenic culture**. Finally, when it is necessary to conduct an investigation with genetically homogeneous populations, single cells or fragments of an organism may be isolated and cultured to multiply into such populations. The latter are known as **clones** and are spoken of as **clonal cultures**. These may or may not be axenic.

In spite of the impressive advances in the cultivation of algae in the laboratory, a few words of caution may be quoted to govern possible excessive enthusiasm. Pringsheim (1967A), an eminent microbiologist and champion of the cultivation of algae in the laboratory, wrote: "Phycologists working mainly in the laboratory ought to give more thought to the ecological implications of their findings."

The cultivation of algae in the laboratory, of course, requires understanding of some of the basic principles of algal physiology and of algal nutrition. It is generally assumed that algae[10] are **photoautotrophic** or **phototrophic**, meaning that by using light energy they are able to synthesize their protoplasm from entirely inorganic sources, and in many instances this assumption has been confirmed, provided the proper concentrations of the proper inorganic compounds were included in the culture medium at proper pH. Such a culture medium, the composition of which, qualitatively and quantitatively, is completely known, is called a **defined medium**. It became evident long ago, however, that a number of algae have additional growth requirements, most often in the form of vitamins. For example, *Euglena gracilis* (p. 259) requires vitamin B_{12} as do some species of *Mougeotia* (p. 230) and a number of other algae. Phototrophic algae with such growth requirements are said to be **photoauxotrophic**. Colorless algae (like most bacteria and the fungi) are obligately, and certain chlorophyllous algae facultatively, **heterotrophic**. By this is meant that they require or can use organic compounds (e.g., sugars, amino acids, etc.) in their nutrition in darkness and/or light. Examples of facultatively heterotrophic algae are *Chlorogonium* (p. 84), *Euglena* (p. 258), and *Ochromonas* (p. 363) among others. A few algae, including species of *Ochromonas* and certain euglenoids and dinoflagellates (p. 417), are **phagotrophic** or

[10]Except those lacking chlorophyll (e.g., *Polytoma*, p. 85).

endocytic; that is, they can ingest solid particles of food. It is of interest that *Ochromonas danica* may be photosynthetic and/or saprophytic and phagotrophic.

In addition to the major elements (C, H, O, P, K, N, S, Ca, Fe, Mg), various others are required in trace amounts. These include Zn, Mn, Mo, Cu, Co, and B. The major elements may be provided in the culture medium in different forms. For example, nitrogen may be supplied as NO_3, and NO_2, NH_4, or in organic compounds for various algae, although NO_3 and NH_4 compounds are most frequently used. A few algae have been shown to be able to use extracellular nitrogen from gelatine (e.g., Archibald and Bold, 1970). Many marine algae thrive in culture media in which seawater has been enriched with nitrogen and phosphorus. Both marine and freshwater media often support better algal growth when they are supplemented with soil extract.

Many algae grow well in Pringheim's (1946) biphasic soil-water medium in which a little soil is introduced into the bottom of the culture vessel that is then filled two-thirds full with deionized, tap, or seawater and steamed for 1 hour on each of three successive days. This method essentially consists of providing the desired alga with a more or less natural environment free from competition. Details of preparation of this and other culture media are presented in the Appendix.

Further references to the nutrition and cultivation of algae are available in the publications of Lewin (1959) and Stein (1973).

Classification

Papenfuss (1955) has reviewed the history of the classification of the major groups of algae and also the history of the discovery of their sexual reproduction. Although Linnaeus (1753) recognized 14 genera of "algae," only 4 of them (*Conferva, Ulva, Fucus,* and *Chara*) were algae as we now define them. By 1836, Harvey had recognized four major groups of algae, the brown, red, and green algae and the diatoms, and color, as a manifestation of differing pigmentation, continues to be of prime importance in classifying the major groups of algae. In the nineteenth century, and in the United States until the publication of G. Smith's *Fresh-water Algae of the United States* in 1933, the algae were grouped as a class coordinate with Fungi under one of the divisions (Thallophyta) of the plant kingdom. Investigations of the first quarter of the twentieth century have subsequently revealed that accompanying differences in pigmentation among the great groups of algae are differences in storage products and cellular organization. It was, therefore, concluded that the former Class Algae did not embrace a closely related, cohesive, natural alliance of organisms. Accordingly, Smith (1933, 1950) recognized 11 major groups of algae that, having abandoned the formal categories Thallophyta and Algae, he grouped in 7 (in 1950) major categories or divisions, coordinate with the Bryophyta and other divisions of the plant kingdom. These he designated, in conformity with the International Code of Botanical Nomenclature, Chlorophyta, Euglenophyta, Chrysophyta, Phaeophyta, Pyrrhophyta,[11] Cyanophyta, and Rhodophyta. Papenfuss (1946), however, pointed out that to use the designation "Chlorophyta," literally "green plants" for the green algae precluded its

---

[11]Sometimes spelled Pyrrophyta.

use for other members of the plant kingdom with identical pigmentation and storage products (Craigie, 1974). He suggested, therefore, that the names for the algal divisions include *phyco*[12], the group names being accordingly Chlorophycophyta, Euglenophycophyta, etc. The inclusion of *phyco* indicates the members of the several groups are at the algal level of organization. Papenfuss' suggestion in this regard is followed in this book and the following divisions of algae will be discussed: Cyanochloronta,[13] Chlorophycophyta, Charophyta, Euglenophycophyta, Phaeophycophyta, Chrysophycophyta, Pyrrhophycophyta, Cryptophycophyta, and Rhodophycophyta. Their distinctive characteristics are set forth briefly in Table 1.1 (see also W. Stewart, 1974).

These divisions of algae differ in their cellular organization, pigmentation, cell wall chemistry, storage products, and flagellation (or its absence), among other respects. Table 1.1 attempts to summarize these characteristics. It must be emphasized that the data in the table should be considered as tentative and probably subject to change because in many of the divisions the data have been educed from only a few representatives.

It has been said often that "nature mocks at human categories" and this certainly seems to be the case when one compares the more-or-less current systems of classification used by various authors. The authors regard the system used in this text, as they do all systems of classification of living organisms, as tentative, and always to be modified in the light of new data. Klein and Cronquist (1967) have reviewed the classification of the algae (and other plants) considering chemical, structural, and functional criteria and have recognized six divisions, the blue-green algae being grouped with the bacteria.

The Fossil Record of the Algae

The algae are ancient organisms, their history possibly extending back about 3.1 billion years into the Precambrian epoch of the earth's history, although this has recently been questioned (Schopf, 1976.) Schopf (1970) has summarized our knowledge of the biota of the Precambrian, and he and his associates, and others, have described a number of Precambrian algae. Seemingly, the most ancient recognizable algae were prokaryotic, and one of these, *Archaeosphaeroides barbertonensis* Schopf and Barghoorn, presumed to be similar to our extant coccoid blue-green algae, is illustrated in Fig. 1.5. Additional evidence of the early occurrence of blue-green algae is provided

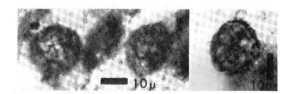

Fig. 1.5 *Archaeosphaeroides barbertonensis* Schopf and Barghoorn, a sphaeroidal, alga-like fossil from Early Precambrian (3.1 × 10⁹ years old). (After Schopf and Barghoorn in Science 156: 508–512, 1967; Copyright by the American Association for the Advancement of Science.)

[12]From the Greek *phykos*, seaweed, hence alga, as in phycology, the study of algae.
[13]Literally, the blue-green ones, a designation that is noncommittal with respect to their possible algal or bacterial affinity.

TABLE 1.1 SUMMARY OF SOME ALGAL DIVISIONS AND THEIR MORE SIGNIFICANT CHARACTERISTICS

Division	Common Name	Pigments and Plastid Organization in Photosynthetic Species	Stored Food	Cell Wall[a]	Flagellar Number and Insertion[b]	Habitat[c]
Cyanochloronta	Blue-green algae	Chlorophyll *a*; C-phycocyanin, allophycocyanin, C-phycoerythrin; β-carotene and several xanthophylls	Cyanophycean granules (alanine and aspartic acid); polyglucose (glycogen-like)	α, ε-Diaminopimelic acid, glucose amine, alanine, etc.	Absent	fw, bw, sw, t
Chlorophyco-phyta	Green algae	Chlorophyll *a*, *b*; α-, β-, and γ-carotenes + several xanthophylls; 2–5 thylakoids/stack[d]	Starch (amylose and amylopectin) (oil in some)	Cellulose in many (= β − 1, 4-glucopyranoside), hydroxy-proline glycosides; xylans and mannans; or wall absent; calcified in some[e]	1, 2–8, many, equal, apical	fw, bw, sw, t
Charophyta	Stoneworts	Chlorophyll *a*, *b*; α-, β-, and γ-carotenes + several xanthophylls; thylakoids variably associated	Starch resembling that of land plants	Cellulose (= β − 1, 4-glucopyranoside); some calcified	2, equal, sub-apical	fw, bw
Eugleno-phycophyta	Euglenoids	Chlorophyll *a*, *b*; β-carotene + several xanthophylls; 2–6 thylakoids/stack, sometimes many	Paramylon (= β − 1, 3-glucopyranoside), oil	Absent	1–3 (−7) apical, sub-apical	fw, bw, sw, t
Phaeophyco-phyta	Brown algae	Chlorophyll *a*, *c*; β-carotene + fucoxanthin and several other xanthophylls; 2–6 thylakoids/stack.	Laminaran (= β − 1, 3-glucopyranoside, predominantly); mannitol	Cellulose, alginic acid, and sulfated mucopolysaccharides (fucoidan)	2, unequal, lateral	fw (very rare), bw, sw
Chrysophyco-phyta	Golden and yellow-green algae (including diatoms)	Chlorophyll *a*, *c* (*c* lacking in some); α-, β-, and ε-carotene + several xanthophylls; including fucoxanthin	Chrysolaminaran (β-1, 3-glucopyranoside, predominantly); oil.	Cellulose, silica, calcium carbonate, mucilaginous substances, and some chitin; or wall absent	1–2, unequal or equal apical	fw, bw, sw, t

		Pigments / thylakoids	Storage product	Flagella[b]	Cell wall[a]	Habitat[c]
Pyrrhophycophyta	Dinoflagellates	Chlorophyll a, c; β-carotene + several xanthophylls; 3 thylakoids/stack	Starch (oil in some)	2, one trailing, one girdling	Cellulose or absent; mucilaginous substances	fw, bw, sw
Cryptophycophyta	Cryptomonads	Chlorophyll a, c; α-, β-, and ε-carotene; distinctive xanthophylls (alloxanthin, crocoxanthin, monadoxanthin); phycobilins; 2 thylakoids/stack	Starch	2, unequal subapical	Absent	fw, bw, sw
Rhodophycophyta	Red algae	Chlorophyll a, (d in some Florideophycidae); R- and C-phycocyanin, allophycocyanin; R- and B-phycoerythrin. α- + β-carotene + several xanthophylls; thylakoids single, not associated	Floridean starch (glycogen-like)	Absent	Cellulose,[f] xylans, several sulfated polysaccharides (galactans) calcification in some	fw (some), bw, sw (most)

(The pigment description continues from the previous page: "in Chrysophyceae, Bacillariophyceae, and Prymnesiophyceae; 3 thylakoids/stack")

[a] In terms of cell wall chemistry, the vegetative cells have received most attention. Spores, akinetes, dormant zygotes, and other resting stages have not been studied, but it is clear that their walls may contain other substances, e.g., waxes and other nonsaponifiable polymers and phenolic substances. See also Parker (1970), Mackie and Preston (1974), Darley (1974), and Hellebust (1974).

[b] In motile cells, when these are produced.

[c] fw = freshwater; bw = brackish water; sw = salt water; t = terrestrial (soil, rocks, etc.).

[d] Based on Gibbs (1970), Dodge (1973).

[e] Others are wall-less or have xylans, mannans, other glucans, some silica, or protein. Also, nearly all skeletal polysaccharides (cellulose, xylans, mannans) are accompanied by one or more mucilaginous substances (e.g., arabino-galactans and sulfated mucopolysaccharides).

[f] Lacking in some Bangiales, which have mannans and/or xylans as the primary wall component.

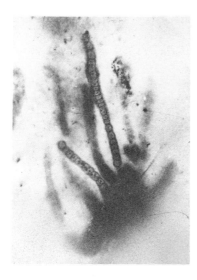

Fig. 1.6 *Langiella scourfieldii* Croft and George. A Middle Devonian blue-green alga. × 158 (all magnifications are approximate). (After Croft and George.)

by the occurrence of stromatolites, which are layered calcareous structures similar to the growth pattern of some modern Cyanochloronta. Stromatolites approximately 2.8 billion years old have been discovered. The Cyanochloronta have persisted from the Precambrian to the present apparently with little modification. Croft and George (1959) found well-preserved representatives of Stigonemataceae in the Devonian chert at Rhynie, Scotland (Fig. 1.6).

During the Middle Precambrian (from 2.5 to 1.7 billion years ago), the diversity of microfossils, including algae, increased. For example, 12 species of microscopic

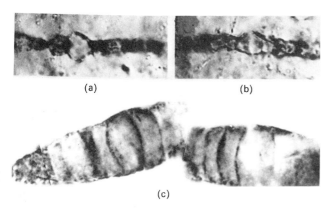

(a)   (b)

(c)

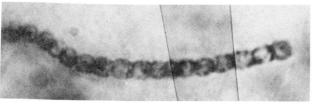

(d)

Fig. 1.7 (*a*), (*b*) *Archaeonema longicellularris* Schopf with heterocysts. (*c*) *Obconicophycus amadeus* Schopf and Blacic. (*d*) *Veteronostrocale amoenum* Schopf. (*a*), (*b*) × 450; (*c*) × 1008; (*d*) × 450. ((*a*), (*b*) after Schopf and Barghoorn, J. Palaeont. 43: 111–118; (*c*), (*d*) after Schopf and Blasic, J. Palaeont., 45: 925–960; courtesy of Soc. Eco. Palaeont. and Mineral.)

plants were recognized by Barghoorn and Tyler (1965), of which several were possibly coccoid forms (Licari, et al., 1969; Cloud, et al., 1975), while others were heterocystous (Licari and Cloud, 1969) and nonheterocystous filaments of Cyanochloronta.

In Late Precambrian black cherts, around 900 million years old, about 30 species of microorganisms were described including blue-green (Fig. 1.7), green, and possibly red algae. The evidence for the occurrence of eukaryotic algae is the presence in the cells of putative nucleus-like bodies and other cellular differentiations. Cloud, et al. (1969) have described as the most ancient green algae several unicellular forms presumed to be chlorococcalean in affinity. These were from strata 1.2 to 1.4 billion years old.

According to Schopf (1970), the geological history of the several groups of algae extends as far back as the period cited in parentheses in the following: Cyanochloronta (Precambrian), Chlorophycophyta (Precambrian), Rhodophycophyta (Cambrian), Phaeophycophyta (Late Precambrian), Charophyta (Silurian), Prymnesiophyceae (Triassic), Xanthophyceae (Cretaceous), Euglenophyceae (Cretaceous), Bacillariophyceae (Cretaceous), and Chrysophyceae (Cretaceous). Schopf (1970) has also summarized the occurrence of fossil green algae; his summary is reproduced as Fig. 1.8. In a more recent publication (1974), he has designated the Precambrian as the age of blue-green algae and discussed their possible origin and evolution.

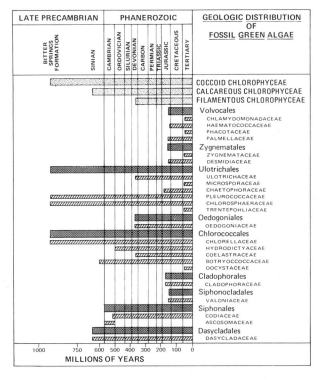

Fig. 1.8 Histogram of geologic distribution of fossil green algae. (After Schopf, J. W. 1970. Pre-cambrian micro-organisms and evolutionary events prior to the origin of vascular plants. Biol. Rev. 45: 319–352, Cambridge University Press.)

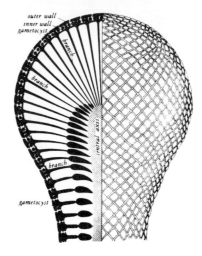

Fig. 1.9 *Ischadites iowensis* Owen, Kesling, and Graham, a dasycladacean green alga, reconstruction. × 5. (After Kesling and Graham, J. Paleont. 36: 943–952; courtesy Soc. Eco. Paleont. and Mineral.)

In a series of major publications, Johnson and Konishi (1956, 1958) Johnson, et al. (1959), and Johnson and Høeg (1961) have reported on algae of the Ordovician, Silurian, Devonian, and Mississippian. As would be expected, calcareous algae, such as certain members of the Codiales and Dasycladales of the Chlorophycophyta and the Solenoporaceae and Corallinaceae of the Rhodophycophyta, are the most frequently and best-preserved representatives of these earlier algal floras. Kesling and Graham (1962) described in considerable detail the Dasycladalean alga *Ischadites;* their reconstruction of it is reproduced in Fig. 1.9. Several calcareous Dasycladacean genera, as reconstructed, are illustrated in Fig. 1.10, as are representative Rhodophycophyta from the Ordovician and Mississippian in Fig. 1.11.

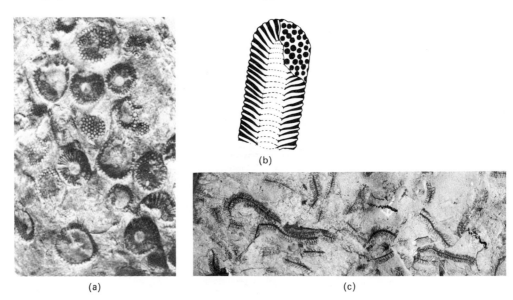

Fig. 1.10 Dasycladacean green algae (Ordovician). (*a*) *Cyclocrinus porosus* Stalley. (*b*) *Anatolepora carbonica* Konishi. (*c*) *Primicorallina trentonensis* Whitfield. (*a*) × 0.4; (*b*) × 23; (*c*) × 0.5. [(*a*) after Stolley in Johnson; (*b*) after Konishi; (*a*), (*b*) courtesy of Colorado School of Mines; (*c*) courtesy Dr. S. Mamay.]

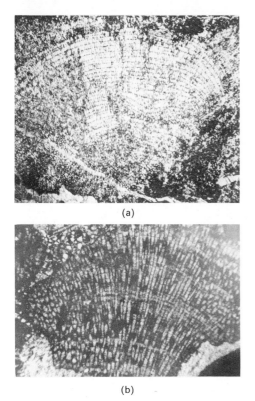

(a)

(b)

Fig. 1.11 Fossil Rhodophyceae. (*a*) Ordovician: *Pterophyton kiaeri* Hoeg. Longisection showing compact tissue. (*b*) Mississippian: *Parachaetetes thomasii* Johnson and Konishi. Longisection. (*a*) × 15; (*b*) × 13.

One of the most interesting and problematical fossil "algae" is *Prototaxites* (Fig. 1.12), described first by Dawson from the Upper Devonian. Some specimens of *Prototaxites* were up to 2.2 m long and 0.9 m in diameter. These axes consisted of large-diameter, longitudinal tubes embedded in a matrix of thin-walled, septate filaments. This type of organization, it has been suggested, is like that of the stipe of kelps, but Arnold (1952), after careful study of well-preserved material, was unwilling to assign *Prototaxites* to any family of recent seaweeds. Schmid (1976), however, on the basis

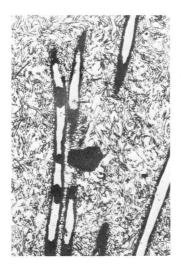

Fig. 1.12 *Prototaxites southworthii* Arnold. Longisection. × 71. (After Arnold.)

Fig. 1.13 *Hungerfordia dichotoma* Fry and Banks, a *Dictyota*-like alga from the Upper Devonian. × 0.6. (After Fry and Banks, J. Paleont. 29: 37–44; courtesy Soc. Eco. Paleont. and Mineral.)

of the nature of the septal pores, has concluded that *Prototaxites* is probably not related to fungi or algae.

Other fossil noncalcareous algae of possible phaeophycophytan affinity have been described from the Upper Devonian of New York by Fry and Banks (1955). One of these, which the authors suggest resembles *Dictyota*, is illustrated in Fig. 1.13. By contrast, Baschnagel (1966) has described the freshwater desmid, *Paleoclosterium leptum* (Fig. 1.14), from Middle Devonian strata of New York.

Microalgae have been reported from much more recent deposits. Sarjeant (1974) has summarized our knowledge of fossil dinoflagellates (see p. 417). Gray (1960) has illustrated *Botryococcus braunii* Kützing and two species of *Pediastrum* (Fig. 1.15)

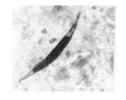

Fig. 1.14 *Paleoclosterium leptum* Baschnagel, a Middle Devonian desmid. × 840. (After Baschnagel.)

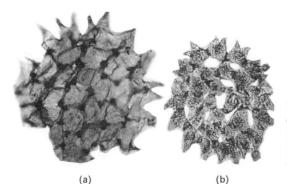

(a)   (b)

Fig. 1.15 *Pediastrum* from the Miocene. (*a*) *P.* cf. *boryanum* (Turp.) Menegh. (*b*) *P.* cf. *duplex*? Meyen. × 240. (After Gray, J. Paleont. 34: 453–463; countesy Soc. Eco. Paleont. and Mineral.)

(a)   (b)   (c)

Fig. 1.16 Coccothiths of Coccolithophoridacean Chrysophycean algae. (*a*) *Colvillea barnesae*, Upper Cretaceous. (*b*) *Discoaster lodoensis*, Lower Tertiary. (*c*) *Braarudosphaera bigelovii*, Lower Tertiary. (a) × 3500; (*b*) × 1000; (*c*) × 2000. (After Black.)

from the Middle Miocene of Oregon, while Evitt (1963A) found well-preserved material of *Pediastrum* in both the Lower and Upper Cretaceous. From Eocene strata, Bradley (1970) has reported and illustrated two blue-green algae, *Stigonema anchistina* and *Symploca hedraia*, and two green algae, *Schizochlamys haywellensis* and *Spirogyra wyomingia*. It should be noted that, like Gray's Miocene and Evitt's Cretaceous algae, Bradley's are referable to extant genera.

Characteristic remains of certain Chrysophycophyta (p. 357), the Coccolithophorids, are abundant in certain Mesozoic and Tertiary rocks. These remains are in the form of small calcareous structures called **coccoliths** (Figs. 1.16, 7.22), which in life comprised ornamentations in the cell surface of flagellates. The latter are widely distributed at present in the upper layers of ocean waters. Black (1965) has illustrated and discussed coccoliths.

The siliceous frustules of diatoms have been preserved in great abundance from the Jurassic through the Mesozoic into the Cenozoic. Great mounds of diatomaceous earth, as deposits of fossil frustules are called, occur at Lompoc, California, and in the Calvert Cliffs of the Patuxent River in Maryland. Figure 1.17 illustrates some species of fossil diatoms from Hungarian deposits.

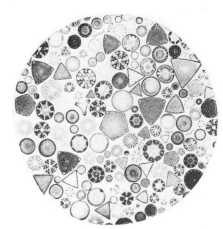

Fig. 1.17 Fossil marine diatoms from Hungary. × 25. (After Mann, from Tiffany, *Algae, the Grass of Many Waters*, 2nd ed., 1958; courtesy Charles C Thomas, Publisher, Springfield, Illinois.)

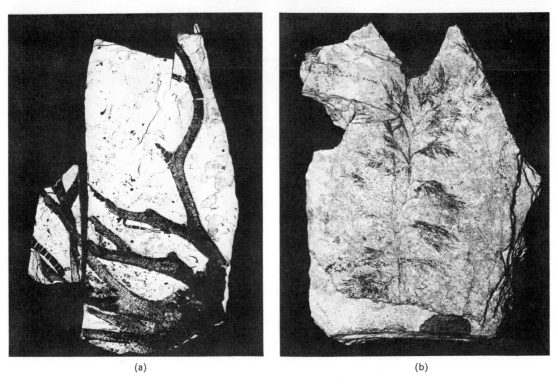

(a)   (b)

Fig. 1.18 (*a*) *Julescraneia grandicornia* Parker and Dawson, a Laminarialean brown alga. (*b*) *Chondrides flexilis* Parker and Dawson, a Ceramialean red alga. (After Parker and Dawson.)

Parker and Dawson (1965) described 10 genera and 22 species of noncalcareous algae from the Miocene deposits in Los Angeles County. Their records include one green alga, four genera of brown algae, and five of red algae. None of the species seem to be extant. Several of their specimens are illustrated in Fig. 1.18.

Phillips, et al. (1972) have discussed several species of the algal genus *Protosalvinia* (Fig. 1.19) from the Upper Devonian. The frequently bifurcate thalli had tetraspores

Fig. 1.19 *Protosalvinia furcata* (Dawson) Arnold. Surface of bifurcate specimen showing reticulate cellular pattern. × 10. (After Phillips, Miklas, and Andrews, Rev. Palaeobot. and Palyn. 14: 171–196; Elsevier Pub. Co.)

in depressions at their apices. The authors consider *Protosalvinia* to have phaeophyco-phytan or rhodophycophytan affinities, probably the former.

Finally, the oogonia of Charophyta have been relatively abundantly preserved as fossils (Peck, 1953; Johnson and Konishi, 1956; Peck and Eyer, 1963; and Grambast, 1974) from the Devonian and Lower Mississippian. The earlier representatives had more than five sheath cells around the oogonia, and in some they were dextrally coiled, unlike those of later and extant Charophyta that have sinistral coiling of the five oogonial sheath cells. Several representative fossil oogonia ("gyrogonites") are illustrated in Fig. 1.20.

Thus, although the fossil record of the algae is a long and ancient one that indicates that the prokaryotic Cyanochloronta preceded the groups of eukaryotic algae, it sheds no light on the origins and relationships of the several extant divisions of algae. For such considerations, the discussion has come to be based entirely on the comparative consideration of the morphology, physiology, and biochemistry of living algae. Flügel (1977) has summarized recent results and developments regarding fossil algae.

(a) (b) (c) (d) (e) (f) (g) (h) (i) (j) (k) (l)

Fig. 1.20 Fossil Charophytan oogonia (zygotes). (*a*), (*e*), (*h*), (*j*): *Catillochara moreyi* Peck. Palaeozoic. (*b*), (*c*), (*d*), (*g*): *Stellatochara prolata* Peck and Eyer. Lower Triassic. (*f*), (*i*), (*k*), (*l*): *Palaeochara acadica* Bell. Pennsylvanian. × 70. (After Peck and Eyer, J. Paleont. 37: 833–844; courtesy Soc. Eco. Paleont. and Mineral.)

Biological and Economic Aspects of Algae

Although the biological and economic significance of bacteria and fungi is widely appreciated, even by laymen, that of algae is less so. The algae play both beneficial and detrimental roles in nature. A number of these are cited in the following brief discussion.

In the first place, algae are highly important as primary producers of organic matter in aquatic environments because of their photosynthetic activities. Were they (along with aquatic angiosperms) to disappear from aquatic environments, animal life would lack its primary source of food and energy. Furthermore, during daylight hours, algae, of course, are oxygenating the water in their immediate vicinity. As examples of the primary dependence of aquatic animals on algae may be cited the association of algae with corals, certain protozoa, sponges, and marine worms. Not only are the structural materials of algal cells and their walls of great importance as primary food sources, but the products of their secretion and excretion are also significant (Trench, 1971). Thus, it will be recalled that polysaccharides, amino acids, DNA, RNA, enzymes, and other proteins are liberated by algae into the surrounding medium (Aaronson, 1971) and that these products become available to organisms associated with the algae. For example, Benson and Muscatine (1974) have recently described the feeding of fish on the slimy exudate of corals, the carbon components of which exudate originates in the photosynthesis of the algae that are symbiotic components of the coral. Of great significance in this relation are the chloroplasts of *Codium* and *Caulerpa* ingested by certain marine opisthobranchs and slugs, which remain viable within the cells of these animals and photosynthesize there (Taylor, 1970; Trench, et al., 1972).

More than 70 species of marine algae (mostly red and brown algae) have been used for food in oriental countries and fewer in the western world as well. Among these are certain red algae and kelps, some of which have been used for centuries. Of the red algae one of the most important is *Porphyra* (p. 477) called "nori" in Japan, "laver" in England and the United States, "sloke" in Ireland, "slack" in Scotland, and "luche" in southern Chile. *Porphyra* is widely cultivated in the shallow parts of bays in the Orient, especially Japan, where it is harvested from brush and/or nets submerged in those waters. *Palmaria* (p. 540), called "dulse" in Canada and the United States, "dillisk" in Ireland, and "sol" in Iceland also is eaten. *Chondrus crispus* (p. 525), the "Irish Moss" or "carragheen" occurs on both shores of the Atlantic from which it has been harvested in great quantities for many years. Its hydrocolloids are used, after appropriate extraction, in a variety of ways as stabilizers and thickenings in ice creams and prepared foods. One of these hydrocolloids is called **carrageenan**. Of the brown algae the stipes and blades of various kelps have long been eaten. More extensive accounts of the uses of algae as food (and in other ways) may be found in Tilden (1935), Newton (1951), Boney (1965), H. Johnson (1965), Dawson (1966), Chapman (1970), Dixon (1973), and Johnston (1965, 1966, 1970, 1976), among others.

In addition to carrageenan, **agar** (from the Polynesian word **agar-agar**) is extracted from various red algae and used in packing canned foods and in the treatment of constipation; and it is of paramount importance in microbiology as a *relatively* inert agent for solidifying culture media in which role it long ago supplanted gelatin. Hydrocolloids from kelps and other brown algae and hydrophilic derivatives of alginic acid, collectively called algins, have uses similar to those of red algal hydrocolloids. Kelps are also important sources of potash.

Mention should also be made of the use of algae as fodder. Cattle and sheep may graze on certain marine algae at low tide in the Scottish islands and in Chile, and preparations of kelps and of certain unicellular green algae have been used as supplements to the food of poultry, cattle, and hogs. Finally, it may be noted that in coastal areas coarser algae (kelps and fucoids) have been used as fertilizers both for improving the texture of rocky soils and their fertility, as in the Aran Islands off the west coast of Ireland.

The role of many blue-green algae in nitrogen fixation is discussed on p. 34 and can scarcely be overemphasized. Their activity in this connection greatly improves the fertility of rice fields.

With increasing awareness of problems of pollution control and waste disposal, the important activities of algae have come to be recognized (Silva and Papenfuss, 1953; Palmer and Tarzwell, 1955; Palmer, 1959). Raw primary sewage or secondary sewage products are in some areas introduced into shallow oxidation ponds or waste stabilization ponds. The oxidation is accomplished by bacteria whose activities are enhanced by the photosynthetic oxygen of associated algae.

The importance of fossil diatoms, as diatomaceous earth, in a number of places has been cited on p. 25. For further reading on the utilization of algae the reader should consult the literature by Zaneveld (1959), Krauss (1962), Boney (1965), Johnson (1965), Diaz-Piferrer (1967), Levring, et al. (1969), Zajic (1970), Michanek (1971), and the Proceedings of the International Seaweed Conferences.

Finally, among beneficial aspects of algae should be cited their increasingly widespread use in biological research. A classic example of this is the use of cultures of *Chlorella*, *Scenedesmus*, *Anacystis*, and other microalgae in investigations of photosynthesis. Our knowledge of sexual reproduction at the cellular and molecular level has advanced greatly through studies of the process in *Chlamydomonas* and other Volvocalean algae.

The paragraphs above have cited briefly some beneficial activities or uses of algae. There are also negative aspects that should be considered. First among these is the fact that certain algae or their products are toxic to animals (Gorham, 1964; Loeblich and Loeblich, 1975; Kadis, et al., 1971). Examples of this, mentioned on p. 47, are the poisoning of livestock by the blue-green alga *Microcystis* and of fish and shellfish (and, as a result, human beings) by the toxins produced by dinoflagellate red tides. Contact dermatitis by the blue-green alga *Lyngbya* has been reported by Moikeha, et al. (1971) and Moikeha and Chu (1971). The role of airborne algae as causative agents in inhalant allergies has been emphasized (McElhenney and McGovern, 1962; Bernstein and Safferman, 1970). Schwimmer and Schwimmer (1964) have discussed the relation of algae to medicine.

Another adverse activity of algae is their propensity for forming water blooms, alluded to on p. 32. Algal blooms, aside from their negative aesthetic effect, are harmful in several other ways. Some algae, when present in large concentrations impart unpleasant taste to drinking water, or their products (during life or upon their decay) may liberate substances deleterious to aquatic animals (e.g., Harris and James, 1974; Loeblich and Loeblich, 1975). Furthermore, when bodies of water become covered with thick mats of filamentous algae, the latter form a barrier between the water and the atmosphere so that, especially at night, an anaerobic condition develops in the water in which the animal life suffers from anoxia.

Another phenomenon that deserves mention at this point, and one that should be investigated further in nature, is autoinhibition and alloinhibition (allelopathy) and allostimulation demonstrated in the laboratory long ago by Pratt and Fong (1940) in *Chlorella* and more recently by Parker and Bold (1961), Fitzgerald (1969), Kroes (1971), D. Harris (1970, 1971), Harris and Parekh (1973), and Harris and Caldwell (1974).

2

Division Cyanochloronta

Introduction

The blue-green algae, here designated by the divisional name Cyanochloronta, are sometimes called Cyanophyta or Cyanophycophyta, the latter emphasizing their algal affinities. By contrast, Stanier, et al. (1971) consider them to be bacteria bacause of their cellular organization and biochemistry. The authors, however, are impressed by the fact that the blue-green algae contain chlorophyll *a*, which differs from the chlorophyll of those bacteria which are photosynthetic, and also by the fact that free oxygen is liberated in blue-green algal photosynthesis but not in that of the bacteria. In light of these considerations, while acknowledging that they have close affinities with the bacteria, the blue-green algae have been retained by the authors of this book among the algae with the noncommittal designation Cyanochloronta (Bold, 1973). Two excellent and comprehensive books on blue-green algae appeared in 1973, one by Fogg, et al. and the other edited by Carr and Whitton; the latter contains chapters by 26 authors; a comprehensive review by Wolk (1973) of the physiology and cytological chemistry of blue-green algae is exceedingly useful. Among the morphological and taxonomic treatments of the group are those of Geitler (1932, 1960), Drouet and Daily (1956), Desikachary (1959), Bourrelly (1970), and Drouet (1968, 1974). The summary of a symposium on blue-green algae edited by Desikachary (1972) contains papers on special topics by more than 60 authors.

The blue-green algae have, through the years, been of great interest in biogeology. Nannofossils of blue-green algae described from layered formations known as stromatolites date back some 1.9 billion years. In this context the blue-green algae are often considered to have been the organisms responsible for the early accumulation of oxygen in the earth's atmosphere.

The origins of the blue-green algae are obscure. It is tempting to assign them to an evolutionary path parallel to that of the photosynthetic bacteria, but as the group that evolved the pathway for evolution of oxygen. No transition forms between these two types of photosynthesis have been found, however, and thus there are no likely candidates for the progenitorship of the blue-green algae.

Occurrence and Habitat

Blue-green algae are ubiquitous in waters of a great range of salinity and temperature, and they occur in and on the soil and also on rocks and in their fissures. Little (1973) reported the occurrence of *Gloeocapsa*, *Nostoc*, and *Lyngbya* in the supralittoral zone of marine shores. A number have been recovered from the atmosphere (R. Brown, 1971). In general, blue-green algae seem to be more abundant in neutral or slightly alkaline habitats, although some (e.g., *Chroococcus*, p. 45) are said to occur in bog waters at pH 4. Brock (1973) reported that blue-green algae were absent from waters whose pH was less than 4 or 5, while certain eukaryotic algae were present. Blue-green algae are both planktonic and benthic. Among the former, several are characteristically members of water blooms (Fig. 2.1), for example, *Microcystis aeruginosa* Kütz., *Anabaena flos-aquae* (Lyngb.) Breb., and *Trichodesmium erythraeum* Ehrenb., the last common in tropical waters, including the Red Sea (which probably was so named because of the color of the alga). At least two blue-green algae, *Microcystis aeruginosa* and *Anabaena flos-aquae*, are responsible for acute poisonings of various animals (Gorham, 1964; May and McBarron, 1973).

Blue-green algae, along with certain bacteria, occur in alkaline hot springs such as those of the Yellowstone and various other parts of the world where they live

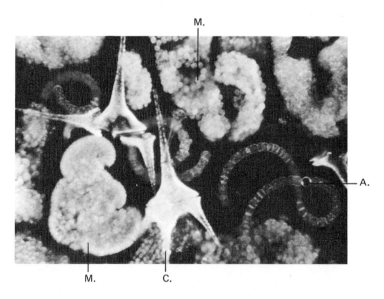

Fig. 2.1   Microscopic view of a drop from a water bloom from a Wisconsin lake. A., *Anabaena* sp.; C., *Ceratium* sp., a dinoflagellate; M., *Microcystis aeruginosa*, a blue-green alga. × 850. (Courtesy of Dr. G. P. Fitzgerald.)

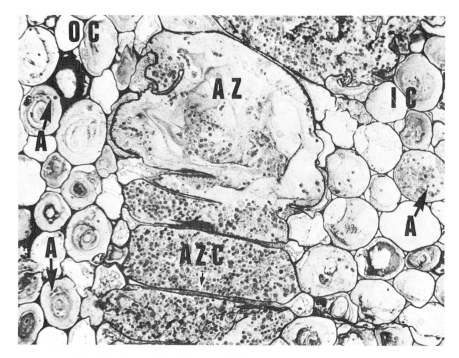

Fig. 2.2 Transection of a root of *Macrozamia communis* L. with endophytic blue-green algae (*Nostoc* or *Anabaena*). Algae (A) inside cells of the inner (IC) and outer (OC) cortex on either side of algal zone (AZ, AZC). × 112. (After Nathanielsz and Staff.)

at maximum temperatures of 73–74°C. Castenholz (1968, 1969) and Brock (1967) have made incisive investigations of the blue-green algae in these thermophilic environments.

Blue-green algae may form mats on the surface of bare soil as primary colonizers and are important in adding organic matter to the soil and in preventing incipient erosion. Some Cyanochloronta of soil have been shown to have remained viable for 107 years (Cameron, 1962) and they have been recovered from house dusts (Bernstein and Safferman, 1970).

A number of blue-green algae grow in association with other organisms. Thus, *Gloeocapsa* and *Nostoc*, among others, are the phycobionts of lichens, while others like *Nostoc* and/or *Anabaena* occur within the plant bodies of certain liverworts, water ferns (Peters, 1975; Duckett, et al., 1975), cycads (Grilli, 1974; Grilli-Caola, 1975; Nathanielsz and Staff, 1975A, B) (Fig. 2.2), and angiosperms (Silvester and McNamara, 1976) where they fix nitrogen. Certain types are associated with Protozoa, where they have been called "cyanelles."

In addition to poisoning animals (p. 56), blue-green algae may be deleterious to human beings. Thus, Moikeha, et al. (1971) and Moikeha and Chu (1971) have isolated a toxic factor from *Lyngbya majuscula* that causes dermatitis.

Nitrogen Fixation

Of great significance biologically is the fact that certain blue-green algae can fix elemental (gaseous) nitrogen. They thus are independent of other combined nitrogen sources (Fogg, 1974; Fogg, et al., 1973; Carr and Whitton, 1973.) Three kinds of blue-green algae have been shown to fix nitrogen: (1) the filamentous heterocystous[1] species, (2) certain unicellular (nonheterocystous) species (Wyatt and Silvey, 1969; Rippka, et al., 1971), and (3) certain nonheterocystous filamentous species, e.g., *Plectonema boryanum* (Stewart and Lex, 1970), albeit only under microaerophilic conditions. The nitrogen-fixing enzyme complex nitrogenase is oxygen-sensitive, so that the highest rate of nitrogen fixation occurs under reduced oxygen tensions. There is evidence that heterocysts reduce the elemental nitrogen and transfer it to the adjacent vegetative cells. Van Gorkon and Donze (1971) obtained evidence for this through the circumstance that phycocyanin is also a nitrogen reserve, so that it declines in nitrogen-starved cultures; evidence for this was presented also by de Vasconcelos and Fay (1974). It reappears first, under aerobic conditions, in the vegetative cells next to the heterocysts. Under anaerobic conditions, the phycocyanin developed evenly in all the vegetative cells. These experiments show again the sensitivity of nitrogenase to oxygen and indicate that less oxygen apparently is present in the heterocysts than in the vegetative cells. This strengthens the hypothesis that the heterocyst is the site of nitrogen fixation, and this has been confirmed by Fleming and Haselkorn (1973), Weare and Beneman (1973), and Bradley and Carr (1976). Postgate (1974), however, suggests that microaerophilic nitrogen fixation may go on also in the vegetative cells of some heterocystous species.

It has been hypothesized that the sheaths of *Gloeocapsa* (Fig. 2.12), a unicellular blue-green alga that fixes nitrogen, may somehow effect microaerophilic conditions in the cells that permit nitrogenase activity (Rippka, et al., 1971).

The nitrogen-fixing capacity of blue-green algae has been made use of in the cultivation of rice in which their growth is encouraged in the rice paddies. Fogg (1974) has summarized our knowledge of nitrogen fixation in blue-green algae.

Cellular Organization

Among living organisms, only the bacteria and blue-green algae are prokaryotic, which means that they lack membrane-bounded nuclei. This is considered by biologists to be of such fundamental significance that they often classify bacteria and blue-green algae in a category "Prokaryota" segregated from all other living things, the "Eukaryota." For many years (ca. 1900–1950), light microscopists debated the organization of blue-green algal cells, but electron microscopy in the last 2 decades has made possible our first real understanding of it. The following account draws on both light and electron-microscopic data. More comprehensive accounts of cellular organization are available in the publications of Lang (1968A), Carr and Whitton (1973), and Fogg, et al. (1973). Of great interest in this connection is the description by Rippka,

---

[1]See p. 42 for an explanation of heterocysts.

et al. (1974) of a prokaryotic organism, *Gloeobacter violaceus*, with chlorophyll *a*, two carotenoids, allophycocyanin, phycocyanin, and phycoerythrin but, unlike other blue-green algae, lacking thylakoids. Also of great evolutionary significance is the report by Lewin and Withers (1975) of the occurrence of a prokaryotic alga which lacks phycobilin segments but which contains chlorophyll *a* and *b*.

The Cell Wall and Other Investments

Considerable confusion exists regarding the nomenclature for investments exterior to the cell wall of blue-green algae. It has been suggested (Martin and Wyatt, 1974) that there may be three types of investment surrounding blue-green algal cell walls: (1) a *sheath* that is immediately adjacent to the cell wall and visible without staining; (2) a slimy, mucilaginous "shroud" that surrounds the organism (with or without a sheath); this has indefinite, not sharply defined, limits; (3) similar shrouds with well-defined limits. The chemistry of the outer investments of blue-green algae (Fig. 2.3) has not been adequately investigated. The available evidence indicates that they are composed of pectic acids and mucopolysaccharides (Dunn and Wolk, 1970). Upon analysis, the latter from various organisms have yielded glucose, hexuronic acids, D-xylose, ribose, galactose, rhamnose, and arabinose (Wang and Tischer, 1973.) The outer investments of blue-green algae, which are somewhat analogous to the capsules of bacteria, are sometimes visible without special treatment, but they become strikingly clear when the organisms are mounted in dilute India ink (Fig. 2.26c). Electron microscopy reveals that the outer investments are fibrillar, the fibrils being embedded in an amorphous matrix. Baker and Bold (1970) reported that the sheaths of some Oscillatoriacean algae vary in thickness and consistency as the environmental conditions change.

The cell wall of blue-green algae, which lies between the plasmalemma and mucilaginous sheath, is a complex, usually four-layered structure (Fig. 2.3). The second layer (centrifugally) is a mucopolymer which is dissolved by lysozyme which also digests the cell walls of Gram-negative bacteria. Upon fractionation, this mucopolymer has been shown to be composed of muramic acid, glucosamine, alanine, and glutamic and α-ϵ diaminopimelic acids. In the filamentous blue-green algae very delicate plasmodesmata or protoplasmic strands effect protoplasmic continuity across the transverse walls. Various pores and depressions in the longitudinal walls have been revealed by electron microscopy; these may be related to the gliding motility of some species (see p. 39).

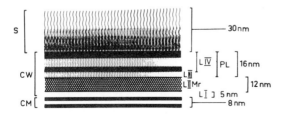

Fig. 2.3 Diagram of the organization of the cellular envelopes of blue-green algae, as seen in transection, based on electron microscopy. Cm, plasmalemma or plasma membrane; CW, cell wall; S, sheath or slime layer; Mr, murein; PL, plastic layers; LI–IV, cell wall layers. (After Carr and Whitton, modified from Jost.)

The Protoplast

The photosynthetic lamellae or thylakoids of blue-green algae (Figs. 2.4, 2.5), unlike those of other chlorophyllose plants, are not enclosed in membrane-bounded groups to form chloroplasts. Instead, they lie free in the cytoplasm, in some species more or less restricted to the periphery, but, depending on the intensity of the incident light and age of the cell, they may be distributed throughout the protoplast. The thylakoids are the site of chlorophyll *a*, and the accessory pigments also occur on their surfaces in the form of small particles, the *phycobilisomes*. The accessory pigments are *c-phycocyanin*, *c-allophycocyanin*, and *c-phycoerythrin*,[2] the two former blue and the

[2]*c* for Cyanochlorontan, to distinguish from *r*, rhodophycophytan.

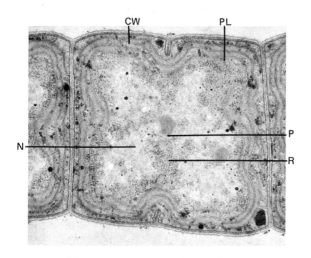

(a)

Fig. 2.4 Ultrastructural organization of a filamentous blue-green alga, *Plectonema boryanum* Gom. (*a*) Early stage in cell division. (*b*) Later stage in cell division. CW, cell wall; N, nucleoplasm; P, polyhedral body; PL, photosynthetic lamellae or thylakoids; R, ribosomes. × 3100. (Courtesy of Professor R. M. Brown, Jr.)

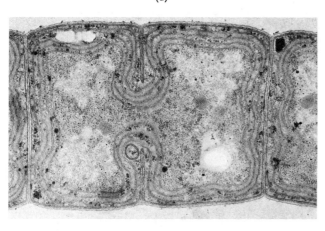

(b)

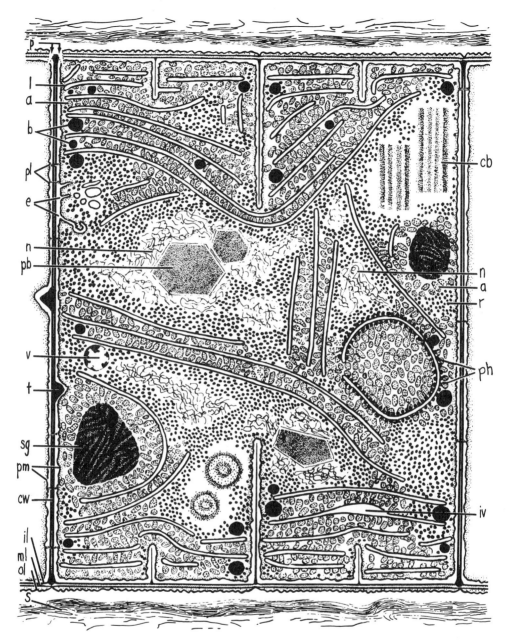

Fig. 2.5 Diagrammatic summary of the organization of the cells of blue-green algae [modified slightly from Pankrantz and Bowen, their terminology brought into conformity with current usage by Lang (personal communication)]. Key to lettering, starting at p (upper left) and proceeding in a counterclockwise direction: p, junctional pores; l, thylakoid; a, polyglycoside (or glycogen) granule; b, lipid body; pl, microplasmodesmata; e, elaborations of plasma membrane; n, nucleoplasm (site of histoneless DNA); pb, carboxysome (probable site of ribulose diphosphate carboxylase); v, polyphosphate body; t, local thickening; sg, cyanophycin granule (composed of arginine and aspartic acid 1:1); pm, plasmalemma; cw, cross wall (composed of LI and LII layers); il, ml, ol: cell wall of four layers (il/ml = peptidoglycan LII; ol = layer IV); s, sheath; iv, intrathylakoidal space; ph, phycobilisomes; r, ribosome; cb, cylindrical body.

latter red. They are biliproteins and are composed of a protein with chromophoric groups (Goodwin, 1974). A variety of carotenoid pigments also is present (Table 1.1, p. 18). The phycobilisomes (Gray, et al., 1973; Wildman and Bowen, 1974; Gantt, 1975) are multiprotein complexes composed of phycocyanin and sometimes of phycoerythrin and allophycocyanin. These accessory pigments apparently transfer light energy they absorb to chlorophyll *a*. It has been suggested that the accessory pigments also have a protective shading function that prevents oxidation in intense light of the other photosynthetic pigments.

The DNA of blue-green algal cells lacks a histone coating and is present as fine fibrils (Figs. 2.4, 2.5) that may appear as a reticulum (Pankratz and Bowen, 1963). In addition to the thylakoids and DNA, Cyanochlorontan protoplasts contain a variety of other bodies including 70S ribosomes, gas vacuoles or vesicles, polyglucan granules, cyanophycin granules, polyhedral bodies, and possibly lipid droplets (Figs. 2.4, 2.5) among others. Polyglucan granules are polymers of glucose similar to animal glycogen, while cyanophycin granules are copolymers of aspartic acid and alanine (Simon, 1971). Polyphosphate bodies, as the name implies, are aggregates of linear polyphosphates; the polyhedral bodies may be carboxydismutase. Andreis (1975) has surveyed the various types of vacuoles in blue-green algae, reporting that some are enlarged thylakoids. Figure 2.5 from the publication of Pankratz and Bowen (1963) summarizes cellular organization in blue-green algae.

Gas vacuoles (Fig. 2.6) are visible in many planktonic blue-green algae as reddish granules under intermediate magnifications of the light microscope. They disappear

Fig. 2.6 *Anabaena flos-aquae.* Micrograph of freeze-etched gas vacuoles. × 12,500. (After Walsby. Micrograph by D. Branton.)

when the cells are subjected to increasing pressure and persist when they are placed in a vacuum. It has been suggested that gas vacuoles, which are seen to be composed of gas vesicles at the electron-microscopic level, aid in cellular buoyancy and in shading the photosynthetic pigments, especially in floating blue-green species; the latter suggestion has been criticized. However, Porter and Just (1976) have reported evidence for the light-shading role of gas vacuoles.

Motility

Many filamentous blue-green algae are not enclosed in firm sheaths; the hormogonia[3] of those that are, and some unicellular species, undergo movement when in contact with the substrate, e.g., agar, a glass slide, or each other. This movement, accomplished without evident organs of locomotion, is called **gliding movement** and occurs also in some filamentous bacteria. The mechanism of gliding movement is not completely understood. It has been suggested that it is a sort of propulsion caused by the secretion of slime and also by contractile waves on the surface of the cell (Halfen and Castenholz, 1971A, B; Halfen, 1973; Fogg, et al., 1973, Chapter 6). In this connection, the minute pores in the walls revealed by electron microscopy have been suggested as the pathways for mucilage secretion. On the other hand, Halfen and Castenholz found fibrils in the wall that they thought to be involved in motility and the sites of waves of propulsion of the trichomes (chains of cells). It is thought that the oscillation of certain trichomes like those of species of *Oscillatoria* (p. 51) is related to these waves of propulsion of the superficial fibrils.

Form

The Cyanochloronta contain unicellular, colonial, and filamentous species. While few of the first are strictly unicellular, a lag in the separation of the progeny of cell division in many results in a temporary colonial organization (Figs. 2.11, 2.12). Groups of cells embedded in polysaccharide matrices comprise colonies. These may be flat or slightly curved plates (Fig. 2.14), spheres (Fig. 2.16), or irregular aggregates. The cells of the colony are similar and undifferentiated from each other, and the colonies are noncoenobic. Cell division restricted to one direction results in an unbranched trichome (Fig. 2.22). Branching trichomes arise by division of certain cells in a different plane (Fig. 2.30) or by false branching (Fig. 2.29). A filament is composed of a chain of cells, the **trichome**, and the enveloping sheath, if one is present. Evans, et al. (1976) have demonstrated clearly that filamentous organization may be lost or modified as the environment changes.

Nutrition

An in-depth discussion of the nutritional physiology of Cyanochloronta is outside the scope of this text. Those with special interests in this subject will find data in the reviews of Holm-Hansen (1968) and Wolk (1973) and in the books by Carr and

[3]Hormogonia: see p. 40.

Whitton (1973), Fogg, et al. (1973), and Stewart (1974). It was at first thought that blue-green algae were obligately photoautotrophic, but heterotrophic growth on glucose, fructose, or sucrose, in darkness, has been documented for several organisms. In addition, in dim light, several species are photoheterotrophic. Several blue-green algae, as is true of other algal groups, require vitamin B_{12} for growth. Certain unicellular species have been shown to have a generation time or cycle of 3–8 hours under optimal conditions.

Reproduction

In the unicellular blue-green algae, reproduction is effected by cell division (Figs. 2.11, 2.13) and subsequent separation of the cellular progeny. The latter may lag so that incipient colonies (*Chroococcus*, Fig. 2.11) or filaments (*Anacystis nidulans*, Fig. 2.13) may develop. In cell division in most blue-green algae, the cell becomes constricted in the median plane and the two inner wall layers grow centripetally until a septum is formed. The invagination of the wall layers is preceded by invagination of the plasmalemma and thylakoids.

Colonial and filamentous Cyanochloronta reproduce by fragmentation in which segments of the organism become separated from the parent, glide or float away, and grow into new individuals. Fragmented sections of trichomes, called **hormogonia** (Fig. 2.7) are motile. They arise by separation of adjacent terminal walls in the trichome or by the death of certain cells that may become biconcave **separation discs** or **necridia.** Lamont (1969) claims that hormogonium formation in *Microcoleus* involves transcellular breakage.

Several kinds of unicellular reproductive agents are produced by many of the filamentous Cyanochloronta. These include akinetes, endospores, exospores, and heterocysts.

An **akinete** develops from a vegetative cell that becomes enlarged and filled with food reserves (cyanophycin granules) and augments its wall externally (Figs. 2.8, 2.9) by an additional complex investment. The latter may be ornamented and colored. After a period of dormancy, the akinete may germinate (Fig. 2.9*b*, *c*) giving rise to a vegetative trichome. Miller and Lang (1968) and Clark and Jensen (1969) studied the ultrastructure of akinete development in *Cylindrospermum*. Ueda and Sawada (1972) reported that akinetes of *Cylindrospermum* at maturity have 30 times as much DNA

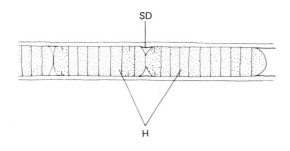

SD

H

Fig. 2.7 Hormogonium formation (diagrammatic). H, hormogonium; SD, separation disc.

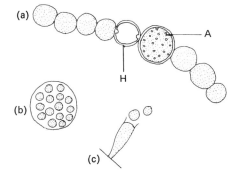

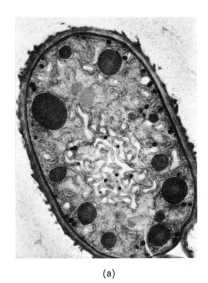

Fig. 2.8 (a) Trichome of vegetative cells with one akinete (A) and one heterocyst (H) (diagrammatic). (b) Endospore formation. (c) Exospore formation.

(a)

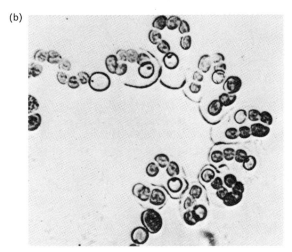

(b)

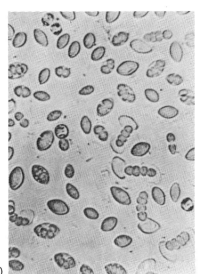

(c)

Fig. 2.9 (a) *Anabaenopsis circularis* (West) Wolozynska and Miller. Section of an akinete. Note prominent storage granules and multilayered wall. (b) *Nostoc linckia*. Germinating akinetes. (c) *Anabaena doliolum* Bharadwaja. Germinating akinetes. (a) × 6013; (b) × 387; (c) × 300. [(a) after Lang, courtesy of The Royal Society; (b), (c) courtesy of Professor R. N. Singh.]

as the vegetative cells. Tyagi (1974) reported that in *Anabaena doliolum* akinete formation begins midway between two heterocysts and proceeds toward the heterocysts. They
found that potassium nitrate and ammonium chloride inhibit akinete formation and
that glucose promotes it.

Heterocysts (Figs. 2.8*a*, 2.10) are cells with homogeneous-appearing contents (as
viewed with the light microscope) and transparent walls. They arise by external augmentation of the walls of vegetative cells by three additional layers (Dunn, et al., 1971).
The role of heterocysts in nitrogen fixation and in relation to akinete formation has
already been mentioned (p. 34). Heterocysts increase in number when nitrogen in the
environment has been depleted (de Vasconalos and Fay, 1974). Heterocysts may be
terminal or intercalary in the trichome and may be evenly distributed among the
vegetative cells. Where they are connected to the latter, the wall is modified in the form
of a **polar nodule** through which there is a channel; the septum across the end of the
pore channel (Fig. 2.10*a*) is transversed by microplasmodesmata that seemingly provide continuity between the heterocyst and vegetative cell. Heterocysts contain much
smaller amounts, perhaps none, of the biliprotein pigment, and their thylakoids are
arranged in a reticulate pattern, as compared with vegetative cells. Lang (1965) and
Lang and Fay (1971) have described electron microscopically the development of the
heterocyst in *Anabaena azollae* Strasburger.

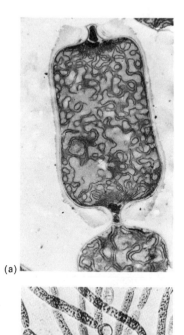

(a)

(b)

Fig. 2.10 (*a*) *Anabaena cylindrica* Lemm. Section of a heterocyst: Note contorted thylakoids,
polar nodules, and thickened wall. (*b*) *Gloeotrichia ghosei* Singh. Germination of heterocysts
in situ. (*a*) × 5559; (*b*) × 302. [(*a*) after Lang;
(*b*) after Singh and Tivari.]

In addition to their probable role in nitrogen fixation, heterocysts have also been reported (Wolk, 1967) to evoke the formation of akinetes adjacent to them. They can also germinate to form trichomes (Wolk, 1965; Singh and Tiwari, 1971) in some cases, as they do abundantly in *Gloeotrichia ghosei* (Fig. 2.10*b*). Tyagi (1975) has summarized much that is known about heterocysts.

Endospores (Fig. 2.8*b*) are formed by endogenous divisions of the protoplast into two or more portions that emerge when the parental wall ruptures. **Exospores** (Fig. 2.8*c*) are cut off *seriatim*, much like fungal conidia, from the cellular apices of certain genera such as *Chamaesiphon* (Fig. 2.21).

In addition to these methods of asexual reproduction, there have been reports of genetic recombination in the Cyanochloronta, although plasmogamy has not been discovered. The evidence has been deduced mainly from investigations of mutants of *Anacystis nidulans*[4] resistant to one or another antibiotic or with a growth requirement. For example, Bazin (1968), working with two strains, one streptomycin-resistant and the other polymixin-B-resistant, grew them together briefly and was then able to obtain colonies on agar containing both streptomycin and polymixin B. More recently, Orkwizewske and Kaney (1974) demonstrated genetic transformation in *A. nidulans*. In one experiment, when DNA extracted from cells of a streptomycin-resistant mutant was added to cultures of the wild type (streptomycin-sensitive), colonies of the latter grew in media containing streptomycin. In another experiment, DNA from the wild type added to phenyl alanine- and ornithine-deficient mutants produced by transformations some individuals in the population capable of growing in the absence from the culture medium of both phenyl alanine and ornithine. It is possible also that blue-green algal viruses are evoking new genetic combinations in blue-green algae in nature.

Brief mention must be made, finally, of viruses that attack blue-green algae; these have been designated phycoviruses or blue-green-algal viruses (BGA viruses). Such a virus was first discovered by Safferman and Morris (1963) as a destructive agent for certain strains of *Lyngbya*, *Plectonema*, and *Phormidium* and was, therefore, called "LPP-1," a reference to the host algae. The virus lyses and destroys the host algal cells. K. Smith, et al. (1966A, B; 1967) have studied the infection cycle and replication of the virus electron microscopically. There is some evidence from field and laboratory tests that phycoviruses like LPP-1 may be effective in controlling unwanted blooms of blue-green algae in nature (Safferman, 1973). The subject of algal viruses has been reviewed by R. Brown (1972).

Classification

The unicellular, colonial, and filamentous blue-green algae have been variously classified by different authors. Fritsch (1942), Desikachary (1959), and Bourrelly (1970) classified them in five orders, while Fott (1971) cites four orders. G. Smith

---

[4]Komarek (1970) is of the opinion that this organism should be designated *Synechococcus leopoliensis* (Racib.) Komarek; Stanier, et al. (1971) also have reviewed problems regarding the name of the organism.

(1950) recognized only three orders, a practice followed in the present volume. These orders are the Chroococcales, the Chamaesiphonales, and the Oscillatoriales. They may be distinguished by the following key:

1. Producing endospores or exospores ....Chamaesiphonales
1. Lacking endospores and exospores 2
   2. Unicellular or colonial ....Chroococcales
   2. Filamentous, the cells of the trichomes contiguous ....Oscillatoriales

Order 1. Chroococcales

In the order Chroococcales are classified unicellular and colonial (noncoenobic) blue-green algae that do not produce endospores or exospores. These organisms are often called "the coccoid blue-green algae." Stanier, et al. (1971) observed phototactic gliding movement of a number of coccoid Cyanochloronta in culture both macroscopically and microscopically.

In the unicellular members of the order, cell division or binary fission results in reproduction. In the colonial members, cell division effects increase in size of the noncoenobic colonies that reproduce by fragmentation. The genera discussed below are members of the family Chroococcaceae and may be distinguished with the aid of the following key:

1. Unicellular, or colonies few-celled, readily dissociating
   into the unicellular state 2
1. Colonial..4
   2. Cells with broad sheaths usually thicker than the cell itself,
      sometimes colored ....*Gloeocapsa*
   2. Cells with relatively thin, colorless sheaths 3
   3. Cells spherical except during or just after division ....*Chroococcus*
   3. Cells short-cylindrical ....*Anacystis*
      4. Colonies spherical or irregular 5
      4. Colonies flat or curved plates, or cubes 6
   5. Colonies small, spheroidal; cells homogeneous in content .....*Coelosphaerium*
   5. Colonies spherical to irregular and clathrate, the cells with
      dark or reddish granules ....*Microcystis*
      6. Colonies flat or curved plates ....*Merismopedia*
      6. Colonies sarcinoid or cubical ....*Eucapsis*

Drouet and Daily (1956) revised the classification of the coccoid Cyanochloronta reducing the number of genera to six on the basis of their extensive studies of herbarium specimens of types. The numerous genera that were described previously in many instances did not have clearly defined characteristics. Stanier, et al. (1971), by contrast, are unwilling to recognize the classification of Drouet and Daily. They argue that, as is the case with bacteria, the organisms must be grown in clonal axenic culture and appraised for their physiological and biochemical characteristics including the base composition of their DNA, if their classification is to be reliable. Kenyon (1972) has grouped 34 species of coccoid blue-green algae on the basis of their fatty acid composition. Stanier, et al. urge that one follow the system of classification of Geitler (1932) for the present. The generic and specific limits in the coccoid blue-green algae as presently conceived are not always clearcut and reliable.

CHROOCOCCUS Nägeli   *Chroococcus* (Gr. *chroos*, color + Gr. *kokkos*, berry) (Fig. 2.11) occurs epipelically and in the plankton. According to Drouet and Daily (1956) it should be included in the genus *Anacystis*. The organism is unicellular or grouped in aggregates of two or four because of the failure of the cellular division products to separate promptly. Stanier, et al. (1971) have pointed out that in *Chroococcus* the products of cell division are hemispheres, while in *Gloeocapsa* they are spheres or they have rounded poles. In *C. turgidus* (Kütz.) Näg. one can usually distinguish a colorless central region from the peripheral pigmented portion of the protoplast. Except in planktonic species, the sheaths are thin, delicate, and colorless.

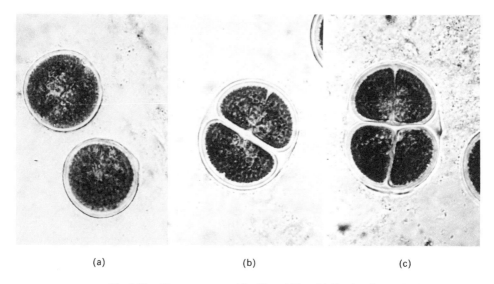

(a)   (b)   (c)

Fig. 2.11   *Chroococcus turgidus* (Kütz.) Näg. (*a*) Single cells, one beginning binary fission. (*b*), (*c*). Further division stages. × 960.

GLOEOCAPSA Kützing   *Gloeocapsa* (Gr. *gloia*, glue + L. *capsa*, a box or case) (Fig. 2.12) also has been classified in the genus *Anacystis* by Drouet and Daily (1956). The cells are ovoid-ellipsoidal and surrounded by copious sheaths (Fig. 2.8) within which several generations of cells may be included. The sheaths sometimes are pigmented. A number of species of *Gloeocapsa* occur on moist rocks, while others are aquatic. The cell poles are rounded after division. Jensen and Sicko (1972) have investigated the fine structure of the cell wall of *G. alpicola*.

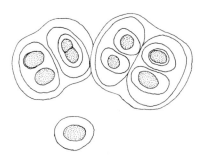

Fig. 2.12 *Gloeocapsa* sp. Note thick sheaths. × 900.

ANACYSTIS Meneghini   *Anacystis* (Gr. *ana*, + Gr. *kystis*, bladder, hence cell) (Fig. 2.13), especially *A. nidulans* (Lyngb.) Drouet and Daily, is a minute blue-green alga widely used in physiological and genetical investigations. The individual cells are ovoid-cylindrical to bacilliform and undergo transverse division. Each is surrounded by a delicate sheath. The cells may occur embedded in a common matrix. Kunisawa and Cohen-Bazire (1970) produced filamentous mutants of *A. nidulans*.

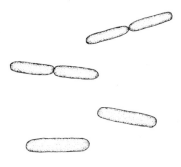

Fig. 2.13 *Anacystis nidulans* (Richt.) Drouet and Daily. × 2250.

MERISMOPEDIA Meyen   In *Merismopedia* (Gr. *merismos*, division + Gr. *pedion*, plain) (Fig. 2.14) the cells are arranged within a matrix in a single flat or curved sheet. This is maintained and grows by cell division in two directions. *Merismopedia* species may be planktonic or epipelic and occur in quiet waters. Reproduction of the colonies is by fragmentation.

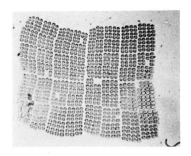

Fig. 2.14 *Merismopedia* sp. × 119.

EUCAPSIS Clements and Shantz   Cell division in three perpendicular planes results in sarcinoid, cubical organization illustrated by *Eucapsis* (Gr. *eu*, perfect + L. *capsa*, a box) (Fig. 2.15). *Eucapsis* grows readily in laboratory cultures and has been isolated from soil. Reproduction is accomplished by fragmentation.

Fig. 2.15   *Eucapsis* sp. × 263.

COELOSPHAERIUM Nägeli   The cells in the spherical or irregular colonies of *Coelosphaerium* (Gr. *koilos*, + Gr. *sphaira*, sphere) (Fig. 2.16) are arranged at the periphery of the colonial matrix. Cells of *Coelosphaerium* are bright blue-green or they may be dark and filled with gas vesicles. *Coelosphaerium* is often present in the plankton. The colonies reproduce by fragmentation.

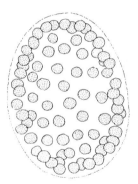

Fig. 2.16   *Coelosphaerium kützingianum* Näg. × 600.

MICROCYSTIS Kützing   The colonies of *Microcystis* (Gr. *micros*, small + Gr. *kystis*, bladder, hence cell) (Fig. 2.17) are spherical or irregular and sometimes clathrate. In contrast to those of *Coelosphaerium*, the cells of *Microcystis* are evenly distributed throughout the colonial matrix. They often have a blackish or reddish appearance because of the numerous gas vesicles they contain. *Microcystis* is strictly planktonic. There is evidence (Vance, 1965) that *Microcystis*, a common cause of water blooms, secretes substances inhibitory to other algae. In addition, *M. aeruginosa* Kützing produces a toxin called the "fast death factor" (Hughes, et al., 1958) that is toxic to animals ingesting the alga. *Microcystis aeruginosa* is included in *Anacystis* by Drouet and Daily (1956).

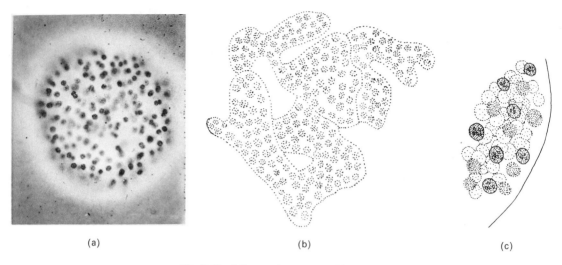

(a)   (b)   (c)

Fig. 2.17 *Microcystis aeruginosa* Kütz. (*a*) Single living colony mounted in India ink. (*b*) Another type of colony. (*c*) Portion of the same, more highly magnified. (*a*) × 34; (*b*) × 135; (*c*) × 750.

Order 2. Chamaesiphonales

The members of the order Chamaesiphonales are unicellular, crustose, or filamentous. Many are marine. Depending on the genus, the cells may form endospores or exospores. Representatives of three families, the Pleurocapsaceae, Dermocarpaceae, and Chamaesiphonaceae are discussed in the following account. The families may be distinguished by means of the following key:

1. Cells undergoing vegetative cell division or desmoschisis . . . . . Pleurocapsaceae
1. Cells not undergoing desmoschisis 2
   2. Reproduction by endospores . . . . Dermocarpaceae
   2. Reproduction by exospores . . . . Chamaesiphonaceae

Family 1. Pleurocapsaceae

The members of this family are epiphytic on other algae and aquatic angiosperms, lithophilic or boring within calcium carbonate shells.

XENOCOCCUS Thuret   The spherical cells of *Xenococcus* (Gr. *Xenos*, + Gr. *kokkos*, berry) (Fig. 2.18) grow epiphytically on filamentous algae. They undergo anticlinal divisions, anticlinal with reference to the host, thus increasing the size of the colonies. Each cell can produce many endospores.

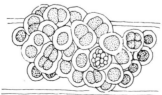

Fig. 2.18 *Xenococcus* sp. × 650.

HYELLA Bornet and Flahault   The branching trichomes of *Hyella* (after *F. Hy*) (Fig. 2.19), which grow by desmoschisis, live within calcareous shells or within other algae. The basal filaments may become pluriseriate. Any of the cells may divide to form endospores.

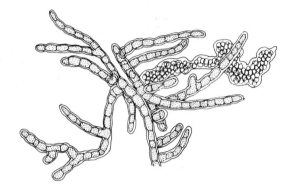

Fig. 2.19 *Hyella caespitosa* Born. et Flah. × 218.

Family 2. Dermocarpaceae

Desmoschisis does not occur in the endospore-forming family Dermocarpaceae.

DERMOCARPA Crouan   The cells of *Dermocarpa* (Gr. *derma*, skin + Gr. *karpos*, fruit) (Fig. 2.20) are spherical to slightly saccate or pyriform and grow attached to the substrate in groups. Reproduction is accomplished solely by endospores which may develop in large numbers within the vegetative cells.

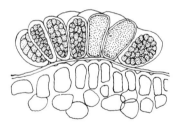

Fig. 2.20 *Dermocarpa* sp. Note endospores in some of the cells. × 263.

Family 3. Chamaesiphonaceae

CHAMAESIPHON Braun and Grunow   *Chamaesiphon* (Gr. *chamai*, on the ground, hence sessile + Gr. *siphon*, a tube) (Fig. 2.21), the only genus in the family, is a widely distributed epiphyte. It occurs on aquatic angiosperms, mosses, and algae, especially on *Cladophora* and *Oedogonium*, and also on nonliving substrates. The cells are attached at their bases. As they mature, the protoplast at the distal pole abstricts a chain of spores called **exospores**, so called because they are soon exposed to the environment and shed. Kann (1972) has summarized the taxonomy of the genus.

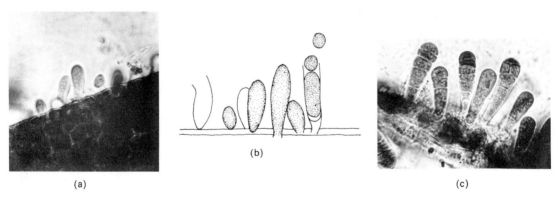

Fig. 2.21 *Chamaesiphon* sp. (*a*) Young cells on the margin of the leaves of an aquatic moss, *Fontinalis*. (*b*), (*c*) Exospore formation. (*a*) × 864; (*b*), (*c*) × 1200.

Order 3. Oscillatoriales

In the Oscillatoriales are classified all the filamentous Cyanochloronta that do not produce endospores or exospores. Reproduction is by hormogonium formation, and in some cases akinetes are formed. Heterocysts also may germinate to form new trichomes. These algae may be grouped in two series: those which produce heterocysts and those which lack them; the two groups are sometimes classified in separate suborders (G. Smith, 1950). The five families belonging to the Oscillatoriales may be distinguished by means of the following key:

```
1. Heterocysts absent ....................................Oscillatoriaceae
1. Heterocysts present ................................................2
    2. Trichomes tapered ...................................Rivulariaceae[5]
    2. Trichomes not tapered ........................................3
3. Trichomes unbranched ...................................Nostocaceae
3. Trichomes branched ..............................................4
    4. Branches originating by rupture of the sheath and
       trichome emergence ............................Scytonemataceae[5]
    4. Branches originating by division of certain cells
       in a new direction .................................Stigonemataceae
```

Family 1. Oscillatoriaceae

Considered to be the simplest of the filamentous Cyanochloronta, the Oscillatoriaceae are undifferentiated except for the terminal cells that may differ in form from the other cells of the trichome. Drouet (1968) has monographed the family on the basis of field and herbarium studies reducing the number of genera to 6 (*Spirulina, Arthro-*

[5]In his revision, Drouet (1973) has included the Scytonemataceae and Rivulariaceae in the Nostocaceae.

spira, Oscillatoria, Schizothrix, Microcoleus, and *Porphyrosiphon*). Bourrelly (1970), however, recognizes 11 genera. The following representative genera of the family are discussed in the account below: *Oscillatoria, Lyngbya, Arthrospira, Spirulina*, and *Microcoleus*. These may be distinguished with the aid of the key that follows:

1. Trichomes helically twisted 2
1. Trichomes straight or slightly curved 3
  2. Transverse walls seemingly lacking or obscure
    with light microscopy ....*Spirulina*
  2. Transverse walls clearly visible with light microscopy ....*Arthrospira*
3. With more than one trichome within a sheath ....*Microcoleus*
3. One trichome within a sheath, if the latter is present 4
  4. Individual sheaths of the trichomes absent or obscure ....*Oscillatoria*
  4. Trichomes with a firm, obvious, hyaline sheath ....*Lyngbya*

OSCILLATORIA Vaucher and LYNGBYA Agardh   The trichomes of *Oscillatoria* (L. *oscillare*, to swing) (Fig. 2.22) are cylindrical and unbranched; they lack a sheath or have only a very delicate one. The trichomes often occur in floating masses or form a shiny mass on moist soil. The individual cells are often shorter than broad, except for the apical cell, which may be capped and attenuated.

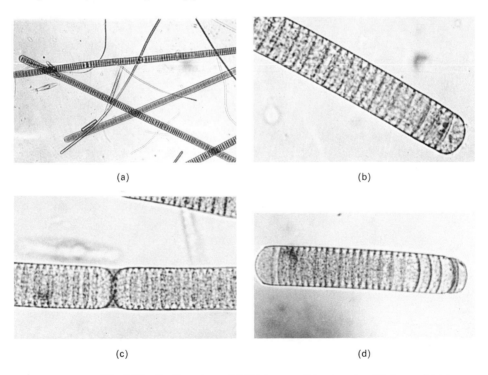

(a)　　　　　　　　　　　　　　(b)

(c)　　　　　　　　　　　　　　(d)

Fig. 2.22 *Oscillatoria* sp. (*a*) Trichomes of two species. (*b*) Apex of trichome, more highly magnified. (*c*), (*d*) Hormogonium formation. In (*c*), note dead cell, the site of fragmentation of the trichome; (*d*) shows a single hormogonium. (*a*) × 77; (*b*)–(*d*) × 450.

The trichomes of *Oscillatoria* exhibit gliding growth, rotation, and (under certain circumstances) an oscillatory motion. Reproduction is accomplished by the formation of hormogonia.

Lyngbya (after *H. C. Lyngbye*, a Danish phycologist) (Fig. 2.23) differs from *Oscillatoria* only in having a distinct sheath. Drouet (1968) regards the sheath as a characteristic affected by the environment, and he included *Lyngbya* in the genus *Oscillatoria*. Baker and Bold (1970), however, found that some species of Oscillatoriacean algae produced well-defined sheaths under all conditions of culture they used and so were inclined to retain *Lyngbya* as a valid genus. *Lyngbya* occurs in both freshwater and marine water and on soil.

(a)

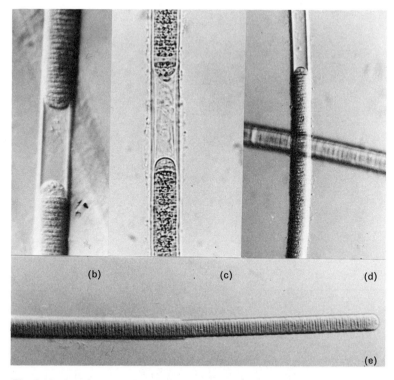

(b)   (c)   (d)

(e)

Fig. 2.23 *Lyngbya* sp. (*a*) Filament showing hormogonium formation. (*b*)–(*e*) Filaments showing sheaths and trichomes. (*a*) × 300; (*b*), (*c*) × 363; (*d*) × 198. [(*b*)–(*e*) after Martin and Wyatt.]

SPIRULINA Turpin and ARTHROSPIRA Stizenberger   *Spirulina* (L. *spirula*, a small coil) (Fig. 2.24*a*) and *Arthrospira* (Gr. *arthron*, joint + Gr. *speira*, a coil) (Fig. 2.24*b*) are helical members of the Oscillatoriaceae. The two genera have been separated in the past because *Spirulina* was thought to lack transverse walls, but staining and electron microscopy (Holmgren, et al., 1971) have revealed that delicate transverse walls are present in *Spirulina*. The latter occurs in both freshwater and marine water and its trichomes are actively motile, the trichomes being rotary and also bending.

Léonard and Compère (1967) reported that algal cakes, sold at the market of Fort-Lamy, Chad Republic, Africa, are composed largely of *Spirulina platensis* (Nordst.) Geitl., which is abundant in small "natron" (Na_2CO_3) lakes north of Lake

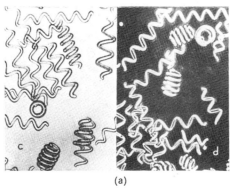

(a)

Fig. 2.24 (*a*) *Spirulina* sp. (*b*), (*c*) *Arthrospira* sp. (*a*) × 70; (*b*) × 80; (*c*) × 800. [(*a*) after Léonard and Compère.]

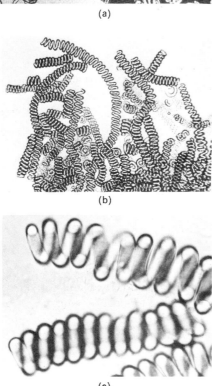

(b)

(c)

Chad and in east Africa. Chemical analysis showed the cakes contain 45–49% protein, dry weight, and it was suggested that this *Spirulina* was a promising food source. Melack and Kilham (1974) found the rates of photosynthesis are exceptionally high in these lakes and that *S. platensis* is frequent in them.

MICROCOLEUS Desmazieres   In *Microcoleus* (Gr. *micros*, small + Gr. *koleon*, sheath) (Fig. 2.25) a bundle of trichomes, the latter sometimes twisted about each other, occurs within the same sheath. Individual trichomes may protrude from the apex of the sheath. The outer walls of the apical cells are often thickened. Baker and Bold (1970) distinguished a number of varieties of three species of *Microcoleus* in culture on the basis of the configuration of the plant mass and its pattern of growth. The various species of *Microcoleus* are both freshwater and marine and some grow on damp sand.

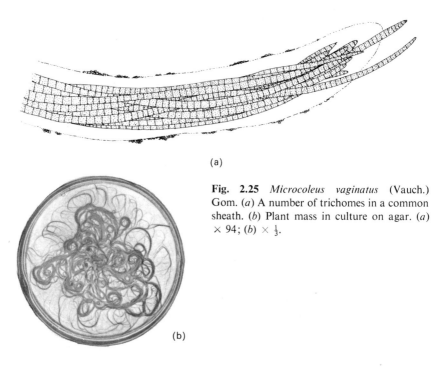

(a)

(b)

Fig. 2.25 *Microcoleus vaginatus* (Vauch.) Gom. (*a*) A number of trichomes in a common sheath. (*b*) Plant mass in culture on agar. (*a*) × 94; (*b*) × ⅓.

Family 2. Nostocaceae

The trichomes of the Nostocaceae are unbranched and form heterocysts and akinetes at maturity. The heterocysts may be terminal or intercalary. Three genera, *Nostoc*, *Anabaena*, and *Cylindrospermum*, are discussed below as representative of the family. They may be distinguished from each other with the aid of the following key:

1. Heterocysts always terminal ....*Cylindrospermum*
1. Heterocysts intercalary, but some may be terminal.......................2
    2. Filaments aggregated at maturity in a plant mass of recognizable form, macroscopically or microscopically ....*Nostoc*
    2. Filaments isolated or in amorphous strata ....*Anabaena*

NOSTOC Vaucher   *Nostoc* (name used by Paracelsus) (Fig. 2.26) is more common as a terrestrial and subaerial alga than as an aquatic one. It is widely distributed

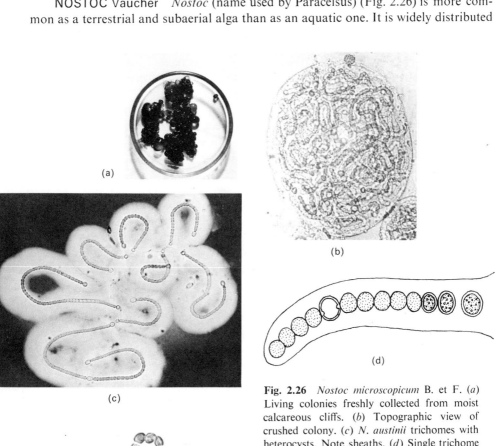

(a)

(b)

(c)

(d)

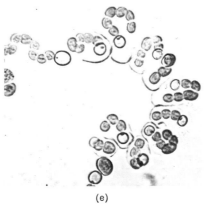

(e)

Fig. 2.26 *Nostoc microscopicum* B. et F. (*a*) Living colonies freshly collected from moist calcareous cliffs. (*b*) Topographic view of crushed colony. (*c*) *N. austinii* trichomes with heterocysts. Note sheaths. (*d*) Single trichome with heterocysts and akinetes. (*e*) *N. linckia* (Roth) Bornet. Germination of akinetes. (*a*) × 0.2; (*b*) × 76; (*c*) × 141; (*d*) × 240; (*e*) × 252. [(*c*) after Martin and Wyatt; (*d*) courtesy of Dr. R. N. Singh.]

on and in alkaline soils and on moist rocks and cliffs. The gelatinous aggregations of filaments have suggested such names as "star jelly," "witches' butter," and "mares' eggs" (often baseball-size in wet meadows) for such species as *N. commune* Vaucher. The trichomes are surrounded with individual sheaths and at maturity occur within a common matrix. The bead-like cells undergo generalized cell division that increases the length of the contorted trichomes, the sheaths of which may be colored yellow or brownish. Lazaroff and Vishniac (1964) and Kantz and Bold (1969) have investigated the morphogenesis of certain strains of *Nostoc* in culture. The latter authors distinguished two types of development in laboratory culture, the *N. piscinale* type and the *N. commune* type. In the latter, development of the motile trichome to form a mature aggregate of filaments takes place entirely within a matrix with a firm surface pellicle; the aggregate thus has a fixed shape that is absent in the *N. piscinale* type. The mature trichomes of most species of *Nostoc* produce akinetes that often occur in chains. Lazaroff (1973) has summarized his own and other investigations of the life cycle of *Nostoc* as related to nutrition and light. Ginsburg and Lazaroff (1973) have described the ultrastructural development of a strain of *Nostoc*.

ANABAENA Bory   *Anabaena* (Gr. *anabaino*, I will go up) (Fig. 2.27) is not easily distinguishable from the *N. piscinale* type of *Nostoc*. Most species of *Anabaena* are aquatic, and a number are planktonic, frequently components of water blooms along with *Microcystis aeruginosa*. *Anabaena* is sometimes said to be distinguishable from *Nostoc* by the criterion that colonies of *Nostoc* have macroscopically recognizable form, while aggregates of *Anabaena* do not. Kantz and Bold (1969), on the basis of investigation in culture of more than 100 isolates from soil of *Nostoc* and *Anabaena*, concluded that the most reliable distinction between the genera was the continuing motility of the trichomes of *Anabaena* as contrasted with the transient motility of those of *Nostoc*. Whether the planktonic species of *Anabaena* also remain motile remains to be seen.

Walsby (1974A, B) has investigated a number of extracellular products of *Anabaena cylindrica* Lemm. These include amino acids (serine and threonine) in a group of complex pigmented and fluorescent compounds.

As in *Nostoc*, mature trichomes of *Anabaena* produce heterocysts and akinetes. The latter may be similar to or different in size and shape from the vegetative cells. Germination into trichomes of both the heterocysts and akinetes of *Anabaena* has been recorded.

Gorham (1964) and Carmichael, et al. (1975) reported on the toxicity of certain species of *Anabaena* to cattle and other animals.

CYLINDROSPERMUM Kützing   *Cylindrospermum* (Gr. *kylindros*, cylinder + Gr. *sperma*, seed) (Fig. 2.28) is similar to *Anabaena* in that its trichomes occur singly or they may be embedded in an amorphous matrix. It differs from *Anabaena* in that the heterocysts are always basal and, under certain conditions, the adjacent cell becomes transformed into a cylindrical akinete. Both aquatic and terrestrial species of *Cylindrospermum* are known.

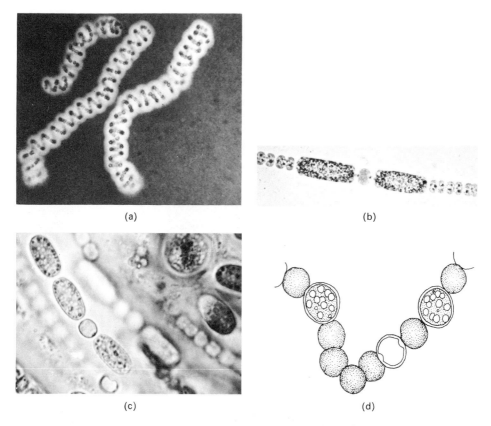

Fig. 2.27 *Anabaena* sp. (*a*) A planktonic species mounted in India ink. Note sheaths. (*b*) *Anabaena* sp. showing a segment of a trichome with vegetative cells, a heterocyst, and two akinetes. (*c*) Heterocysts and three akinetes of another species. (*d*) Mature trichome composed of vegetative cells, akinetes, and a heterocyst. (*a*) × 75; (*b*) × 225; (*c*) × 317; (*d*) × 450.

Fig. 2.28 *Cylindrospermum* sp. Note basal heterocyst and adjacent akinete. × 225.

Family 3. Scytonemataceae

The trichomes of members of the Scytonemataceae are enclosed in firm sheaths that may be colored. The trichomes are characterized by false branching in which, without initiating cell division in a new plane, the trichomes or their hormogonia

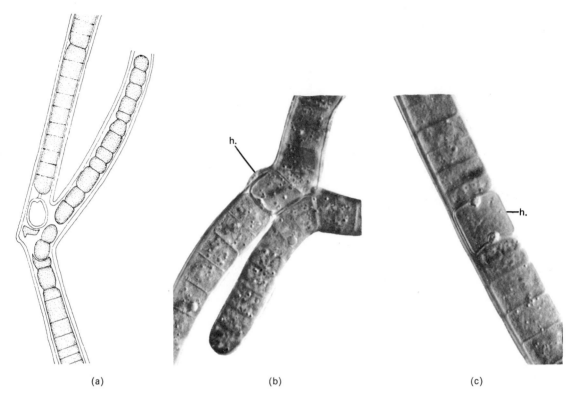

(a)   (b)   (c)

Fig. 2.29   (*a*) *Tolypothrix* sp. (*b*), (*c*) *Scytonema* sp. (*b*) Note double false branching. (*c*) Segment of a filament; note intercalary heterocyst. h. heterocyst. (*a*), × 413; (*b*), (*c*), × 2660. [(*b*), (*c*) courtesy of Mr. Robert Slocum.]

rupture or grow through the sheath (Fig. 2.29). The trichomes are usually heterocystous, but akinetes are rarely produced. Two representatives are discussed and illustrated below, namely, *Tolypothrix* and *Scytonema*.

TOLYPOTHRIX Kützing and SCYTONEMA Agardh   The trichomes of *Tolypothrix* (Gr. *tolype*, woolen ball + Gr. *thrix*, hair) (Fig. 2.29*a*) are of uniform diameter and enclosed in narrow sheaths. The false branches typically arise singly from the vicinity of a heterocyst, but the occasional occurrence of double false branches raises a question as to the validity of number of false branches as a criterion to delimit *Tolypothrix* from *Scytonema* (Gr. *skytos*, leather + Gr. *nema*, thread) (Fig. 2.29*b*). Stein (1963) has investigated this variability in *Tolypothrix* in culture and in field material. While *Tolypothrix* is most often aquatic, *Scytonema* is subaerial or terrestrial. Its sheaths are firm and thick and sometimes lamellated. Its (usually) double false branches arise in a position intercalary to heterocysts. *Scytonema* forms coarse, felty patterns on moist stones, wood, and soil. Drouet (1973) includes *Tolypothrix* in *Scytonema*.

Family 4. Stigonemataceae

The members of the Stigonemataceae differ from other trichomatous Cyanochloronta in exhibiting true branching, the latter initiated by division of certain cells in a new plane. The trichomes of some genera are in part pluriseriate. *Hapalosiphon*, *Stigonema*, and *Fischerella* are here chosen to represent the Stigonemataceae and they may be distinguished with the aid of the following key.

1. Axes different morphologically from the branches . . . .*Fischerella*
1. Axes and branches similar 2
   2. Trichomes mostly uniseriate throughout . . . .*Hapalosiphon*
   2. Trichomes pluriseriate, at least in part . . . .*Stigonema*

HAPALOSIPHON Nägeli   *Hapalosiphon* (Gr. *hapalos*, soft + Gr. *siphon*, tube) (Fig. 2.30) often grows in neutral or slightly acid waters as an epiphyte on other aquatic vegetation. The cells are short-cylindrical. The sheaths are hyaline and intercalary heterocysts and akinetes may be present. Hormogonia form usually from the branches that may arise unilaterally or bilaterally, according to the species.

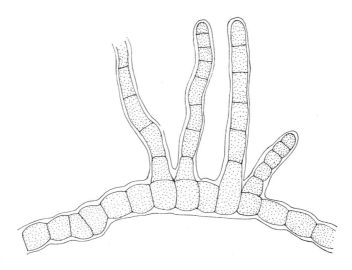

Fig. 2.30 *Hapalosiphon* sp. Note true branching by cell division in a new direction. × 638.

STIGONEMA Agardh   *Stigonema* (Gr. *stigon*, dotted + Gr. *nema*, thread) (Fig. 2.31) is an inhabitant of moist rocks and soil more frequently than it is aquatic. The main trichomes of *Stigonema* are at least in part pluriseriate and encased within a firm sheath that may be colorless or yellow-brown. Growth is largely apical, and the branches are morphologically similar to the main axes. The cells may be spherical or flattened by compression. They seemingly are connected by coarse protoplasmic strands. The generic limits between *Stigonema* and *Hapalosiphon* are not so well

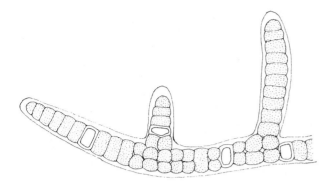

Fig. 2.31 *Stigonema* sp. Note pluriseriate condition of trichome. × 300.

defined as desirable because of the occasional occurrence of pluriseriate segments in the trichomes of *Hapalosiphon*. Hormogonia are produced from the ends of the branches in *Stigonema*.

FISCHERELLA (Born. et Flah.) Gomont   *Fischerella* (from the genus *Fischera*) (Fig. 2.32) differs from *Stigonema* in that it displays a dimorphism of organization, only the main axes being pluriseriate in part. The branches consist of long, slender cells with a narrow sheath, in contrast to the more prominent sheath of the main axis. The heterocysts may be intercalary or lateral, and hormogonia form from the ends of the lateral branches. Species of *Fischerella* occur on moist tree bark and in

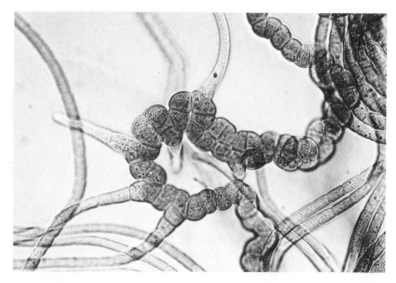

Fig. 2.32 *Fischerella ambigua* (Näg.) Gom. Note dimorphic branches. × 675. (After Thurston and Ingram.)

and on soil. Thurston and Ingram (1971) have investigated development of *F. ambigua* at both the light- and electron-microscope level. Martin and Wyatt (1974) investigated the comparative morphology and physiology of five strains of *Fischerella*.

Family 5. Rivulariaceae

The members of the Rivulariaceae are heterocystous, and their trichomes taper from the base to the apex or from the middle toward both ends. *Rivularia*, *Calothrix*, and *Gloeotrichia* have been chosen as representative of the family and are discussed below. They may be distinguished by use of the following key:

1. Trichomes not in colonies, not planktonic . . . .*Calothrix*
1. Trichomes often in spherical gelatinous colonies or planktonic 2
   2. Lacking akinetes . . . .*Rivularia*
   2. Producing akinetes at maturity . . . .*Gloeotrichia*

CALOTHRIX Agardh   The various species of *Calothrix* (Gr. *kalos*, beautiful + Gr. *thrix*, hair) (Fig. 2.33) occur in both freshwater and marine water and may grow attached to stones, woodwork, or other algae and aquatic angiosperms. The filaments are markedly tapered and either unbranched or falsely branched. In the latter case, the false branches become freed from the parent trichomes. The individual trichomes are ensheathed, but the sheaths may be confluent. Heterocysts are usually basal and the akinetes, if present, are adjacent to the basal heterocysts.

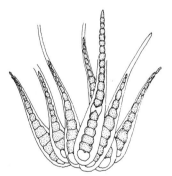

Fig. 2.33   *Calothrix* sp. × 289.

RIVULARIA (Roth) Agardh and **GLOEOTRICHIA Agardh**   *Rivularia* (L. *rivulus*, a small brook) (Fig. 2.34*a*) and *Gloeotrichia* (Gr. *gloia*, glue + Gr. *thrix*, hair) (Fig. 2.34*b*) differ from each other in that no akinetes occur in *Rivularia*. Both genera contain species that are aggregated in gelatinous balls and in both the trichomes taper to hairlike apices. The common matrix is of much firmer consistency in *Rivularia* than it is in *Gloeotrichia*. Species of both genera grow attached to submerged stones, wood, or aquatic plants, but some species of *Rivularia* are subaerial on moist cliffs,

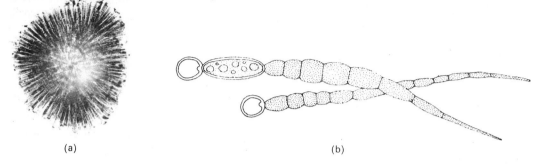

(a)   (b)

Fig. 2.34 (*a*) *Rivularia* sp. (*b*) *Gloeotrichia* sp. Note basal heterocyst and adjacent akinete. (*a*) × 79; (*b*) × 600.

while *Gloeotrichia* is always aquatic. The akinetes of *Gloeotrichia* are prominent and cylindrical adjacent to the basal heterocyst and may occur in chains. In *G. ghosei* Singh, the heterocysts germinate frequently (Fig. 2.9*b*) to form new trichomes (Singh and Tiwari, 1970; Claasen, 1973); *G. natans* (Hedw.) Rabenh. is planktonic. Drouet (1973) includes *Rivularia* and *Gloeotrichia* in *Calothrix*.

3

Division Chlorophycophyta

General Features

The Chlorophycophyta, or green algae, comprise one of the major groups of algae, when one considers the abundance of their species and genera and frequency of occurrence. They grow in waters of a great range of salinity, varying from oligotrophic freshwaters to those that are marine and supersaturated with solutes; a number grow in brackish waters. Several orders of green algae are exclusively marine. Both benthic and planktonic species occur. A number grow in subaerial habitats.

A great range of organization of the plant body occurs in the Chlorophycophyta including unicellular, colonial (coenobic and noncoenobic), filamentous, membranous, and tubular types (Fig. 1.2). Examples of genera illustrating all these are included in the following account.

Cellular organization in the Chlorophycophyta, as in all algae except the Cyanochloronta,[1] is eukaryotic (Dodge, 1973; Bisalputra, 1974; Evans, 1974). Pickett-Heaps (1975A), in a magnificent volume, has summarized much of our knowledge of green algal ultrastructure. The cells are for the most part uninucleate, but the multinucleate (**coenocytic**) condition characterizes several orders (e.g., Caulerpales) and occurs among certain genera of Chlorococcales (e.g., *Hydrodictyon*, p. 125). Nuclear and cell division have been intensively and extensively studied in the green algae at the light-microscopic (Bold, 1951; Godward, 1966) and ultrastructural level, and several different patterns have been recognized. On the basis of these and other criteria, Pickett-Heaps (1972D), Pickett-Heaps and Marchant (1972), Stewart and Mattox (1975), and Pickett-Heaps (1975A, 1976) have proposed important changes in current classifications of the algae. These proposals are discussed on p. 245. Two basic patterns (Fig. 3.1) have emerged from the studies of these investigators regarding nuclear and cell division in the green algae: (1) Intranuclear mitosis, with the nuclear

---

[1]P. 31; the Cyanochloronta are sometimes considered to be bacteria.

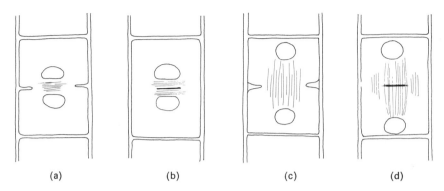

(a)   (b)   (c)   (d)

Fig. 3.1 Patterns of cell division in green algae (diagrammatic) (based on the work of Pickett-Heaps and Floyd, Mattox and Stewart). (*a*), (*b*) Spindle absent, nuclei in close proximity. (*a*) Phycoplast and furrowing present. (*b*) Phycoplast and cell plate present. (*c*), (*d*) Spindle persistent, nuclei remote from each other. (*c*) Note furrowing. (*d*) Note cell plate and phragmoplast-like organization.

envelope closed at metaphase or open at the poles, interrupted by the microtubules of the spindle; other microtubules, transverse to the longitudinal axis of the spindle, are present at telophase and are called the **phycoplast** (Fig. 3.1*a, b*); the latter functions somehow in cytokinesis, by furrowing or by cell-plate formation; the daughter nuclei at telophase are in close proximity. (2) The spindle and nuclear envelope are open and a phycoplast is absent (Fig. 3.1*c, d*); at telophase the microtubules of the intranuclear spindle are persistent and a **phragmoplast**-like structure is organized as cytokinesis by furrowing proceeds. This phragmoplast-like structure suggests cytokinesis in the Charophyta (p. 247) and in the land plants.

The most conspicuous organelle in the algal cell is the **chloroplast**[2] (Fig. 3.7), which occurs in a variety of patterns visible with the light microscope; the form of the chloroplast is an important criterion in the classification of the green algae. It may be massive, in relation to cell size, and parietal and cup-like; or sponge-like, netlike, asteroidal, or axile; organized as separate segments that may or may not be joined in a reticulum; ribbon-like or bar-like. Gibbs (1962A, B, C, 1970), Bisalputra (1974), and Dodge (1973) have reviewed the comparative ultrastructure of algal chloroplasts, which are bounded **thylakoids,** or **photosynthetic lamellae** (flattened sacs), which are the site of the photosynthetic pigments (Table 1.1). Ribosomes and DNA are present in the chloroplasts. Bands of two to six thylakoids are present in many green algal chloroplasts, and dense stacks of them, or **grana,** like those in land plants, occur in some.

In most green algae the plastids contain one or more specially differentiated regions called **pyrenoids** (Fig. 3.7). Gibbs (1962A) and Griffiths (1970) have summarized the occurrence, organization, and functions of these organelles in the green (and other) algae. In many organisms, extensions of the thylakoidal system enter the pyrenoid. In the green algae the pyrenoid is the site of starch formation, one or more

[2]Sometimes called the chromatophore in the older literature.

starch grains forming within the chloroplast closely appressed to the surface of the pyrenoid. It has been suggested (Griffiths, 1970) that the pyrenoid is the region of temporary storage for early products of photosynthesis that, upon overproduction, are converted into starch. An alternate view is that the pyrenoid is the site of production of starch synthetase that polymerizes glucose molecules from the chloroplast into starch on the pyrenoid surface. While in some algae (e.g., *Scenedesmus*, p. 138) it has been shown (Bisalputra and Weier, 1964) that all the starch of the chloroplast arises in association with the pyrenoid, this has not been established for others. Furthermore, pyrenoids are entirely absent from some starch-forming algae (e.g., *Microspora*, p. 147).

In some algae (e.g., *Tetracystis*, p. 129), the pyrenoids divide during cell division (Brown and Bold, 1964), so that the division products of the pyrenoid are transmitted with the divided plastid segment directly to the cellular progeny. In other cases, where cell divisions are repeated in rapid succession and result in a large cellular progeny (as in zoosporogenesis), the pyrenoid disappears in the parental cell, and the young cells apparently develop them *de novo* (Brown and Arnott, 1970).

In addition to pyrenoids, the chloroplasts of motile green algae, and most motile reproductive cells of nonmotile green algae, contain specially pigmented organelles, each of the latter called the **stigma** or **red eyespot** (Fig. 3.7*b*), which, according to one school of thought, is considered to be the site of light perception.[3] Hartshorne (1953) showed that mutants of *Chlamydomonas* lacking eyespots, although still positively phototactic, reacted more slowly than the wild type that had stigmata, and Walne and Arnott (1967) and Arnott and Brown (1967), among others, have investigated the ultrastructure of the stigma. In *Chlamydomonas*, for example, the latter consists of a single layer of 15–45 electron-dense granules mutually compressed; these granules varied between 75 and 100 μm. Arnott and Brown found that in *Tetracystis* the eyespot granules are about 80 Å in diameter and confined within a flat area bounded by the plastid membrane and a thylakoid. Dodge (1969B) has reviewed the fine structure of algal stigmata.

The flagella of motile cells of most[4] green algae that have been investigated are equal in length (isokontan) and smooth. In some organisms, e.g., *Chlamydomonas reinhardtii*, delicate, hairlike extensions arise along the flagellum (Fig. 3.2), while in others, e.g., *Pyramimonas* (Fig. 3.5*g*), *Prasinocladus* and *Trichosarcina*, the flagellar surface is covered with minute scales.

The flagella of algal cells differ in their place of insertion on the cell and in their number, length, and appendages. These variations have been summarized in Fig. 3.2. Flagella may be smooth, that is, lacking hairs except for a delicate terminal fibril; these are referred to as **acronematic** flagella (Fig. 3.2*j*). Some smooth flagella lack the terminal fibril and thus are bluntly terminated. **Pleuronematic** flagella have one or

---

[3]For further discussion of eyespots and phototaxis, see *Euglena*, p. 258; also consult Bendix (1960), Halldal (1962), and Diehn (1969A, B; 1973) and the references therein.

[4]In *Bracteacoccus*, p. 116, Fig. 3.47*b*, the flagella on the zoospores are slightly unequal; also in *Heteromastix*; (the latter is not discussed in this book).

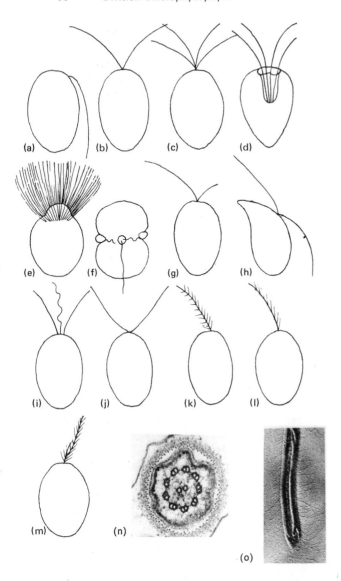

Fig. 3.2 Flagellar number, insertion and organization among the algae. Note apical origin in (*b*)–(*c*) and (*g*); origin at the base of a depression (*d*); subapical origin in (*a*) and (*e*); median origin in (*f*); and lateral insertion in (*a*) and (*h*). (*a*) *Pedinomonas*. (*b*) *Chlamydomonas*. (*c*) *Carteria*. (*d*) *Pyramimonas*. (*e*) *Oedogonium*. (*f*) *Gymnodinium*. (*g*) *Tribonema*. (*h*) *Ectocarpus*. (*i*) *Chrysochromulina*; note haptonema (see p. 360) between flagella. (*j*)–(*o*) Flagellar organization: (*j*) Acronematic flagella: smooth with delicate tips. (*k*) Pleuronematic: flagellum with two rows of radiating mastigonemes, flimmer or fibrils. (*l*) Stichonematic flagellum: flagellar appendages in a single row. (*m*) Pantacronematic; pleuronematic with attenuated, unornamented apices. (*n*), (*o*) *Chlamydomonas reinhardtii* Dangeard Flagellar organization. (*n*) Transection showing 9 + 2 arrangement of component fibrils within flagellar sheath. (*o*) Part of pleuronematic flagellum. [(*n*) and (*o*) After Ringo.]

more rows of lateral hairs (flimmer or mastigonemes). When these hairs arise unilaterally from the flagellum, it is said to be **stichonematic**; if the mastigonemes are arranged in two rows, the term **pantonematic** is applied. If this latter flagellum also has a terminal fibril, it is said to be **pantacronematic.**

The pigments of the chloroplasts of green algae (Table 1.1, p. 18) are chlorophylls *a* and *b* (Meeks, 1974); α, β, and γ carotenes; and several xanthophylls (Goodwin, 1974). The last are oxygenated derivatives of carotenes (Nakayama, 1962). In addition to primary carotenoids that are present during active growth of the cells, some vegetative cells (e.g., those of *Chlorococcum* and *Spongiochloris*) lose their chlorophylls and develop secondary carotenoids among which McLean (1967, 1968) has reported echinenone, canthaxanthin, and astacene and two additional, as yet unidentified pigments. Similar changes in pigmentation probably occur in maturing zygotes and akinetes of certain green algae that turn red.

In addition to the chlorophylls and carotenoid pigments, some green algae (e.g., *Zygnema ericetorum*) contain purple vacuolar pigments that are probably iron-tannin complexes (Alston, 1958).

Green algae, of course, contain such other organelles of eukaryotic cells as Golgi apparatus, mitochondria, and endoplasmic reticulum (Dodge, 1973; Evans, 1974). In many (e.g., *Chlamydomonas*, *Chlorococcum*) the protoplast fills the cell; while in some (e.g., *Spirogyra*, p. 226) a large, central, aqueous vacuole is present within it. The motile cells of green algae contain **contractile vacuoles** in their colorless cytoplasm (Fig. 3.7). These serve an osmoregulatory function. They appear *de novo* in some organisms (e.g., *Spirogyra*, p. 226, a nonmotile alga) and rapidly reduce the fluid content of the cell vacuole. Contractile vacuoles are absent in marine species.

In all the vegetative cells of green algae, except those of the Polyblepharidaceae (p. 71) and some of the motile reproductive cells of others, the protoplast is surrounded by a more or less firm wall that lies just outside the plasmalemma or plasma membrane. In many instances, cells in the senescent state in cultures that have depleted nitrogen levels thicken their walls tremendously (Archibald and Bold, 1970). Many organisms have two distinct wall layers distinguishable at the light-microscopic level: an inner, rather firm layer and an outer, capsular layer (Fig. 2.30), which may be stratified. These layers that bound the algal protoplast, formerly thought to consist entirely of polysaccharides, have been shown to contain from 10–69% protein (Gotelli and Cleland, 1968) and also hydroxyproline glycosides (Miller, et al., 1972). Hanic and Craigie (1969) isolated a chemically resistant cuticular layer from 11 Chlorophycophytan marine algae (among others) and found it contained up to 70% protein. By contrast, Lorch and Weber (1972) report only 0.32% of the cell wall weight in *Pleurotaenium* was nitrogen, 1.7% lipid, and the remainder contained glucose, galactose, xylose, arabinose, and glucuronic and galacturonic acids. Roberts (1974) has discussed the cell walls of algae with special reference to their glycoprotein composition.

It used to be inferred from rather crude chemical tests (sulfuric acid and iodine) that the inner cell wall (inner, if two layers were apparent) was composed of cellulose (polymerized glucose) in the green algae. More recent investigations (Kreger, 1962; O'Colla, 1962; Parker, 1964, 1969, 1970; Preston, 1968; Hanic and Craigie, 1969; Miller, et al., 1972; Siegel and Siegel, 1973; Catt, et al., 1976), unfortunately performed on relatively few green algae, indicate that while cellulose is present in the walls of some (e.g., *Chaetomorpha*, p. 182), it is absent from some others (e.g., *Chlamydomonas*, p. 75; *Acetabularia*, p. 218; and *Bryopsis*, p. 201). Instead, some of them have walls largely composed of polymers of xylose (e.g., *Bryopsis* and *Caulerpa*, p. 199) or mannose (e.g., *Acetabularia* and *Batophora*, p. 223). These polymers occur as fibrils embedded in a nonfibrillar matrix, the latter (at least in some cases), composed of hemicelluloses. The vast majority of green algae have not been analyzed critically with respect to the composition of their cell walls, so that caution is indicated in making generalizations. Siegel and Siegel (1973) and MacKie and Preston (1974) have summarized our knowledge of algal wall chemistry.

Reproduction

Many types of asexual and sexual reproduction, as summarized on pp. 10–14, occur among the Chlorophycophyta. Organisms with H, h; H, d; D^i, h + d, and D^h, h + d are discussed among the illustrative genera that have been included in the following account.

Classification and Key to the Orders

The classification of green algae, as is the case with most other algal divisions, differs with the classifier. In this book the following orders of the single class Chlorophyceae are recognized: Volvocales, Tetrasporales, Chlorococcales, Chlorosarcinales, Chlorellales, Ulotrichales, Chaetophorales, Oedogoniales, Ulvales, Cladophorales, Acrosiphoniales, Zygnematales, Caulerpales, Siphonocladales, and Dasycladales. They may be distinguished with the aid of the following key:

1. Unicellular or colonial; in the former case not composed of mirror-image semicells nor having conjugation of amoeboid gametes 2
1. Filamentous, membranous, tubular; or, if unicellular, the cells composed of mirror-image semicells and/or having conjugation of amoeboid gametes 6
    2. Cells or colonies motile. . . . . .Order Volvocales (p. 69)
    2. Cells or colonies nonmotile. 3
3. Undergoing desmoschisis or division of vegetative cells to form cellular complexes or packets . . . . .Order Chlorosarcinales (p. 127)
3. Not undergoing desmoschisis; cell division forming naked or walled reproductive cells 4
    4. Cellular organization *Chlamydomonas*-like; cells often in gelatinous masses or colonies . . . . .Order Tetrasporales (p. 104)
    4. Cells not *Chlamydomonas*-like 5
5. Cells capable of producing zoospores . . . . .Order Chlorococcales (p. 114)
5. Cells forming only nonmotile autospores or autocolonies . . . . .Order Chlorellales (p. 133)
    6. Filamentous or, if unicellular, with mirror-image semicells and/or conjugation of amoeboid gametes 7
    6. Membranous, tubular, radially symmetrical, or composed of interwoven tubes or siphons 12
7. Lacking zoospores, sexual reproduction by conjugation of amoeboid gametes . . . . .Order Zygnematales (p. 225)
7. Producing zoospores 8
    8. Cells uninucleate 9
    8. Cells large, multinucleate 11
9. Zoospores with ring of numerous flagella . . . . . . .Order Oedogoniales (p. 159)

Order 1. Volvocales

The order Volvocales includes those unicellular and coenobic green algae that are normally actively motile by means of flagella. They exhibit positive phototaxis; that is, they move toward light of proper wavelength and intensity (Bendix, 1960). Temporary nonmotile periods, the so-called "*Palmella* stages," may occur in the lives of these organisms. These stages were designated *Palmella* stages because of their resemblance to the genus *Palmella* (p. 106), which is permanently organized in mucous masses in which the component cells are Volvocalean in organization. Organisms in the *Palmella* stage readily revert to the typical motile condition when environmental conditions change, for example, when fresh culture medium is added to laboratory cultures. Nakamura, et al. (1975) investigated, ultrastructurally, induced *Palmella* stages of *Chlamydomonas*.

Cellular Organization

The unicellular members of the order have one, two, four, or eight flagella, while the coenobic members have two or four flagella. The latter are of the whiplash type; that is, each consists of a sheath (an extension of the cell's plasma membrane or plasmalemma) surrounding a circle of nine double fibrils around two central ones Fig. 3.2*n*). Hairlike extensions of the flagellar sheath are present in some members of the order (Bouck, 1972; Dodge, 1973).

Except for the members of one family (Polyblepharidaceae) and certain other species, the protoplasts of the cells are surrounded by cell walls that may be complex in organization as revealed by electron microscopy. For example, Hills, et al. (1973)

[5]Except in *Derbesia* (p. 206).

reported that the protoplast of *Chlamydomonas reinhardtii*[6] Dangeard is surrounded by seven distinct layers (Fig. 3.8). Later, Catt, et al. (1976) showed that the outermost (smooth) layer was an adsorption product from the culture medium. The major portion of this complex wall is a glycoprotein. It should be noted, however, that few Volvocalean algae have been studied as intensively with reference to their cell walls.

The most conspicuous feature of Volvocalean cells and, indeed, that of most Chlorophycophyta is the chloroplast that varies in position and form in the cell, so that cuplike parietal, asteroidal, lenticular, and lamellate types occur in the various members of the order. The chloroplasts of most species contain at least one pyrenoid, the latter apparently a center of formation of the enzyme starch synthetase, for starch grains originate all over the surface of the pyrenoid. Also in the chloroplast of many members of the order is a stigma or "red eyespot" (Fig. 3.7), the latter designation an allusion to its supposed role in light perception. Volvocalean cells are uninucleate and lack large central vacuoles. Instead, except in marine species, most contain one to many small contractile vacuoles that seemingly have an osmoregulatory role. The contractile vacuoles may be apical, at the flagellar bases, or distributed over the surface of the protoplast. Electron microscopy reveals the presence of such typical organs of eukaryotic cells as mitochondria, Golgi apparatus, and endoplasmic reticulum, as well as the organelles mentioned above.

Reproduction

Both asexual and sexual reproduction occur in the members of the Volvocales.

ASEXUAL REPRODUCTION   Unicellular members of the Volvocales reproduce as a result of mitosis and cellular division, which may be repeated rapidly, so that two, four, eight, or more young cells may arise within and be released by the parental cells (Fig. 3.9). In wall-less species, e.g., *Dunaliella* (Fig. 3.4), division may occur while the cells are motile, but most walled organisms, e.g., *Chlamydomonas*, become transiently nonmotile during mitosis and cell division, which often occur in darkness. In cellular division, some of the organelles (pyrenoid and chloroplast) are directly transmitted to the progeny after they have divided preceding to or during cytokinesis, while others (contractile vacuoles, stigmata) seem to arise *de novo* in the cellular progeny.

Volvocalean colonies are coenobia that reproduce by autocolony formation (Fig. 3.21d). This is effected by repeated bipartitions of all (*Pandorina*, Fig. 3.21d) or some (*Volvox*, Fig. 3.27a) of the cells of the colony to form embryonic colonies that, upon liberation from the parent, gradually grow by increase in cell size (not by additional cell divisions) to that characteristic of adults of the species.

SEXUAL REPRODUCTION   Sexual reproduction of Volvocalean algae is isogamous, anisogamous, or oogamous, depending on the organism. All three types occur among the sexual species of *Chlamydomonas*, while each of the several types

---

[6]Sometimes spelled "*reinhardi*"; the spelling of the specific epithet as here given is that of Dangeard.

characterizes the several genera of the colonial Volvocales. Meiosis is zygotic in all the Volvocales that have been studied cytologically or genetically or both. As far as known, accordingly, all are probably haplobiontic and haploid (H, h) (see, however, Papenfuss, 1955, p. 126).

Occurrence

Volvocalean algae are ubiquitous. Many of the unicellular species occur in soil as well as in fresh water and brackish and marine waters. The coenobic species are widely distributed in bodies of freshwater of various sizes and degrees of permanence. When the latter dry, the organisms persist as dormant zygotes or akinetes; they may be readily induced to becoming vegetative and motile by flooding aliquots of the dry soil with water or culture medium (Starr, 1973). Certain Volvocales have been known to cause water blooms. Among these are *Chlamydomonas, Carteria, Chlorogonium,* and *Volvox.*

Classification

Representatives of six families of Volvocales are included in the following discussion, namely, Polyblepharidaceae, Chlamydomonadaceae, Phacotaceae, Volvocaceae, Astrephomenaceae, and Spondylomoraceae. These families may be distinguished readily by means of the following dichotomous key:

1. Unicellular 2
1. Coenobic 4
  2. Cells lacking a cell wall ....Polyblepharidaceae
  2. Cells bounded by walls or loricas 3
3. Walls delicate to prominent, usually not brown,
   not composed of two portions ....Chlamydomonadaceae
3. Walls or loricas often brown; sometimes obviously
   composed of two parts, often not closely
   appressed to the protoplast ....Phacotaceae
  4. Cells not enclosed in a gelatinous matrix ....Spondylomoraceae
  4. Cells within a gelatinous matrix 5
5. Cells within a common matrix, autocolonies everting
   during ontogeny ....Volvocaceae
5. Cells within individual matrices, autocolonies
   not everting during ontogeny ....Astrephomenaceae

Family 1. Polyblepharidaceae

Once an assemblage that contained a larger number of genera and species, the Polyblepharidaceae has suffered attrition, because a number of its former members have been by some authors transferred to the class Prasinophyceae (p. 73). Three of the remaining members of the family are discussed below, although one of them, *Polyblepharides,* is seemingly an extremely rare alga.

The naked cells of members of the Polyblepharidaceae contain single chloroplasts with eyespots and single nuclei and have contractile vacuoles if they grow in freshwater. They divide while in motion or in the encysted state, the naked cells undergoing longitudinal cleavage. One, two, four, or eight flagella characterize the several genera. All can form smooth-walled gelatinous cysts.

PEDINOMONAS Korschikov   *Pedinomonas* (Gr. *pedinos*, flat + Gr. *monas*, single individual) (Fig. 3.3) is almost unique among the Chlorophycophyta in being uniflagellate with a posteriorly directed flagellum. The latter arises anteriolaterally from the cell near the lateral nucleus. The naked cells are slightly compressed and asymmetric and contain a parietal chloroplast with a stigma and single pyrenoid; one contractile vacuole is present in the colorless cytoplasm.

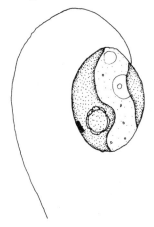

Fig. 3.3 *Pedinomonas minor* Korsch. ×2550. (After Ettl.)

Asexual reproduction is by longitudinal division of the motile cells and sexual reproduction is isogamous.

Pedinomonas occurs in water and soil. Ettl (1967) has monographed the genus and recognizes nine species. He has studied related genera (1966) and is of the opinion that perhaps *Pedinomonas* and certain similar organisms should be segregated from the Volvocales. Ettl and Manton (1964) and Pickett-Heaps (1974D) have described the ultrastructural organization of *P. minor* Korsch; the alga has recently been reported from Ohio (Kalinsky, 1971).

DUNALIELLA Teodoresco   *Dunaliella* (in honor of *M. F. Dunal*) (Fig. 3.4) occurs in extremely saline habitats, such as Great Salt Lake in Utah and in seaside rock pools in which the salt concentration significantly exceeds that of normal seawater. Contractile vacuoles are absent, and stigmata may or may not be present in the cells of the several species that may develop a red pigment, haematochrome (p. 85).

Longitudinal division of the motile cells comprises asexual reproduction. Marano (1976) has followed the process electron microscopically. Sexual reproduction is isogamous and meiosis occurs at zygote germination (Lerche, 1937) that results in the formation of motile individuals. Hyams and Chasey (1974) have investigated the ultrastructural organization of the flagella of *Dunaliella*.

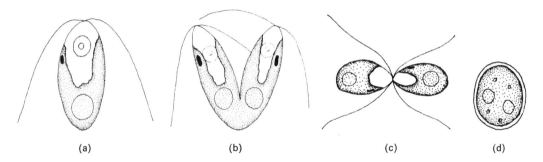

(a)   (b)   (c)   (d)

Fig. 3.4 *Dunaliella salina.* (*a*) Vegetative cell. (*b*) Cell division. (*c*) Union of isogametes. (*d*) Zygote. (*a*), (*b*) × 1331; (*c*), (*d*) × 675. [(*c*) and (*d*) after Lerche.]

PYRAMIMONAS Schmarda   *Pyramimonas* (Gr. *pyramis*, pyramid, + Gr. *monas*, single organism) (Fig. 3.5) differs from *Dunaliella* most obviously in being quadriflagellate.[7] Its species occur in freshwater and brackish and marine waters. The four flagella arise from the base of an apical pit or depression. The cells themselves are four-ridged, at least near the anterior, the four chloroplast lobes corresponding to these ridges. The electron microscope reveals that the cells are asymmetrical (Fig. 3.5*b*) with the nucleus at one side of the apical depression and a large storage body and a reservoir containing flagellar scales on the other. These scales are liberated through a canal to the region of the flagellar bases in the pit. In addition to two kinds of flagellar scales, Manton (1966A, 1968) and Norris and Pearson (1975) have demonstrated that the cell surface is covered with three types of scales that arise within the Golgi vesicles. Moestrup and Thomsen (1974) have shown that in *P. orientalis* there are five kinds of scales produced by the two Golgi bodies in each cell. These scales are stored in a reservoir from which they move to the plasmalemma and flagellar surface. Swale (1973) found that there were three different layers of scales of differing morphology on the surface of *P. tetrarhynchus* Schmarda.

The cells of *Pyramimonas* divide while in motion (Fig. 3.5*e*). They also may undergo encystment. Manton (1966, 1968), Manton, et al. (1963), Belcher (1968, 1969C), Swale and Belcher (1968), Swale (1973), Pearson and Norris (1975), Norris and Pearson (1975), and Pennick and Clarke (1976) made electron microscopic investigations of *Pyramimonas* with special reference to cell division and the origin of the scales of the cell surface (Fig. 3.5*d, f, g*).

Pyramimonas and several other flagellates together with *Prasinocladus* and several coccoid green algae have by some phycologists been removed from the Chlorophyceae and assigned to a separate class Prasinophyceae because of their scale-covered cells and flagella (see Boney, 1970). Not all investigators have recognized this class (e.g., Fott, 1971). Stewart and Mattox (1975) point out that certain filamentous green algae (e.g., *Trichosarcina*, p. 176) have scales on their zoospores, and they question the desir-

[7]*P. amylifera* Cour. however has eight flagella.

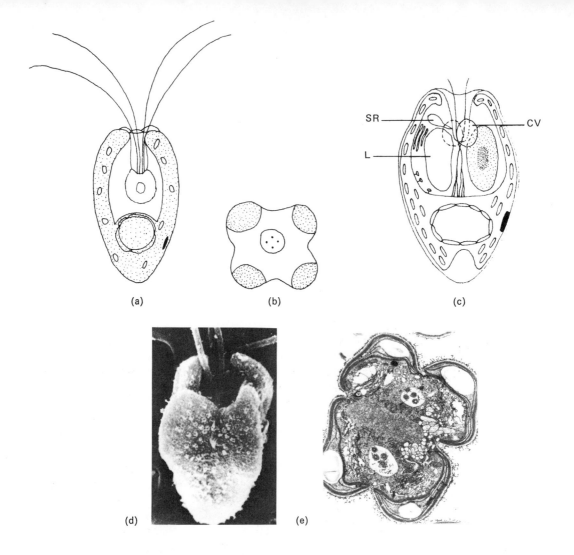

(a)

(b)

(c)

SR

L

CV

(d)

(e)

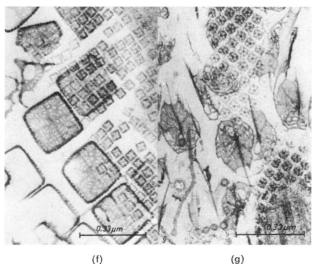

0.33 μm

0.33 μm

(f)

(g)

Fig. 3.5 *Pyramimonas.* (*a*)–(*c*) *P. tetrarhynchus* Schmarda. (*a*) Vegetative individual in optical section. Note apical depression from which flagella emerge. (*b*) Anterior polar view of lobed cell. (*c*) Diagram of ultrastructural organization. (*d*)–(*f*) *P. parkeae* Norris and Pearson. (*d*) Scanning electron micrograph showing flagellar pit and body scales. (*e*) Transection of a cell in early division. Note two flagellar pits, each surrounding four flagella. (*f*) Body scales of two sizes. (*g*) Flagellar scales of two sizes. CV, contractile vacuole; L, liquid food reserve; SR, scale reservoir. (*a*), (*b*) × 750; (*d*) × 6399; (*e*) × 1567; (*f*), (*g*) × 32,676. [(*a*) after Belcher; (*d*), (*f*), (*g*) after Norris and Pearson; (*e*) after Pearson and Norris.]

ability of recognizing the class Prasinophyceae. Sexual reproduction is unknown among these algae.

POLYBLEPHARIDES Dangeard *Polyblepharides* (Gr. *polys*, many + Gr. *blephar*, cilium) (Fig. 3.6) is among the rarest of algae and is mentioned here, not only because of its eight apically inserted flagella but also in the hope that it will be again encountered in nature and taken into laboratory culture. *Polyblepharides fragariiformis* Hazen was observed in a roadside pool in Vermont, while *P. singularis* Dang. has been reported from North Carolina. Other than by longitudinal division, reproduction is unknown for *Polyblepharides*.

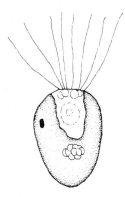

Fig. 3.6   *Polyblepharides fragariiformis* Hazen. ×750.

Family 2. Chlamydomonadaceae

The Chlamydomonadaceae, also unicellular, may be distinguished from the Polyblepharidaceae by their walled cells (Figs. 3.7*b*, 3.8). The flagellate organisms in the Chlamydomonadaceae are distinguished at the generic level by their flagellar number and cell form (whether circular, elliptical, flattened, or angular in transection.)

The ubiquitous, biflagellate *Chlamydomonas* (Gr. *chlamys*, mantle + Gr. *monas*, single organism) (Figs. 3.7–3.11), with several hundred species, is one of the largest algal genera; it occurs in soils and aquatic habitats including brackish and marine ones. *Carteria* (after *H. J. Carter*) (Fig. 3.12), which differs from *Chlamydomonas* only in being quadriflagellate, contains fewer species, and these, seemingly, are less widely distributed but occur in the same habitats as *Chlamydomonas*. The biflagellate *Chlorogonium* (Gr. *chloros*, green + Gr. *gonos*, offspring) has cells that are attenuated at one or both poles. *Haematococcus* and *Polytoma* also are discussed below as representative of the Chlamydomonadaceae.

CHLAMYDOMONAS Ehrenberg   Cellular organization of *Chlamydomonas* (Figs. 3.7, 3.8) has been studied intensively at both the light- and electron-microscopic levels (Lewin, 1952B; Lewin and Meinhart, 1953; Gibbs, et al., 1958; Jones and Lewin, 1960; Walne, 1966; Ringo, 1967; Johnson and Porter, 1968A, B; R. Brown, et al., 1968; Friedmann, et al., 1968; Goodenough, 1970; Schötz, et al., 1972; Witman,

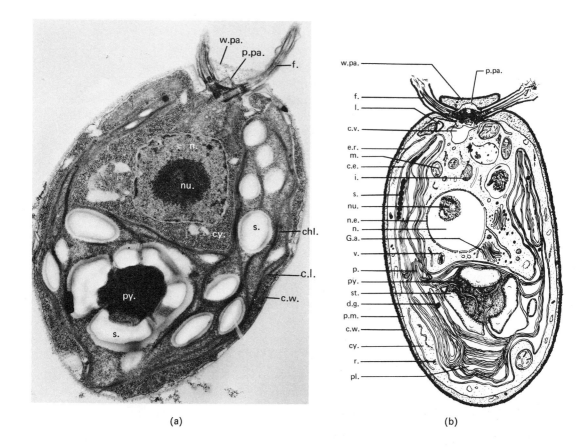

(a)   (b)

Fig. 3.7 Ultrastructure of *Chlamydomonas*. (*a*) Electron micrograph of a section of *C. reinhardtii* Dang. c.l., chloroplast lamella; c.w., cell wall; chl., chloroplast; cy., cytoplasm; f., flagellum; n., nucleus; nu., nucleolus; p.pa., plasma papilla; s., starch; w.pa., wall papilla. (*b*) Diagram of cellular organization of *C. eugametos* Moewus, c.e., chloroplast envelope; c.v., contractile vacuole; c.w., cell wall; cy., cytoplasm; d.g., dense granule; e.r., endoplasmic reticulum; f., flagellum; G.a., Golgi apparatus; i., inclusion; l., lipid body; m., mitochondrion; n., nucleus; n.e., nuclear envelope; nu., nucleolus; p., plastid; p.pa., plasma papilla; pl., chloroplast; p.m., plasma membrane; py., pyrenoid; r., ribosomes; s., stigma; st., starch; v., vesicle; w.pa., wall papilla. (*a*), × 3236; (*b*), × 3996. [(*a*) courtesy of Dr. D. L. Ringo; (*b*) courtesy of Professor P. L. Walne.]

et al., 1972; Chasey, 1974; Bray, et al., 1974; and Osafune et al., 1975) and the details are illustrated and summarized for two species in Fig. 3.7 and its caption.

The chloroplast varies in form among the many species of *Chlamydomonas*. It may be parietal and cup- or urn-shaped, or H-shaped (in optical section), or otherwise. It usually contains one or more pyrenoids. One or two to many contractile vacuoles

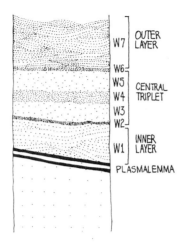

Fig. 3.8 Ultrastructural organization of the cell wall of *Chlamydomonas*, diagrammatic. Note seven layers; the outermost is an adsorption product of the medium. (After Roberts, Gurney-Smith, and Hills, J. Ultrastr. Res. 40: 599–613, Academic Press.)

occur in the colorless cytoplasm; when there are only two, they are anterior; otherwise they may be distributed over the surface of the protoplast.

Walne and Arnott (1967), Nakamura, et al. (1973), and Gruber and Rosario (1974) have investigated electron microscopically the organization of the stigma of *Chlamydomonas*. The stigma is embedded in the chloroplast (Fig. 3.7) and is composed of granules about 75 nm in diameter. Watson (unpublished) studied the stigma of *C. reinhardtii* and found it to have a maximum length of 3 μm. Some aspects of phototaxis in *Chlamydomonas* have been analyzed by Sachs and Mayer (1961) and Stavis and Hirschberg (1973) who showed that positive phototaxis was most intense during the exponential phase of growth. Earlier, Sachs and Mayer (1961) and Marbach and Mayer (1970) investigated the relationship between phototaxis and metabolism in *Chlamydomonas*, while Cain (1965) assayed comparatively nitrogen utilization in 38 chlamydomonad algae. Nultsch and Throm (1975) have ascertained the effect of external factors on phototaxis in *Chlamydomonas reinhardtii*.

The cell wall of *Chlamydomonas* has been studied carefully in only a few species. In *C. reinhardtii* it has been shown to consist of seven layers within which a glycoprotein layer is contained (Fig. 3.8), (Roberts, et al., 1972; Hills, et al., 1973); the outermost layer was later reported to be the product of adsorption from the culture medium. Cellulose is absent from the wall. Hills (1973) has been able to assemble the wall layers from two fractions dissociated in $8M$ lithium chloride when these subunits were dialyzed against water.

Ringo (1967A, B) and Chasey (1974) have investigated in great detail the organization and behavior of the flagella of *C. reinhardtii* (Fig. 3.2*n, o*). They exhibit typical organization within the flagellar sheath, namely, two central fibrils surrounded by nine pairs of fibrils. The central fibrils run to the flagellar tip, while the peripheral ones become single and terminate some distance below it. Bouck (1972), McLean, et al. (1974), and Ringo (1967A) have demonstrated the presence of very fine, hairlike extensions, the **mastigonemes**, on the flagella.

Under conditions unfavorable to the organisms, e.g., desiccation and/or depletion of nitrogen from the environment, many species of Chlamydomonadaceae develop secondary carotenoids, thus becoming orange or reddish, thickening their cell walls, and entering a period of dormancy. These thick-walled, dormant cells are called **akinetes.** Their *Palmella* stages have already been discussed (p. 69).

In asexual reproduction (Fig. 3.9), the cells of most species become nonmotile, and their nuclei undergo one or more mitoses, each followed by a cytokinesis. In this

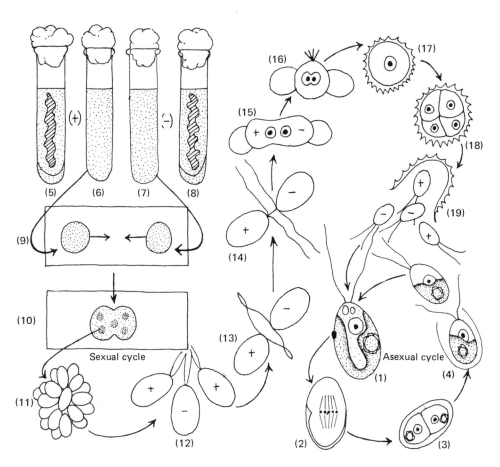

Fig. 3.9 Diagrammatic summary of asexual and sexual reproduction in *Chlamydomonas moewusii*. (1–4) Asexual cycle. (1) Vegetative cell. (2) Mitosis and incipient cytokinesis. (3) Two young vegetative cells within parent cell. (4) Emergence of young vegetative cells from parental wall. (5–19) Sexual cycle. (5, 8) Sexually compatible strains (+ and −) growing separately in the palmelloid state on agar slants. (6, 7) Suspensions of + and − cells in distilled water on low-nitrogen culture medium. (9) The same in drops on a 3 × 1 in. glass slide. Note homogeneous dispersal of maturing gametes. (10) Confluence of + and − droplets. Note agglutination into clusters or clumps. (11) Single cluster. (12) Portion of cluster showing flagellar cohesion. (13) Early pairing. Note cohesion of flagellar tips. (14) Pair in which connecting strand has just been established, the flagella no longer cohesive. (15) Plasmogamy. (16) Karyogamy. (17) Mature zygote. (18, 19) Germination of the zygote and release of meiotic products (meiospores).

way, two, four, or more young cells arise within the parental wall from which they ultimately escape by rupture of the wall and/or its enzymatic breakdown (Schlösser, 1966, 1976). Upon release, the young cells are smaller than their parent but gradually increase to the size characteristic of the species and then repeat the asexual cycle.

Johnson and Porter (1968 B) and Goodenough (1970) have investigated nuclear and cell division in *C. reinhardtii*. The pyrenoid and chloroplast divide during cell division. The basal bodies of the flagella replicate, so that the division products each have two, and microtubules apparently participate in the formation of the cleavage furrow. Centrioles are present at the poles of the spindle (Coss, 1974). Triemer and Brown (1974) have described nuclear and cell division in *C. moewusii* that differs in several respects from that in *C. reinhardtii*. The haploid chromosome number in several species of *Chlamydomonas* has been reported to be 8 (Buffaloe, 1958; Levine and Folsome, 1959) or 16 (Bischoff, 1959). Although Buffaloe and Levine and Folsome reported 8 chromosomes in *C. reinhardtii*, McVittie and Davies (1971) and Loppes, et al. (1972) report the number to be 16. This is of interest, since Hastings, et al. (1965) have reported that 16 linkage groups occur in that organism. Maguire (1976) has recently confirmed the chromosome number as 8 in *C. reinhardtii*, has reported on changes in the life cycle of vegetative cells as related to light intensity, and has confirmed cytologically the occurrence of zygotic meiosis (see p. 12).

Sexual reproduction in *Chlamydomonas* has been investigated in depth by means of light and electron microscopy and biochemically. Clonal cultures, which are populations that arose from single individuals, may or may not, depending on the species, undergo sexual reproduction. If they do, the clones are said to be **self-compatible** or **homothallic**; if they do not, they are designated **self-incompatible** or **heterothallic.** The sexual reproduction of *C. chlamydogama* Bold, *C. eugametos* Moewus., *C. moewusii* Gerloff, *C. reinhardtii* Dang., and *C. pseudogigantea* Korsch. has been studied by a number of investigators (Moewus, 1933; Bold, 1949; Lewin, 1950, 1952A, B, 1953, 1954A, 1956, 1957, 1974A; Sager and Granick, 1954; Bernstein and Jahn, 1955; Tsubo, 1956, 1961A, B; Trainor, 1958, 1959, 1960, 1961; Coleman, 1962; Jones and Wiese, 1962; Wiese and Jones, 1963; Kates and Jones, 1964, 1966; Wiese, 1965; Friedmann, et al., 1968; Wiese, 1969; Wiese and Metz, 1969; Wiese and Shoemaker, 1970; Van Dover, 1972; Wiese and Hayward, 1972; Richards and Sommerfield, 1974; Ishiura and Iwasa, 1973; McLean and Brown, 1974; Triemer and Brown, 1974, 1975A, B; Bergman, et al., 1975; Goodenough and Weiss, 1975; Martin and Goodenough, 1975; Cavalier-Smith, 1975, 1976; Mesland 1976; Snell, 1976A, B; Deason and Ratnasabapathy, 1976; and Weiss, et al., 1977).

Under conditions of proper illuminations (i.e., intensity and quality) and adequate supply of CO_2, low nitrogen content of the culture medium, and temperatures between 18 and 25°C, sexual reproduction (Figs. 3.9, 3.10) can be demonstrated in several species with heterothallic clones, such as *C. eugametos* and *C. moewusii*. Cultures are grown for 3 or 4 weeks on an inorganic solid culture medium[8] and then scraped from the agar and suspended in a harvesting medium that contains 0.1 the nitrogen in the basal medium. Soon after they have been suspended, the cells develop flagella, and many cells of the suspension mature as potential gametes (Figs. 3.9, 3.10*a*). If aliquots

[8]BBM agar, see Appendix, p. 571.

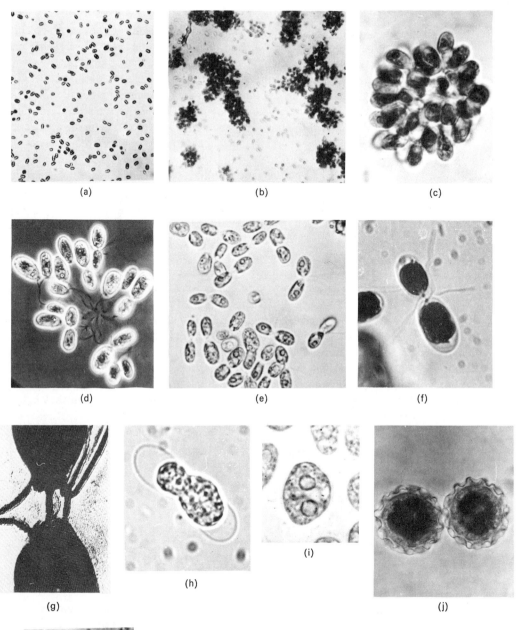

Fig. 3.10 Sexual reproduction in *Chlamydomonas moewusii*. (*a*) Gametes of +
mating type. × 94. (*b*) Cluster formation after addition of − mating type. × 116.
(*c*) Single cluster at higher magnification. × 750. (*d*) Slightly flattened cluster
showing flagellar agglutination. × 936. (*e*) Pairs of gametes. × 465. (*f*) Single
pair; note flagella and plasma papillae. × 1767. (*g*) Members of pair connected
by a protoplasmic strand. × 29,900. (*h*) Plasmogamy. × 1200. (*i*) Recently
formed zygote. × 996. (*j*) Two dormant zygotes. × 1170. (*k*) Germinating
zygote; note four meiotic products. × 1200. [(*e*) after Lewin and Meinhart.]

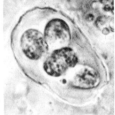

way, two, four, or more young cells arise within the parental wall from which they ultimately escape by rupture of the wall and/or its enzymatic breakdown (Schlösser, 1966, 1976). Upon release, the young cells are smaller than their parent but gradually increase to the size characteristic of the species and then repeat the asexual cycle.

Johnson and Porter (1968 B) and Goodenough (1970) have investigated nuclear and cell division in *C. reinhardtii*. The pyrenoid and chloroplast divide during cell division. The basal bodies of the flagella replicate, so that the division products each have two, and microtubules apparently participate in the formation of the cleavage furrow. Centrioles are present at the poles of the spindle (Coss, 1974). Triemer and Brown (1974) have described nuclear and cell division in *C. moewusii* that differs in several respects from that in *C. reinhardtii*. The haploid chromosome number in several species of *Chlamydomonas* has been reported to be 8 (Buffaloe, 1958; Levine and Folsome, 1959) or 16 (Bischoff, 1959). Although Buffaloe and Levine and Folsome reported 8 chromosomes in *C. reinhardtii*, McVittie and Davies (1971) and Loppes, et al. (1972) report the number to be 16. This is of interest, since Hastings, et al. (1965) have reported that 16 linkage groups occur in that organism. Maguire (1976) has recently confirmed the chromosome number as 8 in *C. reinhardtii*, has reported on changes in the life cycle of vegetative cells as related to light intensity, and has confirmed cytologically the occurrence of zygotic meiosis (see p. 12).

Sexual reproduction in *Chlamydomonas* has been investigated in depth by means of light and electron microscopy and biochemically. Clonal cultures, which are populations that arose from single individuals, may or may not, depending on the species, undergo sexual reproduction. If they do, the clones are said to be **self-compatible** or **homothallic**; if they do not, they are designated **self-incompatible** or **heterothallic.** The sexual reproduction of *C. chlamydogama* Bold, *C. eugametos* Moewus., *C. moewusii* Gerloff, *C. reinhardtii* Dang., and *C. pseudogigantea* Korsch. has been studied by a number of investigators (Moewus, 1933; Bold, 1949; Lewin, 1950, 1952A, B, 1953, 1954A, 1956, 1957, 1974A; Sager and Granick, 1954; Bernstein and Jahn, 1955; Tsubo, 1956, 1961A, B; Trainor, 1958, 1959, 1960, 1961; Coleman, 1962; Jones and Wiese, 1962; Wiese and Jones, 1963; Kates and Jones, 1964, 1966; Wiese, 1965; Friedmann, et al., 1968; Wiese, 1969; Wiese and Metz, 1969; Wiese and Shoemaker, 1970; Van Dover, 1972; Wiese and Hayward, 1972; Richards and Sommerfield, 1974; Ishiura and Iwasa, 1973; McLean and Brown, 1974; Triemer and Brown, 1974, 1975A, B; Bergman, et al., 1975; Goodenough and Weiss, 1975; Martin and Goodenough, 1975; Cavalier-Smith, 1975, 1976; Mesland 1976; Snell, 1976A, B; Deason and Ratnasabapathy, 1976; and Weiss, et al., 1977).

Under conditions of proper illuminations (i.e., intensity and quality) and adequate supply of CO_2, low nitrogen content of the culture medium, and temperatures between 18 and 25°C, sexual reproduction (Figs. 3.9, 3.10) can be demonstrated in several species with heterothallic clones, such as *C. eugametos* and *C. moewusii*. Cultures are grown for 3 or 4 weeks on an inorganic solid culture medium[8] and then scraped from the agar and suspended in a harvesting medium that contains 0.1 the nitrogen in the basal medium. Soon after they have been suspended, the cells develop flagella, and many cells of the suspension mature as potential gametes (Figs. 3.9, 3.10a). If aliquots

[8]BBM agar, see Appendix, p. 571.

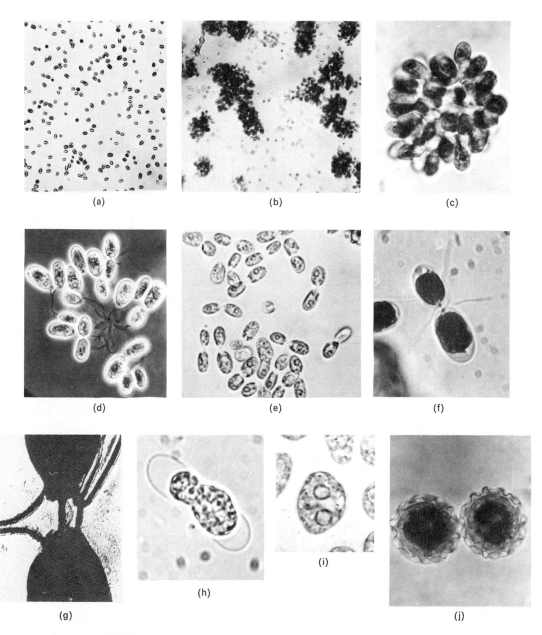

Fig. 3.10 Sexual reproduction in *Chlamydomonas moewusii*. (*a*) Gametes of +
mating type. × 94. (*b*) Cluster formation after addition of − mating type. × 116.
(*c*) Single cluster at higher magnification. × 750. (*d*) Slightly flattened cluster
showing flagellar agglutination. × 936. (*e*) Pairs of gametes. × 465. (*f*) Single
pair; note flagella and plasma papillae. × 1767. (*g*) Members of pair connected
by a protoplasmic strand. × 29,900. (*h*) Plasmogamy. × 1200. (*i*) Recently
formed zygote. × 996. (*j*) Two dormant zygotes. × 1170. (*k*) Germinating
zygote; note four meiotic products. × 1200. [(*e*) after Lewin and Meinhart.]

of compatible suspensions are mixed after they have been illuminated for 1 or 2 hours, sexual reproduction is initiated. The first indication of this is the formation of clusters of gametes (Fig. 3.10*b*, *c*, *d*). These arise because of an attractant ("gamone," Wiese, 1969) that is present on the flagellar tips. Thus, large numbers of compatible gametes become attached to each other (Fig. 3.10*f*, *g*, *h*). It has been shown that the supernatant of a culture or a preparation made from the flagella of one mating type will effect clustering of the cells of the other (Wiese, 1965, 1969). Wiese and Wiese (1975) have educed evidence that in three related taxa one active site on the flagellar tips of the gametes of one mating type is glycosidically bound mannose that in complementary fashion is bound by the mating type substance of the other strain. Speciation in *Chlamydomonas* is postulated by Wiese and Wiese (1977) to proceed from mutation at the mating-type site on the flagella. While the cells are clustered, the flagella twist about each other, thus repeatedly bringing the cellular apices in proximity. Probably both by mechanical abrasion and enzymatic action, at least in *C. moewusii*, the wall papillae of the gametes are perforated, and the plasma papillae come in contact and unite (Brown, et al., 1968). Once this bridge of protoplasm (Fig. 3.10*g*) has become established between two gametes, they are firmly attached, and their flagella no longer remain agglutinated. In *C. moewusii* and *C. eugametos*, the flagella of only one member of the gametic pair beat, so that the pairs emerge from the clusters swimming in only one direction. After several hours of unidirectional movement, the pairs settle down, usually in aggregations, the gametic protoplasts emerge completely from their walls, and plasmogamy and karyogamy follow (Figs. 3.9, 3.10*i*, *j*). Schlösser (1976) and Schlösser, et al. (1976) have reported that the gametes of *C. reinhardtii* become naked through enzymatic dissolution of their walls prior to undergoing plasmogamy.

Immediately after plasmogamy, in which the chloroplasts unite, (Brown, et al., 1968; Bastia, et al., 1969) the zygote secretes a primary wall which is supplemented by a thicker secondary wall (Fig. 3.10*k*) which forms between the primary wall and the protoplast. As the zygote matures, large quantities of starch and oil accumulate and obscure its protoplasmic organization; in some species, secondary carotenoids also develop, so that the zygotes become orange-red.

After a period of dormancy, the zygotes germinate (Fig. 3.10*j*, *l*), each giving rise to four (*C. moewusii*, *C. chlamydogama*, *C. eugametos*) or eight (*C. reinhardtii*) motile cells. Genetic (e.g., Sager, 1955) and cytological (Buffaloe, 1958; Maguire, 1976; and Triemer and Brown, 1975B) evidence reveals that meiosis is zygotic; i.e., it occurs in the first two nuclear divisions of the germinating zygote. *Chlamydomonas*, accordingly, like all Volvocales so far investigated, is haploid. Eight products of zygote germination occur in *C. reinhardtii*[9] because a mitotic division follows the meiosis in the zygote.

Sexual reproduction in *C. reinhardtii* differs in several respects from that in *C. moewusii*, according to Friedmann, et al. (1968) and Triemer and Brown (1975B). The gametes in *C. reinhardtii* lack cell walls, and gametic union is effected through a special "fertilization tubule" that possibly is produced by only one of the compatible strains, but this requires confirmation. Finally, the zygotes in *C. reinhardtii* become motile only after plasmogamy is far advanced, while in *C. moewusii* motility of the connected

[9]Strains with both four and eight zygotic progeny are known (Maguire, 1976).

pairs ceases at plasmogamy. In *C. chlamydogama*, Van Dover (1972) reported motility of the zygotes, as in *C. reinhardtii*.

Certain physiological and biochemical aspects of sexual reproduction have also been investigated. Sager and Granick (1954) were the first to demonstrate the requirement of low nitrogen concentration in the culture medium as an important factor in the sexual process in *C. reinhardtii*, and this has been confirmed for other species. Addition of nitrogen to vegetative cells maturing as gametes reverses this maturation. Trainor (1975), however, has pointed out that the concentration of nitrogen in most culture media is much higher than that even in certain eutrophic waters and he has obtained mating in such natural media. Sager and Granick (1954) also showed that the requirement of light for sexual reproduction could be obviated by growing the organisms in low-nitrogen media in darkness. Although attraction of gametes of *Chlamydomonas* to each other has sometimes been speculated to be the result of positive chemotaxis, the work of Förster and Wiese (1954, 1955) and of Jones and Wiese (1962), among others, clearly demonstrated that gametic clustering was a result of agglutination. They showed further that the agglutinating substances (**gamones**) were present on the flagella and that they were species specific and mating-type specific within the species. Cell-free filtrates, flagellar extracts, and flagella of one mating type cause agglutination of cells of the compatible mating type on a reciprocal basis. Förster, et al. (1956), Wiese (1974), and Wiese and Wiese (1975) reported that the agglutinating agents were glycoproteins that are surface components of the flagella. Wiese (1969) has summarized many aspects of sexual reproduction in *Chlamydomonas* and other algae. Wiese and Hayward (1973) and McLean and Brown (1974) have ingeniously studied the action of various enzymes on the mating capacity of the two compatible strains of several different species of *Chlamydomonas*. They found that two types of interactions affected gametic contact. In one type, a proteinaceous segment of the mating-type substance of one strain reacts with a carbohydrate segment of the compatible strain. The other type seemingly involves interaction between two proteinaceous substances. McLean, et al. (1974) have recently demonstrated that minute vesicles that bleb off the flagellar surface contain the gamone that is effective in gametic

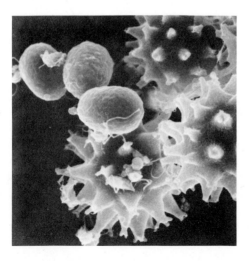

Fig. 3.11 *Chlamydomonas pseudogigantea* Korsch. Oogamous gamete pairs and mature zygotes. × 537. (Courtesy of Dr. John Heimke.)

agglutination. Finally, McLean and Bossman (1975) have presented evidence that the flagellar agglutination in *C. moewusii* is caused by enzyme-substrate binding (see also Wiese and Wiese, 1975).

Isogamy, anisogamy, and oogamy occur among the species of the single genus *Chlamydomonas*. *Chlamydomonas suboogama* (Tschermak-Woess, 1959, 1962) and *C. pseudogigantea* (Van Dover, 1972) are of interest in providing evidence that oogamy was derived from anisogamy because their female gametes may or may not be flagellate at the time of gametic union. In *C. pseudogigantea* Korsh. certain vegetative cells settle to the bottom of the culture vessel and lose their flagella. They undergo divisions to form 16–64 minute sperm. The latter make contact with the cell surface of certain cells that look like vegetative cells and that function later as eggs. After a period of motility, the potential egg becomes motionless and then its protoplast emerges, loses its flagella, and undergoes union with the sperm.

Chlamydomonas has been intensively studied genetically since the pioneering work of Pascher (1916, 1918) who successfully hybridized two species and followed the inheritance and segregation of characteristics in both the haploid vegetative cells and the diploid zygotes. An introduction to the genetic literature concerning *Chlamydomonas* may be obtained by consulting Lewin (1952A, 1953, 1976), Levine and Ebersold (1960), Mattoni (1968), Gillham (1969), Levine and Goodenough (1970), and Sager (1972, 1974). A number of mutants of several species of *Chlamydomonas* have occurred spontaneously, and others have been induced by various mutagenic agents. These mutants were, for the most part, biochemical and related to nutrition (e.g., vitamin- and acetate-requiring, nonphotosynthetic) or to motility (Lewin, 1952, 1974A; Levine and Ebersold, 1960). Hyams and Davis (1972) obtained a wall-less mutant of *C. reinhardtii* which also lacked flagella. Sager (see 1972 for summary, 1974) has described the action of a number of cytoplasmic genes that affect such characteristics as streptomycin resistance. Ettl (1976) has summarized the taxonomy of *Chlamydomonas*.

CARTERIA Diesing   *Carteria* (after *H. J. Carter*) differs from *Chlamydomonas* largely in being quadriflagellate (Fig. 3.12). Lembi (1975A, B) has investigated the

Fig. 3.12 *Carteria crucifera* Korschikov. Vegetative cell. × 1125. (After Akins.)

ultrastructure of its flagellar apparatus, however, and has pointed out some differences in the organization of the pyrenoid. Its asexual reproduction is by cell division and liberation of the two, four, or eight motile division products from the parental cell walls. Van Dover (1972) has recently studied sexual reproduction in a variety of *C. eugametos*, a species in which clonal cultures produce compatible gametes.

CHLOROGONIUM Ehrenberg   *Chlorogonium* (Gr. *chloros*, green + Gr. *gonos*, offspring) (Fig. 3.13) cells are biflagellate but are usually elongate and pointed at one or both poles. They have more than two contractile vacuoles (Ettl, 1958). In the formation of the daughter cells, the first, and sometimes the ensuing, cytokineses are transverse. Both isogamous and oogamous species of *Chlorogonium* are known and meiosis is zygotic. Pringsheim (1969) has reviewed the status of *Chlorogonium*, and Pringsheim and Weisener (1960) showed it grew well when acetate was supplied as a carbon source.

(a)   (b)

Fig. 3.13 *Chlorogonium elongatum* Dang. (*a*) Vegetative cell. (*b*) Parent cell containing progeny. × 694.

HAEMATOCOCCUS C. A. Agardh   *Haematococcus* (Gr. *haema*, blood + Gr. *kokkos*, berry) (Fig. 3.14) is sometimes assigned to membership in the family Haematococcaceae (Droop, 1956A, B) because its protoplast is connected to the wall by radiating strands of protoplasm. Bowen (1967) has devoted especial attention to the ultrastructure of the wall, and Wygasch (1963) also has studied its ultrastructure. *Haematococcus lacustris* occurs in granitic pools, along with *Stephanosphaera* (p. 99), and in bird baths. Droop (1961) has reported that two strains of *H. pluvialis* can use organic nitrogen and acetate and grow heterotrophically in darkness. Pringsheim (1966) investigated the nutrition of 12 strains of five species of *Haematococcus*. Luxurious growth of all occurred when acetate and thiamine were present in the culture medium. Upon desiccation, the cells become transformed into nonmotile, brick-red

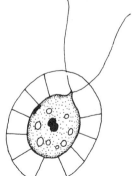

Fig. 3.14 *Haematococcus lacustris* (Girod.) Rostaf. × 694.

akinetes. The red color is called haematochrome or astaxanthin (3 : 3' diketo 4 : 4' dihydroxy-β carotene) (Lang, 1968B). Asexual reproduction is by division of the cells into two or four motile products and isogamous sexual reproduction occurs sporadically. The nomenclatural problems in this genus are staggering (Droop, 1956B).

POLYTOMA Ehrenberg   *Polytoma* (Gr. *polys*, many + Gr. *tome*, section) (Fig. 3.15) is of interest because it lacks chlorophyll and other pigments and, accordingly, is colorless. Lang (1963B) has shown that the cell of *Polytoma* contains a starch-bearing plastid, albeit a colorless one, and this has also been investigated by Scherbel, et al. (1974). Its nutrition is heterotrophic in light and in darkness. It is morphologically similar to *Chlamydomonas*. The cells form motile daughter cells and sexual reproduction is isogamous.

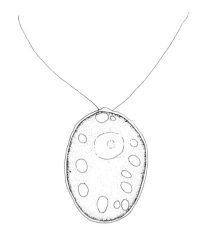

Fig. 3.15 *Polytoma uvella* Ehr. × 900.

Family 3.   Phacotaceae

The unicellular algae in this family have unusual, nonliving surface layers that are usually called **loricas**, although the morphological distinction between them and cell walls is not always clear. The loricas may or may not be impregnated with iron and/or manganese salts and thus may be brown. In some (e.g., *Phacotus*) the lorica is composed of two parts that separate at reproduction, while in others (some species of *Dysmorphococcus*) this characteristic is not evident.

PHACOTUS Perty *Phacotus* (Gr. *phacotos*, lentil-form) (Fig. 3.16) has a *Chlamydomonas*-like protoplast that is in contact with the lorica only at the anterior. Cell division into two, four, or eight motile cells is the only known method of reproduction. The cellular progeny is liberated by the splitting apart into two portions of the parental lorica.

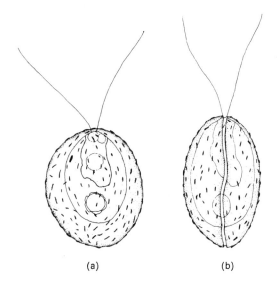

(a)   (b)

Fig. 3.16 *Phacotus lenticularis* (Ehr.) Stein. (*a*) Front view. (*b*) Lateral view. Note lorica. × 900.

PTEROMONAS Seligo The cells of *Pteromonas* (Gr. *pterion*, wing + Gr. *monas*, single organism) (Fig. 3.17) have their loricas extended into a projecting wing around the cell. This wing may be wider near the anterior pole to form shoulder-like protuberances. The lorica is composed of two shell-like portions joined at the wings. The cells are dorsiventral, being compressed in lateral aspect. Electron microscopy (Belcher and Swale, 1967A) reveals that the lorica, which appears homogeneous with light microscopy, is characterized by a hexagonal pattern.

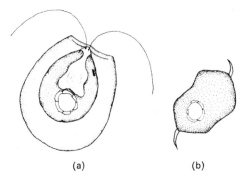

(a)   (b)

Fig. 3.17 *Pteromonas angulosa* (Carter) Lemm. (*a*) Front view. (*b*) Polar view. × 975. (After Belcher and Swale.)

The protoplast is *Chlamydomonas*-like with a parietal chloroplast and pyrenoid, a conspicuous stigma, and two apical contractile vacuoles. At reproduction, two or four division products are organized as immature cells and liberated by opening of the parental lorica in two portions.

DYSMORPHOCOCCUS Takeda   In *Dysmorphococcus* (Gr. *dys*, faulty + Gr. *morphos*, form + Gr. *kokkos*, berry) (Fig. 3.18) the loricas, which may be brown, depending on the age of the cells and the surrounding medium, are separated by some distance from the protoplast. Bold and Starr (1953) described *D. globosa*, which is unusual in that the young cells are liberated from the parental lorica by its posterior rupture and eversion of an inner wall layer as a colorless vesicle.

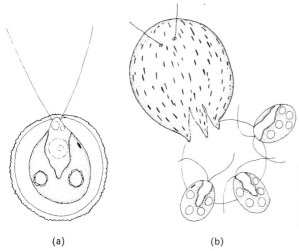

Fig. 3.18 *Dysmorphococcus.* (*a*) *D. variabilis* Takeda. Note lorica. (*b*) *D. globosus* Bold and Starr. Liberation of progeny from posterior tear in lorica (one below vesicle). (*a*) × 1125; (*b*) × 1125.

(a)   (b)

Family 4. Volvocaceae

The family Volvocaceae includes coenobic colonial organisms in which *Chlamydomonas*-like cells are surrounded by a common gelatinous matrix of relatively firm consistency. It is by no means certain that the protoplasts are bounded by cellulosic walls (Lang, 1963A). A **coenobium**, it will be recalled, is a colony in which the number of cells is determined at the time of origin of the colony, and it is not augmented during the existence of the individual colony. Included in the family Volvocaceae is a classic series of organisms long known to both phycologists and protozoologists. Almost all the genera have biflagellate cells with typical Volvocalean organization, and all reproduce asexually by the formation of colonies called **auto-colonies** because they are miniatures of the parental colony. In the development of the autocolonies, except in *Gonium* and *Stephanosphaera*, the hollow autocolonies turn inside out (Fig. 3.29*b*).

The coenobia are slightly curved plates (*Gonium, Platydorina*) or ovoid or subspherical (*Pandorina, Volvulina, Eudorina, Pleodorina, Volvox,* and *Stephanosphaera*).

Figures 3.19 to 3.32 illustrate these organisms and their organization and reproduction. All are haploid with zygotic meiosis. Harris (1971) has reported that a number of Volvocacean algae secrete substances into their surrounding medium, which are self-inhibitory and also inhibitory to some other organisms. A number of genera illustrating the Volvocaceae are discussed in the following account and they may be distinguished by aid of the following key:

1. Coenobia curved or flat, sometimes twisted 2
1. Coenobia spheroids or ellipsoids. 3
   2. Coenobia with posterior mammillate protuberances . . . .*Platydorina*
   2. Coenobia without such protuberances . . . . *Gonium*
3. Cells of the coenobia uniform in size 4
3. Four or up to one-half of the cells at the anterior
   pole smaller in size than the remainder 7
   4. Cells pyriform, anteriorly truncate, or hemispherical 5
   4. Cells spherical or elongate with processes 6
5. Cells contiguous at the center of the
   coenobium, not hemispherical . . . . *Pandorina*[10]
5. Cells not contiguous, hemispherical . . . . *Volvulina*
   6. Cells spherical, arranged in alternating tiers . . . . *Eudorina*
   6. Cells elongate with irregular processes . . . . *Stephanosphaera*
7. Coenobia with a few to 50% small, nonreproductive
   anterior cells . . . . *Pleodorina*
7. Coenobia composed mostly of small, nonreproductive cells;
   relatively few, large, reproductive cells present in the
   posterior hemisphere . . . . *Volvox*

GONIUM Mueller   The genus *Gonium* (Gr. *goneia*, an angle) (Fig. 3.19, 3.20) apparently contains the simplest members of the Volvocaceae. In *G. sacculiferum* Scherffel the coenobium consists of 4 *Chlamydomonas*-like cells (Fig. 3.19), while in

[10]Not contiguous in *P. unicocca* and *P. charkowiensis*, which may belong to the genus *Eudorina*.

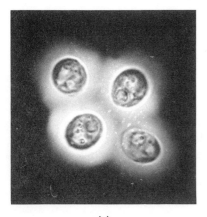

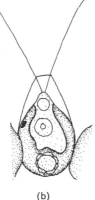

Fig. 3.19 *Gonium sacculiferum* Scherffel. (*a*) Coenobium, India ink preparation showing abundant matrix. (*b*) Cellular organization. (*a*) × 666; (*b*) × 900. [(*a*) courtesy of Prof. R. C. Starr.]

(a)   (b)

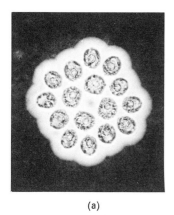

(a)

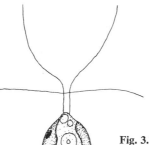

(b)

Fig. 3.20 *Gonium pectorale* Müller. (*a*) Front view of coenobium, India ink preparation. (*b*) Cellular organization. (*a*) × 405; (*b*) × 750. [(*a*) courtesy of Prof. R. C. Starr.]

G. pectorale Muell. (Fig. 3.20) it is composed of 16; other species have 8- or 32-celled coenobia. At maturity, the colonies become quiescent, and all the cells undergo successive mitoses and cytokineses, each forming a new colony. Saito (1972) and Saito and Ichimura (1975) have grown *G. multicoccum* in defined culture medium and demonstrated its requirement for vitamin B_{12}. Shyam and Sarma (1975) investigated mitosis in *Gonium* and later (1976) studied the effects of colchicine on cell division in that organism.

During sexual reproduction, the colonies produce individual, free-swimming gametes that are isogamous (Starr, 1955A; Stein, 1958, 1965, 1966A, B). Stein, in an investigation of 33 sexual populations of *G. pectorale*, found that populations from different geographical areas were mostly sexually compatible and that temperature was one factor in effecting sexual isolation. The individual colonies of most populations of *G. pectorale* are genetically of one or another compatible mating type and only one self-compatible clone has been found. Meiosis is zygotic on the basis of both cytological and genetic evidence. The zygote of *G. pectorale* germinates to form a four-celled coenobium each of whose cells gives rise in the next generation to the typical 16-celled individuals (Stein, 1958A).

PANDORINA Bory   In *Pandorina* (name refers to autocolony liberation) (Fig. 3.21) the 16–32 cells with truncate apices are arranged to form an ellipsoidal coenobium. Polarity is manifested in that the stigmata of the anterior cells of the colony are larger than those at the posterior, a phenomenon that characterizes all the remaining genera in the series, including *Volvox*.

Rayburn and Starr (1974), in an investigation of the nutrition of *P. unicocca*, reported that deficiency of sulfur and nitrogen stimulated sexual reproduction. In addition to autocolony formation (Fig. 3.21*d*), isogamous sexual reproduction occurs in *Pandorina*. Coleman (1959), who demonstrated the existence of 15 pairs of compatible strains among 47 clones collected from various parts of the United States, has studied (1963) the immunological patterns of their flagella and (1975) survival of these clones; she and Zollner (1977) have recently reported polymorphism within *P. morum*.

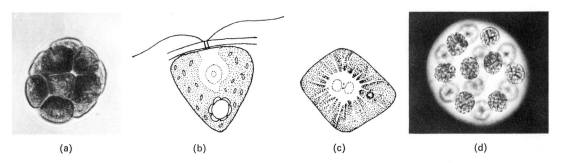

(a)   (b)   (c)   (d)

Fig. 3.21 *Pandorina morum* Bory. (*a*) Coenobium. (*b*), (*c*) Cellular organization: cell wall in (*c*) in anterior-polar view. (*d*) Autocolony formation. (*a*) × 275; (*b*), (*c*) × 900; (*d*) × 180. [(*d*) courtesy of Prof. R. C. Starr.]

VOLVULINA Playfair   *Volvulina* (L. *volvere*, to turn) (Fig. 3.22) differs from *Pandorina* in that its 16 truncate hemispherical cells are not contiguous in the coenobia and their individual envelopes are not confluent. In *V. steinii* Playfair, it is of interest that although the vegetative cells lack pyrenoids, one develops in each zygote (Stein, 1958B.) In *V. pringsheimii* Starr the vegetative cells contain pyrenoids. Carefoot (1966) reported, on the basis of genetic evidence, that meiosis is zygotic in both species. He also reported (1967) that both *V. pringsheimii* and *V. steinii* require vitamin B_{12} and an organic carbon source. Starr (1962) showed that both species are isogamous and heterothallic.

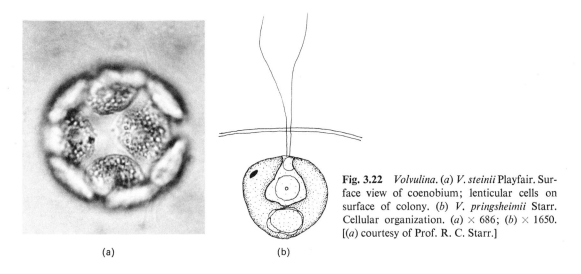

(a)   (b)

Fig. 3.22 *Volvulina*. (*a*) *V. steinii* Playfair. Surface view of coenobium; lenticular cells on surface of colony. (*b*) *V. pringsheimii* Starr. Cellular organization. (*a*) × 686; (*b*) × 1650. [(*a*) courtesy of Prof. R. C. Starr.]

EUDORINA Ehrenberg   In *Eudorina* (Gr. *eu*, well + *dorina*, meaningless) (Fig. 3.23) the cells are arranged in alternating rings to form 16- to 32-celled coenobia. Asexual reproduction is by autocolony formation in colonies that are nonmotile, at least in the later stages.

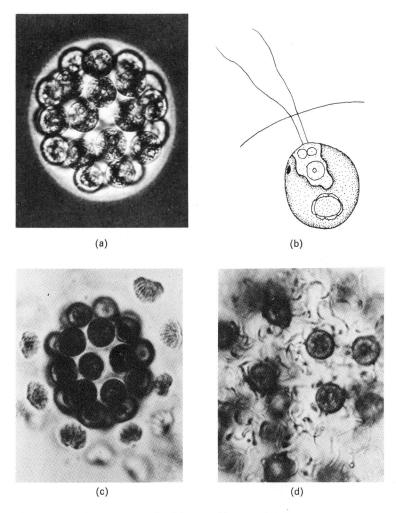

(a)   (b)

(c)   (d)

Fig. 3.23 *Eudorina elegans* Ehr. (*a*) Coenobium; India ink preparation; 32 cells arranged in rings of 4, 8, 8, 8, 4. (*b*) Cellular organization. (*c*) Female colonies surrounded by sperm platelets. (*d*) Female gametes surrounded by sperm. (*a*) × 250; (*b*) × 750; (*c*) × 235; (*d*) × 282. [(*a*) courtesy of Prof. R. C. Starr; (*c*), (*d*) after Goldstein.]

In sexual reproduction (Fig. 3.23*c*), clonal populations may be genetically unisexual or bisexual, and the gametes are anisogamous. In both cases the individual colonies are either male or female. All the cells of a female colony may function as gametes, while all the cells of the male colonies develop into packets of small, sperm-like male gametes. Szostak, et al. (1973) found that increasing light intensity speeds up the formation of sperm packets and also demonstrated the presence of a sperm-packet inducer. The sperm packets are released and swim as units to the female colonies into which individual sperms penetrate. After a period of dormancy, the zygotes germinate producing one (usually) biflagellate cell or occasionally two or three. Genetic evidence (Goldstein, 1964) indicates that meiosis is zygotic.

Goldstein (1964, 1967) studied 73 geographically dispersed clones from 40 natural populations of *Eudorina*, appraising their compatibility and sexual isolation. Toby and Kemp (1975) obtained biochemical mutants in *E. elegans*, while Lee and Kemp (1976) have reported on DNA changes during synchronous growth of that species.

PLATYDORINA Kofoid   The flattened colonies of *Platydorina* (Gr. *platys*, flat + *dorina*, meaningless) (Fig. 3.24) are slightly twisted and somewhat horseshoe-shaped and contain 16 to 32 cells; mammillate projections arise from the colonial matrix at the posterior. Harris and Starr (1969) and Harris (1969) have investigated the morphology and physiology of the single species, *P. caudata* Kofoid.

Asexual reproduction is by autocolony formation (Fig. 3.24*b*). These arise by a series of cleavages identical to those in *Eudorina*, but after eversion the tiny spherical colonies flatten and assume the typical *Platydorina* form. Evidence of this earlier arrangement of the cells in a spheroid is seen in the position of the flagellar poles in the colony, for in a 16-celled colony two of the four central cells face in one direction and two in the other, while along the margin alternate cells face in the same direction.

Harris and Starr (1969) have described sexual reproduction in *Platydorina*. All the strains available to them were dioecious, biflagellate sperms being formed in some colonies and female gametes in the other. The female gametes are similar to vegetative cells and, if unfertilized, can develop into autocolonies. The pattern of inheritance of mating type indicates that meiosis is zygotic. Harris (1969, 1970, 1971, 1972) has reported on the nutrition and sexuality of *Platydorina* as well as demonstrating its secretion of autoinhibitory and alloinhibitory substances, the latter inhibiting the growth of other Volvocacean algae.

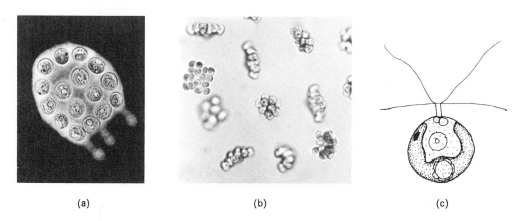

(a)   (b)   (c)

Fig. 3.24 *Platydorina caudata* Kofoid. (*a*) Coenobium, 16-celled with typical posterior projections. (*b*) Autocolony formation; plakeas in various stages of inversion prior to secondary flattening to form typical colonies. (*c*) Cellular organization. (*a*) × 209; (*b*) × 197; (*c*) × 585. [(*a*), (*b*) courtesy of Prof. Richard C. Starr.]

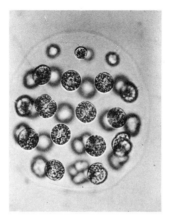

Fig. 3.25 *Pleodorina illinoisiensis* Kofoid. Coenobium, 32-celled, showing early stages in autocolony formation; the anterior tier of four cells is much smaller than the others. × 183. (Courtesy of Prof. Richard C. Starr.)

PLEODORINA Shaw   *Pleodorina* (Gr. *pleon*, more + *dorina*, meaningless) (Figs. 3.25, 3.26) differs from *Eudorina* in the cellular dimorphism of its colonies. In this organism, at least some of the cells toward the anterior pole of the coenobium are smaller than the remainder, and they usually fail to form autocolonies. In *P. illinoisensis* Kofoid (Fig. 3.25), four such nonfertile cells occur, while in the larger *P. californica* Shaw, in which the colonies often contain 128 cells, approximately one-half of the cells, those toward the anterior, are smaller and infertile (Fig. 3.26*a, c*).

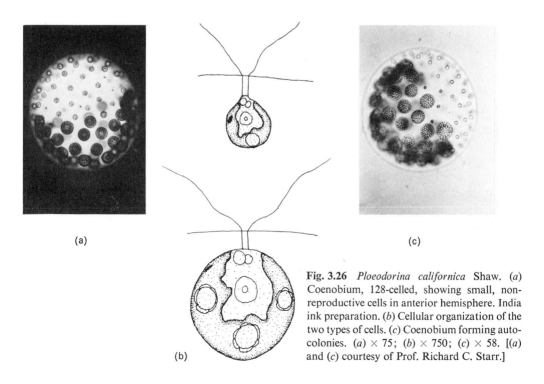

(a)

(b)

(c)

Fig. 3.26 *Ploeodorina californica* Shaw. (*a*) Coenobium, 128-celled, showing small, non-reproductive cells in anterior hemisphere. India ink preparation. (*b*) Cellular organization of the two types of cells. (*c*) Coenobium forming auto-colonies. (*a*) × 75; (*b*) × 750; (*c*) × 58. [(*a*) and (*c*) courtesy of Prof. Richard C. Starr.]

Goldstein (1964) reported that the clonal populations he studied were unisexual, producing either male or female colonies. Sexual reproduction is anisogamous, much like that of *Eudorina*. Goldstein (1964) included *Pleodorina* in the genus *Eudorina* because of intercrossing between several strains of *P. illinoisensis* with several species of *Eudorina*. Gerisch (1959) demonstrated experimentally that at least a few of the anterior cells of *P. californica* had not lost their capacity for reproduction. The fact that these smaller anterior cells are usually present in *P. illinoisensis* and *P. californica*, however, seems to be at least as weighty evidence for retaining them in the genus *Pleodorina*. Hobbs (1971, 1972) has demonstrated differences in the position and orientation of the stigma in *Pleodorina illinoisensis* and has described the organization of the flagellar apparatus, chloroplast, and pyrenoid.

VOLVOX Linnaeus   The spheroidal colonies of *Volvox* (L. *volvere*, to roll) (Figs. 3.27–3.29), species of which are the largest of the Volvocales, have been known since the beginning of the eighteenth century when Leeuwenhoek first described them in January, 1700 (Dobell, 1932). The colonies of most of the species are visible to the naked eye and are strongly phototactic, a phenomenon investigated by Huth (1970).

Each colony is composed of many small (vegetative or somatic) cells and relatively few larger (reproductive) cells, aflagellate **gonidia,** all arranged at the periphery of the coenobium. Cell numbers in the various species range between 1000 and 50,000. The cells are surrounded by mucilaginous envelopes that in most species are not confluent, and thus their boundaries, angular by mutual compression, can be readily demonstrated when stained with methylene blue. Electron-microscopic examination shows that in all species, except those with stellate cells (such as *V. globator L.*), there is a thickened layer of material of unknown composition[11] lying immediately adjacent to the cell's plasma membrane; this layer has been interpreted as a cell wall by some investigators. Burr and McCracken (1973) have reported the occurrence of a fibrillar layer on the surface of the *Volvox* sheath.

The somatic cells in many species (*V. aureus* Ehr., *V. carteri* Stein) have *Chlamydomonas*-like organization (Fig. 3.28); in others (*V. globator*) the construction is more reminiscent of *Haematococcus droebakensis*, inasmuch as the broad intercellular connections give the cells a stellate appearance. In those species having fine connections, or lacking connections, in the mature spheroid, the cell is spherical to subspherical. All species have cytoplasmic connections between the cells during embryonic development, but in some they break as the young individual enlarges. Bisalputra and Stein (1966), Ikushima and Maruyama (1968), and Pickett-Heaps (1970) demonstrated that the cytoplasmic bridges or connections are the stretched remnants of protoplasm of incompletely divided protoplasts.

In addition to the polarity evidenced by gradation in size of the stigmata from anterior to posterior, the asexual reproductive cells (autocolony initials or gonidia) are typically located in the posterior area of the spheroid (Fig. 3.27*a*). The number of gonidia varies from species to species, and also within the species. Spheroids multiply-

---

[11]Kochert (personal communication) says the matrix contains proteins, probably as a glycoprotein.

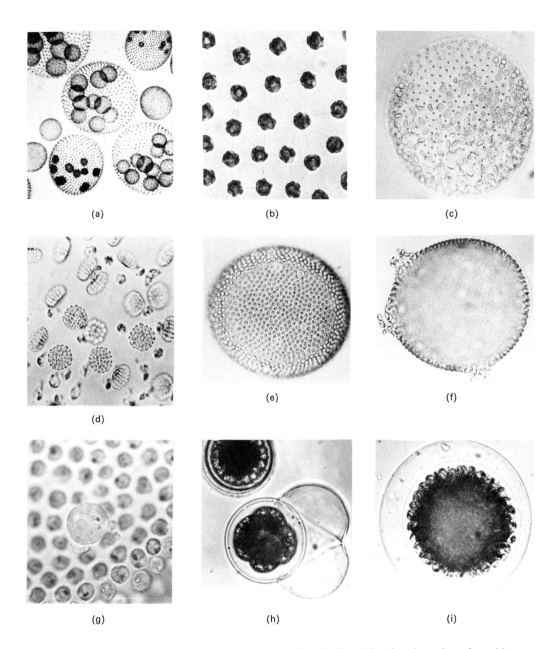

Fig. 3.27 *Volvox aureus* Ehr. (*a*) Coenobia of various sizes, five with auto-colonies. (*b*) Surface view of coenobium showing protoplasmic connections between the cells. (*c*) Male coenobium with sperm packets. (*d*) Sperm packets, more highly magnified. (*e*) Young, asexual, facultative female; the gonidia are undivided and may function as eggs. (*f*) Three sperm packets at surface of a female coenobium. (*g*) Egg with apposed sperms. (*h*), (*i*) Zygote and its germination to form a juvenile colony. (*a*) × 26; (*b*) × 466; (*c*) × 77; (*d*) × 359; (*e*) × 125; (*f*) × 120; (*g*) × 480; (*h*), (*i*) × 330. (Courtesy of Prof. W. A. Darden.)

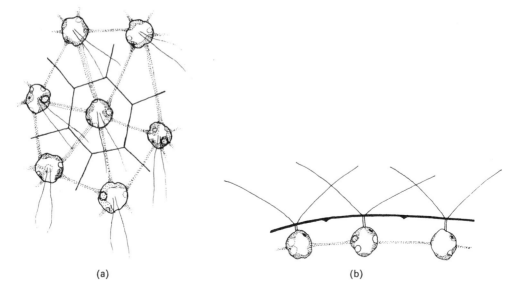

(a)   (b)

Fig. 3.28 *Volvox aureus.* Cellular organization. (*a*) Cells in anterior polar view. (*b*) Cells in lateral view. × 525. (After Bold, *Morphology of Plants*, 3rd ed., 1973, Harper & Row, Publ.)

ing under optimum conditions will be larger and contain greater numbers of reproductive cells than those growing under conditions where light, nutrients, or some other factor is limiting.

New individuals, whether asexual, male, or female, are formed only by a gonidium undergoing a series of cleavages to form an autocolony (Fig. 3.27*a*). Such cleavages may begin when the gonidium is small, each cleavage being followed by a period of cell enlargement, as in *V. aureus*, or a gonidium may undergo extreme enlargement followed by a rapid series of successive bipartitions as in *V. carteri*. Starr (1969) has described the process in *V. carteri* in detail, while Karn et al. (1974) have illustrated it in *V. obversus* (Shaw) Printz. The first cleavage is in a plane parallel to the longitudinal axis of the gonidium and perpendicular to the surface of the parental spheroid in which the gonidium is located. All other divisions are parallel to the longitudinal axis but slightly oblique. This pattern of cleavage is similar to that described as spiral cleavage in some animals and gives the alternating tiers of cells typical of spiral cleavage. Although the eight-celled stage of the *Volvox* autocolony is formed by the same pattern of cleavage that produces an eight-celled stage of *Gonium*, *Pandorina*, or *Eudorina*, it can be observed that in *Volvox* the four cells, not in contact at the base, begin to migrate upward. As they touch, there is formed the typical eight-celled structure with one tier of four cells surrounding a small pore, called the **phialopore**, alternating with a lower tier of four cells. At the same time, the pole of each cell has turned so that the anterior end of each cell projects toward the interior of the cellular mass. Continued divisions of the cells, always in the longitudinal plane of the cell, produce a hollow ball of cells; the latter become increasingly smaller with the succeeding rapid divisions. Apparently a minimal size is reached at which time the cell divisions cease.

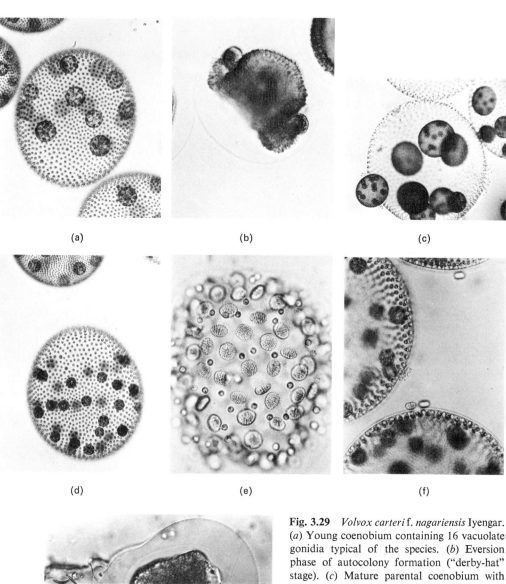

(a)

(b)

(c)

(d)

(e)

(f)

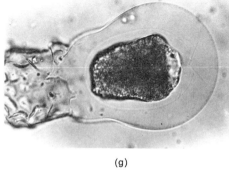

(g)

Fig. 3.29 *Volvox carteri* f. *nagariensis* Iyengar. (*a*) Young coenobium containing 16 vacuolate gonidia typical of the species. (*b*) Eversion phase of autocolony formation ("derby-hat" stage). (*c*) Mature parental coenobium with escaping autocolony; note the gonidia in the latter. (*d*) Young female with approximately 35 eggs. (*e*) Dwarf male coenobium showing the 1:1 ratio of sperm packets to somatic cells. (*f*) Parts of the female coenobia, two with intact sperm packets at their surface and the third with the remains of a dissociated sperm packet. (*g*) Germinating zygote; note single zoospore within the vesicle with two of the three aborted haploid meiotic nuclei at the upper edge of the vesicle. (*a*) × 63; (*b*) × 256; (*c*) × 14; (*d*) × 60; (*e*) × 157; (*f*) × 98; (*g*) × 260. (Courtesy of Prof. Richard C. Starr.)

The phialopore, first apparent in the eight-celled stage, has remained as the one area of the developing hollow sphere in which there are no intercellular connections. Following the last cell division, the phialopore begins to open and the lips to fold back. This continues until the autocolony has completely turned inside out. The mechanics of this eversion are not known, but the results are at once evident, for one sees immediately the development of one flagellum on the anterior pole of each cell, soon followed by the development of a second one.

Mature autocolonies escape from the parental spheroid through rupture of the somatic cell layer at its periphery. The somatic cells of the parental spheroid do not undergo further development in the usual *Volvox*, but Starr (1970B) has described a mutant of *V. carteri* in which the ability of the somatic cells to enlarge and reproduce has been retained. The biochemical action of this mutant locus has not yet been explained.

In most species of *Volvox*, the reproductive cells can be seen in the autocolonies only after eversion, but in *V. carteri* f. *nagariensis* Iyengar (Starr, 1970B) the reproductive cells can be seen to be the result of unequal cell divisions at predictable times and places in the young autocolony, which result in the segregation of a large reproductive cell and a small somatic cell. The former (gonidia) are cut off in a developing autocolony by unequal cleavages of the 16 cells in the anterior half of the autocolony, which becomes the posterior half after eversion. In the developing female coenobium, differentiating divisions occur in the anterior two-thirds of the colony at the division of the 64-celled stage, while in the male colony every cell undergoes an unequal cleavage, usually at the division of the 128- or 256-celled stage. In this case, no further divisions occur, and thus the small cells form the total complement of somatic cells which is equal in number to the large cells which later undergo further cleavage to form packets of sperm.

Sexual reproduction in *Volvox* is strictly oogamous (Figs. 3.27c–i, 3.29d–g) (Starr, 1968; 1969; 1970A, B; 1971A; 1975). Depending on the species and strain, sexual coenobia may contain either eggs or sperm and thus be unisexual or dioecious, or they may contain both and, accordingly, be bisexual or monoecious. In some species, such as *V. aureus*, no specially differentiated eggs are formed, the asexual reproductive cells (gonidia) functioning as such. In some species, individuals containing only eggs and others containing only sperm may be produced in the same clonal population, but in most species the male and female individuals are formed in separate clonal populations.

The egg cells are usually darker and more dense in appearance than gonidia, and in some monoecious spheroids they are decidedly flask-shaped prior to fertilization. The minute, elongate biflagellated sperm are produced by cell divisions that result in the organization of the curved platelets of 16–64 yellowish sperms in some species or of compressed globoids of 256–512 sperms in others. Deason, et al. (1969) have studied the ultrastructural development of the sperms of *V. aureus*.

The groups of sperm are liberated from their parental colonies and, presumably by chemotactic stimulation (in dioecious species), swim to the surface of female colonies (Figs. 3.27f, 3.29f), where they dissolve a hole in the matrix and then dissociate into individual sperms that penetrate the female colonies and fertilize the eggs. In monoecious spheroids, the sperm packets dissociate within the spheroid where they are produced and fertilize the eggs therein. The zygotes very soon develop thick walls and become red (Fig. 3.27*h, i*). Colonies that contain zygotes ultimately disintegrate as the zygotes enter dormancy.

It has been demonstrated genetically and cytologically (Starr, 1970B, 1975) that meiosis is zygotic in *V. carteri* f. *nagariensis*. It is assumed that this is true in all other species. As germination (Figs. 3.27*h, i*; 3.29*g*) of the zygote begins, meiosis occurs, and three of the haploid nuclei are extruded; a similar occurrence has been described by Schreiber (1925) for *Eudorina*. The single motile cell that emerges then swims for a short time, and undergoes successive cleavages to form a small juvenile colony, the gonidia of which initiate a new asexual cycle.

In *Volvox*, as in many other green algae, sexual reproduction is not usually observed as a continuing process throughout the growth of a natural population but rather as a phenomenon that affects a large part of the population on occasion. The change of the mode of reproduction from asexual to sexual in some algae is thought to be associated with the depletion of some nutritive factor such as available nitrogen, and thus a large part of the population can be expected to react in like manner, this resulting in the production of large numbers of gametes. In certain dioecious species of *Volvox*, the factor that ensures that a large number of individuals of both sexes will be produced and mature at the same time has been demonstrated to be a chemical produced by the males (or sperm) of the species involved. A few males appear first in a population and secrete the **inducing factor**; thus, in succeeding generations, most of the individuals produced are sexual. Darden (1966) first described this phenomenon in *V. aureus* where filtrates from a sexual population when added to asexual populations caused the developing autocolonies to mature as males rather than as asexual coenobia. Starr (1968) described similar induction phenomena in a number of species of *Volvox*, detailed studies of which have been published by Darden (1970, 1971, 1973, 1974), Kochert (1968), McCracken (1970), McCracken and Starr (1970), VandeBerg and Starr (1971), Starr (1969, 1970A, B, 1971A, 1972), Pall (1973), Starr and Jaenicke (1974), Kochert and Yates (1974) and Karn, et al. (1974). The inducing substances of each species are species-specific, and their biological activity is only to effect the production of a sexual individual. The sex of the individual produced is genetically determined, the inducer merely initiating the sexual pattern of development rather than the asexual. Starr and Jaenicke (1974) have isolated the inducer from *Volvox carteri* f. *nagariensis* and shown it to be a glycoprotein of *ca.* 30,000 MW with activity at concentrations as low as $(3 \times 10^{-16})M$. Meredith and Starr (1975) have elucidated the genetic basis for the potency of the male-inducing hormone of *V. carteri* f. *nagariensis*.

The potential of *Volvox* for use in studies of development has been demonstrated by those workers mentioned in the preceding paragraph and by Sessoms and Huskey (1973). The latter have with chemomutagenic agents induced and isolated morphogenetic mutants of *V. carteri* f. *nagariensis*.

STEPHANOSPHAERA Cohn *Stephanosphaera* (Gr. *stephanos*, crown + Gr. *sphaira*, sphere) (Fig. 3.30) is monotypic, *S. pluvialis* Cohn being the only known species. The organism occurs in shallow granitic pools. The coenobia are spherical to

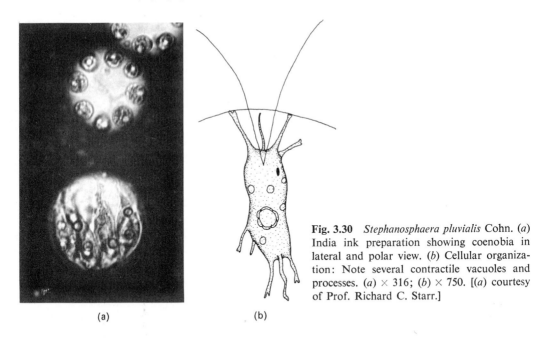

Fig. 3.30 *Stephanosphaera pluvialis* Cohn. (*a*) India ink preparation showing coenobia in lateral and polar view. (*b*) Cellular organization: Note several contractile vacuoles and processes. (*a*) × 316; (*b*) × 750. [(*a*) courtesy of Prof. Richard C. Starr.]

(a)   (b)

ellipsoidal and most frequently contain eight unwalled, elongate, biflagellate cells arranged in a ring. Each cell contains a stigma, several pyrenoids in the chloroplast and several contractile vacuoles. The cell surface is extended into processes that may fork (Fig. 3.30). Droop (1956A) has pointed out the similarities in cellular organization of *Stephanosphaera* and *Haematococcus* and included them in the same family. He also (1961) has investigated the nutrition of the former and reported that it requires vitamin B_{12} and thiamine and that it is facultatively heterotrophic.

Reproduction is by autocolony formation and sexual reproduction is isogamous (Strehlow, 1929), clonal cultures being bisexual (Droop, 1956A). Meiosis is inferred to be zygotic.

Stephanosphaera, along with *Haematococcus*, as noted above, is sometimes classified in a family Haematococcaceae (e.g., Smith, 1950), because of the protoplasmic cellular processes, but as Bourrelly (1966) pointed out, protoplasmic extensions also occur in some species of *Volvox*. Whether the autocolonies of *Stephanosphaera* undergo eversion, as in most other Volvocaceae, is not known.

Family 5. Astrephomenaceae

Bourrelly (1966) does not recognize the family Astrephomenaceae, erected by Pocock (1953) for her genus *Astrephomene*. The organization of the coenobia and their ontogeny are quite different from those of the Volvocaceae, however.

ASTREPHOMENE Pocock   *Astrephomene* (Gr. *a*, negation + Gr. *strephomene*, turning itself) (Fig. 3.31), described first by Pocock (1953), although a motile coeno-

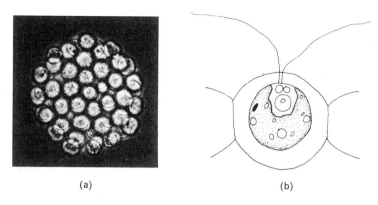

(a)   (b)

Fig. 3.31 *Astrephomene gubernaculifera* Pocock. (*a*) India ink preparation showing individual cellular sheaths (see margin of coenobium) and absence of common
matrix: Note two of four posterior small rudder cells. (*a*) × 223; (*b*) × 750. [(*a*)
courtesy of Prof. Richard C. Starr.]

bium, was classified by her in a separate family because the developing autocolonies
do not undergo eversion.

The spherical coenobia consist of 16–128 cells each surrounded by its own matrix.
The cells are identical except for the posterior two-seven cells, which are smaller, and
they have rather stiff flagella, which seemingly function as a rudder. Hence, they have
been called "rudder cells." The cells of *Astrephomene* lack pyrenoids but have several
contractile vacuoles. They are not enclosed in a common matrix and readily dissociate.

Asexual reproduction is by autocolony formation. Because the anterior poles of
cells are divergent at the end of the first cell division in autocolony formation, although
the future divisions are longitudinal, eversion of the colonies does not occur (Pocock,
1953; Stein, 1958).

Brooks (1966) has studied sexual reproduction in *Astrephomene* and found six
sexually isolated groups among 26 unisexual clones. When compatible coenobia are
brought together, they become agglutinated by their flagellar tips so as to form clusters
of colonies. From these, individual cells are freed and function as isogametes. The
germinating zygote forms a single cell, which, after several hours of motility, gives
rise to a small colony.

Family 6. Spondylomoraceae

The members of this family are coenobic; however, the cells of the coenobia are
not embedded in a gelatinous matrix. They are arranged in alternating tiers of four,
with their long axes more or less parallel to the long axis of the colony. This is a small
family of which only one representative is discussed below.

PYROBOTRYS Arnoldi *Pyrobotrys* (Gr. *pyr*, fire + Gr. *botrys*, bunch of
grapes) (Fig. 3.32) has been found to occur in pools rich in organic matter. The biflagellate cells are attenuated posteriorly, and their chloroplasts lack pyrenoids.

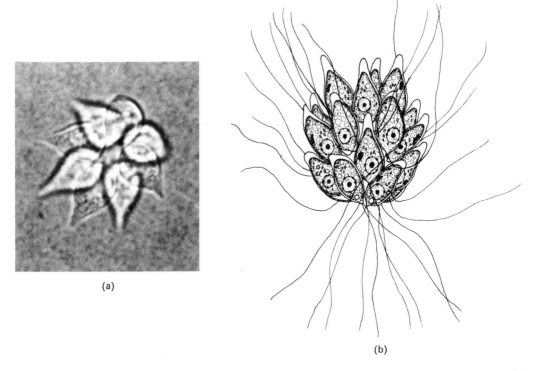

(a)

(b)

Fig. 3.32 *Pyrobotrys* sp. (*a*) Single living coenobium. (*b*) *P. gracilis* (Korsh.) Korsh. Organization of the colony. (*a*) × 952; (*b*) × 825. [(*b*) after Silva and Papenfuss.]

Asexual reproduction is by autocolony formation. The generic name *Pyrobotrys* of Arnoldi is said to be synonymous with *Uva* of Playfair (Bourrelly, 1962A; Dillard and DaPra, 1971).

In sexual reproduction, the isogametes are produced in pairs from the cells of the coenobia. They unite to form quadriflagellate zygotes which may be four-lobed (Fig. 3.32) and which were once classified as a distinct genus and species, *Chlorobrachis gracillima*.

Summary

The discussion of the order Volvocales is longer than that of other chlorophycean orders for several reasons. First, the organisms comprise an introduction to the Chlorophycophyta. Second, they illustrate certain basic biological and phycological generalizations, e.g., cellular organization and both asexual and sexual reproduction, for which they have served as objects of fruitful researches. Third, the production of biflagellate and quadriflagellate reproductive cells in a number of other orders of the green algae suggests that organisms similar to some of the Chlamydomonadaceae were involved in the evolutionary development of the group. Finally, the Volvocales are inherently diversified.

The occurrence of typical Volvocalean cellular organization throughout the group, which organization involves polarized motile cells with apically inserted flagella (except in *Pedinomonas*); contractile vacuoles, for the most part massive chloroplasts with one or more pyrenoids; and the lack of large central vacuoles, represents a cohesive group of characteristics. Volvocalean organization is unicellular or coenobic, reproduction in the latter instance being by autocolony formation. It has been speculated that the coenobic habit arose by mutations that resulted in failure of unicellular parental cells to free their offspring.

Many Volvocales form transitory nonmotile phases in which the cells are embedded in gelatinous matrices, the so-called *Palmella* stages. Their sexual reproduction includes isogamy, anisogamy, and oogamy (all represented among various species of the single genus *Chlamydomonas*). Genetic and cytological evidence has demonstrated that meiosis is zygotic and that it occurs in the early stages of germination of the previously dormant zygotes. Studies of sexual reproduction, especially that of *Chlamydomonas* and *Volvox*, have revealed that this process is under chemical controls genetically determined in the organisms but dependent for expression on the proper environmental conditions in the surrounding medium.

ORDERS 2. TETRASPORALES, 3. CHLOROCOCCALES, 4. CHLOROSARCINALES, AND 5. CHLORELLALES

In addition to the order Volvocales, to which are assigned motile unicellular and colonial (coenobic) green algae, there are other, unicellular and colonial (coenobic and noncoenobic) green algae which are nonmotile, although some of them have flagellate reproductive cells. The latter, being bi- or quadriflagellate and essentially *Chlamydomonas*- or *Carteria*-like in organization, have suggested to many phycologists that the organisms which produce such motile stages have evolved from flagellate ancestors.

Nonmotile unicellular[12] and colonial green algae other than Volvocales are usually classified in two orders, Tetrasporales and Chlorococcales (e.g., Smith, 1950). In this book, however, they have been assigned to four orders: Tetrasporales, Chlorococcales, Chlorosarcinales and Chlorellales. As conceived in this text, the orders of unicellular and colonial green algae may be distinguished with the aid of the following key:

    1. Cells or colonies typically flagellate and actively motile . . . . Volvocales
    1. Cells or colonies typically nonmotile 2
        2. Cells *Chlamydomonas*-like in organization, similar
        in polarity to that organism, often enclosed in
        gelatinous matrices; macroscopic or microscopic . . . . .Tetrasporales
        2. Cells not *Chlamydomonas*-like in form and organization when
        mature; not polarized except in stalked species 3

---

[12]Except for the desmids, which are included in the Zygnematales.

3. Not producing zoospores but only nonmotile spores or
   autocolonies; a few producing biflagellate gametes ....Chlorellales
3. Flagellate cells (zoospores and/or gametes) formed in reproduction..........4
   4. Cell division resulting only in the formation of
      flagellate cells (zoospores or gametes) distinctly
      different from their parental cells ....Chlorococcales
   4. In addition to producing flagellate reproductive cells, the parent
      cells undergoing partition to form two, four, or more cells similar
      to the parent cells; the daughter cells coherent for variable
      periods and often forming cellular complexes ....Chlorosarcinales

Order 2. Tetrasporales

A. General Features

The order Tetrasporales, which takes its name from the genus *Tetraspora* (Fig. 3.41), eloquently illustrates the statement that "nature mocks at human categories." This is so because it is often difficult to decide whether a given organism should be assigned to the Volvocales or to the Tetrasporales. While organisms classified in the latter have *Chlamydomonas*-like cellular organization and polarity, their cells are nonmotile, or only feebly motile, within a matrix, except during the reproductive phases. Although these would appear to be decisive and tangible criteria for separating the Tetrasporales from the Volvocales, two examples of the practical difficulties to be encountered may be cited at this point. First, it will be recalled that many Volvocalean organisms (e.g., *Chlamydomonas*, *Gonium*, and *Pandorina*) grown on agar may become nonmotile and form gelatinous "*Palmella*" stages. It is not easy, without prolonged observation of the algae in culture, to decide whether such stages belong to organisms of the order Tetrasporales or Volvocales. Second, and even more perplexing, are such intermediate types as *Gloeococcus* (Fig. 3.33) in which the flagellate cells move slowly within nonmotile colonies; this occurs also in palmelloid phases of *Chlamydomonas*

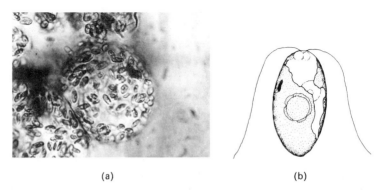

(a)   (b)

Fig. 3.33 *Gloeococcus minutissimus* King. (*a*) Spherical colonies in which the cells are feebly motile (*b*) Single cell: Note *Chlamydomonas*-like organization. (*a*) × 152; (*b*) × 1350.

(e.g., *C. suboogama*, Tschermak-Woess, 1959). Accordingly, the decision where to classify a given organism may be difficult and is often subjective insofar as some tetrasporalean algae are concerned.

In other cases, an additional characteristic, the presence on the cells of nonmotile pseudoflagella[13] as in *Tetraspora* (Fig. 3.41) and *Apiocystis* (Fig. 3.42), is a reliable and distinctive attribute.

Fritsch (1935) included the members of the Tetrasporales as a suborder within the Volvocales, while Smith (1950, 1955), Bourrelly (1966), and Fott (1972) recognize the Tetrasporales as an independent order.

The hydrophilic colloidal matrices in which the cells of Tetrasporalean algae are embedded and which they secrete has not been characterized chemically. King (1971) has discussed the organization of these matrices, and Fig. 3.35 illustrates his proposed terminology.

Asexual reproduction is accomplished by fragmentation in the colonial species and by zoospore formation. Sexual reproduction in the organisms selected to represent the Tetrasporales in this book is isogamous, if it occurs. Meiosis is presumably zygotic, as in the Volvocales, the zygote being the only diploid phase in the H, h life cycle.

Representatives of three of the several families of Tetrasporales are discussed below; namely, the Palmellaceae, Chlorangiaceae, and Tetrasporaceae. The latter differ from the first two families in having cells with pseudoflagella, the structure of which is discussed below.

B. Illustrative Organisms

Family 1. Palmellaceae

As noted above, it is often not easy to distinguish between palmelloid stages of the Chlamydomonadaceae and members of the Palmellaceae. King (1971) attempted to do so by devising a "motility standard." For this purpose he used as Volvocalean organisms of reference such well-known, widely studied and readily available species of *Chlamydomonas* as *C. chlamydogama* Bold, *C. eugametos* Moewus, *C. moewusii* Gerloff, *C. reinhardtii* Dangeard, and *Carteria olivieri* G. S. West, among others, growing them in a variety of culture media, in which they were flagellate and motile for periods up to 4 weeks. By contrast, such palmellacean algae as five species of *Gloeocystis*, one of *Asterococcus*, *Palmella texensis*, and *Palmellopis terrestris* in the same media, produced motile cells only for the first few days after inoculation. Thus, their motile period was transitory, and the motile cells gave rise to gelatinous colonies that enlarged and fragmented because of continuing cell division.

Gloeococcus, Palmella, Palmellopsis, Gloeocystis, Asterococcus, Hormotila, and *Hormotilopsis* have been chosen to illustrate the family Palmellaceae. They may be distinguished with the aid of the following key:

---

[13]Often called "pseudocilia"; because of their length, more appropriately here designated "pseudoflagella."

1. Cells feebly motile within the colonial matrix . . . . *Gloeococcus*
1. Cells nonmotile in the vegetative phase 2
    2. Cells at the apices of gelatinous, stalk-like tubes at maturity 3
    2. Cells in amorphous masses 5
3. Producing biflagellate zoospores . . . . *Hormotila*
3. Producing quadriflagellate zoospores 4
    4. Flagella arising from the bottom of an apical depression . . . . . *Prasinocladus*
    4. Flagella arising from the cell apex . . . . *Hormotilopsis*
5. Cells with an asteroidal chloroplast . . . . *Asterococcus*
5. Cells with a parietal, cuplike chloroplast 6
    6. Cells lacking contractile vacuoles . . . . *Palmella*
    6. Cells with contractile vacuoles 7
7. Individual sheaths not thicker than the protoplasts . . . . *Palmellopsis*
7. Individual sheaths capsular, clearly thicker
    than the protoplasts, firm . . . . *Gloeocystis*

GLOEOCOCCUS Braun   *Gloeococcus* (Gr. *gloia*, glue + Gr. *kokkos*, berry) (Fig. 3.33), here represented by *G. minutissimus* (King, 1973), is perhaps one of the most primitive members of the family. The organism consists of noncoenobic, spherical colonies that contain feebly motile, biflagellate, *Chlamydomonas*-like cells embedded in a gelatinous matrix. The colonies may reproduce by budding off new colonies.

Reproduction thus occurs by fragmentation to form small colonies and by zoospore formation and liberation. The biflagellate zoospores, produced in two's and four's from the parental cells, settle and initiate new colonies. *Gloeococcus minor* A. Braun is aquatic and produces colonies the size of an apple; *G. minutissimus* is edaphic.

PALMELLA Lyngbye and **PALMELLOPSIS** Korschikoff   *Palmella* (Gr. *palmos*, trembling), here represented by *P. texensis* Groover and Bold (1969) (Fig. 3.34), is characterized as follows: (1) The cells are embedded in a common matrix; (2) the cells are *Chlamydomonas*-like in organization, but the vegetative cells lack

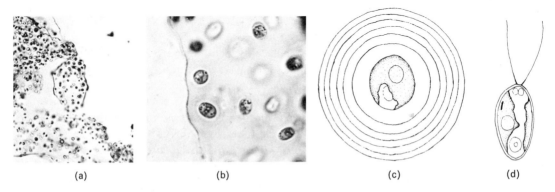

(a)   (b)   (c)   (d)

Fig. 3.34 *Palmella texensis* Groover and Bold. (*a*) Part of plant mass from a stationary-phase culture at low magnification. (*b*) A segment of the preceding at greater magnification. (*c*), (*d*) Cellular organization. (*d*) Zoospore. (*a*) × 44; (*b*) × 280; (*c*), (*d*) × 900. [(*c*), (*d*) after Groover and Bold.]

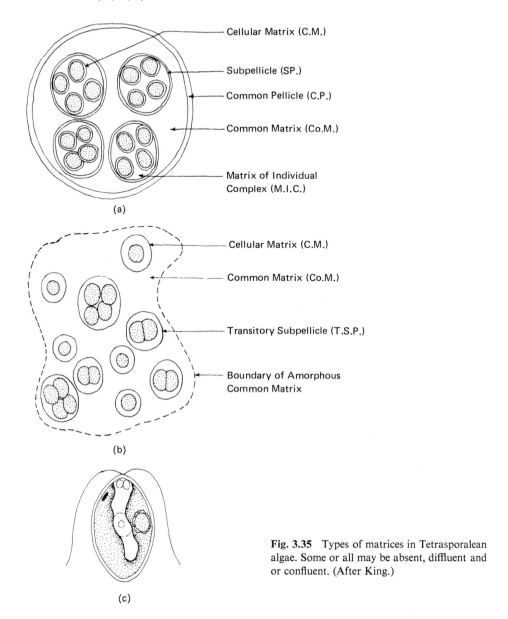

Cellular Matrix (C.M.)

Subpellicle (SP.)

Common Pellicle (C.P.)

Common Matrix (Co.M.)

Matrix of Individual
Complex (M.I.C.)

(a)

Cellular Matrix (C.M.)

Common Matrix (Co.M.)

Transitory Subpellicle (T.S.P.)

Boundary of Amorphous
Common Matrix

(b)

(c)

Fig. 3.35 Types of matrices in Tetrasporalean algae. Some or all may be absent, diffluent and or confluent. (After King.)

flagella and contractile vacuoles; (3) biflagellate, walled zoospores are produced. Apparently the only species available in culture collections, *P. texensis*, often has concentrically stratified individual matrices.

Palmellopsis (Fig. 3.35c), of which one species, as yet unpublished (*Palmella* sp. *of* King), is available in culture, differs from *Palmella* only in having contractile vacuoles in the vegetative cells, according to Korschikoff (1953).

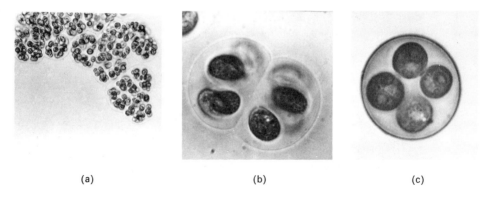

(a)   (b)   (c)

Fig. 3.36 *Gloeocystis.* (*a*) *Gloeocystis* sp. Colonies at low magnification. (*b*) *Gloeocystis* sp. at higher magnification. (*c*) *G. ampla* Kütz. (*a*) × 131; (*b*) × 675; (*c*) × 792. (Courtesy Dr. J. M. King.)

GLOEOCYSTIS Nägeli   The cells of *Gloeocystis* (Gr. *gloia*, glue + Gr. *kystis*, bladder) (Fig. 3.36) are enclosed in thick, capsular sheaths of firm consistency that may be stratified. The cells form discrete colonies, readily separable from the plant mass, and may reproduce by budding. King (1971) studied five species of *Gloeocystis* in culture and compared them with species of *Chlamydomonas* and other genera of palmelloid algae. All five species formed biflagellate zoospores. Fott and Nováková (1971), however, regard *Gloeocystis*, as presently interpreted, as a mixture of palmelloid green algae. They would limit *Gloeocystis* to organisms similar to the type species for which Nägeli did not record zoospores. The resolution of this nomenclatural disagreement is beyond the scope of the present text.

ASTEROCOCCUS Scherffel   *Asterococcus* (Gr. *astros*, star + Gr. *kokkos*, berry), with several species, consists of spherical cells embedded in a common gelatinous matrix (Fig. 3.37). Each cell contains an asteroidal chloroplast, the arms of which radiate from a central pyrenoid. The vegetative cells sometimes contain a stigma and contractile vacuoles. Biflagellate zoospores are produced, but cell division to form aplanospores within the matrix increases the size of the plant mass. Aplanospores are potential zoospores that have failed to become motile. Ettl (1964), Nováková (1964), and King (1971) have studied several species of *Asterococcus* in culture; *A. superbus* (Cienk.) Scherffel and another species are available in culture collections.

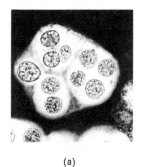
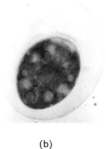

Fig. 3.37 *Asterococcus superbus* (Cienk.) Scherffel. (*a*) Colony beginning to fragment; India ink preparation. (*b*) Cellular organization; note asteroidal chloroplast. (*a*) × 360; (*b*) × 624. (Courtesy Dr. J. M. King.)

(a)   (b)

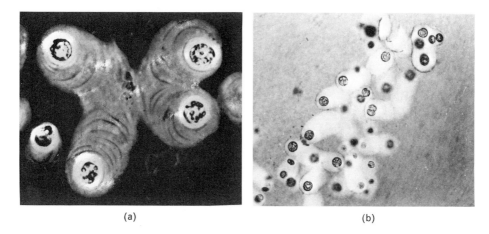

(a)   (b)

Fig. 3.38 (*a*) *Hormotila blennista* Trainor and Hilton. Four cells at the apices of
gelatinous cylinders. (*b*) *H.* sp. (*a*) × 480; (*b*) × 116. [(*a*) after Trainor and
Hilton.]

HORMOTILA Borzi and **HORMOTILOPSIS** Trainor and Bold   *Hormotila*
(Gr. *hormos*, motile + Gr. *tyle*, callus) and *Hormotilopsis* (Gr. *hormotila* + Gr. *opsis*,
appearance of), at maturity, consist of branching, stratified gelatinous stalks in which
the cells are embedded (Figs. 3.38, 3.39). These dichotomously branching stalks
develop in stationary-phase cultures through the unipolar accumulation of a gelatinous secretion. The zoospores of *Hormotila* are biflagellate, while those of *Hormotilopsis* are quadriflagellate. In rapidly growing (log-phase) cultures, the branching
stalks are poorly developed.

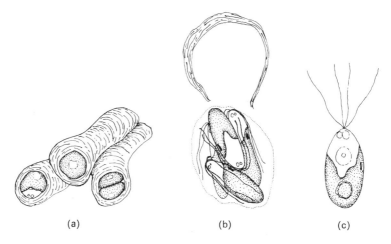

(a)   (b)   (c)

Fig. 3.39 *Hormotilopsis gelatinosa* Trainor and Bold. (*a*) Growth habit. (*b*)
Zoospore release. (*c*) Single zoospore. (*a*) × 375; (*b*), (*c*) × 900. (After Trainor
and Bold.)

Hormotila blenista Trainor and Hilton (1965) was isolated from soil as were Hormotilopsis gelatinosa Trainor and Bold (1953) and H. tetravacuolaris Arce and Bold (1958). Monahan and Trainor (1970, 1971) reported that filtrates of cultures of Hormotila blenista were autostimulatory and that they stimulated growth of two other organisms.

Family 2. Chlorangiaceae

The members of this family are sessile and attached by stalks. Only a single representative, Prasinocladus, is discussed below.

PRASINOCLADUS Kuckuck   The thallus Prasinocladus (Gr. prasios, leek + Gr. klados, branch) (Fig. 3.40) consists of segmented tubes with cells at their apices. The cells are Pyramimonas-like in organization but are walled and Chihara and Hori (1972) and Hori and Chihara (1974A, B) have described the organization, development, and ultrastructure of P. marinus (Cienk.) Waern and of P. ascus Proskauer. The tubes are longer and less frequently branched in P. ascus (Proskauer, 1950; Chihara, 1963) than in P. marinus. According to them, the organism reproduces by Pyramimo-

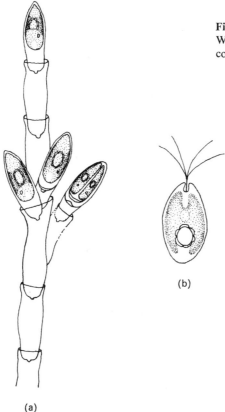

Fig. 3.40 *Prasinocladus marinus* (Cienk.) Waern Kuckuck. (*a*) Portion of dendroid colony. (*b*) Zoospore. (*a*) × 300; (*b*) × 675.

(b)

(a)

nas-like, quadriflagellate zoospores (Fig. 3.40*b*). The flagella of the zoospores arise from the bottom of an apical depression and the eyespot is median. The flagella bear two rows of scales on their surface (Parke and Manton, 1965). The four-lobed, cup-shaped chloroplast contains a single basal pyrenoid into which a lobe of the nucleus protrudes.

After a period of motility, the zoospores attach to the substrate by their flagella, and their protoplasts contract toward the distal portion of the cell lumen. The apical depression later evaginates to form a sort of papilla. The protoplast then secretes a new wall within the original one that is soon ruptured as the new cell wall and protoplast elongate; the original wall of the germling remains as a stalk. This process of formation of new walls within the old is repeated, so that the elongate stalk is composed of cast-off cell walls. Cell division results in dichotomy of the stalk. Under unfavorable conditions, *Prasinocladus marinus* may form spiny-walled cysts.

Prasinocladus, Pyramimonas, and several other green algae, because of their apical depression and scaly flagella, are sometimes grouped in a class, Prasinophyceae, coordinate with Chlorophyceae. Ricketts (1974) has investigated the culture requirements of the group. In light of the discussion of Stewart and Mattox (1974, p. 245) regarding the occurrence of scaly flagella in other green algae, the authors have tentatively retained *Pyramimonas* and *Prasinocladus* in the Chlorophyceae. *Prasinocladus* occurs in marine and brackish water, but Smith (1950) reports it can grow in practically freshwater.

Family 3. Tetrasporaceae

These noncoenobic colonial algae occur either as amorphous, gelatinous masses or as entities of definite form, i.e., spheroidal, saccate, or digitate. Most of the cells produce two long, flagella-like processes, the **pseudoflagella**, at their anterior poles. These are stiff and immobile and this is possibly correlated with their ultrastructural organization. Three representatives of the Tetrasporaceae will be discussed briefly below and may be distinguished with the aid of the following key:

1. Plant mass gelatinous, sometimes cylindrical,
   evident macroscopically . . . .*Tetraspora*
1. Plants microscopic 2
   2. Plant saccate, attached . . . .*Apiocystis*
   2. Plant consistency of spherical, planktonic colonies . . . .*Paulschulzia*

TETRASPORA Link   The several species of *Tetraspora* (Gr. *tetra,* four + Gr. *spora,* spore) (Fig. 3.41) may be of definite form macroscopically (e.g., digitate and up to 15 cm long) or more irregular and amorphous. They are gelatinous, extremely slippery to the touch, and, in the writers' experience, occur in cold, sometimes running, water. Microscopically, at least some of the cells are arranged in groups of four, an indication that two successive bipartitions characterize cell reproduction that results in growth of the organism. The characteristic groups of four cells suggested the generic name. The cells are surrounded by individual matrices that become indistinct as their contained cells secrete additional material to the matrix. Pickett-Heaps (1973A)

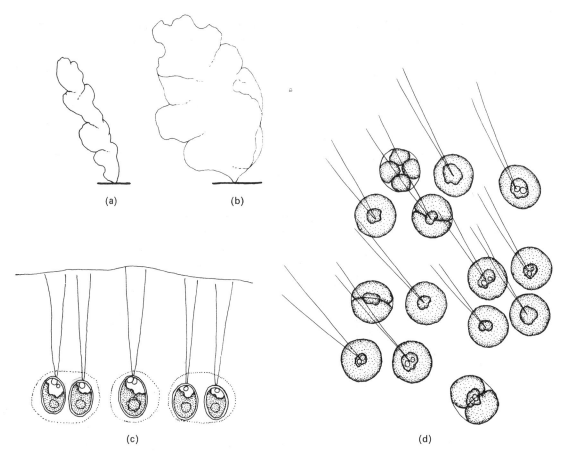

Fig. 3.41 *Tetraspora.* (*a*) *T. lubrica* (Roth) Ag. (*b*) *T. gelatinosa* Kütz. (*c*) Cellular organization, lateral view: Note pseudoflagella. (*d*) Cellular organization, surface view. (*a*), (*b*) × 0.5; (*c*) × 638; (*d*) × 1125.

described cell division in a species of *Tetraspora* in culture and reported that cytokinesis involves formation of a phycoplast.

Lembi and Herndon (1966) and Wujek and Chambers (1966) investigated the ultrastructural organization of the pseudoflagella in *T. lubrica* (Roth) Ag. and *T. gelatinosa* (Vauch.) Desv., respectively. The pseudoflagella lack the two central fibrils of functional flagella, and it has been conjectured that this accounts for their nonmotility.[14]

In addition to fragmentation, zoospores are readily produced when environmental conditions change [e.g., bringing the organisms into the laboratory from cold, running water (Rhodes and Herndon, 1967)]. The biflagellate zoospores, after a short period of motility, initiate new colonies. The flagella of the zoospores are developed anew and are not related to the pseudoflagella.

[14]In the diatom *Lithodesmium*, however, the functional flagellum of the motile male gamete also lacks the central pair of fibrils (Manton and Stosch, 1966).

Isogamous sexual reproduction has been reported for two species of *Tetraspora* (Klyver, 1929; Geitler, 1931). Geitler showed that individual plants of *T. lubrica* were unisexual and designated them as "+" and "−" since they were isogamous. He reported that + (plus) gametes would clump in extracts of − (minus) plants and vice versa. The zygotes, after a period of dormancy, divide to form four or eight nonmotile products, each of which may initiate a colony. It has been inferred, but not proved, that meiosis is zygotic.

APIOCYSTIS Nägeli   The saccate colonies of *Apiocystis* (Gr. *apion*, pear + Gr. *kystis*, bladder) (Fig. 3.42) are always attached, usually to other algae or to aquatic angiosperms, and are composed of several hundred cells at maturity. The cells may contain stigmata. Here, also, pseudoflagella are present and project beyond the common matrix. Biflagellate zoospores initiate new colonies and isogamous sexual reproduction has been recorded.

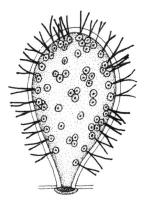

Fig. 3.42 *Apiocystis brauniana* Näg. Mature colony. Note pseudoflagella. × 131.

PAULSCHULZIA Skuja   *Paulschulzia* (after Paul Schulz) (Fig. 3.43) is plank-tonic. Its colonies are spherical or subspherical and composed of cells each with two contractile vacuoles and with two very long pseudoflagella. There is some tendency

Fig. 3.43 *Paulschulzia pseudovolvox* Skuja. Note pseudoflagella. × 263.

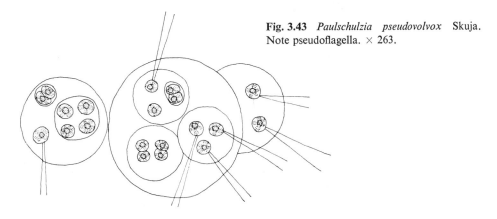

for the four products of cell division to persist as tetrads and for complexes of colonies to form. Reproduction by biflagellate zoospores is also known. *Paulschulzia pseudo-volvox* (Schulz emend. Teil) Skuja occurs in England and continental Europe and is available in culture collections. It is sometimes difficult to demonstrate the pseudo-flagella and the colonial matrices. The former sometimes become more apparent after staining with dilute methylene blue and the latter by mounting colonies in India ink.

Order 3. Chlorococcales

The order Chlorococcales, which takes its name from the genus *Chlorococcum* (Fig. 3.44), contains nonmotile, unicellular, and colonial (coenobic) green algae in which the division of the cells gives rise to zoospores and/or gametes. These may either be naked (unwalled) or have walls which are completely free from those of the parental cells which produce them (Fig. 3.44*b*). The parental cells do not *directly* form

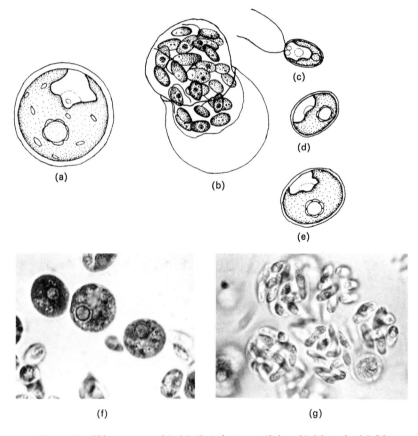

Fig. 3.44   *Chlorococcum.* (*a*)–(*e*) *C. infusionum* (Schrank) Menegh. (*a*) Mature vegetative cell. (*b*) Release of zoospores. (*c*)–(*e*) Increasingly older vegetative cells. (*f*) *Chlorococcum* sp. Cells from 2-week-old culture. (*g*) Zoosporogenesis. (*a*), (*b*) × 638; (*c*)–(*e*) × 750; (*f*), (*g*) × 280.

cellular progeny *similar to themselves* by cell division, as they do in the members of the Chlorosarcinales (p. 127). In other books (e.g., Smith, 1950; Bourrelly, 1966; Fott, 1971), the order is more broadly conceived, so as to include also the nonzoospore-producing unicellular green algae, which the present authors have classified in the order Chlorellales (p. 133). Unlike those of the Volvocales and Tetrasporales, the cells of most Chlorococcales are not polarized, and the special organelles (stigmata, contractile vacuoles) that characterize the cells of Volvocales and some Tetrasporales are usually absent from the vegetative cells of Chlorococcales, although they appear in their motile reproductive cells.

Representatives of four families of Chlorococcales are discussed in the following account. Three of these families (Chlorococcaceae, Protosiphonaceae, and Characiosiphonaceae) contain unicellular genera; the fourth (Hydrodictyaceae) comprises genera that are coenobic. These families include both subaerial and aquatic algae.

Family 1. Chlorococcaceae

Of the family Chlorococcaceae, the large genus *Chlorococcum* (ca. 38 species) and several other genera have been chosen as illustrative organisms.

CHLOROCOCCUM Meneghini and Similar Algae *Chlorococcum* (Gr. *chloros*, green + Gr. *kokkos*, berry) (Fig. 3.44) species are widely distributed in soils but grow luxuriantly when isolated into liquid culture media (Archibald and Bold, 1970). The spherical or ellipsoidal, uninucleate cells contain a single parietal (sometimes cup-shaped) chloroplast that includes one or more pyrenoids.[15]

Reproduction is accomplished by the division of the nonmotile vegetative cells into biflagellate zoospores that are *Chlamydomonas*-like in organization (Fig. 3.44g). After a period of active swimming, these settle down, lose or withdraw their flagella, and (while increasing in size over a period of several days) gradually become spherical. Under certain conditions, e.g., depletion of nutrients and/or of water in the culture medium, the zoospores fail to leave their parental cell walls. Thus, the motile phase is omitted, and the enlarging, potential zoospores are called **aplanospores.** It must be emphasized that aplanospores have the ontogenetic potentiality for being flagellate and motile. Archibald and Bold (1970) have characterized 38 species of *Chlorococcum* with respect to certain supplementary attributes, including configuration of the plant mass on agar, sensitivity to antibiotics, and production of extracellular proteases and amylase and of secondary carotenoids upon senescence.

Isogamous sexual reproduction occurs in several species of *Chlorococcum*. The gametes in such species are morphologically similar to the zoospores but function as gametes. They are thus designated as facultative gametes. The conditions under which zoospores function as gametes have not been precisely defined.

In addition to *Chlorococcum*, there are a number of other spherical Chlorococcaceae (Bold, 1970) that have been classified on the basis of criteria first suggested by

---

[15]For the use of pigmentation, pyrenoid ultrastructure, and isozyme analysis in the taxonomy of *Chlorococcum*, see McLean (1968), Brown and McLean (1969), and Thomas and Brown (1970B).

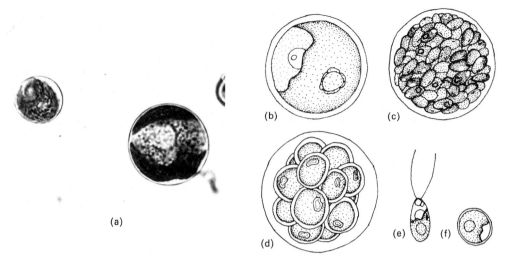

Fig. 3.45 *Neochloris.* (*a*) *N. cohaerens* Groover and Bold. Vegetative cells. (*b*)–(*f*) *N. pyrenoidosa* Arce and Bold. (*b*) Vegetative cell. (*c*) Zoosporogenesis. (*d*) Aplanospore formation. (*e*) Zoospore. (*f*) Young vegetative cell. (*a*) × 1330; (*b*)–(*f*) × 900.

Starr (1955D). These criteria include the position and form of the chloroplast (parietal, asteroidal, reticulate, lens-like, etc.); the presence or absence of a pyrenoid or pyrenoids; and the flagella length of the zoospores and their behavior upon quiescence (whether they become spherical immediately following the cessation of motility or whether they become spherical only gradually, during a several-day period of enlargement, as in *Chlorococcum*). Thus *Neochloris* (Gr. *neos*, new + Gr. *chloros*, green) (Fig. 3.45*a*, *b*) differs from *Chlorococcum* in that its zoospores become spherical immediately upon quiescence. In *Radiosphaera* (L. *radius*, ray + L. *sphaera*, sphere) (Fig. 3.46*a*, *b*), the chloroplast is asteroidal; in *Bracteacoccus* (L. *bractea*, thin metal plate

Fig. 3.46 *Radiosphaera dissecta* (Korsch.) Starr. (*a*) Vegetative cell: Note asteroidal chloroplast. (*b*) Zoospore. × 750.

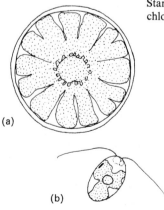

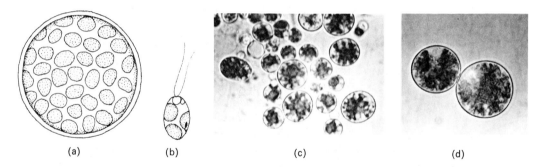

Fig. 3.47 (*a*), (*b*) *Bracteacoccus* sp. (*a*) Vegetative cell: Note lenticular chloroplasts. (*b*) Zoospore. (*c*) *Trebouxia magna* Archibald. (*d*) *T. excentrica* Archibald. In (*c*) and (*d*), note axile chloroplasts. (*a*), (*b*) × 750; (*c*) × 580; (*d*) × 650. [(*c*), (*d*) courtesy of Prof. Patricia A. Archibald.]

+ Gr. *kokkos*, berry) (Fig. 3.47*a*), the cells contain numerous lenticular chloroplasts lacking pyrenoids. *Trebouxia* (after *O. Treboux*) (Fig. 3.47*c, d*), the common phycobiont of lichens, differs from *Chlorococcum* in its axile chloroplast and production of zoospores that rapidly become spherical as the motile period ends. Archibald (1975) has recently monographed the genus *Trebouxia*. Fisher and Lang (1971B) have investigated the comparative ultrastructure of five species of *Trebouxia* that are lichen phycobionts. *Pulchrasphaera* (L. *pulchratudo*, beauty + L. *sphaera*, sphere) (Fig. 3.48*a–d*) has a pari-

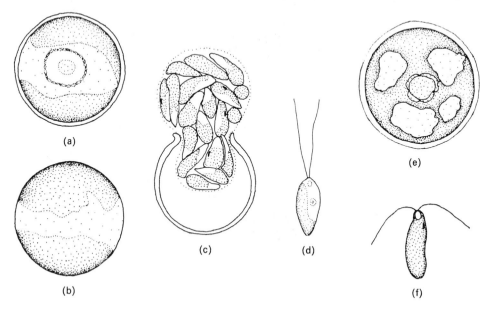

Fig. 3.48 (*a*)–(*d*) *Pulchrasphaera macronucleata* Deason. (*a*), (*b*) Vegetative cells. (*c*) Zoospore release. (*d*) Zoospore. (*e*), (*f*) *Neospongiococcum*. (*e*) *N. punctatum* (Arce and Bold) Deason: Note spongy chloroplast. (*f*) *N. cohaerens* Deason. Zoospore. × 900. (After Deason.)

etal chloroplast and lacks a pyrenoid, and its zoospores become spherical immediately upon quiescence (Deason, 1967). *Neospongiococcum* (Deason, 1971; Deason and Cox, 1971) has sponge-like chloroplasts (Fig. 3.48*e, f*).

CHARACIUM A. Braun and **PSEUDOCHARACIUM** Korsch.   *Characium* (Gr. *characion*, a marsh reed) (Fig. 3.49) is sometimes classified in a family of its own (Smith, 1950) but it is here included in the family Chlorococcaceae by the writers (as by Bourrelly, 1966), because it is *Chlorococcum*-like, but the cells are stalked and attached, or at least attenuated, at one or both poles. Lee and Bold (1974) studied a number of species of *Characium* of which many were epiphytic on other algae and aquatic angiosperms. Several species [e.g., *C. pseudopolymorphum* (Trainor and Bold) Philipose] occur with some frequency in soil. Lee and Bold (1974) have shown that the epiphytic species grow readily when attached to various nonliving substrates, and thus their epiphytism is not a requirement for growth.

Reproduction is by the formation of biflagellate zoospores (Fig. 3.49*c*), and isogamy has been demonstrated in *C. starrii* Fott (Starr, 1953). The zoospores of several species vary in their mode of attachment to the substrate. They may attach at their anterior poles; by their flagella and, by subsequent inversion, at their posterior poles; and directly at their posterior poles (Lee and Bold, 1974).

(a)   (b)

(c)   (d)

Fig. 3.49 *Characium.* (*a*)–(*c*) *C. oviforme* Lee and Bold. (*a*) Vegetative cells attached by their short stalks to a single particle. (*b*) Single cell showing stalk. (*c*) Zoosporogenesis. (*d*) *C. vacuolatum* Lee and Bold. (*a*), (*b*) × 600; (*c*) × 638; (*d*) × 525.

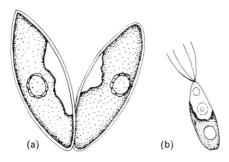

(a)   (b)

Fig. 3.50 *Pseudocharacium* sp. (*a*) Two vegetative cells. (*b*) Zoospore. × 750.

Pseudocharacium (Fig. 3.50) differs from *Characium* only in forming quadriflagellate zoospores.

Family 2. Protosiphonaceae

As herein conceived, the family Protosiphonaceae includes the single genus *Protosiphon*.

PROTOSIPHON Klebs   *Protosiphon* (Gr. *protos*, prior, hence primitive + Gr. *siphon*, a tube) (Figs. 3.51, 3.52) is a widely distributed alga on bare soils where its green, vesicular cells may form extensive patches. As the soil dries, these become orange-red. In nature and sometimes in laboratory culture on agar, the saccate cells achieve a length of 1 mm. One pole of the cell is typically extended as a tubular, presumably absorptive, structure, the rhizoid, while the other is enlarged and bulbous (Fig. 3.51*a*). These saccate cells have a large central vacuole and a reticulate chloroplast, the boundaries of which are difficult to establish with light microscopy. A number of pyrenoids occur in the chloroplast, and all but the youngest cells are multinucleate. Such multinucleate structures are often called **coenocytes.** Coenocytes

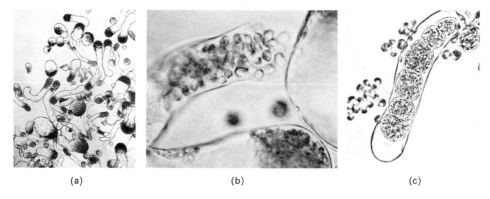

(a)   (b)   (c)

Fig. 3.51 *Protosiphon botryoides* Klebs. (*a*) Vegetative sacs (cells) in various stages of development. (*b*) Zoosporogenesis. (*c*) Sac containing coenocysts. (*a*) × 14; (*b*) × 129; (*c*) × 102.

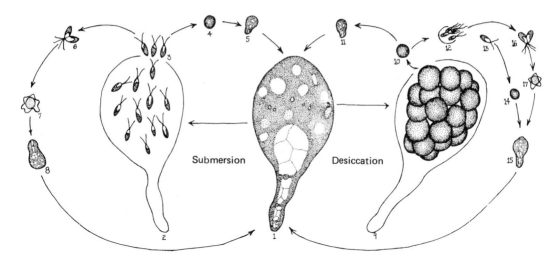

Fig. 3.52 *Protosiphon botryoides* Klebs. Diagrammatic summary of the life cycle. 1. Vegetative sac. 2. Zoosporogenesis. 3–5. Asexual development of vegetative sac. 6. Union of isogametes. 7. Zygote. 8. Germination of zygote to form new sac. 9. Coenocyst formation. 10, 11. Development of coenocyst into vegetative sac. 10, 12–15. Zoosporogenesis in coenocyst and development of new vegetative sacs. 16, 17. Union of gametes from coenocyst.

develop when repeated nuclear divisions are not followed by cytokineses. Such nuclear division is called **free nuclear division.** Berkaloff (1967) has followed ultrastructural changes in organization in *Protosiphon* throughout its development.

The reproductive cycle of *Protosiphon* is greatly influenced by environmental factors as illustrated in Fig. 3.52. When the saccate cells are submerged (as might occur during rain, in nature), they respond by undergoing autonomous plasmolysis and zoosporogenesis. In the latter process, the condensed protoplast undergoes a type of cytokinesis known as **progressive cleavage.** This is so called because the multinucleate protoplast is divided progressively by cleavage furrows into smaller segments until ultimately uninucleate, biflagellate zoospores have been formed. These begin to swim freely within the parental cells from which they are ultimately released. After a period of motility, they lose or withdraw their flagella and immediately become spherical. The zoospores sometimes function as aplanospores. In either case, the minute, spherical cells undergo free-nuclear division,[16] synthesize additional cytoplasm, increase in size rapidly, and become saccate. O'Kelley and Herndon (1961), Stewart and O'Kelley (1966), and Durant, et al. (1968) have investigated some of the factors that influence zoosporogenesis. The latter authors reported that blue light (400–460 nm) inhibits zoosporogenesis, which is enhanced by red light (530–700 nm). Thomas, et al. (1975) and O'Kelley and Hardman (1976) have investigated the photochemistry involved.

[16]Division of nuclei not accompanied by cytokinesis.

The zoospores, under conditions not yet defined precisely, may also function facultatively as isogametes that, upon fusion, develop into polyhedral zygotes (Fig. 3.52). These germinate, after a period of dormancy, into small, vegetative sacs. The site of meiosis has as yet not been determined with certainty but has been speculated to be zygotic.

In populations on soil and on agar, as the substrate dries, the saccate cells undergo cleavage into thick-walled, spherical multinucleate cells (Fig. 3.51c, 3.52) called **coenocysts.** These are at first green but become brick-red if desiccation continues. The coenocysts may become green again and develop into sacs if the substrate is slightly moistened. If they are submerged in water, they undergo progressive cleavage to form zoospores that may behave as gametes.

Thomas and Brown (1970) and Thomas (1971) have isolated 32 races of *Protosiphon* and studied their morphological variation and enzymes in relation to their classification.

Family 3. Characiosiphonaceae.

This family includes the single genus *Characiosiphon*.

CHARACIOSIPHON Iyengar   In contrast to the edaphic habitat of *Protosiphon*, *Characiosiphon* (Gr. *characion*, a marsh reed + Gr. *siphon*, tube) (Figs. 3.53, 3.54) is aquatic. It grows on stones of shallow streams in northern India (Iyengar, 1936), but strains isolated by Starr from Indian soils grow readily in laboratory culture media that contain soil or soil extract. Stewart (1971) has summarized the organization and reproductive cycle of *C. rivularis* Iyengar, the only species.

The coenocytic plant body (cell) is elongate, cylindrical to clavate, and narrowed at the attached pole. In culture the cells reached 0.5 cm in length and 0.1 cm in width, although they attain greater dimensions in nature.

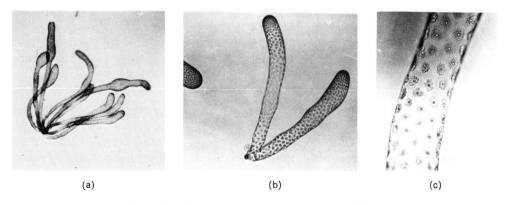

(a)   (b)   (c)

Fig. 3.53 *Characiosiphon rivularis* Iyengar. (*a*), (*b*) Vegetative plants. (*c*) Portion of sac showing chloroplasts, each with a pyrenoid. (*a*) × 10; (*b*) × 16; (*c*) × 81. (Courtesy of Dr. Jeanette Stewart.)

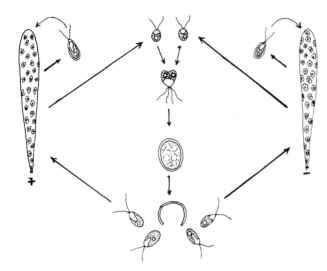

Fig. 3.54 *Characiosiphon rivularis.* Diagrammatic summary of life cycle. The + and − plants reproduce asexually by forming zoospores. They also produce isogametes that form a zygote that, at germination, forms four zoospores. Of the latter, two develop into + plants and two into −.

The cells contain a large central vacuole surrounded by a thin, peripheral layer of protoplasm that contains large (20–25 μm) discoidal plastids each with a single central pyrenoid. Although its organization would suggest an affinity with the Caulerpales (p. 190), *Characiosiphon* lacks siphonein and siphonoxanthin which characterize that order. Numerous contractile vacuoles are present in the cytoplasm, especially in the vicinity of the chloroplasts. A single nucleus lies in the cytoplasm just slightly centripetally to each chloroplast, so that the cells are coenocytic.

In asexual reproduction the sacs produce biflagellate zoospores that may swim for several hours before settling, often in clusters. Aplanospores also occur in *Characiosiphon*.

In sexual reproduction wall-less, biflagellate gametes are produced. These are smaller than the zoospores and may be slightly ameboid. The gametes, like the zoospores, are positively phototactic. It has been established that gametes within a clonal culture will not unite, so that the clones are unisexual. As gametic union proceeds, the flagella are withdrawn, and the zygotes increase in size gradually and become a rusty orange in color. After a short period of dormancy, the zygotes germinate forming four zoospores and meiosis is zygotic. Although *Characiosiphon* produces aplanospores, it does not produce coenocysts like those of *Protosiphon*.

Family 4. Hydrodictyaceae

The members of this family are organized as coenobic colonies (Figs. 3.55, 3.58) and occur in quiet or slowly moving waters. Three genera of the family, *Pediastrum*, *Sorastrum*, and *Hydrodictyon*, will be discussed as representative.

PEDIASTRUM Meyen *Pediastrum* (Gr. *pedion*, plain + Gr. *astron*, star) (Figs. 3.55, 3.56), a genus that contains a number of species, consists of flat coenobia in which the marginal cells, and in some species the internal ones, have one or two

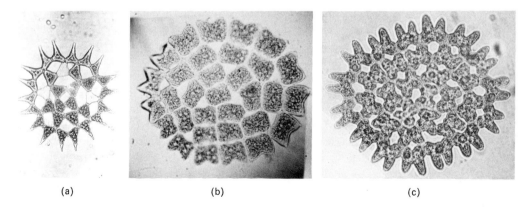

(a)   (b)   (c)

Fig. 3.55   *Pediastrum.* (*a*) *P. simplex* Meyen. (*b*) *P. boryanum* (Turpin) Menegh. (*c*) *P. duplex* Meyen. (*a*) × 201; (*b*) × 231; (*c*) × 256.

horn-like protuberances. Some species of *Pediastrum* are planktonic. Sulek (1969) has summarized the morphology and taxonomy of *Pediastrum* from material collected in nature and also from cultured organisms. Several investigators (Parker, 1969; Gawlik and Millington, 1969; and Millington and Gawlik, 1970) have provided data on the cell walls of *Pediastrum*. According to Parker the walls of three species contain silica. The other investigators showed that the walls consisted of two layers, the inner reticulate and thicker.

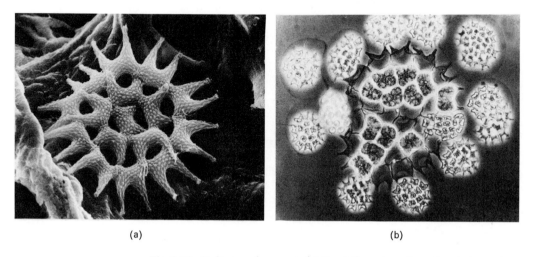

(a)   (b)

Fig. 3.56   *Pediastrum boryanum.* (*a*) Vegetative colony. Scanning electron micrograph. (*b*) Autocolony formation. Note cells in parent colony filled with zoospores and the empty cells from which the zoospores have been liberated to form autocolonies within vesicles. (*a*) × 952; (*b*) × 340. (After Marchant.)

Asexual reproduction takes place by autocolony formation (Fig. 3.56*b*). In this process the protoplast of some or all of the cells of the coenobium undergoes divisions to form biflagellate zoospores. These move freely within a vesicle that is the emergent inner layer of the parental cell wall. After a short period of motility, the zoospores aggregate in one plane and, as they grow, develop the cellular form characteristic oj the species. Marchant (1974A) reported that the zoospores aggregate, as in *Hydrodictyon* and *Sorastrum*, at sites on their surface beneath which microtubules occur. This is borne out by experiments in which colchicine destroyed the microtubules with resulting failure of the cells to aggregate (Marchant and Pickett-Heaps, 1974).

Sexual reproduction, which involves the union of biflagellate gametes, has been observed infrequently (Palik, 1933; Davis, 1967) in *Pediastrum*. The isogametes are smaller than the zoospores and are liberated from the cells that produce them. Davis (1967) illustrated their union as at the posterior poles, which is very unusual, and this is in need of confirmation. The zygotes, after a period of dormancy, form zoospores that rapidly become nonmotile, thick-walled, polyhedral cells, sometimes called **polyeders.** New coenobia arise within the polyhedral cells upon their germination.

SORASTRUM Kützing   The coenobia of *Sorastrum* (Gr. *soros*, a heap + Gr. *astron*, a star) (Fig. 3.57) are spherical, and the component cells have one to four spines on their outer faces. Their basal stalks are joined at the center of the coenobium. The adult cells are multinucleate, and each contains a massive chloroplast with a pyrenoid.

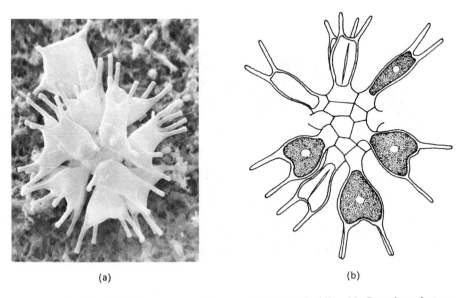

(a)   (b)

Fig. 3.57 *Sorastrum americanum* (Bohlin) Schmidle. (*a*) Scanning electron micrograph of coenobium. (*b*) Flattened coenobium; autocolonies have been formed by three of the cells. (*a*) × 108; (*b*) × 500. [(*a*) after Marchant; (*b*) after Smith.]

Reproduction is accomplished by autocolonies that arise from zoospores produced by each cell much in the manner of *Pediastrum*. Marchant (1974B) has described zoosporogenesis and autocolony formation in *Sorastrum*. Here as in *Pediastrum* and *Hydrodictyon* the microtubules in the zoospores apparently label the sites of cohesion in the formation of the young autocolonies.

HYDRODICTYON Roth   *Hydrodictyon* (Gr. *hydro*, water + Gr. *dictyon*, net), the water net (Fig. 3.58), is widely distributed in lakes and slowly flowing streams. One species, *H. reticulatum* (L.) Lagerh. occurs in the United States. The coenobia of *Hydrodictyon* are typically cylindrical and closed at the poles, the component cells themselves being cylindrical. Nets up to a meter long occur, the cells of larger nets themselves being large. They are arranged end-to-end in polygonal configurations in which six members is a common number, but three to nine cells may be joined. The coenocytic cells contain large central vacuoles, and the protoplasm, with a reticulate chloroplast with pyrenoids and many minute nuclei, is peripheral.

Asexual reproduction is by autocolony formation (Fig. 3.58*b*). The autocolonies arise from biflagellate zoospores that move between the cell wall and vacuolar envelope of the cell. Hawkins and Leedale (1971) and Marchant and Pickett-Heaps (1970, 1971, 1972) have made electron-microscopic investigations of *H. reticulatum*. Persistent centrioles replicate and move to the poles of the spindle during mitosis; the nuclear envelope, except at the poles, remains intact during mitosis. During zoosporogenesis, the pyrenoids disintegrate, and cleavage results in uninucleate zoospores. Prior to cleavage, the vacuole becomes separated from the peripheral cytoplasm by a special, thin layer of homogeneous cytoplasm, the **vacuolar envelope.** This confines the zoospores as a single layer between the cell wall and vacuole. They move freely within this space but ultimately retract their flagella and become joined in groups. Their contacts are made by those sites on their surfaces underlain by microtubules; amorphous material appears between their appressed surfaces, and wall formation follows. Marchant and Pickett-Heaps (1974) reported that treatment of the zoospores of *Hydrodictyon* with colchicine destroyed the peripheral microtubules and with them the ability of the zoospores to aggregate into colonies. The vacuole, vacuolar membrane, and the confining cell wall thus comprise a mold in which the young nets are organized before they are liberated by disintegration of the parental cell wall. This is the origin of the cylindrical form of the net in *H. reticulatum*.

Sexual reproduction in *H. reticulatum* is accomplished by isogamous gametes that swim freely throughout the cell lumen before they are shed into the water through a pore in the parental cell wall. The zygotes enlarge and form four zoospores each (Fig. 3.58*d*), following meiosis. These zoospores settle down to form polyhedral cells (Fig. 3.58*d, e*), which subsequently germinate, liberating flattened circular or saccate nets (Fig. 3.58*f*), the latter two-layered. After the component cells of these have enlarged and become cylindrical, they give rise to typical cylindrical nets.

Marchant and Pickett-Heaps (1972B) have also investigated sexual union with the electron microscope. The gametes differ from zoospores in that at least some have a small, anterior apical cap and they are smaller than zoospores. At the time of gamete union the apical cap elongates to form a fertilization tube that is the agent of initial

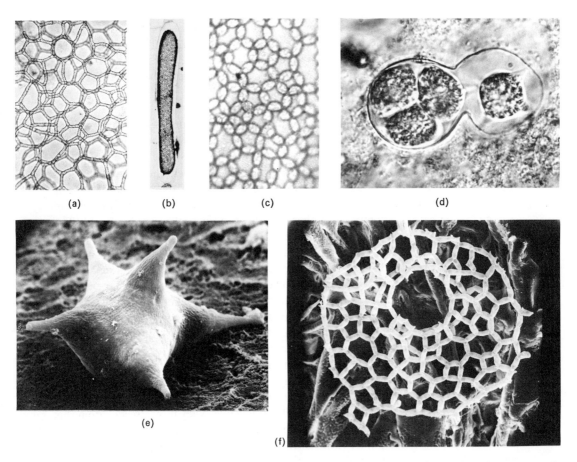

(a)

(b)

(c)

(d)

(e)

(f)

Fig. 3.58 *Hydrodictyon reticulatum* (L.) Lagerh. (*a*) Portion of young coenobium, flattened. (*b*) Young net within parental cell. (*c*) Portion of young net, more highly magnified. (*d*) Germinating zygote (or azygote); note four products (zoospores) that have germinated *in situ* to form polyhedral cells. (*e*) Polyhedral cell, S.E.M. (*f*) Juvenile, circular net derived from polyhedral cell. (*a*) × 59; (*b*) × 22; (*c*) × 350; (*d*) × 570; (*e*) × 1280; (*f*) × 126. [(*d*), (*e*), and (*f*) after Pickett-Heaps and Marchant.]

gametic contact. Gametes that fail to unite may form spherical azygotes. Gametes have been seen conjugating within a single open net cell, but it has not been proved that both were derived from the same cell.

Pocock (1937, 1960) summarized many data regarding the genus *Hydrodictyon*. In addition to *H. reticulatum*, three other species have been described, namely, *H. patenaeforme* Pocock and *H. africanum* Yamanouchi, both natives of Africa, and *H. indicum*. The first two consist of flattened, plate-like nets. The cells of *C. patenaeforme* remain cylindrical, while those of *H. africanum* are barrel-shaped and later become spherical at which time the coenobium breaks up into single spherical cells. Both of these species reproduce only sexually. *Hydrodictyon indicum* Iyengar is not well known.

Order 4. Chlorosarcinales

The members of the small order Chlorosarcinales differ fundamentally from the members of the Volvocales, Tetrasporales, and Chlorococcales (and Chlorellales, p. 133) in their ability to undergo desmoschisis, sometimes called "vegetative cell division," in addition to producing zoospores. Herndon (1958) first proposed the concept on which the order is based, conceiving it to include unicellular green algae that are characterized by desmoschisis and the production of zoospores, and his proposal has been supported by Brown and Bold (1964), Groover and Bold (1969), and Round (1971). Bourrelly (1966) included a number of the genera here classified with the Chlorosarcinales in the order Chaetophorales, because of his opinion that they represent primitive members of that order of filamentous algae, but in 1972 he recognized the order Chlorosarcinales. In the writers' opinion, they share with the Chaetophorales only the attribute of undergoing desmoschisis, which, it will become clear below, they share also with all filamentous and parenchymatous algae and green plants.

Desmoschisis or "vegetative cell division" has been defined by Groover and Bold (1969) as follows: Desmoschisis is a type of cytokinesis with which *all* the following phenomena are associated: (1) a sequence, often repetitive, consisting of karyokinesis (nuclear division), cytokinesis (cytoplasmic division), and cell wall deposition; (2) the division products are neither motile cells, potentially motile, or otherwise specialized cells; (3) the development of the cell walls of the newly divided protoplasts is initiated adjacent to, and continuous with, the parental cell wall; (4) each of the young protoplasts forms a wall over its entire surface; (5) the walls of the young protoplasts remain closely contiguous with the parental cell wall, at least for a short period; and (6) immediate rupture or hydration of the parental cell wall does not occur to liberate the contained division products, although dissociation may occur gradually, after an interval.

The characteristics of desmoschisis summarized above are in marked contrast to those of **eleutheroschisis** (Groover and Bold, 1969) and its results, which occur in the other orders of unicellular Chlorophyceae.[16a] In eleutheroschisis, the products of cell division are either naked (e.g., zoospores of *Neochloris*) or surrounded by completely new walls (e.g., *Chlorococcum* zoospores) that are not intimately related with the parental cell walls. Furthermore, the parental walls are promptly discarded when their enclosed division products (zoospores, aplanospores, hypnospores, or gametes) mature and are liberated. Desmoschisis results in the members of the Chlorosarcinales in the formation of incipient tissues, a step of great evolutionary significance, which was involved in the development of all filamentous and parenchymatous algae and, indeed, in the development of other members of the plant kingdom.

A word of caution is necessary at this point regarding confusion in the use of the term "**vegetative cell division**": because of this confusion Groover and Bold (1969) proposed the term desmoschisis. Smith, 1950, pp. 219–220; 1955) used the term "vegetative cell division" with quite a different meaning than that of Fritsch (1935). Vegetative cell division meant to Smith the formation of *vegetative* cells by division rather than the formation of reproductive cells. The *method* of division and relationship of the parental walls to those of the cellular progeny were not involved in Smith's concept

[16a]Eleutheroschisis occurs, of course, when members of the Chlorosarcinales form zoospores.

of vegetative cell division. Because of this confusion, Groover and Bold (1969) proposed the term **desmoschisis.** The criteria of desmoschisis summarized by Groover and Bold have merely augmented Fritsch's concept of vegetative cell division.

The genera and species of Chlorosarcinales so far described are almost without exception members of the soil algal flora. Bold (1970) has discussed the methods of study and taxonomy of these organisms. Thomas and Groover (1973) have analyzed some of the species of Chlorosarcinaceae electrophoretically and immunologically in an attempt to establish evidence of real relationship among them.

Approximately 10 genera have been assigned to the Chlorosarcinales, of which *Chlorosarcina, Chlorosarcinopsis, Tetracystis, Pseudotetracystis, Borodinellopsis, Axilosphaera,* and *Planophila* are included in the following discussion. As noted previously, Bourrelly (1966) included these organisms in a putative primitive family of his order Chaetophorales but later (1972) recognized the order Chlorosarcinales.

In the classification of the genera of Chlorosarcinalcs, investigators have been guided by the generic characteristics first recognized by Starr (1955) in classifying the spherical genera of Chlorococcaceae, namely, type of chloroplast, presence or absence of a pyrenoid, and zoospore behavior at quiescence. The genera herein chosen for discussion all are members of the family Chlorosarcinaceae and may be distinguished by means of the following key:

1. Zoospores quadriflagellate . . . .*Planophila*
1. Zoospores biflagellate 2
   2. Cells lacking pyrenoids . . . .*Chlorosarcina*
   2. Cells with pyrenoids 3
3. Chloroplast cuplike and parietal 4
3. Chloroplast axile or asteroidal 6
   4. Cells often in cubical, tissue-like complexes, the zoospores
      rapidly become spherical upon quiescence . . . .*Chlorosarcinopsis*
   4. Cells in tetrahedral mounds if aggregated,
      never in cubical packets. 5
5. Zoospores not becoming spherical immediately
   upon quiescence . . . .*Tetracystis*
5. Zoospores becoming spherical immediately
   upon quiescence . . . .*Pseudotetracystis*
   6. Chloroplast axile, pyrenoid excentric . . . .*Axilosphaera*
   6. Chloroplast asteroidal, pyrenoid central . . . .*Borodinellopsis*

Family 1. Chlorosarcinaceae

Seven representatives of this family are discussed below.

CHLOROSARCINA Gerneck and **CHLOROSARCINOPSIS** Herndon   The organisms are characteristic members of the soil algal flora, *Chlorosarcinopsis* (Gr. *chloros,* green + L. *sarcina,* packet + *opsis,* appearance of) being more widely distributed than *Chlorosarcina* in the writers' experience. Pyrenoids are absent from the parietal, cuplike chloroplasts of *Chlorosarcina* (Gr. *chloros,* green + L. *sarcina,* packet) (Chantanachat and Bold, 1962) (Fig. 3.59) but present in those of *Chloro-*

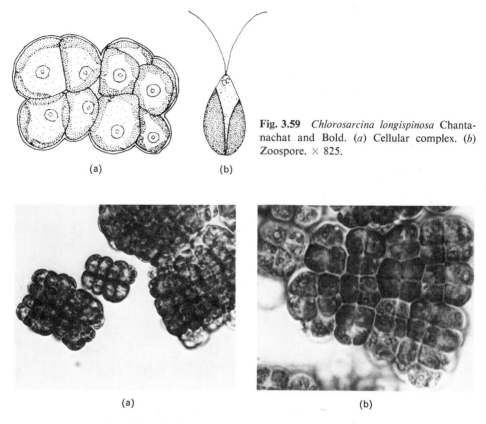

Fig. 3.59 *Chlorosarcina longispinosa* Chanta-nachat and Bold. (*a*) Cellular complex. (*b*) Zoospore. × 825.

(a)   (b)

(a)   (b)

Fig. 3.60 *Chlorosarcinopsis minor* Groover and Bold. (*a*) Young packets. (*b*) Mature packet, more highly magnified. × 618. (After Groover and Bold.)

sarcinopsis (Fig. 3.60). In both organisms, desmoschisis at successively perpendicular planes may build up cubical packets of tissue-like complexes. In some species of *Chlorosarcinopsis* the cells may dissociate before such packets are formed.

In reproduction, the cells produce biflagellate zoospores (Fig. 3.59*b*) that, after a brief period of motility, rapidly become minute, spherical, vegetative cells. These enlarge and initiate the desmoschises by which the typical cellular complexes arise.

Isogamous sexual reproduction was reported to occur in *Chlorosarcinopsis dissociata* by Herndon (1958).

TETRACYSTIS Brown and Bold *Tetracystis* (Gr. *tetra*, four + Gr. *kystis*, bladder) (Fig. 3.61) also is a widely distributed member of the edaphic soil flora. Its cells contain parietal chloroplasts with pyrenoids, and desmoschisis results in tetra-hedral or cruciform tetrads that may build up into noncubical, tissue-like mounds. The biflagellate zoospores of *Tetracystis*, unlike those of *Chlorosarcina* and *Chlorosarcinop-sis*, retain their ovoidal shape at quiescence and become spherical cells only during

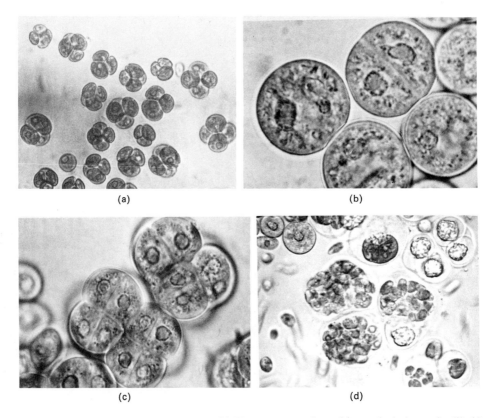

Fig. **3.61** *Tetracystis.* (*a*) *T.* sp., an organism with tetrahedral tetrads. (*b*), (*c*) *T. isobilateralis* Brown and Bold. Note diads and tetrads. (*d*) *T.* sp. [same as (*a*)]. Zoosporogenesis. (*a*) × 516; (*b*) × 860; (*c*), (*d*) × 688.

subsequent growth. Isogamous sexual reproduction has been reported by Brown and Bold (1964) to occur in *T. aggregata, T. excentrica,* and *T. isobilateralis.* These authors have described comparatively the ultrastructural organization of the several species of *Tetracystis.* Arnott and Brown (1967) have investigated electron microscopically the stigma in the zoospores and the fate of the pyrenoid during zoosporogenesis (Brown and Arnott, 1970). The pyrenoid of the parental cell disappears and new pyrenoids arise in the zoospores.

PSEUDOTETRACYSTIS Arneson *Pseudotetracystis* (Fig. 3.62) differs from *Tetracystis* in that its zoospores become spherical immediately upon quiescence. In addition to ovoid zoospores, *P. terrestris* forms elongate isogametes (Arneson, 1973).

BORODINELLOPSIS Dykstra *Borodinellopsis* (*Borodinella* + Gr. *opsis,* similar to) (Fig. 3.63) is similar to *Tetracystis* in that desmoschisis forms noncubical mounds but differs in that its cells contain asteroidal chloroplasts with central pyre-

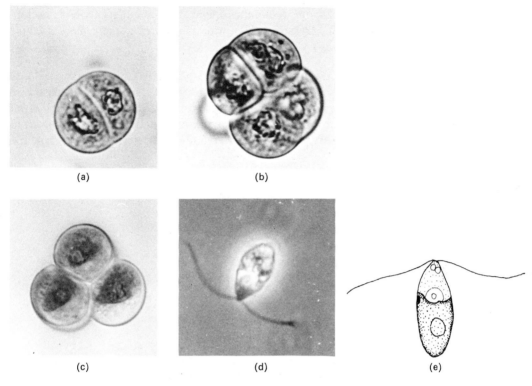

(a)

(b)

(c)

(d)

(e)

Fig. 3.62 *Pseudotetracystis terrestris* Arneson. (*a*)–(*c*) Vegetative cells. (*d*), (*e*) Zoospores. (*a*)–(*c*) × 850; (*d*) × 1000; (*e*) × 1200.

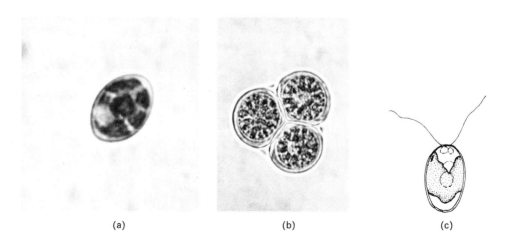

(a)

(b)

(c)

Fig. 3.63 *Borodinellopsis texensis* Dykstra. (*a*) Young vegetative cell, showing asteroidal chloroplast. (*b*) Tetrad of mature vegetative cells. (*c*) Zoospore. (*a*) × 1000; (*b*) × 600; (*c*) × 1200. (After Dykstra.)

noids. The biflagellate zoospores retain their ovoid shape at quiescence and develop in growth into spherical cells. Dykstra (1971), who first described the genus, based it on the single species *B. texensis*.

AXILOSPHAERA Cox and Deason   *Axilosphaera* (Gr. *axon*, axis + Gr. *sphaira*, sphere) (Fig. 3.64) differs from *Tetracystis* only in having an axile plastid with a slightly excentric pyrenoid. In addition to desmoschisis, in which diads and tetrads are produced and often dissociate, the organism produces *Chlamydomonas*-like zoospores, according to Cox and Deason (1968), who described *A. vegetata* from Tennessee soil.

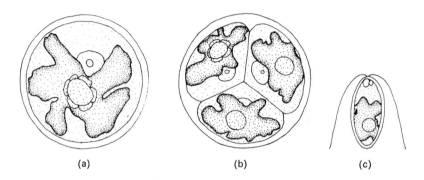

(a)   (b)   (c)

Fig. 3.64 *Axilosphaera vegetata* Deason. (*a*) Vegetative cell in optical section: Note axile chloroplast. (*b*) Tetrad that has arisen by desmoschisis. (*c*) Zoospore. × 1875. (Modified from Cox and Deason.)

PLANOPHILA Gerneck   *Planophila* (L. *planus*, flat + Gr. *philos*, loving) (Fig. 3.65) is somewhat similar to *Chlorosarcinopsis*, but its zoospores are quadriflagellate. Groover and Hofstetter (1969) isolated and described *P. terrestris* from Tennessee soil. In this species they reported that one or two contractile vacuoles may be present in the vegetative cells that have parietal, but perforate, chloroplasts with one to three pyrenoids. The cells are uninucleate.

In addition to quadriflagellate zoospores (Fig. 3.65*b*), aplanospores may also be formed. The zoospores become spherical immediately upon quiescence. Sexual reproduction was not observed.

Fig. 3.65 *Planophila terrestris* Groover and Hofstetter. (*a*) Vegetative cells. (*b*) Zoospore. (*a*) × 612; (*b*) × 1800. [(*a*) after Groover and Hofstetter.]

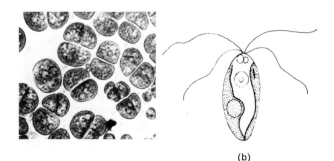

(b)

Order 5. Chlorellales

The unicellular and colonial green algae classified in the order Chlorellales all lack the capacity for zoospore formation, and only a few form flagellate gametes (e.g., *Eremosphaera*, *Golenkinia*, *Scenedesmus*). In the vast majority of members of this rather large order, reproduction is restricted to the formation of nonmotile autospores and autocolonies. The colonial members of the order are often coenobic (e.g., *Scenedesmus*) (Fig. 3.72). The cells of members of the Chlorellales have parietal or lenticular chloroplasts, usually with pyrenoids, and are uninucleate.

The organisms here segregated in the Chlorellales are by other authors (e.g., Fritsch, 1935; Smith, 1950; Bourrelly, 1966; and Fott, 1971) classified in the Chlorococcales; according to their system of classification, this would include both organisms which produce zoospores and those which do not. Although, admittedly, the occurrence of biflagellate gametes in a few species of the several genera cited above would suggest that they are transitional, the vast majority produce no flagellated cells, and so, in this book, they have been assigned to the order Chlorellales. Many species of this order are planktonic and several, like the ubiquitous *Chlorella* itself (which occurs in both marine water and freshwater and in soil), are algal "weeds."

The genera here chosen from among 57 genera (Bourrelly, 1966) to represent the Chlorellales may be classified in two families, the Chlorellaceae and the Scenedesmaceae. The Chlorellaceae contain unicellular genera and the colonial coenobic ones are assigned to the Scenedesmaceae.

Family 1. Chlorellaceae

CHLORELLA Beijerinck   The cells of *Chlorella* (Gr. *chloros*, green + L. *ella*, diminutive) (Fig. 3.66) are small (2–12 μm), spherical, or ellipsoidal and usually occur as isolated individuals. *Chlorella* is probably the first alga to have been grown extensively in axenic culture. The chloroplast is parietal, with a pyrenoid in most species, and the cells are thin-walled. Atkinson, et al. (1972) have studied the ultrastructure and chemistry of the wall of *Chlorella* cells and have reported that **sporopollenin** is present.

Chlorella is ubiquitous in soil and is a common contaminant of containers of water that are undisturbed for long periods. It occurs in both freshwater and marine

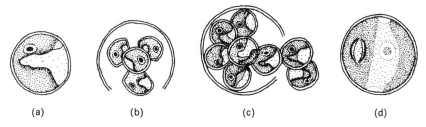

(a)   (b)   (c)   (d)

Fig. 3.66   *Chlorella.* (*a*)–(*c*) *Chlorella* sp. (*a*) Vegetative cell. (*b*), (*c*) Autosporogenesis. (*d*) *Chlorella vulgaris* Beij. var. *vulgaris* Fott. $\times$ 1875. (Modified from Fott and Nováková.)

water (Kessler, et al., 1968). *Chlorella* has for many years been used in research on the process of photosynthesis. Members of the genus are widespread as supposed symbionts within certain animals, e.g., *Paramecium, Hydra,* and *sponges.* Karakashian and Karakashian (1965) investigated the ability of 26 strains of *Chlorella* (and other algae) to infect *Paramecium bursaria;* approximately one-half of them were able to do so. Oschman (1967) studied electron microscopically the *Chlorella* cells growing symbiotically within *Hydra viridis,* while Karakashian, et al. (1968) examined by this technique the symbiosis of *Chlorella* and *Paramecium bursaria.*

In reproduction, which is exclusively asexual, each mature cell divides producing 4, 8, or (more rarely) 16 **autospores** (Fig. 3.66*b, c*) that are freed by rupture or dissolution of the parental walls; the latter may persist in axenic cultures. Bisalputra, et al. (1966), Wanka (1968), and Griffiths and Griffiths (1969) have studied cellular organization and autosporogenesis in *Chlorella* at the ultrastructural level.

Two monographs of *Chlorella* have been published. That of Shihira and Krauss (1964) is based largely on physiological-biochemical characteristics, while, by contrast, that of Fott and Nováková (1969) distinguishes nine species on the basis of cellular morphology. Kessler and associates, in a series of investigations (see Vinayakumar and Kessler, 1975), studied the comparative physiology and biochemistry of various species and strains of *Chlorella* as a basis for elucidating their taxonomy. The citations of most of the reports on these researches are given in Kessler's 1972, 1974, and 1976 papers. DaSilva and Gyllenberg (1973) have considered the taxonomy of *Chlorella* by the technique of continuous classification. They analyzed 41 *Chlorella* strains with reference to 28 morphological and physiological criteria.

PROTOTHECA Krüger   *Prototheca* (Gr. *protos,* primitive + L. *theca,* covering) (Fig. 3.67) is usually considered to represent a colorless *Chlorella.* The spherical or ellipsoid cells reproduce by autospore formation. Nadakavukaren and McCracken (1973) have studied the ultrastructure of *P. zopfii* Krüger and demonstrated the presence of starch-containing plastids, a rather significant evidence of the algal nature

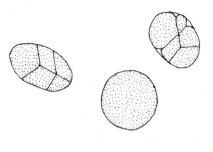

Fig. 3.67   *Prototheca moriformis* Krüg. Note autosporogenesis. × 750. (After Chodat from Fritsch.)

of *Prototheca*. This organism has been found to be pathogenic to humans (e.g., Ashford, et al., 1930). Pringsheim (1963) summarized much of our knowledge of colorless algae. El-Ani (1967) has described the life cycle of *P. wickershamii*, while Cooke (1968A, B) has summarized the relevant literature and taxonomy of *Prototheca*.

GOLENKINIA Chodat   *Golenkinia* (after *M. Golenkin*) (Fig. 3.68) is unicellular, the spherical cells having spines radiating from their walls. Cellular organization is *Chlorella*-like. Two or three species are known. Ellis and Machlis (1968A) studied the nutrition of four strains of *Golenkinia* only one of which did not have a vitamin requirement.

In asexual reproduction, two, four, or eight autospores are formed and liberated from the parental walls. A report that quadriflagellate zoospores are produced by *Golenkinia*, if confirmed, would necessitate its transfer to the Chlorococcales.

Oogamous sexual reproduction has been described for two species. In *G. minutissima* Iyengar and Balakrishnan, according to Starr (1963), the vegetative cells are

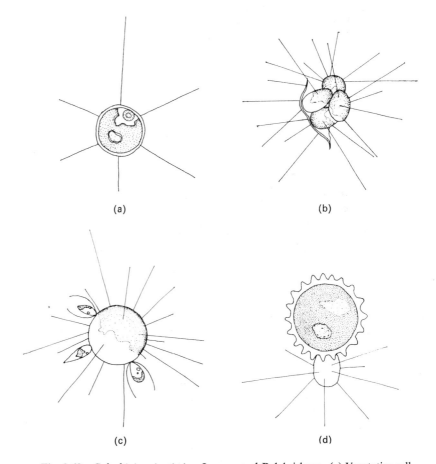

(a)   (b)

(c)   (d)

Fig. 3.68 *Golenkinia minutissima* Iyengar and Balakrishnan. (*a*) Vegetative cell in optical section. (*b*) Autosporogenesis. (*c*) Oogonial cell with sperm. (*d*) Zygote with remains of oogonial wall. × 1010. (After Starr.)

about 8 μm in diameter. Certain cells produce 8–16 biflagellate, naked sperms; the eggs are like vegetative cells, but after the attachment of the sperms their protoplasts escape from their cell walls and gametic union occurs (Fig. 3.68c). The zygote enters a period of dormancy and develops a spiny wall. The site of meiosis has not been demonstrated but may be inferred to be at the germination of the zygote. Møestrup (1972) has investigated the ultrastructure of the vegetative cells and sperms of *Golenkinia minutissima*, while Ellis and Machlis (1968B) found reduced nitrogen and phosphorus levels in the culture medium stimulated sexual reproduction.

OOCYSTIS Nägeli   The ellipsoid cells of *Oocystis* (Gr. *oon*, egg, + Gr. *kystis*, bladder) (Fig. 3.69) may be solitary or in groups confined by the distended parental cell walls of several generations. Polar thickenings occur on the walls of many species. The cells may contain single parietal chloroplasts or several chloroplasts, with or without pyrenoids. *Oocystis* occurs in quiet waters and also in soil.

Reproduction is exclusively by the formation of autospores. Bold and Groover (1968) and Reháková (1969) have investigated the variability of a number of species of *Oocystis* in culture, while Robinson and White (1972) have investigated the ultrastructure of *O. apiculata* W. West.

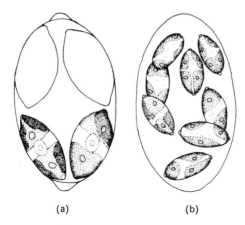

(a)   (b)

Fig. 3.69 *Oocystis marsonii* Lemm. (*a*) Cell with four autospores. (*b*) Cell with eight autospores. × 792. (After Reháková.)

EREMOSPHAERA De Bary   *Eremosphaera* (Gr. *eremos*, solitary + Gr. *sphaira*, sphere) (Fig. 3.70), in contrast to *Chlorella*, is a relatively large organism, its single cells reaching diameters of 200 μm in one species (*E. viridis*). *Eremosphaera viridis* usually occurs in acidic waters among other algal vegetation. The cells contain numerous lenticular chloroplasts, most of them with pyrenoids. Strands of cytoplasm containing fewer chloroplasts radiate from the large central nucleus. The chloroplasts are sometimes joined in reticulate configurations. Fott and Kalina (1962) and Smith and Bold (1966) have made monographic studies of the genus, while Robinson, et al. (1976) have reported on cytokinesis in *Eremosphaera*.

In asexual reproduction of *E. viridis*, each cell forms two (or, more rarely, four) autospores which are liberated by the rupture of the parental walls which persist.

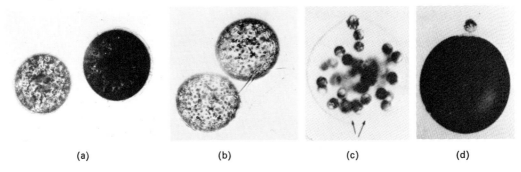

(a)   (b)   (c)   (d)

Fig. 3.70 *Eremosphaera viridis* DeBary. (*a*) Vegetative cells. (*b*) Liberation of two autospores; parental wall remnant visible. (*c*) Gametangium containing sperms. (*d*) Fertilization. (*a*) × 125; (*b*) × 144; (*c*), (*d*) × 160. [(*c*), (*d*) after Kies.]

Kies (1967) discovered the oogamous sexual reproduction of a strain of *E. viridis*. In his bisexual clone, the biflagellate sperms arise by repeated bipartitions of the protoplast of the parental cell; 16, 32, or 64 sperms are produced. The sperms attach to the eggs that resemble autospores (Fig. 3.70*c*, *d*); under this stimulus, the egg develops a fertilization papilla and contractile vacuoles appear in its cytoplasm that plasmolyzes slightly. The zygote then thickens its wall.

ANKISTRODESMUS Corda   Species of *Ankistrodesmus* (Gr. *ankystron*, hook + Gr. *desmos*, bridge) (Fig. 3.71) are common in water and soil and often grow in such abundance as to form water blooms. The individual cells are long and slender-cylindrical and may be tapered toward both ends; they may be spirally twisted. The cells are sometimes aggregated in small groups, and each has a median nucleus and parietal chloroplast without a pyrenoid.

Reproduction is accomplished by the formation of 2–16 autospores which lie in bundles within the parental cell walls which rupture medianly to release them.

Komarkova-Legenerová (1969) has monographed the genus and discussed its relation to the genera *Raphidium* Kützing, *Selenastrum* Reinsch, and *Quadrigula* Printz. She has recognized eight species in the genus *Ankistrodesmus*.

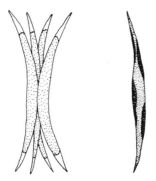

Fig. 3.71 *Ankistrodesmus falcatus* (Corda) Ralfs. Group and single cell showing plastid. × 900. (After Komárková-Legnerová.)

Family 2. Scenedesmaceae

The members of this family are coenobic and occur in the plankton, among benthic algae in quiet bodies of freshwater, or they may be present in soil. In addition to reproduction by autocolony formation, several species of members of the family produce biflagellate gametes.

SCENEDESMUS Meyen   Species of *Scenedesmus* (L. *scena*, stage board + Gr. *desmos*, bond) (Fig. 3.72) are widely distributed in freshwaters and soil. The cylindrical cells, with rounded or pointed ends, are laterally joined in groups of 4 or 8 or (more rarely) 16. The terminal cells, and some of the others, in some species [e.g., *S. quadricauda* (Turp.) Breb.] have spines. Some species have in addition tufts of fine bristles (Trainor and Massalski, 1971, Massalski, et al., 1974) to which has been ascribed the buoyancy of the strains that have them. The cells are uninucleate and have a laminate chloroplast that contains a pyrenoid. Sulek (1975) has investigated mitosis in *S. quadricauda* (Turp.) Breb. and reported the chromosome number to be $n = 11$ or 13. Ultrastructural studies on *Scenedesmus* include those of Bisalputra, et al. (1963, 1964) on the cell wall and pyrenoid and that of Nilshammer and Walles (1974) on cellular differentiation. Komarek and Ludirk (1971, 1972) have used ultrastructural characteristics of the cell wall in differentiating among the species of *Scenedesmus*, while Staehelin and Pickett-Heaps (1975) have investigated species with reticulate patterns on their walls. The latter are extremely complex in organization.

Trainor (1963A, B, C) and Trainor, et al. (1976) found that the composition of the medium influenced the form of one of the strains in his collection of *Scenedesmus* cultures: when an axenic strain of (probably) *S. dimorphus* (Turp.) Kütz. was grown in the presence of yeast extract, the cells of the coenobia were joined only at their apices as in the genus *Dactylococcus* Nägeli; while in a culture medium containing 0.1%

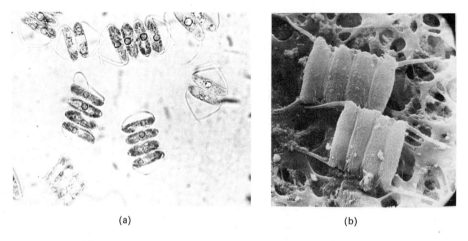

(a)   (b)

Fig. 3.72 *Scenedesmus quadricauda* (Turp.) Bréb. (*a*) Living coenobia. (*b*) Two coenobia, S.E.M. (*a*) × 210; (*b*) × 527. [(*b*), after Pickett-Heaps.]

glucose, the cells were coherent laterally to form typical coenobia. He reported further (Trainor, 1963B) that *S. longus* Meyen resembles the unicellular *Chodatella subsalsa* Lemm. in liquid medium, but typical coenobia developed in the same medium solidified with agar. Trainor (1963C) reported that *S. dimorphus*, which produced typical coenobia in axenic culture, was entirely unicellular in the presence of a soil-inhabiting bacterium. These, and other examples of polymorphism in algae, have been summarized by Trainor, et al. (1971).

Reproduction in *Scenedesmus* is by autocolony formation in which each parental cell forms a miniature colony that is liberated through a tear in the parental wall (Pickett-Heaps and Staehelin, 1975). Trainor and Burg (1965B) surprisingly reported the production of biflagellate gametes by two different clonal strains of *S. obliquus*, both unisexual, from many habitats. The colonies form biflagellate motile cells that are obligate isogametes; i.e., they do not develop parthenogenetically. If they fail to unite sexually, they lyse and disintegrate. To the present, sexual reproduction has been described only for *S. obliquus*.

COELASTRUM Nägeli   Species of *Coelastrum* (Gr. *koilos*, hollow + Gr. *astron*, star) (Fig. 3.73) are less widespread in the writers' experience than those of *Scenedesmus*. The coenobia are in most species hollow spheres composed of from 4 to 128 cells. Depending on the species, the cells may be contiguous or united by extensions

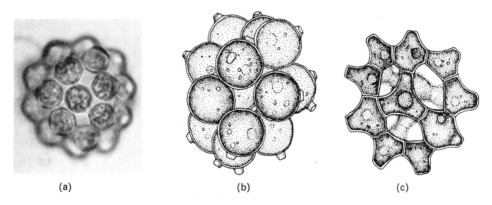

(a)   (b)   (c)

Fig. 3.73   *Coelastrum.* (*a*) *Coelastrum microsporum* Näg. Living coenobium. (*b*) *C. cubicum* Näg. (*c*) *C. probiscideum* Bohlin. (*a*) × 300; (*b*), (*c*) × 600. [(*b*), (*c*) after Skuja.)

of their cell walls. The uninucleate cells are *Chlorella*-like, although the mature cells of one species (*C. probiscideum* Bohlin) have been reported to be multinucleate. Five bristles are present in one species (Reymond, 1974).

Reproduction is by the formation of autocolonies by any or all the cells of a mature coenobium. The pyrenoid undergoes division (Chan, 1974). Trainor and Burg (1965A) have reported the occurrence of biflagellate (and triflagellate and quadriflagellate) cells in cultures of *C. microsporum* Nägeli under conditions of nitrogen starvation;

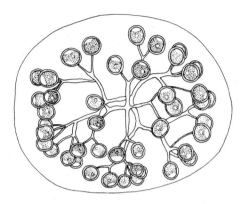

Fig. 3.74 *Dictyosphaerium pulchellum* Wood. × 440. (After Smith.)

Cain (personal communication) reports that motile cells were not observed in two other species of *Coelastrum* under investigation.

Chan (1973) has reviewed the genus *Coelastrum* and relevant literature, while Chan and Ling Wong (1975) have studied *C. reticulatum* electron microscopically.

DICTYOSPHAERIUM Nägeli   The coenobia of *Dictyosphaerium* (Gr. *dictyon*, net + Gr. *sphaira*, sphere) (Fig. 3.74) consist of cells in groups of four borne on stalks. The individual cells are *Chlorella*-like in organization. The stalks are derived from the parental walls at autosporogenesis, which may be repeated to form extensive dendroid systems. These are embedded in a gelatinous matrix. Oogamy has been reported in one species, *D. indicum* Iyeng. and Raman.

The representatives of the Chlorellales discussed in the preceding account illustrate the criteria that characterize the order. The latter contains those unicellular and colonial green algae which are nonmotile and which fail to produce flagellate cells, except for a few species that occasionally produce biflagellate gametes. Reproduction is mainly by the formation of nonmotile autospores and autocolonies.

Order 6. Ulotrichales

In the Ulotrichales are classified filamentous, unbranched green algae with uninucleate cells; many of the genera produce zoospores.[17] They may be aquatic, in freshwater or marine water, or they may be present in soil and on damp surfaces. Some (e.g., *Ulothrix zonata*) exhibit differentiation and polarity in that a specially modified basal cell, the **holdfast**, attaches them to the substrate. Others (e.g., *Klebsormidium*) are free floating and lack holdfasts at maturity. Growth of the filaments is diffuse or generalized in that all the cells, except perhaps for the holdfast, are capable of repeated division. Ramanathan (1964) has summarized the order.

Cellular organization is characterized by a parietal, band-like chloroplast, either a complete cylinder or a curved plate, with one or more pyrenoids. The cells may con-

[17]Those that do not might conceivably be confused with members of the Zygnematales (p. 225) from which they differ in their more simple chloroplasts and lack of amoeboid gametes that conjugate.

tain a large central vacuole (e.g., *U. zonata*) with the result that the cytoplasm is largely peripheral. Floyd, et al. (1971, 1972) have investigated electron microscopically nuclear and cell division in *Ulothrix* and other filamentous algae and have reported that cytokinesis is accomplished by cell-plate formation,[18] as in most land plants, but without formation of a phragmoplast. In another genus of Ulotrichales (*Klebsormidium*, p. 144), they and Pickett-Heaps (1972E) report cytokinesis by furrowing and the absence of a cell plate. These authors consider this difference to be of great significance taxonomically and phylogenetically.

Asexual reproduction in these filamentous algae may involve fragmentation of the filaments and/or zoospore formation. The zoospores are biflagellate or quadriflagellate and have contractile vacuoles and, usually, stigmata. They may be extremely thin-walled or possibly naked. Under unfavorable conditions, e.g., desiccation and others not completely understood, the vegetative cells may thicken their walls and function as akinetes.

Sexual reproduction occurs in many, but not all, of the Ulotrichales and isogamous, anisogamous, and oogamous types have been reported. As far as is known, meiosis is zygotic, although the evidence is inadequate.

By some authors, the order Ulotrichales is conceived as having broader limits than in this text. Thus, Smith (1950) included within the order filamentous algae with multinucleate cells and hair-bearing, branching forms (e.g., *Chaetophora*). The authors have a more narrowly circumscribed concept of the order as is evident from the opening paragraph of this section.

Three families, Ulotrichaceae, Microsporaceae, and Cylindrocapsaceae, of the Ulotrichales are included in the following discussion. They may be distinguished by referring to the following key:

1. Pyrenoids lacking, chloroplast reticulate; cell walls composed of
   overlapping H-shaped (in optical section) segments . . . . *Microsporaceae*
1. Pyrenoids usually present, chloroplast cylindrical or a curved,
   parietal plate or asteroidal, H-shaped segments absent 2
   2. Chloroplasts cylindrical or curved plates; sexual
      reproduction, if present, isogamous . . . . *Ulotrichaceae*
   2. Chloroplasts asteroidal; sexual reproduction oogamous . . *Cylindrocapsaceae*

The representative genera may be distinguished by the following key:

1. Filaments *usually* attached by basal holdfast cells 2
1. Filaments not attached, free-floating at maturity,
   sometimes occurring as single cells or short segments 3
   2. With cylindrical, band-like chloroplasts . . . . *Ulothrix*
   2. With asteroidal chloroplasts . . . . *Cylindrocapsa*
3. Cell walls composed of H-shaped (in optical section) segments . . . .*Microspora*
3. Walls continuous, not composed of H-shaped segments 4
   4. Filaments surrounded by a prominent gelatinuous sheath 5
   4. Filaments lacking a broad sheath 6
5. Cells short-cylindrical, with rounded ends . . . .*Geminella*

[18]Except in *U. zonata* (Stewart, et al. 1972).

5. Cells spherical or ellipsoidal, in the latter case
with the long axes transverse .... *Radiofilum*
6. Filaments elongate.................................. *Klebsormidium*[19]
6. In short segments, often occurring in the unicellular state...... *Stichococcus*

Family 1. Ulotrichaceae

The members of this family are unbranched filaments or bacilliform cells that may be associated in short filaments. The latter may be firmly attached to a substrate, as in *Ulothrix zonata* (Weber and Mohr) Kütz., free-floating or inhabitants of soil. The aquatic species occur in both marine water and freshwater.

Asexual reproduction is accomplished by fragmentation and/or the formation of biflagellate and quadriflagellate zoospores or by aplanospores. Isogamous or anisogamous gametes are produced in some species, and several types of life cycle have been reported in the genus *Ulothrix* itself. Five genera, namely, *Ulothrix*, *Klebsormidium*, *Radiofilum*, *Geminella*, and *Stichococcus*, and discussed below as representative members of the Ulotrichaceae.

ULOTHRIX Kützing   The various species of *Ulothrix* (Gr. *oulos*, wooly + Gr. *thrix*, hair) (Fig. 3.75) are either freshwater or marine. In both habitats they grow on stones and wood to which they are attached by means of unicellular holdfasts (Fig. 3.75*a*, *b*). Growth of the filaments is diffuse or generalized; that is, all the cells (except perhaps the holdfast) undergo divisions of the desmoschisis type. The chloroplasts are parietal, usually with one to several pyrenoids, and may be complete or incomplete short cylinders. The cells of young filaments are shorter than those that are maturing. Floyd, et al. (1971) reported that cell division is by the formation of a cell plate and that plasmodesmata, protoplasmic strands, traverse the transverse cell walls.

In asexual reproduction the cells of the filament form one (*U. fimbriata* Bold, 1958) or more quadriflagellate zoospores (Fig. 3.75*d*, *e*) that swim relatively slowly and ultimately settle with their flagellate poles toward the substrate undergoing cell division to form a basal holdfast cell and an increasingly longer filament. The zoospores are like *Carteria* in their organization, having the organelles, e.g., stigma and contractile vacuoles, which characterize Volvocalean cells. These disappear as the young filaments develop from the zoospores. Zoospores of *Ulothrix* may omit the motile period and function as thin-walled **aplanospores**. The latter (in *U. fimbriata*) thicken their walls and resist desiccation as **hypnospores**. Lokhorst (1974) and Lokhorst and Vroman (1972, 1974A, B) reported that successive asexual cycles reproducing by zoospores occur when freshwater species of *Ulothrix* were grown under short diurnal photoperiods (8 hours of light, 16 hours of darkness). Under longer periods of light, the organisms became yellowish and underwent gametogenesis, producing biflagellate, isogamous gametes (Figure 3.75*f*, *g*) that are smaller than the zoospores. These united to form zygotes that are spherical to pyriform and stalked (*U. tenerrima*) (Fig. 3.75*h*) like *Codiolum* (see p. 186). Lokhorst (1969) discussed the generic criteria in which

---

[19]See p. 246 regarding the classification of *Klebsormidium*.

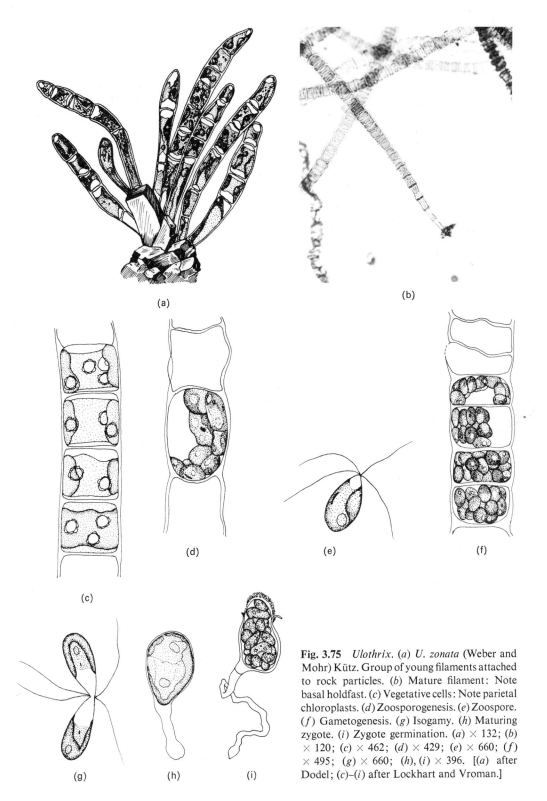

(a)

(b)

(c)

(d)

(e)

(f)

(g)

(h)

(i)

Fig. 3.75 *Ulothrix.* (*a*) *U. zonata* (Weber and Mohr) Kütz. Group of young filaments attached to rock particles. (*b*) Mature filament: Note basal holdfast. (*c*) Vegetative cells: Note parietal chloroplasts. (*d*) Zoosporogenesis. (*e*) Zoospore. (*f*) Gametogenesis. (*g*) Isogamy. (*h*) Maturing zygote. (*i*) Zygote germination. (*a*) × 132; (*b*) × 120; (*c*) × 462; (*d*) × 429; (*e*) × 660; (*f*) × 495; (*g*) × 660; (*h*), (*i*) × 396. [(*a*) after Dodel; (*c*)–(*i*) after Lockhart and Vroman.]

Ulothrix and *Klebsormidium* differ. The zygotes, under short photoperiods, produce four or more quadriflagellate zoospores or aplanospores at germination (Fig. 3.75*i*). Gametes that fail to unite can develop parthenogenetically into new filaments; however, this development is slow. The life cycle of these freshwater species is similar to that described long ago by Gross (1931) for *U. zonata*. In all of these, meiosis is *inferred* to be zygotic, although really convincing cytological and/or genetic confirmation is lacking. Thus, Gross published drawings of *U. zonata* purported to represent zygotic nuclei in meiosis with four pairs of diakinetic chromosomes, while Sarma (1963) reported a chromosome number of $n = 10$ in the same species.

Kornmann (1963A, B, 1964B) and Perrot (1968, 1970, 1971, 1972) have studied the life cycles of several marine species of *Ulothrix*. According to Kornmann, *U. acrorhiza* Kornmann is entirely asexual, producing only by zoospores. *Ulothrix speciosa* (Carm.) Kornmann, *U. subflaccida* Wille, and *U. flexuosa* Kornmann all have filamentous gametophytes and a *Codiolum*-like[20] sporophyte. Perrot (1972) has reported that a diplobiontic isomorphic life cycle occurs in *U. flacca* (Dillw.) Thuret. A form of the latter was reported by her to have an alternate life cycle much like that of the freshwater species. She reported (1972) that *U. pseudoflaccida* Wille undergoes heteromorphic alternation having a filamentous gametophyte and a *Codiolum*-like sporophyte. The cytological aspects of these life cycles have not been elucidated.

KLEBSORMIDIUM Silva, Mattox, and Blackwell[21]   The species of *Klebsormidium* (after *G. Klebs*, German phycologist + *Hormidium*, Gr. *hormos*, chain) (Fig. 3.76), unlike those of *Ulothrix*, occur most frequently in soil (Mattox and Bold, 1962) or on moist substrates, but aquatic species also are known. The filaments are completely undifferentiated, lacking a holdfast at maturity, and the curved, laminate, parietal chloroplasts occupy not more than one-half of the cellular periphery. The cells are uninucleate and their chloroplasts contain only one pyrenoid. Stewart, et al. (1972) have demonstrated the presence of single peroxisomes in the cells of *Klebsormidium*. Peroxisomes are probable sites of catalase activity in cells. Furthermore, in contrast to *Ulothrix*, Floyd, et al. (1972) and Pickett-Heaps (1972E) have shown that cytokinesis in *Klebsormidium* is by furrowing, and no cell plate is present. In some species the filaments fragment readily into few-celled segments or even into single cells.

Reproduction occurs by fragmentation and, in some species, by the infrequent production of biflagellate zoospores and aplanospores. Cain, et al. (1974) reported that zoospores formed in *K. flaccidum* when grown under a diurnal regime of 8 hours of light and 16 hours of darkness. Isogamy has been reported in one species.

GEMINELLA Turpin   The cells of the filaments of *Geminella* (L. *geminus*, twin) (Fig. 3.77) are surrounded by a continuous tubular sheath and are elongate, cylindrical with rounded end walls; the cells are not always closely contiguous. Each contains a laminate chloroplast with a single pyrenoid.

---

[20]*Codiolum*, see p. 189.

[21]The nomenclature of this alga has recently been reviewed and brought into conformity with the International Code by Silva, et al. (1972).

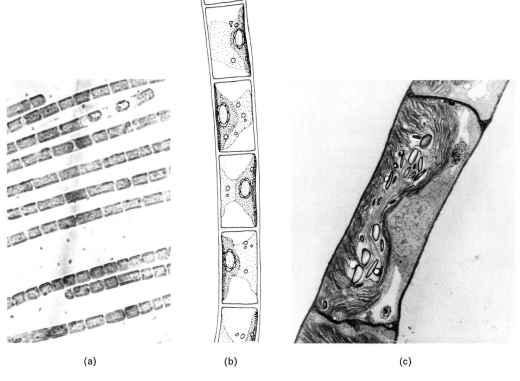

(a)   (b)   (c)

Fig. 3.76 *Klebsormidium flaccidum* (A. Br.) Silva, Mattox, and Blackwell. (*a*) Filaments at low magnification. (*b*) Cellular organization. (*c*) Electron micrograph showing parietal chloroplast filled with starch and lateral nucleus. (*a*) × 150; (*b*) × 594; (*c*) × 1980. [(*c*) after Floyd, Stewart, and Mattox.]

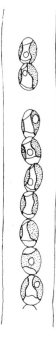

Fig. 3.77 *Geminella* sp. Note broad sheath. × 462.

Reproduction is by fragmentation of the filaments, and the occurrence of zoo-spores has not been confirmed.

RADIOFILUM Schmidle   *Radiofilum* (L. *radius*, + L. *filum*, thread) (Fig. 3.78) is similar to *Geminella* in that the filaments are surrounded by a gelatinous sheath, but the cells of *Radiofilum* are spherical or ellipsoidal, in the latter case with their long axes perpendicular to the long axis of the filament. The cell wall of one species is clearly composed of two overlapping halves.

Reproduction is by fragmentation, zoospores not having been recorded.

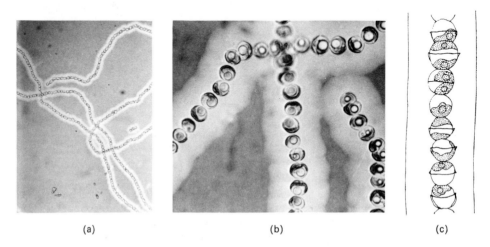

(a)   (b)   (c)

Fig. 3.78   *Radiofilum* sp. (*a*) Filaments at low magnification. (*b*) Segments of filaments, more highly magnified. (*a*), (*b*) India ink preparation. (*c*) Cellular organization. (*a*) × 53; (*b*) × 308; (*c*) × 396.

STICHOCOCCUS Nägeli   *Stichococcus* (Gr. *stichos*, line or series + Gr. *kok-kos*, berry) (Fig. 3.79) occurs either in few-celled filaments or as single cells because of the dissociation of those filaments. Species of *Stichococcus* occur in soils and in both estuarine water and freshwater. The cells are minute in most species and contain parietal, laminate chloroplasts that have pyrenoids; a single nucleus is present. Hayward (1974) studied the growth of *S. bacillaris* Naeg. in culture.

Reproduction is accomplished by fragmentation of the filaments and by division in the unicellular state. Pickett-Heaps (1974C) has described cell division of *S. chloranthus* Krüger at the ultrastructural level on the basis of which he postulated its relationship to *Klebsormidium*. Zoospores and sexual reproduction are absent in *Stichococcus*.

Family 2. Microsporaceae

The family Microsporaceae is herein represented by the single genus *Microspora*, the distinguishing characteristics of which are cited below.

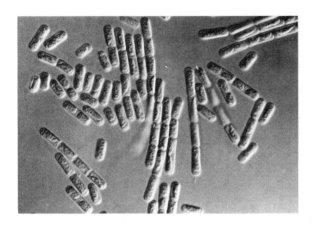

Fig. 3.79 *Stichococcus chloranthus* Krüger. Few-celled filaments and single cells. ×540. (After Pickett-Heaps.)

MICROSPORA Thuret   In the writers' experience, *Microspora* (Gr. *micros*, small + Gr. *spora*, spore) (Fig. 3.80) occurs in neutral or slightly acid waters rather than in alkaline ones. The young filaments, which begin development from zoospores attached by a holdfast, are usually free-floating at maturity. Distinctive characteristics include the cell walls, which are composed of overlapping segments, H-shaped in optical section (Fig. 3.80*b*), and the reticulate chloroplast, which lacks pyrenoids. The cell walls of some species are noticeably thickened.

Pickett-Heaps (1973B) has examined electron microscopically wall structure and cell division in *Microspora*. During nuclear division, centrioles are present at the poles of the spindles, and the nuclear envelope, although perforated, is persistent. Both

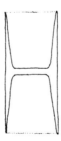

Fig. 3.80 *Microspora loefgrenii* (Nordst.) Lagerh. (*a*) Segment of a filament. (*b*) "H-shaped" wall segment. (*a*) × 297. (After Hazen.)

furrowing and a cell plate are involved in the transverse division of the cells. The long (cylindrical) arms of the H-shaped walls destined to enclose the next pair of cellular progeny are already present in the interphase cell before division. As the new pair of cells elongates, cylindrical (longitudinal) walls, which will function in the products of the next cell division, are secreted.

Reproduction occurs by fragmentation of the filaments and, apparently, also by biflagellate and/or quadriflagellate zoospores (Wichmann, 1937). These may attach by holdfasts or may give rise to free-floating filaments. Both aplanospores and akinetes have also been reported in *Microspora*, and Wichmann has reported that union of isogamous biflagellate gametes occurs.

Family 3. Cylindrocapsaceae

Cylindrocapsa is the only member of this family to be discussed in the text.

CYLINDROCAPSA Reinsch   *Cylindrocapsa* (Gr. *cylindros*, cylinder+Gr. *capsa*, a box) (Fig. 3.81), a rather infrequently encountered alga, differs from other ulotrichalean algae in its asteroid chloroplasts and thick, stratified cell walls, as well

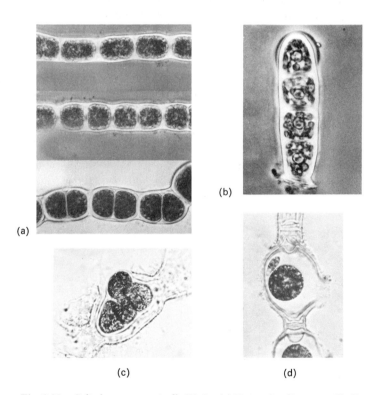

(a)

(b)

(c)

(d)

Fig. 3.81 *Cylindrocapsa geminella* Wolle. (*a*) Vegetative filaments. (*b*) Germling from zoospore. (*c*) Zoospore emergence. (*d*) Oogonium with sperm: Note oogonial pore. (*a*) × 293; (*b*) × 494; (*c*), (*d*) × 293. (After Hoffman and Hofmann.)

as in its specialized oogamous sexual reproduction. Pickett-Heaps and McDonald (1975), however, report that the chloroplasts of *C. involuta* Reinsch are parietal. Hoffman and Hofmann (1975) are doubtful that their organism belongs to the genus *Cylindrocapsa*. The ellipsoid cells, arranged in unbranched filaments, are attached to the substrate by holdfasts.

In addition to fragmentation of the filaments, biflagellate zoospores, and aplanospores arise singly or in two's and four's in each cell (Hoffman and Hofmann, 1975; Hoffman, 1976). These may develop into new filaments. Hoffman (1976) has reported on the fine structure of *Cylindrocapsa* zoospores.

In sexual reproduction (Fig. 3.81*d*) certain cells enlarge and function as female gametangia or oogonia, each having a single pore and surrounding a single egg. The male gametangia, antheridia, are small, often biseriate cells, each of which gives rise to two biflagellate sperms. After fertilization, the zygotes thicken their walls and undergo a period of dormancy. The site of meiosis in *Cylindrocapsa* is unknown.

In summary, the Ulotrichales include unbranched[22] filamentous green algae that form biflagellate or quadriflagellate zoospores, although a few genera (e.g., *Stichococcus*, *Radiofilum*, and *Geminella*) fail to do so. The cells contain cylindrical, laminate, reticulate, or asteroidal chloroplasts, usually with pyrenoids (except *Microspora*), and undergo cytokinesis by furrowing or cell plate formation or both. Sexual reproduction is isogamous, anisogamous, or oogamous, in those genera in which it occurs.

Order 7. Chaetophorales

In the Chaetophorales are classified branching, filamentous green algae with uninucleate cells. The organisms are often differentiated into two systems of filaments, an attached, basal, prostrate system from which erect branching filaments develop. This type of organization is known as **heterotrichy**. In extreme cases, the frequency of branching in the prostrate system may produce a pseudoparenchymatous, disc-like structure (Fig. 3.83*b*). Furthermore, either the prostrate (*Microthamnion*) (Fig. 3.87) or the erect system (*Coleochaete*) (Fig. 3.92) may be poorly developed. The cells are Ulotrichalean in organization with cylindrical or curved parietal plastids, usually with pyrenoids. Unicellular, colorless hairs, multicellular hairlike branches, or sheathed cytoplasmic protuberances may be present. To some extent these characteristics depend for their expression on environmental factors such as nutrients and/or aeration (Cox and Bold, 1966; Tupa, 1974, and Yarish, 1975, 1976).

Authors differ in their concept of the order; Tupa (1974, Table 2) has summarized some of these differences. At one extreme are the systems of Smith (1950), Printz (1964), and Fott (1971): Smith includes the Chaetophorales as a *family* in his order Ulotrichales, while Fott includes them and *all* (except the Zygnematales) the filamentous green algae in a single order Ulotrichales. Printz, by contrast, has classified the Ulotrichaceae and Chaetophoraceae, along with eight other families, in his order Chaetophorales. Fritsch (1935) and Bourrelly (1966, 1972) are more restrictive in their concept of the Chaetophorales. The writers' viewpoint corresponds more closely to

---

[22]*Ulothrix* and *Klebsormidium* occasionally have branches.

theirs. Once again, it is obvious that nature mocks at human categories; the classification, based largely on ultrastructural criteria, suggested by Stewart, et al. (1973) eloquently attests to this conclusion!

The members of the order Chaetophorales, as the order is herein conceived, may reproduce by fragmentation and/or biflagellate or quadriflagellate zoospores and by aplanospores. The vegetative cells of some genera (*Stigeoclonium*) may thicken their walls as akinetes, dissociate, and survive long periods of desiccation.

Sexual reproduction, when it occurs, may be isogamous, anisogamous, or oogamous, and both haplobiontic and diplobiontic life cycles have been reported to occur among the various members of the Chaetophorales; but our knowledge of these is woefully inadequate.

Representatives of three families will be discussed in the following account, namely, the Chaetophoraceae, Aphanochaetaceae, and Coleochaetaceae. These may be distinguished in the following key:

    1. With a slightly or well-developed erect system of filaments ...Chaetophoraceae
    1. Basal system prominent; erect system absent or rudimentary 2
        2. At least some of the cells with elongate,
           unicellular hairs ....Aphanochaetaceae
        2. At least some cells with ensheathed
           cytoplasmic extensions ....Coleochaetaceae

Family 1. Chaetophoraceae

The members of this family are clearly heterotrichous, its genera varying in the degree of development of the prostrate and erect systems. In some genera the tips of the filaments are prolonged as attenuated, multicellular, hairlike branches (Fig. 3.82, 3.84). The family encompasses two lines of variants from the balanced type of heterotrichy evidenced, for example by *Stigeoclonium* (Fig. 3.83*b*). In *Microthamnion*, on the one hand, the erect branching filaments are attached to the substrate by a simple holdfast cell that has a bulbous base. By contrast, the prostrate, rather than the erect, system may dominate, the latter being rudimentary in *Gongrosira* and *Pseudendoclonium* or absent as in *Protoderma*.

Chaetophora, Stigeoclonium, Draparnadlia, Draparnaldiopsis, Microthamnion, Fritschiella, Protoderma, Gongrosira, and *Pseudendoclonium* have been included in the following account to illustrate the characteristics of the family. They may be distinguished by the following key:

    1. Erect filaments well developed...2
    1. Erect filaments rudimentary, plants prostrate, attached ....:...7
        2. Filaments with elongate basal rhizoids, in
           part parenchymatous ....*Fritschiella*
        2. Filamentous, not pluriseriate and parenchymatous 3
        3. Filaments enclosed in gelatinous sheaths that are evident
           macroscopically or demonstrable microscopically with India ink ..*Chaetophora*
        3. Gelatinous sheaths absent or inconspicuous 4
        4. Main axes and branches of successive orders of

CHAETOPHORA Schrank. *Chaetophora* (Gr. *chaetos*, hair + Gr. *phoros*, bearer) (Fig. 3.82) occurs in running or quiet freshwater and is often recognizable macroscopically because the filaments of some species are embedded in globular or elongate gelatinous colonies attached to underwater substrates including aquatic angiosperms. All the species secrete copious gelatinous matrices that are spectacularly apparent when the plants are mounted in India ink preparations. The colonies may be calcified, especially the basal system.

The larger cells contain band-shaped chloroplasts, and the axial filaments may be intertwined and appear pluriseriate. The cells of the branches, which end in attenuate, multicellular, hairlike tips, have massive chloroplasts.

Quadriflagellate zoospores and akinetes are produced, and the biflagellate gametes are isogamous. The life cycle of *Chaetophora* has not been elucidated in culture.

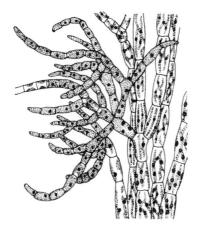

Fig. 3.82 *Chaetophora incrassata* (Hudson) Hazen. Habit of portion of living plant. × 116.

STIGEOCLONIUM Kützing *Stigeoclonium* (L. *stigens*, sharp + L. *clonium*, branch) (Fig. 3.83) is more widely distributed than *Chaetophora* and grows attached to stones and woodwork or epiphytically on the leaves and stems of aquatic angiosperms. Islam (1963) and Cox and Bold (1966) have contributed to our knowledge of *Stigeoclonium*, which differs from *Chaetophora* in lacking copious gelatinous matrices and in the uniseriate axes and their gradual gradation into branchlets. Cox and Bold

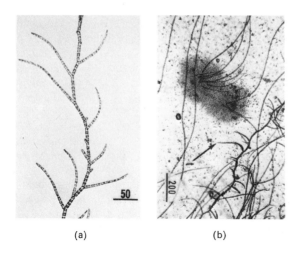

(a)   (b)

Fig. 3.83 (a) *Stigeoclonium variabile* (Näg.) Islam. Segment of aerial branch. (b) *S. farctum* Berthold. Note heterotrichy: basal and erect systems. (Magnification indicated by 50- and 200-μ bars.) (After Cox and Bold.)

showed that the various species of *Stigeoclonium* vary in the degree of development of their basal systems and its branching, and they have reviewed the conflicting and contradictory literature regarding sexual reproduction in the genus. Floyd, et al. (1972) reported that cell division is by the formation of a cell plate, as in *Ulothrix*.

Asexual reproduction has been reported to occur by means of both biflagellate and quadriflagellate zoospores. Several types of life cycle have been described for *Stigeoclonium* (Cox and Bold, 1966), but they should be reinvestigated. The site of meiosis has not been determined, and much remains unknown regarding the variation in life cycle.

DRAPARNALDIA Bory   In *Draparnaldia* (after *J. P. R. Draparnaud*) (Fig. 3.84) the main axes consist of markedly larger cells than those of the branches that

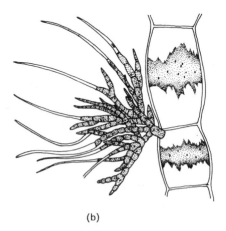

(a)   (b)

Fig. 3.84 (a) *Draparnaldia* sp. Portion of plant showing habit of growth. (b) *D. glomerata* Ag. Segment of main axis with a lateral branch system. (a) $\times$ 66; (b) $\times$ 97. [(b) after Hazen.]

originate at one point near the transverse walls of the axial cells. The chloroplasts of the latter are beautifully fimbriate cylinders containing a number of pyrenoids. The branchlets end in multicellular hairs.

Asexual reproduction is accomplished by quadriflagellate zoospores that arise in the cells of the branches. The fact that the main axial cells do not produce reproductive cells represents a degree of specialization seen more frequently in the brown and red algae. In sexual reproduction, which, seemingly, has not been studied in more than 60 years, quadriflagellate gametes are reported to form thick-walled zygotes that germinate to form two or four young filaments; meiosis has been *inferred* to be zygotic. The life cycles of the six or seven American species of *Draparnaldia* need to be investigated.

DRAPARNALDIOPSIS Smith and Klyver   *Draparnaldiopsis*   (*Draparnaldia* + Gr. *opsis*, appearance of) (Fig. 3.85) differs from *Draparnaldia* in that the main axes are differentiated into alternate long and short cells. The branches arise only from the latter. One species of this rather rare alga, *D. alpina* Smith and Klyver, has been reported in the United States. An Indian species, *D. indica*, has an isomorphic, diplobiontic life cycle according to Singh (1945).

Fig. 3.85 *Draparnaldiopsis alpina* Smith and Klyver. Note smaller axial cells from which branches arise. × 188. (After Smith.)

FRITSCHIELLA Iyengar   *Fritschiella* (after *F. E. Fritsch*) (Fig. 3.86) grows on soil and occasionally on tree bark but is a rather uncommon alga in the United States. A mature plant consists of cushion-like aggregations of branched, *Chlorosarcinopsis*-like packets that terminate in uniseriate branches and of subterranean, colorless rhizoidal cells. This highly differentiated organism is sometimes cited as an example of a possible precursor to land plants because of its parenchymatous habit and its differentiation and especially because its cells undergo cytokinesis by the centrifugal development of cell plates (McBride, 1970).

In reproduction, some plants produce quadriflagellate zoospores and others biflagellate gametes. These may unite to form zygotes that, with or without intervening dormancy, develop into new plants. Meiosis occurs during zoosporogenesis,

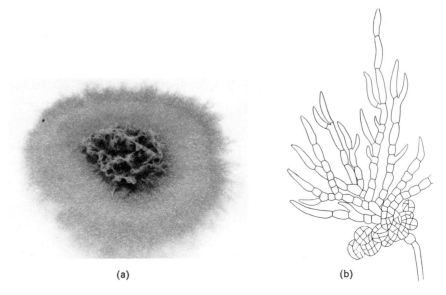

(a)   (b)

Fig. 3.86 *Fritschiella* sp. (*a*) Culture on agar medium. Note prostrate system and erect tufts. (*b*) Outline drawing of a plant showing erect, prostrate (sarcinoid), and rhizoidal systems. (*a*) × 1.5; (*b*) × 225. [(*a*) courtesy of Dr. E. W. Ruf, Jr.; (*b*) after Singh.]

and the life cycle is diplobiontic and isomorphic. Melkonian (1975) has investigated the ultrastructural organization of the zoospores, while Melkonian and Weber (1975) studied the effects of kinetin in *Fritschiella* in axenic mass cultures.

MICROTHAMNION Nägeli   *Microthamnion* (Gr. *micros*, small + Gr. *thamnion*, shrub) (Fig. 3.87) differs from other chaetophoracean algae in that its minute branching filaments are attached to the substrate by single holdfast cells with taper-

Fig. 3.87 *Microthamnion kützingianum* Näg. (*a*) Living plant. (*b*) Plant, more highly magnified, showing chloroplasts and absence of pyrenoids. (*c*) Zoospore. (*a*) × 252; (*b*) × 356; (*c*) × 1125. [(*a*) after Tupa.]

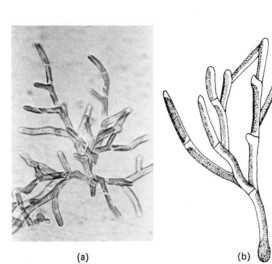

(a)   (b)   (c)

ing bases. The branches always arise at the distal ends of the cells. The parietal, laminate chloroplasts lack pyrenoids. The biflagellate zoospores arise in multiples of two from the vegetative cells. Watson and Arnott (1973) and Watson (1975) have described the ultrastructure of the zoospores.

Sexual reproduction has not been confirmed for *Microthamnion*. Its species occur in soil and in freshwater.

PROTODERMA Kützing, GONGROSIRA Kützing, and PSEUDENDOCLO-NIUM Wille   These genera are all similar in that the prostrate phase of the heterotrichous organism is the dominant one (Fig. 3.88). Tupa (1974) has made an intensive investigation of these organisms in nature and in culture and has critically evaluated the characteristics that are reliable in distinguishing the genera and species.

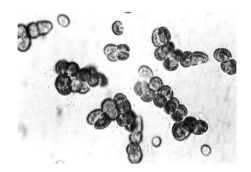

Fig. 3.88   *Protoderma sarcinoidea* (Groover and Bold) Tupa. × 375. (After Tupa.)

In *Protoderma* (Gr. *protos*, first + Gr. *derma*, skin) (Fig. 3.88), the erect phase is entirely lacking, and the prostrate branching filaments grow tightly appressed to the substrate, the latter either rocks or aquatic angiosperms. *Protoderma sarcinoidea* (Groover and Bold) Tupa was isolated from soil. The cells are almost isodiametric. The organism develops from biflagellate zoospores that attach to the substrate and undergo bipolar developments to form few-celled filaments with branches of limited growth all enclosed in a gelatinous sheath. Sexual reproduction has not been observed in *P. sarcinoidea*.

In *Gongrosira* (Fig. 3.89), the loosely branched prostrate filaments ultimately give rise to tufts of short, erect branches forming cushionlike plant bodies. The cells

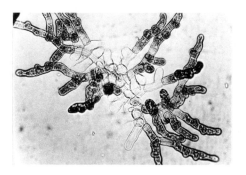

Fig. 3.89   *Gongrosira papuasica* (Borzi) Tupa. Empty cells have liberated zoospores. × 298. (After Tupa.)

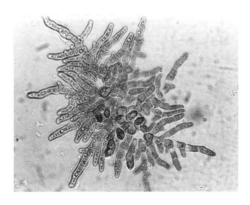

Fig. 3.90 *Pseudendoclonium   basiliense* var. *brandii* Vischer. Note incipient erect filaments at center. × 350. (After Tupa.)

are longer than wide and the chloroplast, which contains a pyrenoid, is shorter than the cell length. *Gongrosira papuasica* (Borzi) Tupa was isolated from material growing epiphytically on a variety of aquatic angiosperms, other algae, and mosses. The organism reproduces by biflagellate zoospores that undergo bipolar germination. Sexual reproduction has not been observed in *Gongrosira*.

Pseudendoclonium (Gr. *pseudes*, false + Gr. innertwig or branch) (Fig. 3.90) is like *Gongrosira* in having both a prostrate system and erect branches of limited growth. The several species investigated by Tupa (1974) were growing epiphytically on various aquatic angiosperms. The genera differ in that the zoospores of *Pseudendoclonium* are quadriflagellate, while those of *Gongrosira* are biflagellate. Mattox and Stewart (1973) have reported the presence of scales on the surface of the zoospores. Sexual reproduction has not been observed in *Pseudendoclonium*.

Family 2. Aphanochaetaceae

The members of the Aphanochaetaceae may be distinguished from other chaetophoralean algae by their production of unicellular hairlike cells (Fig. 3.91*b*). *Aphanochaete* is discussed below as representative of the family (see Tupa, 1974).

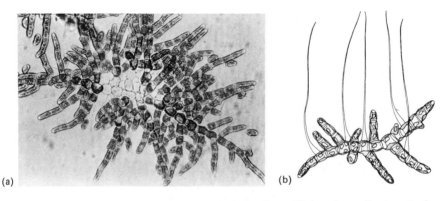

(a)   (b)

Fig. 3.91 *Aphanochaete.* (*a*) *A. elegans* Tupa. Living plant cells at center have liberated zoospores. (*b*) *A. confervicola* var. *major* Tupa. Note hairs. (*a*) × 350; *b* × 350. (After Tupa.)

APHANOCHAETE A. Braun   *Aphanochaete* (Gr. *aphanes*, to appear + Gr. *chaelos*, hair) (Fig. 3.91), like *Protoderma*, is epiphytic on various aquatic angiosperms. The prostrate organism in some species has rather open branching, as in *A. confervicola* (Nag. ex Kütz.) Rabenh., while in *A. magna* Godward the branching is so dense as to form a disc. The vegetative cells of *Aphanochaete* have parietal chloroplasts with pyrenoids. The piliferous cells vary in abundance in accordance with external conditions, and the fragile hair cells are easily broken off.

Reproduction occurs by means of quadriflagellate zoospores. In *A. magna*, germination of the zoospores is tetrapolar, and cruciform young thalli develop. These later undergo intercalary branching and become disc-like.

Sexual reproduction has not been observed often in *Aphanochaete* but has been reported to be anisogamous, the gametes being quadriflagellate. Further investigation to confirm this is indicated.

Family 3. Coleochaetaceae

The members of this family, of which *Coleochaete* (Fig. 3.92) is here chosen as the sole representative, are characterized by the basally ensheathed cytoplasmic processes that protrude from some or all of the cells. All are epiphytic primarily on aquatic angiosperms, although they attach readily in field and laboratory on nonliving substrates.

COLEOCHAETE Brébisson   In *Coleochaete* (Gr. *koleon*, sheath + Gr. *chaetos*, hair), the several species differ in respect to the abundance of their branching; abundant branching results in orbicular thalli (Fig. 3.92) closely appressed to the substrate. Growth in these occurs by the division of the marginal cells in both a tangential and radial direction, a process studied in *C. scutata* Bréb. by Marchant and Pickett-Heaps (1973). When the cells undergo radial divisions, they develop a cell plate and phragmoplast similar to those in the land plants, and the protoplasts are interconnected by plasmodesmata. In circumferential division of the cells a modified phragmoplast is involved. The vegetative cells contain parietal laminate chloroplasts with pyrenoids. McBride (1974) and McBride, et al. (1974) have studied thallus ontogeny and the seta-bearing cells. The latter contain a revolving chloroplast and usually do not divide. If the surrounding vegetative cells are destroyed, however, the setiferous cells dedifferentiate and regenerate new thalli. The ensheathed hairs are delicate and readily broken off near their bases. The electron microscope reveals that the seta is entirely ensheathed in the plasma membrane and that an extension of the chloroplast is present in the seta. The base of the latter is surrounded by a collar (McBride, 1974). Marchant (1977) also has studied the setiferous cells electron microscopically.

In asexual reproduction, single biflagellate zoospores are produced by the vegetative cells; aplanospores also are known.

Sexual reproduction in *Coleochaete* is oogamous (Fig. 3.92*e, f*). The oogonia are scarcely distinguishable from vegetative cells except in a few species (e.g., *C. pulvinata* A. Br.) in which they have an elongate protuberance, the **trichogyne** (Fig. 3.92*f*), a structure characteristic of the female gametangium of red algae. The antheridia are small cells, each of which produces a small, biflagellate sperm.

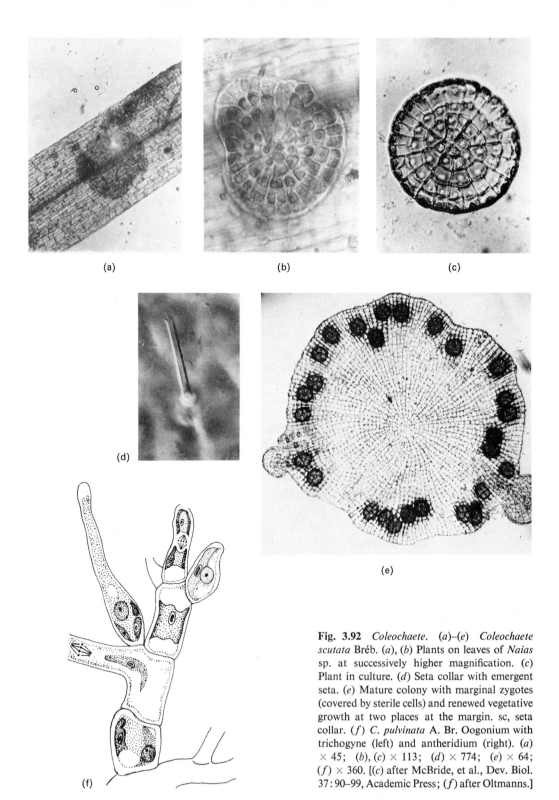

(a)

(b)

(c)

(d)

(e)

(f)

Fig. 3.92 *Coleochaete.* (*a*)–(*e*) *Coleochaete scutata* Bréb. (*a*), (*b*) Plants on leaves of *Naias* sp. at successively higher magnification. (*c*) Plant in culture. (*d*) Seta collar with emergent seta. (*e*) Mature colony with marginal zygotes (covered by sterile cells) and renewed vegetative growth at two places at the margin. sc, seta collar. (*f*) *C. pulvinata* A. Br. Oogonium with trichogyne (left) and antheridium (right). (*a*) × 45; (*b*), (*c*) × 113; (*d*) × 774; (*e*) × 64; (*f*) × 360. [(*c*) after McBride, et al., Dev. Biol. 37: 90–99, Academic Press; (*f*) after Oltmanns.]

The zygote enlarges after fertilization and becomes covered by overgrowth of the surrounding vegetative cells. After a period of dormancy, meiosis occurs, followed by mitosis and cytokinesis, so that 8–32 biflagellate zoospores are produced and released. Meiosis is zygotic and the life cycle, accordingly, is haplobiontic and haploid. Hopkins and McBride (1976) have recently confirmed this by microspectrophotometric analyses of the DNA cycle.

In summary, chaetophoralean algae are heterotrichous, branching Chlorophyceae with uninucleate cells. In their heterotrichy both the erect and prostrate portions of the plant may be well developed (e.g., *Stigeoclonium*) or one or the other may be partly or entirely lacking (e.g., *Microthamnion, Coleochaete*). Asexual reproduction is by biflagellate or quadriflagellate zoospores and sexual reproduction varies from isogamous to oogamous. Many chaetophoralean algae are epiphytic on aquatic angiosperms and on other aquatic algae and aquatic mosses.

Order 8. Oedogoniales

While it is sometimes difficult to characterize incisively the higher taxonomic categories, such is not the case with the order Oedogoniales, for its members have clearly distinctive characteristics. The plants are branched or unbranched filaments composed of cells with parietal nuclei and parietal, reticulate chloroplasts containing pyrenoids, all surrounding one or more central vacuoles (Fig. 3.93*b*). In addition, they have an unusual pattern of cell division, both basal and intercalary, to be described below. All have oogamous sexual reproduction and all are seemingly haplobiontic and haploid. Both the sperms and zoospores have a subapical ring of flagella, up to 120 in the case of the zoospores, so that they have sometimes been classified in a separate taxon, the Stephanokontae (Gr. *stephanos*, crown + Gr. *kontos*, oar). Finally, in many species, sexual reproduction is characterized by the production of dwarf male plants.

Family 1. Oedogoniaceae

The Oedogoniales contain a single family, the Oedogoniaceae, in which are classified three genera, *Oedogonium, Bulbochaete*, and *Oedocladium*. These may be distinguished by the following key:

1. Filaments unbranched ....*Oedogonium*
1. Filaments branched 2
   2. Some of the cells bearing bulbous-based bristles ....*Bulbochaete*
   2. Bristles with bulbous bases absent ....*Oedocladium*

Tiffany (1930, 1937) and Gauthier-Liévre (1963, 1964) have made taxonomic summaries of the Oedogoniaceae.

OEDOGONIUM Link   *Oedogonium* (Gr. *oedos*, swelling + Gr. *gonos*, reproductive structure) (Fig. 3.93–3.97) is a large genus with several hundred species that grow in freshwater and are attached to various substrates including stone, wood, and, most frequently, the stems and leaves of aquatic angiosperms. The unbranched fila-

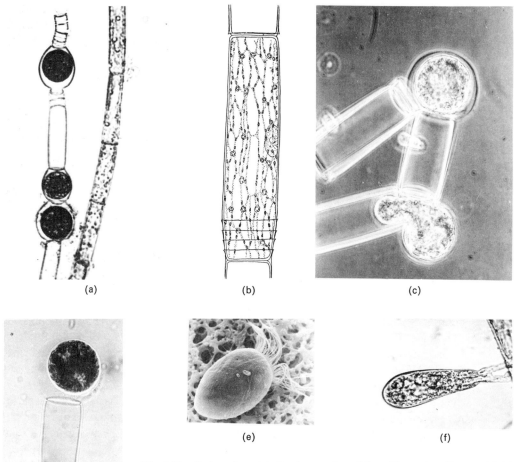

(a) (b) (c)

(e) (f)

(d)

Fig. 3.93 *Oedogonium*. (*a*) *Oe. intermedium* Wittr. Vegetative filament (right) and sexually mature filament (left); in the latter, note three oogonia and (empty) antheridia above. (*b*) *Oe. crassum* (Hass.) Wittr. Vegetative cell. Note parietal nucleus and wall rings. (*c*), (*d*) *Oe. cardiacum*. Zoosporogenesis and liberation. (*e*) Zoospore. Note crown of flagella. (*f*) Germinating zoospore showing protuberant holdfasts. (*a*) × 343; (*b*) × 320; (*c*) × 265; (*d*) × 700; (*e*) × 306; (*f*) × 500. [(*b*) after Smith; (*c*) and (*e*) after Pickett-Heaps.]

ments are usually attached by a basal holdfast cell the proximal portion of which may be simple or multilobed (Fig. 3.93*f*). The short or elongate cylindrical cells, which may be nodulose, contain a reticulate parietal chloroplast with a number of pyrenoids, a large central vacuole or vacuoles, and a prominent parietal nucleus.

Growth and cell division in the mature filament are largely intercalary, only certain of the cells undergoing division. These may be recognized by the cell division scars or "rings" that occur in their walls near one end of the cell (Fig. 3.93*b*, 3.95*b*). The process of cell division has been described carefully on the basis of both light- and electron-microscopic studies by Hill and Machlis (1968), Pickett-Heaps and Fowke

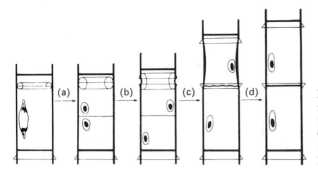

Fig. 3.94 *Oedogonium.* Diagrammatic summary of cell division in *Oedogonium.* (*a*) Telophase of mitosis, splitting of wall ring. (*b*) Cytokinesis. (*c*), (*d*) Rupture and stretching of wall ring as protoplasts enlarge. (After Pickett-Heaps.)

(1969), and Coss and Pickett-Heaps (1974A, B). Figure 3.94 summarizes this process. Cell division is preceded by mitosis, the nucleus lying at this time nearer to one end of the cell at which a ringlike thickening (Fig. 3.94*a*) has developed. After mitosis, a strand of cytoplasm extends across the vacuole between the daughter nuclei that ultimately move to the equators of the new daughter cells as the cytoplasmic septum is completed. At this point, the outer wall layer splits in the vicinity of the ringlike thickening, and the latter becomes elongated into a cylinder as the protoplasts of the two daughter cells elongate. The protoplasts elongate so that the cytoplasmic septum within which the transverse wall is deposited comes to lie at the level of the ruptured end of the longer cell. The expanding ring forms the outer wall of the other cell at first; but as elongation ceases, a second layer is deposited between the expanded ring and the protoplast. Repeated divisions of this cell in similar fashion cause the accumulation of ringlike scars near one end.

Asexual reproduction in *Oedogonium* is both by fragmentation and zoospore formation, the latter process and its products studied at the light-microscopic level by Retallack and von Maltzahn (1968) and at the electron-microscopic level by Hoffman (1968, 1970), Hoffman and Manton (1962), and Pickett-Heaps (1971, 1972A, B, C). In zoosporogenesis, centrioles appear and multiply on the surface of the nuclear envelope forming two rows. The nucleus then moves to the midpoint of the lateral wall of the vegetative cell, and the centrioles function as basal bodies, each generating a flagellum. The protoplast of the potential zoospore secretes a hyaline layer which becomes the vesicle (3.93*c*, *d*) within which the zoospore is released and a basal mucilage which causes contraction of the protoplast and probably functions in extrusion of the zoospore. The latter emerges through rupture of the vegetative cell wall near the lateral wall and may have as many as 120 flagella (Hoffman and Manton, 1962) attached to a colorless, anterior cytoplasmic dome (Fig. 3.93*e*). The zoospores (and gametes, see below) thus have a ring or crown of flagella and are said to be **stephanokontan.** The zoospores contain a stigma and numerous contractile vacuoles.

After a period of motility, the zoospores attach to the substrate at their colorless poles and shed their flagella (Pickett-Heaps, 1972A). The dome sends out one or more rhizoidal protuberances that spread over, or wrap around, the substrate; a wall is secreted around the zoospores, and development into a germling filament is initiated. Pickett-Heaps (1972B, C) has reported on organization of the germling and on its patterns of cell division.

Sexual reproduction in *Oedogonium* is oogamous, and the eggs and sperms are produced, respectively, in oogonia and antheridia (Fig. 3.95*b*, *c*). The oogonia (Fig. 3.95*b*) arise by division of an oogonial mother cell into a supporting cell and oogonium; in some species the oogonia occur in series (e.g., *Oe. bengalense* Hirn), whereas in others [*Oe. cardiacum* (Hass.) Wittr.] they occur singly, separated by vegetative cells. The antheridia are minute, short-cylindrical cells, each of which produces two (or four, Coss and Pickett-Heaps, 1974) multiflagellate pale-green or yellowish sperms (Figs. 3.95*c*, *d*; 3.96); these bear about 30 flagella (Hoffman and Manton, 1963). Hoffman (1960), Hill and Machlis (1970), and Coss and Pickett-Heaps (1973) have shown that availability of adequate CO_2, low nitrogen content of the medium, and light evoke gametogenesis.

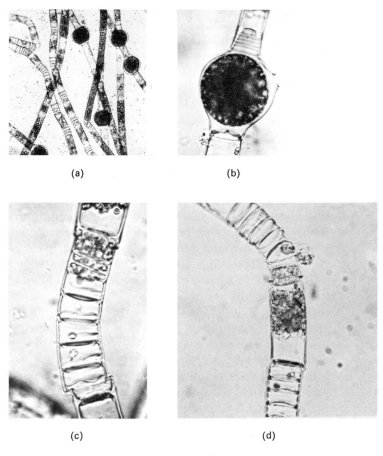

(a)

(b)

(c)

(d)

Fig. 3.95 *Oedogonium cardiacum* (Hass.) Wittr. (*a*) Note male (left) and female (right) filaments. (*b*) Oogonium containing egg and oogonial pore; note rings of cell division above. (*c*) Portion of male filament: Note empty antheridia and those containing two sperms each. (*d*) Liberation of sperms from antheridia. (*a*) × 75; (*b*) × 428; (*c*) × 936; (*d*) × 553.

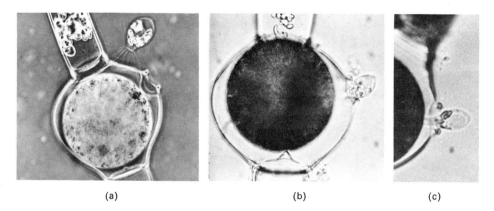

(a)   (b)   (c)

Fig. 3.96 *Oedogonium cardiacum.* Fertilization. (*a*) Sperm approaching oogonium. (*b*), (*c*) Stages in penetration of the sperm. (*a*) × 300; (*b*) × 469; (*c*) × 389. (After Hoffman.)

Several patterns of distribution of the sex organs occur in *Oedogonium.* Individuals of some species (e.g., *Oe. intermedium* Wittr.) are bisexual, antheridia and oogonia occurring on the filament (Fig. 3.93*a*). Other species (e.g., *Oe. cardiacum*) have unisexual filaments that produce either antheridia or oogonia (Fig. 3.95*a*). In still other species there is a dimorphism between the male and female filaments, the former being few-celled dwarfs and epiphytic on the female filaments (Fig. 3.97), the latter, the usual size. Such species are said to be **nannandrous** in contrast to the **macrandrous** species with large male filaments. The dwarf male filaments arise from special zoospores, intermediate in size between sperms and zoospores, and termed **androspores** (Fig. 3.97). These occur singly in special **androsporangia** that may be present in the same filaments as the oogonia, in which case the organism is said to be **gynandrosporous**; or the androspores may arise from other filaments, in which case the organism is said to be **idioandrosporous.**

Hoffman (1960; 1973A, B; 1974) has investigated fertilization in the dioecious macrandrous *Oe. cardiacum* (Fig. 3.96). He showed (1960) that the sperms are chemotactically attracted to the oogonia, which in that species open by pores (while in others the oogonial wall may have a fissure), and that (1973B) polyspermy may occur. Machlis, et al. (1974) have investigated the nature of the sperm attractant in the dioecious *Oe. cardiacum.* The molecular weight of the substance is between 500 and 1500.

After fertilization, the zygote wall thickens, and its contents often become brown-red as dormancy begins. After a variable period, the zygote germinates to form four zoospores. Hoffman (1965) demonstrated cytologically the occurrence of meiosis in the germinating zygote of *Oe. foveolatum* Wittr.

Rawitscher-Kunkel and Machlis (1962) investigated the chemical control of sexual reproduction in an idioandrosporous species, probably *Oe. borisianum* (LeCl.) Wittr., and reported that the androspores are chemotactically attracted to the female filaments in the vicinity of the oogonial mother cells. Only if, and after, the andro-

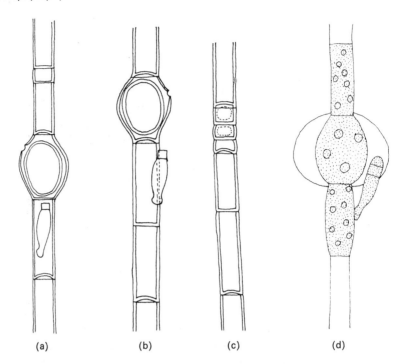

(a)   (b)   (c)   (d)

Fig. 3.97 *Oedogonium.* (*a*) *Oe. crassiunculum* Wittr., a gynandrosporous species. (*b*), (*c*) *Oe. crassiunculum* var. *idioandrosporum* Nord. and Wittr. (*b*) Female filament with attached dwarf male. (*c*) Androspore-forming filament with androsporangia. (*d*) *Oe. borisianum* (Le Cl.) Wittr. Oogonial filament. Oogonial secretion engulfing antheridia of dwarf male. (*a*)–(*c*) × 132; (*d*) × 225. [(*a*)–(*c*) after Tiffany; (*d*) modified from Rawitscher-Kunkel and Machlis.]

spores have attached and begun to grow into dwarf males does the oogonial mother cell divide into an oogonium and supporting cell. The oogonium then develops a massive gelatinous sheath that encloses the sperm-bearing apices of the dwarf males (Fig. 3.97*d*). The sperm are released into the gelatinous sheath and move at random until a papilla extrudes through the oogonial pore. The sperms then move toward this and when one becomes attached to the papilla, the latter is withdrawn carrying a single sperm into the oogonium. Rawitscher-Kunkel and Machlis (1962) also reported that the dwarf males and androspores both could develop into androsporic filaments.

Machlis (1962, 1973) and Hill and Machlis (1970) have investigated the nutrition of *Oe. cardiacum* in defined media but later reported that their cultures had been contaminated with a species of *Corynebacterium.* Upon freeing the alga from the bacterium, Machlis (1973) reported that a marked reduction in the production of antheridia occurred. This could be increased by introducing the original bacterium or another, *Pseudomonas putida.* Whether this is a genuine example of a coevolutionary relationship remains to be elucidated.

BULBOCHAETE C. Agardh   The branching filaments of *Bulbochaete* (Gr. *bulbos*, a bulb + Gr. *chaito*, a bristle) (Figs. 3.98, 3.99) grow attached to stones and woodwork and, most frequently, are epiphytic on aquatic angiosperms and occasionally on other filamentous algae. Most of the cells have long, bulbous-based, colorless bristles or hairs which result from the unequal division of the cells which bear them. The bristles lose their protoplasts at maturity. The chloroplast is parietal and reticulate and contains a number of pyrenoids. Cook (1962) showed that cell division is not restricted to the basal holdfast cell, as previously reported, but that intercalary divisions also occur. Pickett-Heaps (1974A, B) has provided an ultrastructural account of cell division and bristle formation, while Fraser (1975) has described the role of the Golgi apparatus and microtubules in the formation of the phycoplast.

Asexual reproduction is accomplished, as in *Oedogonium*, by zoospores (with 37–49 flagella) that arise singly from the vegetative cells.

Sexual reproduction is oogamous, and among the different species occur those with bisexual filaments, macrandrous forms with unisexual filaments, and those with dwarf males arising from androspores.

Cook (1962) presented a rather complete account of the ontogeny and asexual and sexual reproduction of *B. hiloensis* (Nordst.) Tiffany in culture, his work indicating the meiosis is probably zygotic, since the zygotes germinate to form zoospores. Retallback and Butler (1970A, B; 1972), Fraser and Gunning (1969, 1973), and Pickett-Heaps (1974A, B), using electron microscopy, have contributed much to the understanding of organization and reproduction in *Bulbochaete*.

Fig. 3.98 *Bulbochaete hiloensis* (Nordst.) Tiffany. × 50. (After Pickett-Heaps.)

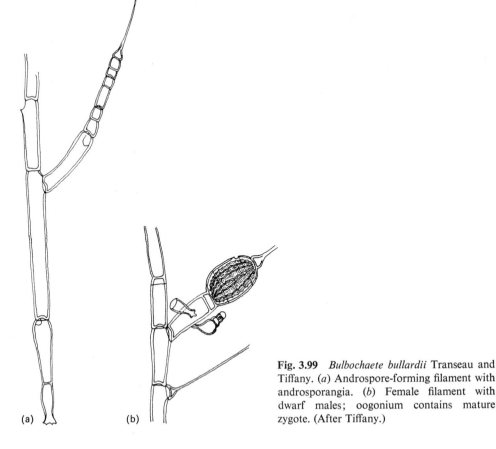

(a)　　　　(b)

Fig. 3.99 *Bulbochaete bullardii* Transeau and Tiffany. (*a*) Androspore-forming filament with androsporangia. (*b*) Female filament with dwarf males; oogonium contains mature zygote. (After Tiffany.)

OEDOCLADIUM Stahl *Oedocladium* (Gr. *oedos*, swelling + Gr. *klados*, branch) (Fig. 3.100) has both aquatic and terrestrial species. The branching filaments lack bulbous hairs but have elongate, slender rhizoidal branches. The other vegetative cells are short-cylindrical with parietal chloroplasts and pyrenoids. Cell division is largely restricted to the apical cell. Both thick-walled akinetes and multiflagellate zoospores effect asexual reproduction in *Oedocladium*.

Sexual reproduction is oogamous and the species may be monoecious or dioecious, macrandrous or nannandrous. *Oedocladium* is less widely distributed than *Bulbochaete* and *Oedogonium*. Beaney and Hoffman (1968) described two new species, *Oe. carolinianum* from near Raleigh, N.C. and *Oe. cirratum* from near Bastrop, Tex. Both of these are gynandrosporous and undergo sexual reproduction in laboratory culture.

In summary, the three genera of Oedogoniales differ clearly from each other but have in common their method of cell division that involves ring formation and their characteristic sexual reproduction in which they differ from other green algae. Their multiflagellate zoospores occur elsewhere in the green algae only in *Derbesia* (p. 206) and in *Bryopsis* (p. 201), in which genera biflagellate gametes are also formed, so that the discontinuity between the Oedogoniales and other green algae may be more apparent than real.

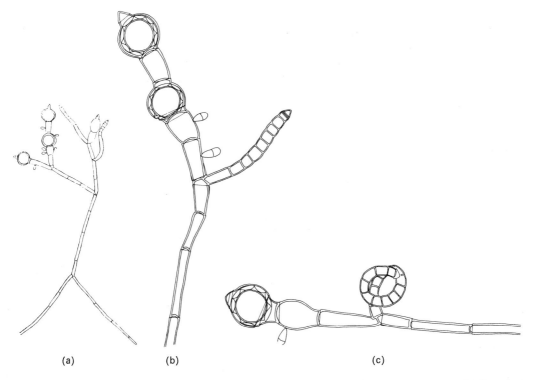

(a)   (b)   (c)

Fig. 3.100   *Oedocladium*. (*a*), (*b*) *Oe. carolinianum* Beaney and Hoffman. (*a*) Habit of growth showing oogonia and oogonial branches and dwarf males. (*b*) Portion of a plant enlarged. The oogonia contain zygotes. (*c*) *Oe. cirratum* Beaney and Hoffman. (*b*), (*c*) × 220. (After Beaney and Hoffman.)

Order 9. Ulvales

The members of the Ulvales illustrate in their organization various patterns of cell division with resultant variation in the form of the organism. Thus, the mature plants are biseriate filaments (*Percursaria*, Fig. 3.101), monostromatic (*Monostroma*, Fig. 3.102) or distromatic (*Ulva*, Fig. 3.104*b*) membranes, hollow tubes (*Enteromorpha*, Fig. 3.103), or solid cylinders (*Schizomeris*, Fig. 3.108). All except *Prasiola* have ulotrichalean cellular organization including laminate, parietal chloroplasts with pyrenoids and single nuclei in each cell, and many have erect thalli that arise from a prostrate basal system. Most produce biflagellate or quadriflagellate zoospores and biflagellate gametes. *Prasiola* is exceptional in its axile or asteroidal chloroplasts and lack of zoospores.

Sexual reproduction in the Ulvales is isogamous, anisogamous, or oogamous; haplobiontic and diplobiontic life cycles, the latter either isomorphic or heteromorphic, have been reported among the various genera; some lack sexual reproduction entirely.

Most of the Ulvales are marine, occurring also in inland waters of high salinity, whereas a few (*Schizomeris, Trichosarcina*) are restricted to freshwater.

The organisms herein included in the order Ulvales have been variously classified by others. Thus, Smith (1950) and Bourrelly (1966, 1972) recognize them, as the present authors do, as a distinct order, while Fritsch (1935), Papenfuss (1960), and Fott (1971) have included them as a family in the Ulotrichales. One of the leading students of the group (Bliding, 1963, 1968) considered the Ulvales as a distinct order.

Five families of Ulvales are included by the authors in the following account, namely, the Percursariaceae, Monostromataceae, Ulvaceae, Schizomeridaceae, and Prasiolaceae. These families may be distinguished by the following key:

1. With axile or stellate chloroplasts . . . . .Prasiolaceae
1. With parietal, laminate chloroplasts 2
   2. Filamentous, biseriate or pluriseriate 3
   2. Tubular or membranous 4
3. Filaments biseriate . . . . .Percursariaceae
3. Filaments pluriseriate . . . . .Schizomeridaceae
   4. Plants membranous or tubular; diplobiontic and isomorphic. . . . . .Ulvaceae
   4. Plants membranous; sporophyte is the enlarged zygote . .Monostromataceae

The genera to be discussed in the following account, as representative of the Ulvales, may be distinguished by the following key:

1. With axile or stellate chloroplasts . . . .*Prasiola*
1. With parietal, laminate chloroplasts 2
   2. Filamentous, biseriate or pluriseriate 3
   2. Tubular or membranous 5
3. Filaments biseriate . . . .*Percursaria*
3. Filaments pluriseriate, at least in part 4
   4. Filaments dissociating into sarcinoid packets . . . .*Trichosarcina*
   4. Filaments not so dissociating . . . .*Schizomeris*
5. Mature plants tubular, sometimes constricted . . . .*Enteromorpha*
5. Mature plants membranous 6
   6. Plant body distromatic . . . .*Ulva*
   6. Plant body monostromatic 7
7. Life cycle diplobiontic and heteromorphic (in some cases
   the enlarged zygote representing the only diploid cell). . . . .*Monostroma*
7. Life cycle diplobiontic and isomorphic . . . .*Ulvaria*

Family 1. Percursariaceae

This family contains the single genus *Percursaria*.

PERCURSARIA In ontogeny, the quadriflagellate zoospores of the marine *Percursaria* (L. *percursum*, run through) (Fig. 3.101) develop into a prostrate disc from which biseriate ribbons arise at the margins; the zoospores may also develop directly into biseriate ribbons with a rhizoidal base (Kornmann, 1956B). *Percursaria percursa* (C. Agardh) Bory is diplobiontic, isomorphic, and anisogamous; the male and female gametes are produced on different individuals, which are unisexual. Kornmann re-

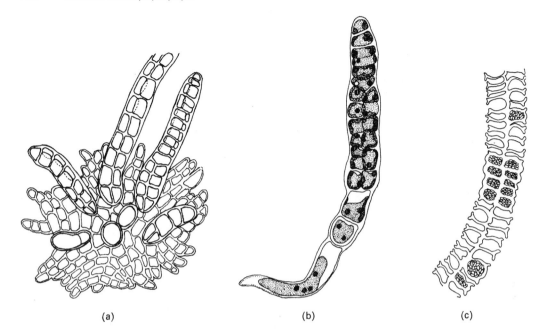

(a)   (b)   (c)

Fig. 3.101 *Percursaria percursa* (Ag.) Bory. (*a*) Habit of growth showing heterotrichy. (*b*) Young plant; initiation of biseriate condition. (*c*) Portion of plant with zoosporangia. (*a*) × 132; (*b*), (*c*) × 198. (After Kornmann.)

ported that biflagellate motile cells can function as gametes or develop parthenogenetically into filaments (presumably both sporophytes and gametophytes) that produce both biflagellate and quadriflagellate motile cells. *Percursaria* is marine and grows attached to stones and woodwork in the intertidal zone.

Family 2. Monostromataceae

The genus *Monostroma* is classified by some authors (Smith, 1950; Bourrelly, 1966, 1972) in the family Ulvaceae. Apparently the genus formerly contained two series of species, all monostromatic, which differed in ontogeny and life cycle, among other respects. Following Bliding (1968) and, in part, Gayral (1965), the authors have included certain diplobiontic isomorphic species, formerly classified in *Monostroma*, in the genus *Ulvaria*, which is classified in the Ulvaceae. Tatewaki (1972) has summarized the various types of life cycle that occur in *Monostroma* (*sensu lato*) and their bearing on classification. Finally, Hori (1972B, 1973) was able to distinguish ultrastructurally eight different kinds of pyrenoids among nine species of *Monostroma* (*sensu lato*) and among several Ulvacean algae.

MONOSTROMA Thuret   *Monostroma* (Gr. *monos*, one + L. *stroma*, layer) (Fig. 3.102*a*, *b*), as its name implies, consists of unistratose (Fig. 3.102*c*), cellular sacs or blades that are attached to the substrate by a basal disc; rhizoids are absent. In some

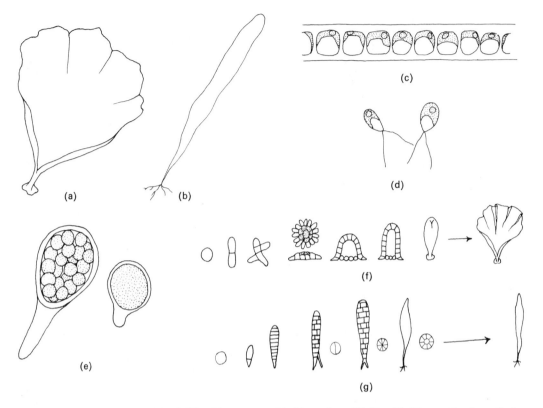

Fig. 3.102 *Monostroma.* (*a*) *M. angicava* Kjellm. (*b*) *M. groenlandicum* J. Ag. (*c*), (*d*) *M. grevillei* (Thuret) Wittr. (*c*) Transection. (*d*) Anisogamous gametes. (*e*) Zygotes (sporophytes), one forming zoospores. (*f*) Ontogeny of *M. angicava*. (*g*) Ontogeny of *M. groenlandicum*. [(*a*), (*b*), (*f*), and (*g*) after Tatewaki; the remainder after Kornmann.]

species the plants persist in the adult stage as sacs (Fig. 3.102*a*), while in others they are membranous (Fig. 3.102*b*). The plants may reach a size of 25 cm. Some species of *Monostroma* are able to tolerate great changes in the concentration of solutes in their environment.

The investigations of Kornmann (1962A, B; 1964A), Kornmann and Sahling (1962), Bliding (1963, 1968), Hirose and Yoshida (1964), and Tatewaki (1972) have greatly augmented our knowledge of the reproductive cycle in *Monostroma*. Tatewaki (1972) has summarized the types of ontogeny and life cycles that occur in five different species. The adult, frond-like or tubular plant bodies may arise by the upwelling of a group of cells in the center of a prostrate disc (Fig. 3.102*f*) (e.g., *M. angicava*), or they may develop directly as erect fronds or tubes through uniseriate and pluriseriate filamentous stages (e.g., *M. groenlandicum*) (Fig. 3.102*g*).

With respect to life cycle, several species of *Monostroma* (e.g., *M. arcticum* Wittrock and *M. undulatum* Wittr.) have asexual cycles only, in laboratory cultivation. *Monostroma angicava* Kjell and *M. groenlandicum* J. Ag. have what Tatewaki (1972) interprets to be heteromorphic life cycles as does *M. grevillei* (Thuret) Wittr. (Kornmann and Sahling, 1962). In these, the plants are gametophytes (dioecious and anisogamous in *M. angicava* and monoecious and isogamous in *M. groenlandicum*). Their gametes unite forming zygotes that become enlarged and *Codiolum*-like (Fig. 3.102e). At low temperatures (5–14°C) these reproduce after a period of vegetative development to form quadriflagellate zoospores (*M. angicava*) or aplanospores (*M. groenlandicum*); these grow into adult gametophytes to complete the cycle. In the authors' opinion it is quite plausible to consider the life cycle of these two species to be haplobiontic and haploid (H, h), since the only alternate in the life cycle is a quiescent zygote, as in *Chlamydomonas*, for example. In *M. zostericola*, by contrast, the leafy frond is the sporophyte, the zoospores of which produce discoid gametophytes (Fig. 3.102f); here the life cycle is clearly heteromorphic. In other species the vesicular stage develops without disc formation (Fig. 3.102). The species of *Monostroma* occur in freshwater and in brackish waters; one that grows near Woods Hole, Mass., is twice daily subjected to both freshwater and to the marine waters of Buzzards Bay as the tide rises.

Family 3. Ulvaceae

The members of the family Ulvaceae are tubular or membranous; in the latter case they pass through a tubular stage. They are almost exclusively marine. Many of the species have a diplobiontic, isomorphic life cycle, while a few are entirely asexual. All reproduce by means of biflagellate or quadriflagellate zoospores. All are anchored to the substrate (rocks, woodwork, or larger algae) by a basal disc or by rhizoidal filaments that are capable of regenerating new plants. They often grow in the intertidal zone so that they are periodically exposed to air. *Enteromorpha*, *Ulva*, and *Ulvaria* are discussed below as representative of the family.

ENTEROMORPHA Link   *Enteromorpha* (Gr. *enteron*, intestine + Gr. *morphe*, form) (Fig. 3.103) is widely distributed in marine habitats where it grows on rocks, woodwork, shells and other algae, often in the intertidal zone. The tubular thalli, which may be constricted at various loci, are attached to the substrate by rhizoids that may or may not be multinucleate. In ontogeny the agents of reproduction (zoospores, parthenogenetic gametes, or zygotes) attach to the substrate and by cell division in one direction form at first uniseriate filaments; these subsequently become pluriseriate and distromatic. The cells of the two layers separate along their contiguous walls to give rise to the tubular type of organization. In this connection, the morphology of *E. linza* (L.) J. Ag. is of interest, being intermediate between *Enteromorpha* and *Ulva*. In the flattened *E. linza* the two layers of the plant are contiguous in the center and separated at the margins. Eaton, et al. (1966) have reported that excised segments of the *Enteromorpha* plant have polarity; rhizoids, upon regeneration, arise only from the proximal (basal) cut surface and photosynthetic papillae from the distal (upper) cut surface.

Fig. 3.103 *Enteromorpha   intestinalis* (L.) Grev. × ⅓.

Depending on the species, the life cycle may be (1) diplobiontic and isomorphic with quadriflagellate zoospores and biflagellate gametes, meiosis occurring during zoosporogenesis, or (2) only asexual, by means of biflagellate or quadriflagellate zoospores. Kapraun (1970), in field and cultural studies of six species of *Enteromorpha* of the Texas coast, reported these types of life cycle among them and an additional type in which alternate generations of sporophytes produced by biflagellate and quadriflagellate motile cells. Kapraun reported that these six sympatric species did not produce hybrids in culture.

A number of investigators have reported the parthenogenetic development of both the male and female (anisogamous) gametes of *Enteromorpha* into new thalli. Isogamy also occurs in a few species. Lersten and Voth (1960), Christie and Shaw (1968), and Jones and Babb (1968) have studied the factors that affect discharge and motility of the zoospores of *Enteromorpha*.

ULVA (L.) Thuret   *Ulva* (Latin name for marsh plant) (Figs. 3.104, 3.105), commonly known as "sea lettuce," is widely distributed in ocean and estuarine waters where it grows attached to rocks, woodwork, and kelps. As in *Enteromorpha* and *Ulvaria*, the reproductive agent (zoospore, parthenogenetic gamete or zygote) passes through uniseriate, pluriseriate, and tubular stages in ontogeny. The adult plants in *Ulva* are distromatic through failure of two cellular layers to separate (which they do in *Enteromorpha* and *Ulvaria*). The plants are anchored to the substrate by multinucleate rhizoidal protuberances at the base. Growth of the plant body is generalized or diffuse. Løvlie and Bråten (1970) investigated mitosis in *U. mutabilis* Føyn and reported that the nuclear membrane was persistent except at the poles.

Most species of *Ulva* are diplobiontic and isomorphic with quadriflagellate zoospores and biflagellate gametes. A few species seemingly lack sexual reproduction and reproduce exclusively by biflagellate or quadriflagellate zoospores. Most of the sexual species are anisogamous. Smith (1947) studied the reproductive cycle of five species of *Ulva* on the California coast and reported that the gametes and zoospores were released only at regular 2-week intervals during the spring tides of the lunar months. The gametophytes shed gametes early in the series of spring tides, while the sporophytes shed zoospores toward the end. Bråten (1971) described gametic union and zygote development in *Ulva mutabilis* Føyn. The four flagella of the young zygote are absorbed and one of the gametic chloroplasts disintegrates (Bråten, 1973). A special attaching substance is secreted by the attaching zygote (Fig. 3.105*d*) (Bråten, 1975A, B). Bråten and Løvlie (1968), Bråten and Nordby (1973), and Løvlie and Bråten (1968) have described additional aspects of the ultrastructure of *Ulva mutabilis* Føyn.

Kapraun (1970) investigated in the field and laboratory the life cycles of *U. lactuca* L. and of *U. fasciata* Delile. He failed to obtain hybrids between the two species. Sarma and Chaudhary (1975) confirmed Kapraun's report of the chromosome number as $n = 10$ for *U. fasciata*. Føyn (1958) reported that parthenogametes of *U. mutabilis* Føyn could grow into diploid sporophytes that produced zoospores by meiosis. Føyn and his associates have made extensive genetic and development studies of *U. mutabilis* Føyn. These have been summarized by Løvlie (1968).

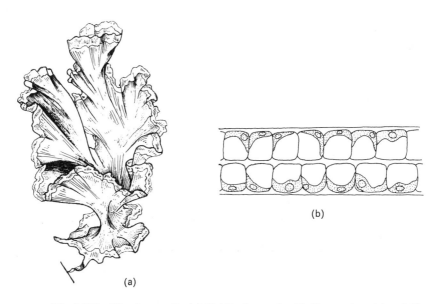

(a)

(b)

Fig. 3.104   *Ulva lactuca* L. (*a*) Habit of growth. (*b*) Transection. (*a*) × 0.13; (*b*) × 264. [(*a*) after Thuret.]

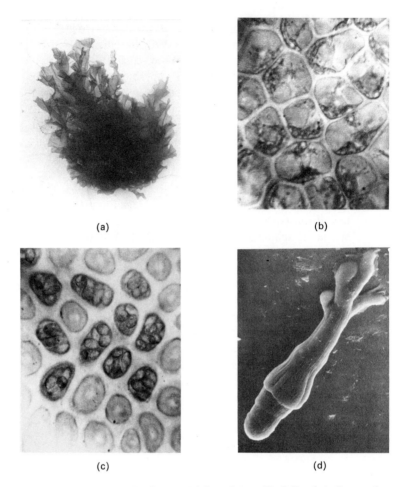

(a)

(b)

(c)

(d)

Fig. 3.105 *Ulva.* (*a*) *Ulva lactuca.* Living plants. (*b*) Cells of thallus, surface view: Note parietal chloroplasts. (*c*) Zoosporogenesis; some cells have liberated zoospores through pores. (*d*) *U. mutabilis* Føyn. Young germling (from zygote) showing rhizoidal protuberances surrounded by attachment material. (*a*) × 525; (*b*) × 470; (*c*) × 425; (*d*) × 210. (After Bråten.)

ULVARIA Ruprecht   The thalli of *Ulvaria* (Fig. 3.106) are monostromatic like that of *Monostroma.* In *Ulvaria* the reproductive agent (zoospore, parthenogenetic gamete, or zygote) develops a uniseriate filamentous stage that becomes successively pluriseriate, distromatic, tubular, and vesicular. The small vesicle opens during early ontogeny to give rise to the monostromatic adult. Unlike *Monostroma*, *Ulvaria* produces basal rhizoids, and its life cycle is diplobiontic and isomorphic [see, however, Kapraun and Flynn (1973)].

Dube (1967) has described the organization and diplobiontic, isomorphic life cycle of the northern Pacific coast species *Ulvaria obscura* var. *blyttii* (Aresch.) Bliding

(Fig. 3.106) as *Monostroma fuscum*. The gametophytes of his organism were unisexual and anisogamous, and meiosis occurred at the inception of zoosporogenesis. Partheno-genetic development of the gametes also occurred. Kapraun and Flynn (1973), by contrast, reported the life cycle of *U. oxysperma* (Kütz.) Bliding to be entirely asexual, reproduction occurring by biflagellate and quadriflagellate zoospores.

Family 4. Schizomeridaceae

The members of this family are inhabitants of freshwater and are pluriseriate filaments, at least in part. Two genera are here chosen to represent the family, namely, *Trichosarcina* and *Schizomeris*.

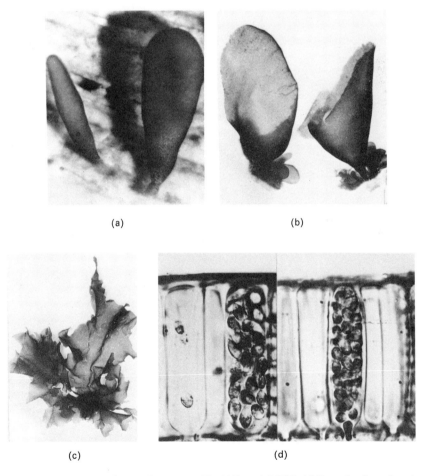

(a) (b)

(c) (d)

Fig. 3.106 *Ulvaria obscura* var. *blyttii* (Aresch.) Blid. (*a*) Juvenile plants in culture. (*b*) Rupture of vesicles. (*c*) Mature plant from nature. (*d*) Left, transection of gametophyte showing gametangia and gametes. (*a*) × 17; (*b*) × 46; (*c*) × 0.06; (*d*) × 506. (After Dube, as *Monostroma fuscum*.)

TRICHOSARCINA Nichols and Bold   *Trichosarcina* (Gr.   *thrix*,   hair + L. *sarcina*, a packet) (Fig. 3.107) is ulvacean in its ontogeny; that is, quadriflagellate zoospores develop into uniseriate and late pluriseriate filaments. The cells in these filaments remain coherent until late in development when they dissociate into sarcinoid packets. The pluriseriate filaments and sarcinoid packets may undergo zoosporogenesis. The zoospores have scales on their surfaces (Mattox and Stewart, 1973). Mattox and Stewart (1974) have investigated cell division in *Trichosarcina* and find it similar to that in *Ulva*.

Trichosarcina has been isolated from granitic pools in Texas and from soil at Coral Gables, Florida. It was first described by Nichols and Bold (1965). Nichols (personal communication) has ascertained that the Florida organism carries on anisogamous sexual reproduction.

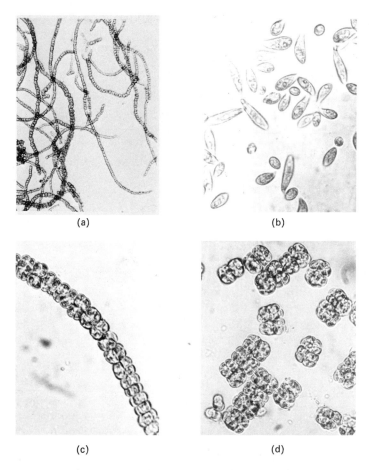

(a)   (b)

(c)   (d)

Fig. 3.107 *Trichosarcina polymorpha* Nichols and Bold. (*a*) Habit of growth: Note uniseriate and pluriseriate filaments. (*b*) Zoospores and germlings. (*c*) Pluriseriate filament. (*d*) Dissociation into sarcinoid packets. (*a*) × 66; (*b*) × 360; (*c*) × 339; (*d*) × 241. (After Nichols and Bold.)

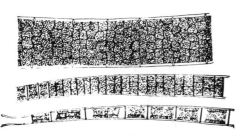

Fig. 3.108 *Schizomeris leibleinii* Kütz. Portions of thallus showing uniseriate and pluriseriate condition. × 200. (After Smith.)

SCHIZOMERIS Kützing   The mature, unbranched, solid-cylindrical thalli of *Schizomeris* (Gr. *schizo*, cleave + Gr. *meros*, part) (Fig. 3.108) grow in running freshwaters. The plant is attached to the substrate by a basal holdfast cell that may be lobed. In the early, uniseriate filament, the apical cell is pointed. The pluriseriate portions of the thallus display a bricklike arrangement of cells and may be constricted at intervals. Ringlike transverse walls, remnants of the uniseriate phase, persist in the pluriseriate one; the cells between the ringlike transverse walls undergo intercalary division to form the parenchymatous plant body. Mattox, et al. (1974) report that unlike *Ulva*, in which cell division is by furrowing, in *Schizomeris* it is accomplished by a cell plate with associated microtubules (the latter as a **phycoplast**, see p. 64). Furthermore, plasmodesmata, or protoplasmic connections, occur in the transverse walls. These investigators consider *Schizomeris* to be related to the Chaetophorales on the basis of these considerations.

   In addition to reproduction by fragmentation, which results in free-floating plants, zoospores develop singly in the vegetative cells. Prasad and Srivastava (1963) reported great variability in the flagellation of the zoospores of *S. leibleinii* Kütz., two to eight or more flagella being common on the zoospores in clonal cultures. McBride (1968) investigated the conditions that evoke zoosporogenesis in the same species and found similar variation in the flagellation of the zoospores. Birkbeck, et al. (1974) have investigated the quadriflagellate zoospores electron microscopically. Sexual reproduction has not been confirmed in *Schizomeris*.

   Sarma and Chaudhary (1975B) have reviewed the literature of *Schizomeris* and, in contrast to other investigators, report the chromosome number to be $n = 14$ in their strain of *S. leibleinii*.

Family 5. Prasiolaceae

PRASIOLA Meneghini   The single genus of this family, *Prasiola* (Gr. *prasios*, green) (Fig. 3.109), was originally classified in an order of its own because of its stellate or axile chloroplasts and its supposed lack of motile cells. Inasmuch as biflagellate sperms have been described (Friedmann, 1959) and in view of the fact that other orders of Chlorophyceae (e.g., Chlorococcales) have members with a variety of chloroplast types, *Prasiola* is here included with the Ulvales, a practice followed also by Bourrelly (1966).

   The thalli of *Prasiola* are monostromatic and have the appearance of miniature Ulvas or Monostromas. They are attached to the substrate by marginal rhizoids or

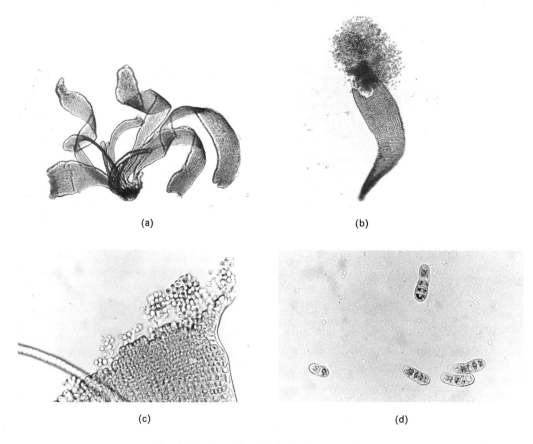

(a)

(b)

(c)

(d)

Fig. 3.109 *Prasiola calophylla* (Carm. et Grev.) Menegh. (*a*) Habit of growth. (*b*) Fertile plant in culture, liberation of aplanospores. (*c*) Apex of fertile plant, more highly magnified. (*d*) Germinating aplanospores. (*a*) × 25; (*b*) × 130; (*c*) × 702; (*d*) × 273. (After Kornmann and Sahling.)

those originating at the base of a stalk. *Prasiola* species occur in both freshwater and marine water, in the former in rapidly running streams; some grow on damp soil. Some species grow on substrates with a high concentration of nitrogen from the droppings of birds. The cells are arranged in geometric, pseudosarcinoid groups (Fig. 3.109*c*) and in some species give the appearance under low magnification of plurilocular gametangia or bryophytic antheridia.

The thallus of *Prasiola* is diploid as in *P. stipitata* Suhr (Friedmann, 1959), and mature individuals form spores or gametes near their apices. The fertile regions become two-layered as a result of paradermal divisions. The protoplasts of the cells in this region are liberated from their walls as spherical spores, essentially similar to the monospores of red algae (p. 464), and remain in a common chamber formed between the persistent outer walls of the upper and lower surfaces of the plant. They are gradually liberated from this cavity and may develop into plants similar to those from which they came or into gametophytic plants (Fig. 3.110*b*).

In the latter, certain distal cells undergo meiosis followed by mitoses and cytokineses in a plane parallel to the thallus surface. The resultant haploid cells undergo

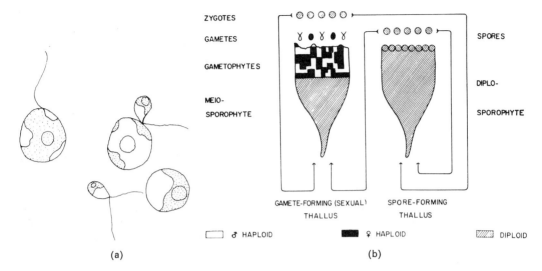

Fig. 3.110 *Prasiola stipitata*. (*a*) Eggs and sperm and zygote with one emergent sperm flagellum. (*b*) The possible courses of the life cycle. [(*a*) based on Friedmann; (*b*) after Friedmann.]

further mitoses and cytokineses in three perpendicular directions so that sarcinoid packets of cells result (Fig. 3.110*b*). These packets differ in color; those that are lighter (because of the smaller size of their plastids) become male gametes and those that are darker, female gametes. This segregation reflects the fact that the sex genes are segregated in meiosis I. Like the spores, the gametes are liberated into a common cavity between the surface layers of the plant body.

Friedmann and Manton (1960) have described in detail the oogamous sexual reproduction in *P. stipitata*. The male gametes are biflagellate, while the female gametes are somewhat irregular in form and lack flagella. The male gamete or sperm glides over the surface of the egg remaining in contact by means of one of its flagella (Fig. 3.110*a*); this causes rotation of the egg. The union becomes more durable when one of the sperm's flagella is absorbed into the egg cytoplasm; the egg then becomes motile through the activity of the second flagellum of the sperm. The latter and the body of the sperm are ultimately incorporated into the egg and the zygote becomes walled. Zygotes presumably grow into new diploid individuals that may produce spores or gametes.

The life cycle of *P. stipitata* may be interpreted in two different ways. One could consider, as Friedmann (1959) did (Fig. 3.110*b*), that the haploid tissue that arises after meiosis represents male and female "gametophytes" borne on diploid thalli. Alternatively, it is possible to consider the life cycle as consisting of individuals that under certain conditions undergo meiosis to form gametes as they do in such diploid plants as some species of *Bryopsis*. The life cycle would then be H, d. Bravo (1965) reported that a life cycle essentially similar to that of *P. stipitata* occurs in *P. meridionalis* S. and G. on the West Coast of North America.

More recently Kornmann and Sahling (1974) have reported on the reproductive cycles of *Prasiola* at Helgoland. Those species subjected to submersion by the sea grow best in marine media, while others sprayed only occasionally with seawater grow better in freshwater media. At Helgoland *P. calophylla* is entirely asexual but heteromorphic, the spores of the leafy thallus growing into unicellular aplanosporangia. *Prasiola furfuracea* reproduces only by spores.

In summary, the Ulvales contain a diverse assemblage of green algae, which are filamentous (biseriate or pluriseriate), tubular, or membranous (monostromatic or distromatic). The cells of all, except the rhizoidal cells of some, are uninucleate. A few lack sexual reproduction, but the majority that have it exhibit a variety of life cycles including haplobiontic and haploid (*Monostroma*),[23] diplobiontic and isomorphic (*Percursaria, Enteromorpha, Ulva,* and *Ulvaria*), and the unusual type demonstrated in *Prasiola*. In all, the zoospores are biflagellate or quadriflagellate, if produced, and the gametes, biflagellate. Sexual reproduction is isogamous, anisogamous, or oogamous. Except for *Trichosarcina, Schizomeris,* and some species of *Prasiola,* the group is largely marine.

Order 10. Cladophorales

With the exception of *Protosiphon* and *Hydrodictyon,* the green algae discussed up to this point had uninucleate cells.[24] By contrast, the members of the Cladophorales all have cells with more than one nucleus, the number varying from few (in some species of *Rhizoclonium*) to many (as in *Cladophora*). The multinucleate condition arises ontogenetically in that mitosis and cytokinesis are not closely correlated in time as they are in algae with uninucleate cells. Thus, the agents of reproduction (zoospores, parthenogenetic gametes, or zygotes) attach to the substrate, and their single nuclei undergo a series of mitoses without ensuing cytokineses to give rise to multinucleate cells.

The reproductive cells are biflagellate or quadriflagellate in the asexual individuals, while the gametes are biflagellate. In those with sexual reproduction, the life cycle is isomorphic, D^i, h + d.

The members of the Cladophorales cited below in illustration of the order are *Rhizoclonium, Chaetomorpha, Cladophora, Pithophora,* and *Sphaeroplea.* The first four are classified in the family Cladophoraceae and the last in the family Sphaeropleaceae.

Family 1. Cladophoraceae

The members of this family are branched or unbranched. In their sexual reproduction, if it occurs, the gametes are isogamous and the life cycle is D^i, h + d.

RHIZOCLONIUM Kützing   The species of *Rhizoclonium* (Gr. *rhizo*, root + Gr. *klonos*, branch) (Fig. 3.111) occur in freshwater and brackish and marine waters. The

---

[23]Sometimes interpreted as diplobiontic and heteromorphic by those who consider the zygote a diploid alternate.

[24]In a few cases (e.g., *Pediastrum*) they become multinucleate just before zoosporogenesis.

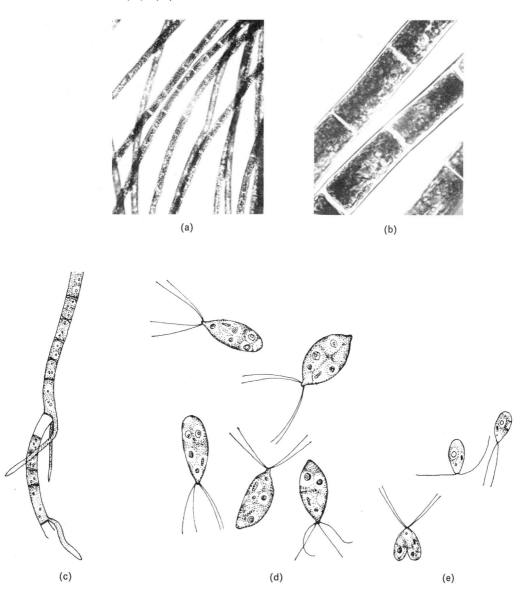

Fig. 3.111 *Rhizoclonium.* (*a*), (*b*) *R. hieroglyphicum* (Ag.) Kütz. (*a*), (*b*) Filaments at successively higher magnification. (*c*)–(*e*) *R. kochianum* Kütz. (*c*) Segment of filament with rhizoidal branches. (*d*) Zoospores. (*c*) Gametes and isogamous gametic union. (*d*) Zoospores. (*e*) Gametes and isogamous gametic union. (*a*) × 56; (*b*) × 263; (*c*) × 113; (*d*), (*e*) × 563. [(*c*)–(*e*) after Bliding.]

filaments may or may not be attached with a holdfast cell with basal lobes. *Rhizoclonium* species sometimes resemble sterile plants of *Oedogonium* from which they may be distinguished by the absence of cell division rings in *Rhizoclonium*. The latter may occur intermingled with other algae, e.g., *Enteromorpha*. The cells are several to many times longer than broad and contain a reticulate, parietal chloroplast with pyrenoids and have more than one nucleus. The filaments are unbranched or one- or few-celled rhizoidal branches (Fig. 3.111c) may be present.

In the freshwater species, reproduction occurs by fragmentation, and biflagellate zoospores seemingly are produced but rarely. In two marine species [*R. kochianum* Kütz. and *R. riparium* (Roth) Harvey], a diplobiontic isomorphic life cycle involving quadriflagellate zoospores and biflagellate gametes was reported by Bliding (1957). Nienhuis (1974) found that sexual reproduction was suppressed in *R. riparium* by reduced salinity and increased concentration of nitrogen. *Rhizoclonium* species are sometimes difficult to distinguish from *Chaetomorpha* or sparingly branched species of *Cladophora*.

CHAETOMORPHA Kützing   *Chaetomorpha* (Gr. *chaeto*, hair + Gr. *morphe*, form) (Fig. 3.112) may be free-floating or attached to rocks and shells, and its species are widely distributed throughout the world in marine waters. The cells are barrel-shaped and in some species are large enough to be visible to the unaided eye. The filaments grow singly or in tufts and are anchored by an elongate holdfast cell with a prominent attaching basal portion; this distinguishes *Chaetomorpha* from *Rhizoclonium*. The cells contain a reticulate chloroplast composed of many segments and containing many pyrenoids; a large number of nuclei is present in each cell. Kornmann (1968, 1969) has demonstrated that growth is generalized in various species of the genus and has characterized taxonomically (1972A) a number of the European species.

In some species, the life cycle is diplobiontic and isomorphic (D^i, h + d), the plants producing either quadriflagellate zoospores or biflagellate gametes (Köhler,

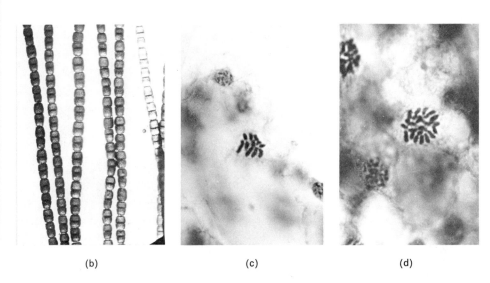

(a)   (b)   (c)   (d)

Fig. 3.112   *Chaetomorpha aerea*. (*a*), (*b*) Filaments at successively higher magnification. (*c*), (*d*) Plates of metaphase chromosomes of the gametophyte and sporophyte, respectively. (*a*) × 2.7; (*b*) × 7.5; (*c*), (*d*) × 748. (After Kornmann.)

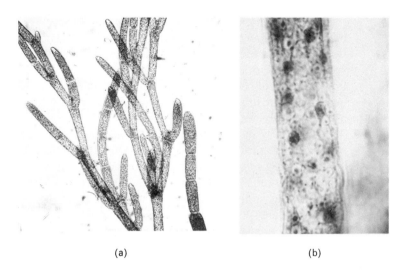

(a)   (b)

Fig. 3.113 *Cladophora* sp. (*a*) Habit of growth. (*b*) Cellular organization: Note many nuclei (darkly staining structures). (*a*) × 72; (*b*) × 360.

1956). The differences in chromosome complement in sporophyte and gametophyte are illustrated in Fig. 3.112*c, d.* Kornmann (1969) had confirmed this for the Australian *C. darwinii.* Parthenogenetic development of gametes occurs to repeat the gametophytic phase.

CLADOPHORA Kützing   The various species of the genus *Cladophora* (Fig. 3.113) occur in freshwater or brackish and marine waters. Jónsson (1962B) has investigated the morphology and cytology of *Cladophora* and other genera in the family, and van den Hoek (1963) has summarized knowledge about the European species. Most species of *Cladophora* are attached to the substrate (rocks, woodwork, or coarser algae) by rhizoidal cells.

The pattern of growth among the various species of *Cladophora* is exclusively apical, exclusively intercalary, or a mixture of both, depending on the species. The cell walls consist of cellulose I microfibrils (Nicolai and Preston, 1959). Hanic and Craigie (1969) reported that the surface cuticle contained up to 70% protein. Cell division is, of course, independent of nuclear division. The transverse walls arise by annular furrowing and deposition of wall material in the furrow.

The large cells of *Cladophora* are multinucleate (Fig. 3.113*b*), the nuclei lying beneath the chloroplast in the colorless cytoplasm, which may be frothy and reticulate. Wik-Sjöstedt (1970) has studied mitosis and meiosis in nine species of *Cladophora*; she found that the basic chromosome number was six but that polyploid series occurred in most species. The chloroplast, which contains numerous pyrenoids, gives the appearance of a reticulum or at times that of a network composed of jigsaw-like segments but is not completely continuous. McDonald and Pickett-Heaps (1976) have described cell division with the aid of electron microscopy.

In several marine species of *Cladophora* the life cycle has been shown to be diplobiontic and isomorphic (D^i, h + d), the individual gametophytes being unisexual and isogamous. The zoosporangia and gametangia may have the appearance of ordinary vegetative cells, or, prior to the fertile period, cell divisions may result in the production of chains of shorter cells. The zoospores, which arise after meiotic divisions, are quadriflagellate in the diplobiontic species and grow into gametophytes that produce biflagellate gametes.

Apparently in some species of *Cladophora* reproduction occurs only asexually by biflagellate or quadriflagellate zoospores. The freshwater *C. glomerata* was long ago reported (List, 1930) to be diploid, meiosis occurring in the formation of haploid gametes. This requires confirmation. In a few species no motile reproductive cells have been recorded, so that their reproduction seemingly is accomplished only by fragmentation. In some species thick-walled akinetes may form, and the basal rhizoids also are resistant to adverse conditions and capable of regenerating. Moore and Traquair (1976) reported that silicon, present in the cell walls of *Cladophora glomerata* (L.) Kütz., is an essential nutrient for that species.

PITHOPHORA Wittrock *Pithophora* (Gr. *pithos*, small cask + Gr. *phora*, bearer) (Fig. 3.114) has the gross appearance of *Cladophora* but is restricted to freshwater. The cellular organization is like that of *Cladophora*, the cells being multinucleate and having a parietal chloroplast with pyrenoids. Other than by fragmentation, reproduction occurs exclusively through terminal and intercalary akinetes, zoospores not being produced. Under favorable conditions, the akinetes germinate to form new filaments. *Pithophora oedogoniana* (Mont.) Wittr. is fairly common in lily pools and quiet waters in the southern United States.

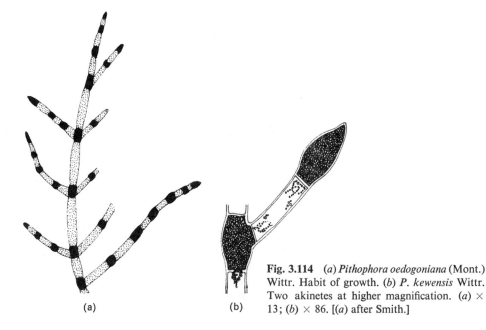

(a)

(b)

Fig. 3.114 (*a*) *Pithophora oedogoniana* (Mont.) Wittr. Habit of growth. (*b*) *P. kewensis* Wittr. Two akinetes at higher magnification. (*a*) × 13; (*b*) × 86. [(*a*) after Smith.]

Family 2. Sphaeropleaceae

This family includes the single genus *Sphaeroplea*.

SPHAEROPLEA C. Agardh   *Sphaeroplea* (Gr. *sphaira*, sphere + Gr. *pleon*, many) (Figs. 3.115, 3.116), with four species, is the sole member of the family, the relationships of which are obscure. It is here classified with the Cladophoraceae because of its multinucleate cells. The latter are unique among the green algae in being 15–60 times as long as broad and in having extremely thin walls that lack a gelatinous sheath. The cells contain a series of large vacuoles segregated from each other by protoplasmic septa (Fig. 3.115) containing a portion of the chloroplast and nuclei. Pyrenoids are present in the chloroplast. The filaments lack holdfasts throughout their development. *Sphaeroplea* is not a common alga in the writers' experience. It has been found in flooded ditches and gravel pits.

Asexual reproduction occurs solely by fragmentation, zoospores not being produced by the vegetative filaments. In sexual reproduction the gametes are anisogamous or oogamous, depending on the species. Whether the various species are monoecious or dioecious has not been determined with certainty. Certain cells undergo increase in nuclear number, and their protoplasts cleave to form very numerous biflagellate male gametes (Fig. 3.116*a*). In other cells, which function as female gametangia or oogonia, the protoplast cleaves to form a large number of eggs or female gametes that are uninucleate at maturity (Fig. 3.116*b*). In *S. cambrica* Fritsch, at least some of the female gametes are biflagellate, so that anisogamy may be said to occur. In others the female gametes are nonflagellate eggs. Depending on the species, fertilization may occur within the oogonial cell or outside. In the former case, the sperms enter through a lateral pore in the oogonial wall.

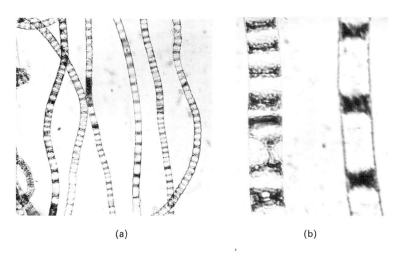

(a)   (b)

Fig. 3.115   *Sphaeroplea annulina* (Roth) Ag. (*a*) Filaments at low magnification. (*b*) Cellular organization. (*a*) × 19; (*b*) × 180.

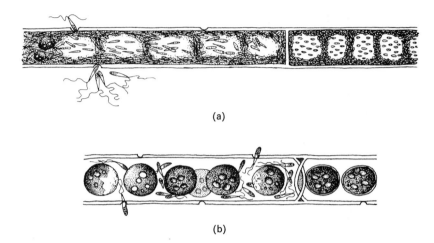

(a)

(b)

Fig. 3.116 *Sphaeroplea annulina.* (*a*) Male filament with sperms escaping. (*b*) Female filament; sperms surrounding eggs. × 185. (After Cohn and Klebahn from Oltmanns.)

After a period of dormancy, the zygotes germinate, forming four zoospores. These germinate into cells pointed at both ends and lacking a holdfast. The life cycle of *Sphaeroplea* is thought to be the H, h type.

Order 11. Acrosiphoniales

The Acrosiphoniales contains a single family, the Acrosiphoniaceae, with the characteristics of the order. The members of this family differ from those of the Cladophorales in having single, perforate chloroplasts in their cells, and many have a heteromorphic life cycle in which the haploid gametophyte is succeeded by a saclike unicellular sporophyte classified in the genus *Codiolum* (Fig. 3.117*e*). This type of life cycle does not always occur (see *Acrosiphonia*, p. 189). Kornmann (1965) considers the operculate gametangia (Fig. 3.120*b*) as the most distinguishing characteristic of the order, while Jónsson (1962B) emphasized plastid organization and the heteromorphic life cycle. *Codiolum* was described first as an independent genus, its connection with the Acrosiphonialean phase in the life cycle not having been established until about 1959. The sporophytes and gametophytes differ with respect to their requirements of temperature (and photoperiod?) and thus often occur in a given locality at different times of the year. The three representative genera of Acrosiphoniales discussed below, namely, *Urospora*, *Spongomorpha*, and *Acrosiphonia*, are all marine and occur mostly in colder waters.

Family Acrosiphoniaceae

UROSPORA Areschough   (= *pro parte Codiolum* A. Braun) The rather coarse filaments of *Urospora* (Gr. *oura*, tail + Gr. *spora*, spore) (Figs. 3.117, 3.118) are unbranched and grow attached to solid substrates by means of basal rhizoidal cells that

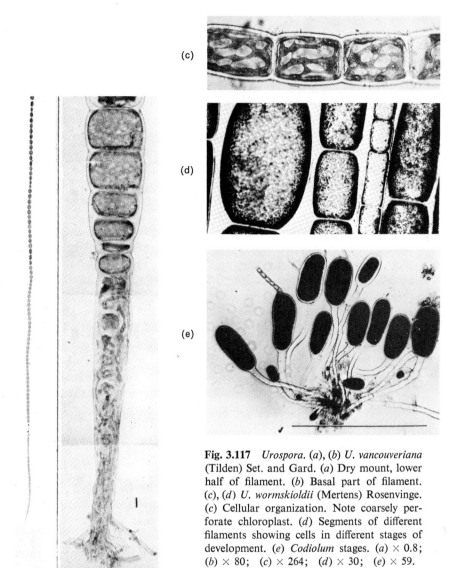

(c)

(d)

(e)

Fig. 3.117 *Urospora.* (*a*), (*b*) *U. vancouveriana* (Tilden) Set. and Gard. (*a*) Dry mount, lower half of filament. (*b*) Basal part of filament. (*c*), (*d*) *U. wormskioldii* (Mertens) Rosenvinge. (*c*) Cellular organization. Note coarsely perforate chloroplast. (*d*) Segments of different filaments showing cells in different stages of development. (*e*) *Codiolum* stages. (*a*) × 0.8; (*b*) × 80; (*c*) × 264; (*d*) × 30; (*e*) × 59. (Courtesy of Prof. Louis Hanic.)

(a)   (b)

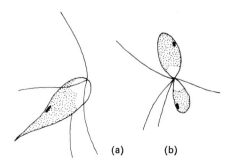

(a)   (b)

Fig. 3.118 *Urospora wormskioldii.* (*a*) Zoospore. (*b*) Anisogamous gametes. × 750. (Courtesy of Prof. Louis Hanic.)

augment the primary basal holdfast of the germling. *Urospora* grows in colder marine waters (e.g., Puget Sound, Hudson River at New York, Cape Cod, Helgoland) where it is present during certain months of the year (Kornmann, 1961B; Hanic, 1965).

The cells are large and multinucleate and contain a parietal perforate chloroplast with pyrenoids (Fig. 3.117c) that characterizes the family. Many nuclei migrate to that region of the cell at which division is occurring. Cell division and hence growth are generalized in the young filament, but growth later becomes basal (Kornmann, 1966).

The taxonomy of the several species of *Urospora* requires further investigation, and several types of reproductive cycle have been reported. Kornmann (1961B) and Hanic (1965) have presented the most detailed accounts of reproduction for *U. wormskioldii*. The former studied the species from Helgoland and the latter from Puget Sound.

According to Hanic the life cycle is probably D^h, h + d, although the chromosome cycle has not been elucidated. The elongate filaments are unisexual gametophytes that produce anisogamous biflagellate gametes (Fig. 3.118b) on different individuals, according to Hanic. The resulting zygotes develop into *Codiolum gregarium* A. Br. (Fig. 3.117e). These grow vegetatively for long periods but finally become fertile (at sufficiently low temperatures) and liberate quadriflagellate zoospores (Fig. 3.118). Both the male and female gametes also develop without union into *Codiolum* plants, according to Hanic (1965).

Although seemingly identical morphologically to Hanic's alga, the two varieties of *U. wormskioldii* studied by Kornmann (1961B) undergo entirely asexual life cycles in which dwarf filaments and *Codiolum* plants are produced. The reproductive agents of one variety are biflagellate swarmers and in the other they are quadriflagellate; sexual reproduction is absent.

Because the generic name *Codiolum* antedates *Urospora*, Kornmann (1961) has used it on occasion for the diplobiontic plant.

SPONGOMORPHA Kützing   *Spongomorpha* (Gr. *spongos*, sponge + Gr. *morphe*, form) (Fig. 3.119) superficially resembles *Cladophora* and *Acrosiphonia* but differs in its uninucleate cells. The branching filaments occur in dense tufts. Growth is apical, and cell elongation is confined to the apical cells of the branches (Kornmann, 1967).

Jónsson (1962B) and Kornmann (1961A, 1964D) have investigated the life cycle of *Spongomorpha*. In *S. lanosa* the filamentous plants are dioecious and form biflagellate isogametes. These are liberated from operculate gametangia (Kornmann, 1965) (Fig. 3.119d). After gametic union, the zygotes develop into a unicellular, obligately endophytic alga, *Chlorochytrium inclusum* Kjellman (Jónsson, 1962B), which lives within the red alga *Polyides rotundus* (Huds.) Grev., or into *Codiolum petrocelidis* (Fig. 3.119e), which lives within the thallus of the red alga *Petrocelis hennedyi* (Harv.) Batters or in *P. cruenta* J. Ag. At maturity, these unicellular sporophytes (*Codiolum* or *Chlorochytrium*) liberate quadriflagellate zoospores that reestablish the gametophytic phase. On the basis of his investigations, Kornmann (1972B) concluded that *Chlorochytrium inclusum* and *Codiolum petrocelidis* were different phenotypes of the same organism.

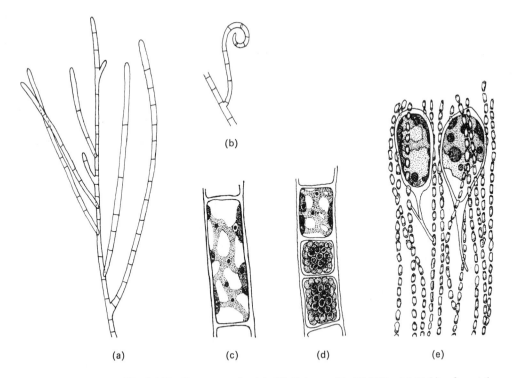

(a)   (c)   (d)   (e)

Fig. 3.119 *Spongomorpha.* (*a*)–(*d*) *S. lanosa* (Roth) Kütz. (*a*) Habit of growth. (*b*) Hooklike branch. (*c*) Chloroplast organization. (*d*) Gametogenesis. (*e*) *S. aeruginosa* (L.) van den Hoek. Sporophyte (= *Codiolum petrocelidis* Kuckuck in *Petrocelis hennedyi*). (*a*), (*b*) × 44; (*c*), (*d*) × 83; (*e*) × 119. (After Kornmann.)

ACROSIPHONIA   *Acrosiphonia* (Gr. *acros*, apex + Gr. *siphon*, tube) (Fig. 3.120) differs from *Spongomorpha* in its multinucleate cells. Its growth is strictly apical as in *Spongomorpha*. Hudson and Waaland (1974) have investigated mitosis and cytokinesis in *A. spinescens* (Kutz.) Kjellm. As cytokinesis is about to occur, a bridge of cytoplasm with chloroplast intrudes across the vacuole. Certain of the numerous nuclei migrate to that site and undergo mitosis, while the remaining nuclei fail to divide. The nuclei are distributed mostly to the upper (new apical) cell at cytokinesis. No phragmoplast or phycoplast is present during cytokinesis.

Four types of life cycle occur in *Acrosiphonia* (Fan, 1959, as *Spongomorpha coalita*; Kornmann, 1970A, B, 1972B). These include (1) a D^h, h + d cycle in which *Codiolum* is the sporophyte (as in *A. spinescens*); (2) no alternation but the incipient *Codiolum* phase grows directly into a gametophytic filament (*A. grandis*); (3) a succession of gametophytes (*A. arcta*); and (4) a succession of asexual plants produced by biflagellate swarmers (*A. sonderi*). The *Codiolum* stage (Fig. 3.120*c*) occurs endophytically in thalli of the red algae *Petrocelis hennedyi*, *P. cruenta*, and *Polyides*

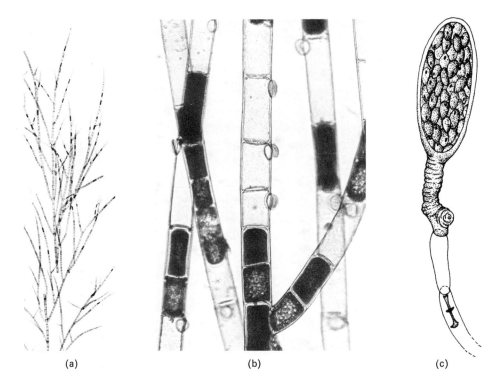

(a)   (b)   (c)

Fig. 3.120 *Acrosiphonia arcta* (Dillw.) J. Ag. (*a*) Habit of growth; dark cells are fertile. (*b*) Fertile filaments: Note empty, operculate gametangia. (*c*) Sporophyte. (*a*) × 38; (*b*) × 66; (*c*) × 356. [(*a*) after Kornmann and Sahling; (*b*), (*c*) after Kornmann.]

rotundus. These alternate types of life cycle and the *Chlorochytrium* phase have caused a great deal of confusion in our understanding of *Acrosiphonia*.

Order 12. Caulerpales

The Caulerpales, along with the Siphonocladales and Dasycladales, are those green algae with a siphonous or coenocytic construction. This type of construction consists of tubular filaments, lacking cross walls (or septa) except for delimiting reproductive structures. It may occur as single, uniaxial systems (e.g., *Derbesia* and *Bryopsis*) or as consolidated, multiaxial systems (e.g., *Codium* and *Halimeda*). Members of the Caulerpales may be distinguished from the other siphonous green algae by the distinctive pattern of cell division (so-called "segregative cell division") in the Siphonocladales and the radially symmetrical organization present in the Dasycladales.

Our interpretation of the Caulerpales is that of a broad assemblage encompassing a number of families that other authors have split off into independent orders, such as the Dichotomosiphonales, Derbesiales, Bryopsidales, and Codiales. In some instances such taxonomic fragmentation was done prematurely, either on the basis of isolated life-history information or other criteria that later proved to be not so restricted in their occurrence. The present account recognizes and will treat the following families: Codiaceae, Udoteaceae, Derbesiaceae, Bryopsidaceae, Dichotomosiphonaceae, Caulerpaceae, and Phyllosiphonaceae. A key to these families is presented below:

1. Uniaxial construction (single siphon) 2
1. Multiaxial construction (many siphons consolidated into
   more massive thalli) 6
    2. Endophytic or endozoic habitat ....Phyllosiphonaceae
    2. Free-living existence 3
3. Thalli differentiated into creeping rhizome-like portions bearing
   rhizoids and erect, often elaborate photosynthetic portions; cell
   walls bearing inwardly projecting extensions (trabeculae) ....Caulerpaceae
3. Thalli not so differentiated; trabeculae absent 4
    4. With oogamous sexual reproduction; growing
       in freshwater ....Dichotomosiphonaceae
    4. With anisogamous sexual reproduction, marine 5
5. With alternation of heteromorphic phases, typically
   involving a branched, tubular asexual stage and a saccate
   sexual stage..Derbesiaceae
5. Alternation of phases either lacking or in some instances a filamentous
   protonemal stage may alternate with the more conspicuous,
   branched, tubular sexual stage ....Bryopsidaceae
    6. Producing specialized compound reproductive structures with predetermined
       discharge tubes; two types of plastids (chloroplasts and
       leucoplasts) present.....................................Udoteaceae
    6. Producing simple reproductive structures; one type of plastid
       (chloroplasts) present ....Codiaceae

In addition to the traits mentioned previously, members of the Caulerpales seemingly have distinctive xanthophyll pigments, namely, **siphonein** and **siphonaxanthin**, the occurrence of both pigments being almost entirely restricted to this order, although siphonaxanthin alone does occur in some Siphonocladales (Goodwin, 1974). Cell walls may be composed of mannan, as in *Codium* and *Derbesia*, or of xylan, as in *Caulerpa*, *Udotea*, and *Halimeda*. The taxonomic utility of cell wall chemistry in the siphonous green algae and in other groups has been explored by Parker (1970). Zoospores are largely absent, except for *Derbesia* and for some populations of *Bryopsis*. It should be pointed out, however, that the reproductive biology of many marine Caulerpales is still little understood.

Family 1. Codiaceae

The reproductive trait of simple gametangia and the cytological trait of only one type of plastid serve to distinguish the Codiaceae from the Udoteaceae, while the multiaxial construction seen in the Codiaceae quickly separates it from the Bryopsidaceae with its uniaxial construction. Large plant sizes may be reached by the interweaving of the siphons into multiaxial systems, which are anchored below by a mass of rhizoids. The gametangia are of definite form, and reproduction is anisogamous.

CODIUM Stackhouse   (Gr. *kodion,* the skin of an animal) The thalli of this rather widespread marine genus, containing about 80 species (Silva, 1962A), exhibit tremendous variation (Fig. 3.121) ranging from prostrate crusts to hollow spheres to erect systems, consisting possibly of branched cylindrical axes or unbranched laminate blades. All share a spongy, noncalcareous texture and are composed of interwoven siphons, (Fig. 3.122*a*) arranged in a colorless, vertically aligned, medullary region and a photosynthetic, horizontally aligned, cortical region. As the coenocytic filaments approach the periphery of the tightly congested multiaxial system, each tip is dilated into an enlarged **utricle.** The densely arranged utricles form a compact, palisade-like surface layer. The shape of the utricle is an important characteristic in the classification of the species (Silva, 1951). From the sides of these utricles are produced the reproductive structures (Fig. 3.122*b*), usually regarded as gametangia. Chloroplasts are small and discoidal and lack pyrenoids (Hori and Ueda, 1967).

Sexual reproduction is by means of biflagellate anisogametes, with a marked difference in the size of the male and female gametes. Thalli are usually dioecious, although monoecious plants have been reported. Mature gametangia are readily distinguishable (Borden and Stein, 1969) by the dark green color and the lumpy,

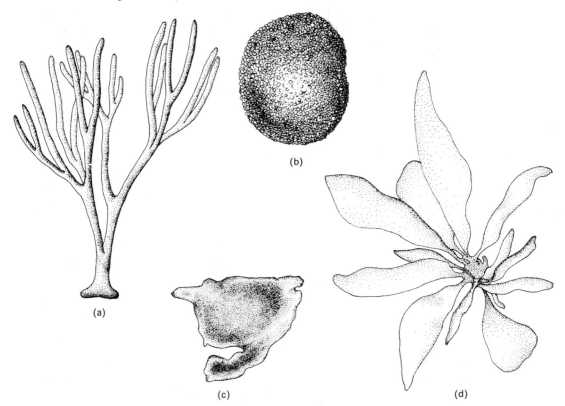

(b)

(a)

(c)   (d)

Fig. 3.121   *Codium.* (*a*) *C. fragile* (Suring.) Har. (*b*) *C. mamillosum* Harv. (*c*) *C. setchellii* Gardner. (*d*) *C. latum* Suring. (*a*) × 0.4; (*b*) × 1.2; (*c*) × 0.4; (*d*) × 0.3.

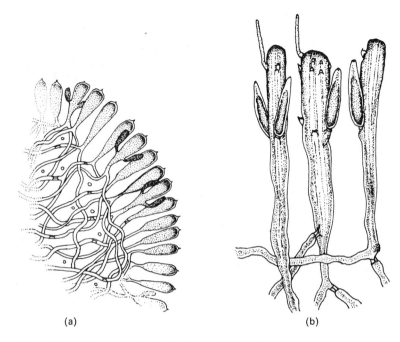

(a)   (b)

Fig. **3.122**   *Codium*. (*a*) *C. fragile* (Suring.) Har., diagrammatic transverse section of a thallus branch. (*b*) *C. prostratum* Levr., utricles bearing gametangia and
hairs. (*a*) × 20; (*b*) × 53. [(*a*) after Smith; (*b*) after Silva.]

uneven appearance of the mass of gametes within the female gametangia and the bright
yellow color and the even appearance of the mass of gametes within the male gametangia. In Pacific Coast populations of *Codium fragile* (Sur.) Hariot, such as from the
Monterey Peninsula, California (Smith, 1955) and from Vancouver Island, British
Columbia (Borden and Stein, 1969), the fusion of gametes has been reported, and the
absence of parthenogenetic development of female gametes was suggested. In Atlantic
populations of *Codium fragile* subsp. *tomentosoides* (van Goor) Silva, such as those
from Long Island, New York (Churchill and Moeller, 1972) and from Connecticut
(Ramus, 1972A), only one type of motile cell was observed, equivalent to the female
gametes of the west coast, and these germinated directly, apparently without any
fusion. It is noteworthy that this latter alga was inadvertently introduced into waters
of the Atlantic coastline and is presently distributed from New Jersey to Maine. It
has reached what might be considered to be epidemic proportions, interfering both
with normal swimming activities of man and also displacing shellfish from their
attachment to the substrate (see Wassman and Ramus, 1973A). Recent reports have
examined this "nuisance" seaweed both in the field (Churchill and Moeller, 1962;
Malinowski and Ramus, 1973; Wassman and Ramus, 1973B) and in the laboratory
(Ramus, 1972A). Ramus (1972A) determined that shear forces generated by a shaker
culture were required for the transformation of the unconsolidated filaments into the
differentiated multiaxial axes and for their maintenance.

Family 2. Udoteaceae

This family is comprised of a number of tropical to subtropical macroscopic genera, many of which have the ability to deposit calcium carbonate in their thalli. The multiaxial construction relates this family to the Codiaceae, but the presence of specialized compound reproductive structures is a distinguishing characteristic. Another difference is the condition of **heteroplastidy**, in which two types of plastids are present: chloroplasts and amylogenic leucoplasts. Several genera of this family have been grown in seawater aquaria under laboratory conditions, and their development has been followed (Colinvaux, et al., 1965).

1. Thalli calcified, consisting of siphonous filaments consolidated
   into massive, compact structures....................................2
1. Thalli noncalcified, consisting of loosely branched,
   unconsolidated filaments ....*Chlorodesmis*
   2. Axes divided into flattened or cylindrical segments separated
      by short, flexible zones ....*Halimeda*
   2. Axes lacking a segmented organization 3
3. Thallus consisting of a cylindrical stalk of consolidated filaments
   terminated by a tuft of unconsolidated filaments ....*Penicillus*
3. Thallus consisting of a stalk portion gradually expanding into a
   broad fanshaped or funnelshaped blade ....*Udotea*

HALIMEDA Lamouroux   This genus is easily recognized by its articulated pattern of flattened, lime-encrusted portions alternating with flexible noncalcareous joints (Fig. 3.123). Thalli of *Halimeda* (Gr. *halimos*, marine) are usually anchored in sandy substrate by a massive, fibrous holdfast. The flattened outer faces of the utricles (Fig. 3.124*a*) present a polygonal aspect of the thallus surface (Fig. 3.124*b*). Aragonite crystals have been shown (Wilbur, et al., 1969) to be developed on the wall surfaces in the interutricular spaces at about 36 hours' age. Eventually the space becomes filled with clusters of randomly arranged aragonite crystals. Clusters of grape-like

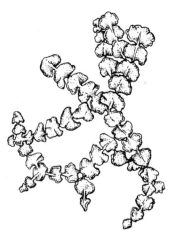

Fig. 3.123 *Halimeda opuntia* (L.) Lamour. Portion of plant. $\times$ 0.9.

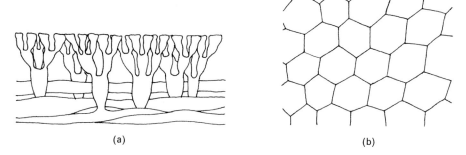

(a)   (b)

Fig. 3.124 *Halimeda tuna* (Ellis and Solander) Lamour. (*a*) Sagittal section through the cortex. (*b*) Surface view of the peripheral utricles. (*a*) $\times$ 19; (*b*) $\times$ 38. (After Hillis.)

gametangia (Fig. 3.125) are borne by **gametophores** arising marginally or super-ficially from the flattened lobes (Graham, 1975), giving a densely fringed aspect. Sexual reproduction is anisogamous (Meinesz, 1972C). About 2 dozen species are currently recognized. The genus has been monographed (Hillis, 1959), with some additional species later described (Colinvaux and Graham 1964; Goreau and Graham, 1967; Taylor, 1962B, 1973). The accumulated limestone remains of *Halimeda* are stated to contribute significantly to the gradual buildup of coral reefs.

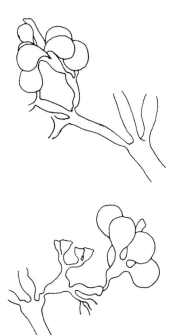

Fig. 3.125 *Halimeda cryptica* Colinv. and Graham. Gametophores bearing clusters of gametangia. $\times$ 75. (After Graham.)

Fig. 3.126 *Udotea conglutinata* (Ellis and Solander) Lamour. × 0.4.

UDOTEA Lamouroux The thalli of *Udotea* (Gr. *udor*, water) are stipitate, fan-shaped or funnel-shaped blades (Fig. 3.126), arising from either a single holdfast or from a branched rhizoidal system. The filaments are laterally joined to form a pseudoparenchymatous structure. Sexual reproduction has been recently described in *Udotea* (Meinesz, 1969), as involving anisogamous gametes (Fig. 3.127*a*) from dioecious plants. The method is described as **holocarpic** in that the entire contents of the siphonous filaments are transformed into gametes (Fig. 3.127*b*). From the zygote to the mature thallus three successive stages (Fig. 3.128) are passed through (Meinesz, 1972B): a uninucleate **protosphere** lacking leucoplasts followed by a juvenile filament-

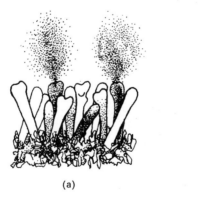

(a)

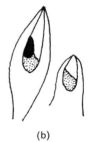

(b)

Fig. 3.127 *Udotea petiolata* (Turra) Børg. (*a*) Margin of thallus showing papillae at the moment of gamete emission. (*b*) Anisogametes. (*a*) × 75; (*b*) × 2900. (After Meinesz.)

196

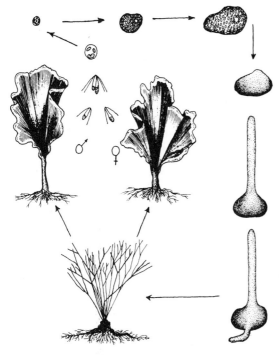

Fig. 3.128 *Udotea petiolata* (Turra) Børg. Scheme representing the life history, showing development of zygote into uninucleate protosphere, followed by juvenile filamentous phase, which is followed by the adult phase. (After Meinesz.)

ous stage, which is both coenocytic and heteroplastidic, and finally the more typical mature phase of the alga, composed of the consolidated siphons and provided with utricles.

PENICILLUS Lamarck Familiarly known as "Neptune's shaving brush," *Penicillus* (L. *penicillus*, a painter's brush) consists of a simple stalk arising from a basal cluster of rhizoids and terminating in a tuft of loose, dichotomously branched filaments, which are often constricted above the points of forking (Fig. 3.129). Early in the growing season the terminal tufts are bright green, but with age they become superficially impregnated with limestone and thus become whitened. Reproduction is unknown in this genus, although the early development has been recently described (Meinesz, 1972A).

Fig. 3.129 *Penicillus capitatus* Lamarck. × 0.5.

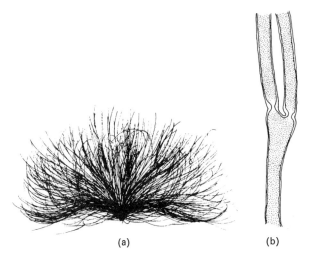

(a)

(b)

Fig. 3.130 *Chlorodesmis.* (*a*) Habit of *C. comosa* Harv. and Bailey. (*b*) *C. hildebrandtii* A. and E. S. Gepp, photosynthetic filaments. (*a*) × ⅓; (*b*) × 34. (After Ducker.)

CHLORODESMIS Harvey et Bailey   Unlike the previous representative genera of this family, *Chlorodesmis* (Gr. *chloros*, green + Gr. *desmos*, a bond) consists of noncalcified tufts of free, dichotomously branched siphons (Fig. 3.130*a*). Thalli may be present commonly and conspicuously in coral reefs and are of pantropical distribution (Ducker, 1967). Constrictions are characteristically present above the point of branching (Fig. 3.130*b*). *Chlorodesmis* has been subjected (Ducker, et al. 1965) to a numerical taxonomic analysis, and four species emerged from this study. A fifth species was later described (Ducker, 1969). Reproduction by means of elaborate compound reproductive structures (Fig. 3.131), whose biflagellated swarmers drain into a common central siphon and are discharged through an apical pore, has been described (Ducker, 1965).

Fig. 3.131 *Chlorodesmis baculifera* (J. Ag.) Ducker. Compound reproductive organ with apical discharge tube. × 29. (After Ducker.)

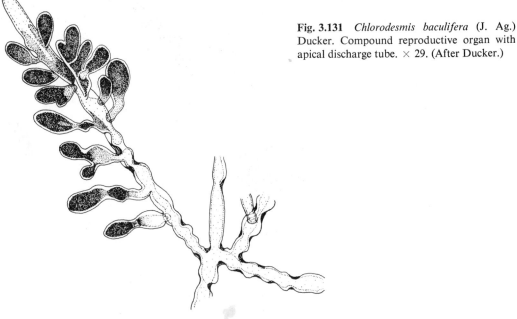

Fig. 3.132 *Caulerpa.* Cross section of axis showing trabeculae. × 13.

Family 3. Caulerpaceae

The family Caulerpaceae is monotypic, containing the single large genus *Caulerpa.* The family is distinguished on the basis of trabeculae (Fig. 3.132), which are branching, cylindrical ingrowths of wall material, traversing the central lumen, or vacuome, of the coenocyte and providing for mechanical support. The Caulerpaceae, like the Udoteaceae, is characterized by the occurrence of two types of plastids: chloroplasts and amyloplasts.

CAULERPA Lamouroux   The general habit of the thalli of *Caulerpa* (Gr. *kaulos*, a stem + Gr. *herpo*, to creep) conforms to a pattern of a horizontal rhizome-like portion, anchored by means of periodic rhizoidal outgrowths, and erect photosynthetic portions. The latter are modified into a great variety of shapes (Fig. 3.133), and this diversity accounts for the fairly large number of species and varieties occurring throughout tropical and subtropical seas of the world. The inherent limitations of the coenocytic habit are seemingly overcome by the combination of especially firm walls and the wall ingrowths, and thus the thalli of some species reach a size of 1 meter in length. On the basis of morphological evidence, early workers speculated that bilateral Caulerpas are more recent and have been derived from older radial Caulerpas. Recent ultrastructural studies (Calvert, *et al.,* 1976) of the chloroplast structure of 28 of the 73 species in the genus support this phylogenetic theory. Evidence has been gained that indicates a large chloroplast with a pyrenoid and starch grains was replaced by a small chloroplast lacking them.

In a study (Dawes and Rhamstine, 1967) of *Caulerpa prolifera* (Forssk.) Lamour., fine structural similarities were observed between the blade and the rhizome, except for the thicker nature of the walls and the trabeculae of the rhizome and the more numerous trabeculae. The rhizoid, however, was distinctive from the other two organs in its lack of chloroplasts. Blade proliferation in this same species has been demonstrated (Dawes, 1971) to be significantly enhanced by the application of low concentrations of indole-3-acetic acid, although assays for the existence of endogenous IAA proved negative. In reference to chloroplasts, it might be inserted that experimentation is underway (Giles and Sarafis, 1972) in which isolated chloroplasts have been shown to survive for more than 2 weeks under *in vitro* conditions and also to divide.

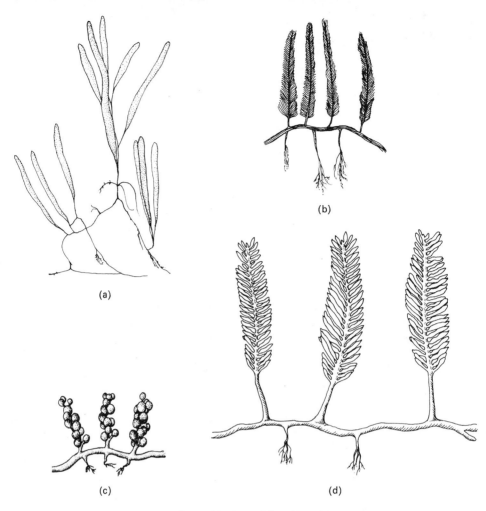

Fig. 3.133 *Caulerpa.* (*a*) *C. prolifera* (Forssk.) Lamour. (*b*) *C. sertularioides* (Gmelin) Howe. (*c*) *C. racemosa* (Forssk.) J. Ag. (*d*) *C. floridana* Taylor. (*a*), (*b*) × 0.2; (*c*) × 0.8; (*d*) × 0.43.

Sexual reproduction is known for a few species. Rather than differentiated gametangia appearing, a broad area of cytoplasm becomes differentiated into gamete production, narrow discharge papillae (Fig. 3.134) (about 2 mm long) being formed. Meiosis occurs at gametogenesis, reproduction is anisogamous, and the life cycle is, accordingly, the H, d type. Zygote development in *Caulerpa serrulata* (Forssk.) J. Ag. was followed (Price, 1972), and stout axes resembling the original plant were obtained after 5 months, refuting the possible occurrence of a heteromorphic life history suggested by Goldstein and Morrall (1970).

The production of a deadly poison by species of *Caulerpa*, referred to as **caulerpicin** (Doty and Aguilar-Santos, 1966), has been recognized in Hawaii and in the Philippines and is known to enter into marine food chains (Doty and Aguilar-Santos, 1970). Its chemical structure has been characterized (Santos and Doty, 1971). Since freshly collected plants of *Caulerpa* are eaten in salads in some areas of the Pacific, the health hazard is obvious.

Fig. 3.134 *Caulerpa prolifera* (Forssk.) La-mour. Fertile blade with discharge papillae. × 0.8. (After Schussnig.)

Family 4. Bryopsidaceae

The construction of members of this family consists of a uniaxial system, lacking trabeculae and the branches being produced pinnately or radially from the main axes. The family cytologically resembles the Codiaceae in that only one type of chloroplast is present, and they are provided with a pyrenoid and accumulate starch. Unlike the Codiaceae, the siphons demonstrate no aggregation. The two genera included in this account can be separated on the basis of the unspecialized gametangia produced in *Bryopsis* and by the specialized, or modified, gametangia in *Pseudobryopsis*.

BRYOPSIS Lamouroux   The generally delicate, feather-like fronds of *Bryopsis* (Gr. *bryon*, a moss + Gr. *opsis*, an appearance) (Fig. 3.135) are to be found in relatively protected waters, ranging from temperate to tropical seas. One species in Japan (*B. maxima* Okamura) reaches a spectacular length of over 40 cm, but most species are less than 10 cm long.

An interesting wound-healing response has been observed (Burr and West, 1971B), which perhaps compensates for the delicate nature of the coenocytic thalli. If the siphonous tubes are punctured, preexisting protein bodies in the cytoplasm immediately migrate to the site of the injury and are involved in the formation of a plug, preventing further loss of protoplasm. This reaction has also been studied cytochemically (Burr and Evert, 1972). It is also possible (Tatewaki and Nagata, 1970) to obtain extracellular protoplasts by crushing *Bryopsis* plants between glass slides, and new cell walls will be synthesized *in vitro* and develop into normal plants.

The life history of *Bryopsis* has recently come under close scrutiny, and some remarkable facts concerning the variable nature of the life history have been brought to light. Although on the basis of older work *Bryopsis* was long contended to be haplobiontic and diploid with gametic meiosis, the culturing investigations of Rietema

Fig. 3.135 *Bryopsis maxima* Okam. Portion of plant showing arrangement of pinnae. × 0.5.

(1969, 1970) on populations of *B. plumosa* (Hudson) C. Ag. from a variety of European localities revealed two alternative life histories. In the first, the familiar plants are unisexual, their anisogametes fusing in pairs to produce zygotes that germinate to directly return the familiar macroscopic plants. In the second, the difference occurs in the germination of the zygote. It forms a creeping filamentous germling, whose cytoplasmic contents cleave up to produce numerous stephanokontan swarmers, which settle down and develop into new gametophytes, approximately 50% male and 50% female. The latter pattern is more typical of populations from northern latitudes (Rietema, 1970), while from a locality such as Naples both patterns exist within the same population. Quite parallel variations exist in the related species, *B. hypnoides* Lamx. (Rietema, 1971; Diaz-Piferrer and Burrows, 1974). Again, culturing results indicate that geographical differences exist in respect to the life-history pattern. A variation was demonstrated (Bartlett and South, 1973) in a population from Newfoundland, in which the zygotes derived from the fusion of anisogametes released from bisexual plants produced prostrate germlings that either formed new plants directly or else released more anisogametes rather than the stephanokontan zoospores observed in some European populations. It was speculated that the site of meiosis might be prior to gamete formation in either phase: larger plant or the prostrate germling. No sexual reproduction was observed (Diaz-Piferrer and Burrows, 1974) in this species in Caribbean populations.

The ultrastructure of the vegetative plant and gametangial formation in *Bryopsis hypnoides* from California were reported on by Burr and West (1970). It is interesting that the cytoplasm peripheral to the large central vacuole has two layers: an outer zone with all the organelles except chloroplasts and an inner zone with the chloroplasts. Many vacuolar profiles extend into the cytoplasm. The migration of the chloroplasts during the dark phase and their aggregation at the bases of branches were also described. Pinnae destined to become gametangia (Fig. 3.136) would first have a plug formed in the pore separating the cytoplasm in the branch from that of the main axis. The cytoplasm in the pinna becomes fenestrate, and soon the smaller male gametes become metamorphosed in the distal half and larger female gametes in the proximal half. It should be interjected that this positioning may be reversed, as, for example, the population of this species observed by Bartlett and South (1973). The eventual discharge of the gametes is light-triggered, with an almost explosive discharge occurring about 6–8 minutes following the light exposure. Within about 1 minute almost all the gametes have been released. The male gamete is of interest in its having a giant mitochondrion, occupying the greater volume of the cytoplasm; the female gamete has many small mitochondria and a larger chloroplast with a conspicuous eyespot. A final point to make is the fact that the germling produced from the zygote of *B. hypnoides* contains a very large primary nucleus (Burr and West, 1971A) comparable to the so-called giant nucleus occurring in *Acetabularia*. The division of this giant nucleus has been observed (Neumann, 1969A). Chromosomes in a mitotic configura-

Fig. 3.136 *Bryopsis hypnoides* Lamour. Axis with pinnae. × 16.

tion were reported, followed by the dissolution of the giant nucleus and the appearance of two small nuclei, which proceed to undergo mitotic divisions.

Further evidence of some relationship between *Bryopsis* and *Derbesia*, an alga forming stephanokontan zoospores (see p. 206), is the observation of Hustede (1964) that a species of *Bryopsis* from the Mediterranean (*B. halymeniae* Berthold) alternates with *Derbesia neglecta* Berth. This exciting discovery has been confirmed (Rietema, 1972). An interesting difference in regard to staining properties of the cell walls has been pointed out (Huizing and Rietema, 1975). A positive reaction to Congo red and zinc-chlor-iodine is given by the gametophytic stage (*Bryopsis halymeniae*), but a negative reaction is given by the sporophytic stage (*Derbesia tenuissima*). This difference can be correlated with the report (Preston, 1968) of the absence of xylan and cel-

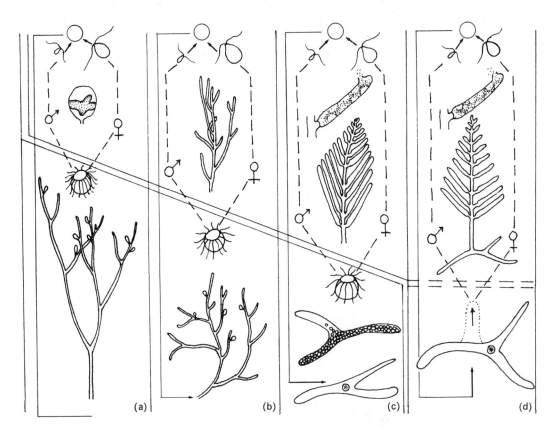

Fig. 3.137 Comparative representation of life histories in *Bryopsis* and *Derbesia*. (*a*) *Derbesia marina* (Lyngb.) Sol. as sporophyte and *Halicystis ovalis* (Lyngb.) Aresch. as gametophyte. (*b*) *Derbesia neglecta* Berth. as sporophyte and *Bryopsis halymeniae* Berth. as gametophyte. (*c*) *Bryopsis plumosa* (Huds.) C. Ag. and *B. hypnoides* Lamour., in which the creeping phase produces zoospores. (*d*) *Bryopsis plumosa* (Huds.) C. Ag. and *B. hypnoides* Lamour., in which the creeping phase functions as a protonema. (After van den Hoek, et al. and Neuman.)

lulose in the walls of *Derbesia* and the apparent presence of cellulose in the walls of *Bryopsis*. A similar staining difference occurs in the *Derbesia tenuissima-Halicystis parvula* (p. 207) (Rietema, 1973), suggesting an interesting correlation between the chromosome number of these algae and the composition of their cell walls. The *Derbesia*-like sporophytic phases seem to lack cellulose, but the haploid phases (*Bryopsis* or *Halicystis*) seem to have it. Van den Hoek, et al. (1972) presented a schematic analysis of the interrelationships of *Bryopsis* and *Halicystis* (as gameto-phytes) and *Derbesia* (as sporophyte), which is seen in Fig. 3.137.

PSEUDOBRYOPSIS Berthold   This alga vegetatively resembles *Bryopsis*, with the difference that in *Pseudobryopsis* (Gr. *pseudes*, false + *Bryopsis*, *q.v.*) the pinnules are often densely arranged and more slender in relation to the diameter in the main axes (Feldmann, 1969; Diaz-Piferrer, 1965). Septa are typically laid down at the base of pinnules once they have reached their full length. The main distinction from *Bryopsis* is the modified appearance of the gametangia (Fig. 3.138); they are usually ovoid and pedicellate and have a terminal papilla. Feldmann (1969) noted that in *P. myura* (Agardh) Berth., the type species, the plants are bisexual but with separate male and female gametangia.

The mode of gamete extrusion has been described (Feldmann, 1969), in which the gamete mass is discharged due to internal pressure; the gametes become motile only after the entire contents have been extruded as is the situation in *Codium* and unlike that of *Bryopsis*, in which the gametes are fully motile within the gametangium. The zygote has been shown (Mayhoub, 1974) to develop directly into a creeping fila-ment (Fig. 3.139) morphologically and cytologically resembling *Ostreobium quecketti* Bornet and Flahaut (see p. 209).

The 10 described species of this genus have been characterized by Diaz-Piferrer (1965); subsequently a small planktonic species has been added to the list from New

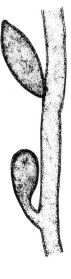

Fig. 3.138 *Pseudobryopsis blomquistii* Diaz-Piferrer. Gametangia. × 317. (After Diaz-Piferrer.)

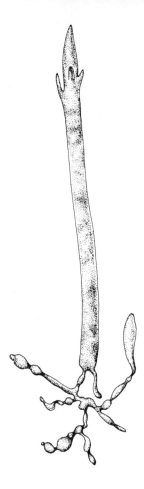

Fig. 3.139 *Pseudobryopsis myura* (Ag.) Berth. Young axis arising from a prostrate, creeping system resembling *Ostreobium* (p. 209). × 66. (After Mayhoub.)

Zealand (Cassie, 1969). Although Taylor (1962C) has pointed out that the generic name *Trichosolen* Montagne, 1860, antedates *Pseudobryopsis* of Berthold, 1904, Diaz-Piferrer (1965) has presented a case for conserving the latter name.

Family 5. Derbesiaceae

The basis for the recognition of this monotypic family is the unusual hetero-morphic life history, involving a branching siphonous sporophyte and a vesicular gametophyte, the so-called *Halicystis*-phase (Fig. 3.137a). The name *Derbesia* Sol., 1847, antedates *Halicystis* Areschoug, 1850, and thus takes priority.

DERBESIA Sol. The initial discovery of the existence of an alternation of heteromorphic generations in *Derbesia* (named after the French phycologist *Alphonse Derbes*) occurred when Kornmann (1938) followed in culture the development of zoospores released from the sporangia (Fig. 3.140a, b) of *Derbesia marina* (Lyngbye) Sol. These zoospores are of the stephanokontan type and are thus reminiscent of those produced in the Oedogoniales. From the zoospores germinated vesicular stages that enlarged into spherical plants identifiable as *Halicystis ovalis* (Lyngbye) Aresch. (Fig. 3.140c). Such a startling relationship between these two genera, previously classified

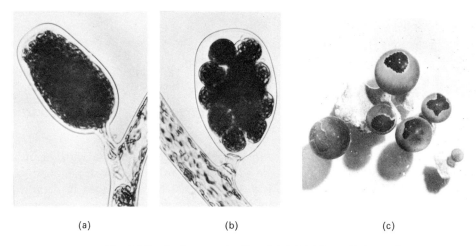

(a)   (b)   (c)

Fig. 3.140 *Derbesia marina* (Lyngb.) Sol. (*a*) Laterally borne, immature sporangium; (*b*) Mature sporangium containing zoospores; (*c*) Vesicular gametophytes. (*a*), (*b*) × 264; (*c*) × 1.4. [(*a*), (*b*) courtesy of Carolina Biological Supply Company; (*c*) courtesy of Dr. James Sears.]

remotely because of their dissimilar appearances, was subsequently confirmed by Feldmann (1950), who linked together *Derbesia tenuissima* (De Not.) Crn. fr. and *Halicystis parvula* Schmitz in the same pattern of a branched, tubular sporophyte and a saccate gametophyte, or D^h, h + d pattern. Feldmann obtained *Derbesia* from the zygotes of the *Halicystis* stage.

The fact that more species of *Derbesia* are recognized than species of *Halicystis* (Gr. *halos*, the sea + Gr. *kystis*, a bladder) is perhaps indirect evidence that the life-history scheme discussed above will not be consistent for all examples. Evidence has been presented (Sears and Wilce, 1970) that for *Derbesia marina* occurring in New England there is a direct development of sporophytes from the zoids of sporophytes. Another anomalous situation is the observation that *Derbesia neglecta* Berth. alternates with a species of *Bryopsis* (see p. 204). Such interrelationships would suggest that the establishment of a separate order of Derbesiales (Feldmann, 1954) would be premature. In a review article, Neumann (1974) has stressed similarities existing in many of the Caulerpales, including a "giant nucleus" present in the young germling of several members.

The *Halicystis* stage is of widespread occurrence but is not generally abundant. On one occasion the authors observed hundreds of the small balloon-like thalli (about the size of marbles) of *H. ovalis* washed ashore on a sandy Monterey beach after a storm. But more typically it will take the alert eyes of an underwater diver to spot the dark green sacs attached to their substratum, which is often a calcareous red alga. The base of the sphere is elongated into a colorless peg-like portion, which is embedded in the substratum. Beneath the rather thick wall of the sac is a thin peripheral layer of protoplasm, with numerous nuclei and chloroplasts. The greater volume of the sac is the vacuole, the properties of which have been the subject of many physiological

Fig. 3.141 *Derbesia marina* (Lyngb.) Sol. Differentiated gamentangial region, from which gametes are explosively released. × 3. (After Hollenberg, Amer. J. Bot. 22: 782–812.)

studies (Blinks, 1955). Plants are separately sexed and are easily distinguished by the dark green color of the female gametangia and the yellowish-tan color of the male gametangia. The process of gametogenesis has been described in detail by Hollenberg (1935). The region of the protoplasm destined to become a gametangium (Fig. 3.141) becomes cut off from the surrounding protoplasm by a cell membrane and gradually becomes much thicker. The contents are cleaved into a mass of biflagellate gametes, of which the female gametes are relatively larger (Fig. 3.142).

An electron-microscope study (Wheeler and Page, 1974) of gametangial differentiation indicated a thick accumulation of protoplasm in the region of the incipient gametangium by migration from vegetative regions. A gametangial membrane separates the gametangium from the rest of the coenocyte. Cleavage of the protoplasm into gametes occurs such that by 0.5 hour before release mature gametes are present, and the pore in the wall is capable of rupture. Gelatinization of the outer wall layer and apparent enzymatic digestion of the inner wall layer combined with increased turgor pressure effect wall dissolution. Under normal conditions the onset of the light period triggers a spectacular discharge of gametes. Cytoplasmic debris plugs the pore following gamete discharge. Periodicity in both the formation of gametangia and the release of gametes has been investigated (Page and Kingsbury, 1968). Formation of gametangia occurred usually every 4–5 days for cultured plants (unlike náturally occurring plants, which had a periodicity of 2–3 weeks), and the events of gamete formation were synchronous for the population. Plants in continuous light did not

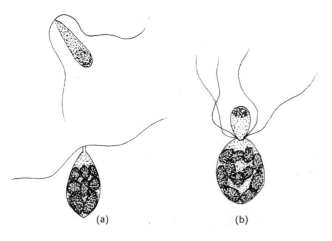

Fig. 3.142 *Derbesia marina* (Lyngb.) Sol. (*a*) Anisogametes released from the *Halicystis*-phase. × 600. (*b*) Fusion of gametes. × 615. (After Hollenberg, Amer. J. Bot. 22: 782–812.)

(a)   (b)

demonstrate so great a degree of synchrony of formation and release. An endogenous physiological cycle with a period of about 4 days (Page and Sweeney, 1968) interacting with environmental conditions maintains this periodicity in sexual reproduction.

Interestingly the development of zoids from the sporophyte is under the influence of daylength, with zoids maintained in long-day conditions developing into complete, vesicular "*Halicystis*" plants and zoids maintained in short-day conditions developing into slender rhizoidal plants lacking a spherical part (Rietema, 1973). Similarly, naked protoplasts extruded from the "*Halicystis*" stage behave as the zoids and are influenced by the daylength, with long-day conditions promoting the development of normal spherical plants and short-day conditions favoring the development of rhizoidal plants.

The formation of zoospores by the sporophyte has been investigated (Neumann, 1969B). From 15 to 40 nuclei in the young sporangium undergo meiosis, and a double wall separates the cytoplasm of the sporangium from the thallus after meiosis. Only one nucleus from each series of meiotic divisions proceeds toward spore formation, the other three nuclei aborting. Mitotic divisions occur resulting in multinucleate zoospores.

Family 6. Phyllosiphonaceae[25]

The endophytic or endozoic habitat is characteristic of members of this family, as well as their relatively microscopic size. The filaments are usually irregularly branched and may be swollen. Chloroplasts lack pyrenoids. Two genera will be discussed: the first, *Phyllosiphon*, is a parasite within the leaves or stems of flowering plants; the second, *Ostreobium*, occurs within the calcareous substrates of shells, corals, and other marine hosts.

PHYLLOSIPHON Kühn   The infected leaves of aroid hosts may display discoloration and irregular blotches due to the presence of the irregularly or dichotomously branched, endophytic filaments (Fig. 3.143*a*) of *Phyllosiphon* (Gr. *phyllon*, a leaf + Gr. *siphon*, a tube). Aplanospores are the only reproductive structures known, and they directly return new thalli.

OSTREOBIUM Born. and Flah.   The filaments of this alga (Fig. 3.143*b*) tend to be very irregular in form, with a penetrating endolithic existence. The correct affinity of *Ostreobium* (Gr. *ostreon*, an oyster + Gr. *bios*, life) to the Chlorophyceae has been confirmed (Jeffrey, 1968) on the basis of its pigment composition. The problems encountered when working with this alga are obvious, since decalcification of the material is necessary to free the filaments from their calcareous host (shells, corals, and coralline algae) and this results in the alteration of the filament size and their spatial relationships. Scanning electron microscopy has been used (Lukas, 1974)

---

[25]The Phyllosiphonaceae (including *Phyllosiphon* but not *Ostreobium*) has been transferred by Bourrelly (1968) to the Vaucheriales (p. 389) of the Xanthophyceae because of the alleged presence of food reserves comparable to chrysolaminaran rather than starch.

Fig. 3.143 *Ostreobium duerdenii* Weber-van Bosse. Filaments filled with starch granules. × 79. (After Weber-van Bosse.)

in better visualizing the intact alga. A recent account (Lukas, 1974) recognizes five species of this pantropical genus.

Family 7. Dichotomosiphonaceae

The primary distinction of this family from the remaining groups of the Caulerpales is the oogamous reproduction. The family is characterized as heteroplastidic, with leucoplasts bearing starch grains and chloroplasts lacking pyrenoids.

DICHOTOMOSIPHON Ernst   This freshwater alga, with the single species *D. tuberosus* (A. Br.) Ernst recognized, is of sporadic occurrence in the United States. It might be confused with the superficially similar alga *Vaucheria*, its siphonous counterpart in the Xanthophyceae, but the presence of occasional constrictions in *Dichotomosiphon* (Gr. *dichotomos*, divided equally + Gr. *siphon*, a tube) (Fig. 3.143) serves to distinguish these two algae. The presence of unusual structures termed "striated tubules" within both chloroplasts and leucoplasts has been reported (Moestrup and Hoffman, 1973). Spherical oogonia and smaller antheridia are borne at the terminal ends of filaments, often in groups of two, three, or four, with the antheridia subtending the oogonia (Fig. 3.144). Moestrup and Hoffman (1975) have described the ultrastructural organization of the minute sperm of *Dichotomosiphon*, which are discharged rapidly by explosive rupture of the tip of the antheridium. Large akinetes, with a tuberous appearance, are also produced.

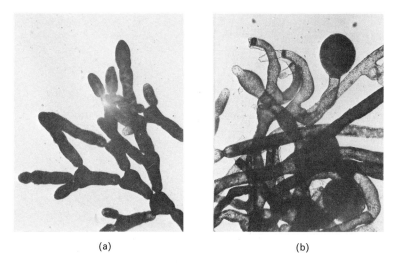

(a)   (b)

Fig. 3.144 *Dichotomosiphon.* (*a*) Dichtomously branched axes with characteristic constrictions. × 228. (*b*) Enlarged oogonia and discharged antheridia. × 125. (Courtesy of Dr. Eugene Shen.)

Order 13. Siphonocladales

The most distinctive trait of this exclusively marine and tropical order is the manner in which cell division occurs. Termed **segregative cell division**, this process involves the cleavage of the protoplast into a few to many protoplasmic portions of varying size, each of which rounds up and secretes an enveloping membrane. Each mass of protoplasm then expands until it comes in contact with adjacent segments. Expansion of the new segments may be endogenous or exogenous. In the endogenous type, as in *Dictyosphaeria* Decaisne, the new segments simply swell within the parental vesicle, pressing against one another and forming a pseudoparenchymatous tissue (Fig. 3.145). In the exogenous type, as in *Siphonocladus* Schmitz, expansion of the new

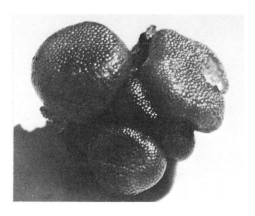

Fig. 3.145 *Dictyosphaeria cavernosa* (Forssk.) Børg. Thallus composed of endogenously developed segments contained within the parental vesicle. × 1.2.

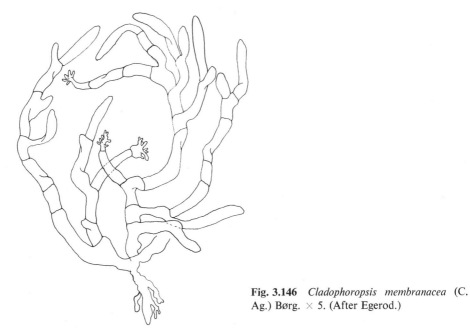

Fig. 3.146 *Cladophoropsis membranacea* (C. Ag.) Børg. × 5. (After Egerod.)

segments causes them to protrude outward from the parental vesicle, with the parental wall being incorporated into the lateral outgrowth (Fig. 3.146).

Chloroplasts are lobed and arranged in a network; pyrenoids are absent. There are no special reproductive structures. Biflagellated swarmers have been observed in several of the genera, but it is not certain whether they are sexual or asexual. Sexual reproduction by biflagellated gametes has been reported for *Valonia utricularis* (Roth) C. Ag., and the life history is alleged to be H, d, but this requires confirmation.

The status of the Siphonocladales has been discussed by various authors (Fritsch, 1947; Egerod, 1952; Chapman, 1954; Nizamuddin, 1964; Jónsson, 1965). In this account three families will be treated.

Family 1. Siphonocladaceae

Thalli of this family are generally loosely branched and lack the small lenticular cells seen in the family Valoniaceae. Following segregative cell division the segments mature exogenously, such that the segments bud out laterally to form branches.

CLADOPHOROPSIS Børgesen   Although it bears a resemblance to *Clado-phora* (p. 183), *Cladophoropsis* (an appearance similar to *Cladophora*, *q.v.*) is readily distinguishable from that genus by its pattern of segregative cell division. Usually there is only a single outgrowth per parental cell, and the emergent cell is continuous with the parental cell, remaining nonseptate at the base. A septum may be laid down to separate off the branch, but it occurs secondarily. A loose system of uniseriate axes (Fig. 3.146) is thus brought about.

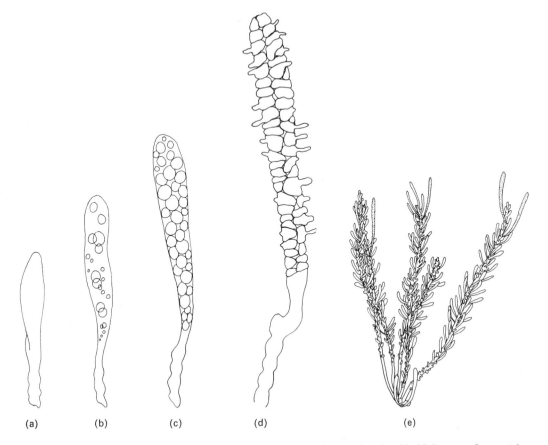

(a)   (b)   (c)   (d)   (e)

Fig. 3.147 *Siphonocladus tropicus* (Crouan) J. Ag. (*a*)–(*d*) Pattern of segregative cell division and exogenous development of the primary vesicle. (*e*) Later stage of development. (*a*)–(*d*) × 8; (*e*) × 1.7. (After Egerod.)

SIPHONOCLADUS Schmitz   Parent cells in *Siphonocladus* (Gr. *siphon*, a tube + Gr. *klados*, a branch) cleave into a large number of vesicles (Fig. 3.147) with a multiseriate arrangement. The cylindrical axes can be branched repeatedly in this way, with branching out to the fourth order. Tufted, densely branched thalli result.

Family 2. Valoniaceae

Plants in this family lack a central axis, and the aspect of a mat or aggregation of vesicles is commonly effected. Small, lenticular cells are frequently observed in thalli of this family; these cells may serve to form hapteroid cells on the lower surface of the alga to attach it to the substratum.

VALONIA C. Ginnani   The most familiar species of the genus *Valonia* (after *Valoni*, an Italian botanist) is *V. ventricosa* J. Agardh, which is alleged to be the

Fig. 3.148 *Valonia ventricosa* J. Ag. Specimens growing in a clump of *Galaxaura*. × 1.3. (After Kodachrome transparency courtesy of Dr. W. Randolph Taylor.)

largest plant cell, a single spherical vesicle (Fig. 3.148) at times reaching about 10 cm in diameter. In reality, however, this alga is multicellular due to formation by segregative division of many small cells in the peripheral zone (Egerod, 1952). Most of these cells remain unchanged and are essentially continuous with the original parental wall; however, the cells on the lower surface of the enlarged vesicle exogenously protrude to form rhizoids (Fig. 3.149) which attach the alga. The cell walls of *V. ventricosa* have been used in many investigations (Gardner and Blackwell, 1971).

Family 3. Anadyomenaceae

This family, which is considered by some to be more aligned with the Cladophorales, includes foliose thalli, which are constructed of a branching system of filaments whose cells anastomose into a monostromatic thallus. The individual cells may be closely contiguous, filling up the interstices with small cells, or it may be a looser arrangement with an open reticulate system being produced. There are no special organs for attachment such as haptera or tenaculae. The absence of segregative cell division, the apparent alternation of isomorphic phases with meiosis at sporogenesis, and septation by the ingrowth of lateral walls all point to a relationship of this family with the Cladophorales.

Fig. 3.149 *Valonia trabeculata* Egerod. Exogenously produced rhizoids at base of plant. × 38. (After Egerod.)

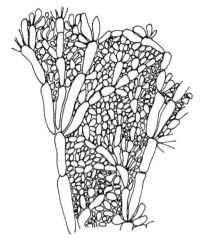

Fig. 3.150 *Anadyomene saldanhae* Joly and Oliveira-Filho. Portion of blade showing larger "rib" cells and small cells filling in the interstices. × 20. (After Joly and Oliveira-Filho.)

ANADYOMENE Lamour. Blades of *Anadyomene* (Gr. *anadyomene*, arising from the sea) are recognized microscopically to be composed of closely contiguous cells in a branching filamentous system (Fig. 3.150). The larger cells give a riblike appearance, and the fan-shaped thallus (Fig. 3.151), which may become lobed or divided, is attached to the substratum by means of a short stalk and rhizoids. Some species of this genus have been collected at great depths (Joly and Oliveira Filho, 1968; Humm, 1956), *A. menziesii* Harv. being dredged from approximately 200 m in the Gulf of Mexico. Although specimens of *A. stellata* (Wulfen) C. Ag. reach a maximum size of 10 cm height or in width, specimens of *A. menziesii* measured up to 45 cm long and 25 cm wide (Humm, 1956).

Swarmers are formed in the small cells. An alteration of isomorphic generations has been observed for populations of *Anadyomene stellata* (Wulf.) Ag. in the Indian Ocean and in the eastern Mediterranean Sea (Mayhoub, 1975), the monoecious

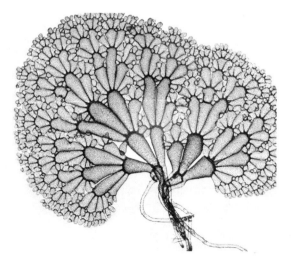

Fig. 3.151 *Anadyomene stellata* (Wulf.) C. Ag. × 19. (After J. Rosowski and B. C. Parker, Selected Papers in Phycology, University of Nebraska.)

gametophytes producing anisogametes. In a different population of this species in the Mediterranean a direct reproduction by quadriflagellate zoospores was observed (Jonsson, 1962A). Plants became fertile at 25°C but remained sterile when maintained at 15°C.

Order 14. Dasycladales

This well-defined order[26] of green algae contains about 8 extant genera, but more than 50 fossil genera have been described. The fossil record extends back to the Ordovician, with 3 living genera recognized from the Cretaceous. Since the Triassic their distribution has been centered in the tropics and subtropics. The construction of extant forms is more elaborate than that of fossil forms.

The order is clearly demarcated on the basis of the following characteristics:

1. Radial symmetry based upon a nonseptate primary axis with whorls of laterals; branching may extend to several orders.
2. Reproduction is by operculate cysts (except in *Dasycladus*); these cysts function as gametangia and release isogametes.
3. A process termed **diaphysis** occurs (Fig. 3.152), whereby new growth pushes through older growth, rupturing it and leaving a scar.

Fig. 3.152 *Acetabularia.* Enlarged, semidiagrammatic representation of diaphysis, with emergence of new axis and scar remaining. × 10. (After Egerod.)

The superficial deposition of limestone is common throughout this order but is absent in *Batophora* and barely present in *Dasycladus*. This trait is responsible for permitting such a good fossil record.

The whorls of laterals borne by the central axis may be loosely arranged or densely aggregated, depending on the genus. In genera such as *Bornetella* and *Cymopolia* the

---

[26]A systematic revision and a discussion of the phylogenetic interrelationships have been recently offered (Valet, 1969). Observations from scanning electron microscopy have clarified the relationships of *Acetabularia* with certain other extant genera (Bailey, et al., 1976).

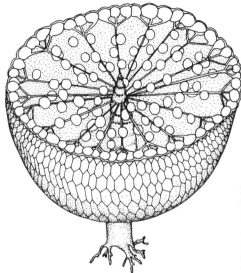

Fig. 3.153 *Bornetella sphaerica* (Zanard.) Solms-Laubach. Oblique aspect of sectioned thallus showing radial arrangement of primary laterals and terminal whorls of secondary laterals, which are laterally coherent to form the cortex. Gametangia are borne along primary laterals. × 5. (After Egerod.)

ultimate or penultimate laterals are inflated and become appressed laterally to form a compact surface of hexagonal facets (Fig. 3.153). Despite the extensive branching in some of these forms and the constricted appearance where branches arise, a thallus is essentially unicellular, with the laterals in all ranks in cytoplasmic communication with the primary axis. A small opening or pore lies between laterals of different ranks.

Thalli remain uninucleate until they become fertile. An enormous (so-called "giant") nucleus is situated in a basal rhizoid (Fig. 3.154), and at the time that the thallus becomes reproductively mature this large primary nucleus undergoes fragmentation, with numerous secondary nuclei resulting and being moved upward into the

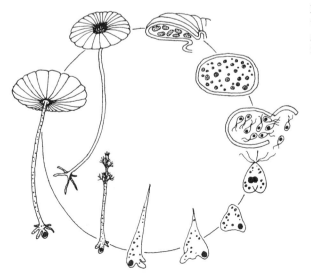

Fig. 3.154 *Acetabularia*. Schematic representation of life history, with production of isogametes from the germinating cysts. (Modified from Brachet.)

(a)   (b)

Fig. 3.155 *Acetabularia crenulata.* (*a*) Portion of cap with rays containing cysts; (*b*) Emergence of rays to form a cap. (*a*) × 100; (*b*) × 300. [(*a*) courtesy of Richard Rezak; (*b*) courtesy of G. Bailey, E. Cox, and R. Rezak.]

gamentangia by cytoplasmic streaming. Gametangia arise as terminal or lateral outgrowths usually on laterals of the first order (Figs. 3.153, 3.164); in *Batophora* the gametangia occur on laterals of both the first and second orders (Fig. 3.162).

   The protoplasm within the gametangia cleaves into uninucleate portions, each forming a cyst (Fig. 3.155*a*). It is noteworthy that the cyst walls in *Acetabularia* have been demonstrated (Herth, et al., 1975) to be composed of cellulose, whereas in other parts of the plant the only structural polysaccharide is a β-1-4-mannan. This condition of the composition of the cell wall varying at different times in the life history is also a feature of *Derbesia* (p. 206). These operculate cysts, which may be calcified, are then shed as the parent plant distintegrates. Eventually the cyst geminates (Fig. 3.154), the single nucleus undergoing many mitotic divisions. It has been maintained that the final series of divisions of the nuclei within the cysts are meiotic. Recent studies have suggested, however, that this notion is incorrect. One piece of evidence (Green, 1973) comes from isolating single cysts of *Acetabularia*. Although motile gametes are released, new diploid vegetative cells do not appear in these cultures started from single cysts. But cultures containing combinations of two or more cysts are productive in that new diploid thalli are developed. It would thus appear that the gametes produced by any one cyst are all of the same mating type, indicating that meiosis must have occurred prior to cyst formation. More recently, Koop (1975) has demonstrated by a spectrophotometric method that only the primary nucleus is diploid and that all other nuclei including those within the cyst are haploid. This would signify that meiosis occurs before or during the fragmentation of the primary nucleus.

   Valet (1969) has recognized two families, Dasycladaceae and Acetabulariaceae, the latter containing the two genera *Acetabularia* and *Halicoryne*, which are regarded as more advanced in that alternating fertile and sterile whorls are produced or only one whorl of fertile laterals. This distinction is not present in the Dasycladaceae.

Fig. 3.156 *Acetabularia*. Group of plants. × 2.2.

A third family, the Receptaculitaceae, comprised of fossil genera only, has been recognized (Nitecki, 1971) to include forms in which the primary laterals are spirally arranged rather than in whorls. But some doubt has been expressed (Rietschel, 1977) that "receptacultids" are truly Dasyclads.

Three representative genera of the Dasycladales will be discussed in greater detail.

ACETABULARIA Lamour.   The distinctive shape of *Acetabularia* (L. *acetabulum*, vinegar cup) (Figs. 3.156, 3.157) is well known to even first-year biology students, since this alga is often included in introductory textbooks as a classical organism involved in studies of morphogenesis. It is familiarly known as "mermaid's wine glass."

Fig. 3.157 *Acetabularia*. Cultured specimens, lacking calcification. × 1.9. (Courtesy of Carolina Biological Supply Company.)

Acetabularia has been the focus of numerous articles as well as several recent books and symposia (Brachet, 1965; Gibor, 1966; Schweiger, 1969; Brachet and Bonotto, 1970; Puiseux-Dao, 1970, 1975). The reason for its being particularly well suited for studies of morphogenetic mechanisms is its peculiar morphology. Some species are only a few millimeters in height, but others are several centimeters tall. The larger thalli can be subjected to experimental **merotomy**, which is the cutting or sectioning of portions of cells, with or without nuclei. Isolated cytoplasm proceeds to differentiate new wall material (Werz, 1968). By such enucleation experiments followed by intraspecific and interspecific grafting of nucleate or enucleate portions, basic information concerning nuclear control of morphogenesis has been gained.

Beginning in the 1930's Hämmerling demonstrated that the rhizoid containing the nucleus can be easily removed and the axial cell can be sectioned into two, three, or many pieces, each of which can survive for several months. It was also realized that it was possible to graft severed portions of two different species to each other before these cut sections have healed. Because of its very large size and other useful attributes the merits of *Acetabularia major* Mart. for experimental work have been extolled (Schweiger, et al., 1972, 1974).

The two species that have been employed most extensively in these graft experiments have been *Acetabularia crenulata* Lamour. and *A. mediterranea*[27] Lamour. The cap of *A. crenulata* has about 30–35 gametangial rays, which are pointed (or spurred) and joined only at the base (Fig. 3.155b), whereas the cap of *A. mediterranea* has about twice that number of gametangial rays, which are not pointed and are fused along their edges. Another difference is that multiple fertile whorls, alternating with sterile whorls of hairs, may be produced in *A. crenulata*, which is reminiscent of the related genus *Halicoryne*. If nucleate portions of these species are grafted, a cap of intermediate appearance is formed, but the graft is sterile when interspecific. It is also possible to graft enucleate portions of these two species and obtain a viable graft, with a cap of usually intermediate morphology being produced.

If a nucleate portion of one species is grafted to an enucleate portion of the other species, a cap is formed that shows a greater resemblance to the species of the portion containing the nucleus, although there is still some indication of the influence of the species of the enucleate portion. If that first cap is removed prior to maturation and a second cap is allowed to be formed, this second cap will be a pure expression of the species of the portion containing the nucleus (Puiseux-Dao, 1970). These facts support the contention originally postulated by Hämmerling in the 1930's that morphogenetic substances of nuclear origin reside in the cytoplasm and are influential in the formation of the particular shape of the cap. These factors will be used up in the production of the cap but will be continually supplied to the cytoplasm if the nucleus is still present; in an enucleate portion, their supply will be used up after the initial cap is formed.

Although the fertile whorl of laterals that compose the cap is the conspicuous feature of *Acetabularia*, sterile whorls are also formed prior to maturation of the alga.

---

[27]The correct name of this species is *A. acetabulum* (L.) Silva (Silva, 1952), but *A. mediterranea* is apparently ingrained in the literature.

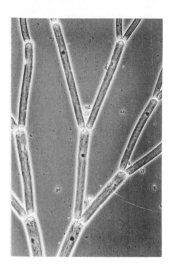

Fig. 3.158 *Acetabularia*. Sterile whorls of living plant. × 93.

These sterile whorls (Fig. 3.158) branch several times, forming progressively finer branches. Only the youngest of these branches are present, with older whorls being shed. These sterile whorls have been demonstrated (Gibor, 1973A) to function in the uptake of solutes and increase the surface area of the cell. Light is required to maintain these sterile whorls of hairs, and the hairs seem to influence the elongation of the main cell axis (Gibor, 1973B).

In some species of *Acetabularia* a pectic sheath, called a **velum** (Valet, 1967, 1968), is developed as a protective covering over the emerging whorls of laterals (Fig. 3.159). It also occurs in other genera of this order. The fine structure of the gametes and zygotes of *Acetabularia* (Crawley, 1966) and of the morphology of the interface between nucleus and cytoplasm, relating structural to functional changes during both the vegetative and reproductive stages, has also been described (Franke, et al., 1974; Berger, et al., 1975). A wide variety of nucleocytoplasmic interactions is now recognized both physiologically and biochemically (Hämmerling, 1963; Schweiger, 1969; Spring, et al., 1974).

The nuclear cycle in *Acetabularia* seems to be of general occurrence throughout the order (Valet, 1969). From the time of germination of the zygote to the time right before final maturation of the cap, a single nucleus is maintained in the rhizoidal portion of the cell, and this so-called primary nucleus (Fig. 3.160) becomes greatly

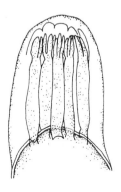

Fig. 3.159 *Acetabularia clavata* Yamada. Formation of a whorl of sterile laterals covered by a velum. × 116. (After Valet.)

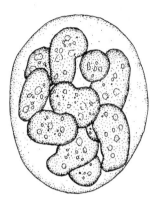

Fig. 3.160 *Acetabularia.* Enlarged primary nucleus. (After Schwieger.)

increased in volume. Its fine structure and biochemistry have been investigated in detail (Burr and West, 1971; Berger, et al., 1975). The nucleus is surrounded by a cytoplasmic layer that offers protection when it is isolated or manipulated (Crawley, 1963; Zerban, et al., 1973). Many invaginations in the nuclear envelope give it an irregular aspect.

Lampbrush-type chromosomes have been observed (Spring, et al., 1975) in the primary nucleus, and there is some evidence (Puiseux-Dao, 1970) suggesting that it is polyploid. Membrane components synthesized in the cytoplasm are incorporated into the expanding primary nuclear envelope (Franke, et al., 1975), and an interesting correlation between nuclear pore density and total pore number with the phase in the life cycle has been pointed out (Zerban and Werz, 1975). Minimal pore densities and total pore number occur in the gamete cell nucleus, whereas maximal values were observed in the mature primary nucleus of the vegetative phase.

Once the gametangial rays have been elaborated, the large primary nucleus undergoes a process of fragmentation (**amitosis**), a tremendous number of nuclei being moved upward into the gametangial rays by cytoplasmic streaming. These secondary nuclei are then fairly uniformly dispersed in the rays and incorporated with some surrounding cytoplasm in the formation of operculate cysts (Fig. 3.155). Mitotic divisions by the secondary nucleus result in about 20–40 nuclei being present within each cyst, prior to the dormancy of the latter.

Germination of the cysts has been observed (Koop, 1975) to depend on a period of maturation, maximal germination being achieved after 12–15 weeks. Although the actual germination is not light-dependent, light does influence germination. Increased temperature accelerates germination up to 21°C, but beyond that a significant retardation occurs. At germination additional divisions occur, which also seem to be mitotic, and cleavage of the protoplasm within the cyst takes place, with isogamous gametes of one sex being released by the abscision of the lid of the cyst. Evidence also suggests that cysts sometimes form zoospores (Hämmerling, 1964).

Circadian rhythmicity in regard to photosynthetic activity in *Acetabularia* has received attention (Van den Driessche, 1966). The peak of photosynthesis that occurs during the day under a normal light-dark cycle continues to be expressed under constant light conditions. This oscillation, or circadian rhythm, will continue for 6–7 weeks, which is remarkably stable (Schweiger, et al., 1964). Enucleate cells will also demonstrate (Sweeney and Haxo, 1961) such a rhythm in photosynthesis.

BATOPHORA J. Agardh   The noncalcified, moderate-sized plants (Fig. 3.161) of *Batophora* (Gr. *bato*, a bramble, prickly bush + Gr. *phoreo*, I bear) tend to be gregarious and bear loosely arranged whorls of laterals that are repeatedly branched. Large spherical gametangia are produced at the nodes of the laterals of the first and second order (Fig. 3.162), and the cysts give rise to biflagellate gametes. Although typically marine, *Batophora oerstedii* J. Ag. has also been recorded (Proctor, 1961) far inland, such as from sinkholes in New Mexico, where the salinity was measured to be about one-third that of normal seawater.

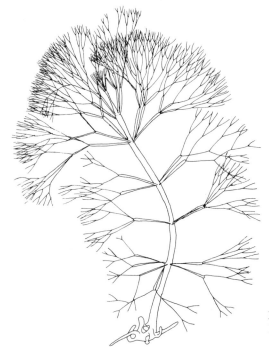

Fig. 3.161   *Batophora oerstedii* J. Ag. General habit of plant. × 14.

CYMOPOLIA Lamour.   The bushy plants (Fig. 3.163) of *Cymopolia* (Gr. *cyma*, surf + Gr. *polos*, grey) are heavily encrusted with limestone. The plants have a segmented, or jointed, construction, paralleling that of the Codialean genus *Halimeda* (p. 194) and of articulated coralline algae (p. 509), providing the algae with flexibility. The cortex is composed of the secondary laterals, which are terminally inflated and closely contiguous. Limestone is deposited in the interstices of these laterals. Gametangia are borne singly (Fig. 3.164) at the tip of the primary laterals, surrounded by the inflated secondary laterals. This genus occurs in the Gulf of Mexico (Florida and Mexico) and the Caribbean.

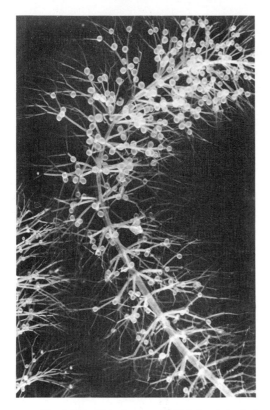

Fig. 3.162 *Batophora oerstedii* J. Ag. Spherical gametangia produced at the nodes of primary and secondary laterals. × 3.9.

Fig. 3.163 *Cymopolia barbata* (L.) Lamour. General habit of plants. × 0.26.

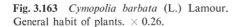

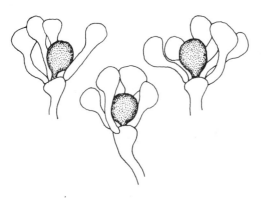

Fig. 3.164 *Cympolia barbata* (L.) Lamour. Gametangia borne at tips of primary laterals and surrounded by secondary laterals. × 35. (After Taylor.)

Order 15. Zygnematales

Like the Oedogoniales, the Zygnematales may be clearly distinguished from other Chlorophycophyta. Their characteristics are so unifying and so striking that they are sometimes assigned to a special class of algae, the Conjugatophyceae (Fott, 1971) or Zygnemaphyceae (Round, 1971), as distinct from other groups of Chlorophyceae.

The cells of Zygnematalean algae are uninucleate and characterized by their strikingly prominent chloroplasts with pyrenoids. The plastids may be bar- or board-shaped, asteroid, or ribbonlike, the last often twisted, among other varieties of plastids. Many of the Zygnematales secrete copious gelatinous materials, presumably pectinaceous in composition, which may be readily visible or spectacularly demonstrable upon immersing the cells in diluted India ink. In many Zygnematales, the colorless cytoplasm may be observed to be in a state of rotating motion called *cyclosis*. Mix (1975) has investigated comparatively the organization of the cell walls in Zygnematales. These are composed of three layers. The outer layer is mucous and amorphous, while the two inner layers are fibillar. Pores are present in the walls of the Desmidiaceae but not in those of the other two families.

Members of the Zygnematales are either unicellular (Fig. 3.175) and called desmids (*sensu lato*) or filamentous[28] (Fig. 3.165). The desmids are grouped into two different families on the basis of the organization of their cell walls and method of cell division and sexual reproduction. A few desmids are permanently filamentous, while others may be temporarily so. Furthermore, some species of filamentous Zygnematales may undergo dissociation into segments composed of one or more cells. Fragmentation results in asexual reproduction in many filamentous species. Vegetative cells may also thicken their cell walls and function as akinetes.

The Zygnematales lack flagellate motile cells, their gametes usually being amoeboid in sexual reproduction. In the latter process, the naked, amoeboid protoplasts in some members of the order emerge from their walls and unite in pairs to form zygotes. In others, the gametangia develop tubular protuberances or **conjugation papillae** that meet and, their terminal walls having dissolved, the protuberances form a **conjugation tube** into which both gametic protoplasts flow and unite; or one of them passes across the tube into one of the gametangia. After a period of dormancy, the zygotes germinate to form one, two, or four products that re-initiate the vegetative phase. All the Zygnematales that have been studied are haplobiontic and haploid with zygotic meiosis.

The Zygnematales are exclusively freshwater algae. The filamentous members may form slimy, frothy masses on the water surface and are called "pond scums." Certain desmids, like species of *Cosmarium* and *Staurastrum*, may form water blooms.

The Zygnematales are usually divided into three families, the Mesotaeniaceae, Desmidiaceae, and Zygnemataceae. The North American species of the first two families have been summarized taxonomically by Prescott, et al. (1972, 1975). The

---

[28]A few desmid genera, e.g., *Desmidium* (Fig. 3.180) and *Hyalotheca* (Fig. 3.179), are filamentous.

first two families are desmids, *sensu lato*. These families may be distinguished by the following key:

    1. Filamentous, but sometimes dissociating even to the unicellular
       state; walls not equatorially indented or incised . . . . .Zygnemataceae
    1. Unicellular; or filamentous with cell walls equatorially incised or indented . .2
       2. Cell walls continuous, daughter cells arising by enlargement
         of partitioned halves of parental cells . . . . .Mesotaeniaceae
       2. Cell walls of the cellular halves or semicells of different age;
         daughter cells arising as buds from the partitioned
         parental cells . . . . .Desmidiaceae

Family 1. Zygnemataceae

The members of the Zygnemataceae, of which between 534 and 580 species have been recognized (Hoshaw, 1968), have short or long cylindrical cells and are unbranched, except for occasional short rhizoidal branches that attach some species to the substrate. The walls are continuous and often surrounded by slimy sheaths. Transeau (1951), Randhawa (1959), and Gauthier-Lièvre (1965) have monographed the family on the basis of field and herbarium studies; Yamagishi (1963) has discussed the classification of the family. The unpublished work of Allen (1958) indicates that taxonomic studies supported by field and laboratory investigation of cultures would be profitable in elucidating the taxonomy of this group of organisms. Four genera of Zygnemataceae have been chosen here to exemplify the family, namely, *Spirogyra*, *Sirogonium*, *Zygnema*, and *Mougeotia*. These may be distinguished readily with the aid of the following key:

    1. Chloroplast consisting of one or more parietal, undulate-margined bands 2
    1. Chloroplast stellate or a simple bar 3
       2. Chloroplast(s) spiral . . . . *Spirogyra*
       2. Chloroplast(s) not spiral . . . . *Sirogonium*
    3. Each cell with two stellate chloroplasts . . . . *Zygnema*.
    3. Each cell with a simple, rotatory, bar-shaped chloroplast . . . .*Mougeotia*

Hoshaw (1968) has written an excellent and rather comprehensive review of the Zygnemataceae including data and literature references regarding their ecology, physiology, reproduction, and genetics.

SPIROGYRA Link *Spirogyra* (Gr. *speira*, a coil + Gr. *gyros*, twisted) (Figs. 3.165, 3.166) is probably the best known and most widely distributed of all the Chlorophycophyta. The silky masses of this alga occur resting on the bottom or floating on a variety of bodies of freshwater. More than 275 species of *Spirogyra* have been described (Transeau, 1951), taxonomic criteria being the size and shape of the cells, the number of chloroplasts, and details of sexual reproduction, especially the size and shape of the zygote and color and type of ornamentation of its walls.

The cylindrical cells of *Spirogyra* contain a large central vacuole in which the nonspherical nucleus is suspended by threads of cytoplasm. The cytoplasmic strands

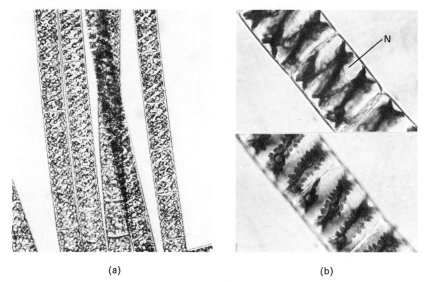

(a)

(b)

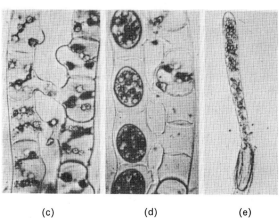

(c)   (d)   (e)

Fig. 3.165 *Spirogyra* sp. (*a*) Group of vegetative filaments. (*b*) Cellular organization; cell above in optical section, nucleus (N) visible; lower cell in surface view. (*c*), (*d*) Early and late stages of conjugation, the latter showing zygotes (Z). (*e*) Germinating zygote (Z). (*a*) × 465; (*b*) × 350; (*c*)–(*e*) × 375. [(*c*)–(*e*) after Bold, *Morphology of Plants*, 3rd ed., 1973, Harper & Row, Publ.]

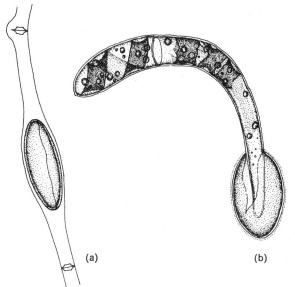

(a)   (b)

Fig. 3.166 (*a*) *Spirogyra groenlandica* Rosenvinge. Terminal or "lateral" conjugation. (*b*) *S. inflata* (Vauch.) Kütz. Germination of the zygote. (*a*) × 185; (*b*) × 396.

are connected with the thin parietal layer of cytoplasm within which the spiral, band-like chloroplast or chloroplasts are contained (Fig. 3.165b). The parietal cytoplasm undergoes cyclosis in the cell. The chloroplasts vary in number from one to several in the various species. Their margins are beautifully scalloped and they contain numerous, prominent pyrenoids. In many species of *Spirogyra* the cell wall is invested with a broad gelatinous sheath visible to the unaided eye or readily demonstrable when the filaments are mounted in India ink. The transverse walls are, according to the species, plane or replicate. In the latter case there occur corresponding infoldings of the terminal walls of adjacent cells that are H-shaped in optical section. Species with replicate terminal walls dissociate readily upon eversion of the replication, although nonreplicate species also may dissociate. The filaments of at least some species of *Spirogyra* undergo gliding motion (Yeh and Gibor, 1970). Fowke and Pickett-Heaps (1969A, B) have investigated mitosis and cytokinesis in *Spirogyra* at the electron-microscopic level. The nuclear envelope persists into anaphase. Early cytokinesis is by ingrowth of an annular septum, but a phragmoplast apparatus also takes part in cytokinesis.

Asexual reproduction occcurs by fragmentation or dissociation of the filaments and their subsequent growth.

Sexual reproduction, a process long ago first called **conjugation,** occurs after an extended prior period of vegetative development. Fowke and Pickett-Heaps (1971) have investigated the ultrastructural aspects of sexual reproduction in *Spirogyra*. Conjugation (Fig. 3.165c, d) most frequently involves aggregation of the filaments in a common gelatinous matrix and establishment of tubular connections between pairs of cells by elongation of protuberances, the conjugation papillae, in the lateral walls by the addition of new wall material to the latter. The contiguous ends of the papillae are dissolved to form the **conjugation canal**, and the protoplasts of both vegetative cells, now the gametangia and gametes, lose large quantities of water and undergo autonomous plasmolysis through the action of numerous contractile vacuoles that have developed, apparently, *de novo*. One of the gametic protoplasts (inferred to be male) then migrates across the canal and unites with that of the connected cell forming a zygote that subsequently develops a thick, dark wall and enters a period of dormancy. This type of sexual union is often interpreted as physiological anisogamy because the motile gamete and the filament that produces it are inferred to be male and the other filament female. The space between the zygote and gametangial walls is filled with a hyaline substance so that the zygotes are fixed in position. Grote and Pfautsch (1977) studied early conjugation by scanning electron microscopy.

Nuclear union follows soon after plasmogamy and, in one species, *S. crassa* Kütz., Godward (1961) has shown that the zygotic nucleus undergoes meiosis before the onset of dormancy, three of the four meiotic nuclei degenerating. The life cycle of *Spirogyra* thus is haplobiontic and haploid. After a period of dormancy, the zygote germinates (Fig. 3.165e) to form a new filament. Pessoney (1968) has shown that, in contrast to some populations of Zygnematales that withstand desiccation by means of thick-walled akinetes, it is the dormant zygotes that are the agents of survival in *Spirogyra*.

Pessoney (1968) also investigated the role of nitrogen concentration (and of light and aeration) on the abundance of conjugation in *Spirogyra* (and other Zygnematacean algae). He found that low concentration of nitrogen enhanced the degree of conjugation as did high light intensity.

In addition to the scalariform conjugation described above, terminal conjugation may occur in some species and in those that also undergo scalariform conjugation. In terminal conjugation, adjacent cells of the same filament conjugate by establishing conjugation canals near the terminal walls of the cells (Fig. 3.177a).

SIROGONIUM Kützing   The filaments of *Sirogonium* (Gr. *siera*, series, + Gr. *gon*, reproductive structure) (Fig. 3.167) superficially resemble those of *Spirogyra* in that the cells contain 2 to 10 ribbonlike chloroplasts, but these tend to be oriented parallel to the long axis of the cells. Furthermore, the filaments lack the gelatinous investments characteristic of those of *Spirogyra*.

The most important difference from *Spirogyra* is apparent at the time of sexual reproduction for the gametangia conjugate directly without forming conjugation canals (Fig. 3.167b). Hoshaw (1965) investigated the sexual reproduction of *S. melanosporum* Transeau in culture and found that clonal cultures undergo sexual reproduction. The male gametes were regularly smaller than the female. After a period of dormancy, the zygotes germinated into new filaments.

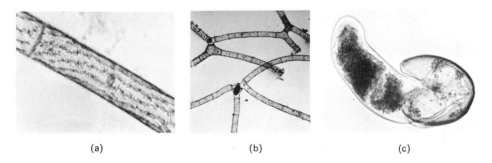

(a)   (b)   (c)

Fig. 3.167   *Sirogonium melanosporum* (Randhawa) Transeau. (*a*) Vegetative cell stained with I_2KI. (*b*) conjugation; note zygote below; differentiating gametes upper left. (*c*) Germinating zygote with two-celled germling. (*a*) × 147; (*b*) × 19; (*c*) × 181. (After Hoshaw.)

ZYGNEMA C. A. Agardh   The cells of *Zygnema* (Gr. *zygon*, yoke + Gr. *nema*, thread) (Fig. 3.168) are short-cylindrical and contain two stellate chloroplasts with radiating protuberances with a nucleus between them. The cells are surrounded with gelatinous sheaths of varying thickness, depending on the species. The filaments of *Zygnema* are but a few inches long in contrast to the much longer filaments of *Spirogyra*. Approximately 100 species of *Zygnema* have been described.

Pessoney (1968) and McLean and Pessoney (1971) have demonstrated that a number of species of *Zygnema* apparently survive desiccation by means of akinetes as well as by their thick-walled zygotes. They dried a species of *Zygnema* with akinetes on soil, agar, and filter paper and found that the akinetes germinated rapidly when flooded with culture medium.

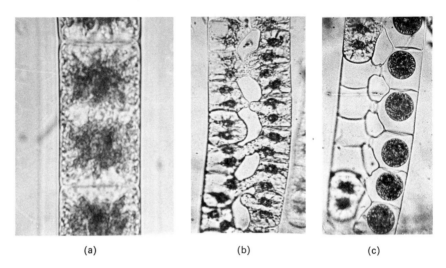

(a)   (b)   (c)

Fig. 3.168 *Zygnema* sp. (*a*) Vegetative cell. Note two stellate chloroplasts with nucleus between (arrow) and broad sheath. (*b*) Early and (*c*) late stages in conjugation. Note conjugation papillae, conjugation tubes, and zygotes. (*a*) × 298; (*b*), (*c*) × 149.

Sexual reproduction by conjugation (Fig. 3.168*b, c*) occurs in most species of *Zygnema*, and both scalariform and terminal conjugation have been observed. The zygotes in some species are formed in the conjugation canals (physiological isogamy) and in others in one of the filaments, as in *Spirogyra*. The walls of the dormant zygotes are blue or brown, depending on the species. At germination, the zygote gives rise to a single vegetative filament. The life cycle is presumed to be haplobiontic and haploid with zygotic meiosis.

MOUGEOTIA C. A. Agardh   The filaments of *Mougeotia* (after *A. Mougeot*) (Figs. 3.169, 3.170) are unbranched and composed of cylindrical cells with one (rarely two) bar-like chloroplast containing pyrenoids. The plastid rotates on its long axis

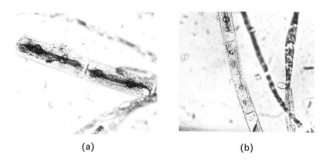

(a)   (b)

Fig. 3.169 *Mougeotia* sp. Vegetative cells showing board-like chloroplasts. (*a*) Face view. (*b*) Edge view. (*a*) × 116; (*b*) × 149.

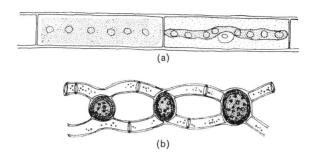

Fig. 3.170   (a) *Mougeotia* sp. Cells with chloroplasts in two views. (b) *M. calcarea* (Clev.) Wittr. Conjugation. (a) × 248; (b) × 109. [(b) after Wittrock from Oltmanns.]

orienting with its edge to white light of low intensity, phytochrome being the mediative pigment; in red light, the plastid presents its broad surface to the incident rays (Wagner, et al., 1972). Beck-Hansen and Fowke (1972) studied mitosis in a species of *Mougeotia* electron microscopically.

In sexual reproduction (Fig. 3.170b) almost all the species are isogamous, the zygote being formed in the conjugation canal. Both scalariform and terminal conjugation occur, the former most frequently. In all species of *Mougeotia* there is a cytoplasmic residue in the gametangia after conjugation has been completed. The germinating zygote of *Mougeotia* develops a germling filament and meiosis is presumed to be zygotic.

Family 2. Mesotaeniaceae

The members of the Mesotaeniaceae are often referred to as the **saccoderm desmids** because their walls, like those of most algal cells, are complete investments composed of a homogeneous unit, insofar as they are observable with light microscopy. This is in contrast to the members of the Desmidiaceae, often called **placoderm desmids**, the walls of which are composed of two half components of different age and origin (see p. 236). The term *desmid* itself is derived from the Greek word *"desmos,"* a bond, which was inspired probably by the union of two semicells in many genera (e.g., Fig. 3.177). Mollenhauer (1975) has edited a volume summarizing various aspects of desmids.

The family Mesotaeniaceae is a smaller assemblage than the Desmidiaceae and is represented in this book by four genera: *Mesotaenium*, *Spirotaenia*, *Cylindrocystis*, and *Netrium*. Key to the genera representing the Mesotaeniaceae is as follows:

1. Chloroplast a simple, bar-like structure . . . . *Mesotaenium*
1. Chloroplast different 2
  2. Chloroplast a spirally twisted ribbon . . . . *Spirotaenia*
  2. Chloroplast otherwise 3
3. Chloroplasts stellate . . . . *Cylindrocystis*
3. Chloroplasts axial, ridged . . . . *Netrium*

MESOTAENIUM Nägeli   *Mesotaenium* (Gr. *mesos*, middle + Gr. *taenia*, band) (Fig. 3.171) might readily be mistaken for a chlorococcalean or chlorellalean alga, if one did not observe its sexual reproduction. The cells are short- or long-cylindrical, depending on their age and the frequency of cell division. Each contains a single nucleus and a bar-like chloroplast with one or more pyrenoids; the cells look much like those of *Mougeotia*.

Asexual reproduction is accomplished by transverse division of the cells after which they separate and elongate to the size characteristic of the parental cells (Fig. 3.171*b*, *c*).

Starr and Rayburn (1964) and Biebel (1973) have studied in laboratory culture sexual reproduction of *M. kramstai* Lemm. in which the clones are unisexual. In conjugation two compatible cells become enveloped in a gelatinous matrix and develop

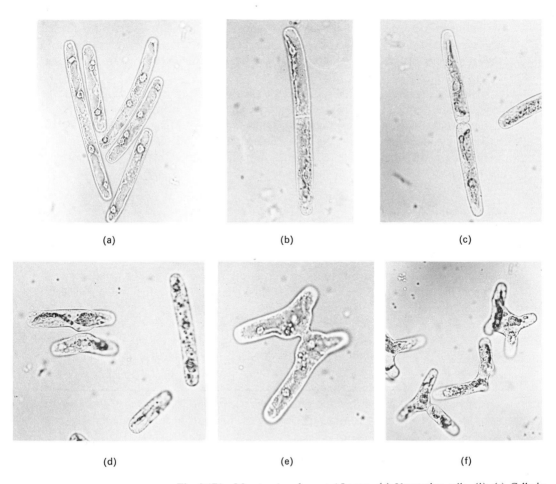

(a)   (b)   (c)

(d)   (e)   (f)

Fig. 3.171 *Mesotaenium kramstai* Lemm. (*a*) Vegetative cells. (*b*), (*c*) Cells in division. (*d*), (*e*), (*f*) Stages in conjugation. (*a*) × 291; (*b*) × 358; (*c*) × 347; (*d*) × 341; (*e*) × 429; (*f*) × 259. (After Starr and Rayburn.)

a broad conjugation canal between them. The zygote is formed in this (Fig. 3.171*d–f*). Gelatinous material is secreted in the gametangia by the migrating gametic protoplasts. Germinating zygotes of other species produce two or four young cells.

SPIROTAENIA Brébisson   The cells of *Spirotaenia* (Gr. *speira*, spiral + Gr. *taenia*, band) (Fig. 3.172) are straight or slightly curved cylinders with rounded apices. Most species have a single, spirally twisted, ribbonlike chloroplast with pyrenoids and their cells contain single nuclei. Asexual reproduction is by transverse cell division.

In sexual reproduction the zygote develops between two paired individuals (Hoshaw and Hilton, 1966; Biebel, 1973). Potthoff (1928) reported that karyogamy occurs only at the end of zygotic dormancy and that this is followed by meiosis and germination of the zygote into four young *Spirotaenia* cells (Fig. 3.172*d*).

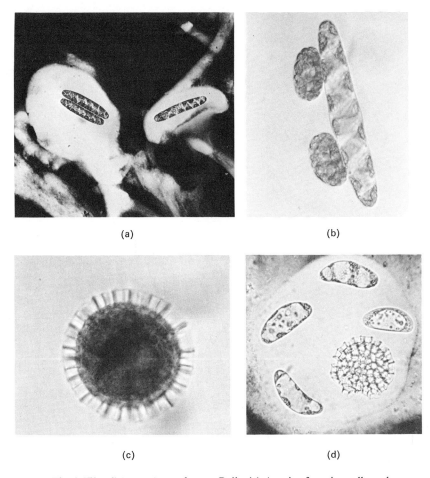

(a)   (b)

(c)   (d)

Fig. 3.172 *Spirotaenia condensata* Bréb. (*a*) A pair of mating cells and an unmated cell. India ink preparation: Note copious gelatinous matrices. (*b*) Gametogenesis. One cell (walls not visible) has formed two gametes; other cell in early phase. (*c*) Mature zygotes. (*d*) Zygote germination: Note four meiotic products. (*a*) × 84; (*b*) × 240; (*c*) × 400; (*d*) × 195. (After Hoshaw and Hilton.)

CYLINDROCYSTIS Meneghini   *Cylindrocystis* (Gr. *kylindros*, cylinder + Gr. *kystis*, bladder) (Fig. 3.173) is *Zygnema*-like in that its cells contain two axile chloroplasts each with a single pyrenoid. The cells are short- or long-cylindrical and undergo transverse division in asexual reproduction.

In sexual reproduction a short, broad conjugation canal is established between pairs of compatible cells and the gametes move into this isogamously to form a zygote (Biebel, 1975) (Fig. 3.173c, d).

After a period of dormancy, the zygote germinates to form four short-cylindrical cells that are liberated within a matrix. Meiosis has not been demonstrated in *Cylindrocystis* but is presumed to be zygotic.

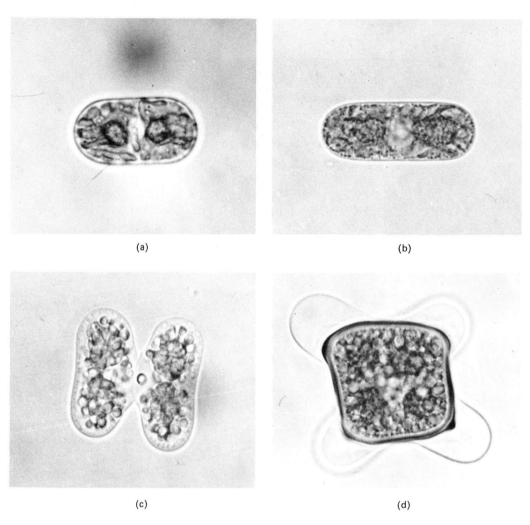

(a)

(b)

(c)

(d)

Fig. 3.173   *Cylindrocystis.* (*a*) *C. crassa.* Vegetative cell. (*b*)–(*d*) *C. brebissonii.* (*b*) Vegetative cell. (*c*) Early conjugation. (*d*) Mature zygote. × 833. (After Biebel.)

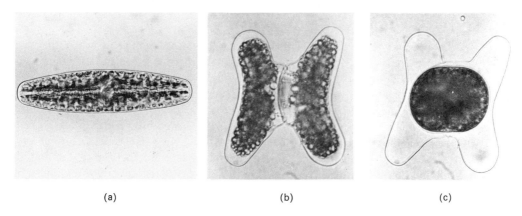

(a)   (b)   (c)

Fig. 3.174   (*a*) *Netrium digitus* var. *lamellosum*. Vegetative cell. (*b*), (*c*) *N. digitus* var. *digitus*. (*b*) Early conjugation. (*c*) Immature zygote. (*a*) × 300; (*b*), (*c*) × 276. (After Biebel.)

NETRIUM Nägeli   The cells of *Netrium* (origin unknown) (Fig. 3.174) are elongate with rounded or flattened ends. They contain two massive axial chloroplasts, each with an elongate pyrenoid, and in transection may be seen to have six or more radiating chloroplast ridges. The nucleus lies near the equator of the cells.

Biebel (1964) studied asexual and sexual reproduction in 87 clonal cultures of *Netrium*. Grown on an 18-hour light 8-hour dark cycle, the cells divided abundantly between 11 and 13 hours after the inception of the light period. Cell division follows mitosis in *Netrium* and is accomplished by an annular ingrowth of the cell wall.

In the sexual reproduction (Fig. 3.174*b*, *c*) of *Netrium digitus*, Biebel (1964, 1973) found that one variety was unisexual in clonal culture and one bisexual. Depletion of or omission of nitrogen from the culture medium and high light intensity evoked sexual reproduction in which pairs of cells become contiguous, develop broad conjugation canals into which the gametic protoplasts move, and unite isogamously. Meiosis is zygotic and the germinating zygote, after dormancy, may give rise to two or four progeny. Biebel and Reid (1965) demonstrated that segregation of mating type and of zygote wall genes occurs at meiosis in *N. digitus* var. *lamellosum*.

Family 3. Desmidiaceae

Members of the Desmidiaceae, the placoderm desmids, are unicellular, colonial, or filamentous (Figs. 3.175, 3.179). Their cell walls are composed of two halves of different age because of the manner of cell division, and, furthermore, they are interrupted by pores through which mucilaginous substances are secreted. The wall itself, which in many cases consists of two layers, is surrounded by a gelatinous sheath.

The cells are composed of two mirror-image counterparts or semicells that contain conspicuous chloroplasts with pyrenoids. These chloroplasts vary in number and organization in the genera discussed below. The conspicuous nucleus lies at the equator of the cell, the region of the **isthmus**, where the cell may be deeply incised to form a sinus in some species (Fig. 3.177).

The desmids exemplify several different patterns of symmetry when they are viewed in polar aspect. In lateral aspect all are bilaterally symmetrical. Some have forms of beautiful complexity (e.g., *Micrasterias*, Fig. 3.177).

The cell division of members of the Desmidiaceae is their most distinguishing characteristic and the process is illustrated in Fig. 3.176c, d. Cell division is preceded by mitosis, which occurs in the isthmus. Mitosis is followed by cytokinesis after which each semicell then buds forth a new semicell, which in some genera (*Cosmarium*, *Micrasterias*) is at first colorless. This gradually enlarges to the size of the mature semicell and becomes provided with its own chloroplast(s) by the migration and subsequent division of the parental plastid(s).

Pickett-Heaps (1973C, 1974E) has published beautiful micrographs prepared with the scanning electron microscope illustrating in three dimensions the symmetry and wall organization of certain desmids.

Asexual reproduction, of course, is accomplished by cell division of the unicellular and by fragmentation of the filamentous types.

Sexual reproduction is by conjugation in which a conjugation canal may or may not develop. In the filamentous genera conjugation canals are characteristically present. In most species the amoeboid gametes unite in the conjugation canals and develop dark zygotes with ornamented or smooth walls that may be spinose. Starr (1955C) has described methods for isolating sexual strains of Desmidiaceae, while Coesel and Texeira (1974) have investigated the occurrence of sexual reproduction in nature and in laboratory culture. They suggest that it is lack of sexual potential rather than ecological factors that accounts for the infrequent incidence of sexual reproduction in nature.

Meiosis occurs in the germination of the zygote, but unlike those of many Mesotaeniaceae the zygotes of the Desmidiaceae usually give rise to only one or two progeny at germination. Brandham (1965) found that polyploid individuals of desmids appeared spontaneously in culture.

Members of the Desmidiaceae seem to be most abundant in slightly acid waters (pH 5–6); however, water blooms of *Cosmarium* have been observed by the writers in waters of pH 9.1.

Six genera of the family have been chosen as its representatives in the following account, namely, *Closterium*, *Cosmarium*, *Micrasterias*, *Staurastrum*, *Desmidium*, and *Hyalotheca*. These may be distinguished with the aid of the following key to some genera of Desmidiaceae:

1. Unicellular or sometimes attached temporarily after cell division 2
1. Filamentous 5
   2. Cells elongate, sometimes arcuate . . . .*Closterium*
   2. Cells not elongate or arcuate 3
3. Cells incised only at the isthmus, apices of semicells
   without processes . . . .*Cosmarium*
3. Semicells variously incised 4
   4. Cells flat in end view . . . .*Micrasterias*
   4. End views of cells radiate . . . .*Staurastrum*
5. Cells with scarcely perceptible median constrictions . . . .*Hyalotheca*
5. Cells with obvious median constrictions . . . .*Desmidium*

CLOSTERIUM Nitzsch   Species of *Closterium* (Gr. *kloster*, spindle) (Fig. 3.175) are elongate unicells, narrowed toward both poles, sometimes slightly tumid at the equator and slightly arcuate. The semicells contain single axial chloroplasts that may be radially ridged. The prominent central nucleus lies between the two plastids at the equator of the cell. There is a vacuole near each pole of the cell in which one or more granules of calcium sulfate may be observed in Brownian movement. The cell walls of *Closterium* contain pores (Pickett-Heaps and Fowke, 1970) through which mucilaginous material is secreted. Mix (1969) studied nine species of *Closterium* electron microscopically and reported that the cell wall is three-layered, consisting of an amorphous outer layer and a fibrillar primary and secondary wall. The outer wall is smooth or longitudinally striate. Pores in longitudinal rows, through which mucilage is secreted,

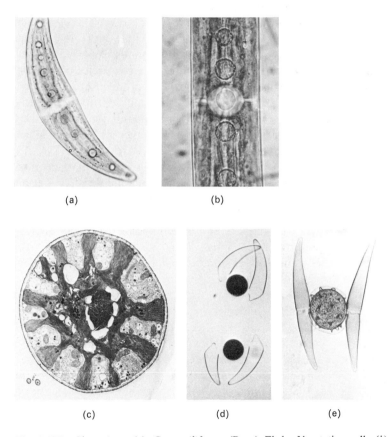

(a)   (b)

(c)   (d)   (e)

Fig. 3.175 *Closterium*. (*a*) *C. moniliferum* (Bory) Ehrb. Vegetative cell. (*b*) *Closterium* sp. Central region of cell. Note large nuclei and four (of many) pyrenoids. (*c*) *C. littorale*. Transection of cell. Note axile chloroplast with radiating ridges. (*d*) *C. moniliferum*. Twin zygotes formed by recently divided and paired cells. (*e*) *C. calosporum* Wittr. var. *maius* West and West. Dormant zygote between two empty cells. (*a*) × 192; (*b*) × 390; (*c*) × 1800; (*d*) × 138; (*e*) × 105. [(*a*), (*d*) after Lippert; (*c*) after Pickett-Heaps and Fowke; (*e*) after Cook.]

are present in all species. Chardard (1975) has reported on organization of the micro-fibrils of the wall in *C. acerosum* (Schrank) Ehrenb. Mucilage secreted from the poles of the cell results in movement across the substrate (Kies, 1964; Yeh and Gibor, 1970).

Asexual reproduction is accomplished by cell division. In this process the nuclei undergo mitosis in the center of the cell and this is followed by cytokinesis. The daughter nuclei migrate to the equator of the new semicells in which each plastid divides into two, the nucleus assuming a position between the two halves. Meanwhile, each gradually develops a new semicell and these secrete new cell walls. Pickett-Heaps and Fowke (1970) have studied the process in *C. littorale* Gay electron microscopically. According to them, a microtubular center precedes the daughter nuclei as they migrate and this center probably has some role in the division of the chloroplast of each semi-cell just after cytokinesis.

A number of investigators have studied sexual reproduction in *Closterium*. Cook (1963) investigated 18 clones of the *C. venus-C. dianae* complex belonging to seven species. Both unisexual and bisexual clones were among them. The zygotes varied in shape and in wall ornamentation. Ichimura and Watanabe (1976) reported on mor-phological variation among three related species of *Closterium*. Kies (1964) investi-gated the factors that evoke sexual reproduction in *C. acerosum* (Schrank) Ehr. The bisexual clones of this species undergo cell division preceding conjugation, which occurs 24 hours later between the immature semicells of the compatible pairs. Conju-gation occurred only in sufficiently aged cultures in which the medium had been depleted (such as in 30–35-day-old cultures).

In the process of conjugation (Fig. 3.175d, e), the compatible individuals become associated in pairs and develop conjugation papillae that merge and enlarge to form a broad canal within which the zygote is formed in some species, but in others the canal is evanescent. The zygote wall thickens and becomes brown and variously ornamented. Lippert (1967, 1975) investigated sexual reproduction in *C. moniliferum* (Bory) Ehrb. and in *C. erhenbergii* Meneghini. In a bisexual clone of the former, cells pair and then divide and conjugate so that twin zygotes are formed (Fig. 3.175d). Both unisexual and bisexual clones were found in this species. Meiosis is zygotic and the zygote germinates to form two new individuals. Ichimura (1972) reported that although nitrogen depletion and light were necessary for conjugation papillae forma-tion in *C. strigosum*, light was not necessary for the completion of conjugation.

Pickett-Heaps and Fowke (1971) and Dubois-Tylski (1973, 1975) investigated conjugation in *C. littorale* and *C. moniliferum*, respectively, electron microscopically. They reported that the conjugation papillae develop by augmentation of the wall material at the point of contact of two individuals and that the zygote, formed within the united papillae, secretes six layers of wall material as it matures.

Fox (1958) and Kies (1964) have described zygote germination in *Closterium*. Only zygotes that have matured for several months germinate when transferred to fresh culture media. Meiosis occurs during germination, and two of its products degenerate. A protoplasmic vesicle emerges as the zygote germinates and its contents divide into two small Closteria. The life cycle, as in all Zygnematales, is haplobiontic and haploid. Starr (1958A) reported an unusual type of resistant asexual spore in *C. didymotocum* Ralfs var. *pygmaeum* Starr.

COSMARIUM Corda   The cells of *Cosmarium* (Gr. *cosmina*, an ornament) (Fig. 3.176) are deeply incised at the isthmus to form a **sinus** and thus conspicuously divided into semicells. These may be variously lobed and their cell walls ornamented in different patterns, according to the species. Each semicell contains an axile chloroplast with radiating ridges or plates; the pyrenoids occur in the axial portions of the plastids. Mix (1966) investigated wall organization in *Cosmarium* and found the walls consisted of fibrils within a structureless matrix. Starr (1958B) showed that increasing ploidy resulted in radial symmetry in *C. turpinii* Bréb.

Pickett-Heaps (1972F) has investigated electron microscopically cell division in *Cosmarium botrytis* Meneghini (Fig. 3.176c, d). He reported that during early prophase the isthmus elongates slightly and that new wall material in the form of a girdle is deposited around the isthmus. Cytokinesis occurs during the telophase of mitosis, and a wall layer continuous with the girdle is deposited between the two cells. As the daughter semicells enlarge, new wall material is synthesized for augmentation of the walls of the semicells; the plastid arms from the old semicells protrude into the new ones and are finally severed. Just as the new semicells reach maximum size, a secondary wall layer is deposited between the plasma membrane and primary cell wall, and the pattern of ornamentation characteristic of the species develops. Mucilage pores are present in the secondary wall.

Starr (1954A, B, 1955B, C) was one of the first to master the control of sexuality in laboratory cultures of desmids, particularly *Cosmarium turpinii* Bréb.[29] He established for the first time that unisexual clones occur in desmids. In conjugation (Fig. 3.176f), pairs of compatible cells become surrounded by mucilaginous material and lie with their long axes perpendicular to each other. The cells of each conjugant then open at the isthmus and the protoplasts escape and unite to form a spherical zygote that soon secretes a spiny wall (Fig. 3.176f). After a period of dormancy, during which a thick, olive-brown wall is secreted within the spiny one, the zygote protoplast emerges in a hyaline vesicle and divides to form two daughter cells. Genetic evidence (Starr, 1954B, 1955C) indicates that meiosis is zygotic, mating type segregating at meiosis I, the two zygotic progeny being of opposite mating type. Starr (1954B) also demonstrated the inheritance of a lethal factor causing zygote abortion in certain gametic combinations.

MICRASTERIAS C. A. Agardh   The species of *Micrasterias* (Gr. *micros*, small + Gr. *asterias*, star) (Fig. 3.177) are among the most beautiful of the desmids. The bilaterally symmetrically incised semicells are perhaps unrivaled for beauty among algal cells. The cells of many species are circular in one view but quite flattened in lateral aspect. One species of *Micrasterias* is filamentous. Each semicell contains a single massive chloroplast with pyrenoids and the nucleus lies in the isthmus. Lacalli (1973) investigated cytokinesis and primary wall deposition in *M. notata*, while Ueda and Yoshioka (1976) studied wall development in isotonic and hypertonic solutions.

Cell division in *Micrasterias* is a spectacular process as the semicells of elaborately incised species develop after cytokinesis.

[29]As *C. botrytis* var. *subtumidum.*

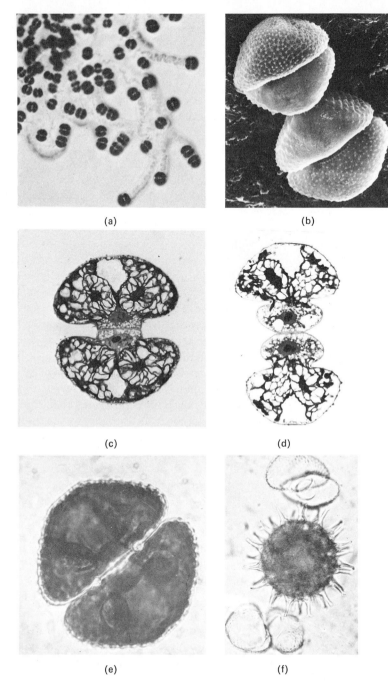

Fig. 3.176 *Cosmarium.* (*a*)–(*d*) *C. botrytis* Menegh. (*a*) Cells growing on agar. Note trails left by motile cells. The paired cells have recently undergone division. (*b*) Pair of cells after division. Note immature walls on adjacent semicells. (*c*), (*d*) Stages in cell division and formation of new semicells. (*e*) *C. turpinii* Bréb., living cell. (*f*) Sexual reproduction. Note empty gametangia with zygote between them. (*a*) × 12; (*b*) × 540; (*c*), (*d*) × 640; (*e*) × 191; (*f*) × 48. [(*a*)–(*d*) after Pickett-Heaps.]

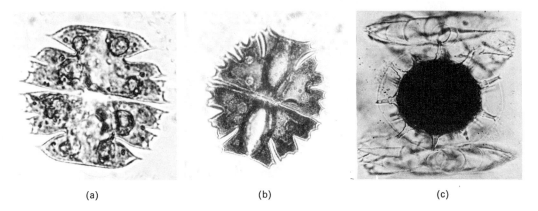

(a)   (b)   (c)

Fig. 3.177   *Micrasterias.* (*a*) *M. truncata* (Corda) Breb. (*b*) *M. americana* (Ehr.) Ralfs. (*c*) *M. papillifera* Breb. Zygote with empty gametic walls. (*a*) × 345; (*b*) × 189; (*c*) × 400. [(*c*) after Coessel and Texeira.]

Kies (1968) investigated in the laboratory the sexual reproduction of a bisexual clone of *M. papillifera* Breb. In this process, pairs of cells become aligned with their flat surfaces parallel. Soon a denser gelatinous drop develops between the partners within which the conjugation papillae of the two cells develop and ultimately unite. The zygote forms within the conjugation canal and ultimately develops a spiny wall (Tassigny, 1971). Conjugation and the complicated zygote wall and spine formation have been studied in detail by Kies (1970A, B; 1975). Two cells emerge from the germinating zygotes of *Micrasterias* (Lenzenweger, 1968, 1975). Coesel and Texeira (1974) have illustrated zygote formation in *Micrasterias papillifera* (Fig. 3.177*c*).

STAURASTRUM Meyen   The cells of *Staurastrum* (Gr. *stauron*, cross + Gr. *astron*, star) (Fig. 3.178) are bilaterally symmetrical in lateral view and triradially or hexaradially symmetrical in polar view instead of being flattened like those of *Micrasterias*. The numerous species differ in the ornamentation of their walls.

Asexual reproduction is by cell division and isogamous conjugation to form spiny-walled zygotes has been observed (Winter and Biebel, 1967) (Fig. 3.178*d–m*). These have been reported to give rise to from one to four cells at germination.

HYALOTHECA Ehrenberg   *Hyalotheca* (Gr. *hyalos*, colorless + Gr. *theka*, case) (Fig. 3.179) exemplifies the filamentous members of the Desmidiaceae. The unbranched chains of cells, each inconspicuously indented at the isthmus, are surrounded by a copious gelatinous sheath. The individual cells are short-cylindrical or discoidal, each at maturity containing two chloroplasts with pyrenoids.

Cell division is typically desmidiacean and results in growth of the filaments; their reproduction is by fragmentation.

Conjugation occurs between compatible individuals, and conjugation canals are formed. The zygotes may develop within the latter or in one or another of the gametangia. Meiosis occurs at zygote germination and two young individuals emerge.

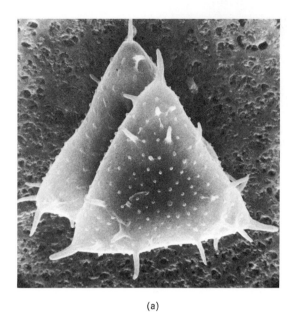

(a)

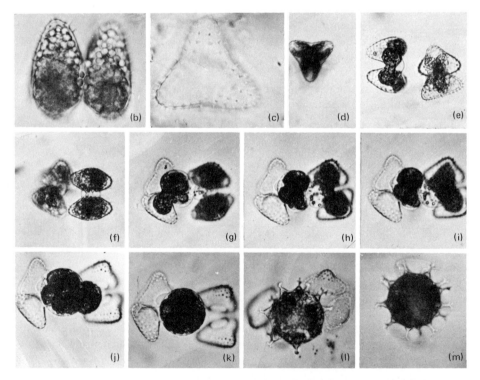

(b)   (c)   (d)   (e)

(f)   (g)   (h)   (i)

(j)   (k)   (l)   (m)

Fig. 3.178 *Staurastrum.* (*a*) *S. cristatum* (Naeg.) Arch. Scanning electron micrograph. (*b*)–(*m*) *S. gladiosum* Turner. Vegetative cells (*b*), (*c*) and sexual reproduction (*d*)–(*m*). [(*a*) × 1000; (*b*), (*c*) × 550; the remainder × 205] [(*a*) after Pickett-Heaps; the remainder after Winter and Biebel]

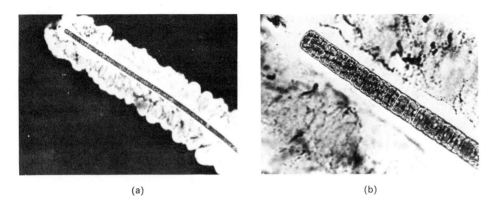

(a)   (b)

Fig. 3.179 *Hyalotheca dissiliens* (J. E. Smith) Bréb. India ink preparation showing sheaths. (*a*) × 29; (*b*) × 171.

DESMIDIUM C. A. Agardh   The filaments of *Desmidium* (Gr. *desmos*, bond)
(Fig. 3.180) are spirally twisted and usually enclosed in a prominent mucilaginous
sheath. The indentation (sinus) between the cells is shallow and groove-like. Each cell
contains two axile chloroplasts with pyrenoids. Here, again, cell division results in
growth of the filaments that multiply by fragmentation.

Conjugation occurs between compatible filaments, and conjugation canals are
formed. The zygotes develop in the canals or (in one species) in the gametangia of one
of the filaments.

This brings to a conclusion discussion of the order Zygnematales, the three
families of which seem clearly to represent a cohesive group. They share such common
attributes, among others, as rather elaborate chloroplasts with prominent pyrenoids,
the frequent presence of copious gelatinous sheaths around the cells and filaments,
the absence of flagellate gametes, and reproduction involving the union of usually
amoeboid gametes. All are haplobiontic and haploid with zygotic meiosis. It has been
suggested that the Desmidiaceae are somewhat isolated from the other two families,
which are similar in the organization of their cell walls and in the method of cell division. This supposed relationship is more striking when one observes the dissociation
of Zygnematacean filaments into unicellular components that are similar to meso-
taeniacean desmids.

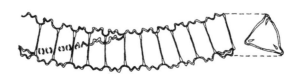

Fig. 3.180 *Desmidium schwartzii* C. A. Agardh. × 218. (After Britton and
Tiffany.)

Classification and Phylogeny of the Green Algae

Papenfuss (1953) and Christensen (1971), among others, have reviewed the history of the classification of the algae in general, including that of the Chlorophycophyta, and Round (1963, 1971) and Christensen (1971) have discussed the classification of the latter. Papenfuss recognized 11 orders of green algae, while Bourrelly (1966, 1972) has classified genera with freshwater species in 19 orders, the majority of which are in the class Chlorophyceae; in addition, Bourrelly recognizes the classes Zygophyceae, Prasinophyceae, and Charophyceae.

Round (1971) was among the most recent phycologists to consider the classification of the green algae. Like the authors, he classified the Charophyta in a separate division from the Chlorophycophyta but, unlike the authors, he erected a division Prasinophyta. Under his division Chlorophyta (=Chlorophycophyta of the present book), Round recognized four classes: the Zygnemaphyceae (containing the members of the writers' Zygnematales); the Oedogoniophyceae (with the single order Oedogoniales); the Bryopsidophyceae (containing siphonous and many other coenocytic Chlorophyceae of the present text), and the Chlorophyceae (with 18 orders).

About three-quarters of a century ago Blackman (1900) suggested that evolution in the green algae proceeded from ancestral green flagellates along three different pathways: the **volvocine**, the **tetrasporine**, and the **chlorococcine** (siphonous). In the volvocine pathway, it was postulated, flagellate precursors retained their motility but became grouped in motile colonies (coenobia). Tetrasporine algae, it was hypothesized, lost their motility but retained the capacity for their cells to divide directly into somatic cells similar to the parents. In the chlorococcine line, the primitive ancestors lost both their motility and capacity to divide into somatic cells similar to the parental cells. These three lines of development are reflected, respectively, in the currently accepted orders Volvocales (volvocine) and Tetrasporales (tetrasporine), while the Chlorococcales and coenocytic orders (Caulerpales, Dasycladales, and Siphonocladales) represent the chlorococcine trend.

In addition to these, another important development, namely, the capacity for **desmoschisis** (vegetative cell division, in part, *sensu* Fritsch), discussed by Groover and Bold (1968) (see p. 127), must be taken into account in considering algal evolution. Desmoschisis makes possible the formation of tissue complexes and the type of multicellular body observable in many filamentous and membranous algae.

The origin of filamentous algae has been suggested to have followed two different pathways. According to Smith (1950), the filamentous green algae may have arisen from tetrasporalean ancestors in which cell division became restricted to one plane and the gelatinous matrix at the cellular poles became restricted, so that the cells became contiguous, end-to-end. Fritsch (1935), however, was impressed by the ontogenetic behavior of the zoospores and motile zygotes of many filamentous algae that directly develop into filaments after they settle by restriction of cell division to one direction. The division of the latter is also by desmoschisis, so that chlorosarcinalean precursors also should be considered.

While it has often been postulated that the filamentous, membranous, and tubular green algae evolved from biflagellate and quadriflagellate precursors, because their zoospores and/or gametes are similarly flagellated, the origin of the Oedogoniales with their multiflagellate (stephanokontan) motile cells is often considered less certain because stephanokontan green flagellates are unknown. The occurrence of both isokontan and stephanokontan motile cells, however, in the life cycle of the same organism, e.g., in *Derbesia* and *Bryopsis*, seems to offer an alternative explanation.

The multinucleate (coenocytic) condition in the green algae seems to have evolved repeatedly in different groups. For example, some species of Chlorococcales (*Spongiochloris*, *Characium*, and *Hydrodictyon*) are coenocytic, as are *Sphaeroplea*, *Cladophora*, Caulerpales, Dasycladales, and Siphonocladales. The coenocytic or partially coenocytic condition arises ontogenetically, of course, by the repetition of mitosis without ensuing cytokinesis, and it has been suggested that this, too, was its phylogenetic origin.

Finally, what of the precursors of the Zygnematales? Because *Chlamydomonas* and Chlamydomonads have a variety of chloroplasts and because their gametes at plasmogamy may be amoeboid, it has been suggested that members of the Zygnematales also had a green flagellate ancestry.

Recently, a new system of classification of the green algae has been proposed on the basis of ultrastructural organization and comparative biochemistry by Pickett-Heaps and Marchant (1972), Stewart, et al. (1973), Stewart and Mattox (1975), and Pickett-Heaps (1975, 1976). These authors, as a result of their investigations of the course of nuclear and cell division in a number of green algae, on the basis of the organization of their zoospores and/or gametes and in consideration of biochemical differences, have postulated that the Chlorophycophyta contain two distinct classes, the Charophyceae and the Chlorophyceae.

Nuclear and cell division in members of these classes are especially distinctive at telophase (Fig. 3.1). In the Charophyceae, the telophasic nuclei are relatively far apart, and the microtubular spindle between them is persistent until the completion of telophase. In some of the members of the class (e.g., *Chara*) and also in *Coleochaete*, the spindle persists and a cell plate is organized from a **phragmoplast**, as in more complex green land plants, and stages in the evolution of that structure have been interpreted to occur in other members. In other green algae, e.g., *Klebsormidium*, cleavage (without cell plate formation) divides the cytoplasm in the presence of a persistent spindle (Fig. 3.1). By contrast, in the Chlorophyceae, the telophasic nuclei are in close proximity, the microtubules of the spindle disappear, but microtubules perpendicular to the axis of the spindle, designated as **phycoplast**, are organized at the site of cytokinesis, which may be by cell plate or furrowing (Fig. 3.1). A phycoplast is absent in the Charophyceae.

Correlated with these differences between the classes are differences in the organization of their motile cells (Pickett-Heaps, 1975). In the Charophyceae, the motile cells contain a flat band of microtubules at the flagellar bases, while the motile cells of Chlorophyceae have four cruciately arranged flagellar roots. Furthermore, flagellar insertion in the Charophyceae is slightly lateral, rather than anterior as in the Chlorophyceae. Finally, the Charophyceae, like more complex green land plants, produce glycolate oxidase, while the Chlorophyceae lack it (Frederick, et al. 1973).

The Charophyceae, according to the proposals of Stewart and Mattox (1975) include the orders Klebsormidiales, Zygnematales, Coleochaetales, and Charales, rather strange bedfellows on the basis of traditionally recognized criteria! The Volvocales, Microsporales, Ulvales, Chaetophorales, and Oedogoniales are retained in the Chlorophyceae. Were this classification to be widely accepted, certain groups of green algae currently considered to be homogeneous or natural alliances would have to be dismembered; for example, *Klebsormidium*, now classified as a member of the Ulotrichaceae, would have to be transferred to the family Klebsormidiaceae, order Klebsormidiales, of the Charophyceae. It should be noted also that phylogenetic schemes that postulated chlamydomonad and carterial ancestry for the nonvolvocalean groups of green algae would have to be modified, because the organization of the motile cells of these putative ancestors differs from that of some of the more complex green algae (and land plants). For discussion of further details on this classification, the reader is referred to the summary paper by Stewart and Mattox (1975).

Kornmann (1973) and Fott (1974) have also made recent proposals bearing on the classification of the green algae. The former, impressed with the significance of *Codiolum* (Fig. 3.117*e*, p. 187) stages in certain Chlorophycophyta, proposed that organisms having them in their life cycles be assigned to a new class, Codiolophyceae. This class, according to Kornmann, would include members of the Ulotrichales, Monostromatales, Codiolales, and Acrosiphonales. Fott (1974), discussing the phylogeny of eukaryotic algae, recognized nine basic classes of algae, three of them members of the division Chlorophycophyta, namely the Chlorophyceae, Conjugatophyceae, and Charophyceae.

4

Division Charophyta

The divisional heading, which lacks the root *phyco*, is an indication of the authors' uncertainty that these plants, the stoneworts and brittleworts, are, in fact, algae. They have been variously classified by others. Some, impressed by the presence in their cells of chlorophylls *a* and *b* and starch stored in the chloroplasts, have disregarded morphological differences and have assigned them, as an order or class, to the Chlorophycophyta.[1] At the other extreme, perhaps, are those who would assign them to the Bryophyta. The authors, while conceding that their pigmentation and storage products are similar to those of the green algae, recognize that these are shared also with the chlorophyllous land plants. They are, therefore, impelled by the differences in vegetative organization and reproduction of the Charophyta to classify them as a separate division of uncertain affinity. Grambast (1974) has reviewed the fossil history of the Charophyta and concluded that they occupy an isolated position between green algae and bryophytes. Most phycologists agree that a single class, Charophyceae; order, Charales; and family, Characeae should be included in the group. Wood and Imahori (1964, 1965) in a monumental and profusely illustrated work have monographed the Charophyta on a worldwide scale.

The four genera of Charophyta which occur in North America (Tindall, et al., 1965) and which are included in the following discussion may be distinguished with the aid of the following key:

```
1. Oogonial crown cells 5, in a single tier ...............................2
1. Oogonial crown cells 10, in a double tier ..............................3
   2. Stipulodes[2] present, regular and prominent........................Chara
   2. Stipulodes normally absent; if present,
      irregular and rudimentary ...............................Nitellopsis[3]
3. Axes of the branchlets strongly percurrent and unbranched .........Tolypella
3. Branchlets forked or branched ....................................Nitella
```

[1]Pickett-Heaps (1975A, B) has emphasized that the organization of the sperm of the Charophyta is unlike that of motile cells of most Chlorophycophyta.

[2]Single-celled processes subtending branchlets.

[3]Found only once in the United States (Tindall, et al., 1965).

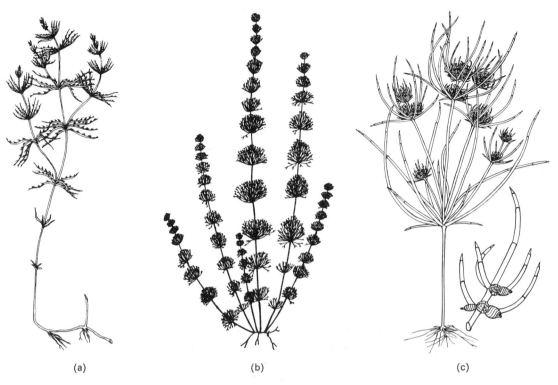

(a)   (b)   (c)

Fig. 4.1 Charophyta, habit of growth. (*a*) *Chara foetida* A. Br. (*b*) *Nitella tenuiissima* (Desv.) Kütz. (*c*) *Tolypella prolifera* (Ziz.) Leonhardi; detail of branchlet with sex organs at the right. (*a*) × $\frac{1}{3}$; (*b*) × $\frac{2}{3}$; (*c*) × $\frac{2}{3}$. [(*a*) after Groves and Bullock-Webster; (*b*), after Prescott; (*c*) courtesy of Professor Takashi Sawa.]

Wood (1968) has prepared a helpful summary guide to the North American species of Charophyta.

These genera are illustrated in Fig. 4.1 and are not discussed separately because of basic similarities in their organization and reproduction.

The Charophyta are widely distributed in freshwater and some occur also in brackish water. They often grow in waters with sandy or muddy bottoms in which they are anchored by their rhizoids. Several grow in waters that run over limestone rocks to which they also are attached by their rhizoids. The common names "stonewort" and "brittlewort" were suggested by the circumstance that many species, especially of *Chara*, become encrusted with calcium carbonate. Forsberg (1963, 1965) and Shen (1971) have succeeded in growing *Chara* in defined culture media in the axenic state.

Vegetative Morphology

Like the Ulvales and some of the Caulerpales, Siphonocladales, and Dasycladales of the Chlorophycophyta, the members of the Charophyta are recognizable macroscopically. They are erect, in quiet waters, or bend with the current in running water and may be 30 cm or more in length. The intact plant of charophytes comprises two physically continuous, but ontogenically distinct, structures. One of these, the **pro-**

tonema, however, is rather inconspicuous at the base of the predominant adult shoot that displays important taxonomic characteristics. The most striking feature of the adult shoot is the whorled pattern of its laterals of limited growth arising at the main axial nodes. The whorls consisting of a number of laterals of limited growth precisely alternate with the internodes, which are single and multinucleate cells, which may attain a length of several centimeters. In many species of *Chara* the internodes are covered with corticating cells that in ontogeny grow from the nodes adjacent to a given internode and meet. The branching rhizoids that anchor the plants to the substrate primarily arise from the protonematal rhizoidal node, but adventitious rhizoids may develop from any nodes of the adult shoot. The rhizoids undergo apical growth and may proliferate, giving rise to additional adult shoots.

The growth of the main axis of the adult shoot is apical and unlimited. A prominent dome-shaped apical cell divides transversely periodically to form a short-cylindrical subapical cell (Pickett-Heaps, 1967A). By a single transverse division, the subapical cell produces two cells, the upper of which, the nodal initial, develops into a nodal complex after a series of divisions, whereas the lower becomes an internode without further division. This mode of growth results in the very regular geometric construction of the adult shoot, as observed in longitudinal sections of the apex (Fig. 4.2). The internodal cells elongate gradually; as they develop very large central vacuoles, the original nucleus divides repeatedly amitotically (Shen, 1966B, 1967A; Roberts and Chen, 1975), so that the twelfth internodal cell from the apex may contain as many as 1370 crescentic nuclei. The nuclear divisions in the nodal cells are mitotic, and at cytokinesis a phragmoplast and cell plate are present (Pickett-Heaps, 1967A, B).

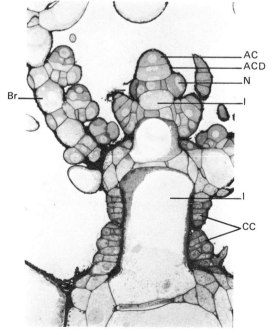

Fig. 4.2 *Chara* sp. Median longitudinal section of the apex. AC, apical cell; ACD, apical cell derivative with two nuclei; Br, branch with sex organs; CC, corticating cells from adjacent nodes; I, internode; N, node. × 2000. (Courtesy of Prof. J. D. Pickett-Heaps.)

Nodal cells divide longitudinally (vertically) into two and then undergo a series of longitudinal divisions by curved walls so that ultimately the two central cells become surrounded by 6–20 pericentral cells. The two central cells in some genera undergo further divisions (Frame and Sawa, 1975). The pericentral cells at the nodal complex protrude and function as apical cells repeating the method of segregation into nodal and internodal cells. The apical cell, unlike that of the main axis, ceases to divide after a certain number of divisions. Cellular progenies by this limited activity of the apical cell, particularly the internodal cells, simply increase in size until the lateral growth reaches to a certain length. Whorls of laterals of limited growth, or branchlets, which are sometimes spoken of as "leaves," are thus formed. The main axis may divide repeatedly to give rise to branches of unlimited growth. They arise from the basal nodes of branchlets. Although cytoplasmic streaming occurs in most of the cells, it is especially marked in the internodal cells. The cytoplasm adjacent to the central vacuole displays active rotating cyclosis. The peripheral cytoplasm and chloroplasts are stationary.

Reproduction

Fragments of the charophyte plant body are able to produce rhizoids and adventitious shoots from their nodes and can become established. Furthermore, the rhizoids themselves may spread and give rise to colonies of erect, photosynthetic adult shoots. Other than by these methods, reproduction is sexual and oogamous.

Chara itself may be monoecious or dioecious (McCracken, et al., 1966; Proctor, 1971A, B). The sex organs of the Charophyta are characteristic and unique in the plant kingdom. Because of this they have been readily recognized as fossils (p. 27). The sex organs develop from the nodes of branchlets or "leaves," the antheridia and oogonia being closely associated except in dioecious species. Both antheridia and oogonia arise laterally (Fig. 4.2) from certain pericentral cells of the branchlet nodes in all the taxa of charophytes except in *Nitella* and some species of *Tolypella* (Sawa, 1974) whose antheridia develop from the apical cells of the branchlets. In the latter case the antheridium is terminal on the branchlet.

The globose antheridia (Figs. 4.3, 4.4) are green when they are young but become orange-red as the chloroplasts of their wall cells, the **shield cells**, develop secondary

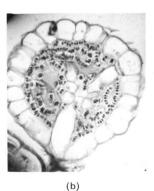

Fig. 4.3 *Chara seijuncta* A. Br. (*a*) Branchlet with oogonium (note five crown cells) and antheridium. (*b*) Section of an antheridium. Note, in centripetal order, the shield cells, spermatogenous filaments, manubria, primary capitula, and pedicel. (Compare with Fig. 4.4.) (*a*) × 48; (*b*) × 34.

(a)   (b)

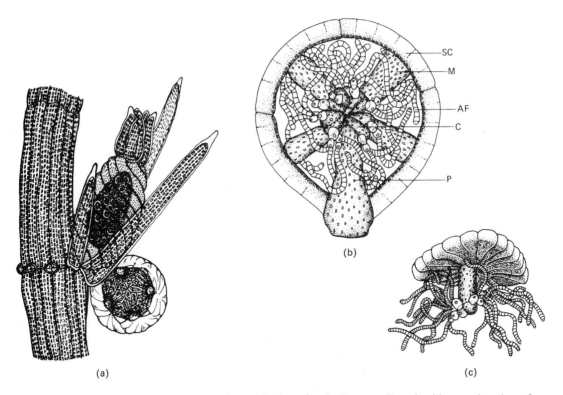

(a)

(b)

(c)

Fig. 4.4 *Chara.* (*a*) *Chara fragilis* Desvaux. Branch with oogonium (note five crown cells) and antheridium. (*b*) Median section of an antheridium (diagrammatic). (*c*) Shield cell, manubrium, capitula, and antheridial fragments. AF, antheridial filaments; C, capitulum; M, manubrium; P, pedicel or stalk; and SC, shield cells. (*a*) × 66; (*b*) × 110; (*c*) × 70. [(*a*) after Sachs; (*b*), (*c*) from *The Plant Kingdom* by William H. Brown, Copyright © 1935 by William H. Brown. Used by permission of the publisher, Ginn and Co. (Xerox Corporation).]

carotenoids. The shield cells are incompletely compartmentalized. Pickett-Heaps (1968A) has studied their development electron microscopically. The antheridium (Figs. 4.3, 4.4) is supported by a columnar stalk or **pedicel**. This bears eight cells, each of which is a **primary capitulum**, connected to the wall or shield cells by an elongate cell, the **manubrium**. The primary capitula cut off secondary (or even tertiary or additional) capitular cells, each of which develops a colorless **spermatogenous** ("antheridial") **filament** of thin-walled, boxlike cells. Each of the latter produces a single biflagellate sperm. Pickett-Heaps (1968B) and Turner (1968) have investigated spermatogenesis in *Chara* and *Nitella*, respectively. At maturity the eight shield cells separate in varying degrees, exposing the spermatogenous filaments to the water and permitting escape of the biflagellate sperm. The sperms have subapically inserted flagella (Pickett-Heaps, 1968B; Turner, 1968) (Fig. 4.5).

The oogonia (Figs. 4.3*a*, 4.4*a*) also arise from nodal cells of the branchlet; these undergo transverse divisions to form a row of cells of which the most distal differentiates into a single egg or oosphere and a certain number of small sterile cells (Sawa and Frame, 1974) and the most proximal becomes the pedicel. The intermediate

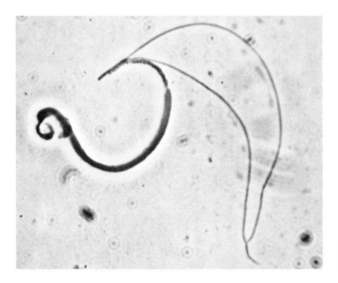

Fig. 4.5 *Nitella* sp. Mature sperm. × 2700. (After Turner.)

cell, which seemingly is a nodal cell, divides to form a central cell surrounded by five pericentral cells that comprise a sheath of tube cells. These elongate in helical fashion as the oogonium grows, and each of them cuts off one or two small cells at the apex, which results in the formation of the **corona** of five crown cells in *Chara* and *Nitellopsis* and one of 10 cells in *Nitella* and *Tolypella* (Fig. 4.6).

At maturity the helical tube cells develop intercellular slits at the level just under the corona and thus provide a pathway for sperms to enter the oogonium. The fertilized egg is filled with food storage products that give it a dense appearance. After fertilization the walls of the tubular sheath cells thicken differentially and the wall of the zygote thickens, becoming dark brown or black. The zygote of *Chara* and of all oogamous algae is sometimes called an "oospore."

Germination follows a period of dormancy. At germination (Fig. 4.7) the nucleus lies at the coronal end of the zygote where it undergoes two successive divisions that have been interpreted to represent meiosis, although the evidence is not conclusive. Sawa (1965) investigated nuclear divisions in different parts of mature *Nitella opaca*

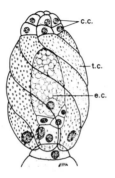

Fig. 4.6 *Nitella opaca* Ag. Living, immature oogonium. cc, crown cells; ec, egg cells; tc, tube cells. × 289. (Courtesy Prof. Takashi Sawa.)

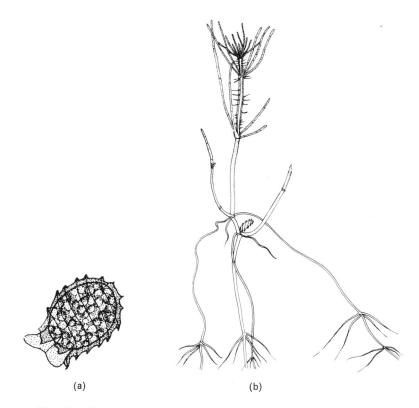

Fig. 4.7 *Chara* sp. Germination of the zygote. (*a*) Early stage. (*b*) Later stage. (*a*) × 30; (*b*) × 3. (After DeBary and N. Pringsheim.)

Ag. and *N. flexilis* (L.) Ag. for the purpose of karyotype determination and observed that all nuclear divisions are mitotic except for amitoses in the internodal cells. Shen (1966; 1967A, B, C) studied oospore germination in *Chara zeylanica* Kl. ex Willd. and measured nuclear DNA at various times in its development. He showed by micro-spectrophotometric means that the sperms contain the same amount of DNA as the somatic cells. This indicates that meiosis is not gametic, as had been suggested by one investigator.

Ross (1959), Proctor (1960, 1967), Carr and Ross (1963), Forsberg (1965), Shen (1966A, B), and Takatori and Imahori (1971) all have investigated germination of the zygote of *Chara*. Shen showed that the dormant zygotes of *Chara contraria* Kütz. and *C. zeylanica* require a period of dormancy before germination but that storing them at low temperature (5–7°C) shortened the dormancy requirement.

At germination (Fig. 4.7) the zygote wall splits at the region of the former corona and a filament, the primary protonema, with a slender basal rhizoid, emerges. The protonema is at first colorless with an apical nucleus. It undergoes cell division to form four to five cells. The second cell from the basal colorless cell undergoes divisions to form an internodal cell with a nodal cell at both ends of it. The lower nodal cell, by further divisions, gives rise to rhizoids, while the upper organizes a nodal complex in which one of the pericentral cells produces an apical cell to initiate development of the adult shoot. Adventitious protonemata and adult shoots may develop from the rhizoids and thus increasingly large colonies may form. Germination is similar to the pattern just described, except for minor differences, in all species.

Griffin and Proctor (1964), McCracken, et al., (1966), Proctor (1967; 1970; 1971A, B; 1972), Proctor and Wiman (1971), and Grant and Proctor (1972) have investigated the occurrence of reproductive isolation in various geographical races of *Chara*.

5

Division Euglenophycophyta

The relatively few chlorophyllose members of the Euglenophycophyta share with the Chlorophycophyta and Charophyta the presence of chlorophylls *a* and *b* in their chloroplasts. Other than this, they differ in many respects even though it might appear superficially that they should be classified in the same group. Leedale (1967) has prepared an excellent treatise on Euglenoids.

The Euglenophycophyta differ from the green algae in cellular organization and biochemistry. With respect to the latter, they store their excess photosynthate as **paramylon**, a β 1-3 polymer of glucose (Meeuse, in Leedale, et al., 1965) that is present in the cytoplasm and not in the chloroplasts like the α 1-4 linked polymer of glucose, the starch of the Chlorophycophyta.

Euglenoids are widely distributed, occurring in freshwater, and brackish and marine waters and also on moist soils and mud. They are often abundant and may form water blooms in ponds, tanks, and puddles, especially those to which livestock have access, and this may be correlated with the facultatively heterotrophic existence of some of them. About 800 species of Euglenoids (including both chlorophyllose and colorless species) have been described (Leedale, 1967.) Lackey (in Buetow, 1968) has discussed the ecology of *Euglena* itself.

Euglenoid cells are bounded by the plasmalemma exteriorly and just within it by a proteinaceous layer, the **pellicle**[1] (Fig. 5.4b); the latter is helical in organization and composed of overlapping pellicular strips. Euglenoid cells thus are naked except in those genera (e.g., *Trachelomonas*, Fig. 5.6) in which a lorica is present. The pellicle may be pliable or rigid, depending on the species, and the overlapping and interlocking edges of the pellicular strips are visible as helical or longitudinal markings on the cell surface. Leedale (1967), Arnott and Walne (1967), and Guttman and Ziegler (1974) have investigated the ultrastructure of the pellicle. The latter is composed of up to 80% of proteins and the remainder of lipids and carbohydrates. Some Euglenoids undergo flowing, contracting, and expanding "euglenoid movement" or **metaboly** when not swimming, the mechanism of which is not understood.

[1]Sometimes called the periplast.

Just underneath the pellicle and parallel to each of its strips are rows of mucilage-producing bodies that are in communication with the exterior of the cell by canals through which mucilage may be extruded.

Except when they are encysted or in a *Palmella* phase, Euglenoids are flagellate, having two or several flagella. When there are two, one may be nonemergent from the **anterior invagination** (Fig. 5.2), which consists of a **canal** and a **reservoir**. Euglenoid flagella are rather coarse as compared with those of Chlorophycophyta. They have the usual (9 + 2) fibrillar arrangement and in addition a **paraflagellar rod** (Fig. 5.2). The flagella emerge from basal bodies located just beneath the reservoir wall. They bear one or more rows of fine hairs. As the emergent flagellum passes from the reservoir to the canal, it may be thickened laterally by a **flagellar swelling**.

In the anterior of chlorophyllose euglenoid cells, the prominent eyespot or stigma occurs in the colorless cytoplasm, in contrast to that of many other algae in which it is located within the chloroplast. The stigma lies at about the level of the flagellar swelling. Walne and Arnott (1967), Leedale (1967), and Walne (1971), among others, have investigated the organization of euglenoid stigmata and have found that they are prominent in some species. Thus, in *Euglena granulata* they may be 7×8 μm and consist of 50–60 orange-red granules contained singly or in groups within unit membranes. Absorption spectra of concentrated eyespot granules are in the range of 360–520 μm with a peak also at 660–675 (Batra and Tolin, 1964; Bartlett, et al., 1972). Astaxanthin and/or echinenone have been reported to be pigments in the euglenoid eyespot together with other carotenoid pigments.

Beginning with Engelmann (1882), many investigators have suggested that the eyespot of euglenoids (and of other flagellate algal cells) was involved in receiving light stimuli. These organisms are positively phototactic to light of low intensity and negatively phototactic with respect to bright light and darkness. A number of investigators (Cobb, 1963) have shown that the action spectrum of phototaxis and the absorption spectrum of the eyespot pigments coincide. There are two different suggestions regarding the role of the eyespot in light perception. According to one point of view, the eyespot is the site of light perception, and it controls cellular movement as to direction and hence as to phototactic responses. According to the other interpretation, the real site of light perception is thought to be the flagellar swelling, and the eyespot is considered to function as a shading organelle for it. Benedetti and Chechucci (1975) reported that fluorescent microscopy suggests the paraflagellar body contains flavin. Pagni, et al., (1976) recently demonstrated the presence of flavins in the eyespots of *E. gracilis* var. *bacillaris*. Much has been written regarding the role of the eyespot and the phenomenon of phototaxis in euglenoids and other organisms (Bendix, 1960; Halldal, 1962, 1964; Batra and Tolin, 1964; Walne and Arnott, 1967; Leedale, 1967, 1971; Jahn and Bovee, 1968; Wolken, 1967; and Diehn, 1969A, B, 1973). Bartlett, et al., (1972) isolated into pure preparations the eyespot granules of *E. gracilis* var. *bacillaris* Pringsheim and found that the absorption spectra of the cell-free granules corresponded to those in intact cells. Later, Hilenski, et al. (1976) reported that the eyespot granules contained 6% lipids including wax esters, triacylglycerols, free fatty acids, and phospholipids.

The chloroplasts of Euglenophycophyta vary in form among the different species and genera. They may be small, simple discs; large and platelike with entire or dissected margins; or ribbonlike and arranged in stellate fashion. Pyrenoids of various types may be present within the chloroplasts and are seemingly centers for paramylon formation in some species, but paramylon also develops without relationship to the pyrenoid or plastid. Paramylon, as noted above, is a β 1-3 glucose polymer that occurs outside the chloroplast in the colorless cytoplasm. The chloroplasts contain chlorophylls a and b, β carotene, and several xanthophylls (Table 1.1, p. 18).

A contractile vacuole lies at the anterior of the euglenoid cell and discharges periodically into the reservoir. Smaller, accessory vacuoles coalesce to form new contractile vacuoles after they have discharged into the reservoir.

The nucleus is prominent in euglenoid cells (Fig. 5.4a) and often readily visible in living individuals in the center or posterior of the cell. Mitosis is largely intranuclear and characterized by a persistent nucleolus (endosome) that divides during mitosis and by the arrangement of the chromosomes with their long axes parallel to the long axis of the division figure. The chromatids separate gradually along their long axes and migrate to the poles. Microtubules are present in a spindle-like arrangement within the nucleus. Cytokinesis occurs by longitudinal cleavage of the protoplast.

Certain genera of euglenoids have the capacity to encyst and thus to withstand unfavorable environmental conditions. Cell division may occur in the swimming phase or in the palmelloid condition, and one recently described species (*E. myxocylindracea*, Bold and MacEntee, 1973) is seemingly permanently palmelloid. Reproduction in euglenoids is entirely asexual by cell division. Reports of the occurrence of meiosis in a species of *Phacus* and in a colorless euglenoid, *Hyalophacus*, require confirmation and further elucidation.

Nutrition

Very few euglenoids have been grown successfully in axenic culture, and thus we are insufficiently well-informed regarding the nutrition of the chlorophyllose forms, most investigators having studied one species, *E. gracilis* Klebs and its varieties that grow readily in the axenic state. In spite of having chlorophylls a and b, chlorophyllose euglenoids are not absolutely photoautotrophic but are rather photoauxotrophic because they require one or more vitamins. Although they can use organic compounds as nitrogen sources, there is not unequivocal evidence that they require them. None of the chlorophyllose euglenoids is phagotrophic, but some of the colorless ones are. Some of the green species are facultatively heterotrophic; that is, they are able to grow in darkness if organic carbon is provided and, indeed, their growth in light may be stimulated by such carbon compounds. Kivie and Vesk (1974) reported the pinocytotic uptake of protein from the reservoir of *Euglena*.

Strains of *Euglena gracilis* have been "bleached" by growing them at temperatures of 32–35°C (Pringsheim and Pringsheim, 1952). The flagellar swelling and eyespot may be retained. At these high temperatures cell division continues, but chloroplast division falls behind, with colorless individuals (lacking plastids) resulting. These are unable to become green again because they lack plastids. Treatment of *E. gracilis* with ultraviolet light or radiation (Lyman, et al., 1959) and with streptomycin (Provasoli, et al. 1948) also resulted in bleaching. Kronstedt and Walles (1975) bleached *E. gracilis* with the antibiotic porfiromycin and showed that colorless plastids, the stigma, and flagellar swelling persisted. Ehara, et al. (1975) have studied the degeneration of the chloroplasts electron microscopically, while Palisano and Walne (1976) have investigated two permanently bleached strains of *E. gracilis*. Leedale and Buetow (1976) reported on the ultrastructure of carbon starvation in bleached *E. gracilis*. Diamond and Schiff (1974) have obtained streptomycin-resistant mutants of *E. gracilis*. The development of colorless euglenoids from chlorophyllose precursors is thought to be of phylogenetic significance in that it might well explain the morphological similarity between such green genera as *Euglena* (below) and *Phacus* (Fig. 5.10), on the one hand, and the colorless *Astasia* and *Hyalophacus*, on the other.

Although there are a number of colorless euglenoids, only a few representatives are included in the following discussion. Leedale (1967) has prepared an account of both green and colorless euglenoids, and the works of Wolken (1967) and Buetow (1968) may be consulted for a more extensive discussion of some of the topics referred to herein.

Members of three orders of Euglenophycophyta, Class Euglenophyceae are discussed in the following account. These include *Eutreptia* of the order Eutreptiales; *Euglena, Astasia, Trachelomonas, Phacus, Hyalophacus,* and *Colacium* of the Euglenales; and *Peranema* of the Heteronematales.

Order 1. Eutreptiales

The Eutreptiales differ from the Euglenales in having two emergent flagella of equal length and in its members undergoing very active euglenoid movement. *Eutreptia* is the sole member of the order discussed here; it is a member of the family Eutreptiaceae.

EUTREPTIA Perty   *Eutreptia* (Gr. *eu*, well, true + Gr. *trepein*, to turn) (Fig. 5.1) is a *Euglena*-like organism with two long, equal, active emergent flagella. The cells are markedly variable in form due to their euglenoid movement. According to the species, the chloroplasts may be disc-like or ribbon-like, in the latter case radiating from a parmylon center (Fig. 5.1). *Eutreptia* occurs mostly in marine and brackish waters.

Order 2. Euglenales

Of the two flagella at the anterior of the cells of members of the Euglenales, only one emerges from the reservoir and canal (Fig. 4.3). The genera of Euglenales discussed below belong to the single family Euglenaceae.

EUGLENA Ehrenberg   The genus *Euglena* (Gr. *eu*, good + Gr. *glene*, eyeball) (Figs. 5.2–5.4) is a familiar and large one, approximately 152 taxa having been described (Gojdics, 1953) although a number of these are probably synonymous. Pringsheim

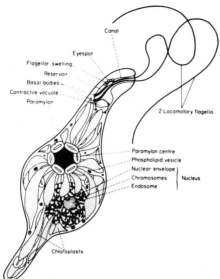

Canal
Eyespot
Flagellar swelling
Reservoir
Basal bodies
Contractile vacuole
Paramylon

2 Locomotory flagella

Paramylon centre
Phospholipid vesicle
Nuclear envelope
Chromosomes   Nucleus
Endosome

Chloroplasts

Fig. 5.1 *Eutreptia pertyi* Pringsheim. Cellular organization. × 878. (After Leedale, G. F., *Euglenoid Flagellates,* © 1967. Reproduced by permission of Prentice-Hall, Inc., Englewood Cliffs, New Jersey.)

Fig. 5.2 *Euglena gracilis* Klebs. Cellular organization. × 1000 (After Leedale, G. F., *Euglenoid Flagellates,* © 1967. Reproduced by permission of Prentice-Hall, Inc., Englewood Cliffs, New Jersey.)

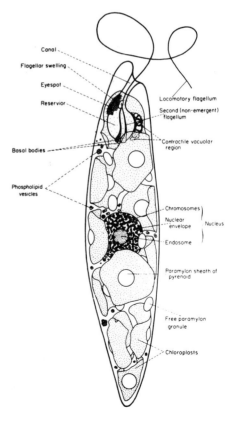

Canal
Flagellar swelling
Eyespot
Reservoir

Locomotory flagellum
Second (non-emergent) flagellum
Contractile vacuolar region

Basal bodies

Phospholipid vesicles

Chromosomes
Nuclear envelope   Nucleus
Endosome

Paramylon sheath of pyrenoid

Free paramylon granule

Chloroplasts

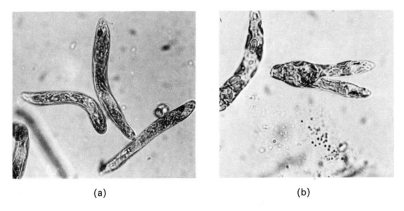

(a)   (b)

Fig. 5.3 *Euglena mesnili* Deflandre and Dusi. Living individuals. Note stigmata and one individual in division. (*a*) × 350; (*b*) × 298.

(1956) published an incomplete monograph of the genus based on his investigation of organisms in the field and in the laboratory. He discussed the reliable taxonomic criteria for delimiting the species and emphasized the indispensability of growing the organisms in unialgal culture for taxonomic determinations. He classified the various species in six subgenera based on the degree of rigidity of their pellicles and the nature of the chloroplasts and pyrenoids. *Euglena* is widely distributed in freshwater and occurs also on mud (Lackey, in Buetow, 1968).

The pellicular strips, which converge at the cell extremities, are arranged in helical fashion, their overlapping giving the appearance of helical striae on the cell surface

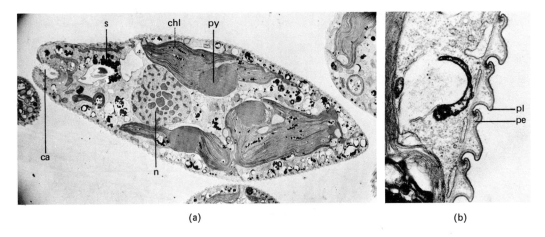

(a)   (b)

Fig. 5.4 *Euglena granulata* (Klebs) Schmitz. (*a*) Electron micrograph of thin section. (*b*) Section of cell surface. ca, canal; chl, chloroplast; n, nucleus; pe, periplast; pl, plasmalemma; py, pyrenoid; s, stigma. (*a*) × 1590; (*b*) × 16,500. [(*a*) courtesy of Prof. H. J. Arnott; (*b*) courtesy of Prof. H. J. Arnott and P. L. Walne, from Bold.]

(Fig. 5.4*b*). Some species (e.g., *E. spirogyra*) have helically arranged, wartlike ornamentations on the cell surface. In most species, the orifice of the canal is slightly lateral to the cell apex. The chloroplasts are discoid-lenticular, shield-shaped, or ribbonlike and may or may not contain pyrenoids. *Euglena oxyuris* Schmarda is the largest species in the genus, reaching dimensions of 95 to 530 μm, while *E. minuta* Prescott, $12 \times 15 \mu$m, is among the smallest. When they encyst, the cells become spherical and surrounded by a gelatinous sheath within which they may undergo movement and revolve.

Cell division (Fig. 5.3) is preceded by mitosis of the typical euglenoid type. Cytokinesis is longitudinal, beginning at the anterior end. Boasson and Gibbs (1973) reported synchronous division of the chloroplasts at cytokinesis in *E. gracilis*.

ASTASIA Dujardin   *Astasia* (Gr. *astatos*, unsteady) (Fig. 5.5) is a colorless counterpart of *Euglena* with respect to cellular form but differs in several respects. Its nutrition, of course, is saprophytic, and chloroplasts, eyespots, and the flagellar swelling of *Euglena* are absent, although paramylon grains are present. Both marine and freshwater species of *Astasia* have been described. The experimental bleaching of *Euglena* (p. 257) by high temperature and streptomycin suggested a derivation of *Astasia* from *Euglena*. *Astasia longa* Pringsheim, for example, is very similar to bleached *Euglena gracilis* (Pringsheim and Pringsheim, 1952). However, some morphological and biochemical differences have been reported (Blum, et al., 1965). The colorless races of *E. gracilis* may not lose their eyespots and flagellar swellings. Loss of the chloroplasts is permanent, and colorless strains lacking eyespots and flagellar swellings are seemingly identical to *Astasia longa*.

Fig. 5.5 *Astasia fritschii* Pringsheim and Hovasse. $\times$ 627. (After Pringsheim.)

TRACHELOMONAS Ehrenberg   In *Trachelomonas* (Gr. *trachelos*, neck + Gr. *monas*, single organism) (Figs. 5.6, 5.7), the naked cells are enclosed in a nonliving investment called the **lorica**. The latter has a short neck or collar through which the emergent flagellum passes; the lorica of the numerous species is variously ornamented, and these ornamentations have been used as taxonomic criteria, but Leedale (1967) suggested that, as in *Euglena*, other criteria of cellular organization might result in a more natural classification. The lorica is said to be pectic and impregnated with iron

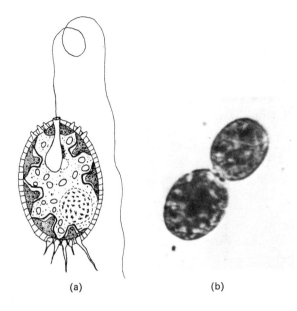

(a)   (b)

Fig. 5.6 *Trachelomonas.* (*a*) *T. armata* (Ehrbg.) Stein. Cellular organization. (*b*) *T. grandis* Singh. Cell division. (*a*) × 525; (*b*) × 280. [(*a*) courtesy of Dr. Kamala Prasad Singh.]

and/or manganese salts and is brown or yellow. Among the many species of *Trachelo-monas* occur various types of chloroplasts and pyrenoids as in *Euglena.*

After mitosis and cytokinesis, the latter frequently diagonal, one or both of the daughter protoplasts emerges through the neck of the lorica, increases there in size, and then forms a lorica of its own (Singh, 1956). The exact mechanism of the formation of a lorica and its chemical composition are imperfectly known. Furthermore, the differences between a lorica and a cell wall are by no means clear. Leedale (1975B) has reported that in lorica formation, a delicate "skin-like" layer is first formed on the surface of the naked cell. The lorica is formed between this layer and the protoplast;

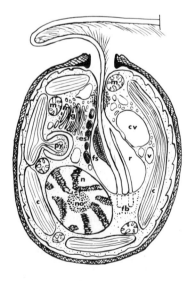

Fig. 5.7 Cellular organization of *Trachelo-monas*, diagrammatic: c, chloroplast; cv, contractile vacuole; es, eyespot; fb, flagellar base; fs, flagellar swelling; g, Golgi apparatus; m, mitochondrion; n, nucleus no, nucleolus; py, pyrenoid; r, reservoir; note mastigonemes on flagellum. × 660. (After Dodge.)

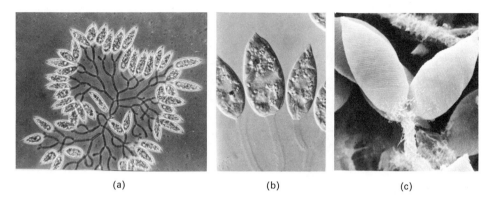

(a)   (b)   (c)

Fig. 5.8 *Colacium mucronatum* Bourrelly et Chadefaud. (*a*) Phase-contrast photomicrograph showing dichotomously branched stalks. (*b*) Nomarski photomicrograph showing stalked cells; the largest cell has just undergone nuclear division. (*c*) SEM micrographs showing surface of cells and stalks. (*a*) × 109; (*b*) × 270; (*c*) × 1000. (Courtesy of Prof. J. Rosowski.)

the "skin-like" outer layer ultimately disappears. Dodge (1975C), too, has reported on the ultrastructural organization of *Trachelomonas*.

COLACIUM Ehrenberg   The cells of *Colacium* (Gr. *kolak*, parasite) (Figs. 5.8, 5.9) are usually sessile and attached to the substrate by an anterior, mucilaginous stalk that is striate and ridged. Rosowski and Krugens (1973) report that *Colacium* species grow on a variety of aquatic animals including the protozoan *Vorticella* and on rotifers and copepods. Rosowski and Willey (1975) described a new species, *C. libellae*, that lives in the rectums of larval damselflies. The free-swimming-stage organism has one

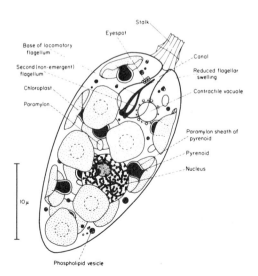

Stalk
Eyespot
Base of locomotory flagellum
Second (non-emergent) flagellum
Chloroplast
Paramylon
Canal
Reduced flagellar swelling
Contractile vacuole
Paramylon sheath of pyrenoid
Pyrenoid
Nucleus
10 μ
Phospholipid vesicle

Fig. 5.9 *Colacium mucronatum*. Attached cell, lateral aspect. (After Leedale, G. F., *Euglenoid Flagellates*, © 1967. Reproduced by permission of Prentice-Hall, Inc., Englewood Cliffs, New Jersey.)

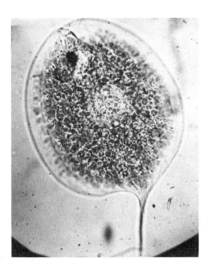

Fig. 5.10 *Phacus pleuronectes* (O.F.M.) Duj. Living individual. Note canal, central nucleus, and stigma. (Flagellum not visible.) × 1800. (After Bold.)

emergent flagellum and attaches itself to the substrate at the anterior pole and secretes a stalk. Cell division results in the formation of branching colonies (Fig. 5.9*a*). Individual cells may be freed from the colony, undergo euglenoid movement, and then swim by means of the emergent flagellum. In the sessile cells, both flagella are nonemergent and reported to beat within the reservoir. At cell division each of the cellular progeny secretes a new stalk that is attached to the original one, and thus dendroid colonies develop. Rosowski and Willey (1977) have described in detail the stalks of *C. mucronatum* Bourr. and Chad.

In addition to their habitat on aquatic animals, species of *Colacium* have been found on mud and in freshwater attached to aquatic plants. *Colacium* is sometimes classified in a separate family from the other euglenoids discussed above because of its sessile habit.

PHACUS Dujardin   The cells of *Phacus* (Gr. *phakos*, lentil) (Fig. 5.10) are free-swimming, more or less flattened, and rigid. Most species grow in freshwater but a few occur in marine waters. The cells are marked with longitudinal or helical striae and the cells of many species terminate in a spine. All have small, discoid chloroplasts lacking pyrenoids. One long flagellum is emergent through the canal, while, as in *Euglena*, a second shorter flagellum is nonemergent. Dynesius and Walne (1975) have described the ultrastructure of the reservoir and flagella in *Phacus*; the flagella have hairs.

Krichenbauer (1937) reported the union of nuclei in single individuals followed by meiosis, but this requires confirmation.

HYALOPHACUS Pringsheim   *Hyalophacus* (Gr. *hyalinos*, glassy, clear + Gr. *phakos*, lentil) (Fig. 5.11) is similar to *Phacus* but lacks chloroplasts. Two species are known, one with eyespot and flagellar swelling and the other lacking them (Leedale, 1967). *Hyalophacus* may well represent a *Phacus* that, like *Euglena gracilis* upon special treatment, has lost its chloroplasts.

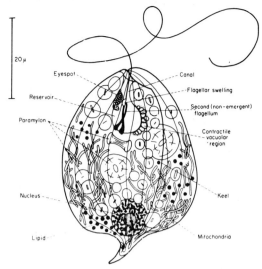

Fig. 5.11 *Hyalophacus ocellatus* Pringsheim. Living cell. × 660 (After Leedale, G. F., *Euglenoid Flagellates*, © 1967. Reproduced by permission of Prentice-Hall, Inc., Englewood Cliffs, New Jersey.)

Order 3. Heteronematales

The members of the Heteronematales are colorless and phagotrophic. They lack both eyespots and flagellar swellings. Although they have euglenoid characteristics, they are decidedly different from the chlorophyllose genera in nutrition and have special organelles for the ingestion of particles and other organisms. *Peranema* of the family Heteronemataceae is the only representative of the order included in the present volume.

PERANEMA Dujardin   The colorless cells of *Peranema* (Gr. *pera*, beyond + Gr. *nema*, thread) (Fig. 5.12) are *Euglena*-like in form and contain abundant paramylon.

Fig. 5.12 *Peranema trichophorum* (Ehr.) Stein. × 660 (After Leedale, G. F., *Euglenoid Flagellates*, © 1967. Reproduced by permission of Prentice-Hall, Inc., Englewood Cliffs, New Jersey.)

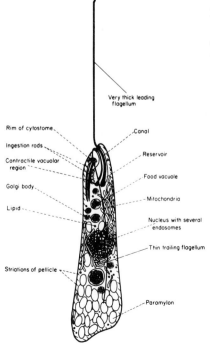

Peranema trichophorum (Ehrenb.) Stein is biflagellate, like *Eutreptia*. However, one flagellum, the more slender of the two, is directed posteriorly and appressed to the cell body. The anteriorly directed flagellum is especially actively motile near its tip. Lateral to the apex of the cell and opening of the canal, a special organelle of ingestion, which contains ingestion rods, is present. The rods may be temporarily protruded, become attached to the prey (bacteria, algae, yeasts, and *Euglena*), and then, with the latter, pulled into the cell body where the prey is digested in food vacuoles.

In summary, the Euglenophycophyta, although some of its members have chlorophyll *a* and *b* like the Chlorophycophyta, differ in many other respects. Thus, they store paramylon instead of starch; they have proteinaceous pellicles, canals and reservoirs, and stigmata; and they have stored food (paramylon) lying free in the cytoplasm. They seemingly have not evolved beyond a flagellate level of organization, except perhaps for *Colacium*.

The *Euglenoid* group contains a heterogeneous assemblage of chlorophyllose and colorless genera. Some of the colorless type like *Astasia* and *Hyalophacus* seem clearly to have been derived from chlorophyllose predecessors; but others, like *Peranema*, with their phagotrophic nutrition and special ingestion apparatus, seem to be less closely related.

6

Division Phaeophycophyta

General Features

The division Phaeophycophyta, the brown algae, constitutes an important assemblage of plants, classified in more than 250 genera and over 1500 species. They are almost exclusively of marine occurrence; only about 5 genera have been recorded[1] from freshwater. On temperate rocky shores in both northern and southern hemispheres brown algae are a conspicuous intertidal component, extending from the upper littoral zone into the sublittoral zone, sometimes to depths of 220 m in clear tropical waters.[2] They seem to flourish in temperate to subpolar regions, exhibiting the greatest diversity in regard to species and to morphological expressions in these colder ocean waters. They predominate in the lower littoral to upper sublittoral zones, and many forms, particularly members of the order Fucales, thrive in a regime of long periods of daily emersion during periods of low tide.

A combination of physiological and structural features ties the brown algae into a distinct taxonomic grouping. Any one criterion taken separately might be observed in other algal groups, but the combination of these traits is the distinction that enables us to recognize them as a separate division. The photosynthetic pigments include chlorophylls *a* and *c*, *β* carotene, violaxanthin, and fucoxanthin (Goodwin, 1974), with occasional traces of diatoxanthin and diadinoxanthin. The active participation of fucoxanthin as an accessory pigment in photosynthesis was demonstrated (Haxo and Blinks, 1950) by the marked activity in the region 500–540 nm, in which range fucoxanthin has absorptive properties.

[1]E.g., see reports of *Heribaudiella* (Smith, 1950) and *Sphacelaria* (Thompson, 1975) both in the United States and *Pseudobodanella* (Gerloff, 1967) and *Lithoderma* (Bourrelly, 1968A) both in Europe. *Pleurocladia lacustris* A. Br. is a euryhaline species, inhabiting both freshwater and marine habitats (Waern, 1952; Wilce, 1966).

[2]Attached plants of the Dictyotalean alga *Lobophora variegata* (Lamour.) Womers. were gathered in 1968 by divers who emerged from the submersible "Deep Diver" at this depth near the "Tongue of the Ocean" in the Bahamas (Earle, personal communication).

The food reserve is known as laminaran, which is a soluble polysaccharide reserve composed primarily of β, 1–3–linked glucans, with a variable degree of β, 1–6–linkages (Craigie, 1974), chemically similar to the reserves in the Chrysophycophyta and Euglenophycophyta. The laminaran content ranges from less than 2 to 34% of the algal dry weight (Powell and Meeuse, 1964). Mannitol is also of universal occurrence in the Phaeophycophyta; sucrose and glycerol have also been reported.

Cell walls are composed of an inner cellulosic layer and an outer slimy or gummy layer. Alginic acid, a polymer of 5-carbon acids (D-mannuronic and L-guluronic acid), is localized both in the cell walls and in the intercellular spaces (Evans and Holligan, 1972). The alginates (salts of alginic acid) are reported (Mackie and Preston, 1974) to have both structural and ion-exchange roles. The proportion of these two uronic acids can be quite variable, with reports (Haug, et al. 1969) of the mannuronic content ranging from about 30% [in the cortex of *Laminaria hyperborea* (Gunn.) Fosl.] to 97% (in the intercellular fluid of *Ascophyllum*). The alginic acid may constitute up to 24% of the dry weight of the alga, and alginates are being used extensively in a variety of commercial purposes because of the emulsifying or stabilizing properties. These are discussed in Chapter 1. In addition to alginic acid, sulfated polysaccharides are present in water-soluble extracts of brown algae, and the term *fucoidan* has been applied. L-Fucose (2-deoxy-L-mannose) is apparently the main component of fucoidan, but acid hydrolysis of fucoidan also results in various amounts of D-xylose, D-galactose, and uronic acid (Mackie and Preston, 1974).

Brown algae range in size from microscopic epiphytes to the largest of marine plants, namely, *Macrocystis* (p. 335), which reaches 60 m or more in length (Womersley, 1954). Structurally, the simplest brown algal forms are composed of erect, branched, or unbranched filaments arising from a prostrate, filamentous basal system, i.e., with a heterotrichous organization. Representative of these are members of the order Ectocarpales and of the family Myrionemataceae in the order Chordariales. Other **haplostichous** forms include loose collections of branched filaments held together by mucilage in rather amorphous masses (e.g., *Leathesia*, p. 288); multiaxial, branching systems (e.g., *Cladosiphon*, p. 291); or compactly arranged, pseudoparenchymatous aggregations of filaments into prostrate crusts (e.g., *Ralfsia*, p. 283); erect, branched axes from a prostrate crust (e.g., *Analipus*, p. 285); or blades (e.g., *Desmarestia*, p. 293). The parenchymatous condition is characteristic of several orders. These **polystichous** forms, derived from the ability of cells to divide in various planes to effect a true tissue, may be terete, solid axes, branched (e.g., *Stictyosiphon*, p. 309) or unbranched (e.g., *Myriotrichia*, p. 308), solid blades (e.g., *Punctaria*, p. 310), or more massively constructed thalli with differentiation of cells into tissues (orders Laminariales and Fucales).

Growth may occur by various methods in the brown algae. **Diffuse growth** is seen in most Ectocarpales and many Chordariales. **Trichothallic growth**, in which cell divisions are localized at the base of one or several filaments, is expressed in the Desmarestiales, Cutleriales, and some Chordariales. The orders Sphacelariales, Dictyotales, and Fucales have **apical growth**, with a single apical cell, groups of apical cells, or marginal apical cells cutting off segments proximally. An intercalary meristem is present in the Laminariales. In addition, a **meristoderm**, which consists of a layer of

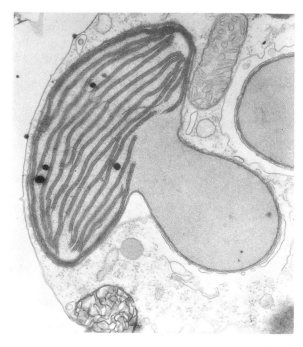

Fig. 6.1 *Pilayella littoralis* (L.) Kjellm. Section through a swimming zoospore, showing chloroplast with projecting pyrenoid, the latter surrounded by a transparent cap of metabolite. Note occurrence of thylakoids arranged in groups of three within the chloroplast. × 16,650. (After L. V. Evans in J. Cell Science 1: 449–454, 1966.)

superficial meristematic cells which can divide and add cells centripetally to the thallus, is characteristic of Laminariales and Fucales.

Chloroplasts in the brown algae (Fig. 6.1) may be present singly, a few, or many per cell, and their number is a dependable taxonomic criterion. They may be discoid, platelike, or ramified; in *Bachelotia* (of the Ectocarpales) they are arranged in a distinctive stellate configuration (Fig. 6.2a). In the Phaeophyceae examined, photosynthetic lamellae are comprised of groups of three thylakoids with some interconnections

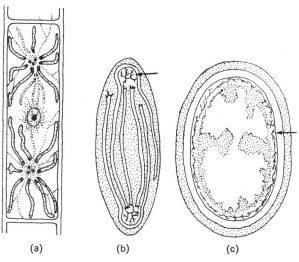

(a)   (b)   (c)

Fig. 6.2 (a) *Bachelotia antillarum* (Grun.) Gerloff. Stellate configuration of chloroplasts; (b), (c) *Sphacelaria*: diagrammatic representations of sectioned chloroplast, showing the ringlike arrangement of the genophore (arrow) in vertical (b) and horizontal section. (a) × 726. [(a) after Blomquist; (b) and (c) after Bisalputra and Bisalputra.]

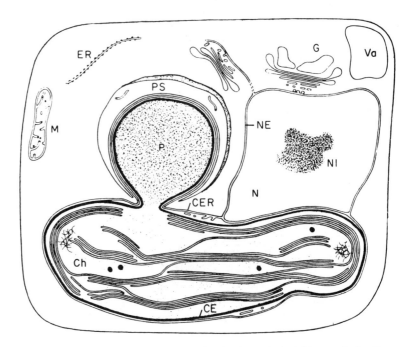

Fig. 6.3 Diagrammatic representation of a hypothetical brown algal cell to illustrate the relationships of the organelles and various membrane associations. The chloroplast endoplasmic reticulum (CER) is derived from the outer membrane of the nuclear envelope and surrounds the chloroplast (C) with its projecting pyrenoid (P). (Permission of G. B. Bouck and Rockefeller University Press for J. Cell Biol. 26: 523–537, 1965.)

running between lamellae (Dodge, 1973). **Girdle lamellae** are present, lying just within the chloroplast envelope. A DNA zone, or **genophore**, has been clearly demonstrated to be (Bisalputra and Bisalputra, 1969) located just beneath the girdle lamella in *Sphacelaria* (Fig. 6.2*b*, *c*).

A **chloroplast endoplasmic reticulum** (CER) has been noted (Gibbs, 1962C; Bouck, 1965) in the brown algae, as for several other algal classes[3] (Dodge, 1973). The CER is an evagination of the outer nuclear membrane, extending to encompass the chloroplast(s) and also the pyrenoid(s), if present (Fig. 6.3).

Pyrenoids in the brown algae are of the single-stalked type (Fig. 6.3) (Dodge, 1973) in that they project outward from the chloroplast but yet are continuous with the chloroplast matrix. Usually a close-lying sheath of polysaccharide reserve surrounds the emergent pyrenoid. Some phylogenetic significance has been attached to the presence or absence of pyrenoids (Evans, 1966; Hori, 1971, 1972A), their apparent absence being taken as a trait associated with the more advanced orders, *viz.*, Dictyotales, Sphacelariales, Laminariales, and Fucales. However, the detection of their

[3]These include the Xanthophyceae, Chrysophyceae, Bacillariophyceae, Prymnesiophyceae, Cryptophyceae, and some Eustigmatophyceae.

presence, albeit rudimentary, in the eggs (but not in the vegetative cells) of some Fucalean genera (Evans, 1968), their poorly developed presence in *Sphacelaria* (Hori, 1972A), and their presence in members of Dictyotales and Laminariales (Chi, 1971) would effectively lessen the phyletic value of this cytological feature.

Reproduction

Essentially, two types of reproductive structures can be encountered in the brown algae. One type consists of a multicellular or **plurilocular** organ (Fig. 6.4*a*), each cell of which produces a single motile cell. The entire structure is derived by mitotic divisions, and the label **mitosporangium** is sometimes applied. This structure can function as a gametangium producing haploid sexual cells when it is present on a haploid individual, or it can function as a sporangium producing diploid asexual cells when present on a diploid individual. Parthenogenetic development of unfused gametes may also occur.

The second type of reproductive structure, the **unilocular sporangium** (Fig. 6.4*b*), is a single cell, usually spherical and enlarged. It is the usual site of meiosis, thus a **meiosporangium**, although many examples of **apomeiosis** have been reported. Synaptonemal complexes, which constitute evidence of meiotic crossing-over (Heywood and Magee, 1976), have been observed (Toth and Markey, 1973) in the unilocular sporangia of *Pilayella* (p. 281) and *Chorda* (p. 323). Following meiosis the products may be

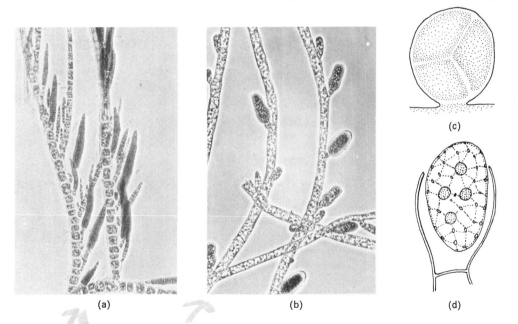

(a)   (b)   (d)

Fig. 6.4 Reproductive structures. (*a*), (*b*) *Ectocarpus*. (*a*) Plurilocular organs. (*b*) Unilocular organs. (*c*) *Dictyota*: unilocular sporangium containing four spores. (*d*) *Tilopteris*: quadrinucleate monosporangium. (*a*), (*b*) × 73; (*c*) × 113; (*d*) × 188. [(*d*) based on Reinke.]

released as nonmotile spores or "tetraspores" (in most Dictyotales), or the haploid nuclei may undergo mitotic divisions to produce a large number of nuclei, the protoplasm becoming cleaved into 16, 32, 64, 128, or more haploid motile cells, which are then released.

All brown algae release motile cells at some time in their life history, either as gametes or zoospores. The typical motile cell (Fig. 6.5a) is somewhat asymmetric in shape, slightly reniform, with two dimorphic flagella inserted laterally or subapically. The forward-directed flagellum is longer and is of the tinsel type, i.e., covered with hairlike appendages called **mastigonemes**. The basally directed flagellum is relatively short and smooth. Some exceptions to this generalization exist. The sperm of the Dictyotales is uniflagellate (Fig. 6.5b), with a single tinsel flagellum pointed in an anterior direction. Sperm in the Fucales have a shorter anterior flagellum and a longer posterior flagellum. One other distinction about their sperm (Fig. 6.5c) is the presence of a proboscis-like process projecting from the forward end of the cell (in *Fucus*, p. 350) or a variety of barbs or appendages on the tinsel flagellum (in *Himanthalia*). The proboscis in sperm of *Fucus* is considered (Manton and Clarke, 1956) to facilitate attachment to the large egg.

Multicellular **propagules** are accessory means of asexual reproduction in *Sphacelaria* (p. 298). Spores are produced within the unilocular sporangia of members of the Dictyotales, with usually four or eight such spores formed per sporangium (Fig. 6.4c). A different type of spore is observed in the Tilopteridales, where a single, large, quadrinucleate protoplast is released per sporangium, and it is referred to as a **monospore** (Fig. 6.4d).

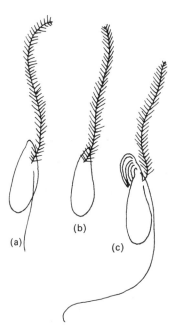

Fig. 6.5 Motile cells in the brown algae. (*a*) Typical zoospore. (*b*) Sperm of Dictyotales. (*c*) Sperm of *Fucus*.

(a)   (b)   (c)

Classification and Key to the Orders

The bases for categorizing the brown algae have been somewhat modified in recent years, although the fundamental criteria remain. Within the Phaeophycophyta usually a single class is recognized (Papenfuss, 1955), the Phaeophyceae, although as many as five other classes have been included (Maekawa, 1960). On the other hand, the brown algae have also in recent years been regarded as a class within the division Chromophyta (Christensen, 1962, 1964) or the division Chrysophyta (Dodge, 1974A), this latter division being a more restricted grouping than Christensen's Chromophyta.

Although the three classes based upon life history introduced by Kylin (Isogeneratae, Heterogeneratae, and Cyclosporeae) are generally no longer maintained (*cf.* reasons offered by Papenfuss, 1951B), the pattern of the life history remains a significant ordinal criterion. Kylin's Cyclosporeae has been reinstated as the subclass Cyclosporidae (Scagel, 1966) on the basis of the different flagellation in the Fucales (and presumably Durvilleales), the other orders of brown algae being placed in the subclass Phaeophycidae.

The characteristics that are of greatest utility in distinguishing the orders of brown algae include the pattern of life history, type of sexuality, manner of growth, construction (filamentous versus parenchymatous), and (to a lesser extent) the presence or absence of pyrenoids with the chloroplasts. The following key employs these traits and others:

1. Thalli haplobiontic (with a single free-living phase, the diploid phase) ....*Fucales* (p. 343)
1. Thalli diplobiontic (with alternate forms of free-living gametophyte and sporophyte) 2
2. Thallus of haplostichous construction (i.e., filamentous or pseudoparenchymatous organization 3
2. Thallus of polystichous construction (i.e., parenchymatous organization) at least in one phase of the life history..................7
3. Life history essentially isomorphic (may be slightly heteromorphic) 4
3. Life history heteromorphic (with a relatively small gametophyte and a relatively larger sporophyte) 5
4. Germlings with a loosely branching, creeping development; multiple chloroplasts per cell and pyrenoids present ..Ectocarpales (p. 274)
4. Germlings with a discoid-type development (i.e., closely coherent filaments); single chloroplast per cell and lacking pyrenoids ....Ralfsiales (p. 282)
5. Sexual reproduction isogamous ....Chordariales (p. 286)
5. Sexual reproduction oogamous 6
6. Growth trichothallic with a single filament terminating each axis (i.e., uniaxial organization).................Desmarestiales (p. 292)
6. Growth trichothallic, with conspicuous tuft of filaments terminating each axis and with a meristematic zone at the base of each filament ....Sporochnales (p. 292)
7. Life history of isomorphic phases (except for *Cutleria*, whose heteromorphic life history represents a derived condition); both phases of parenchymatous organization (in Tilopteridales this condition is seen in the lower polysiphonous portions) 8

Order 1. Ectocarpales

It is generally accepted that the Ectocarpales contain the most primitive forms of brown algae in regard to construction, sexuality, and life history. The members are little specialized, consisting of uniseriate branched filaments with a heterotrichous organization. Growth is typically diffuse; however, the genus *Feldmannia* is distinguished by localized zones of growth. The habit of these algae is that of loose tufts of filaments, ranging from less than 1 to 10 cm in height. Some species reach 25 cm (Cardinal, 1964). Sometimes the alga is simply an inconspicuous endophytic filament of microscopic size.

The classical interpretation of the life history (Papenfuss, 1935) is that of an alternation of isomorphic generations. This concept has been subjected to modifications in recent years, and several instances of somewhat divergent morphologies between gametophyte and sporophyte have been described for species in several genera (in *Ectocarpus* by Kornmann, 1956A, and Müller, 1972A; in *Giffordia* by Kornmann, 1954; and in *Feldmannia* by Kornmann, 1953). The difference in size between gametophyte and sporophyte, however, does not approach the strikingly dissimilar nature of these two phases occurring in the Chordariales.

There has been some controversy over whether members of the Ectocarpaceae such as *Ectocarpus* and *Giffordia* might have stages in their life history representative of the family Myrionemataceae (p. 286) in the Chordariales. The alleged occurrence of these myrionematoid stages (Baker and Evans, 1971) has been refuted by Clayton 1972) and Müller (1972A), who regard such stages as contaminants. These myrionematoid stages are not to be equated with the rhizoidal systems expressed when axenic cultures of *Ectocarpus* are grown on defined medium lacking kinetin. Pedersen (1973) discovered that the production of the erect axes occurs when kinetin above a critical level is added to the medium. Less than half of the genera of the Ectocarpaceae are included in the present account. They may be distinguished by the following key:

1. Reproductive organs (both unilocular and plurilocular) with an
   intercalary, catenate arrangement ....*Pilayella*
1. Reproductive organs laterally or terminally located but
   not in an intercalary arrangement.....................................2
   2. Few chloroplasts per cell, ribbon-shaped or
      irregularly lobed ....*Ectocarpus*
   2. Several to many chloroplasts per cell, discoid 3
3. Thallus endophytic, not evident or forming small spots;
   colorless hairs often present ....*Streblonema*
3. Thalli basically epiphytic or free-living; usually macroscopic;
   colorless hairs absent 4
   4. Localized zones of cell division, with no branching occurring
      distal to these zones; reproductive organs usually pedicellate
      and not in secund series ....*Feldmannia*
   4. Cell divisions diffuse throughout thallus; reproductive organs
      sometimes occurring in secund series and often sessile
      on the axes ....*Giffordia*

ECTOCARPUS Lyngbye   Thalli of *Ectocarpus* (Gr. *ektos*, external + Gr. *karpos*, fruit) (Fig. 6.6) are epiphytic or epilithic tufts of branched uniseriate filaments, of variable size, and with cells containing relatively few branching, ribbon-shaped chloroplasts (Fig. 6.7). Although the genus formerly contained a very large number of species, recent workers (Hamel, 1939; Kuckuck and Kornmann, 1955, 1956; Cardinal, 1964; Russell, 1966, 1967) have restricted the definition of *Ectocarpus* to include relatively fewer species. Nevertheless, the genus has a widespread distribution throughout the oceans of the world, and one might expect to encounter it on almost any shore.

Ectocarpus siliculosus (Dillw.) Lyngb. has been studied by several investigators over the years. In the early 1880's Berthold observed this species in the vicinity of Naples and confirmed that the plurilocular organs (Fig. 6.4*a*) functioned as gametangia, since the swarmers from these organs fused as gametes; he also was able to see the fusion of the gametic nuclei. If gametes did not fuse, they were capable of parthenogenetic germination. Knight (1929), studying this same species at the Isle of Man in the Irish Sea, contributed another fact about the life history, namely, that plurilocular organs were present on both haploid and diploid plants. She also studied cytologically

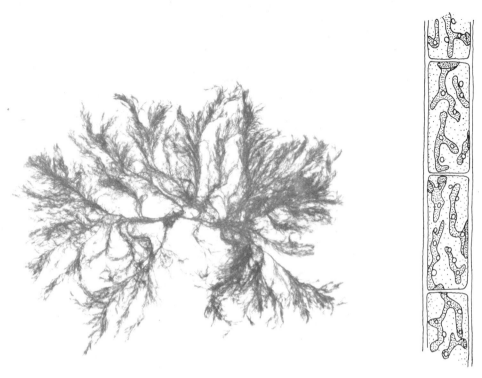

Fig. 6.6 (*Left*) *Ectocarpus*. Habit. × 1.2.

Fig. 6.7 (*Right*) *Ectocarpus*. Vegetative cell containing ribbon-shaped chloroplasts. × 264.

the unilocular sporangia (Fig. 6.4*b*) and recognized that they were the site of meiosis.

Next, Papenfuss (1935), working with material from Cape Cod, Massachusetts, established that the life history was of the D^i, h + d type (Fig. 6.8). Some thalli, epiphytic on the brown alga *Chordaria* (p. 290), bore plurilocular organs only, while other thalli were epiphytic on the brown alga *Chorda* (p. 323) bore both unilocular and plurilocular organs. The thalli on *Chordaria* proved to be dioecious gametophytes,

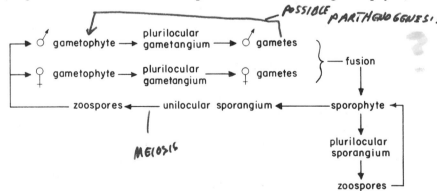

Fig. 6.8 Pattern of life history in *Ectocarpus siliculosus* (Dillw.) Lyngb.

with physiologically anisogamous motile cells being produced. The female gametes settled down first after a brief period of motility and attracted usually a single male gamete but at times several male gametes (Fig. 6.9). The actual fusion of the gametes took place rapidly (less than 20 seconds). About 5% of the female gametes were observed to germinate parthenogenetically. Thalli epiphytic on *Chorda* turned out to be the diploid phase, bearing both plurilocular and unilocular organs on the same individual or one type only. The zoospores from the plurilocular organs germinated directly into other asexual plants, whereas the zoospores from the unilocular organs germinated into haploid sexual plants. A slight difference in the overall stature of the plants, the cell size, and the size of the plurilocular organs was noted by Papenfuss between haploid and diploid phases, with the diploid phase being the more robust. The finding by Knight (1931) that zoids from unilocular sporangia can fuse has recently been refuted (Müller, 1975). Occasional "twin" or aggregated zoids are the result of incomplete separation during sporogenesis.

This traditional concept of the life history of *Ectocarpus siliculosus* was altered by Müller (1972B), who studied a clone originating from Naples. The gametophytic and sporophytic phases could be morphologically distinguished, and Müller reported that gametophytes could be haploid or diploid and sporophytes could be haploid, diploid, or tetraploid. The haploid sporophytes arose from the parthenogenetic germination of unfused male or female gametes; their unilocular sporangia were apomeiotic. A process of spontaneous multiplication of chromosome number also occurred, zoids from the plurilocular sporangia of haploid sporophytes occasionally developing into tetraploid sporophytes. Zoids from the unilocular sporangia of these latter plants yield both diploid sporophytes and diploid gametophytes. Müller's interpretation of the various interconnected pathways possible in this species is obviously much more complicated than the scheme presented in Fig. 6.8.

The gametes of *Ectocarpus siliculosus* have been demonstrated (Müller, 1967A) to be isogamous and are produced from dioecious gametophytes, but behaviorally they are anisogamous, the female gametes settling on the substrate first. A pleasant, sweet odor is noticeable from a suspension of female gametes. It has been shown (Müller, 1967B, 1968) to be a volatile substance, which modifies the swimming behavior of the male gametes, attracting them to the settled female gametes, with which they

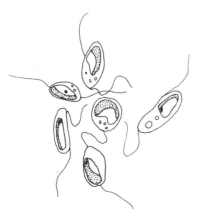

Fig. 6.9 *Ectocarpus siliculosus*. Cluster of male gametes around a settled female gamete. × 1254. (After D. Müller.)

fuse. This substance was the first "erotactin" to be chemically identified (Müller, et al., 1971) in the algae and was later designated (Machlis, 1972) **ectocarpin**.

Ectocarpus has been isolated into axenic culture (Boalch, 1961A, 1961B). Such pure cultures have been analyzed and determined to have requirements both for iodine (Pedersen, 1969; Woolery and Lewin, 1973) and kinetin (Pedersen, 1968). *ALGAE FROM*

The ultrastructure of *Ectocarpus* has been reported in regard to vegetative cells (Oliveira and Bisalputra, 1973) and unilocular and plurilocular organs (Baker and Evans, 1973A, 1973B; Lofthouse and Capon, 1975).

GIFFORDIA Batters   One of the original reasons for the recognition of *Giffordia* (named for Miss *I. Gifford*) was the alleged occurrence of small-loculed (male?) and large-loculed (female?) plurilocular gametangia, with a third type of plurilocular organ with locules of intermediate size. This alga typically occurs as small to large tufts, made up of much branched uniseriate filaments, whose cells contain a large number of discoid chloroplasts (Fig. 6.10a). Erect axes may be anchored to the substrate by a mass of rhizoids issuing from basal cells (Fig. 6.10b) or there may be a

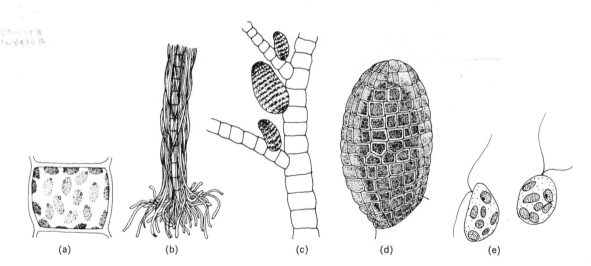

(a)   (b)   (c)   (d)   (e)

Fig. 6.10   *Giffordia.* (*a*) Vegetative cell with many chloroplasts. (*b*) Base of axis with rhizoidal investment. (*c*) Sessile attachment of the plurilocular organs. (*d*) Plurilocular organ. (*e*) Zoospores. (*a*) × 264; (*b*) × 33; (*c*) × 66; (*d*) × 264; (*e*) × 1056. [(*a*), (*b*), (*c*), and (*d*) after Kuckuck-Kornmann; (*e*) after Clayton.]

heterotrichous organization with prostrate filaments. The plurilocular organs are usually sessile (Fig. 6.10c) and, depending on the species, may be cylindrical or asymmetric in shape (Fig. 6.10d). The plurilocular organs and unilocular organs in some species occur in close, secund series (Fig. 6.11). In most species there is no evidence of sexuality (Edwards, 1969A; Cabrera, 1970; Clayton, 1974). Plurispores (the zoids released from plurilocular organs) have an eyespot and several chloroplasts (Fig. 6.10e).

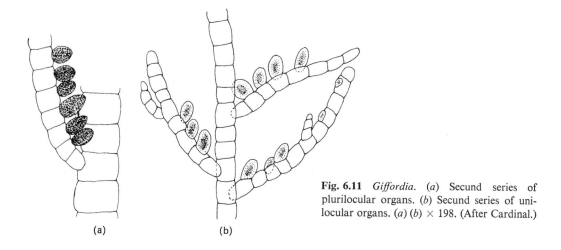

Fig. 6.11 *Giffordia.* (*a*) Secund series of plurilocular organs. (*b*) Secund series of unilocular organs. (*a*) (*b*) × 198. (After Cardinal.)

(a)   (b)

FELDMANNIA Hamel Hamel (1939) segregated *Feldmannia* (named for the French phycologist Professor *Jean Feldmann*) from *Giffordia* because of several differences, including the smaller size of the tufts, the greater degree of branching near the base, the pedicellate and nonseriate nature of the reproductive organs, and the localized zones of cell division (Fig. 6.12*a*) above which branching does not occur. Cells are short in the meristem region (Fig. 6.12*b*) and elongate above and below this zone

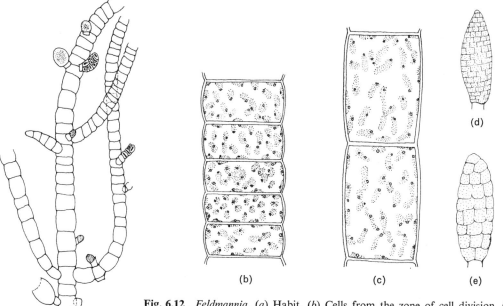

(a)   (b)   (c)   (d)   (e)

Fig. 6.12 *Feldmannia.* (*a*) Habit. (*b*) Cells from the zone of cell division. (*c*) Cells above the region of cell division. (*d*) Microsporangium. (*e*) Megasporangium. (*a*) × 66; (*b*)–(*e*) × 264. [(*a*), (*b*), and (*c*) after Kuckuck-Kornmann; (*d*) and (*e*) after Cardinal.]

(Fig. 6.12c). Plants typically are epiphytic, but the base may become endophytic, as in the case of *F. padinae* (Buffham) Hamel growing on the loose tissue of its host *Codium*. *Feldmannia* shares with *Giffordia* the characteristic of having cells with a large number of discoid chloroplasts.

For some species three different types of plurilocular organs have been reported (Cardinal, 1964): **microsporangia** or antheridia (Fig. 6.12d), **megasporangia** (Fig. 6.12e), and **meiosporangia**, this last category being those with locules of intermediate size. However, sexuality has apparently not yet been observed. A cycle has been deduced by Kornmann (1953) involving *Feldmannia* as the apparent gametophyte and *Acinetospora*[4] as the apparent sporophyte. Subsequent culturing studies by Knoepffler-Peguy (1974) showed that a *Feldmannia–Acinetospora* relationship did exist but was independent of an alternation of nuclear phases.

Like the two previous genera, *Feldmannia* is of widespread occurrence, and there is a great degree of morphological plasticity expressed in these algae. Recent investigators (Knoepffler-Peguy, 1970; Clayton, 1974), however, would argue that these genera have integrity and that generic boundaries can be drawn, in contrast to Ravanko's (1970) conclusion from her culturing studies that the taxonomic criteria used were inconstant and that generic differences represented merely developmental or environmental stages.

STREBLONEMA Derb et Sol.   The primarily endophytic habit of *Streblonema* (Gr. *streblos*, twisted; Gr. *nema*, thread) is what distinguishes it from the other Ectocarpaceaen genera being discussed. Its filaments may be undetectable to the unaided eye, with just the reproductive organs emerging from the host tissue. Or they may produce more conspicuous spots or discolorations, particularly when on a red or green algal host. In some species plurilocular organs (Fig. 6.13a) are multiseriate, resembling those of *Ectocarpus*, whereas in other species they are uniseriate. True hairs are frequently encountered in this genus. These structures (termed "true" or "phaeophycean" hairs) are multicellular colorless filaments with a basal meristem and often with a basal collar (Fig. 6.13b).

It is reasonable to believe that some algae described as species of *Streblonema* are in reality microscopic (= adelophycean) stages of members of the Chordariales (p. 286) and Dictyosiphonales (p. 308), whose macroscopic (delophycean) stages often alternate with such minute phases in the course of their life history. This fact is attested to by the discovery by Loiseaux (1970A) that a Pacific coast species of *Streblonema* produces in culture a small *Scytosiphon* (p. 315).

Although some authors (e.g., Hamel, 1939) would place *Streblonema* in the family Myrionemataceae of the Chordariales on account of its terminal or subterminal growth, most recent authors (e.g., Parke and Dixon, 1964; Kuckuck-Kornmann, 1954) continue to regard this genus as belonging to the Ectocarpaceae.

---

[4]*Acinetospora* was thought to be a distinct genus on the basis of the production of short laterals or hooklike branches above the localized zones of cell division, but its status is now doubtful (Knoepffler-Peguy, 1974).

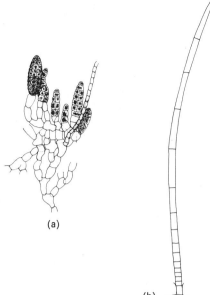

(a)

(b)

Fig. 6.13 *Streblonema.* (*a*) Portion of thallus with endophytic filaments and emergent plurilocular organs and true hair. (*b*) True hair, with basal meristem. (*a*) × 210; (*b*) × 420. (After Kuckuck-Kornmann.)

PILAYELLA Bory   The most common species of *Pilayella* (named for *M. de la Pylaie*) is *P. littoralis* (L.) Kjellm, an alga exhibiting great morphological variation (Russell, 1963). It can reportedly attain 50 cm in overall length and is often epiphytic on *Fucus* and *Ascophyllum* (p. 352) and less commonly epilithic. It can be collected year-round in many localities but is not found in tropical or subtropical water. Its distinction is the arrangement of both types of reproductive organs (unilocular and plurilocular) in intercalary catenate series (Fig. 6.14). Terminal and lateral sporangia may rarely occur (Jorde and Klavestad, 1959). Two other features are noteworthy: Opposite branching is frequent and cells contain many discoid chloroplasts.

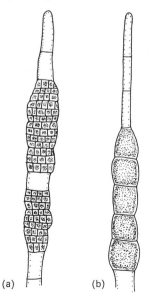

Fig. 6.14 *Pilayella littoralis.* (*a*) Series of plurilocular organs. (*b*) Series of unilocular organs. (*a*), (*b*) × 224.

(a)   (b)

In the 1920's Knight recognized that there were two kinds of plurilocular organs in *P. littoralis*, one occurring on the haploid plant and the other occurring on the diploid plant. She also demonstrated that the unilocular sporangium was the site of meiosis. The D^i, h + d, type of life history was indicated but with modifications. Fusion of gametes was observed only in early spring; during the rest of the year there was parthenogenetic development of gametes. Zoids produced by plurilocular sporangia on diploid plants served to recyle that phase asexually. Zoids from the unilocular sporangia could also recycle the diploid phase due to apparent apomeiosis. A final complication, the reported occasional fusion of zoids from the unilocular organs functioning as gametangia, has been reported in other brown algae (cf. Caram, 1972, p. 153, for a list of such plants), which would signify that both plurilocular and unilocular organs can behave as gametangia. This remains a controversial issue among those who culture brown algae. The occurrence in nature and in culture of plants bearing only unilocular sporangia has been reported in *Pilayella littoralis* f. *rupincola* by West (1967).

The presence of occasional vertical cell walls in *Pilayella*, thus resulting in an incipient parenchymatous organization, has led certain authors (e.g., Christensen, 1962) to place this genus in the Dictyosiphonales (p. 308). Russell (1964) has pointed out that additional cases are known of species normally placed in nonparenchymatous orders in which rare vertical walls are produced. It was his opinion that the logical step was to place the Dictyosiphonales as a suborder in the Ectocarpales. If this were to be followed, the Ectocarpales would be a broadly defined order. The present authors prefer to retain separate orders despite the occasional exception such as *Pilayella*.

Order 2. Ralfsiales

This order of brown algae was recently established (Nakamura, 1972) to include mostly crustose forms (with the exception of *Analipus*). The justification of the setting up of this order was the fact that, in addition to the apparent D^i, h + d, type of life history present, the members demonstrated a discoid-type of germination and early development (Fig. 6.15) rather than forming loosely branching, creeping filaments as in the Ectocarpales. A single platelike chloroplast lacking a pyrenoid is present in each cell.

The only convincing demonstration of sexuality in this order has been in *Nemoderma tingitana*, in which Kuckuck (1912) observed the fusion of anisogametes issuing

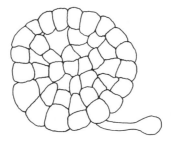

Fig. 6.15 *Ralfsia*. Discoid germling with germination tube of zoospore still remaining. × 383.

from the lateral plurilocular organs of two types. In three other European members' studies by Loiseaux (1968), a direct type of development was observed. For example, in *Ralfsia verrucosa* if one started with crusts with unilocular organs or crusts with plurilocular organs, the next generation of crusts obtained in culture would have sporangia similar to those of the parent, implying an apomeiotic cycle.

The Ralfsiales has been recognized (Nakamura, 1972) to contain three families, which are differentiated by the following key:

1. Reproductive organs are terminally borne on the
   assimilatory filaments ....Lithodermataceae
1. Reproductive organs in intercalary or lateral positions
   but not terminal..2
   2. Unilocular sporangia lateral; plurilocular
      organs intercalary ....Ralfsiaceae
   2. Unilocular sporangia intercalary; plurilocular
      organs lateral ....Nemodermataceae

Family Ralfsiaceae

Although this family is made up of mostly crustose members, the genus *Analipus* has also been assigned to it (Nakamura, 1972; Wynne, 1972B) due to some strong similarities. Members of this family are typically epilithic, forming barely noticeable spots on rocks to very broad expanses covering boulders or intertidal rock. They can be in the intertidal zone or range into the sublittoral. In a monographic treatment of this family on the California coast, Hollenberg (1969) described a total of 11 representatives classified among seven genera. His circumscription of the Ralfsiaceae included the Lithodermataceae.

RALFSIA Berkeley   Thalli of *Ralfsia* (named for *J. Ralfs*, British phycologist) are composed of closely coherent filaments forming a hypothallial layer of horizontally directed rows and an epithallial layer of assurgent or vertically directed rows (Fig. 6.16a), a compact pseudoparenchymatous organization thus being brought about. Normally only unilocular or plurilocular organs are produced on a given thallus, this implying an alternation of isomorphic phases. Unilocular organs (Fig. 6.16b), are produced laterally from photosynthetic filaments arising somewhat loosely above the epithallial layer. Plurilocular organs (Fig. 6.16c) are produced directly by the transformation of a somewhat thickened epithallial layer. Each row of cells in a plurilocular organ terminates with a sterile cell.

One of the most distinctive species of this genus is *Ralfsia fungiformis* (Gunn.) S. et G. (Fig. 6.17), inhabiting the colder waters of the North Atlantic and North Pacific. For this species both types of reproductive organs have been reported (Edelstein, et al., 1968) on the same thallus.

Culturing studies in recent years have indicated that *Ralfsia* is a heterogeneous collection of species, some actually constituting the alternate phase of some Scytosiphonaceae and thus having no relationship at all to *Ralfsia* in the restricted sense. This is discussed in more detail on p. 315.

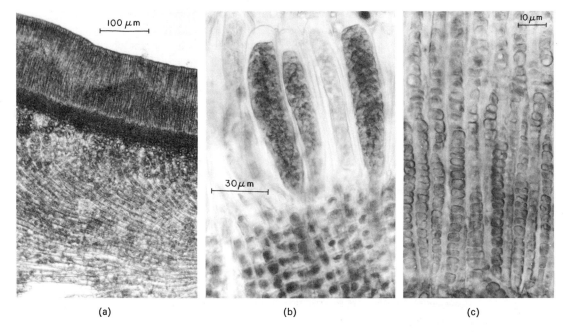

(a) (b) (c)

Fig. 6.16 *Ralfsia fungiformis* (Gunn.) S. et G. (*a*) Vertical section through crust showing lower hypothallial zone and upper epithallial zone. (*b*) Unilocular organs. (*c*) Plurilocular organs terminated by sterile cells. (After Edelstein, Chen, and McLachlan; Journal of Phycology 4: 157–160, 1968.)

Fig. 6.17 *Ralfsia fungiformis.* Habit. (After Edelstein, Chen, and McLachlan; Journal of Phycology 4: 157–160, 1968.)

Fig. 6.18 *Analipus japonicus* (Harv.) Wynne. Habit. × 0.3.

ANALIPUS Kjellman   *Analipus* (Gr. *analipos*, barefoot) is a genus with two species occurring in the North Pacific. *Analipus japonicus* (Harvey) Wynne (Fig. 6.18) is of common mid-littoral occurrence on both sides of the Pacific Ocean and consists of erect axes thickly covered with nonbranched laterals. The erect axes arise from a well-developed prostrate portion, the latter being a means to perennate this species, while the erect axes are seasonal.

Thalli bear either plurilocular organs or unilocular organs. When the cortex of a fertile lateral becomes transformed into plurilocular organs (Fig. 6.19), the entire photosynthetic tissue is shed, leaving behind the colorless medullary tissue. Despite their disparate morphological appearances, *Analipus* shares several attributes with *Ralfsia*, including the cytological feature of a single plastid per cell without a pyrenoid (Hori, 1971), the presence of a sterile cell terminating every row of plurilocular organs (Wynne, 1971), and the discoidal germination pattern. The occurrence of an isomorphic life history reported by Abe (1935, 1936) needs verification.

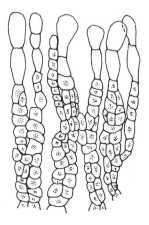

Fig. 6.19 *Analipus japonicus.* Plurilocular organs terminated by sterile cells. × 355.

Order 3. Chordariales

This order is a fairly diverse assemblage of brown algae, including some forms that are microscopic, although the majority are macroscopic. It is conceded to be a primitive group, parallelling the Ectocarpales in several respects. In fact, some current treatments (Russell, 1964; Parke and Dixon, 1964) continue to include this order within the Ectocarpales, a circumscription proposed by Oltmanns (1922) and followed by Fritsch (1945). We would maintain the Chordariales as a distinct order on the basis of the heteromorphic life history (D^h, h + d), in which the haploid microthallus (the gametophyte) alternates with a diploid macrothallus (the sporophyte). This concept has been corroborated by recent reports (Caram, 1961, 1965; Dangeard, 1969B; Loiseaux, 1967B, 1970B). This typical cycle can be altered by the fusion of zoids issuing from the unilocular sporangia, however, as reported for *Leathesia* (Dangeard, 1965, 1969A), *Myrionema* (Loiseaux, 1964, 1967B), and other genera. Another alternative scheme is the cycle in which the **unispores** (i.e., the zoids from unilocular sporangia) may return the macrothallus without fusion, as in *Elachista* species (Blackler and Katpitia, 1963; Edelstein, et al., 1971) and *Chordaria* (Kornmann, 1962C). Other anomalous cycles have been described (see p. 291).

Thallus construction may range from epiphytic discs with heterotrichous organization to loose or compact pseudoparenchymatous aggregations of filaments derived by either uniaxial or multiaxial consolidation. Growth may occur by means of divisions of an apical cell or a subapical meristem. Trichothallic or diffuse growth patterns are also common.

Where sexuality has been witnessed in these algae, isogamy is the rule. The zygote may germinate into a stage which persists indefinitely as a microscopic filamentous system, termed the **plethysmothallus**, which produces plurilocular organs, to asexually recycle the diploid stage until suitable conditions exist to induce the expression of the macroscopic sporophyte. This plethysmothallus has also been referred to as a **protonema** when the macroscopic sporophyte is observed to arise directly from it.

Only a representative selection of families of this order is discussed in the present account. A key to selected families of Chordariales follows:

1. Plants large, erect, typically branched ....Chordariaceae
1. Plants small to moderate-sized, discoid, tufted, or pulvinate 2
  2. Plants minute, consisting of a basal layer giving
     rise to erect, unbranched filaments ....Myrionemataceae
  2. Plants relatively larger, not constructed as above 3
  3. Plants tufted, consisting of a basal portion embedded in the host tissue
     and an upper portion of extended filaments ....Elachistaceae
  3. Plants pulvinate or globose, organized into a cortex and a
     medulla and a hollow central portion ....Leathesiaceae

Family 1. Myrionemataceae

This family comprises small epiphytic algal discs composed of a basal layer of closely adherent, radiating filaments and vertical, unbranched photosynthetic filaments usually about six cells long. Every cell in the basal layer can produce such a vertical filament, a true hair, or a reproductive organ (unilocular or plurilocular sporangium).

Members of this family occurring on the coasts of France and California have been cultured by Loiseaux (1964, 1967A, 1967B, 1970B). Some species were found to have life histories typical for the order, while others had atypical patterns or accessory patterns supplementing the typical D^h, h + d scheme. The genus *Myrionema* is used as an example for this family.

MYRIONEMA Greville   Epiphytic plants of *Myrionema* (Gr. *murios*, numberless + Gr. *nema*, thread) are commonly encountered on a variety of algal and sea grass hosts. Every cell of the monostromatic basal layer gives rise to some type of erect process (Fig. 6.20). Even though the sporophyte falls into a microscopic size range, it is still relatively larger than the gametophyte, which Loiseaux (1967A, 1972) described as a loosely branched filamentous system producing plurilocular organs in *Myrionema strangulans* Grev. and *M. feldmannii* Loiseaux. She also confirmed that the zoids from these plurilocular organs were sexual and fused to return the sporophyte.

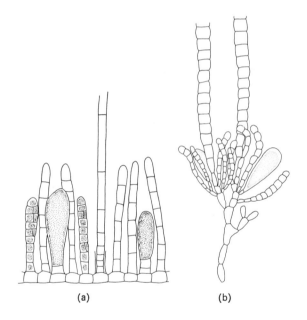

(a)   (b)

Fig. 6.20 (*a*) *Myrionema orbiculare* J. Ag. Vertical section of disc showing unilocular sporangia, plurilocular sporangia, true hair, and photosynthetic filaments arising from the monostromatic basal layer. (*b*) *Elachista*. Portion of thallus epiphytic on another alga. (*a*) × 240; (*b*) × 80.

Family 2. Elachistaceae

These algae are usually small pulvinate growths, often epiphytic but of larger dimensions than the members of the preceding family.

ELACHISTA Duby   Thalli of *Elachista* (Gr. *elachistos*, very small) are differentiated into a colorless, congested basal portion, which may be endophytic in the host tissue, and a photosynthetic portion of erect free filaments, branched at the base only

(Fig. 6.20*b*). Shorter clavate filaments and unilocular sporangia are also produced later in the season.

Early culturing experiments on *Elachista* [*E. fucicola* (Velley) Aresch. and *E. stellaris* Aresch.] by Kylin (1937) indicated that zoids from the unilocular sporangia produced filamentous growths that would directly generate the normal sporophyte, implying that the unilocular sporangia were apomeiotic. A new interpretation was more recently advanced (Wanders, et al., 1972; van den Hoek, et al., 1972) for *E. stellaris*.[5] These authors demonstrated for this species a heteromorphic life history without the intervention of sexuality. The culture studies revealed that the expression of the **macrothallus** or the **microthallus** was dependent on such conditions as temperature, daylength, and the composition of the medium. The diploid macrothallus reproduced directly by **plurispores** or else formed unilocular sporangia, which apparently were meiotic. Unispores from the latter sporangia formed haploid microthalli, which reproduced themselves by plurilocular organs of a different size than those of the macrothalli. Interestingly, when these microthalli were subjected to warm temperatures and long daylengths, they budded off macrothalli, these arising by a process of apparently spontaneous diploidization.

Family 3. Leathesiaceae

More massive aggregations of filaments characterize members of this family. Thalli have spongy to globular or even amorphous shapes. Abundant mucilage production is an additional feature. Thalli are sufficiently large to show an anatomical differentiation into a colorless medulla made up of large-celled filaments and a pigmented small-celled cortex, these two zones blending into each other. The pseudo-parenchymatous nature of these algae, such as *Leathesia*, can be readily appreciated in the field by squeezing a thallus between the fingers and noting its easy disaggregation.

LEATHESIA S. F. Gray   A ubiquitous species of *Leathesia* (named for *Rev. G. R. Leathes*) is *L. difformis* (L.) Aresch, an alga first described by Linneaus as belonging to *Tremella*, a jelly fungus. Its convoluted, brain-like appearance (Fig. 6.21*a*) can be observed in the mid-littoral zone on both the Atlantic and Pacific North American coasts. The mature thallus becomes hollow. The larger colorless medullary cells gradually merge with the smaller pigmented cells of the cortex (Fig. 6.21*b*). Close to the periphery can be detected both unilocular and plurilocular organs. Zoids from the plurilocular organs have been shown (Dangeard, 1965) to recycle the diploid phase, whereas unispores germinated and grow into filamentous stages with plurilocular organs. These were called "prothalli" by Dangeard (1965, 1969A) and were presumably gametophytes that both asexually produced additional gametophytes by parthenogenesis or returned the sporophyte by apparent fusion of gametes. Dangeard (1965) has also observed the fusion of unispores to return the diploid phase directly in an abbreviated cycle.

---

[5]This species is at times placed in its own genus *Symphoricoccus*.

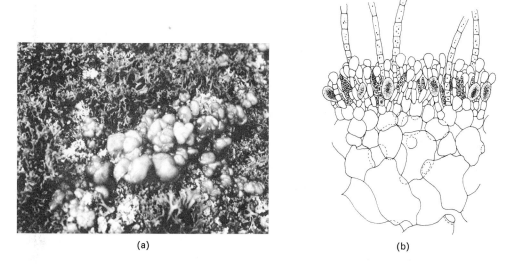

(a)   (b)

Fig. 6.21 *Leathesia difformis* (Linn.) Aresch. (*a*) Habit. (*b*) Section of thallus, showing pseudoparenchymatous nature of construction, both plurilocular and unilocular organs, and true hairs. (*a*) × 0.4; (*b*) × 150. [(*b*) after Kuckuck.]

Family 4. Chordariaceae

This final family of the order contains larger algae with cylindrical, usually branched axes. Some have a sharp distinction between the photosynthetic cortex and a colorless medulla; in others there is a gradual transition of these two tissues. True hairs are common. The sporophyte may have only unilocular sporangia or both unilocular and plurilocular organs. Depending on the genus, the construction may be according to a monopodial or sympodial plan. In the former the primary region of growth is maintained and continues to function; in the latter there is a continual displacement of the region of growth, that region being shifted to the side and a new one replacing it at least temporarily. One must make very careful examinations of the growing tip, usually by slight squashing, to be able to distinguish these two modes. Both uniaxial and mutiaxial genera occur in this family; in some cases cell division is localized in a subapical meristem.

The three genera to be described in this account can be distinguished by the following key:

1. Terete axes relatively firm and dark-brown to black; slippery but
   not excessively mucilaginous, with monopodial development ....*Chordaria*
1. Terete axes very soft and mucilaginous in texture; golden
   brown; with sympodial development 2
   2. Central axis more or less hollow; outer parts with cells
      pseudoparenchymatously compacted together, with a rather
      sharp demarcation between the cortical and
      medullary regions ....*Cladosiphon*

Fig. 6.22 *Chordaria flagelliformis* (O. F. Müll.) C. Ag. Habit. × 0.2.

2. Central axis not hollow; the thallus easily pressed out to show tufts of assimilatory filaments arising from loosely arranged central filaments; the demarcation between medulla and cortex not sharp .... *Eudesme*

CHORDARIA C. Agardh *Chordaria flagelliformis* (O. F. Müll.) C. Ag. is the most familiar form of the genus *Chordaria* (L. *chorda*, a cord), a moderate-sized alga recorded from the North Atlantic and the far North Pacific (i.e., Alaska and the Bering Sea). Often the branching is confined to the base of the plant (Fig. 6.22), and the primary branches usually are longer than the primary axis. Thalli are dark brown to almost black and have a slippery but firm texture. The internal organization is very compact, the cortex having short, multicellular assimilatory filaments with laterally produced unilocular sporangia (Fig. 6.23).

Fig. 6.23 *Chordaria flagelliformis.* Portion of transection of thallus, showing unilocular sporangia and periphery. × 125. (After Kylin.)

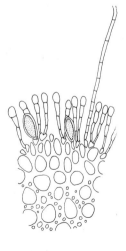

Kornmann (1962C) cultivated *C. flagelliformis* and discovered that a hetero-morphic cycle occurred but that it did not involve sexuality. Unispores from the macrothallus gave rise either to a new crop of macrothalli or to a "dwarf generation," which was filamentous and bore plurilocular organs at lower temperatures. The pluri-spores from this microthallus returned the macrothallus. Since *Chordaria* is seasonal in the North Sea where it was investigated, being found only from April to November, Kornmann believed that it persisted as the microthallus during the remaining months of the year.

EUDESME J. Agardh   Though less frequently encountered than the previous genus, *Eudesme* (Gr. *eu*, well + Gr. *desmos*, bond) can be collected during the summer months on rocks or on other algae and is an excellent genus to portray this family due to the ease with which one can squash out a portion of an axis to demonstrate the loose aggregation of the multiaxial system (Fig. 6.24). *Eudesme virescens* (Carm. ex Harv. in Hook.) J. Ag. has a North Atlantic distribution.

CLADOSIPHON Kütz.   The texture of *Cladosiphon* (Gr. *klados*, a branch + Gr. *siphon*, a tube) approaches the soft, slimy condition of *Eudesme*. *Cladosiphon occidentalis* Kylin has a warm-water distribution and is common in the waters of Florida and the Gulf of Mexico during the winter and spring (Earle, 1969; Edwards and Kapraun, 1973). The sporophyte produces both unilocular and plurilocular organs in the cortical region (Fig. 6.25).

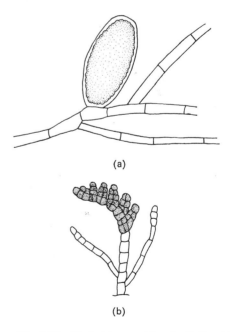

Fig. 6.24 *Eudesme virescens* (Carm. ex Harv. in Hook.) J. Ag. Longitudinal section of thallus. × 65.

Fig. 6.25 *Cladosiphon occidentalis* Kylin. (*a*) Unilocular sporangium. (*b*) Plurilocular spo-rangia. × 248.

Order 4. Sporochnales

This order is relatively small, with about six genera, only one of which (*Sporochnus*) has a distribution that includes North America, namely, from Beaufort, North Carolina, southward to Florida and the Gulf of Mexico, and also Southern California. One distinctive feature of this order is the presence of a tuft of hairs terminating each axis (Fig. 6.26). An intercalary meristem is situated at the base of each of these hairs, and a solid pseudoparenchymatous tissue results from divisions of this meristematic zone.

The original interpretation of a D^h, h + d life history for this order by the French phycologist Camille Sauvageau has recently been confirmed for *Sporochnus pedunculatus* (Huds.) C. Ag. by Caram (1965). She described an alternation of heteromorphic phases, with meiosis occurring in the unilocular sporangia of the diploid macrothallus. Microthalli were either monoecious or dioecious gametophytes. Eggs and sperm were observed, but actual fertilization was not seen. Caram noted apertures at the apex of the oogonial wall, which apparently permitted the entry of sperm. Magne (1953) earlier showed that an alternation of cytological generations was superimposed on this alternation of morphological generations.

Order 5. Desmarestiales

Although containing only a few genera, this order is of interest because the genus *Desmarestia* is frequently encountered in the colder waters of both hemispheres and

Fig. 6.26 *Sporochnus*. Conspicuous tuft of hairs terminate each axis. × 0.24.

Fig. 6.27 *Desmarestia*. Trichothallic growth near apex of thallus. × 73.

can constitute a dominant element in the sublittoral vegetation of the Antarctic (Reinbold, 1928). It is a reasonably clear-cut group as an order, characterized by a D^h, h + d life history, with a microscopic gametophyte (monoecious or dioecious) producing eggs and sperms and a larger sporophyte showing trichothallic growth and firmly pseudoparenchymatous construction. Each apex (Fig. 6.27) is terminated by a single filament with an intercalary meristem, producing laterals both above and below it. Most of the segments are cut off proximally, and a rhizoidal cortication below the zone of cell division rapidly invests the thallus in a compact cortex. The same type of meristem occurs at the apices of all the laterals, and thus the entire thallus is fringed with the pigmented hairs emerging distal to the meristems. These hairs are conspicuous when the thallus is young and rapidly growing but are usually shed by the time the plant is older.

DESMARESTIA Lamouroux *Desmarestia* (named for *A. G. Desmarest*, a French naturalist) is a genus inhabiting cold waters of both the northern and southern hemispheres. A surprising recent range extension was the collection (Diaz-Piferrer, 1969) of two species of *Desmarestia* in deep waters of the Caribbean. The sporophyte is always macroscopic, reportedly reaching 5 m in length in some species. A long list of species ranging from narrow, terete, much branched forms to unbranched ligulate forms has resulted from the great variation displayed by this genus (Fig. 6.28). But when the morphological characteristics have been subjected to rigorous graphical analyses, the results have indicated that the differences are clinal and largely reflect

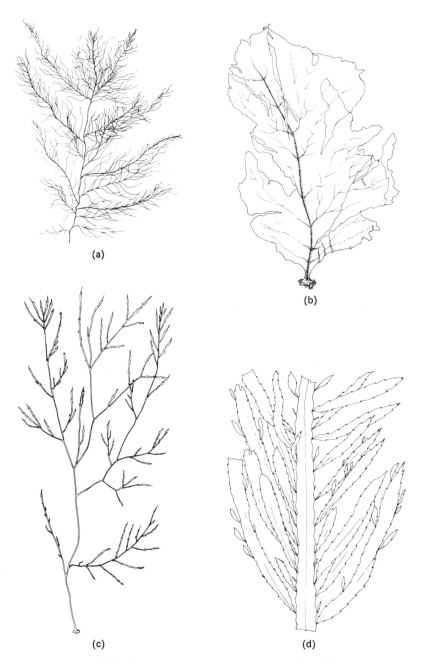

Fig. 6.28 *Desmarestia*. Examples of morphological variation in the genus. (*a*) *D. viridis* (Müller) Lamour. from the Atlantic coast. (*b*) *D. tabacoides* Okam. from Japan. (*c*) *D. latifrons* (Rupr.) Kuetz. from the Pacific coast. (*d*) *D. ligulata* (Lightf.) Lamour. from the Pacific coast. (*a*) × 0.25; (*b*) × 0.19; (*c*) × 0.28; (*d*) × 0.18.

seasonal or gradual geographical changes. In studying the ligulate members occurring on the west coast of North America, Chapman (1972) concluded that the 10 previously recorded taxa of this genus actually represented only 3: two species, one of which had two varieties. The extreme polymorphic nature of this complex is obvious.

Another taxonomic opinion offered by Chapman (1970) was that the three sections of the genus recognized by Setchell and Gardner (1925), namely, Virides (terete, oppositely branched), Aculeateae (terete, alternately branched), and Herbaceae (ligulate, oppositely branched), were untenable, since two species occur (*viz.*, *D. tortuosa* Chapman from British Columbia and *D. kurilensis* Yamada from the Kuril Islands in the western North Pacific) that are transitional between sections Virides and Herbaceae. These two species have a combination of terete and ligulate axes.

On the other hand, a basic distinction seems to exist between those species with alternate branching and those species with opposite branching in that culturing studies indicate that the members of the former group have dioecious gametophytes, whereas members of the latter group have monoecious gametophytes. Chapman and Burrows (1971) confirmed that *D. aculeata* (L.) Lamour. with alternate branching, has separate male and female gametophytes. Kornmann (1962D) demonstrated bisexual gametophytes for *D. viridis* (O. F. Müll.) Lamour., and Nakahara and Nakamura (1971) reported similar findings for *D. tabacoides* Okam., which would be placed in the same subgenus with *D. viridis*.

The gametophytes (Fig. 6.29a) bear antheridia that are produced as short, lateral branches or at the tips of short laterals. Oogonia are produced at the growing tips of a main filament (Kornmann, 1962D) or at the tips of young branches. Each antheridium releases a single sperm, while the oogonium liberates its protoplast as a spherical egg that is fertilized while attached to the oogonial wall, the sporophyte developing (Fig. 6.29b) while remaining firmly attached. A similar pattern of development is seen

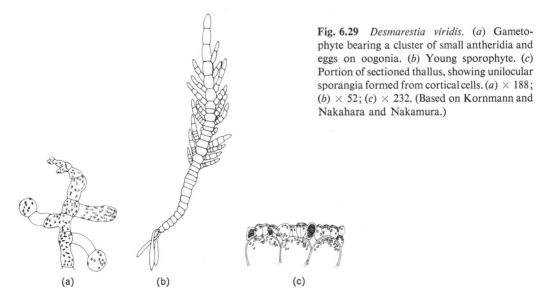

Fig. 6.29 *Desmarestia viridis.* (*a*) Gametophyte bearing a cluster of small antheridia and eggs on oogonia. (*b*) Young sporophyte. (*c*) Portion of sectioned thallus, showing unilocular sporangia formed from cortical cells. (*a*) × 188; (*b*) × 52; (*c*) × 232. (Based on Kornmann and Nakahara and Nakamura.)

(a)   (b)   (c)

in the Laminariales (see p. 319). Kornmann (1962D) noted that in his cultures sporophytes seldom developed from unattached zygotes.

The requirement for cold temperature for maturation of the gametophytes has been indicated by investigators (Kornmann, 1962D; Nakahara and Nakamura, 1971). Kornmann noted that gametophytes grew vegetatively indefinitely at 15°C but rapidly became sexual in the range of 3–5°C. Other evidence (Chapman and Burrows, 1970) revealed that a critical quantity of light energy (involving a combination of light intensity and daylength) was necessary for maturation of the gametophytes; values less than the critical level inhibited maturation.

The sporophytes bear unilocular sporangia as little modified cortical cells (Fig. 6.29c). This is the only type of reproductive organ occurring on the sporophyte.

An interesting physiological property of some of the oppositely branched species of *Desmarestia* deserves our attention. It has been demonstrated (Meeuse, 1956) that free sulfuric acid (at a concentration of $0.44N$) is present in the vacuoles, resulting in a pH value of from 0.8 to 1.8 (Blinks, 1951; Eppley and Bovell, 1958; Schiff, 1962). Malic acid is also present in less concentration. Those who collect specimens of *D. viridis* or *D. ligulata* soon discover that not only do these specimens rapidly deteriorate but that other algae placed in the same bucket or container are also spoiled. Better success is obtained by keeping the *Desmarestia* isolated and in chilled seawater.

Order 6. Cutleriales

This brown algal order, including only three genera, is clearly delimited from other orders by the following combinations of features: trichothallic growth, parenchymatous construction, and anisogamous reproduction. The life history varies according to the genus: *Cutleria* has an alternation of heteromorphic phases, while *Zanardinia* undergoes an alternation of isomorphic phases. Despite this difference in life history, the many other shared traits indicate that these two genera should be placed in the same order. Some phycologists, such as Fritsch, have suggested that in this order the isomorphic life history is primitive and the heteromorphic pattern seen in *Cutleria* represents a derived condition.

Only the single genus *Cutleria* will be discussed in detail in the present account.

CUTLERIA Greville   The conspicuous phase of *Cutleria* (named for *Catherine Cutler*, a British phycologist) is represented by the gametophyte. The form of the thallus may be flattened, ribbon-shaped axes, cylindrical axes, or fan-shaped blades, depending on the species. In contrast, the diploid phase consists of a prostrate, encrusting, lobed thallus, which may attain a surface area of several square centimeters but usually is fairly small. This alternate phase was originally regarded as a separate genus, *Aglaozonia;* the custom of designating it as the "Aglaozonia-stage" persists.

The gametophyte of *Cutleria* grows by trichothallic growth, with a conspicuous tuft of hairs arising distal to the meristem (Fig. 6.31b). A true parenchyma is formed proximally to the meristem, due to the longitudinal and transverse segmentation of the cells cut off. In older thalli an inner medulla of large cells and small-celled cortical layers are present.

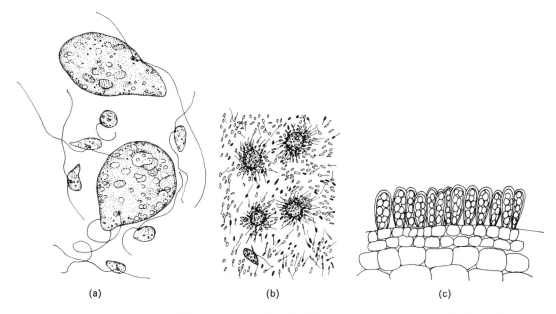

(a)   (b)   (c)

Fig. 6.30 *Cutleria multifida* (Smith) Grev. (*a*) Anisogametes. (*b*) Field of cultured gametes. (*c*) Vertical section of "*Aglaozonia-stage*" showing unilocular sporangia over upper surface. (*a*) × 132; (*b*) × 660; (*c*) × 231. [(*a*) and (*b*) after Kuckuck; (*c*) after Smith.]

The gametophytes produce two types of plurilocular organs, one type with large locules and the second with small locules. Anisogametes (Fig. 6.30*a*) are released, and the clustering of the smaller male gametes around the larger female gametes has been depicted (Fig. 6.30*b*). A sex attractant involved in this behavior has recently been demonstrated (Müller, 1974). It was determined to be a highly volatile compound with a low molecular weight and causes chemotaxis in the male gametes (Jaenicke, 1977).

The sporophyte of *Cutleria* grows by means of a marginal meristem rather than a trichothallic pattern. It also is parenchymatous, and at maturity clusters of unilocular sporangia are formed over the surface (Fig. 6.30*c*).

Cutleria hancockii Dawson (Fig. 6.31*a*) occurs commonly in the spring in the Gulf of California and is the first representative of this order to be recorded (Dawson, 1944) from North America.

Order 7. Sphacelariales

Members of this order are distributed both in warm and cold seas throughout the world. Thalli are generally small, filamentous tufts, but some species may reach up to 15 cm in length. Both epiphytic and epilithic forms occur, with rhizoidal attachment to the substrate. *Sphacelaria bipinnata* (Kutz.) Sauv. is invariably epiphytic on the Fucalean genera *Halidrys* and *Cystoseria*, anchored by penetrating endophytic fila-

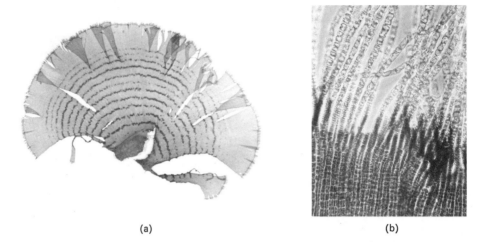

(a)   (b)

Fig. 6.31 *Cutleria hancockii* Dawson. (*a*) Habit. (*b*) Margin with trichothallic growth. (*a*) × 0.75; (*b*) × 100.

ments (Goodband, 1973). All forms of Sphacelariales have a characteristically prominent apical cell, and the segments that are cut off from this apical cell undergo both transverse and longitudinal division (Fig. 6.32). Thus a solid parenchyma results. In some genera, extensive cortication adds to the girth of the filaments.

Other ordinal traits include an essentially isomorphic (or slightly heteromorphic) life history. Sexuality ranges from isogamy [as in *Cladostephus spongiosus* (Huds.) C. Ag. in which the gametes are morphologically indistinguishable but behaviorally anisogamous in that the female gametes become immobile first and settle down before being fertilized] through anisogamy (as in *Sphacelaria furcigera* Kütz. in which gametes are clearly distinguishable into two sizes) and oogamy (as in several species of *Halopteris* from New Zealand, according to Moore, 1951).

The cells in this order contain a large number of lenticular chloroplasts with no obvious pyrenoids. Hori (1972) did detect rudimentary pyrenoids in material of *Sphacelaria* from Japan. The chloroplast of *Sphacelaria* has been examined intensively with the electron microscope (Bisalputra and Bisalputra, 1969; Bisalputra and Burton, 1970). An intimate association between all the photosynthetic lamellae and a ringlike DNA-containing body has been described (Fig. 6.2*b, c*). The term "genophore" has been used for this gene-containing body. It is speculated that this close contact between the genophore and the photosynthetic lamellae may facilitate the development of the chloroplast membrane system.

SPHACELARIA Lyngbye   Representatives of *Sphacelaria* (Gr. *sphakelos*, gangrene) are likely to be collected in any littoral habitat. The many species in this genus range from polar to tropical seas. A prominent apical cell (Fig. 6.32*a*) terminates each axis, and this cell contains an abundance of fucosan vesicles, causing the tips of the plants to have a dark, burnt appearance. The apical cell divides to produce primary segment cells proximally. This primary segment cell elongates and then undergoes a

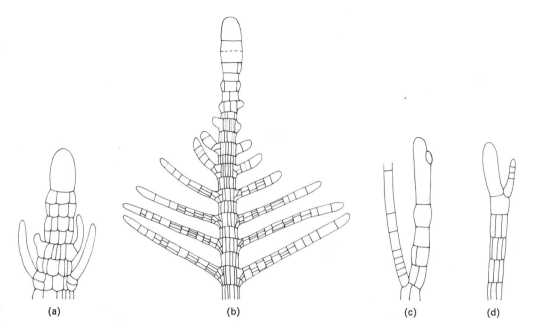

(a)   (b)   (c)   (d)

Fig. 6.32 *Sphacelaria*. (*a*) Apex of axis, showing conspicuous apical cell. (*b*) Pattern of segmentation, with hemiblastic branching. (*c*), (*d*) Hair formation, involving holoblastic branching. (*a*) × 89; (*b*) × 50; (*c*), (*d*) × 63. (After Sauvageau.)

transverse division, which results in two secondary segment cells (one "superior" and the other "inferior"). These cells undergo vertical divisions to produce up to about 20 cells in tiers corresponding to the original secondary segment cells. The cells produced by this septation do not expand in size, which means that the filament remains more or less the same diameter throughout. In some species of *Sphacelaria* and in other genera cortication occurs, involving an upward and downward growth of rhizoidal filaments issuing from the surface cells. An increase in girth then results.

Branches arise by several different methods. In *Sphacelaria* the typical lateral branches (whether determinate or indeterminate) usually arise from a cell occupying the entire height of a superior segment (Fig. 6.32*b*), thus half the length of a primary segment. This method is termed **hemiblastic**. True hairs are developed in *Sphacelaria* by an unequal division of the apical cell. The small hair-initial is considered as the actual apical cell, which is converted into the true hair (or a lateral), and the larger lower cell bends upward to continue as the apex (Fig. 6.32*c*, *d*). This sympodial type of branching is termed **holoblastic** in that the whole apical cell is involved in the formation of the lateral. The third type of branching, which is more prevalent in the heavily corticated genus *Cladostephus*, is designated **meriblastic** and involves the development of a lateral from only a part of a secondary segment rather than the entire segment. Other modes of branching have been described. In the uncorticated species of *Sphac-*

elaria the transverse septa corresponding to the original segmentation of the primary cell are detectable at a distance from the apex.

The development of **propagula** is common in many species of *Sphacelaria*. These are multicellular structures which represent modified branches which are shed and asexually propagate the plant. They vary in shape from club-shaped to biradiate and triradiate structures (Fig. 6.33*a, b*), the latter showing resemblance to conidia produced by a group of aquatic Fungi Imperfecti designated the Hyphomycetes.

A recent report on *Sphacelaria furcigera* from Holland (van den Hoek and Flinterman, 1968) will be discussed as an example to illustrate the life history of this genus. A slightly heteromorphic, diplobiontic life history was demonstrated, the dioecious gametophytes being somewhat less robust than the sporophytes. Plurilocular organs on the gametophytes are of two types: plurilocular macrogametangia (Fig. 6.33*c*), which are dark brown, and plurilocular microgametangia (Fig. 6.33*d*), which are yellowish white. Instead of an apical rupturing of the entire gametangium, as occurs in *Ectocarpus*, pores are developed in almost all the peripheral locules. The larger female gametes have several brown chloroplasts, while the male gametes have a rudimentary pale yellow chloroplast; both have a stigma. Although the unfused male gametes do not survive, female gametes are able to germinate parthenogenetically into female gametophytes or into haploid sporophytes, these latter representing "dead ends" in the life history. The diploid sporophytes produce only unilocular sporangia, which are the site of meiosis. About half of the unispores develop into male gametophytes, the other half into female gametophytes.

The production of sporangia, gametangia (plurilocular organs), or propagula was shown (van den Hoek and Flinterman, 1968) to depend on temperature. Under a 12-hour photoperiod, cultures of gametophytes produced gametangia at 4 and 12°C

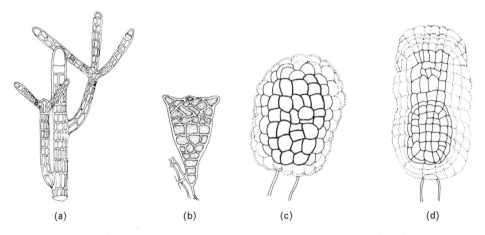

(a)   (b)   (c)   (d)

Fig. 6.33   *Sphacelaria.* (*a*), (*b*) Types of propagules. (*a*) *S. subfusca* S. and G. (*b*) *S. californica* Sauv. (*c*), (*d*) Gametangia. (*c*) Female gametangium. (*d*) Male gametangium. (*a*), (*b*) × 84; (*c*), (*d*) × 430. [(*a*) and (*b*) after Setchell and Gardner; (*c*) and (*d*) after van den Hoek and Flintermann.]

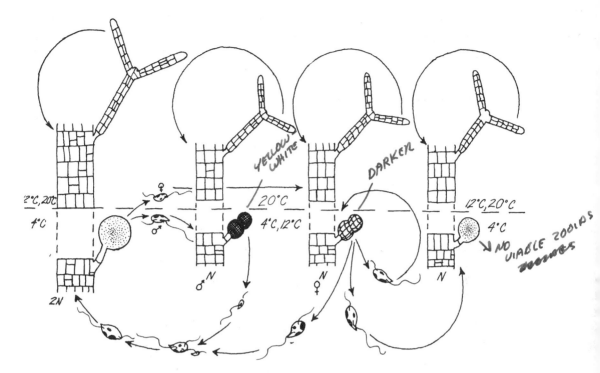

Fig. 6.34 *Sphacelaria furcigera* Kuetz. Life-history pattern involving influence of temperature on sexual or asexual reproduction. (After van den Hoek and Flintermann.)

but formed only propagula at 20°C (Fig. 6.34). Sporophytes formed unilocular sporangia at 4°C but only propagula at 12 and 20°C. In a subsequent publication (Colijn and van den Hoek, 1971) determining the role of photoperiod in the formation of gametangia and propagula, it was observed that cultures grown in higher temperatures (12 and 17°C) produced propagula exclusively under 12- or 16-hour daylengths. Propagula were not formed at lower temperatures (4°C) or under short-day conditions (8-hour daylengths) at elevated temperatures (12 and 17°C). The conclusion is that the formation of propagula was regulated by both temperature and daylength. Plurilocular organs were formed under long-day conditions (16-hours of light) at 4 and 12°C; however, at 17°C they were initially produced but soon were replaced by propagulum production. Unilocular sporangia appeared in these experiments at 4 and 12°C, with some evidence to suggest that short-day conditions were also involved.

Although many species of *Sphacelaria* have been described, it would appear that a great plasticity exists among the forms and that more detailed studies (e.g., Haas-Niekerk, 1965; Irvine, 1956; Goodband, 1971) will reduce some of these taxa to synonymies. In a statistical analysis of several species of the genus, Goodband (1971) concluded that *S. fusca* (Huds.) C. Ag. and *S. furcigera* were distinct species, whereas Haas-Niekerk (1965) and van den Hoek and Flinterman (1968) had regarded them as synonyms. Goodband justified his attitude by stating that *S. furcigera* bears dimorphic plurilocular organs, produces large numbers of hairs, and is partially endophytic, whereas *S. fusca* bears monomorphic plurilocular organs, does not produce hairs, and is epiphytic or lithophytic. Formerly, it was thought that biradiate propagula occurred in *S. furcigera* and triradiate propagula occurred in *S. fusca*. Haas-Niekerk (1965) demonstrated that the latter species can produce both biradiate and triradiate propagula both in field material and in culture. Goodband (1971) observed that the proportion of the two types of propagula produced in culture changes with time. Over a 1-year period of culture the percentage of triradiate propagula in *S. fusca* decreased from 81 to 25%, with a corresponding increase in percentage of the biradiate type.

Order 8. Tilopteridales

A few genera of small filamentous algae (Fig. 6.35*a*), resembling thalli of *Ectocarpus* or *Giffordia* (p. 278), are contained in this order. Although uniseriate in the upper portions, cells in the lower axes become partitioned by longitudinal divisions and take on a polysiphonous appearance. Trichothallic zones of cell division also characterize this order. An important criterion for recognizing these algae as belonging to the same order is the production of monosporangia. Monosporangia (Fig. 6.35*b*) borne on apparent gametophytes contain a single large nucleus (Papenfuss, 1951A; South and Hill, 1971), whereas those on apparent sporophytes (Fig. 6.4*d*) are quadrinucleate (Smith, 1955); cytological evidence (Dammann, 1930) indicates that they are the products of meiosis.

Before the turn of the last century it was suggested from indirect evidence that an isomorphic life history occurs for *Haplospora globosa* Kjellm., a species occurring on

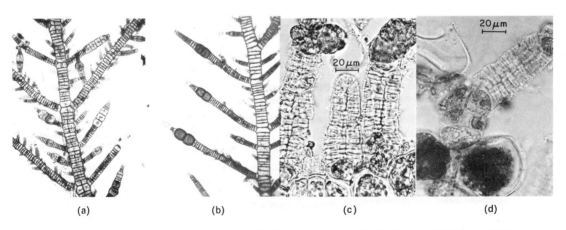

(a)   (b)   (c)   (d)

Fig. 6.35 (*a*), (*b*) *Tilopteris mertensii* (Smith) Kuetz. (*a*) Habit of a field specimen, showing longitudinal divisions and pattern of opposite branches. (*b*) Paired monosporangia. (*c*), (*d*) *Haplospora globosa* Kjellm. (*c*) Gametophyte with antheridia. (*d*) Gametophyte with pair of oogonia and antheridium. [(*a*) and (*b*) courtesy of Dr. James Sears; (*c*) and (*d*) after T. Edelstein and J. McLachlan and permission of the National Research Council of Canada, from the Canadian Journal of Botany 45: 203–210, 1967.)

both sides of the North Atlantic. The gametophyte and sporophyte are vegetatively similar; the gametophyte is apparently oogamous. More recent studies by Sundene (1966) support this idea, and antheridia and oogonia have been reported (Edelstein and McLachlan, 1967A) in Canadian material (Fig. 6.35c, d).

In the only other representative of this order in North America, *Tilopteris mertensii*, which occurs from Massachusetts to Nova Scotia, asexual reproduction by monospores seems to be the only pattern (South and Hill, 1971). Plurilocular organs have been observed in this species, but the function of their zoids remains undetermined.

Order 9. Dictyotales

This clearly demarcated group of brown algae, composed of a single family containing about 20 genera, is best expressed in tropical and subtropical waters. Members exhibit an alternation of isomorphic phases (D^i, h + d). This pattern was recently confirmed (Gaillard, 1972) for two species, *Dictyota dichotoma* (Huds.) Lamour and *Dilophus ligulatus* (Kütz.) Feldm.; yet for *Padina pavonica* (L.) Thivy. a succession of diploid plants was obtained in culture, suggesting apomeiosis in the sporangia. In natural collections disproportionately higher percentages of diploid plants in some species have been recorded (Mathieson, 1966; Liddle, 1972).

Apical growth is the rule in the Dictyotales, with either a single large apical cell as in *Dictyota* (Fig. 6.36a) or a margin of apical initials as in *Padina* (Fig. 6.40). Thalli

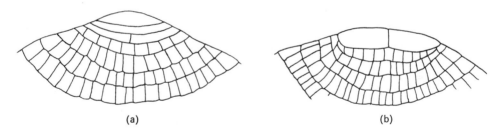

(a)　　　　　　　　　　　　　　　(b)

Fig. 6.36　*Dictyota dichotoma* (Huds.) Lamour. (*a*) Apex of a branch. (*b*) Formation of a dichotomy by the equal longitudinal division of an apical cell. × 136.

are usually a few cell layers in thickness (the range is from two to about eight, the anomalous genus *Dictyotopsis* being monostromatic). The cortical layers, with cells containing a large number of discoid chloroplasts, are clearly distinguished from the colorless medulla.

The order is characterized by pronounced oogamous reproduction, the gametophytes typically being unisexual. Oogonia and antheridia are often located in definite sori, sometimes with spatial relationship to rows of hairs. Sperm are uniflagellate (Manton, 1959), the smooth flagellum being absent. Diploid plants have sporangia whose contents are released in most genera as four large immobile spores (often called "tetraspores"); in *Zonaria*, eight spores are produced per sporangium.

Three genera are included in the present treatment:

1. Growth by means of a single apical cell ....*Dictyota*
1. Growth by means of a row of initials 2
    2. Distal margin circinately inrolled; thalli of some species covered with a light to heavy deposition of calcium carbonate; sporangia with four spores ....*Padina*
    2. Distal margin not circinately inrolled; thalli never covered with deposition of calcium carbonate; sporangia with eight spores.......*Zonaria*

DICTYOTA Lamouroux Plants of *Dictyota* (Gr. *diktyotos*, net-like) are much branched, bushy erect systems (Fig. 6.37), attached by a tuft of rhizoids. The flattened axes are often dichotomously branched but pinnately branched in some species. A prominent apical cell cuts off transverse segments proximally (Fig. 6.36*a*). Divisions of these segments produce walls parallel to the surface, resulting in a primary medullary cell and a primary cortical cell on each surface. The cortical cells undergo further division perpendicular to the surface. The thallus in *Dictyota* is then three cells thick, the smaller cortical cells containing numerous discoid chloroplasts and the larger medullary cells being devoid of chloroplasts. True hairs arise in tufts scattered over the surface. Thalli of some species of *Dictyota* are described as producing a blue-green iridescence when submerged. When examined with the electron microscope, so-called iridescent globules are seen to be clustered around the nucleus. They are membrane-bounded vesicles and contain a larger number of ringlike inclusions (Berkaloff, 1962).

The type species of this genus, *D. dichotoma*, has a widespread distribution, ranging from tropical to temperate waters. It has been the classic example of true dichotomous branching ever since Nägeli first worked out the details of its anatomy in the mid-nineteenth century. The apical cell divides equally with a vertical longitudinal wall. These two cells proceed to cut off segments proximally. Eventually the newly formed segments no longer make contact with the original vertical wall, and the two apical cells separate to form a perfect dichotomy (Fig. 6.36*b*).

Fig. 6.37 *Dictyota dichotoma.* Habit. × 0.28.

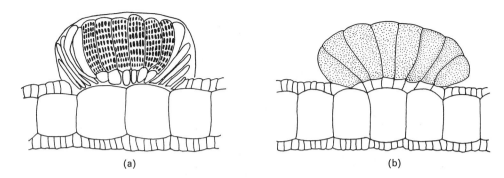

Fig. 6.38 *Dictyota dichotoma.* (*a*) Antheridia. (*b*) Oogonia. × 135.

Gametophytes are typically unisexual. Male plants have much divided plurilocular organs, or antheridia, arranged in sori on both surfaces of the blades. These antheridia develop from cortical cells. The original cortical cell divides periclinally to produce a **stalk cell** and a **primary spermatogenous cell**, the latter undergoing divisions to produce a gametangium (Fig. 6.38*a*) composed of perhaps over 1000 locules. A single uni-flagellate sperm (Fig. 6.5*b*) is formed within each locule. Sterile cells surrounding an antheridial sorus may form a sterile, protective covering called an **involucre** over the antheridia.

An oogonium is likewise initiated by the division of a cortical cell, with the lower cell becoming the stalk cell and the surface cell enlarging into the oogonium (Fig. 6.38*b*), whose contents become metamorphosed into a large egg. The dissolution of the ooginial wall liberates the egg. Usually an oogonial sorus consists of about 25 to 50 oogonia.   *TECHNICALLY   PLURILOCULAR*

The morphologically similar sporophyte produces unilocular sporangia (Fig. 6.4*c*) singly or in small groups. The four large immobile spores from each sporangium are the products of meiosis (Yabu, 1958) and germinate into four gametophytes, two of which are female and two male.

Populations of *Dictyota dichotoma* have been studied in diverse localities, and it has been determined that the production of gametangia is often correlated with the tidal cycle. The precise phasing may vary from locality to locality. For example, in North Wales (Williams, 1905) the development of gametangia is initiated during the neap tide periods of the summer months. Liberation of the gametes occurs about a week later for several days following the highest spring tide. The production of sporangia shows no correlation with the tidal periodicity.

The periodicity demonstrated by this same species in North Carolina (Hoyt, 1927) differed from that in Wales in that gametangia were formed at monthly intervals rather than every 2 weeks. A population from Jamaica was also investigated (Hoyt, 1927) and observed to have a steady production of gametangia due to the overlap of successive crops and their prolonged development. Another factor is the negligible effect of the tide due to an amplitude of only about 15 cm.

It was formerly recognized that the endogenous rhythm in the production of the sex organs could be maintained in the laboratory for some months. It has been

Fig. 6.39 *Padina*. Habit. × 0.5.

demonstrated (Müller, 1962) that the release of gametes can be regulated under *in vitro* conditions. Plants are sensitive to as low an intensity of light as 3 lux (so-called "artificial moonlight"). This intensity of light when administered during a dark period will trigger the production of gametangia, gamete production reaching a maximum 10 or 26 days later. Another species of *Dictyota*, *D. binghamiae* J. Ag., which is widespread on the Pacific coast of North America, shows no obvious periodicity of gamete formation and release (Foster, et al., 1972).

PADINA Adanson   *Padina* (Gr. *padinos*, belonging to a plain) is a familiar and easily recognized alga occurring in tropical and subtropical seas throughout the world. *Padina vickersiae* Hoyt is of common occurrence in the Gulf of Mexico (Edwards, 1970A) and ranges northward in the Atlantic to North Carolina (Taylor, 1960). A distinctive feature of this genus is the deposition of calcium carbonate (= limestone) over the surface of the fan-shaped blades (Fig. 6.39). Depending on the species, this deposition may be light or heavy and may be present on both surfaces or confined to a single surface. Some species lack the deposition entirely. A second noteworthy characteristic is that the distal margin is circinately rolled in (Fig. 6.40), protecting the margin with its apical cells.

Often a conspicuous zonation results from the presence of concentric rows of hairs. The reproductive organs usually occur between these hair bands. Most species

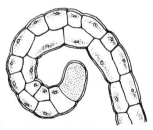

Fig. 6.40 *Padina*. Vertical section of margin showing a single initial from the marginal row of apical cells. × 99. (After Oltmanns.)

prove to be unisexual; however, *Padina pavonica* (L.) Thivy in the Mediterranean was observed (Ramon and Friedmann, 1965) to be influenced by depth and season. Dioecious gametophytes occurred in shallow waters (0–0.5 m), whereas monoecious gametophytes prevailed in depths of 2.5 m and greater. At the intermediate depths (1.0–1.5 m) monoecious plants were found in July but were replaced by dioecious plants in August. An ultrastructural investigation of the two phases (gametophyte and sporophyte) by Fagerberg and Dawes (1973) revealed no substantial differences between the two isomorphic phases, except minor differences reflecting variations in growth activity.

ZONARIA J. Agardh   In contrast to the tristromatic thallus of *Dictyota*, the fan-shaped blades (Fig. 6.41) of *Zonaria* (Gr. *zoni*, belt, girdle) are about eight cells thick. The plant is attached by a felted rhizoidal holdfast, and branching occurs simply by the tearing of the frond. Unilocular sporangia, located in sori between bands of hairs, contain eight spores, which sometimes germinate prior to release. Growth takes place by means of a marginal row of apical cells (Fig. 6.36c). The apical cells reveal a distinctive polar pattern, with four different cytoplasmic zones distinguished by Neushul and Dahl (1972A): apical, nuclear, compound vacuolar, and vegetative. These zones correspond to an intracellular gradient in regard to the distribution of organelles. The nuclei in "dormant," active, and dividing cells have also been described by these authors (Neushul and Dahl, 1972B). The dividing nucleus is of interest in that the nuclear envelope is largely persistent, centrioles are present, and microtubules extend centrifugally from the centriole to form an aster as well as course peripherally outside of the nuclear envelope.

The responses of the apical meristem in *Zonaria* to drug treatment have been investigated (Walker, et al., 1975). Colchicine caused continued elongation of the apical cells without division; after cessation of treatment, abnormal cytokinesis and cell deaths were observed, accompanied by a disruption of the coordinated growth of the marginal meristem. Phenobarbital halted apical cell elongation, and the mitotic cycle was arrested.

Fig. 6.41 *Zonaria*. Habit. × 0.66.

A population of *Zonaria farlowii* S. et G. was observed (Liddle, 1968) in the vicinity of Santa Barbara, California, in reference to the timing of the release of gametes. The initiation and development of the oogonia required a lunar month. Sporangia did not develop according to this regime. Light stimulated mature sporangiate plants to release spores. *Zonaria* has also been analyzed (Dahl, 1971) from the standpoint of providing a morphogenetic system. It was proposed that the apical cells were sensitive to a relatively few internally specified pieces of information, and their behavior (e.g., whether they would divide transversely or longitudinally) was directed by such input.

Order 10. Dictyosiphonales

This somewhat heterogeneous assemblage of brown algae share the following traits: a D^h, h + d life history, in which the sporophyte is the macroscopic phase; parenchymatous construction; and isogamous sexuality. Growth may be apical or diffuse; in some species apical growth in the young plant may be replaced by diffuse growth in older stages. The sporophytes are practically microscopic in some genera, as in *Myriotrichia* (below), but they are still larger than the gametophyte. Life histories are flexible in the sense that the sporophyte often bears plurilocular sporangia in addition to the unilocular sporangia; the zoids from the former organs serve to recycle the sporophyte asexually. Plethysmothalli (see p. 286) are also known to occur in many members, as in the Chordariales, again contributing to a flexible life history. Cells typically contain numerous discoid chloroplasts per cell, usually with pyrenoids. Four families are included in the present account. They are distinguished by the following key:

> 1. Plants of minute size, consisting of filaments that are uniseriate
>   below, becoming multiseriate above .... Myriotrichiaceae
> 1. Plants of larger size, simple or branched 2
>   2. Apices of plants with an extensive uniseriate portion before
>   longitudinal divisions appear .... Striariaceae
>   2. Parenchymatous development occurring very close to apex;
>   uniseriate portion near apices lacking 3
> 3. Growth by means of an apical cell, at least in
>   early stages .... Dictyosiphonaceae
> 3. Growth diffuse .... Punctariaceae

Family 1. Myriotrichiaceae

The members of this small family are included in the Punctariaceae in some treatments, but they are maintained here as a separate group on the basis of their simple construction. They are small and unbranched, often uniseriate in part but becoming multiseriate by vertical wall formation.

MYRIOTRICHIA Harvey *Myriotrichia* (Gr. *murios*, numerous + Gr. *thrix*, hair) includes some minute, filiform tufted epiphytes. Some species show terminal growth, while in others an intercalary meristem is developed at the base, such that the basal portion is uniseriate and the upper portion becomes multiseriate (Fig. 6.42). Its

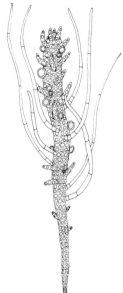

Fig. 6.42 *Myriotrichia.* Habit. × 18.

affinity with the Dictyosiphonales is thus evident by the formation of a parenchyma, despite the small stature (often less than 1 cm in length). True hairs are produced in abundance and are the reason for the generic name.

Cultures of *Myriotrichia filiformis* Harv. were established by Dangeard (1966) from plurispores of the sporophyte, and these—as would be expected—formed plethysmothalli, which produced both plurilocular and unilocular organs as well as directly gave rise to new erect filaments.

An interesting connection with an apparent member of the order Chordariales (p. 286) was contributed by Loiseaux (1969). She obtained thalli of *Myriotrichia* when zoids released by a species of *Hecatonema* were cultured. Clayton (1974) also remarked that other workers have obtained the same type of *Hecatonema* growths in their cultures of members of this order.

Family 2. Striariaceae

These algae are usually branched, cylindrical systems that terminate in long uniseriate portions. Usually at some distance from the apex there is a gradual septation by vertical cell walls (Fig. 6.43*a*), resulting in a parenchyma. A large-celled colorless medulla is surrounded by a cortex of smaller, highly pigmented cells (Fig. 6.43*b*). Most of these forms are highly branched. The sporophyte bears unilocular sporangia, plurilocular sporangia, or both at the same time.

STICTYOSIPHON Kützing   Representatives of *Stictyosiphon* (Gr. *stiktos*, spotted, + Gr. *siphon*, tube) are rather infrequently encountered on the east and west coast of North America, but the presence of a few species has been noted (Taylor,

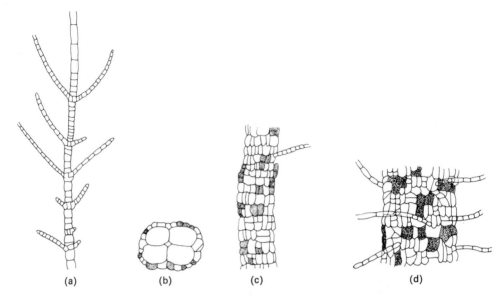

(a)   (b)   (c)   (d)

Fig. 6.43 *Stictyosiphon.* (*a*) Part of plant near apex with septation in progress in the initially uniseriate axis. (*b*) Transection of mature region of thallus, with small cortical cells and large medullary cells. (*c*) Portion of a thallus bearing unilocular sporangia. (*d*) Portion of a thallus bearing plurilocular sporangia and true hairs. (*a*) × 75; (*b*)–(*d*) × 100. (After Kuckuck.)

1962A; Wynne, 1972A; South and Hooper, 1976). In this genus reproductive organs in the sporophyte are formed by a simple conversion of cortical cells into either unilocular or plurilocular organs (Fig. 6.43*c, d*). They are scattered randomly over the surface.

   Stictyosiphon adriaticus Kütz has been investigated (Caram, 1965) in the Mediterranean and observed to have a life history as characteristic of the order, i.e., a heteromorphic pattern.

Family 3. Punctariaceae

   This family includes a large number of genera all with unbranched thalli. The forms of the sporophytes demonstrate a considerable variety. *Punctaria*, with blade-forming thalli, and *Soranthera*, with hollow globose thalli, will be used to exemplify this family.

   PUNCTARIA Greville   Blades of *Punctaria* (L. *punctum*, a point) are most apt to occur as summer annuals, epiphytic on other algae. Some are also epilithic. The thickness of these solid blades varies considerably, depending on the species. Some forms may be only two to four cells thick, as *P. latifolia* Grev., whereas others may be four to seven cells thick, as *P. plantaginea* (Roth) Grev. Upon sectioning the blade

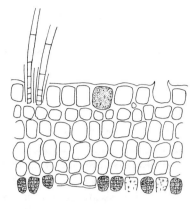

Fig. 6.44 *Punctaria*. Transection of blade with unilocular and plurilocular sporangia. × 110.

(Fig. 6.44*a*), it is noteworthy that there is not a marked difference in size between the cortical and medullary cells. As in *Stictyosiphon*, reproductive organs (plurilocular and unilocular) are formed by the conversion of cortical cells into these structures, and they may protrude from the cortex or remain at the same level (Fig. 6.44*a*). True hairs are often produced in profusion.

Two recent studies (Dangeard, 1963; Clayton and Ducker, 1970) of *Punctaria latifolia*, a wide-ranging species, demonstrated in culture a direct return of foliose plants from unispores of the sporophyte, suggesting apomeiosis in the unilocular sporangia. This is in contrast to the heteromorphic pattern originally reported (Sauvageau, 1929).

SORANTHERA Postels et Ruprecht   The saccate thalli (Fig. 6.45*a*) of *Soranthera* (Gr. *soros*, a heap + Gr. *antheros*, blooming) are distinctive in appearance in that they are balloon-shaped, about 1–2 cm in diameter, and dotted with sori of emergent

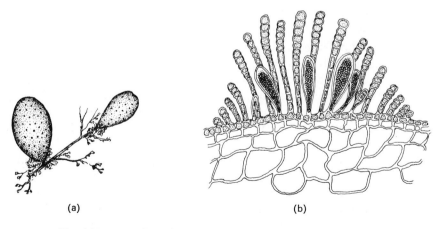

(a)   (b)

Fig. 6.45 *Soranthera ulvoidea* Post. et Rupr. (*a*) Habit of plants epiphytic on a red algal host. (*b*) Sectional view of sorus containing unilocular sporangia and multicellular filaments. (*a*) × 0.8; (*b*) × 248. [(*b*) after Smith.]

unilocular sporangia. The only species, *S. ulvoidea* P. et R., is restricted to the Pacific coast of North America but ranges from the Bering Sea to southern California. Invariably the sacs are epiphytic on the red algae *Odonthalia* or *Rhodomela*, both members of the family Rhodomelaceae (p. 559).

The sporophyte bears only unilocular sporangia, and these are clustered among multicellular photosynthetic filaments (Fig. 6.45b) arising from the continuous cortical layer of small cells. The complete life history was followed in culture (Wynne, 1969A), and the pattern conformed to that typical of the order Dictyosiphonales. The microscopic filamentous gametophytes produced narrow plurilocular organs, the gametangia. When the gametes were followed to the subsequent generation, two types of germlings were observed. The great majority were narrow, twisted filaments and were fated to reach only a few cells in length before dying; these were regarded as derived from unfused gametes. A small percentage of robust, loosely prostrate filamentous growths also appeared (regarded to be derived from zygotes). This second type of germling continued its development, eventually producing parenchymatous knobs or solid packets of cells that expanded into the hollow sacs characteristic of the sporophyte.

Family 4. Dictyosiphonaceae

Two rather dissimilar looking genera are placed in this family, *Dictyosiphon* and *Coilodesme*. In both cases growth is apical at least initially; in *Coilodesme* it would seem to be quickly replaced by diffuse growth. One important common trait is the sunken position of the unilocular sporangia.

DICTYOSIPHON Greville   The type species of *Dictyosiphon* (Gr. *dictyon*, a net + Gr. *siphon*, a tube), *D. foeniculaceus* (Huds.) Grev., is a good example of the formation of a solid parenchyma by the transverse and vertical segmentation of segments from the apical cell (Fig. 6.46a). This occurs close to the apex in contrast to the more remote segmentation characteristic of *Stictyosiphon* (Fig. 6.43a). This species is abundantly branched and forms an entangled mass. At maturity unilocular sporangia are produced from cortical cells, and they become somewhat sunken.

The life history of *D. foeniculaceus*, which occurs on both sides of the North Atlantic, was determined (Sauvageau, 1929) to be heteromorphic, the microscopic gametophyte producing uniseriate plurilocular gametangia and isogametes. Thus, the life-history characteristic of the order is present in the type species of *Dictyosiphon*.

COILODESME Strömfelt   In contrast to the much branched, generally terete thalli of the previous genus, *Coilodesme* (Gr. *koilos*, hollowed + Gr. *desme*, a tuft) includes saccate forms (Fig. 6.46c) tubular, flattened, or greatly swollen and always hollow. Unilocular sporangia are borne sunken or subsurface (Fig. 6.46b). Plurilocular sporangia, although rare, have been described in at least one species (Wynne, 1972A). A direct return to the saccate form of the sporophyte by the germination of unispores was described (Wynne, 1972A) for two species of this genus. The half dozen species of *Coilodesme* are distributed in the North Pacific.

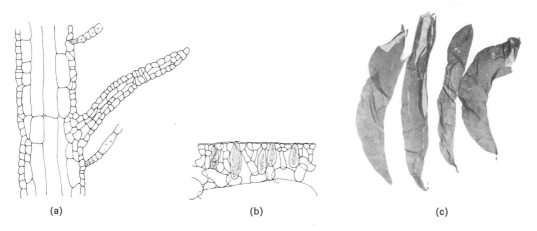

(a)   (b)   (c)

Fig. 6.46 (*a*) *Dictyosiphon foeniculaceus* (Huds.) Grev. Vertical section of an axis showing parenchymatous construction and apical growth in lateral branch. True hairs also occur. (*b*) *Coilodesme bulligera* Stroem. Section of thallus showing immersed unilocular sporangia at periphery. (*c*) *Coilodesme californica* (Rupr.) Kjellm. Habit. (*a*) × 150; (*b*) × 225; (*c*) × 0.26. [(*a*), (*b*) after Kuckuck.]

Order 11. Scytosiphonales

This order includes about eight genera that were formerly placed in the previous order, the Dictyosiphonales, but ample evidence has been accumulated in recent years to warrant their classification as a distinct order of brown algae. Two of the criteria used by Feldmann (1949) in recognizing this order are the presence of a single chloroplast with a conspicuous pyrenoid in each vegetative cell and the presence of only plurilocular organs in the macroscopic plant. A more recent contribution was made by Nakamura (1965) followed by Tatewaki (1966), who demonstrated that an alternative crustose phase resembling the genus *Ralfsia* (p. 283) is present in the life history of some of the members of this order. The precise relationship between the macroscopic phase and the smaller crustose or discoid phase remains controversial. Tatewaki (1966) described for *Scytosiphon lomentaria* (Lyngb.) Link an alternation between a macroscopic sexual stage (the gametophyte) and a crustose sporophyte, bearing unilocular sporangia. The occurrence of a crustose stage was subsequently recorded for the related species *Petalonia fascia* (O. F. Muell.) Kuntze by several authors (Wynne, 1969A; Hsiao, 1969; and Edelstein, et al., 1970) but with no evidence for a sexual cycle. Further discussion is presented below.

Family Scytosiphonaceae

Members of this family are of common occurrence in littoral to sublittoral habitats, often being present for only a few months and then disappearing. In many cases they are winter annuals, as *Petalonia fascia* along the Texas coast (Edwards, 1969A) and *Scytosiphon dotyi* Wynne on the California coast (Wynne, 1969A). The discovery, therefore, of an alternative crustose stage in the life history of some of these algae adds a dimension to our understanding of their seasonal appearance.

Some members are more characteristic of tropical and subtropical waters, as the net-like *Hydroclathrus*, the convoluted balloon-shaped *Colpomenia sinuosa* (Roth)

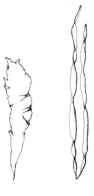

Fig. 6.47 (*a*) *Petalonia fascia* (O. F. Muell.) Kuntze. Habit. (*b*) *Scytosiphon lomentaria* (Lyngb.) Link. Habit. (*a*) × 0.25; (*b*) × 0.3.

Derb. et Sol., and the branched *Rosenvingea*. The two genera to be discussed in greater detail here are both prevalent in temperate waters. They are the foliose *Petalonia* and the unbranched tubular *Scytosiphon*.

PETALONIA Derb. et Sol. Blades of *Petalonia* (Gr. *petalon*, leaf) may be epiphytic or epilithic and usually have a tapered appearance (Fig. 6.47*a*). The sterile blade in cross section (Fig. 6.48*a*) reveals monostromatic cortical layers and a medulla a few cells thick. Unlike *Punctaria* in cross section (Fig. 6.44*a*), there is a size differ-

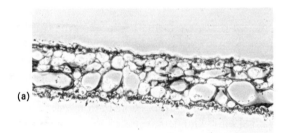

(a)

Fig. 6.48 *Petalonia fascia*. (*a*) Transection of sterile blade. (*b*) Transection of fertile blade. (*c*) Vertical section of fertile crustose stage. (*a*), (*b*) × 170; (*c*) × 118.

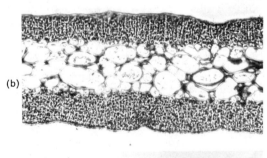

(b)

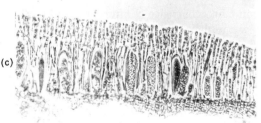

(c)

ence between cortical and medullary cells in *Petalonia*. When fertile, the cortical cells undergo periclinal divisions to develop thickened areas of cells that are converted into files of cells representing plurilocular organs. The entire cortical layer of each blade surface becomes reproductive (Fig. 6.48b), and large numbers of zoids are released.

The alternate phase of *Petalonia* has been described as a crustose stage (Wynne, 1969A; Edelstein, et al., 1970; Lüning and Dring, 1973) resembling *Ralfsia*. This crust consists of a consolidated **hypothallium** and an **epithallium** of multicellular filaments from the bases of which unilocular sporangia are laterally produced (Fig. 6.48c).

Wynne (1969A) found evidence that environmental conditions play a determining role whether the foliose or crustose phase of *Petalonia* is produced. Culturing experiments indicated that formation of the crust (the **microthallus**) was favored by long photoperiod (16:8 hours) and relatively warm temperature (18–19C), whereas the formation of the blade (the **macrothallus**) was favored by a short photoperiod (8:16) and relatively cool temperature (10°C). Intermediate conditions of a 12:12-hour photoregime and 13°C resulted in approximately equal numbers of the blade expression and the crustose expression. No evidence of a sexual cycle between the two expressions was obtained in the California material studied (Wynne, 1969A), and recent further investigations (Nakamura and Tatewaki, 1975) of *P. fascia* in Japan also reported that sexuality was absent. It was observed that parthenogenetically developing gametes of either sex undergo two different pathways, either erect fronds with plurilocular organs or crusts with unilocular organs, the former occurring under relatively cool conditions and the latter under relatively warm conditions.

This problem was analyzed in greater detail by Lüning and Dring (1973) in regard to the possible effect of different qualities of light on the expression of the blade or the crustose stage. They observed a definite morphogenetic effect mediated by light in *Petalonia fascia* and in *Scytosiphon lomentaria*. When these algae were grown exclusively under red light, the erect system (i.e., the blade in *Petalonia* and the tube in *Scytosiphon*) was produced, yet with blue light or white light still being required for release of zoospores. When these algae were grown under blue light, both species developed only the crusts or prostrate filamentous stage. This experimentation was followed up with *Scytosiphon*, and a genuine photoperiodic response was demonstrated (see below).

Finally, *Petalonia fascia* was studied in The Netherlands (Roeleveld, et al., 1974) on a year-round basis in nature and in laboratory. The evidence presented again showed that both temperature and photoperiod influenced the expression of macrothalli or microthalli. Culture experiments closely corroborated the phenology of this species in the field, with only minor discrepancies in the time of optimal blade development being noted. Blades of *P. fascia* were found to be induced by short-day conditions, with apparently three generations of blades being produced during the growing season for this species in The Netherlands, namely, November to May.

SCYTOSIPHON C. Agardh   The type and most widespread species of *Scytosiphon* (Gr. *scutos*, a whip + Gr. *siphon*, a tube), *S. lomentaria*, is distinctive in that the unbranched, tubular thalli are regularly constricted along their length (Fig. 6.47b). Again, as in *Petalonia*, only plurilocular organs are produced by fertile specimens,

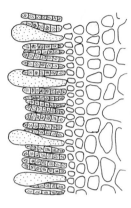

Fig. 6.49 *Scytosiphon lomentaria*. Portion of sectioned fertile thallus, showing plurilocular organs and scattered unicellular paraphyses. × 150.

and they extend over the entire surface. Scattered among the plurilocular organs are conspicuous unicellular paraphyses (Fig. 6.49), which have a clear, refractive appearance and which might be mistaken for unilocular organs.

It was in Japanese material of *S. lomentaria* that Tatewaki (1966) reported a heteromorphic life history consisting of the familiar cylinder as the sexual macrothallus and a smaller prostrate microthallus as the asexual stage, resembling the genus *Ralfsia*.[6] Sexual plants were separately sexed, with slightly anisogamous gametes reported. Gametes of either sex developed parthenogenetically, their type of development depending on the culture conditions of temperature and photoperiod (Nakamura and Tatewaki, 1975). Relatively cool, short-day conditions brought about the development of additional cylindrical plants; relatively warm, long-day conditions brought about crustose discs or tufted plants with unilocular sporangia, which, however, were apomeiotic. These results corroborate those of Wynne (1969A) with *S. lomentaria* from California in regard to the influence of culturing conditions on the pattern of development, except for the apparent absence of sexuality in the California material.

For this same species a light-quality effect was demonstrated (Lüning and Dring, 1973), such that cultures grown under red light produced the macrothallus, whereas cultures grown under blue light produced the microthallus. This two-dimensional growth as well as the formation of true hairs in *S. lomentaria* were directly dependent on blue light (Dring and Lüning, 1975A) and occurred independently of photosynthesis or growth. Finally, these same authors (Dring and Lüning, 1975B) conducted experiments, demonstrating that a genuine photoperiodic response occurred in *S. lomentaria* and that the critical daylength was sharply defined. Macrothalli were produced by 8-hour daylengths, and microthalli were produced by 16-hour daylengths. Between 12- and 13-hour daylengths, differences in the photoperiod of only 15 minutes brought about significant differences in response. They also discovered that the response to

[6]Some workers (Lund, 1966; McLachlan, et al., 1971B) have observed a *Microspongium*-stage, which is a gelatinous, more loosely organized crustose alga than *Ralfsia*. Also, a small unnamed *Scytosiphon* has been reported (Loiseaux, 1970A) to have *Streblonema* and *Compsonema* stages, the latter being a genus of Myrionemataceae (p. 286).

short-daylengths is completely inhibited by a 1-minute light break with a low irradiance of blue light that is administered in the middle of 16-hour dark periods. It was concluded, however, that the photoreceptor pigment is not phytochrome (see p. 476) because the inhibitory effects of the blue light are not reversed by subsequent irradiation with other wavelengths.

The significance of the interest in members of this order is that the perplexing state of our knowledge of their life history has stimulated various approaches to unraveling their biology in various parts of the world, and the aftermath of this heightened research activity has been to develop a better appreciation of the subtle interactions of environmental influences such as photoperiod, quality of light, temperature, and nutrient concentrations upon the morphogenesis of rather simply constructed brown algae.

Order 12. Laminariales

This order includes the largest and most structurally complex of all the algae. They are collectively referred to as the "kelps" and constitute a significant floristic component of the lower littoral and sublittoral zones on almost any rocky coast in temperate or polar seas. On the west coast of North America they extend from the Arctic Circle (Mohr, et al., 1957) southward to Baja California of Mexico (but not into the Gulf of California), being sporadically present in regions of upwelling at the southern extreme of their range (Dawson, 1945, 1951). On the east coast of North America they range again north of the Arctic Circle (Wilce, 1959; Ellis and Wilce, 1961) southward to Long Island Sound, New York (Taylor, 1962A). Two species of *Laminaria* are also known (Joly and Oliveira Filho, 1967) from the coast of Brazil.

The order Laminariales is characterized as follows:

1. A D^h, h + d pattern of life history, involving a macroscopic sporophyte and a microscopic gametophyte.
2. Oogamous reproduction by dioecious gametophytes.
3. Parenchymatous construction.
4. Growth in length by an intercalary meristem, with additional growth in girth[7] contributed by a superficial meristem called the **meristoderm.**

Some kelps such as *Chorda* and *Nereocystis* are annuals, all of their growth taking place in one growing season. Rates of elongation in plants of *Nereocystis*, which can commonly reach 25 m in length, have been measured to be about 3 cm/day (Nicholson, 1970) and 6 cm/day (Scagel, 1947). Other kelp genera such as *Macrocystis*, *Pterygophora*, and *Egregia* are perennials, with one specimen of *Pterygophora* reported (MacMillan, 1902) to be 17 years old on the basis of annual increments of growth discernible in a transection of the stipe. Most species of *Laminaria* are perennial. The perennial species of *Laminaria* simply have the blade slough off after reproduction, the stipe

---

[7]Another secondary meristem, termed a *medullary meristem* and located between the cortex and the medulla, has been described (Clendenning, 1964) in *Macrocystis*.

Fig. 6.50 *Laminaria.* Example of a simple species. × 0.04.

and meristem persisting. The meristem produces a new blade at the beginning of the next growing season.

The sporophyte of the Laminariales (Fig. 6.50) is typically constructed of three parts: the blade (lamina or phylloid), the stipe (cauloid), and the holdfast (consisting of haptera that anchor the alga firmly to the substrate, which is usually rock but may be wood or rarely another alga). The intercalary meristem, or **transition zone**, is located at the juncture of blade and stipe. The blade may be divided or undivided. Such divisions of splits in the blade may or may not extend down into the transition zone.

The only type of reproductive structures present on the sporophyte are the unilocular sporangia. These arise laterally from the base of columnar paraphyses (Fig. 6.51*a*), and the somewhat dilated apices of these paraphyses form a protective layer over the sporangia below. Usually from 16 to 64 zoospores are produced within each sporangium, and they are haploid cells, resulting from the original single nucleus within the young sporangium undergoing meiosis followed by mitotic divisions. Only *Chorda* has zoospores with eyespots.

The sori of unilocular sporangia may be located on any portion of the blade, or there may be special blades to which the fertile areas are confined; these blades are termed **sporophylls**. These traits are employed in distinguishing the four families of the order. A key to these families is presented herewith:

1. Differentiation into stipe and blade lacking..................Chordaceae
1. Differentiation into stipe and blade present...........................2
  2. Longitudinal splits extending through transition zone, dividing up the original blade into several blades and increasing the number of intercalary meristems ....Lessoniaceae
  2. Longitudinal splits not extending into transition zone 3
    3. Sori of unilocular sporangia occurring any place on the lamina ....Laminariaceae
    3. Sori of unilocular sporangia restricted to only certain blades, the sporophylls ....Alariaceae

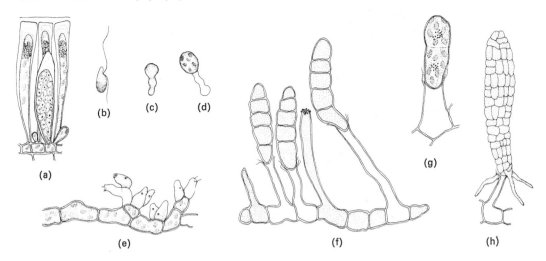

Fig. 6.51 Stages in the life history of a kelp (*Eisenia arborea* Aresch.). (*a*) Unilocular sporangium among longer paraphyses. (*b*) Zoospore released from a unilocular sporangium. (*c*) Early germination stage. (*d*) Germling that has evacuated the original germ wall of settled zoospore. (*e*) Cluster of antheridia borne on a male gametophyte. (*f*) Female gametophyte bearing two incipient oogonia and three discharged oogonia, attached to which are young sporophytes. (*g*) Two-celled sporophyte. (*h*) Young sporophyte. (After Hollenberg.)

The exciting discovery that a heteromorphic life history occurs in the order Laminariales was made in 1915 by Camille Sauvageau in France. He made the observation that a microscopic gametophytic stage alternated with the large sporophyte of *Saccorhiza*. The sequence of events in the reproduction of *Eisenia* is outlined in Fig. 6.51. Zoospores (Fig. 6.51*b*) from unilocular sporangia (Fig. 6.51*a*) germinate (Fig. 6.51*c, d*) into two types of gametophytes: large-celled female gametophytes, which may be of limited growth and little branched, and small-celled male gametophytes, which are abundantly branched. Clusters of antheridia are produced on the male gametophyte (Fig. 6.51*e*), and each produces a single laterally biflagellate sperm. Intercalary or terminal oogonia are developed in the female gametophyte, with the oogonial cell protruding outward and then extruding the protoplast as a spherical egg on the lip of the oogonium. The egg is fertilized and begins its germination still attached to the wall of the oogonium (Fig. 6.51*f, g*).

In an ultrastructural study of the zygote (Bisalputra, et al., 1971), it was shown that the zygote maintains its attachment to the oogonial wall by an attachment cushion, which possibly results from a localized gelatinization of the oogonial wall. Although the chloroplasts in the mature sporophyte lack pyrenoids, chloroplasts in the zygote have them.   *PHYLOGENETIC IMPORTANCE ?*

Rhizoids issue from the young sporophyte, and transverse divisions divide the germling into an elongate structure (Fig. 6.51*h*), which soon also undergoes longitudinal and periclinal divisions, resulting in a polystromatic blade, a shape typical of all members of the order at this juvenile state (Fig. 6.52). Once the blade, stipe, and holdfast have been initiated, growth in length becomes localized in the juncture of the

Fig. 6.52 *Postelsia palmaeformis* Rupr. Field of 20-day-old sporophytes. × 67.

stipe and blade. This transition region becomes responsible for growth in both blade and holdfast in the mature plant.

In all portions of the thallus there are basically three types of tissue: the superficial meristoderm, which is both photosynthetic and meristematic; a parenchymatous cortex; and the central medulla, consisting of a loose network of longitudinally directed cells interconnected by transverse files of cells. The peripheral meristem produces rows of cells inward, and they become separated by mucilage. This mucilage between adjacent cells corresponds to the middle lamella and the primary walls, which are markedly thickened and composed of alginic acid (Roelofsen, 1959; van Went and Tammes, 1973), with great swelling capacity. The secondary wall is much thinner and composed of cellulose.

The cells of the cortex produce papillae from their sides; these papillae are opposite one another and fuse to form cross-connecting filaments. A dense reticulum thus arises, more and more mucilage being deposited in the intervening spaces (Fig. 6.53a). From the innermost cells of the cortex are derived two types of cells: sieve elements and hyphal cells (Schmitz and Srivastava, 1975). They contribute to the formation of the medulla. These cells derived from the inner cortex are initially much longer than wide but undergo transverse divisions to yield vertical files of cells, which separate from each other by swelling of the wall material. Some of these cells differentiate into sieve elements and others into hyphal cells (Schmitz and Srivastava, 1974). The sieve elements greatly elongate, although their diameter at the cross walls remains the same, resulting in a trumpet shape (Fig. 6.53b). This distinctive trumpet shape may be due to the loss of turgor when the tissue is sectioned for the microscope, and thus artifactual, in the opinion of van Went and Tammes (1973); but others (Schmitz and Srivastava, 1974) disagree with this interpretation. The innermost cortical cells continually produce additional sieve elements, which are superimposed upon each other to form "sieve tubes" running mainly longitudinally in the thallus but also running radially, as in wings of *Alaria* (p. 339) (Schmitz and Srivastava, 1975). Older sieve elements become stretched and eventually nonfunctional, but new crops of sieve elements are being continually added to maintain continuity between the blade and the stipe.

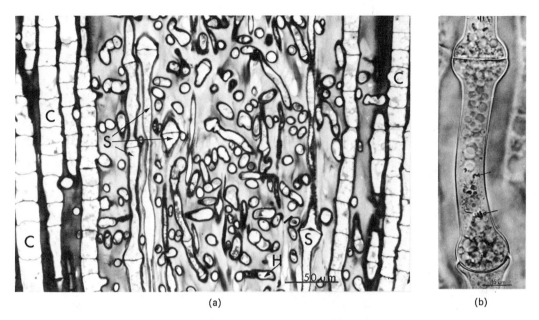

(a)   (b)

Fig. 6.53 (*a*) *Alaria marginata* P. et R. Radial longitudinal section through the midrib at the intercalary growing region. Vertically aligned inner cortical cells (C) become separated from each other along their longitudinal walls by mucilage production. Sieve cells (S) and hyphal cells (H) are present in the medulla. (*b*) *Laminaria groenlandica* Rosenv. A young sieve element *in vivo*. The nucleus is evident as well as vacuoles and chloroplasts (arrows). [(*a*) after K. Schmitz and L. M. Srivastava and permission of the National Research Council of Canada, from the Canadian Journal of Botany 53: 861–876, 1975; (*b*) after Schmitz and Srivastava and permission from Cytobiologie 10: 66–86, 1974.]

The cross walls between adjacent sieve elements contain fields of pores termed *sieve areas*. In *Laminaria* a sieve area contains numerous small pores (Fig. 6.54) but in the structurally more advanced genera *Macrocystis* and *Nereocystis* pores in a sieve area are both larger and fewer (Ziegler and Rück, 1967; Parker, 1971A, B; Nicholson, 1976). These sieve areas have been demonstrated (van Went, et al. 1973; van Went and Tammes, 1972) to have open plasmodesmata that facilitate the flow of fluids between sieve elements. The flow of exudate from a sectioned kelp, particularly the stipe, has been recognized, and translocation velocities have been measured in the giant kelps. For *Nereocystis*, rates of translocation averaging 37 cm/hour were recorded (Nicholson and Briggs, 1972) in laboratory aquaria; for *Macrocystis*, translocation rates in the stipe were measured (Parker, 1965, 1966) at 65–78 cm/hour, after the blades were labeled. Experiments by Nicholson and Briggs (1972) clearly demonstrated that translocation did take place in the medulla, the tissue to which the sieve tubes were confined. To rule out the possibility of the flow of photosynthate through the mucilage ducts, which are located in the cortex, careful girdling of *Nereocystis* blades was carried out, and it was determined that the radioactively labeled photosynthate

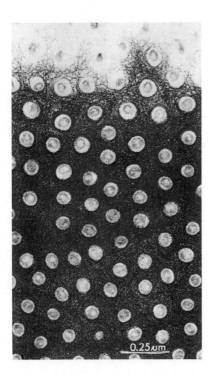

Fig. 6.54 *Laminaria groenlandica*. Oblique surface view through a sieve area. A meshwork of fibrils is seen to compose the secondary wall (upper portion). (After Schmitz and Srivastava and permission from Cytobiologie 10: 66–86, 1974.)

was located in the exudate from the medulla. Results also indicated that this flow does depend on living tissue and the primary product of photosynthesis is the alcohol mannitol.

Various investigators have focused attention on the induction of sexuality in this order. For *Nereocystis*, Vadas (1972) concluded that light was the single most critical factor in gametogenesis. Low light (161 lux) and short daylengths inhibited or retarded sexual maturity, with gametogenesis occurring at 5, 10, and 15°C but not at 20°C. On the other hand, the number of oogonia produced in *Laminaria hyperborea* (Gunn.) Foslie was affected by temperature (Kain, 1964). Axenic cultures of gametophytes have been obtained (Druehl and Hsiao, 1969) by a process of surface sterilization of the sorus prior to spore release and by using a medium containing antibiotics. Optimal conditions for sexuality have been described (Hsiao and Druehl, 1971). The levels of nitrate and phosphate concentrations necessary for gametogenesis in *Laminaria saccharina* have been determined (Hsiao and Druehl, 1973A). The female gametophytes were more limited by low nutrient levels than male gametophytes, oogonial production occurring over a narrower range of nitrate and phosphate concentrations. Field studies (Hsiao and Druel, 1973B) indicated that variations in light were more influential in the rate of gametogenesis than were nutrient levels.

Light quality has been analyzed (Lüning and Dring, 1972, 1975) in respect to gametogenesis in *Laminaria saccharina* (L.) Lamour. It was recognized that 6 to 12-hour irradiance with blue light induces egg formation by female gametophytes at

15°C, while they remained sterile in red light. At lower temperatures, however, some gametophytes became fertile in red light, but blue light greatly increased the percentage of eggs formed. Male gametophytes were similarly induced to form antheridia, and this response, as for the female gametophyte, was independent of photosynthesis and growth.

That sex is genotypically determined in the gametophytes was confirmed by the observation (Evans, 1965) of a large X chromosome in the female gametophyte and a small Y chromosome in the male gametophyte of *Saccorhiza polyschides* (Lightf.) Batt. These sex chromosomes were seen to pair during meiosis. A similar large X chromosome was also noted in four other Laminarialean algae.

Although the life history of the members of this order would appear to be rigidly defined, the expression of gametophyte or sporophyte is not strictly tied to the ploidy level, as shown in *Alaria*. Nakahara and Nakamura (1973) recorded that as many as 13% of the unfertilized eggs developed parthenogenetically into haploid sporophytes indistinguishable from diploid sporophytes. Likewise, haploid sporophytes were developed apogamously from the terminal cells of male gametophytes. Furthermore, by a process of apospory diploid gametophytes could be obtained from vegetative cells of a diploid sporophytes, and by sexual reproduction tetraploid sporophytes could next be obtained.

Representative genera from the four families are discussed briefly in the following account.

Family 1. Chordaceae

This family includes the single genus *Chorda*.

CHORDA Stackh.   The lack of a distinction between stipe and blade is what separates *Chorda* (L. *chorda*, a cord) from other kelps. The unbranched cylindrical axes of *Chorda* may reach 8 (to 12) m in length. This genus is fairly common in cold-temperate waters of the North Atlantic, the North Pacific, the Arctic Sea, and the Bering Sea. This distribution is true for both species, *C. filum* (L.) Stackh., which is the type, and *C. tomentosa* Lyngbye, which differs from the type in having a dense covering of pigmented hairs. One other apparent difference between the two species is that *C. filum* is reported (South and Burrows, 1967) to have dioecious gametophytes, whereas *C. tomentosa* is reported (Sundene, 1963) to have monoecious gametophytes.

The sporophyte of *C. tomentosa* is a cold-water alga, appearing in the winter and disappearing by early summer (Sundene, 1963) in the southern part of its range but persisting until late summer in its northern range. Gametophytes also grew best at low temperatures, not becoming fertile at temperatures above 10–12°C. Gametophytes of *Chorda filum* were able to produce sporophytes at salinities as low as 5‰. Norton and South (1969) expressed the opinion that this tolerance to low salinity was seemingly correlated with the distribution of this species in the Baltic Sea.

In field studies of *C. filum*, South and Burrows (1967) showed a seasonal cycle to occur, with a succession of primary sporophytes (small axes without a localized zone of growth) in February through April, followed by secondary sporophytes with meristems, and these followed by secondary sporophytes lacking meristems. Fruiting of the sporophytes began in midsummer and continued through the winter.

Family 2. Laminariaceae

This family is well represented on both the east and west coasts of North America but is absent in the Gulf of Mexico and the Gulf of California. All genera share the trait of having blades that are either simple or longitudinally split but never with the splits extending into the transition zone. Characteristics of the blade are used as a means of distinguishing a large number of genera. Blades may lack midribs, as in *Laminaria* and *Hedophyllum*, or be provided with them, as in *Costaria*, *Agarum*, and *Pleurophycus*. Perforations of the blade may be a constant feature, as in *Agarum* and *Thalassiophyllum*. Unilocular sporangia are produced in usually large areas on the blade, which are referred to as sori; unilocular sporangia are the only type of reproductive organ produced by the sporophyte, which is true of the order.

LAMINARIA Lamour.   Blades of *Laminaria* (L. *lamina*, blade) are a conspicuous feature of the lower littoral and upper sublittoral zones of most temperate to boreal rocky shores (Fig. 6.55), and a large number of species have been described. All species have in common a holdfast, a stipe, and a blade lacking a midrib. Many variations exist upon that general morphological scheme. The blades of some species are always undivided (Fig. 6.56), whereas those of other species are always divided

Fig. 6.55 *Laminaria* zone. Specimens of this genus populate the lower intertidal zone and extend into the upper sublittoral zone. (Courtesy of Fred Weinmann.)

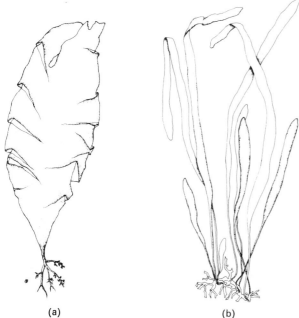

(a)   (b)

Fig. 6.56 (*a*) *Laminaria saccharina* (L.) Lamour. (*b*) *Laminaria sinclairii* (Harv. ex Hook. et Harv.) Farl., Anders., et Eaton. (*a*) × 0.16; (*b*) × 0.2.

or sometimes divided. Most species are perennial, the blade being eroded away above the transition zone and only the holdfast and stipe persisting during the nongrowing season. A new blade is then developed from the tip of the stipe with the onset of the new growing season. In some situations the blade persists until the new growth causes the previous year's blade to be sloughed away (Fig. 6.57). Some annual species are known, such as *L. ephemera* Setchell in Puget Sound.

The holdfasts in the majority of *Laminaria* species are in the form of masses of haptera, which are branched cylindrical processes firmly adhering to the substratum. Holdfasts in the form of a simple disc are also known. In a few species a branching system of rhizomatous horizontal axes is established, which gives rise to new crops of stipes and blades. This expanded rhizome-like system in the west coast *Laminaria sinclairii* (Harvey) Farlow, Anderson, et Eaton (Fig. 6.56*b*) facilitates its survival in exposed lower littoral localities periodically buried in sand (Markham, 1973). Its greatest growth occurs in spring prior to burial, and vegetative proliferation of new blades from the rhizomatous haptera is apparently more significant than the sexual reproduction by the gametophytes. It has been shown (Markham, 1968) that detached portions of haptera in this species, as short as 2.5 mm in length, will produce outgrowths that develop into entire plants.

The blades of some species have a blistered or puckered aspect due to the presence of bullations; these are well developed in the west coast *L. farlowii* Setch., in which they consistently cover the blade, but in the more common *L. saccharina* (L.) Lamour., which occurs on both the east and west coasts, the bullae are variable in their occurrence, being influenced by temperature. In a comparison of two European populations

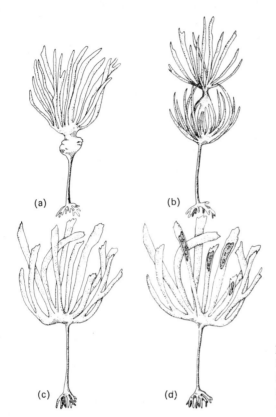

(a)

(b)

(c)

(d)

Fig. 6.57 *Laminaria*. (*a*) Sequence of new blade formation in transition zone. (*b*) Sloughing off of previous year's growth. (*c*) Vegetative growth. (*d*) Reproductive maturation. (After Baardseth.)

of *L. saccharina* (Lüning, 1975), specimens growing at the Isle of Man in the Irish Sea had deep bullations, whereas specimens growing at Helgoland in the North Sea had smooth blades. When sporophytes derived from transplanted Isle of Man gametophytes and hybrid sporophytes between the two populations were grown at Helgoland, bullations were formed. However, the Isle of Man form did not tolerate the water temperatures above 14°C that occur during the summer at Helgoland and thus did not survive. Thus, ecotypic differences within a species are observable.

In general a great deal of morphological variation is recognized in the kelps. The criteria used in the classification of species of *Laminaria* have been reviewed (Burrows, 1964; Wilce, 1965; Druehl, 1968; Markham, 1972), and such traits as gross morphology of the blade, stipe, and holdfast and mucilage duct distribution are usually employed. Recent crossing experiments conducted by Chapman (1975) demonstrated, however, that the presence or absence of mucilage ducts is not a valid taxonomic criterion at least for the nondigitate species he examined. Complete crossability between all mucilage canal types was obtained, and there was evidence that environmental factors were influential in the phenotypic variability of the degree of development of mucilage ducts.

Transplant experiments have been conducted to gain a better understanding of the effect of environmental factors on blade morphology. When plants with very narrow blades (f. *stenophylla*) of *Laminaria digitata* (Huds.) Lamour. were transferred from an extremely exposed locality and allowed to grow in a sheltered habitat, the plants developed into broad-leaved forms with widths eight times greater than those obtained in the exposed locality. This experiment accordingly demonstrates the influence of wave motion (Sundene, 1964A). Another set of such transplant experiments (Svendsen and Kain, 1971) revealed that a common form may be achieved in two different species of *Laminaria* when these species are grown in sites of little water motion because of depth or extreme shelter. The two species, *L. hyperborea* and *L. digitata*, normally do not occur in such sheltered sites; but when they do, a distinctive form is independently developed, which had been called *L. cucullata*. The transplant experiments, supported by hybridization investigations, confirmed that this form represents merely an ecad (a phenotypic variant environmentally induced) for these two distinct species.

Transplant experiments were also involved in the study by Druehl (1967) to analyze the local distribution of two species of *Laminaria* around Vancouver Island, Canada. He determined that the distribution of these species was related to environmental factors. *Laminaria saccharina* was both eurythermal and euryhaline but could not tolerate surf conditions and thus was absent in areas subjected to surf. *Laminaria groenlandica* Rosenv., on the other hand, was tolerant of surf conditions but was sensitive to high temperature and low salinity. Two forms were observed in this latter species; a short-stipe form was prevalent in areas of moderate surf, while a long-stipe form was collected in areas of heavy surf. The results of laboratory growth of the gametophytes and sporophytes revealed a correlation with the field observations in regard to tolerance or lack of tolerance to variations in salinity and temperature.

One of the most intensively studied species of *Laminaria* is *L. hyperborea*. Several European workers (Kain, 1967, 1969, 1975; Larkum, 1972; Lüning, 1970, 1971) have contributed significantly to the autecology of this kelp, the effects of light on the growth both of sporophytes and gametophytes, differences at varying latitudes, and the role of competitors.

The cultivation and harvest of *Laminaria* (Fig. 6.58) are a well-developed science, particularly in the western Pacific [China and northern Japan (Cheng, 1969)]. *Laminaria japonica* Aresch., an especially desirable species, has been grown (Hasegawa, 1972) under experimental conditions in 1 year's time to a size equal to that of 2-year-old specimens in nature with no loss of quality. A method for obtaining axenic cultures of *Laminaria* gametophytes and other kelps has been devised (Druehl and Hsiao, 1969).

The synthesis of sulfated polysaccharides has been demonstrated (Evans, et al., 1973) by means of autoradiography to take place in the secretory cells of three different species of *Laminaria* and then to be passed to the outside via an intricate system of mucilage ducts. It should be emphasized that not all species of *Laminaria* have such mucilage ducts. The translocation of ^{14}C-labeled assimilates, which included mainly mannitol but also several amino acids, was demonstrated (Lüning, et al., 1972) to be

Fig. 6.58 Harvesting *Laminaria japonica* Aresch. on Hokkaido Island, Japan.

from the distal parts of the blade to the basal part and to a smaller extent to the hold-fast. So translocation does occur in the Laminariaceae as it does in the Lessoniaceae even though the plants of *Laminaria* are smaller and lack enucleate sieve tubes and microscopically visible pores.

Another study (Hellebust and Haug, 1972A) indicated that the amino acids are the most likely sources of carbon for alginic acid synthesis and respiration in the dark, since the mannitol seems to be relatively unavailable. The outermost cortical layer carries on the greatest amount of photosynthesis (Hellebust and Haug, 1972B). Although the underlying cortical layer and medulla have significant photosynthetic capabilities, the darkly pigmented outermost cortex strongly shades these tissues, minimizing their photosynthetic activity under normal circumstances. The ultra-structure of the meristoderm cells in haptera has been examined (Davies, et al., 1973).

HEDOPHYLLUM Setchell Blades of *Hedophyllum* (Gr. *hedos*, a seat + Gr. *phyllon*, a leaf) are easily distinguished from other members of the Laminariaceae by the absence of a stipe, which disappears when the alga is in an early stage of growth.

The simple to deeply divided blade thus has a sessile appearance (Fig. 6.59a), with a mass of haptera attaching the alga to the substratum. Additional haptera are formed from the margins of the proximal end of the blade. *Hedophyllum sessile* (C. Ag.) Setch., the single variable species of this genus, is distributed from Alaska to northern California, occupying the lower part of the littoral zone. On Amchitka Island in the Aleutians (Lebednik, et al., 1971) it constitutes a major component of the intertidal flora, occupying a considerable horizontal distance perpendicular to the gradually sloping shoreline but with a relatively small vertical distribution, occupying a narrow zone just above the MLLW line (i.e., the mean lower low water line; *cf.* Doty, 1946). In an investigation of populations of *Hedophyllum* occurring along the Strait of Juan de Fuca, it was observed (Widdowson, 1965) that direct exposure to sunlight causes bullation, or blistering, of the blades and that wave action brings about splitting of the blades. Another study (Paine and Vadas, 1969) in the same area revealed that *Hedophyllum* became the dominant species in lower intertidal areas from which sea urchins had been removed. The fact that *Hedophyllum* was not nearly so conspicuous in control areas was attributed primarily to the grazing effect of herbivores such as sea urchins.

AGARUM Bory   The longitudinal median rib and the numerous perforations in the broad, undivided blade (Fig. 6.59*b*) distinguish *Agarum* (from *agar-agar*, a Malayan word for an edible seaweed) from the other genera of Laminariaceae. The perforations in the blades of *A. fimbriatum* Harv., which occurs in the quiet waters of Puget Sound, may be few or absent. The holdfast is usually fibrous, and the blade often has a ruffled appearance. The circular holes appear in the proximal portion of

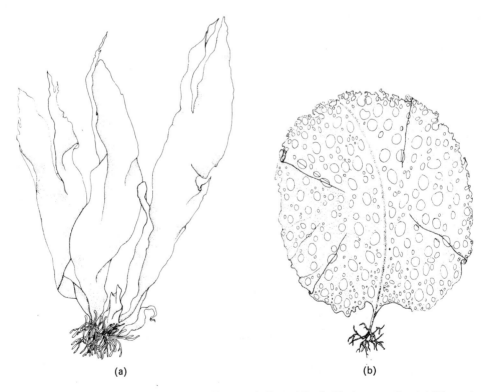

(a)   (b)

Fig. 6.59   (*a*) *Hedophyllum sessile* (C. Ag.) Setch. (*b*) *Agarum cribrosum* (Mertens) Bory. (*a*) × 0.25; (*b*) × 0.2.

the blade, near the transition zone, and enlarge to about 1.0–1.5 cm in diameter, apparently functioning to permit water currents to flow through the blade without pulling it from the substratum. The fruiting areas develop as irregular thickened blotches over the blade surfaces.

Species of this genus are present on both the east and west coasts of North America, restricted to colder waters. *Agarum* occurs on the Washington coast and northward to the Bering Sea and on the coast of northern Massachusetts northward to Ellesmere Island in Baffin Bay. At sites near San Juan Island, Washington, Vadas (1969) attempted to explain the factors that enable *Agarum* to constitute the dominant understory kelp, despite the fact that it was considered competitively inferior to the other kelps because it was not a member of the successional series in colonization. It was determined that the preferential feeding behavior of sea urchins is the primary cause of the dominance of *Agarum*. The urchins most preferred *Nereocystis; Agarum* rated least preferred of the various algal species tested in laboratory studies with urchins. Another factor for the maintenance of large beds of *Agarum* is that algal colonization beneath an *Agarum* canopy is minimal due to competition for light.

At a site in Nova Scotia, Canada, *Agarum cribrosum* Bory was observed (Mann, 1972) to constitute a sublittoral zone ranging from about 10 to 30 m in depth, the lower limit being determined by the hard substratum giving way to soft substratum. In its upper range there was some overlap with species of *Laminaria*, and the *Agarum* and *Laminaria* species together accounted for as much as 83 % of the total biomass of the littoral and sublittoral vegetation.

COSTARIA Grev.   Five prominent longitudinal ribs, which alternately project from either side of the blade, characterize the undivided blade (Fig. 6.60) of *Costaria* (L. *costa*, a rib). Three of the ribs emerge from one blade surface, and the other two ribs emerge from the opposite surface between the other ribs. A groove or depression is present on the side opposite a rib, and bullations are evident in the regions between the ribs. The stipe is generally flattened, becoming cylindrical only where it merges into the holdfast region. The stipe, which may be up to 50 cm in length, is

Fig. 6.60   *Costaria costata* (Turner) Saunders. × 0.17.

marked with numerous delicate grooves, running parallel to the long axis. *Costaria costata* (Turn.) Saund. is a commonly encountered kelp ranging from southern California northward to the Bering Sea.

Family 3. Lessoniaceae

This family, including the largest of all algae, *Macrocystis*, is characterized by the splitting of the original single blade into two, the division extending through the transition zone. This splitting of the original blade may stop with the formation of a single dichotomy (as may be the situation in *Dictyoneurum*, below) or each division may undergo an additional four to six divisions (as in *Pelagophycus*) or undergo an indefinite number of divisions, producing hundreds of new blades (as in *Macrocystis*, p. 335). The stipe assumes diverse morphological shapes and functions. In *Nereocystis* and *Pelagophycus* the terminal end of the stipe is inflated into an enlarged spherical float, or pneumatocyst, that may be up to 20 cm in diameter in *Pelagophycus*. In *Dictyoneurum* the stipe becomes flattened and attached to the substratum by means of marginal haptera and functions as a rhizome, with splits developing and continuing on up through the blade. In *Postelsia* the stipe is an erect, hollow axis bearing a terminal dense cluster of blades; its extreme resiliency allows the alga to withstand the onslaught of the waves.

In North America the Lessoniaceae is restricted in its occurrence to the Pacific Coast, with members such as *Macrocystis*, *Nereocystis*, and *Pelagophycus* contributing to the formation of conspicuous kelp beds lying offshore. These kelp bed communities have been studied in some detail (Dawson, et al., 1960; Neushul and Dahl, 1967; North, 1971).

DICTYONEURUM Rupr. As are so many other kelp genera, *Dictyoneurum* (Gr. *dictyon*, a net + Gr. *neuron*, a nerve) is a monotypic genus, occurring on the Pacific coast of North America. The long, linear blades (Fig. 6.61) of *D. californicum*

Fig. 6.61 *Dictyoneurum   californicum* Rupr. × 0.16.

Rupr. have a distinctive reticulate pattern of irregular narrow ridges crisscrossing both surfaces of the blade and giving a corrugated appearance. A midrib is lacking. The cylindrical, erect stipe of the young blade is soon modified by the stipe becoming flattened and prostrate and functioning as a rhizomatous axis with lateral haptera. The growing end of this rhizome may occasionally undergo the usual longitudinal splitting characteristic of the family, with the split extending outward to divide the blade into two portions. Thalli of *Dictyoneurum* are perennial, with new blades being continually formed with each new growing season.

POSTELSIA Rupr.   One of the most easily recognized marine algae, *Postelsia* (named for A. Postels, German naturalist), occurs in gregarious stands in the mid- to lower-littoral zone of areas exposed to very heavy surf action on the Pacific coast from central California to Vancouver Island. Individually, the plant has a palm-tree-like aspect (Fig. 6.62) with a congested cluster of deeply corrugated blades with dentate margins terminating the very resilient tubular stipe. At the base of the stipe is a massive hemispherical mound of haptera firmly securing the alga to the rock substratum. More than 100 narrow blades formed by the continued splitting of the tightly congested transition zones are borne at the apex. The sori of unilocular sporangia lie within the longitudinal grooves of the ridged blades.

Fig. 6.62 *Postelsia palmaeformis.* Population on a surf-swept rocky shore in central California. (Courtesy of Kimon T. Bird.)

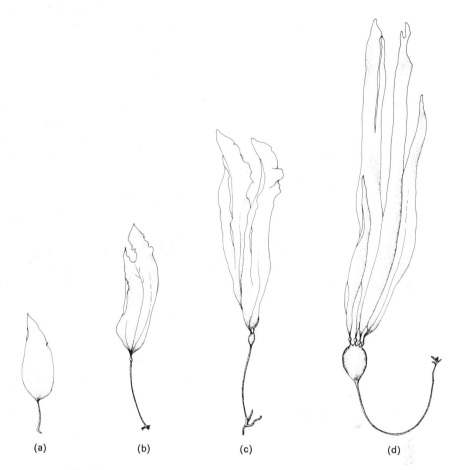

(a)   (b)   (c)   (d)

Fig. 6.63 *Nereocystis luetkeana*(Mert.)P. et R. (*a*)–(*d*) Stages in the development of the mature thallus. × 0.32.

NEREOCYSTIS Post. et Rupr.   Young stages of *Nereocystis* (Gr. *Nereus*, god of the sea + Gr. *kystis*, bladder) have a simple *Laminaria*-like appearance (Fig. 6.63*a*). By the progressive splitting of the original blade and a modification of the stipe (Fig. 6.63) the distinctive form of *Nereocystis luetkeana* (Mert.) Post. et Rupr., the only species, is assumed. The stipe of the mature plant may be 25–30 m or more in length and arises from a hemispherical mound of congested, branched haptera. The final few meters of the stipe gradually enlarges into a hollow portion, which is terminated by a spherical pneumatocyst, which is up to about 15 cm in diameter. The **pneumatocyst**, which serves as a float to maintain the distal portion of the alga at the water's surface, contains a mixture of gases, carbon monoxide being one of the more noteworthy of them (Foreman, 1976).

Upon the pneumatocyst are borne four short, flattened extensions, which represent the transition zones; these are dichotomously branched about four or five times.

Fig. 6.64 *Nereocystis luetkeana*. The terminal end of the stipe is hollow and allows the blades to lie at the surface of the water, as seen in this dense bed in Puget Sound, Washington. (Courtesy of Dr. Susan D. Waaland.)

A blade terminates each of these small branches, about 50–100 blades being produced on a single plant (Fig. 6.64). Individual blades are a few meters in length, and at maturity localized patches of the blades differentiate into sori, progressive increments of this maturation process occurring in a basipetal fashion. The fertile patch of unilocular sporangia (Fig. 6.65) is abscised from the blade, leaving the sterile sur-

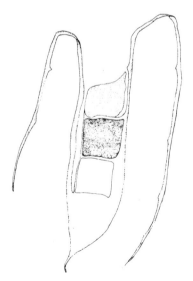

Fig. 6.65 *Nereocystis luetkeana*. Progressive development of sori on a single blade. × 0.15.

rounding tissue as a remnant, and the patch of fertile tissue can be disseminated by the water currents, releasing the zoospores over a considerable distance.

Nereocystis is prevalent to the Pacific coast of North America, its range extending from central California to the Aleutian Islands (Druehl, 1970). The thalli are annuals, and since they may reach a maximum length of 40 m at maturity, they have remarkably rapid rates of growth, as noted on p. 317. Sieve tubes are present in the medulla (Nicholson, 1976).

MACROCYSTIS C. Ag.   The basal portion of *Macrocystis* (Gr. *macro*, large + Gr. *kystis*, bladder) is perennial and capable of regenerating additional stipes. The result is that a tremendously large, entangled mass of stipes and blades may be developed from the original single blade. The early ontogeny of this alga is demonstrated by Fig. 6.66. The *Laminaria*-like blade undergoes a longitudinal division with the split developing acropetally. Following this primary division each segment divides

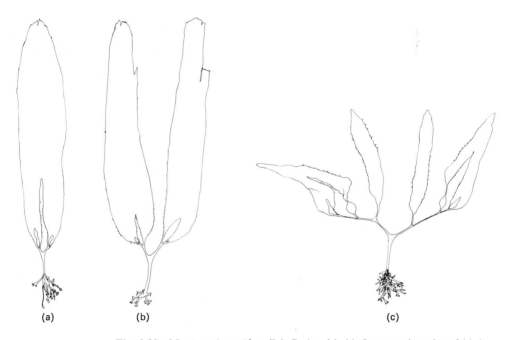

(a)   (b)   (c)

Fig. 6.66 *Macrocystis pyrifera* (L.) C. Ag. (*a*)–(*c*) Ontogenetic series of blade splitting and differentiation into the two lateral series of blades and the two central series of eventually fertile blades. × 0.32.

again to produce a pair of inner blades, which are destined to give rise to the sporophylls (sporangia-bearing blades), and an outer pair of blades, which are destined to continue to split in a unilateral manner, effecting a long trailer of vegetative blades (Fig. 6.67a). The inner pair of blades also continues to divide but in a dichotomous manner, and the resulting blades, which usually lack basal pneumatocysts, eventually produce unilocular sporangia on both surfaces.

Fig. 6.67 *Macrocystis pyrifera*. Terminal blade showing manner of splitting of the transition zone. × 0.27. (After H. B. S. Womersley; published in 1969 by the Regents of the University of California and reprinted by permission of the University of California Press.)

Plants of *Macrocystis* have been observed to bear sori throughout the year, although sporophyte colonization seems greatest in the spring (Foster, 1975A). Reproductive maturity of sporophytes is reached at the age of 9 to 12 months. The mean annual values for the release of zoospores ranged from 1000 to 5000 spores per minute per square centimeter for a sporophyll in a healthy, normal kelp bed (Anderson and North, 1967). A maximum value for spore release of 76,000 spores per minute per square centimeter was also recorded.

The vegetative blades have a relatively small basal pneumatocyst where they are attached to the stipe, and the individual blades (Fig. 6.67*b*), which may reach 40 cm in length, are coarsely rugose with dentate margins. Each mound of *Macrocystis* may have two to many such unilaterally developed "trailers" of blades, and the pneumatocyst in each blade serves to keep these long stipes at the water's surface.

Sieve tubes, which occupy a zone (Fig. 6.68) separating the cortical parenchyma from the medulla (Parker, 1971A), serve to translocate photosynthates to the lower portions of the plant, which may be at depths of 10 or more meters and shaded by the dense upper canopy of the so-called "kelp forest." These sieve tubes have been demonstrated (Parker, 1965, 1971B) to be the major channels of transport for ^{14}C-labeled organic products and fluorescein dyes. Productivity of kelp beds has been the focus of attention of some investigators (Clendenning, 1971). The *in situ* incorporation of ^{14}C in various parts of *Macrocystis* has been measured (Towle and Pearse, 1973) by means of polyethylene bags. The rate of carbon fixation was observed to decline very rapidly below the dense kelp canopy.

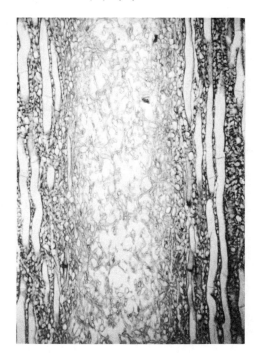

Fig. 6.68 *Macrocystis pyrifera*. Longitudinal section of stipe showing zone of sieve tubes between central medulla and peripheral cortical region. × 56. (After Bold, *Morphology of Plants*, 3rd ed., 1973, Harper & Row, Publ.)

Succession and regulation of community development within a *Macrocystis* bed have been investigated (Foster, 1975A, B), and observations based upon colonization of concrete blocks fastened to the bottom indicated that differences in growth rate and success in interspecific competition for space and light were the significant determinants of which algal species colonized and persisted rather than a strict ecological succession of species. Algal diversity and number of species consistently reached a peak within 100 to 200 days regardless of the time the blocks were placed on the bottom. This peak was then followed by a decrease in species diversity as ephemerals disappeared. The upper canopy of the *Macrocystis* contributed in reducing diversity below (Foster, 1975B).

Large submarine beds of *Macrocystis* are familiar sights (Fig. 6.69) off the coast of southern California. How extensive these beds have been in the past 50 years has been a variable matter, dependent on changing water currents and the abundance of foraging sea urchins and their predators, the sea otter. With the near extinction of the sea otter by the turn of the century, the ecological balance between plant and animal life was altered, and the sea urchins were no longer kept under the same control as before. It was noticed that from 1940 on certain beds of *Macrocystis* seemed to lose their vigor. In the late 1950's a shift in the coastal currents adversely affected the kelp beds by bringing in relatively warmer waters, which *Macrocystis* could not tolerate. The kelp beds off LaJolla and Palos Verdes virtually disappeared; the sea urchins no longer had a sufficient amount of plant debris to feed upon and thus were forced to feed directly upon the plants themselves (Ashkenazy, 1975). The underwater observa-

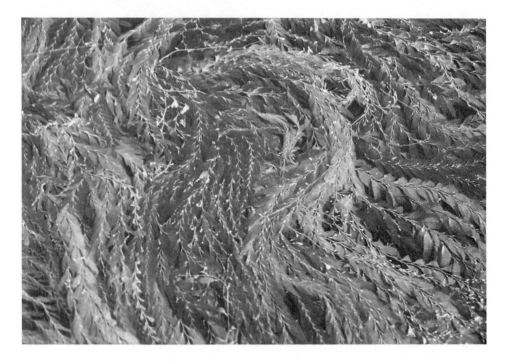

Fig. 6.69 Bed of *Macrocystis* off the coast of Southern California. (Courtesy of Kelco Company, San Diego, California.)

tions made by Dr. Wheeler J. North of the California Institute of Technology were instrumental in recognizing the damage done by the population explosion of sea urchins. The large-scale application of quicklime in controlled amounts from ships has been effective in reducing the number of sea urchins and affording *Macrocystis* the opportunity to recolonize regions (North, 1971), with a resultant beneficial effect on the algin industry (p. 29). Likewise, personnel of The California Department of Fish and Game have been pleased with the increased numbers of fish, lobster, and abalone, which make these kelp beds their home. The comeback of the sea otter has been an additional factor in restoring the natural balance among the components in these kelp communities (Branning, 1976).

Macrocystis is recognized (Womersley, 1954) to have three species, two of which are encountered on the Pacific coast of North America and overlapping in their ranges on the central California coast. The holdfast of *M. pyrifera* (L.) C. Ag., which has the more southerly range of the North Pacific, is characterized by a terete, erect axis, from which issue dichotomously branched haptera (Fig. 6.70a); the holdfast of *M. integrifolia* Bory, which has the more northerly range, is distinguished by a rhizomatous, strongly flattened axis, with marginally produced haptera (Fig. 6.70b). It has also been recognized (North, 1972), on the basis of culturing and transplantation techniques, that plants of *M. pyrifera* from southern Baja California, Mexico, have a greater tolerance to elevated temperature than plants of the same species from La-Jolla, California.

(a)   (b)

Fig. 6.70 *Macrocystis* holdfasts. (*a*) *M. pyrifera*. (*b*) *M. integrifolia*. (*a*) × 0.5; (*b*) × 0.46. (After H. B. S. Womersley; published in 1969 by the Regents of the University of California and reprinted by permission of the University of California Press.)

Family 4. Alariaceae

The members of this family are distinguished by restriction of the unilocular sporangia to special sporophylls, which may arise in two rows in acropetal succession from the transition zone (as in *Alaria*) or from marginal outgrowths of the original blade (as in *Eisenia*). The stipe in some genera may also produce vegetative outgrowths, both simple and branched, as in *Egregia*, in which the original blade is replaced by these secondary lateral blades.

The only member of this family known from the Atlantic coast of North America is *Alaria*, which also is present on the Pacific coast. Several other genera are included in the Pacific coast flora.

ALARIA Grev.   The undivided blade of *Alaria* (L. *ala*, a wing) has a conspicuous flattened midrib extending from the transition zone to the terminal end of the alga (Fig. 6.71). In colder waters, species of this genus (as *A. fistulosa* Post. et Rupr.) may have blades up to 25 m long. The lamina is frequently torn and shredded by the ravages of the sea; indeed, in *A. crispa* Kjellm. the blade portion may be so eroded that only a dense tuft of sporophylls remain, giving a superficial resemblance to *Postelsia palmaeformis* Rupr.

Mucilage glands of two types have been shown (Kasahara, 1973) to be produced in some species of *Alaria*. "Primary mucilage glands" originate from epidermal cells

Fig. 6.71 *Alaria esculenta* (Linn.) Grev. × 0.24.

and are first noticeable because of their enlarged size and refractive contents. The initiation of this type of superficial gland cell ceases when the blade is about 50 cm in length. A second type of mucilage gland is initiated by cortical cells located beneath the epidermis, and this type is produced during the life of the alga from the margin of the transition zone. This second type of gland cell is similar to the secretory cells observed in *Laminaria*, in which mucilage is secreted into a duct or canal.

Alaria is a relatively large genus in terms of number of species described, both in the North Atlantic and the North Pacific. The most recent taxonomic treatment for the temperate areas of the North Pacific was made by Widdowson (1972A), in which a total of 14 species was recognized. Widdowson suggested that the stability of the internal organization indicates that the plant is somewhat independent of environmental influences. In a separate statistical analysis of populations of *Alaria* (Widdowson, 1972B), 5 of the 10 species subjected to the study were distinguishable; 2 species intergraded with each other through a series of apparent ecotypes; and 3 species were indistinct from each other.

A method for growing *Alaria esculenta* (L.) Grev. in tanks supplied with filtered, free-flowing seawater and artificial illumination has been described (South, 1970). Although the sporophytes reached 30 cm in length, they did not produce sporophylls and remained sterile.

PTERYGOPHORA Rupr. A certain resemblance to *Alaria* is evident in *Pterygophora* (Gr. *pteryx*, a wing + Gr. *phora*, a carrying), a monotypic genus of kelp ranging from British Columbia to Baja California. The difference lies in the single terminal blade, which lacks a midrib and is approximately the same size as the lateral

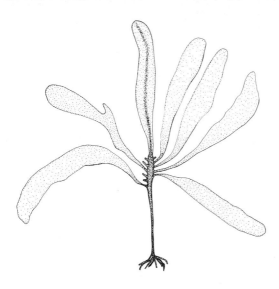

Fig. 6.72 *Pterygophora californica* Rupr. × 0.15.

sporophylls (Fig. 6.72). Usually about six or eight sporophylls are present in two vertical rows, produced basipetally from the transition zone. A new crop of sporophylls and a new terminal blade are developed with each growing season, and these parts are gradually eroded away at the conclusion of the growing season. The tough, woody stipe is capable of surviving for many years, as evidenced by the "annual rings," or increments of growth, revealed in a transection of the stipe. The life history of *Pterygophora californica* Rupr. has been demonstrated (McKay, 1933) to be typical of the order Laminariales.

EGREGIA Aresch.   The juvenile blade of *Egregia* (L. *egregius*, remarkable) has very much the typical appearance of a member of the Alariaceae, with a terminal blade and laterals arising from the region of the transition zone. The primary blade is later replaced by the stipe developing into a compressed, branching rachis, up to 8 m in length. Numerous foliar outgrowths are produced laterally from this rachis system, giving the sporophyte a densely fringed appearance (Fig. 6.73). Some of these lateral

Fig. 6.73 *Egregia*. An herbarium specimen of a young plant. × 0.2.

blades develop small pneumatocysts; others are ligulate and simple or branched. One small ovate to spatulate type of lateral blade bears the unilocular sporangia. The mature form of *Egregia* is anchored to the substratum, which is typically the lower littoral to upper sublittoral zone, by a massive conical holdfast of branched haptera. New axes arise near the base by the metamorphosis of lateral flattened blades, which in turn develop into a rachis system similar to the original blade.

The ecology of *Egregia* in southern California has been the subject of recent investigations (Chapman, 1962; Black, 1974). A large colonization of *Egregia laevigata* Setch. occurs in the spring in the intertidal zone, but a great mortality of these young algae ensues (Black, 1974) due to a combination of intraspecific competition from older plants, which apparently scrape away young ones as their stipes abrade the substratum during the rising and falling tides, and interactions with other plants of the same "year-class," resulting in a self-thinning process due to a density-dependent mortality. A uniform distribution of individual kelps is brought about, with relatively minor reduction due to grazers.

The gametophytes of *Egregia menziesii* (Turn.) Aresch. were shown (Myers, 1928) to be very reduced, the female gametophyte consisting usually of only one or two cells and the male gametophyte of two to four vegetative cells bearing up to a dozen antheridia.

EISENIA Aresch.   Groves of this interesting kelp are of sporadic distribution on the Pacific coast of North America, ranging from Vancouver Island to Baja California (Druehl, 1970). The populations of *Eisenia* (named for Dr. *G. Eisen*) are rather disjunct and may be either subtidal, as at Point Lobos in central California (Hollenberg and Abbott, 1966), or in the lower littoral, as at Laguna Beach in southern California. The primary blade of this alga achieves its mature form by a unique manner of growth. The original blade is largely sloughed away, but the lower portions persist and function as two meristem regions, from which sporophylls are marginally developed. A cluster of up to 50 corrugated sporophylls may be attached to each side of the bifurcation from the erect, woody stipe. The mature alga (Fig. 6.74) is often 1 m high

Fig. 6.74 *Eisenia arborea* Aresch. A preserved specimen. × 0.13.

and is attached by means of a massive hapterous holdfast. The stipe of *E. arborea* Aresch. is stiff enough to hold the plant in an upright position in calm water but is elastic enough to allow the plant to bend through an angle of nearly 90° and return to its upright position (Charters, et al., 1969).

Order 13. Fucales

This order of Phaeophycophyta is unique in the pattern of the life history, namely, type H, d, in which the alga is diploid and meiosis occurs at the time of gametogenesis. Pronounced oogamous sexual reproduction is the rule for most species. Unattached forms are known for several Fucalean genera, such as *Sargassum, Fucus, Ascophyllum,* and *Pelvetia* (Fritsch, 1945). In the Sargasso Sea, lying off the African coast between 20 and 35° latitude north, certain pelagic species of *Sargassum* have lost their capacity for sexual reproduction and rely entirely upon vegetative fragmentation. Evidence suggests these immense tracts of drifting *Sargassum* (as *S. natans* L. and *S. fluitans* Børg.) are derived from occasionally detached littoral forms.

This is a large and diversified order, with a great amount of morphological diversity exhibited by the members. The genus *Fucus* is one of the most common genera of rocky coastlines in the temperate northern hemisphere but is entirely absent in the southern hemisphere. The abundance of *Fucus* and the related *Ascophyllum,* both genera being referred to as "rockweeds" or "wracks," may be so great as to warrant their harvesting for utilization in agriculture in some parts of the world. In tropical waters *Sargassum* often constitutes a dominant element of the flora, and a large number of species have been described. The morphologically distinctive *Durvillea,* occurring in New Zealand and Australia, resembles some of the kelps with its blade-like thalli reaching lengths of 10 m. Its anomalous pattern of growth, which is diffuse rather than apical, has led to some workers (Petrov, 1965; Nizamuddin, 1968) separating it into its own order.

The normal manner of growth in the Fucales is by means of an apical cell or a group of apical cells. The apical cell[8] may be three-sided or four-sided in transection and has the shape of a truncate pyramid (Fig. 6.75) with either three or four lateral cutting faces and a basal cutting face. Parenchymatous construction is effected by the segmentation of the apical cell (Moss, 1967A). The latter is always located within a groove or pit at the terminus of each axis, and a very precise thallus structure is another feature of this order. Perfect dichotomous branching is the result of the division of an apical cell, as in *Fucus.* Thalli are either bilaterally or radially branched and may be flattened or terete. Evidence has been presented (Jensen, 1974) that the more primitive condition is a terete, triradiately branched axis with a three-sided apical cell.

---

[8]This classical notion of a single enlarged apical cell may have to be reassessed in light of McCully's (1966) observations that from four to eight apical initials of equivalent size were invariably observed to occur in the apical groove of *Fucus* rather than a single enlarged apical cell, as portrayed, for example, by Moss (1967A). Moss (1974) stated that the apical cell may function as a quiescent center, an hypothesis applied to apical cells in some vascular cryptogams. Moss (1967A) admitted that the apical cell divides infrequently.

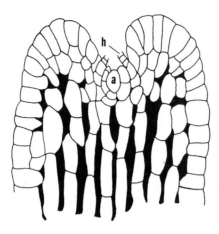

Fig. 6.75 *Fucus*. Median longitudinal section of a young germling with apical cell, a, and bases of hairs, h. (After Oltmanns.)

Therefore, *Fucus*, which has flattened bilaterally branched axes with four-sided apical cells, would be derived, or relatively advanced.

Reproduction in members of this order is preceded by the formation of receptacles, which are fertile areas of branches. Receptacles may be the enlarged, swollen distal ends of branches, as in *Fucus*, or they may be a much branched system of small terete terminal axes, as in *Sargassum*. Scattered over the surface of these **receptacles** are minute openings, which lead into cavities, termed **conceptacles**, in which are produced the eggs and the sperm. An ostiole is the small opening from the interior of the conceptacle to the surrounding water. In some species, a single plant may produce eggs and sperm and is thus a bisexual individual; in other species, eggs and sperm may be restricted to separate plants. Another variation is that bisexual individuals may have eggs and sperm formed within the same conceptacle or they may be formed in different conceptacles (Fig. 6.76).

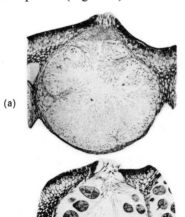

(a)

(b)

Fig. 6.76 Conceptacles of *Fucus vesiculosus* Linn. (*a*) Antheridial conceptacle in sectional view. (*b*) Oogonial conceptacle in sectional view. (*a*) × 33; (*b*) × 20. (After Bold, *Morphology of Plants*, 3rd ed., 1973, Harper & Row, Publ.)

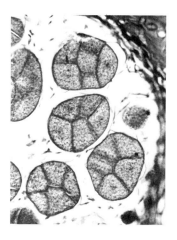

Fig. 6.77   Oogonia of *Fucus*, containing eight eggs. × 124.

A conceptacle is derived from a single originally superficial cell. It becomes *SIMILAR IN MORPHOLOGY TO A RALFSIAN CRUST OR A SORUS* differentiated while still lying close to the apex undergoing a transverse division to produce an upper cell, which does not divide again, and a lower cell, which does redivide to form a layer of cells called the **fertile sheet** (Smith, 1955). This fertile sheet eventually lines the interior of the sunken chamber of the conceptacle, and cells of this fertile layer may function as oogonial initials. An individual cell may divide transversely to form a lower stalk cell and an upper cell that enlarges to form an oogonium (Fig. 6.77). The first two divisions of the nucleus of the oogonium are meiotic, and these four haploid nuclei undergo a mitotic division to produce a total of eight haploid nuclei per oogonium. Various alternatives exist for these eight nuclei depending on the genus.

These alternative patterns of egg formation are outlined in Fig. 6.78. Each of the eight nuclei becomes incorporated into an egg in *Fucus*, the protoplast being cleaved into eight eggs (Fig. 6.78*a*); in *Ascophyllum*, four functional eggs are formed and four supernumerary nuclei are discarded (Fig. 6.78*b*); two eggs are produced in *Pelvetia*, six supernumerary nuclei being extruded between them (Fig. 6.78*c*); in *Hesperophycus* only a single functional egg is cleaved from the protoplast, and the remaining seven nuclei remain behind in some residual cytoplasm (Fig. 6.78*d*); a single egg is also formed in *Cystoseira*, but the seven supernumerary nuclei are extruded peripherally (Fig. 6.78*e*); the seven supernumerary nuclei simply degenerate in *Sargassum*, in the cytoplasm of the functional egg (Fig. 6.78*f*). A gradual progression of progressively fewer eggs is implied in this sequence, and *Fucus* was thought to represent the ancestral type. More recently, however, *Fucus* has been suggested (Manton, 1964C) as a derived type on the basis of ultrastructural details concerning the flagella of the sperm; Manton suggested that *Cystoseira* is more primitive with reference to this characteristic.

Antheridia are produced either directly from the fertile layer or on branched paraphyses arising from the floor of the conceptacle. The development of an antheridium, as is that of an oogonium, is completely the same as that of a unilocular sporangium. The original single nucleus within the antheridium undergoes meiosis, and the

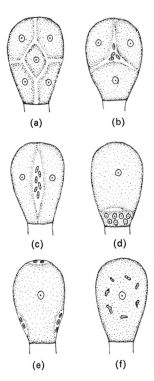

(a)   (b)

(c)   (d)

(e)   (f)

Fig. 6.78 Scheme of alternative patterns of egg production in the Fucales. (*a*) *Fucus*. (*b*) *Ascophyllum*. (*c*) *Pelvetia*. (*d*) *Hesperophycus*. (*e*) *Cystoseira*. (*f*) *Sargassum*. (Based on G. M. Smith.)

4 haploid nuclei undergo a series of mitotic divisions, resulting usually in 64 haploid nuclei, which are incorporated into the 64 sperm. The biflagellate sperm are distinctive in having a longer posteriorly directed flagellum, unlike the situation in other Phaeophycean gametes and zoospores, in which the longer flagellum is the anteriorly directed one. In addition, a variety of flagellar appendages, such as spines or barbs, has been described in Manton and Clarke (1956), Manton, et al., (1953), and Cheignon (1974) at the ultrastructural level for the Fucalean sperm, causing Scagel (1966) to classify this order as a distinct subclass, the Cyclosporidae, within the Phaeophycophyta.

The timing of the release of gametes is correlated with the tidal cycle. When the plants are exposed at low tide, desiccation of the frond causes shrinkage and the extrusion of slime outward through the ostioles of the conceptacles. Each oogonium has a wall consisting of three layers: exochite, mesochite, and endochite. When mature, the outermost layer, the exochite, bursts, releasing the egg or eggs still enclosed by two wall layers. The swelling of mucilage within the conceptacle and its passage through the ostiole passively moves the oogonia outward to the surface of the alga. Once the incoming tide washes over the alga, the endochite, which is the innermost gelatinous layer, imbibes water, swells, and causes the overlying mesochite to rupture, exposing the endochite. The continued swelling of the endochite causes the enclosed egg or eggs to become spherical and eventually to float away. An equivalent process of release of sperm occurs in the antheridia. After an initial rupture of the exochite, the mass of sperm, surrounded only by an endochite, is discharged. With the gelatinization of the endochite the sperm are freed to swim in all directions. A prominent orange eyespot is assembled *de novo* within the single chloroplast of each sperm, and it has been demonstrated (Bouck, 1970) that a complex of fibers and microtubules arises near the posterior basal body and seems to be oriented with the first formed eyespot granules. It is possible that the parabasal body and the eyespot mediate light imping-ing on a photoreceptor bound to the plasma membrane.

In an investigation (Pollock, 1970) of *Fucus distichus* L., it was shown that a period of about 20 minutes was required for the release of the eggs from the freed oogonium and the release of the sperm from the freed antheridia. This process is illustrated in Fig. 6.79. It has long been recognized in *Fucus* that there was an attrac-tion and clustering of sperm around eggs, and this chemotaxis of the sperm was due to a volatile substance released by the eggs (Cook and Elvidge, 1951; Müller, 1972B). This female sex attractant in *Fucus serratus* L. has been called (Müller and Jaenicke, 1973) "fucoserraten."

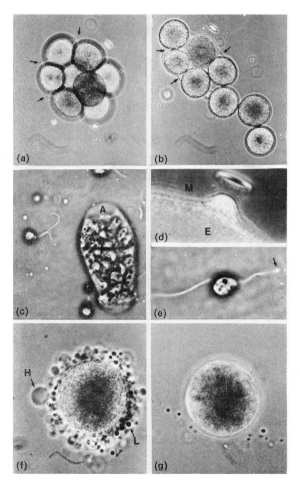

Fig. **6.79** *Fucus distichus* Linn. Sequence of liberation of eggs and fertilization. (*a*) Freshly liberated eggs still enclosed by the oogonial membrane (arrows). (*b*) Eggs leaving their packet. (*c*) Freshly liberated antheridium (A). (*d*) Protuberance on an egg opposite the pore in the oogonial membrane. E, egg; M, oogonial membrane. (*e*) Motile sperm, with shorter anterior flagellum (arrow). (*f*) Unfertil-ized egg with blebs (H, L) on egg surface. (*g*) Fertilized egg. (*a*), (*b*) × 293; (*c*) × 2001; (*d*) × 1573; (*e*) × 3450; (*f*), (*g*) × 776. (Permis-sion from E. G. Pollock and Springer-Verlag, Heidelberg, for Planta 92: 85–99, 1970.)

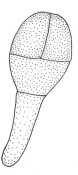

Fig. 6.80 Four-celled stage of germling of *Fucus*, showing polarity of apical and rhizoidal portions. × 198.

The zygote of Fucalean algae has long served as a useful system in studying environmental influences on the establishment of polarity (Moss, 1974; Quatrano, 1974; Nakazawa, 1975). Cell walls are lacking in *Fucus* eggs, but wall synthesis begins within 40 minutes after fertilization (Novotny and Forman, 1975), with a rapid deposition of alginic acid during the first 2 hours. Although there is a homogeneous distribution of cytoplasm prior to fertilization (Quatrano, 1972) and no evidence prior to germination to suggest polarity, about 14 hours after fertilization germination commences with the production of a rhizoidal outgrowth.[9] Several hours later a cross wall is formed at right angles to the rhizoidal outgrowth (Fig. 6.80), and the two cells that are separated are differentiated both in regard to structure and function. Numerous external stimuli have been demonstrated to cause an orientation of the rhizoidal outgrowth. These parameters include unilateral illumination (both visible and ultraviolet), temperature, acidity, and auxin (Jaffe, 1968), although it has been demonstrated (Sussex, 1967) that the external stimuli are not absolutely required for the initiation of polar growth. The processes of rhizoid formation and cell division were separated (Torrey and Galun, 1970) by the growth of *Fucus* zygotes in a seawater-sucrose medium with unilateral illumination. Spherical, multicellular embryos lacking rhizoids were formed.

The requirement for ribonucleic acid formation for rhizoid formation has been demonstrated (Quatrano, 1968), such that the RNA synthesis takes place several hours prior to the production of the proteins that are used for germination. The cytoplasmic area destined to be the site for the emergence of the rhizoid becomes irreversibly fixed from 1 to 3 hours before germination is detectable. Since this timing corresponds to the time when proteins, not RNA, required for rhizoid formation are synthesized, it seems that it is the synthesis of proteins which irreversibly determines a specific region of the cytoplasm to undergo the events which culminate in a polarized cell. Cytoplasmic differentiation observable with the electron microscope occurs in the rhizoidal hemisphere before the rhizoid emerges but at the time that the site has been irreversibly fixed (Quatrano, 1972). Another biochemical distinction in the two-celled embryo is that sulfated polysaccharides are localized only in the rhizoidal cell.

[9]Although a single primary rhizoid is formed in *Fucus* and *Ascophyllum*, germlings in other genera produce four or more primary rhizoids by the longitudinal division of the potential primary rhizoid (Moss, 1974).

Fucoidan appears during rhizoid formation (Quatrano, 1973), the evidence indicating that sulfate becomes incorporated into a carbohydrate fraction that was unsulfated prior to fertilization.

An argument has been made (Caplin, 1968) to designate the reproductive cells of this order megasporangia and microsporangia instead of oogonia and antheridia, respectively. The logic of this position is to conform to the usage of this terminology in discussing the alternation of generations in land plants. Since the thallus in the Fucales is diploid, it is referred to as a sporophyte that produces megasporangia and microsporangia. The four haploid nuclei produced by meiosis within these so-called sporangia would constitute four megaspores and four microspores, respectively. Mitosis by the four megaspores results in four binucleate female gametophytes, which in the case of *Fucus* would be entirely fertile and give rise to eight functional eggs by the cleavage of the protoplast within the original megasporangium. Further mitotic divisions of the four microspores bring about what are interpreted as four 16-celled male gametophytes, which also are entirely fertile, giving rise to a total of 64 sperm. This interpretation has been recently supported (Jensen, 1974) by an account of the South African genus *Bifurcariopsis* Papenfuss. Instead of having a syncytial type of development, as is present in *Fucus* and the great majority of Fucales, a tetrasporic type of development occurs in the so-called oogonia of *Bifurcariopsis*. The unique feature is that two concentric walls enclose each of the four binucleate endosporic megagametophytes. One nucleus degenerates within each unit, and four functional eggs are produced per sporangium, causing Jensen to speculate that this pattern represents the primitive pattern in the order. It is evident that this is largely a semantic issue. Traditionally the terms *oogonia* and *antheridia* are more commonly employed in describing the reproductive structures of the Fucales.

The distribution of Fucalean genera on a worldwide basis has been reviewed (Nizamuddin, 1962, 1970), some genera, as *Fucus* and *Pelvetia*, being widely separated geographically but confined to the northern hemisphere. *Hesperophycus* Setch. et Gardn. and *Pelvetiopsis* Gardn. are examples of genera endemic to the west coast of North America. An example of a very large genus restricted to Australia and New Zealand is *Cystophora* J. Ag. in the Cystoseiraceae. Womersley (1964) assigned 23 species to this genus, almost all of which are sublittoral. Two of the most widely distributed Fucalean genera, occurring in both hemispheres and ranging from tropical to temperate regions, are *Sargassum* and *Cystoseira*. *Turbinaria* Lamour. (Taylor, 1963) is equally widespread but is restricted to tropical and subtropical regions, particularly reef floras.

Family 1. Fucaceae

Plants in this family may have flattened, strap-shaped axes, as in *Fucus*, or they may have cylindrical axes bearing somewhat compressed laterals, as in *Ascophyllum*. Branching may be dichotomous, irregularly pinnate, or radial. Vesicles are present in some species (Fig. 6.81). Some axes become metamorphosed into fertile regions, the receptacles, which may terminate the main branches or be developed from short

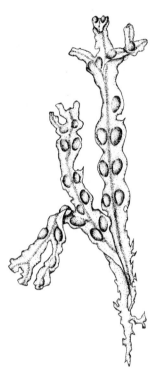

Fig. 6.81 *Fucus vesiculosus.* × 0.65.

lateral branches. The primary distinguishing feature of this family is the initiation of growth by a single four-sided apical cell in the adult stage. (See footnote 8 on p. 343, however, for a word of caution concerning this statement.) Three fairly common genera are discussed below.

FUCUS Linn.   In a monographic treatment by Powell (1963) *Fucus* (L. *fucus* derived from the Gr. *phykos*, seaweed) is regarded as a highly plastic genus, the species exhibiting a wide range of form in response to environmental conditions (Jordan and Vadas, 1972) and in reference to their geographical range. The relative ease, however, in which hybrids may be formed (Burrows and Lodge, 1951, 1953) may be responsible in part for this phenotypic variability. Although well over 100 taxa of this genus have been described, Powell (1963) expressed a conservative judgment in accepting only 6 of these as species. Infraspecific taxa are common; for example, a number of sub-species are recognized (Powell, 1957A, 1957B) for *F. distichus* L., the type species.

The thallus of *Fucus* is attached by a broad discoidal holdfast, from which arise usually bilaterally branched flattened fronds with a fairly distinct midrib (Fig. 6.82*a*). Paired air bladders, or vesicles, may be present, as in *Fucus vesiculosus* L. (Fig. 6.81). Scattered cryptostomata, which are small sterile cavities containing tufts of hairs, are common over the surface of the frond, but especially on the wings rather than on the midrib. The plant body of *Fucus* is highly differentiated and has provided a rich source for study of apical meristems (Moss, 1965, 1966, 1967A), histology of vegetative and reproductive tissues (McCully, 1966, 1968A; Moss, 1968), and regeneration (Moss, 1964, 1967B; Fulcher and McCully, 1969A).

The thallus is organized into a surface layer of columnar cells, which are rich in chloroplasts and **fucosan vesicles,** several underlying layers of cortical parenchyma, and an innermost medulla consisting of a network of filamentous cells. This network is made up of two types of cells: branched **primary filaments** and elongated **secondary fibers** (McCully, 1966), which are all embedded in a very thick matrix. This massive, biphasic matrix is comprised of rigid, oriented fibers of cellulose and alginic acid embedded in an amorphous suspension of the sulfated polysaccharide, **fucoidan,** which fills the interstices. The abundant presence of this hydrophilic polysaccharide is of ecological importance in reducing the effects of desiccation during exposure. This matrix also serves as an ion-exchange resin that buffers against abrupt changes in the osmotic environment (McCully, 1966).

By means of the histological stain toluidine blue, which is metachromatic, and the periodic acid–Schiff's test, four matrix components could be distinguished (McCully, 1966): alginic acid, fucoidan, cellulose, and laminaran. The following indicates their staining properties:

	Toluidine Blue	Periodic Acid–Schiff's
alginic acid	+	+
fucoidan	+	−
cellulose	−	+
laminaran	−	−

In an investigation of oogonia and antheridia, McCully (1968A) demonstrated that the production of alginic acid and fucoidan is separated in time and space. For oogonia, alginic acid synthesis is detectable in the earliest stage of the oogonial initial, the sites of production being scattered throughout the cell, as evidenced by the staining reaction. On the other hand, fucoidan synthesis does not commence until the eight-nucleate stage of the oogonia, with a definite localization of production in the perinuclear region.

The highly polarized nature of the epidermal cells has been examined (McCully, 1968B). The nucleus and chloroplasts are aggregated at the basal end of the cell; a hypertrophied perinuclear dictyosome and numerous vesicles containing fucoidan, alginic acid, and polyphenols are present. The outermost surface of these specialized epidermal cells is characterized by a convoluted plasma membrane, and it is thought that this surface has dual roles of absorbing sulfate and carbonate from the environment and of secreting alginic acid, fucoidan, and polyphenols to the outside. The chloroplasts are regarded (Willenbrink and Kremer, 1973) as the only sites of primary mannitol biosynthesis during photosynthesis.

Four species of *Fucus* have been grown in the laboratory (McLachlan, et al., 1971A) in unialgal conditions, starting from zygotes. *Fucus distichus*, which is a species permanently submerged in high, littoral tide pools, was the only one of the four to complete its life history by producing gametes. For *F. vesiculosus*, a species that becomes exposed in the intertidal region, a tidal cycle of alternating submergence and exposure has been suggested (Fulcher and McCully, 1969B) as a requirement for growth and maintenance, but reasonably good growth under submerged conditions was obtained in this species (McLachlan, et al., 1971A). So-called adventive embryos were developed from rhizoidal filaments in the sporelings of several species of *Fucus* (McLachlan and Chen, 1972). This response appeared to be spontaneous under variable conditions of temperature, light, and nutrition and not due to any wounding.

The four species cultured in the laboratory maintained their characteristic morphologies, even though under identical conditions. This fact lends support to the contention that the extensive phenotypic variation observable in the field is perhaps the result of interspecific hybridization, as has been artificially obtained (Burrows and Lodge, 1951, 1953), and not the result of environmental influences.

A seasonal rhythm in reproduction is usually expressed in *Fucus*. For the two most common species of *Fucus* in England, *F. vesiculosus* L. and *F. serratus* L., their seasonal rhythms stand opposed to one another; the vegetative season of one coincides with the reproductive season of the other (Knight and Parke, 1950). The peak of reproduction is reached in the spring and summer by *F. vesiculosus* and in the autumn and winter by *F. serratus*. The initiation of receptacles continues over a 5-month period, although only 3 months are needed for the development of a receptacle from its initiation up to the time of gamete release.

ASCOPHYLLUM Stackhouse   The sole species of *Ascophyllum* (Gr. *askos*, a wine skin + Gr. *phyllon*, a leaf), *A. nodosum* (L.) LeJolis, is commonly encountered (Fig. 6.82) in great profusion in the mid-littoral zone on both sides of the North Atlantic. In North America it extends from New Jersey to Baffin Island (Taylor, 1962A), both on rocky shores and in sheltered inlets.

The plant consists of linear, compressed axes lacking a distinct midrib and attached by a basal disc. Regular dichotomous branching in one plane is demonstrated; lateral branching is subsequently superimposed upon the original branching pattern. Surgical experiments (Moss, 1970) strongly suggest that the apical cell exerts the controlling influence within the apex. New apical meristems can be regenerated from wound-healing tissue, as in *Fucus* and *Pelvetia*. Large vesicles, or air bladders, are present as occasional swellings of the cylindrical axes. When fertile (usually in the winter months), short stalked lateral receptacles are produced by the unisexual plants. Four eggs are produced by each oogonium in the female plants.

The occurrence of a free-living form referred to as *A. nodosum* ecad *mackaii* (Turn.) Cotton has been documented for Britain (Gibb, 1957) and Newfoundland (South and Hill, 1970). This variant arises from fragments of attached plants, and in the advanced stages of this free-living condition the thalli become globular with markedly dichotomous branching, air-bladders are absent, and plants are sterile. Experiments by Moss (1971) demonstrated that the percentage of eggs that germinate in this unattached form was very low and germination was very slow, although a small percentage of the germlings retained the ability to attach. The normal pattern of development seen in attached plants of this species was apparently irreversibly lost in this morphologically very different free-living form.

(a)   (b)

Fig. 6.82 (*a*) *Fucus.* Dichotomously branched thalli exposed by the outgoing tide. (*b*) *Ascophyllum nodosum* (L.) LeJol. Exposed clumps of thalli on the coast of Rhode Island. (Courtesy of John Harlin.)

PELVETIA Decaisne et Thuret   Species of *Pelvetia* (named after Dr. *Pelvet*, a French naturalist) are widely separated in the northern hemisphere (Nizamuddin, 1970), with *P. canaliculata* (L.) Decaisne et Thuret, the type species, common along the European coast of the Atlantic Ocean, *P. fastigiata* (J. Ag.) DeToni common along the west coast of North America, and *P. wrightii* (Harv.) Okam. common in the western Pacific (Japan and Korea). All three species occur in the mid- to upper-littoral regions of rocky shores.

The perennial plants of *Pelvetia* are very firm in texture, well adapted for the long periods of desiccation. Axes (Fig. 6.83) are dichotomously branched, terete to compressed, with the receptacles developing at the tips of the ultimate dichotomies.

Fig. 6.83 *Pelvetia fastigiata* (J. Ag.) DeToni. × 0.19.

Plants are bisexual, with antheridia and oogonia occurring within the same conceptacles. Only two functional eggs are produced within each oogonium, and six supernumerary nuclei are extruded between these eggs.

Family 2. Sargassaceae

This fairly small family, in terms of number of genera, includes one of the largest
genera of brown algae, *Sargassum*. The distinction from the family Cystoseiraceae is
the location of branches in the axils of subtending leaves. Like the Cystoseiraceae, each
axis is terminated by a single apical cell, which is three-sided in transection, and a
single egg is produced by each oogonium. Only the genus *Sargassum* will be treated in
this account.

SARGASSUM C. Agardh   Morphologically, *Sargassum* (Sp. *sargazo*, seaweed,
used by navigators to describe floating algae) includes some of the most specialized of
the algae (Fig. 6.84), with such differentiated parts as holdfast; cylindrical main axes;
flattened, sterile leaf-like laterals, in the axils of which are developed solitary, spherical
air bladders; variously modified receptacles; or additional indeterminate cylindrical
axes. The main axis of *Sargassum* is perennial and slow growing, giving rise to yearly
crops of radially arranged primary laterals. These laterals are flattened and resemble

Fig. 6.84   *Sargassum filipendula* C. Ag. × 0.64.

leaves, with a midrib and serrated margins. The branches bearing conceptacles are usually like vegetative axes but more condensed and with narrower and shorter segments (Jensen, 1974).

Sargassum is a large genus with more than 150 species described, occurring in tropical, subtropical, and temperate zones of both hemispheres (Nizamuddin, 1970). It is the most conspicuous brown alga in tropical and subtropical waters, ranging from mid-littoral to sublittoral zones. Some of the species of *Sargassum* grow attached, but others are known to have a pelagic existence, floating in tremendous masses as in the Atlantic Ocean west of Africa (the "Sargasso Sea").

The loss of sexual reproduction is associated with the pelagic mode of existence. Simple fragmentation effectively propagates these forms. Two pelagic species of *Sargassum* are recognized (Taylor, 1960; Earle, 1969) for the eastern coast of North America and the Gulf of Mexico: *S. fluitans* Børg. and *S. natans* (L.) Meyen. Cryptostomata are generally absent in these two species but present in the attached species.

A good example of a "weedy" seaweed is provided by *Sargassum muticum* (Yendo) Fensholt, an alga originally known only in Japan[10] but now widespread on the west coast of North America and in British waters due to accidental introductions. Plants, which may attain lengths of 7 m, consist of monopodially branched, cylindrical axes with air bladders and leafy laterals. This species first received attention (Scagel, 1956) when it appeared in the northeastern Pacific, apparently having accompanied an introduction of spat of the Japanese oyster (*Crassostrea gigas*). Its range in the eastern Pacific presently extends from British Columbia to southern California. Its habitat is eelgrass (*Zostera marina*) beds, and it may be displacing this ecologically important sea-grass.

More recently *Sargassum muticum* has appeared at sites on the Isle of Wight in southern England (Farnham, et al., 1973; Jones and Farnham, 1973), its introduction probably also resulting from the importation of the Japanese oyster from Japan or British Columbia. An ecologial study (Fletcher and Fletcher, 1975A) of this species of the Isle of Wight revealed that it out-competed the indigenous Fucalean species in shallow low-tide-level lagoon habitats, where plants formed thick swards covering several square centimeters of shore and also successfully colonized the floating jetties. Antheridial and oogonial conceptacles are produced together on fleshy receptacles, and excised segments, containing apical or lateral meristems, are able to continue growth in culture (Fletcher and Fletcher, 1975B).

Family 3. Cystoseiraceae

This family, recognized (Jensen, 1974) to contain as many as 17 genera, is close to the Sargassaceae but distinguished by the fact that branches do not arise in the axils of subtending "leaves." Additional characteristics include the presence of a single apical cell terminating each axis, this apical cell being three-sided in transection, and the production of one egg per oogonium. The only exception to this latter trait is the South African genus *Bifurcariopsis*, in which four eggs are produced. The forma-

---

[10]The name *Sargassum kjellmanianum* Yendo continues to be used for this species in Japan.

Fig. 6.85 *Cystoseira osmundacea* (Menzies) C. Ag. × 0.26.

tion of two to four primary rhizoids by the embryo is regarded (Nizamuddin, 1962) as an additional diagnostic trait, in contrast to the tuft formed in embryos of Sargassaceae, but this embryological character has been observed in relatively few species and may prove unreliable.

CYSTOSEIRA C. Ag.   A highly differentiated, perennial alga is exemplified by *Cystoseira* (Gr. *kystis*, a bladder + Gr. *seira*, a chain), with a conical holdfast, a woody, angular stipe bearing old leaf scars in the basal part, and a series of highly modified branches of various orders (Fig. 6.85). In the lower portion of the thallus the lateral branches are flattened and present a leaf-like appearance, as in the common west coast species *C. osmundacea* (Menzies ex Turner) C. Ag. In the upper portion of the thallus the primary and subsequent orders of branches are cylindrical and terminate in chains of small vesicles.

Although a few species occur in the southern hemisphere, *Cystoseira* is a genus predominantly of the northern hemisphere, with some species on the west coast of North America and one species on the Florida coast. Roberts (1967) evaluated the morphological criteria used for specific separation in this genus, concluding that five species were present in British waters. Whether plants were solitary or caespitose, whether axes were radial or flattened, whether cryptostomata were present or absent, and whether the swollen bases of shed laterals persisted were all useful in separating the species.

7

Division Chrysophycophyta

Introduction

The present account includes six classes within the circumscription of the division Chrysophycophyta, a group of algae that is quite diversified in regard to pigment composition, cell wall, and type of flagellated cells (see Table 1.1). Yet they share certain features, such as a food reserve composed of a β-linked glucan, chrysolaminaran (Craigie, 1974), formerly called leucosin, which closely resembles the laminaran of the Phaeophycophyta. Another common characteristic is the predominance of carotenoids over chlorophylls, affording the cells of this division a hue other than "grass green" of the Chlorophycophyta. Thus, we refer to the "golden algae" (the Chrysophyceae) and the "yellow-green algae" (the Xanthophyceae). The diatoms (the Bacillariophyceae) and the members of the Prymnesiophyceae are equally golden in appearance, and the members of the Eustigmatophyceae are more yellow-green, this class having been recently segregated (Hibberd and Leedale, 1971B, 1972) from the Xanthophyceae. All the classes placed in this division share the feature of the occurrence of chlorophylls a and c and the absence of chlorophyll b. The Chloromonadophyceae (= Raphidophyceae), a small group of seldom encountered, distinctive flagellates, are included in this division because of pigment similarities to the other members (Guillard and Lorenzen, 1972).

Carotenoids are the accessory pigments that are responsible for this distinctive golden to yellow-green shade. Fucoxanthin, which is also present in the Phaeophycophyta, is a characteristic xanthophyll (oxygen-containing carotenoid) occurring in three of the six classes: the Chrysophyceae, Prymnesiophyceae, and Bacillariophyceae. It may account for as much as 75% of the total pigment component of the algal cells and thus is more noticeable than the chlorophyll(s). Throughout this division, β carotene (nonoxygenated carotenoid) is of ubiquitous occurrence, being the only carotene present in some species or being accompanied by trace amounts of α and γ carotenes (Allen, et al., 1960; Dales, 1960).

Xanthophylls such as violaxanthin, diatoxanthin, and diadinoxanthin are also present in some members of this division (Goodwin, 1974), the last pigment also occurring in the Pyrrhophycophyta. A useful pigment difference has been detected between members of the Eustigmatophyceae and Xanthophyceae, namely, the presence of violaxanthin as the major xanthophyll pigment in the Eustigmatophyceae (Whittle and Casselton, 1975A) and the presence of diadinoxanthin as the major xanthophyll pigment in the Xanthophyceae (Whittle and Casselton, 1975B). Chlorophylls *a* and *c* are present in most members of this division (Meeks, 1974), although the latter pigment is absent in some species. It was long thought that chlorophyll *c* was absent in the class Xanthophyceae (including the Eustigmatophyceae), but some careful recent investigations (De Greef and Caubergs, 1970; Guillard and Lorensen, 1972) have verified its presence in some Xanthophyceae. Chlorophyll *b*, however, is never present in this algal division.

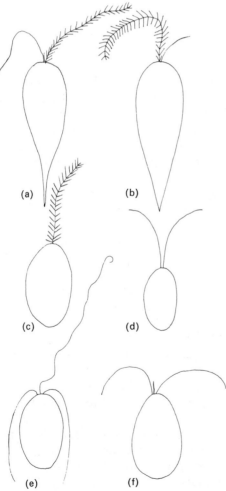

(a)

(b)

(c)

(d)

(e)

(f)

Fig. 7.1 Flagellar patterns in the Chrysophycophyta. (*a*) *Synura*. (*b*) *Ochromonas*. (*c*) *Chromulina*. (*d*) *Isochrysis*. (*e*) *Chrysochromulina*. (*f*) *Prymnesium*.

A variety of flagellated cells characterizes this assemblage, and the differences are of taxonomic utility both at the class and ordinal levels. In the Chrysophyceae the **monad** is basically a biflagellated cell (Fig. 7. 1*a*) with two **heterodynamic**[1] **flagella**, one bearing hairlike appendages called mastigonemes and termed **pleuronematic** and the other flagellum devoid of appendages and termed **acronematic**. The pleuronematic flagellum is usually directed in an anterior direction, and the acronematic flagellum is usually directed in a posterior direction. The two flagella may be approximately equal in length or slightly unequal, such as in an individual cell (Fig. 7.1*a*) in a motile colony of *Synura* (p. 367), or unequal in length, such as in *Ochromonas* (p. 363), the pleuronematic flagellum being relatively longer (Fig. 7.1*b*). In some members of this class, the acronematic flagellum has entirely disappeared (Fig. 7.1*c*), such as in *Chromulina* (p. 369). Interestingly, in some species of *Chromulina* (Fauré-Fremiet and Rouiller, 1957; Belcher and Swale, 1967B) a second vestigial flagellum is present within an invagination or pocket.

The only type of flagellated cell that occurs in the Bacillariophyceae is the sperm, which has been observed in some members of the Centrales. The sperm has a single anterior pleuronematic flagellum. This single flagellum is unusual in that fine-structural examinations (Manton and von Stosch, 1966; Heath and Darley, 1972) revealed it to lack the central doublet microtubules (Fig. 7.2*b*) that are invariably present in typical eucaryotic flagella (Fig. 7.2*a*). It is noteworthy that the occurrence of such a motile stage in the diatoms was not recognized until relatively recently, when von Stosch (1951) studied the centric diatom *Melosira* (p. 406) and confirmed oogamous sexual reproduction. An affinity with the Chrysophycean flagellated cell is evident as well as links between these two groups and the Phaeophycophyta.

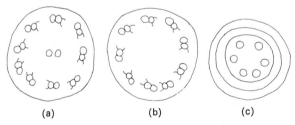

(a)   (b)   (c)

Fig. 7.2 Diagrammatic representation of cross sections through flagella and haptonema as observed with the electron microscope. (*a*) Typical eucaryotic flagellum showing a $9 + 2$ pattern. (*b*) Flagellum of sperm of centric diatom, showing a $9 + 0$ pattern. (*c*) Haptonema.

Motile cells in the Xanthophyceae, which may be expressed in the vegetative stage, such as in *Heterochloris*, or in zoospores, such as those produced by the fila-mentous genus *Tribonema* (p. 388), bear two anteriorly attached flagella, one of which is pleuronematic and the other is acronematic. They are heterodynamic in behavior and unequal in length. The name Heterokontae was the original term used by Luther (cf. Papenfuss, 1955) in describing this group as a class distinct from the Chlorophy-ceae, "heterokont" referring to the unequal length of their flagella. Two flagellar patterns are present in the genus *Vaucheria* (p. 390) of this class. Its zoospores are large, compound structures bearing many pairs of flagella, which are acronematic and

[1]Heterodynamic flagella have independent patterns of beat, in contrast to homodynamic flagella, such as in *Chlamydomonas* (p. 75), in which there is a coordinated pattern of beat.

slightly unequal in length; the antherozoids, however, bear a pair of flagella, one pleuronematic and the other acronematic, the latter pattern being more typical of the class Xanthophyceae.

Motile cells in the class Eustigmatophyceae bear mostly a single pleuronematic flagellum, arising at the anterior end of the cell. Two genera (*Ellipsoidion* and *Pseudocharaciopsis*) of this class have a pair of flagella, the second flagellum being relatively short and acronematic.

The Prymnesiophyceae stands as the most disjunct class of this division in reference to the type of motile cell. Basically, the monad of this class bears two anteriorly located acronematic flagella of equal length, such as in *Isochrysis* (Fig. 7.1*d*), or they may be slightly unequal in length, such as in *Pleurochrysis* (p. 376) but both still acronematic. A variation in this class is the presence of a unique appendage called the **haptonema**, which has an ultrastructure and a behavior quite alien to that of a flagellum. A haptonema in cross section (Fig. 7.2*c*) has been shown (Manton, 1964B) to consist of an outer sheath of three concentric membranes and an inner circle of six or seven microtubules, in contrast to the single sheath surrounding the axoneme of a flagellum and the usual circle of nine paired microtubules and a central pair, i.e., the "9 + 2" arrangement of microtubules (Fig. 7.2*a*) almost universally present in eukaryotic flagella and cilia. Rather than waves being propagated down its length, as is the situation in flagella, the haptonema can be observed to function as a mechanism of attachment, especially when it is long (see *Chrysochromulina*, p. 379). The haptonema may be many times the length of the cell (Fig. 7.1*e*), such as in *Chrysochromulina*, or it may be a very short, stiff appendage, such as in *Prymnesium* (Fig. 7.1*f*) and some of the coccolithophorids (p. 381).

Most of the morphological trends that are present among the members of the Chlorophycophyta can be cited as occurring within the division Chrysophycophyta, particularly within the classes Chrysophyceae and Xanthophyceae. Thus, motile stages (both unicellular monads and colonial aggregations), coccoid stages, palmelloid formations, filaments (simple or branched), and parenchymatous, thalloid structures are known among the Chrysophyceae and Xanthophyceae, these expressions paralleling comparable morphological developments in the Chlorophycophyta. This phenomenon might be referred to as **convergent evolution**.

Two morphological tendencies that are not known among the Chlorophycophyta are expressed within the Chrysophycophyta, namely, the formation of amoeboid or rhizopodial vegetative stages, such as in *Chrysamoeba* (p. 369), and the formation of **plasmodial** colonies, as expressed in the European genus *Myxochloris* (Bourrelly, 1968A). The existence of such amoeboid and plasmodial types, linked with a high prevalence of **facultative heterotrophy**, has stimulated speculation (Bourrelly, 1962B) that the Chrysophycophyta may have possibly served as an ancestral stock from which emerged some of the Protistan groups, such as the Sarcomastigophora (Honigberg, et al., 1964). Various biochemical data have also been cited as evidence of some phylogenetic affinities between these two assemblages.

The six classes of this division are treated in the following account by focusing on the general attributes of each class, listing the orders comprising each class, with a brief description of each order, and then including the descriptions of some representative genera within each class.

CLASS CHRYSOPHYCEAE

A great plasticity in regard to both morphology and nutrition is expressed by the members of this algal class. Freshwater forms are more common than marine forms, and many seem to prefer unpolluted systems and relatively cool- to cold-water temperatures. Photosynthesis is well developed in most members of this class, but colorless forms are not infrequent. Both osmotrophic and phagotrophic modes of heterotrophic nutrition occur, some forms, such as *Ochromonas* (p. 363), having both patterns as well as autotrophic nutrition in the same species.

Although many Chrysophycean algae have naked protoplasts, others have various cell coverings, including scales, loricas, and close-fitting cell walls. The scales are composed of silica in *Synura* (p. 367) and *Paraphysomonas*, the latter a colorless genus in which the scales have been shown (Manton and Leedale, 1961B) to originate within vesicles of complex dictyosomes, whereas in *Synura* the scales are developed within special vesicles derived from cisternae of the endoplasmic reticulum (Schnepf and Deichgräber, 1969). The walls and loricas (Fig. 7.3) can be of a cellulosic-pectic nature with or without a siliceous impregnation. The loricas become calcified in some genera, such as in *Chrysococcus* and *Pseudokephyrion* (Bourrelly, 1957, 1963). Internal skeletons composed of silica characterize one distinctive group in this class, namely, the silicoflagellates.

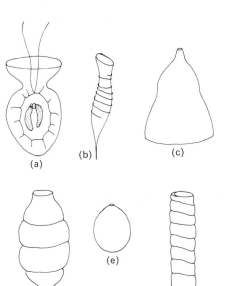

Fig. 7.3 Examples of loricas in the Chrysophyceae. (*a*) *Derepyxis crater*, containing protoplast. (*b*) *Dinobryon suecicum*. (*c*) *Lagynion triangularis*. (*d*) *Pseudokephyrion undulatum*. (*e*) *Chrysococcus tesselatus*. (*f*) *Kephyrion bacilliforme*. (After Bourrelly.)

(a) (b) (c) (d) (e) (f)

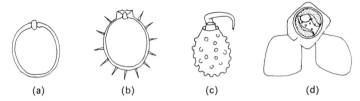

Fig. 7.4 (*a*)–(*c*) Examples of statospores in different species of *Uroglena*. (*d*) Sexually formed statospore of *Kephyriopsis* enclosed in an envelope with loricas of gametes still attached. (*d*) × 2442. [(*a*), (*b*) after Bourrelly; (*c*) after Conrad; (*d*) after Fott.]

One of the most distinctive features of this class is the formation of a characteristic cyst, or **statospore**, that constitutes a resting stage formed endoplasmatically, i.e., internally. The statospore (Fig. 7.4*a–c*) consists of two pieces of different size. The larger portion is first developed within the cytoplasm, and the cytoplasm external to this bottle-shaped piece migrates into it, a plug being finally formed at the mouth. Unlike the loricas of vegetative cells, which may be siliceous, cellulosic-pectic, or (rarely) calcareous in composition, the wall of the statospore is always siliceous in nature (Bourrelly, 1957). It is noteworthy that in some Chrysophycean genera the vegetative phase may have a calcareous lorica, whereas the cyst has a siliceous wall. Thus, the two processes of calcification and silicification occur at different phases in the same alga. The plug of the cyst, upon germination, dissolves, and one or more motile spores are released.

In addition to cell division in unicellular forms or the fragmentation of multicellular forms, such as *Synura* and *Phaeothamnion*, asexual reproduction can also be accomplished by the production of zoospores. Whether the zoospore is uniflagellated or biflagellated is useful in placing the alga into the appropriate subclass of the Chrysophyceae.

Sexual reproduction has been observed in a relative limited number of genera, and it seems to be invariably isogamous. It has been recorded (Bourrelly, 1957; Kristiansen, 1960, 1961, 1963A, 1963B: Fott, 1964) especially in loricate members due to the fact that zygote formation is easily recognized (Fig. 7.4*d*). The loricate cells attach at their openings, and the protoplasts migrate from the loricas to fuse and form the zygote. The empty loricas usually remain attached to the spherical zygote. It is generally believed that the vegetative phase of members of this class is haploid, meiosis being zygotic in those forms undergoing sexual reproduction, thus, a type H, h life history. Autogamous sexual reproduction has been described (Gayral, et al., 1972) in a species of *Ochromonas*.

Classification

In the following account based in part upon Bourrelly's (1968A) treatise the Chrysophycean orders are subdivided into three series based upon the presence or absence of flagella as well as the number of flagella. The biflagellate series include those forms with a motile stage bearing two flagella, which are heterodynamic and of unequal length. The motile stage may be the predominant phase in the life history of the alga or merely a transitory phase, such as in the formation of zoospores. Orders falling into this category are the following:

Ochromonadales: monads, colonies, or palmelloid masses.
Chrysapiales: coccoid, solitary or colonial.
Phaeothamniales: filamentous or thalloid.

In the uniflagellate series, which comprises the second category, a single pleuronematic flagellum is present on the motile cells. The group contains the following orders:

Chromulinales: monads or colonies;
    rhizopodial, palmelloid, or plasmodial aggregations.
Craspedomonadales: monads or colonial, attached or free-swimming; with a ring
    of rhizopodia or a cytoplasmic collar surrounding the flagellum.
Dictyochales: monads, with spongy, amoeboid protoplasm, and a siliceous
    skeleton.
Chrysosphaerales: coccoid forms, solitary or colonial.
Thallochrysidales: filamentous or thalloid forms.

A third series includes forms in which motile cells do not occur or at least have not yet been observed. In some of these forms reproduction is by means of amoeboid cells.

Chrysococcales: palmelloid forms, the naked cells being embedded in mucilaginous envelopes.
Rhizochrysidales: rhizopodial forms, solitary or colonial, naked or loricate.
Stichogloeales: coccoid forms, with a cell wall, solitary or colonial.
Phaeoplacales: forms filamentous or thalloid, with a cell wall.

Selected genera from these orders are discussed in greater detail in the account that follows.

REPRESENTATIVE GENERA

Order 1. Ochromonadales

OCHROMONAS Wyssotzki   The naked, biflagellate monads of *Ochromonas* (Gr. *ochros*, yellow, pale + Gr. *monas*, single organism) have one or two (rarely several) brownish to yellowish chloroplasts, usually a stigma, and contractile vacuoles. Auxotrophy for various vitamins is common among the species of this genus (Provasoli, 1958). *Ochromonas danica* Prings. is auxotrophic for thiamine and biotin, whereas *O. malhamensis* Prings. requires these two vitamins and also B_{12}, i.e., cyanocobalamin (Pringsheim, 1952). Heterotrophic nutrition, both by saprophytic and phagotrophic means, accompanies the photosynthetic mode (Aaronson, 1973A, 1973B; Duboursky, 1974).

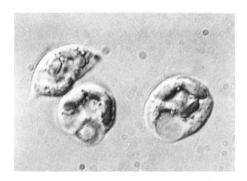

Fig. 7.5 *Ohromonas danica* Pringsheim. Interference light micrographs of cells with elongate shape when not containing engulfed food or with spherical shape when containing matter in their food vacuole. × 1800. (Courtesy of Dr. Garry T. Cole.)

Cells of *Ochromonas danica* have a characteristic shape (Fig. 7.5), the posterior end gradually tapering to a point. A complex system of microtubules lying just beneath the cell surface maintains this shape (Bouck and Brown, 1973). A single, lobed chloroplast is present at the anterior end of the cell, and this chloroplast is enclosed by a membranous sac called the **chloroplast endoplasmic reticulum**, which is in continuity with the outer membrane of the nuclear envelope (Gibbs, 1962D). This configuration, namely, an association of the chloroplast with the endoplasmic reticulum, has been demonstrated (Gibbs, 1962C; Bouck, 1965) to be present in several algal classes, including all five classes of this division as well as the Phaeophyceae and the Cryptophyceae (Dodge, 1973).

Another interesting feature of the chloroplast of *Ochromonas danica* is the presence of chloroplast DNA arranged in a ring-shaped structure, lying just inside the girdle thylakoid (Gibbs, et al., 1974) and resembling that observed in the brown alga *Sphacelaria* (see Fig. 6.2*b*). Such **genophores**, or chloroplast "nucleoids," have been observed in five different algal classes, the Chrysophyceae, Xanthophyceae, Bacillariophyceae, Phaeophyceae, and the Chloromonadophyceae (p. 396), in which classes girdle lamellae are present just within the chloroplast envelope (Gibbs, et al., 1974). In other algal classes, those lacking girdle lamellae, plastid DNA is present but not in this peripheral ring-shaped configuration.

A vacuole is located in the more posterior portion of the cell of *Ochromonas danica*. When particulate matter is contacted at the anterior end of the cell (Cole and Wynne, 1974), this "food" is immediately engulfed into a primary food vacuole, which migrates to the posterior portion of the cell and is incorporated there into the larger food vacuole, where digestion proceeds (Fig. 7.6). The teardrop or obpyriform shape of the cell is altered by endocytosis, the posterior end becoming distended with engulfed food and the cell becoming spherical (Fig. 7.5).

Cells of *Ochromonas* have one long and one short flagellum, and both seem to extend in a forward direction rather than one of them trailing (Bouck, 1971). Cinéphotomicrographic studies (Jahn, et al., 1964) revealed that the cell essentially pulls itself forward by the activity of the long flagellum. Flagellar waves are propagated distally by means of planar undulations. The two rows of mastigonemes are arranged in the same plane as the undulations. The relatively stiff mastigonemes extend outward and function to reverse the thrust of the forward flagellum. The cell also rotates around

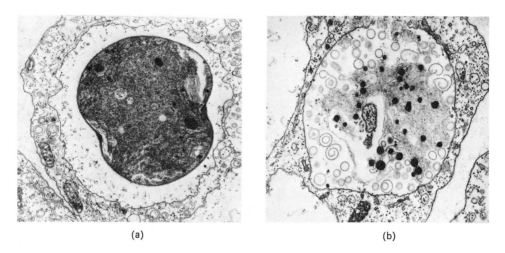

(a)   (b)

Fig. 7.6 *Ochromonas danica* Pringsheim. Thin sections of food vacuoles containing blue-green algal cells in early (*a*) and later (*b*) stage of digestion. (*a*) × 10,080; (*b*) × 13,776. (Courtesy of Dr. Garry T. Cole.)

its long axis, when it has its elongate, obpyriform shape but does not rotate when spherical. A sequence involving the apparent origin of mastigonemes from within the perinuclear space, the initial attachment of the mastigonemes to the membrane within this perinuclear space, the transfer of the presumptive mastigonemes to Golgi cisternae, and the assembly of the mastigonemes onto the base of the flagellum has been eluci-dated by Bouck (1971).

Mitosis and cytokinesis have also been described using the electron microscope (Slankis and Gibbs, 1972). One interesting feature is that the poles of the spindle are occupied by **rhizoplasts**, which are fibrous organelles extending from the flagellar basal bodies. The microtubules of the spindle make direct connections with these rhizo-plasts. The nuclear envelope largely disappears during mitosis, only small portions connected to the chloroplast persisting.

The formation of siliceous cysts occurs in *Ochromonas*. Since the Ochromonada-ceae is usually defined (Bourrelly, 1968A) as being comprised of naked forms and the Synuraceae as containing scaly forms, the inclusion of a pigmented monad bearing body scales in *Ochromonas*, namely, *O. diademifera* Takahashi (1972), has caused other workers (Rees, et al., 1974; Thomsen, 1975) to call attention to how such an interpretation of *Ochromonas* would complicate the separation between the Ochro-monadaceae and Synuraceae. Most species of *Ochromonas* are of freshwater occur-rence (Huber-Pestalozzi, 1941), but marine types are also known. The colorless counterpart of *Ochromonas* is the genus *Spumella*.

MALLOMONAS Perty   The free-swimming cells (Fig. 7.7) of *Mallomonas* (Gr. *mallos*, lock of wool + Gr. *monas*, single organism) have a cell membrane covered with delicately sculptured siliceous scales. Long needle-like projections are often borne

Fig. 7.7 *Mallomonas*. Diagrammatic representation of a longitudinal section of a cell. × 858.

by the scales at the posterior end and sometimes at the anterior end. The systematics within the genus is based upon the fine structure of these scales (Fig. 7.8) (Asmund, 1959; Harris and Bradley, 1960; Fott, 1962; Takahashi, 1975). More than 80 species have been described from freshwater habitats (Bourrelly, 1957). Most forms are photosynthetic, with one or two golden chloroplasts, but colorless forms are also included in the genus.

The cells bear two flagella, but one is reduced to a short peduncle that has a photoreceptor role. A recent proposal has been made (Belcher, 1969A) to include the genus *Mallomonopsis* within *Mallomonas*, the only difference being that the second flagellum is not reduced to a short peduncle in *Mallomonopsis*.

Sexual reproduction has been described (Wawrik, 1960; Kristiansen, 1961) in *Mallomonas*. It is always isogamous and may involve the fusion of cells at their anterior or posterior ends, depending on the species.

Fig. 7.8 *Mallomonas*. Scales. × 16,000. (Courtesy of Sara Marquis.)

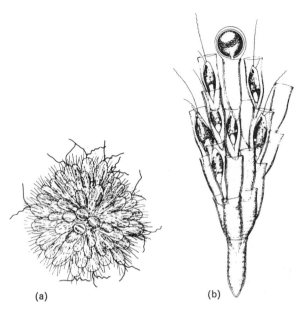

Fig. 7.9 (a) *Synura splendida* Korsh. A colony. (b) *Dinobryon sertularia* Ehr. A colony of monads and one cyst. (a) × 425; (b) × 462. [(a) after Kristiansen; permission of Botanisk Tidsskrift; (b) after H. Skuja in Nova Acta Regiae Soc. Sci. Upsal., Ser. IV, Vol. 18, No. 3, Uppsala, 1964; published by the Royal Society of Sciences of Uppsala. Distributed by Almqvist and Wiksell International, Stockholm, Sweden.]

(a)

(b)

SYNURA Ehrenberg   The free-swimming colonies (Fig. 7.9a) of *Synura* (Gr. *syn*, with, together + NL. *ura*, tail) are spherical to ellipsoidal and may be composed of a few to numerous cells that are held together by their elongated posterior ends. The colony increases in size by the division of individual cells, and occasionally groups of cells pinch off to form new colonies. Each cell bears two heterodynamic flagella of unequal length (Fig. 7.1a). Tiny scales of various shapes (annular, linear, semicircular, clavate, etc.), depending on the species (Schnepf and Deichgräber, 1969; Hibberd, 1973), are borne on the flagella. These scales seem to be restricted to the family Synuraceae. The mechanism of flagellar movement has been investigated (Jarosch, 1970). The two basal bodies are connected by a bridge, and the basal body of the smooth flagella is connected to the stigma region of the chloroplast by a structure arising from two of the triplets (Kristiansen and Walne, 1976). It is possible that this connection might serve to conduct impulses from the stigma to the flagellum (or flagella), thus a photoreceptor-photoresponse mechanism.

Species are distinguished by the form of the chloroplast(s) and the shape in ornamentation of the statospore. But, as in the case of *Mallomonas*, ths most useful taxonomic criterion is the fine-structural appearance of the siliceous scales that cover the cells (Manton, 1955; Petersen and Hansen, 1958; Takahashi, 1975). Isogamous sexual reproduction has been described in one species, *S. petersenii* Korschikov (Wawrik, 1970B).

DINOBRYON Ehrenberg   The loricate, biflagellate cells of *Dinobryon* (NL. from Gr. *dinos*, whirling + L. *bryon*, moss) form free-swimming, arborescent colonies (Fig. 7.9b). Rarely, solitary cells occur. It may be the dominant component of freshwater phytoplankton and may impart a foul, fishy odor to the water, as has happened at the southern end of Lake Michigan during several summers (Palmer, 1959). Loricas are broad and open vase-shaped structures, and the cell is attached to the inside wall of the lorica by a narrow cytoplasmic strand. When cell division occurs, two to four cellular progeny are produced, and one or more of them settle and become attached to the inner margin of the parental lorica. A new lorica, composed of homogeneously laid down fibrils, is formed by the protoplasm, new material being deposited at the outer edge (Hilliard, 1971; Franke and Herth, 1973). An arborescent habit is thus achieved.

In a study of populations of *Dinobryon sertularioides* Ehr., the formation of statospores was observed (Wujek, 1969) to frequently take place within the dendroid colonies. The presence of food vacuoles, indicating phagotrophic ingestion, was also noted.

Order 2. Phaeothamniales

PHAEOTHAMNION Lagerheim in Wittrock et Norstedt   The branched, uniseriate axes (Fig. 7.10*a*) of *Phaeothamnion* (Gr. *phaios*, dusky + Gr. *thamnion*, shrub, bush) are attached by a single basal cell and are usually epiphytic on other algae in freshwater habitats. It has been recorded (Smith, 1950) from California and Wisconsin. The cells, other than the basal cell, contain one to a few chloroplasts. Individual cells may produce usually four or eight zoospores, which resemble *Ochromonas*, with or without a stigma (Bourrelly, 1968A). Siliceous cysts are produced. Under certain conditions of growth the cells of the filament dissociate into a palmelloid condition. The cells are rounded and are surrounded by a thick mucilage. This palmella stage (Fig. 7.10*b*) resembles *Sphaeridiothrix* of the order Phaeoplacales (p. 363), and the independent status of the latter genus has accordingly been questioned (Geitler and Schiman-Czeika, 1970).

Closely related to *Phaeothamnion* is *Apistonema*, a genus of both marine and freshwater species in which the filaments tend to be creeping rather than erect and the pattern of branching is irregular. It has become evident (Gayral and Lepailleur, 1971) that *Apistonema* is a heterogeneous collection of species in that some represent benthic phases of coccolithophorids (p. 381) of the Prymnesiophyceae and others represent

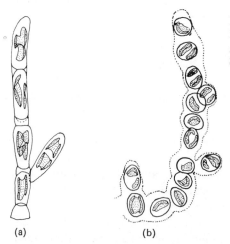

Fig. **7.10** *Phaeothamnion confervicola.* (*a*) Filamentous organization. (*b*) Palmelloid organization.   × 1056.   (After   Geitler   and Schiman-Czeika.)

(a)   (b)

genuine chrysophycean algae, such as *A. submarinum* P. A. Dangeard (Bourrelly, 1968A).

Another alga showing similarity to *Phaeothamnion* is *Chrysomeris*, a marine genus consisting of erect, branched, uniseriate filaments. The zoospores of this alga were observed (Gayral and Haas, 1969) to be biflagellated, with one pleuronematic flagellum and one acronematic, and it was subsequently transferred from the Chromulinales, where it had been classified, to the Ochromonadales.

Order 3. Chromulinales

CHROMULINA Cienkowski   The free-swimming, solitary cells of *Chromulina* (Gr. *chroma*, color + Gr. *lina*, made of) have a single pleuronematic flagellum (Fig. 7.1c), which apparently is a derived condition, in that in at least two species a rudimentary second flagellum has been observed (Fauré-Fremiet and Rouiller, 1957; Belcher and Swale, 1967B) to be present within an invagination or pocket near the exserted pleuronematic flagellum. Cells are naked and lack a well-defined form. Usually one or two chloroplasts are present, lying on opposite sides of the cell.

At least 120 species of *Chromulina*, both freshwater and marine in distribution, have been described. Such criteria as presence or absence of stigma and pyrenoids, structure of cysts, and number and form of chloroplasts are used in distinguishing the species (Conrad, 1931; Bourrelly, 1968A). Cysts have been encountered in relatively few species.

CHRYSAMOEBA Klebs   The solitary, amoeboid cells of *Chrysamoeba* (Gr. *chrysos*, gold + Gr. *amoibe*, change) have delicate, narrow rhizopodia radiating from the roughly spherical protoplast (Fig. 7.11). The cells are generally embedded in a very thin mucilage but are otherwise naked. The cells do not have any relationship with each other unlike the situation in *Chrysarachnion*, a genus in which networks of cells result from the interconnections of their rhizopodia.

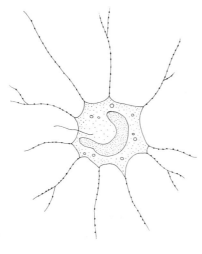

Fig. 7.11 *Chrysamoeba radians* Klebs. Representation of an amoeboid cell bearing filiform rhizopodia and a single obvious flagellum. × 2000. (After Hibberd.)

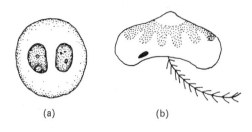

Fig. 7.12 *Phaeaster pascheri* Scherffel. (*a*) Palmelloid stage. (*b*) Motile stage, side view. × 1320. (After Belcher.)

(a)   (b)

An investigation of *Chrysamoeba radians* Klebs revealed (Hibberd, 1971) the presence of a pair of flagella, one short pleuronematic type and the other merely a rudimentary structure, in the amoeboid cells. As in *Chromulina*, the evidence of two flagella, even though one is greatly reduced, implies the artificiality of the distinction between the Ochromonadales and Chromulinales. Amoeboid cells occasionally metamorphose into free-swimming cells, the pleuronematic flagellum becoming relatively elongate. In a culture of *C. radians* a small percentage of cells in the flagellated conditions can always be observed.

PHAEASTER Scherffel   This genus is placed in the family Chrysocapsaceae, which is characterized by a palmelloid vegetative organization, which on occasion produces zoospores of the *Chromulina* type. The palmelloid phase lacks a definite zone of growth, which distinguishes it from the Hydruraceae, in which such a localized zone of growth is a feature, and it also lacks mucous hairs, the distinctive trait of the Chrysochaetaceae. *Phaeaster* (Gr. *phaeo*, dusky + Gr. *aster*, a star) is a freshwater organism (Whitford and Schumacher, 1969), occurring as small nonmotile colonies of cells embedded in mucilage (Fig. 7.12*a*) or as solitary free-swimming cells (Fig. 7.12*b*).

Both stages have been examined (Belcher, 1969B) in cultures of *P. pascheri* Scherffel. The palmelloid groups may contain up to 16 cells, but the mucilaginous vesicles tend to become diffluent as cell division progresses, the cells becoming dissociated. Although flagella are not obvious in the palmelloid cells, they are probably present in a functional condition to judge from the rapidity with which cells can swim away from the matrix (Belcher, 1969B). The motile cell is flattened in the anterior-posterior axis and has one conspicuous flagellum and apparently a second very reduced flagellum. A single parietal chloroplast with radiating lobes is present in each cell. (Fig. 7.12*b*).

Unlike the fairly amorphous gelatinous matrix present in *Phaeaster*, the closely related genus *Gloeochrysis* has a distinctly stratified mucilaginous layer, recalling that of *Gloeocystis* (p. 108) in the Chlorophycophyta. The vegetative cells of *Gloeochrysis* contain contractile vacuoles and a single chloroplast and eyespot per cell (Geitler, 1967; Kalina, 1969).

Order 4. Craspedomonadales

Although treated by Bourrelly (1968A) as a distinct subclass of the Chrysophyceae,[2] namely, the Craspedomonadophycidae, the present account regards the

---

[2]Some authors (Christensen, 1964; Parke and Dixon, 1964) have removed the choanoflagellates to a separate class, the Craspedophyceae. The unusual flagellar structure caused Hibberd (1975) to state that this assemblage does not appear to belong either with the algae or in the plant kingdom.

choanoflagellates as comprising an order paralleling the Chromulinales in the uniflagellate series of Chrysophyceae. This group of flagellates is characterized by the presence of a truncate cytoplasmic collar extending out at the anterior end of the cell (Fig. 7.13*d, e*) and surrounding the single flagellum, which bears a delicate bilateral winglike process or vane (Hibberd, 1975), the winglike process extending along the flagellum from inside the collar to a point about two-thirds along the length of the flagellum where it abruptly ends.

One of the reasons given by Bourrelly in separating off this order as a distinct subclass of the Chrysophyceae is that statospores are still unknown for the choanoflagellates. Nonetheless, a progression of forms can be outlined (Fig. 7.13) that seems to link together the choanoflagellates with members of the Chromulinales. The family Pedinellaceae, in which uniflagellate monads bear an anterior crown of transitory or permanent "tentacles," or rhizopodia, was placed by Bourrelly (1968A) in the Chromulinales, distinct from the choanoflagellates of his subclass Craspedomonadophycidae; yet he recognized that a close relationship exists between this family and the choanoflagellates.

A different interpretation was given by Norris (1965), who regarded the Pedinellaceae as belonging to the order Craspedomonadales, differing from the other families in that the anterior rhizopodia, when present, were not densely aggregated into a discrete collar. Members of the Pedinellaceae are photosynthetic. Cells of *Pedinella* and *Pseudopedinella*[3] have six chloroplasts arranged around the long axis, resulting in

---

[3]Javornicky (1967) expressed the idea that *Pseudopedinella* should be merged into *Pedinella*, but Swale (1969) was inclined to retain the two genera, outlining two groups of morphological traits. Anterior rhizopodia are always present in *Pedinella* but are absent in *Pseudopedinella*, except for *P. ambigua* Bourrelly (Fig. 7.13*c*), which has transitory anterior rhizopodia. Both genera have a posterior, branched or unbranched peduncle, in contrast to *Apedinella* (Fig. 7.13*a*), which lacks such a peduncle (Thronsden, 1971).

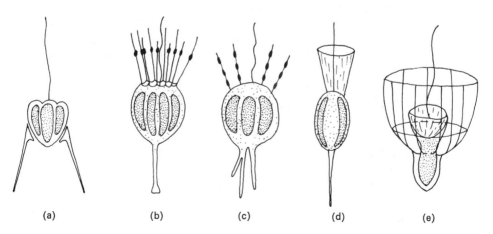

(a)   (b)   (c)   (d)   (e)

Fig. 7.13 (*a*) *Apedinella spinifera* (Throndsen) Throndsen. (*b*) *Pedinella hexacostata* Vysot. (*c*) *Pseudopedinella ambigua* Bourrelly. (*d*) *Stylochromonas minuta* Lackey. (*e*) *Pleurasiga reynoldsii* Throndsen. × 910. [(*a*) and (*e*) after Throndsen; (*b*) after Swale; (*c*) after Bourrelly; (*d*) after Lackey.]

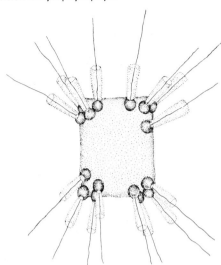

Fig. 7.14 *Proterospongia haeckelii* Kent var. *gracilis.* × 163. (After Skuja.)

a radial symmetry. Food particles, such as bacteria, can be ingested, however, in these photosynthetic genera (Swale, 1969) either by means of the anterior rhizopodia or by a posterior **peduncle**, which is a cytoplasmic protrusion used both in attachment or to "catch" particulate matter. All the choanoflagellates, on the other hand, are colorless, with the exception of *Stylochromonas* (Fig. 7.13*d*), a marine choanoflagellate with two chloroplasts.

Choanoflagellates are members of the **nanoplankton**, inhabiting the **neuston** of tidal pools and comparable calm-water systems (Norris, 1965; Throndsen, 1970; Leadbeater, 1972A; Manton, et al., 1975). Loricas are produced by some choano-flagellates (Fig. 7.13*e*), the posterior end of the lorica at times being expanded or structurally elaborated seemingly to provide a greater surface area for adhesion of the cell to the water-air interface. The composition of the loricas of freshwater choano-flagellates is probably cellulose or chitin. Elaborate siliceous loricas, made of variously arranged ribs, are produced in some marine choanoflagellates (Manton, et al., 1975; Leadbeater, 1975).

The resemblance of choanoflagellates to choanocytes of sponges has been noted (Rasmont, 1959; Fjerdingstad, 1961; Tuzet, 1963; Simpson, 1968), and the similarity of the flagella in both groups has been pointed out (Afzelius, 1961; Feige, 1969; Brill, 1973; Hibberd, 1975). The possibility of the choanoflagellates serving as an ancestral stock to the phylum Porifera has been speculated upon (Bourrelly, 1962B). Several freshwater colonial genera of choanoflagellates, such as *Proterospongia* (Fig. 7.14), *Sphaeroeca*, and *Codosiga*, are known (Bourrelly, 1968A; Hibberd, 1975).

Order 5. Dictyochales

This order includes the **silicoflagellates**, a relatively small group of marine orga-nisms (Fig. 7.15) bearing a single flagellum and a siliceous skeleton composed of a network of tubular elements. Although the fossil record is relatively rich (Deflandre,

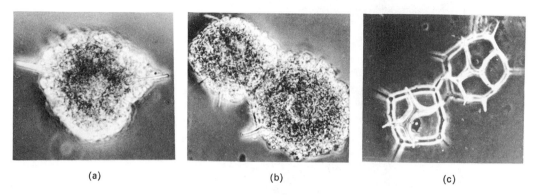

(a)   (b)   (c)

Fig. 7.15 *Dictyocha fibula* Ehr. (*a*) A motile cell containing a skeleton. (*b*) Cell division. (*c*) Skeletons. (*a*), (*b*) × 364; (*c*) × 420. (After Van Valkenburg and Norris; permission of Journal of Phycology.)

1950; Loeblich, et al., 1968; Mandra, 1968; Glezer, 1970), they are relatively rare in present-day seas and may be represented by only a few taxa (Deflandre, 1952; Van Valkenburg and Norris, 1970).

The cytoplasm has a spongy or frothy appearance, with a central dense region containing the nucleus and perinuclear dictyosomes and a less dense ectoplasm containing numerous chloroplasts, mitochondria, and other organelles (Van Valkenberg, 1971B). Isolated portions of ectoplasm radiate from the central region, connected by narrow channels of cytoplasm. Fine pseudopodia may also extend outward from the central region. Although such an amorphous, rhizopodial protoplast would appear to be naked, a delicate cell covering can be detected by using certain stains. This outer boundary is viscous and deformable in swimming cells. In cultures of silicoflagellates, nonswimming stages also occur, usually lacking skeletons; these nonswimming, amoeboid protoplasts have a firmer periplast than those of swimming cells (Van Valkenberg and Norris, 1970).

Whether the skeleton is actually external or internal in relation to the cytoplasm has been variously interpreted over the years. Skeletons are three-dimensional structures, and the nucleus is situated in the central basket. The boundary of the cytoplasm is elusive, flowing continually and changing in shape. Portions of the skeleton do extend beyond the cytoplasm. It has been demonstrated (Van Valkenberg, 1971A) that the skeleton is developed sequentially. It consists of a system of branched tubular elements bearing spinose endings. When the spinose endings make contact with another portion of the skeleton, fusion apparently occurs, resulting in an interconnected pattern (Fig. 7.16).

The first successful cultivation of a silicoflagellate (Van Valkenberg and Norris, 1970) resulted from using a medium with decreased salinity (23–26‰ instead of around 33–35‰) and a reduced enrichment. Optimal temperature for growth was 10°C. The systematics of fossil and extant silicoflagellates has been based upon the structure of the skeleton (Gemeinhardt, 1930). It is noteworthy that in clonal cultures

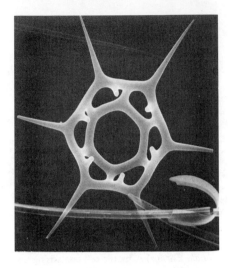

Fig. 7.16 *Distephanus speculum* (Ehr.) Haeckel. Skeleton. × 1480. (Courtesy of Dr. Greta Fryxell.)

of the silicoflagellate identified as *Dictyocha fibula* Ehr. a random collection of 200 skeletons was examined by Van Valkenberg and Norris (1970). On the basis of a standard monograph for this group the skeletons of this clone would be classified into at least six taxa of three genera, some of the skeletons not fitting into any known taxa. The degree of variation possible in skeletal pattern is evident.

Order 6. Chrysosphaerales

Within the uniflagellate series of Chrysophyceae this order contains coccoid forms, solitary or colonial, with a definite wall, thus paralleling the Chrysapiales (p. 363) of the uniflagellate series. The genus, *Chrysophaera*, may have single cells or clusters of cells, producing by vegetative cell division or the production of autospores or zoospores (Whitford and Schumacher, 1969). The zoospores are uniflagellate (Starmach, 1972).

Order 7. Thallochrysidales

Several of the marine filamentous genera, including *Chrysomeris* and *Giraudyopsis*, that had earlier been regarded as belonging to this order in the uniflagellate series, proved to have biflagellated zoospores (Gayral and Haas, 1969; Gayral and Lepailleur, 1971) and accordingly were transferred from this order to the Phaeothamniales (p. 368) of the biflagellate series. Only three genera remain in this order: *Thallochrysis*, with marine and brackish species, and two poorly known monotypic genera occurring in freshwater habitats, *Phaeodermatium* and *Chrysoclonium* (Bourrelly, 1968A).

CLASS PRYMNESIOPHYCEAE

This class is comprised of organisms that produce motile cells bearing two equal, subequal, or unequal **acronematic** flagella, with homodynamic or heterodynamic motion. In addition, some of these motile stages have a **haptonema** arising close to the pair of flagella. The fine structure of the haptonema (Fig. 7.2c) is unlike that of the flagellum (Fig. 7.2a). Its behavior is also dissimilar to that of a flagellum. By means of ciné-photography Leadbeater (1971A) observed that the haptonema is capable of autonomous movement and has an innate mechanism controlling the direction of coiling. In *Chrysochromulina* (p. 379) the haptonema may vary from a highly coiled appendage to an elongated structure. Although it had been suggested that the haptonema is an organelle for attaching or concerned with phagotrophy, Leadbeater discounted these functions on the basis of his observations. Similarly, he pointed out that there was no evidence to support the idea that the haptonema is a tactile organelle or an aid in flotation. In some genera, such as the flagellate *Prymnesium* (Fig. 7.1f) and the motile cells produced by *Phaeocystis*, the haptonema is relatively short and stiff. In *Hymenomonas roseola* Stein the haptonema is reduced to a small bulbous emergence.

Superficial scales may cover both the motile and the nonmotile stages of these algae. The composition of the scales is organic, cellulosic in the case of *Pleurochrysis* (Brown, et al., 1969). A deposition of calcite may also be present on the surface of a second type of scale produced in the **coccolithophorids**, which constitute a large group within this class. Silica, however, is never produced.

The existence of this assemblage within the Chrysophyceae *sensu lato* was recognized by Parke (1961), and the class was subsequently established[4] (Christensen, 1962) to accommodate these organisms.[5] They are more common in the sea than in freshwater (Hibbard, 1976), although this in part reflects the fact that more marine investigations have focused attention upon them. They may occasionally bloom in profusion. An example is *Phaeocystis*, in which the floating gelatinous colonies can form blooms so dense that the sea is locally discolored and the migration patterns of such fish as herring are adversely affected. Coccolithophorids are also known to produce blooms, including *Gephyrocapsa huxleyi* (Lohmann) Reinhardt,[6] which is normally an **oceanic** species but may also be **neritic** in its distribution (Birkenes and Braarud, 1952). Recent evidence has been presented (Gallois, 1976) that blooms of coccolithophorids in the past might have been responsible for the origin of the oil in the North Sea. It was suggested that such blooms might have occurred in bodies of water presenting an environment intermediate between the open ocean and enclosed marine basins. Such conditions would have favored the production of blooms followed by deoxygenation and poisoning of the water. The bottom conditions would have been suitable at least temporarily for the formation of organic-rich sediments, and indeed a coccolith band was observed to be composed of almost entirely a single species, an extinct form.

Most haptophytes are photosynthetic, but heterotrophic growth is possible, either saprophytic (Blankley, 1969; Rahat and Spira, 1967) or phagotrophic (Parke, et al.,

[4]Originally designated the Haptophyceae by Christensen, the class has more recently been assigned the name Prymnesiophyceae by Hibberd (1976).

[5]Bourrelly (1968A) continues to retain this group as a subclass (the Isochrysophycidae) within the Chrysophyceae.

[6]More familiarly known in the literature as *Coccolithus huxleyi* (Lohm.) Kamptn., this taxon has been transferred (Reinhardt, 1972) to *Gephyrocapsa*. It has also been called *Emiliania huxleyi* (Lohm.) Hay et Mohler.

1956; Parke and Adams, 1960; Manton, 1972B). The heterotrophic mode of nutrition would enable these organisms to survive periods when they sink below the **euphotic zone**.

The systematics of the Prymnesiophyceae at present is in an active state of flux. In general, two orders are recognized:

Isochrysidales: motile cells with two acronematic flagella but lacking an obvious haptonema.

Prymnesiales: motile cells with two acronematic flagella and a haptonema, which may range in length from short to long.

The distinction above appears to be unstable, since some of the flagellated cocco-lithophorids (in the order Prymnesiales) seem to lack a haptonema and the suggestion of a haptonema may occur in some members of the Isochrysidales, such as the type species of *Isochrysis, I. galbana* Parke (Parke, 1971).

Cell Covering

It now appears that all members of this class have at least some type of cell cover-ing, even if the monads appear to be naked at the light-microscope level. For example, the biflagellate cells of *Isochrysis* have very fine organic scales detectable with the electron microscope (Parke, 1971). Cells of *Chrysochromulina* may be covered by scales of more than one morphology (Manton, 1972A, 1972B; Leadbeater, 1972B), the simple platelike scales usually underlying the more elaborate types. The walls of some mem-bers, such as *Pleurochrysis*, may represent thick accumulations (Fig. 7.17a) of closely appressed organic scales (Fig. 7.17b) (Brown, 1969; Brown, et al., 1969, 1970, 1973; Leadbeater, 1971B). In some of the filamentous or packet-forming prymnesiophycean genera, such as *Apistonema* and *Chrysotila*, the cell covering changes with age of the

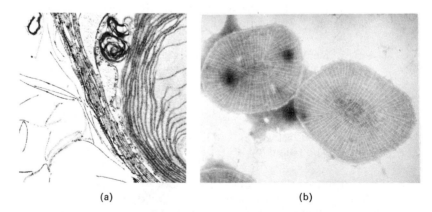

(a)   (b)

Fig. 7.17 *Pleurochrysis scherffelii* Pringsheim. (*a*) Transmission electron micro-graph of section through cell wall showing accumulation of scales. (*b*) Surface view of wall scales. (*a*) × 22,825; (*b*) × 25,875. (Courtesy of Dr. R. Malcolm Brown, Jr.)

culture. A thin layer of mucilage is present on cells in young cultures; this mucilaginous investment thickens with age and may become lamellated (Green and Parke, 1975). In yet older cultures (6 months and older) mineralized elements may become deposited in the mucilaginous layer (Parke, 1971; Green and Parke, 1975). In addition to the usual circular scales covering the cell body, smaller oval scales may be present apparently on the haptonemal surface (Leadbeater, 1970). Scale production within Golgi vesicles has been demonstrated for several different haptophytes (Manton, 1966C, 1967A; Brown, 1969).

The production of special scales that become encrusted with calcium carbonate in the form of calcite, rarely aragonite, crystals is the primary attribute of the **coccolithophorids** (p. 381). The individual scales, called **coccoliths**, are produced within vesicles of the Golgi body (Outka and Williams, 1971) and collectively form a rigid outer skeleton around the cell.

Reproduction and Life Histories

The usual pattern of asexual reproduction is effected by binary cell division or the production and release of several motile or nonmotile cells. Some prymnesiophycean algae form multicellular aggregations, having diverse morphologies: filaments, such as in *Apistonema* and *Crysotila*; palmelloid colonies, such as in *Ochrosphaera*; cuboidal packets (Fig. 7.18), such as in *Sarcinochrysis*; or coccoid unicells and clusters, such as

Fig. 7.18 *Sarcinochrysis*. A typical cuboidal colony. × 500. (Courtesy of Dr. Jimmy T. Mills.)

in *Pleurochrysis*. It is now recognized that these benthic forms exhibit a high degree of **pleiomorphism** and transition from one expression to another. Depending on the age of the culture, a clone might pass through various morphological appearances (Boney and Burrows, 1966; Boney, 1967B; Green and Parke, 1974, 1975).

The status of our understanding of the life histories in the Prymnesiophyceae is still rather incomplete, but a few patterns have been recognized. One of the earliest known cycles is that of *Ochrosphaera neapolitana* Schussnig, in which the fusion of *Ochromonas*-like gametes was reported (Schwarz, 1932). These gametes and also zoospores were released from coccoid cells that bore coccoliths. This benthic phase was the dominant expression of this species, and the gametes and zoospores were ephemeral.

An unusual cycle was reported (Parke and Adams, 1960) in which a motile form identified as *Crystallolithus hyalinus* Gaarder et Markali alternated with a nonmotile form identified as *Coccolithus pelagicus* (Wallich) Schiller. Either phase is capable of indefinite vegetative reproduction of itself, and it is not known what causes a change in the expression or whether there is a corresponding change in the ploidy. Both phases bear coccoliths, and the interesting observation is that holococcoliths are present in *Crystallolithus* and heterococcoliths are present in *Coccolithus*. Therefore, two very distinct types of coccoliths can occur within one life history, which points out the rather artificial nature of the traditional basis of classifying these organisms, namely, the structure of their coccoliths.

The third type of life history characterizes several coccolithophorids in which a motile diploid phase alternates with a haploid benthic phase. The example of *Hymenomonas carterae* (Braarud et Fagerl.) Halldal et Markali is illustrated in Fig. 7.19 (Stosch, 1955, 1967; Rayns, 1962; Leadbeater, 1970, 1971B; Gayral, et al., 1972).

The *Hymenomonas* stage, which bears coccoliths, multiplies by mitosis but under certain conditions undergoes meiosis to produce four products, which germinate into the filamentous benthic stage lacking coccoliths, the *Apistonema* stage. In addition to normal vegetative cell division, two types of motile cells are produced by the *Apistonema* stage. Both types of motile cells from the *Apistonema* stage have two acronematic flagella and scales covering the cell body. One type of motile cell has long, sometimes unequal, flagella and lacks a haptonema, whereas the other has long, equal flagella and a short haptonema, with a second type of scale covering the haptonema. Although Leadbeater (1970) saw no indication of fusion for either type of motile cell, von Stosch (1967) reported that the one type of motile cell was sexual, the zygote developing into the *Hymenomonas* stage, and the other type was asexual, regenerating the *Apistonema* stage.

The life history of *Gephyrocapsa huxleyi*, undoubtedly the most studied of the coccolithophorids in regard to its physiology, remains somewhat uncertain, although patterns have been described (Klaveness, 1972B). Three different cell types are recog-

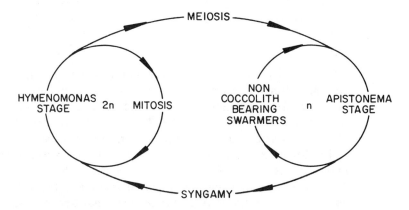

Fig. 7.19 Life history pattern of *Hymenomonas carterae*. (Based upon von Stosch and Leadbeater.)

nized, each capable of indefinite vegetative reproduction: the coccolith-forming type (C cell), the completely naked cell type (N cell), and the scaly motile cell (S cell). Each of these three main cell types reproduces asexually by simple constrictions.

In a pure culture established from any of the three types, other cell types occasionally appear, indicating that the cell types are interconvertible (Paasche and Klaveness, 1970). The factors causing these transitions are not understood, but the transition from C cell to N cell or S cell involves the loss of the coccolith-forming apparatus (Klaveness, 1972A). The motile cell produces noncalcified organic scales. The C cells and the N cells have the same amount as DNA per cell (Paasche and Klaveness, 1970), and it was suggested (Klaveness, 1972B) that the S cells may have a smaller chromosome number and thus represent a possible sexual stage. But such a cycle has not yet been proved. A haptonema has not yet been observed in *Gephyrocapsa huxleyi*.

REPRESENTATIVE GENERA

Order 8. Isochrysidales

ISOCHRYSIS Parke   This genus of marine monads was originally characterized (Parke, 1949) as having two equal, acronematic flagella, a haptonema being absent (Fig. 7.1*d*). The type species of *Isochrysis* (Gr. *isos*, equal + Gr. *chrysos*, gold), *I. galbana* Parke, was later examined (Parke, 1971) in greater detail and observed to have motile cells covered with circular body scales in a mucilaginous matrix and a very reduced haptonema in some but not all of the strains. The presence of a haptonema points out the need for a revision of these two orders of algae. It has been suggested (Green and Parke, 1975) that the Isochrysidales be retained for forms in which a rudimentary haptonema may or may not be present and, if present, undetectable with the light microscope. This criterion would necessitate the transfer of some coccolithophorids (p. 381) to the Isochrysidales, such as *Gephyrocapsa huxleyi* (Klaveness, 1972B), *Ochrosphaera verrucosa* (Lefort, 1971), and *O. neapolitana* (Gayral and Fresnel-Morange, 1971). The last-named species has a small intracellular haptonema.

Benthic phases are known to produce *Isochrysis*-type cells (Parke, 1971; Billard and Gayral, 1972; Green and Parke, 1975). Calcareous elements may become deposited in the thick mucilaginous matrix of these benthic stages, especially in older cultures. These rodlike or cruciform calcareous structures resemble a nanofossil form known as *Tetralithus* (Parke, 1971). They lack the elaborate structure of coccoliths, but their presence is a feature indicating further relationship with the coccolithophorids of the Prymnesiales.

Order 9. Prymnesiales

CHRYSOCHROMULINA Lackey   Originally based upon a small, supposedly triflagellated monad, *C. parva* Lackey, described from a river in Ohio (Lackey, 1939), *Chrysochromulina* (Gr. *chrysos*, gold + Gr. *chroma*, color + Gr. *lina*, made of) includes small marine and freshwater flagellates. The third, relatively long "flagellum" of members of this genus has since been recognized to be a haptonema (Fig. 7.1*e*). Many species have been described from European coastal waters (Parke, et al., 1955, 1958; Manton and Leedale, 1961A; Parke and Manton, 1962; Leadbeater and Manton, 1969A, 1969B, 1971; Leadbeater, 1972B; Green and Leadbeater, 1972).

The most useful characteristic in delimiting species is the morphology of the body scales. Only rarely are these scales discernible with the light microscope, such as the large scales of *C. mactra* Manton (Manton, 1972A). More often, examination with the electron microscope is required. Scales may be simple, rimless plates or have more elaborate shapes, such as cylindrical in *C. megacylindra* (Manton, 1972B), tub-shaped in *C. mactra* (Manton, 1972A), tapered spines in *C. ericina* (Manton and Leedale, 1961A), and other variations. More than one kind of scale may occur on a cell. Three types, including long spine scales, are present on *C. mantoniae* Leadbeater (1972B); four types (three kinds of plate scale and one kind of spine scale) are present in *C. parkeae* Green et Leadbeater (Green and Leadbeater, 1972). The interesting architectural details of these scales have resulted in *Chrysochromulina* being the most intensively studied genus in this class in reference to ultrastructure. The scales are produced in Golgi vesicles (Manton, 1967A, 1967B; Allen and Northcote, 1975).

The pair of flagella and the haptonema arise close together, usually on the ventral surface of the cell. In some species the length of the haptonema, when uncoiled, may be 12 to 18 times the cell body length (Parke, et al., 1959). Considerable metaboly may be exhibited by these monads. Usually two lateral chloroplasts are present per cell, and an immersed or bulging pyrenoid is present, depending on the species (Parke, et al., 1958; Manton, 1966B, 1972B; Leadbeater and Manton, 1971). Phagotrophic feeding can occur (Parke, et al., 1955, 1956, 1959; Manton, 1972B).

PRYMNESIUM Massart   The monads of *Prymnesium* (Gr. *prymn*, the hindmost; stern of a ship) have relatively simple organic scales (Manton, 1966C) arranged in one layer and have two flagella arising near a small, relatively stiff haptonema (Fig. 7.1*f*) (Manton, 1964B). Mitosis has been described (Manton, 1964A) in *Prymnesium*. The nuclear envelope disappears during nuclear division, and the pairs of flagellar bases seem to orient spindle formation.

Members of this genus are **euryhaline**, tolerating a broad range of salinities. Blooms can be produced by *Prymnesium*, the significance being that an extracellular, hemolytic toxin may be associated with these blooms. *Prymnesium parvum* N. Carter has been intensively studied (McLaughlin, 1958; Padilla, et al., 1968; Shilo, 1970) because of its fish-killing capabilities and its interference with pisciculture activities in such countries as Israel. Toxin formation, which is prevented by continuous light, can be controlled in the field by the addition of ammonium sulfate to the fish ponds, causing lysis of the *Prymnesium*.

PLATYCHRYSIS Geitler   The predominant phase of this alga is a nonmotile attached form, in which the two flagella and relatively short haptonema of the transitory motile phase may be retained. The haptonema may serve to attach the cell to the surface film (Norris, 1967A), since this organism is generally **neustonic**, occurring in tidal pools along the Pacific coast. Cells of *Platychrysis* (NL. from Gr. *platys*, flat

+ Gr. *chrysos*, gold) are dorsiventrally flattened and amoeboid and may creep about under the surface film by means of pseudopodia. Both phagotrophic nutrition and photosynthesis are known. Two types of body scales are formed (Chretiennot, 1973), cup-shaped scales lying above thin plate scales. One to four chloroplasts are present in *P. pigra* Geitler, the type species. A temporary motile phase also occurs, these cells being nonflattened and metabolic, which distinguishes these cells from the laterally flattened, nonmetabolic motile cells of *Prymnesium parvum*.

PHAEOCYSTIS Lagerheim   Forming palmelloid colonies in marine plankton, *Phaeocystis* (Gr. *phaios*, dusky + *kystis*, bladder) can produce gelatinous blooms in great abundance, especially in the North Sea. Though more common in temperate waters, this genus extends into the tropics. The gelatinous colonies (Fig. 7.20*a*) contain a large number of cells, each with usually two chloroplasts. Motile cells (Fig. 7.20*b*)

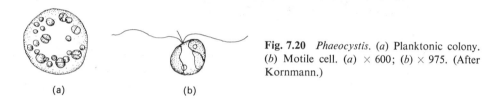

(a)   (b)

Fig. 7.20   *Phaeocystis.* (*a*) Planktonic colony. (*b*) Motile cell. (*a*) × 600; (*b*) × 975. (After Kornmann.)

have two equal, heterodynamic flagella and a very short, noncoiling haptonema with a bulbous tip (Parke, et al., 1971). Both phases, palmelloid colonies and flagellated cells, can be simultaneously present in the same culture and can reproduce independently (Kornmann, 1955). A second category of motile cell, called **microzoospores** because of their relatively smaller size, appeared in older cultures of *Phaeocystis*, but their function is uncertain.

Coccolithophorids

Because of the distinctive nature of the coccoliths that surround these cells, the coccolithophorids are generally placed in the suborder Coccolithophorineae (Papenfuss, 1955); several families are included of both fossil and recent forms. More than 150 genera have been described, and a listing of taxa has been compiled (Loeblich and Tappan, 1966). It is thought that perhaps as many as 250 taxa of extant coccolithophorids exist. New taxa continue to be described, especially fossil forms. Extending into the lower Jurassic of the Mesozoic Period, they have great stratigraphic value, owing to the precision of their skeletal remains (Bramlette, 1958). Coccoliths have often survived in marine sediments in which other fossils were not preserved. Some of the present-day families were in existence early in the Jurassic (Black, 1971), and these families are still flourishing today. Coccoliths have also been employed (McIntyre, 1967) as indicators of climatic conditions during the Pleistocene glaciation.

The great bulk of coccolithophorids are marine, most being strictly oceanic in their occurrence and the minority being neritic (Gaarder, 1971). It is the neritic species, such as *Hymenomonas carterae*, that may have an alternative benthic phase (p. 368).

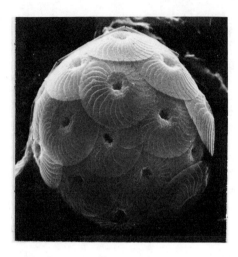

Fig. 7.21 *Calcidiscus* sp. Coccosphere comprised of coccoliths. × 2800. (Courtesy of Dr. Greta Fryxell.)

The coccolithophorids have received relatively less attention than diatoms and dino-flagellates as a phytoplankton group, but they nonetheless constitute an important source of primary productivity, at times outnumbering diatoms in oceanic waters (Gaarder, 1971). Several recent surveys of living coccolithophorids deserve mention: McIntyre and Bé (1967) for the Atlantic Ocean, Gaarder and Hasle (1971) for the Gulf of Mexico, Throndsen (1972) for the Caribbean sea, and Okada and Honjo (1970, 1973) for the Pacific Ocean.

Electron-microscopic examination is essentially a prerequisite for elucidating the more detailed analysis of coccolith structure (Black, 1963, 1965). Many coccolithophorids have organic scales in addition to their coccoliths. Coccoliths are of two basic types. **Holococcoliths** have crystals all of the same size and shape, simple rhomdohedral calcite crystals forming a covering over an organic template. **Heterococcoliths** have morphologically diverse crystals, usually of calcite but occasionally aragonite. The coccoliths of any one coccolithophorid may fit together into a coherent collection termed the **coccosphere** (Fig. 7.21).

A further subdivision of coccoliths is possible on the basis of their shape (Braarud, et al., 1955). The following are some examples:

placolith: coccolith consisting of two layers separated by a cylindrical or tubular
    central piece (Fig. 7.22*a*).
rhabdolith: coccolith bearing a spine or projection (Fig. 7.22*d, f*).
discolith: coccolith bearing a small protuberance.
calyptrolith: dome-shaped coccolith (Fig. 7.22*b*).
scapholith: diamond-shaped coccolith (Fig. 7.22*e*).
pentalith: five-sided coccolith, made up of radially arranged, wedge-shaped
    crystals (Fig. 7.22*c*).

The artificial nature of this system of classification is emphasized by such discoveries discussed previously (p. 378) as that by Parke and Adams (1960), who

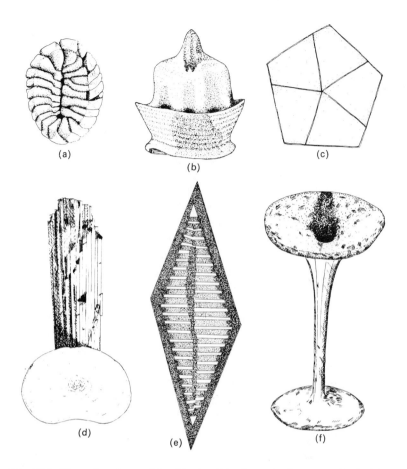

Fig. 7.22   Examples of coccoliths. (*a*) Placolith of *Actinosphaera sera* Black. (*b*) Calyptrolith of *Calyptrosphaera pirus* Kamptner. (*c*) Pentalith of *Braarudosphaera bigelowi* (Gran et Braarud) Defl. (*d*) Rhabdolith of *Rhabdosphaera clavigera* Murray et Blackman. (*e*) Scapholith of *Calciosolenia murrayi* Gran. (*f*) Placolith of *Discosphaera tubifera* (Murr. et Blackm.) Ostenf. [(*a*), (*c*), and (*d*) after Black; (*b*) after Throndsen; (*e*) and (*f*) after Gaarder and Hasle.]

demonstrated in culture that *Coccolithus pelagicus*, which bears placoliths (in the heterococcolith category) fitting together into a coccosphere, has an alternate motile form that had been identified as *Crystallolithus hyalinus*, which bears calyptroliths (in the holococcolith category), not forming a coccosphere. Furthermore, dimorphic coccoliths may be present on a single coccolithophorid.

The reaction of coccoliths to polarized light has also been valuable in categorizing them. Crystallographic analysis reveals whether a coccolith is composed of a single calcite crystal, relatively few, or many. A positive or negative reaction to polarized light may be produced, or in some cases alternating bands of light and dark might be produced when the coccoliths are exposed to polarized light.

The mechanism of coccolith production has been reviewed by Paasche (1968A). The assembly of the coccoliths occurs internally. In *Gephyrocapsa huxleyi* crystal growth begins near the base of the central tubular part of the coccolith (Wilbur and Watabe, 1963), but it is possible that this process begins at the periphery of coccoliths in other species (Black, 1963). The coccolith retains a membranous covering after it has been extruded from the cell. Coccoliths are not formed if the pH of the medium drops below 7 (seawater being slightly alkaline with a pH of 7.8). Coccolithophorids may thus be decalcified by lowering the pH, but the naked cells can produce a new covering of coccoliths within 15 hours in the case of *G. huxleyi* if a more favorable pH is restored (Paasche, 1966).

An interesting relationship of coccolith formation and the nutritional status of the organism was noted (Wilbur and Watabe, 1963) in a strain of *G. huxleyi* that did not form coccoliths under normal conditions of growth. When placed into a medium lacking nitrate, however, this strain formed coccoliths. This observation should be considered in light of the life-history pattern accepted for *Hymenomonas carterae* (Fig. 7.19), in which the coccolith-bearing phase is regarded as the diploid phase and the alternate, noncoccolith-bearing phase as haploid.

Although experiments by Paasche (1964, 1968B) indicated that coccolith formation was a light-dependent process, other work (Blankley, 1969) claimed that calcification and sustained growth could occur in darkness in cultures supplied with glycerol as a carbon source. The formation of coccoliths was demonstrated (Watabe and Wilbur, 1966) to be temperature-dependent, the proportion of naked cells increasing at the extremes of the temperature growth range. But this result might simply reflect the fact that the permanently naked strains of the species being tested (*G. huxleyi*) were favored by the temperature extremes. In a species in which all cells formed coccoliths no temperature influence on calcification was observed (Paasche, 1968B).

The relative uniformity in the capacity of *Gephyrocapsa huxleyi* to form coccoliths was stressed by the work of Paasche (1968B). Although a seven-fold variation in growth rate could be brought about by the manipulation of the parameters of cultivation, the correlated variation in coccolith production per unit of cell volume amounted to less than a two fold variation.

CLASS XANTHOPHYCEAE

Related to the other classes in this division by the absence of chlorophyll *b*, the Xanthophyceae differ from the Chrysophyceae, Prymnesiophyceae, and Bacillariophyceae by the lack of the accessory pigment fucoxanthin. The pigmentation of these algae is consequently yellow-green in appearance rather than the golden hue characteristic of these other classes. Members of the Xanthophyceae are more likely to be confused with the chlorophycean algae because of their greenish to yellow-green aspect. Testing for the presence of starch (by adding iodine to detect the black staining reaction of starch) would result in a negative reaction, indicating that the alga in question is xanthophycean rather than chlorophycean. For most members of the class the nature of the food reserves has been little investigated, although chrysolaminaran (= leucosin) is usually reported (Dodge, 1973). Oils and fats are also present. Analyses of the water-soluble and alkali-soluble carbohydrates in *Tribonema* (p. 388) demonstrated (Cleare and Percival, 1972, 1973) the presence of free mannitol, glucose, a mixture of oligopolysaccharides, a 1,6-linked glucan, and 1,4- and 1,3-linked xylans. The cell wall was cellulosic.

Although older studies did not detect chlorophyll *c* in the Xanthophyceae, more recent investigations of some members have presented evidence that chlorophyll *c* is present (De Greef and Caubergs, 1970; Guillard and Lorenzen, 1972). The fact that chlorophyll *c* does occur in this class as well as in the Chloromonadophyceae (p. 396) lessens the desirability of separating these two classes into a distinct division (the Xanthophyta), as suggested by recent authors (Scagel, et al., 1966). The major xanthophyll present in the Xanthophyceae is diadinoxanthin (Whittle and Casselton, 1975A).

Statospores are produced in some xanthophycean algae, and they are unlike those occurring in the Chrysophyceae in that they consist of two parts of about equal size, which overlap. These resting stages are also formed internally. Many genera, particularly filamentous forms, have walls that are made of so-called H-pieces such that the ends of adjacent H-pieces overlap to enclose the protoplasm. *Tribonema* (p. 388) has this type of wall construction.

Classification

The orders of Xanthophyceae are distinguished on the basis of their vegetative development. Many of the same vegetative tendencies observable in the Chlorophyceae and Chrysophyceae are paralleled by members of this class. The following list presents a brief characterization of these orders.

> Heterochloridales: naked, unicellular, anteriorly biflagellated monads, some with a capacity to be temporarily amoeboid.
>
> Rhizochloridales: dominant amoeboid stage, unicellular or multicellular; biflagellated zoospores produced.
>
> Heterogloeales: immobile, coccoid forms, solitary or colonial, with or without a gelatinous envelope; zoospores with contractile vacuoles.
>
> Mischococcales: immobile, coccoid forms, solitary or colonial, not palmelloid; uni- or biflagellated zoospores produced, lacking contractile vacuoles.
>
> Tribonematales: simple or branched filaments with cross walls; uni- or biflagellated zoospores produced.
>
> Vaucheriales: multinucleate vesicles or filaments, lacking cross-walls; simple or compound zoospores produced.

Some selected genera are discussed in this book. Works by Pascher (1938), Ettl (1956), and Bourrelly (1968A) contain many additional genera and further information.

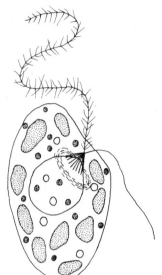

Fig. 7.23 *Olisthodiscus luteus* Carter. × 2400. (After Leadbeater.)

Order 10. Heterochloridales

OLISTHODISCUS Carter   The monads of *Olisthodiscus* (Gr. *olisthos*, slippery + Gr. *diskos*, quoit) contain several chloroplasts and have two subequal, hetero-dynamic flagella. Cultured extensively as food for the growth of oysters and other shellfish, *O. luteus* Carter is a brackish-water species, which has been shown (Hellebust, 1965) to release mannitol and glycolic acid into the medium. The cells have a large number of spherical bodies lying in a subsurface layer (Leadeater, 1969). The two flagella arise in a subapical position and are associated with an elaborate flagellar root system, which in turn is closely connected with the nucleus (Fig. 7.23). The hairs of the pleuronematic flagellum have been observed (Leedale, et al., 1970) to arise in vesicles derived from the perinuclear space, and the vesicles migrate to the base of the flagellum, where the hairs are deposited on the flagellum. Such a process of flagellar hair formation also happens in other classes of this division (Dodge, 1973).

Order 11. Mischococcales

BOTRYDIOPSIS Borzi   This genus consists of nonmotile spherical cells with a thin wall and of variable size. Young stages of *Botrydiopsis* (resemblance to *Botrydium*, q.v.) contain only a few chloroplasts, but the cell gradually enlarges (Fig. 7.24*a*) and eventually contains numerous peripheral discoid chloroplasts and many nuclei. Reproduction is by the formation and release of autospores or zoospores, the latter having two unequal flagella, one pleuronematic and the other acronematic.

CHARACIOPSIS Borzi   Placed in a different family than the free-living *Botrydiopsis*, cells of *Characiopsis* (resembling *Characium*, q.v.) are attached to the sub-

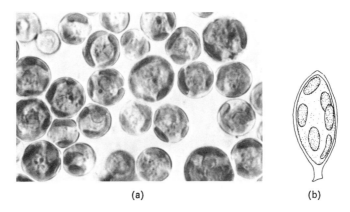

(a)   (b)

Fig. 7.24 (*a*) *Botrydiopsis arhiza* Borzi. Mature cell. (*b*) *Characiopsis obovoidea* Pascher. Vegetative cell. (*a*) × 800; (*b*) × 1320. [(*a*) after Bold, *Morphology of Plants*, 3rd ed., 1973, Harper & Row, Publ.; (*b*) after Dr. K. Lee.]

stratum by a small pad or stalk (Fig. 7.24*b*). The cells are solitary and may be spherical, ovoid, or elongate, the distal end being drawn out into a narrow tip in some species. One trough-shaped chloroplast or a few to many discoid chloroplasts may be present, but pyrenoids are absent.

Reproduction occurs by the formation of zoospores, aplanospores, or auto-spores. In *Characiopsis obovoidea* Pascher the mature cells are uninucleate but become multinucleate just prior to zoosporogenesis, and 16–32 naked zoospores are liberated from a reproductive cell (Lee and Bold, 1974). Resembling the chlorophycean genus *Characium*, *Characiopsis* may be distinguished by its lack of starch and by the zoospores with flagella of unequal length. The monographic treatment of Pascher (1938) included a total of 46 species in this taxonomically difficult genus.

OPHIOCYTIUM Naegeli   This genus can be distinguished from the previous two by the nature of the cell wall, which consists of two pieces joined together. Cells are capable of continued elongation by the addition of intercalary, cupshaped pieces. Cells of *Ophiocytium* (Gr. from *ophis*, snake + Gr. from *kystis*, bladder) may be solitary or grouped into clusters, the latter pattern resulting from the release of zoo-spores by the detachment of the distal wall piece and the attachment and germination of these zoospores into the parental wall.

The mature, multinucleate cells may be straight, curved, or spiraled; spinelike processes often terminate the free ends (Fig. 7.25). The chloroplasts within vegetative cells and zoospores of *Ophiocytium majus* Naegeli contain the three-thylakoid lamellae and peripheral girdle lamellae that are characteristic for many Chrysophycophyta (Hibberd and Leedale, 1971A). Girdle lamellae, which are thylakoids (= photosyn-thetic membranous sacs) forming a concentric arrangement lying within the chloro-plast envelope, are absent in the Prymnesiophyceae (p. 374), Eustigmatophyceae (p. 393), and rare Xanthophyceae (Massalski and Leedale, 1969; Dodge, 1973). In addi-

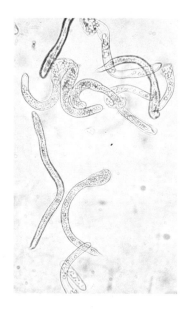

Fig. 7.25 *Ophiocytium*. Vegetative cells. × 380.

tion to zoospore production, aplanospores may be liberated. About a dozen species have been recognized. They are all freshwater in their occurrence and cosmopolitan in their distribution.

Order 12. Tribonematales

TRIBONEMA Derbès et Solier The unbranched uniseriate filaments of *Tribonema* (Gr. from *tribein*, to rub + Gr. *nema*, thread) are composed of uniformly cylindrical to slightly barrel-shaped cells (Fig. 7.26), the walls of which consist of two equal, slightly overlapping halves. When the filaments become dissociated, the

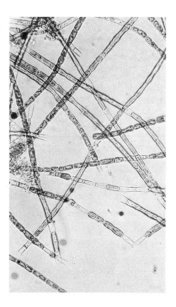

Fig. 7.26 *Tribonema*. Filament. × 129.

H-pieces can be detected (Fig. 7.27), and the ends of filaments appear as the open ends of a cylinder. Each time cell division occurs, a new H-piece is added to separate the two resultant cells. Cells contain one to several chloroplasts, pyrenoids apparently being absent (Massalski and Leedale, 1969), except for *T. pyrenigerum* Pascher. Girdle lamellae are present (Falk and Kleinig, 1968).

In addition to fragmentation of the filaments, reproduction is effected by the formation of zoospores (Fig. 7.28), one or two per cell. They are liberated by the separation of the H-pieces. Sexual reproduction is also known and involves the isogamous union of gametes; a dormant zygote is produced. Statospores and aplanospores are also known to occur in this freshwater genus, which has a cosmopolitan distribution.

Two closely related filamentous genera deserve mention. In *Bumilleria* the characteristic H-pieces are formed less regularly, perhaps for every two to four cells. The walls of *Heterothrix*, also of widespread distribution as a soil alga, are not formed of H-pieces but are single structures.

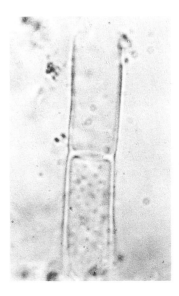

Fig. 7.27 *Tribonema*. So-called "H-piece" consisting of cross wall and portions of wall of two adjacent cells. × 1850.

Fig. 7.28 *Tribonema*. Zoospore. × 1320. (After Massalski and Leedale.)

Order 13. Vaucheriales

BOTRYDIUM Wallroth   Of widespread occurrence on damp soil, the vesiculate thalli (Fig. 7.29) of *Botrydium* (Gr. *botrydion*, in clusters) consist of a conspicuous spherical aerial portion and a colorless rhizoidal portion, the latter being simple or branched and embedded in the soil. The protoplasm is a thin peripheral layer with numerous nuclei and numerous discoid chloroplasts. Plants in nature frequently develop a superficial coating of calcium carbonate.

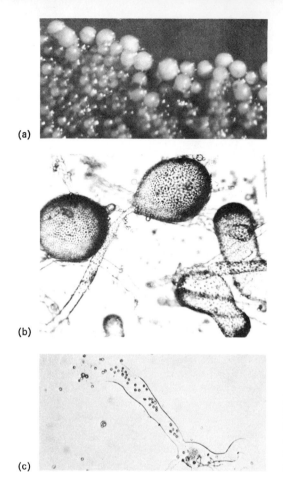

(a)

(b)

(c)

Fig. 7.29 *Botrydium granulatum* Grev. (*a*) Sacs growing on agar surface. (*b*) Individual sacs from an agar-grown culture. (*c*) Release of zoospores from a sac. (*a*) × 4.5; (*b*) × 15; (*c*) × 27. [(*a*) and (*b*) after Bold, *Morphology of Plants*, 3rd ed., 1973, Harper & Row, Publ.]

Reproduction of these saccate plants is carried out by zoospores or aplanospores. Zoospore production is more likely to happen if the thallus is flooded with water. The sac is transformed into a large sporangium, and biflagellated zoospores, each with a single nucleus and two chloroplasts, are released. These zoospores may function as gametes, their isogamous or anisogamous fusion (depending on the species) resulting in zygotes. In some species the rhizoidal portions can be metamorphosed into thickened cysts, which germinate directly or by zoospore production.

VAUCHERIA De Candolle   This branching, filamentous alga consists of tubular coenocytic axes, septa being laid down only when gametangia or zoosporangia are formed. Long regarded as belonging to the green algae, the affinities of *Vaucheria* (named for the Swiss phycologist *M. Vaucher*) with the Xanthophyceae were recognized by analyses of its photosynthetic pigments and food reserves, which agree with those of the Chrysophycophyta rather than those of the Chlorophycophyta. In addition, ultrastructural investigations of the sperm revealed (Moestrup, 1970) the presence of a pair of heterodynamic flagella, including one pleuronematic type.

The peripheral cytoplasm of the tubes contains a large number of discoid chloroplasts and many small nuclei. The filaments extend in length by tip growth, and a

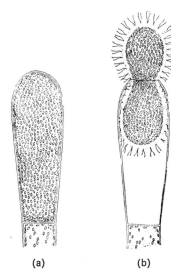

Fig. 7.30 *Vaucheria.* (*a*) Formation of a zoosporangium. (*b*) Release of a single zoospore from a zoosporangium. × 273. (After Bold, *Morphology of Plants*, 3rd ed., 1973, Harper & Row, Publ.)

(a)   (b)

cytoplasmic zonation is evident near the apex. An apical zone, subapical zone, and zone of vacuolation have been described (Ott and Brown, 1974A). The apical zone of actively growing filaments contains numerous vesicles that are thought to be responsible for the deposition of wall material, which is primarily cellulose (Maeda, et al., 1966). Nuclei and chloroplasts are lacking in this apical zone, but many associated pairs of mitochondria and Golgi bodies are present. The ultrastructure of mitosis in *Vaucheria* has been reported (Ott and Brown, 1972). The nuclear envelope remains intact during mitosis, surrounding the spindle apparatus. This "closed" type[7] of nuclear division has been reported in other algae (Dodge, 1973), but gaps or fenestrae occur at the poles of these other algae. This nuclear envelope is without any gaps in *Vaucheria*.

Reproduction is both asexual and sexual. The asexual mode involves the production of zoosporangia (Fig. 7.30*a*) at the tips of the filaments, a septum separating off the protoplasm of this region. The protoplasm is then metamorphozed into a single large zoospore (Fig. 7.30*b*) with numerous pairs of slightly unequal flagella covering the surface. The multinucleate zoospore, which is subsequently liberated, is regarded as a compound zoospore, since it seems to represent a composite structure of many nuclei with flagellar pairs. Zoosporogenesis, which is controlled by an **endogenous circadian rhythm** (Rieth, 1959), has been examined (Ott and Brown, 1974B) in *Vaucheria fontinalis* (L.) Christensen and observed to involve initially a migration and accumulation of nuclei at the tip destined to differentiate into a zoosporangium. The pair of centrioles associated with each nucleus initiates two flagella, which are located in an internal vesicle. The flagella of several nuclei become associated within common vesicles and are called "flagellar pools" (Ott and Brown, 1974B). These groups migrate to the periphery of the incipient zoospore, and the vesicle fuses with the plasma mem-

[7]The completely "closed" type of mitosis is known to occur in some fungi and protozoa. It is regarded (Leedale, 1970) as a primitive condition.

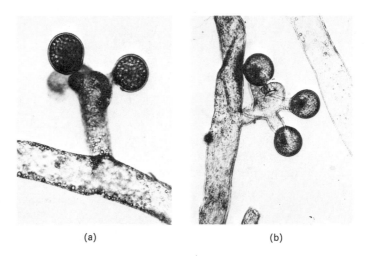

(a)   (b)

Fig. 7.31 *Vaucheria*. Cluster of oogonia and antheridia borne on a lateral branch. × 185.

brane. The numerous pairs of flagella are thus situated on the surface of the cytoplasm within the mature zoosporangium. The multiflagellated zoospore is allowed to escape by the gelatinization of the apical tip of the zoosporangium (Ott and Brown, 1975). When the naked zoospore settles down, the flagella are withdrawn, and a wall layer is laid down seemingly by peripheral vesicles containing fibrillar material.

Oogamous sexual reproduction is easily observed in *Vaucheria* because of the conspicuous nature of the antheridia and oogonia. Depending on the species, the sex organs (Fig. 7.31 and 7.32*a*) may be distributed singly or in clusters on special lateral branches (either on a common branch or on adjoining branches) or be sessile on the main filaments. Significant variation is shown by the antheridia, and the large number of sections that have been established within the genus are distinguished on the basis of criteria related to the antheridia (Blum, 1972). The antheridium, which

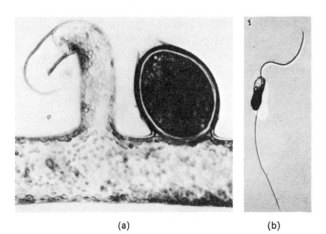

(a)   (b)

Fig. 7.32 *Vaucheria*. (*a*) *V. sessilis* (Vauch.) DeCand. var. *clathrata*. Oogonium containing zygote and discharged antheridial branch. (*b*) *V. sescuplicaria* Christensen. Sperm. (*a*) × 41; (*b*) × 58. [(*a*) after Bold, *Morphology of Plants*, 3rd ed., 1973, Harper & Row, Publ.; (*b*) after Møstrup; permission of Cambridge University Press for Journal Marine Biological Association U. K.]

is multinucleate, produces a large number of small, almost colorless sperm (Fig. 7.32*b*). Spermatogenesis has been described (Moestrup, 1970) at the ultrastructural level. The sperm bears a pair of laterally inserted flagella, one of which has two rows of stiff hairs. An emergent process called a *proboscis*, similar to that in the sperm of *Fucus* (p. 350), is located at the anterior end of the cell.

The oogonium also is initially multinucleate, but all the nuclei except one migrate out from the differentiating oogonium into the subtending filament. The uninucleate protoplasm is then separated by a septum. A pore is formed in a receptive area of the oogonial wall to permit entry of the sperm. The zygote that results is thick-walled and resting. Meiosis occurs at the time of germination of the zygote, and a new siphonous filament emerges from the zygote wall.

Widespread in freshwater, brackish, and marine habitats, *Vaucheria* is often amphibious, living on mud that is periodically immersed in water and then exposed to the air. This genus continues to generate much interest regarding its ecology (Christensen, 1956; Nienhuis and Simons, 1971), systematics and nomenclature (Blum, 1972; Christensen, 1969; Polderman, 1973; Rieth, 1974), cytology (Schulte, 1964), and physiology. Chloroplast distribution is random in the dark but oriented in the light, low light intensity causing the chloroplasts to move into position to receive maximum illumination and high light intensity causing them to orientate themselves to receive minimal illumination (Nultsch, 1974). The photoreceptors of the light-orientated movement of chloroplasts are located in the cytoplasm rather than in the chloroplasts (Fischer-Arnold, 1963). The action spectrum for this movement is essentially identical to that measured for similar phenomena in other plant groups (Haupt, 1963), suggesting a flavin. It is interesting that the chloroplasts of germlings of *Vaucheria woroniniana* Heering have projecting pyrenoids (Marchant, 1972), whereas embedded pyrenoids have been observed in the chloroplasts of mature filaments for other species.

CLASS EUSTIGMATOPHYCEAE

Differences regarding ultrastructural characteristics and pigment composition have led to the separation of some organisms that had formerly been placed in the Xanthophyceae into the recently established class Eustigmatophyceae (Hibberd and Leedale, 1970, 1971B). One of the most distinctive features of this group is the eyespot in the zoospores (Fig. 7.33). Independent of the chloroplast, it consists of an irregular group of globules located at the anterior end of the cell adjacent to the flagellum (or flagella) and is not surrounded by a membrane either individually or as a group. Such an extrachloroplastidic eyespot is reminiscent of that in the Euglenophycophyta.

The naked, amoeboid zoospores are elongate and usually uniflagellate, the flagellum being anteriorly directed and bearing two rows of stiff hairs. A second smooth flagellum is present in two genera, *Ellipsoidion* and *Pseudocharaciopsis*. The pleuronematic flagellum has a distinctive flagellar swelling at its base, where the flagellar membrane is dilated or flared out and lacking the rows of hairs.

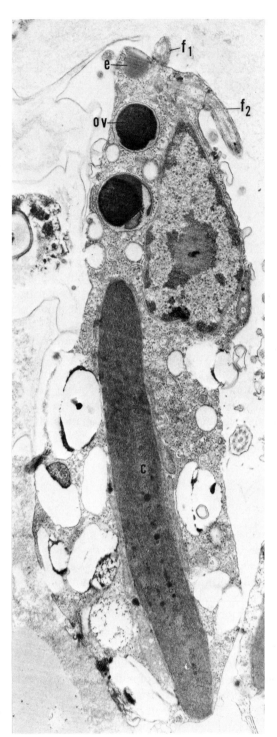

Fig. 7.33 *Pseudocharaciopsis texensis* Lee et Bold. Transmission electron micrograph of longitudinally sectioned zoospore. Note eyespot (*e*) located outside of chloroplast (*c*), the bases of two emergent flagella (f_1, f_2), and osmiophilic vesicles (*ov*). × 22,250. (After Lee and Bold; permission of British Phycological Journal.)

Fig. 7.34 *Pseudocharaciopsis texensis.* Vegetative cell. × 960. (After K. Lee; permission of British Phycological Journal.)

Other ultrastructural differences include the fact that Golgi bodies are absent from the motile cells. Furthermore, girdle lamellae running in a concentric arrangement just within the chloroplast envelope are also absent, which is unlike the situation in other classes of this division except for the Prymnesiophyceae. The single elongate chloroplast in the zoospore lacks a pyrenoid, but the chloroplast of the vegetative cell contains a distinctive polyhedral pyrenoid, projecting from the inner face of the chloroplast by a stalk and not penetrated by membranes (Hibberd and Leedale, 1972). Although the chloroplast is associated with endoplasmic reticulum, there is no relationship to the nuclear envelope. In agreement with other classes in the Chrysophycophyta, chloroplasts contain thylakoids arranged in groups of three.

The pigment composition of the Eustigmatophyceae serves as an additional means of distinguishing this class from the other classes of the division. All members that have been examined (Whittle and Casselton, 1975A) contain violaxanthin as their major xanthophyll pigment, which is lacking in the Xanthophyceae. In addition, only chlorophyll *a* is present as well as *β* carotene (Fork and Brown, 1975).

Only a few coccoid or attached genera have been related with certainty to the Eustigmatophyceae. But it is apparent that many genera, especially coccoid forms, that have been placed in the Xanthophyceae need to be reexamined in light of the information that has emerged from observations of the zoospores. Species assigned to this class include the coccoid forms *Pleurochloris commutata* Pascher, *P. magna* Boye Petersen, *Ellipsoidion acuminatum* Pascher, *Vischeria stellata* (Poulton) Pascher, *V. punctata* Vischer, and *Polyedriella helvetica* Vischer et Pascher. *Pseudocharaciopsis texensis* Lee et Bold differs from the others in being a sessile, fusiform alga (Fig. 7.34), usually affixed to the substratum by a small, discoid pad, paralleling the chlorophycean *Characium* and *Pseudocharacium* (p. 118) and the xanthophycean *Characiopsis* (p. 386). Older vegetative cells become multinucleate and may contain several chloroplasts (Lee and Bold, 1973). The zoospores are naked and somewhat amoeboid, conforming to the pattern in this class.

CHLOROMONADOPHYCEAE

This rather small assemblage of unicellular flagellates has in the past been variously classified, as a group of uncertain position (Smith, 1950), with the Xanthophyceae in a separate division (Scagel, et al., 1966), or with the Cryptophyceae (p. 566).
The detection of small amounts of chlorophyll c along with chlorophyll a (Guillard
and Lorenzen, 1972) is evidence that a relationship exists between the Chloromonadophyceae (also called the Raphidophyceae) and the Chrysophycophyta. Several similarities in regard to ultrastructure can also be listed as reflecting an affinity with the
chrysophytes, namely, heterokontan flagella, chloroplast thylakoids arranged in groups
of three, and the presence of girdle lamellae. Significant differences include the presence
of **trichocysts,** which are known to occur in the Pyrrhophycophyta, and the lack of
direct connections between the chloroplast endoplasmic reticulum and the outer
nuclear membrane. Such direct connections characterize most members of the Chrysophycophyta excepting the Eustigmatophyceae. The presence of **kinetochores** (Heywood
and Godward, 1972), which are of very rare occurrence among the algae (Dodge,
1973) and more characteristic of the chromosomes of animals, and an intranuclear
spindle, which recalls that observed in *Vaucheria* (p. 390) of the Xanthophyceae, are
somewhat distinctive ultrastructural features. Because of this combination of similarities and dissimilarities the chloromonads were regarded by Taylor (1976) to represent a line leading toward the chrysophytes but occupying an isolated position.

These forms are always naked, solitary, and free-living, with two flagella of unequal length, the trailing flagellum being situated close to the cell surface and consequently easily overlooked. Only the anteriorly directed flagellum is pleuronematic,
and its hairs seem to originate within cisternae of the endoplasmic reticulum and accumulate in vacuoles near the flagellar base before being added (Heywood, 1972). Both
colorless and pigmented genera are recognized, the latter having numerous discoid,
grass-green to yellow-green chloroplasts peripherally located. Only freshwater members are known, and contractile vacuoles are present (Schnepf and Koch, 1966). The
colorless members, such as *Reckertia* and *Thaumatomonas*, both of which have been
recorded for the United States (Bourrelly, 1970), are biflagellate and produce pseudopodia. Their membership in this class is indicated by the presence of trichocysts and
a large nucleus, which is characteristic of the class. The classification of the chloromonads has been recently treated by Fott (1968). Only the genus *Gonyostomum* is
discussed below.

Order 14. Chloromonadales

GONYOSTOMUM Diesing   Cells of *Gonyostomum* (Gr. *gony*, knee + Gr.
stoma, mouth) tend to be rather large (up to 100 μm) and are oval to obpyriform in
front view and compressed in side view. Two heterodynamic flagella arise at the anterior end, the trailing flagellum being relatively shorter or longer, depending on the
species. The periplast is thin, flexible in some instances; numerous discoid chloroplasts
and narrow trichocysts are located just beneath the cell surface (Fig. 7.35). These

Fig. 7.35 *Gonyostomum semen* (Ehr.) Diesing. Front view. × 378. (After G. M. Smith.)

trichocysts are easily ejected as thin threads when the cell is adversely influenced by heat, pressure, or other stresses.

Reproduction occurs by longitudinal division, the dividing cell remaining in motion during the process. Acidophilic *Gonyostomum semen* (Ehr.) Diesing, which occurs in bogs and ponds, has been adapted to grow on a defined medium (Heywood, 1973), which included the vitamins thiamine, biotin, and vitamin B_{12} and had a pH of 5.5–5.8.

CLASS BACILLARIOPHYCEAE

The members of this class, referred to as the **diatoms,** are essentially unicellular, although pseudofilamentous and colonial aggregations may occur. They are ubiquitous in their distribution, in aquatic (freshwater, marine, and brackish) and terrestrial or subaerial habitats, requiring at least periodic moisture. Both free-floating (planktonic) and attached (benthic) modes of existence are prevalent among the approximately 200 genera. A significant number of fossil diatoms is also known, the earliest types being marine species dating from the Cretaceous (Patrick and Reimer, 1966). Freshwater-type species appeared in the Oligocene. As a group of organisms there is a greater percentage of extant species of diatoms present in older sediments than other groups of organisms (Wornardt, 1969). For example, about 15% living species of diatoms in the Eocene and 6% in the Upper Cretaceous can be contrasted with about 3.5% living species of mollusks in the Eocene and 0% in the Cretaceous.

The presence of diatoms in the fossil record is the result of the resistant nature of their cell wall, which is composed of silica with an organic coating (Hecky, et al., 1973). The degree of silicification may range from very heavy to extremely light, the genus *Subsilicea* (von Stosch and Reimann, 1970) being an example of the latter condition. The classification of diatoms is almost entirely based upon the structure and ornamentation of the cell wall, which is termed the **frustule**. Since the pattern and markings of the frustule are most readily visualized with specially prepared and cleaned cells, routine examination and identification of diatoms involve nonliving material. With rare exceptions the siliceous components of the wall enable the diatoms to be assigned to particular taxonomic groupings, without one having to depend on living specimens.

Cells of diatoms are uninucleate,[8] the nucleus being either suspended in the center by cytoplasmic strands or displaced to one side of the cell. The chloroplasts, whose shape and number show great variation, are the conspicuous feature of the living cell. They may be numerous discoid structures randomly distributed, such as in *Coscinodiscus* (Fig. 7.36a); or in a stellate configuration, such as in *Striatella* (Fig. 7.36b); or one or two platelike axile or peripheral structures, such as in *Pinnularia* (Fig. 7.36c); or a single, lobed, H-shaped structure, such as in *Gomphonema* (Dawson, 1973A) and *Cocconeis* (Taylor, 1972).

The color of the chloroplasts is typically golden brown but may be yellowish green or dark brown. Photosynthetic pigments include chlorophylls *a* and *c*, β carotene (the principal yellow pigment), fucoxanthin (a brown pigment), and small amounts of diatoxanthin and diadinoxanthin as well as other carotenoids. The cell sap of *Navicula ostrearia* (Gaillon) Bory, a diatom commonly found in oyster beds, is distinctly blue because of the presence of a special vacuolar pigment (Neuville and Daste, 1972). Pyrenoids are present in some species; their presence can be variable within a single genus. They usually appear as bright, refractive bodies, lying within the chloroplast and with thylakoids coursing through them. An orderly paracrystalline arrangement of subunits within the pyrenoid has been reported in some diatoms (Holdsworth, 1968; Taylor, 1972). Even though these organisms are autotrophic, a high incidence of auxotrophy occurs in marine littoral diatoms. Investigations (Lewin and Lewin, 1960; Lewin, 1972) have shown that approximately half of the clones isolated required either cobalamin or thiamine or both for growth.

A few species of colorless diatoms are known, living saprophytically on such substrata as the mucilage of thalli of *Fucus* (p. 350) (Pringsheim, 1967B). It seems likely that these forms arose as nonphotosynthetic mutants from pigmented forms; for example, *Nitzschia alba* Lewin et Lewin might have originated from *N. laevis* Hust. since the frustules have a common fine structure (Lewin and Lewin, 1967). Leucoplasts are present in *Nitzschia alba* (Schnepf, 1969).

In addition to the similarities of pigment composition to other classes in the division Chrysophycophyta, the Bacillariophyceae show other shared characteristics, such as the food reserve, which is chrysolaminaran. Oil droplets are also present.

---

[8]The presence of an unusual bilobed nucleus in *Lauderia annulata* cleve, a marine centric diatom, was reported (Holmes, 1977).

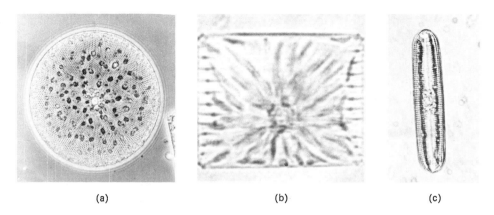

(a)   (b)   (c)

Fig. 7.36 (*a*) *Coscinodiscus.* Valve view, with numerous chloroplasts evident. (*b*) *Striatella.* Girdle view, with chloroplasts in a stellate arrangement. (*c*) *Pinnularia.* Valve view, with two platelike chloroplasts. (*a*) × 391; (*b*) × 363; (*c*) × 435.

Although the presence of flagellated cells was not convincingly demonstrated until relatively recently (Stosch, 1951), their structure is further evidence of the relationship of this class with the Chrysophycophyta. A single, anterior pleuronematic flagellum has been observed on the sperm of some diatoms of the class Centrales (Stosch, 1954, 1958A; Drebes, 1966, 1974; Manton and Stosch, 1966; Manton, et al., 1970B; Schultz and Trainor, 1970; Heath and Darley, 1972; Stosch, et al., 1973). An unusual feature of this single flagellum is the presence of nine peripheral doublet tubules but no central pair (Fig. 7.2*c*) (Manton and Stosch, 1966). Although the derivation of such a uniflagellate cell from a heterokontan biflagellate ancestor is suggested by such a process in the Chrysophyceae, a second vestigial basal body of a flagellum was not detected in the two genera (*Lithodesmium* and *Biddulphia*) investigated.

One departure from the other classes in the Chrysophycophyta is that the vegetative phase of diatoms is diploid, meiosis occurring at the time of gametogenesis. This pattern (an H, d life cycle) thus differs from that of a dominant haploid phase of most other Chrysophycophyta. Almost all diatoms examined thus far are homothallic (Stosch, 1954), sexual reproduction being capable of completion within a single clone. An exception is *Rhabdonema adriaticum* Kuetz., a pennate species that is reportedly heterothallic (Stosch, 1958B).

Structure of the Diatom Frustule

Although observations made with the light microscope have provided the foundation upon which present knowledge of diatom wall structure is based, the innovation of electron microscopy, particularly scanning electron microscopy, has greatly

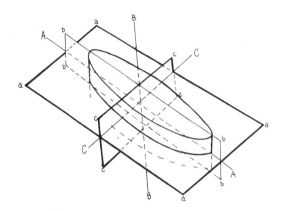

Fig. 7.37 Axes and planes of symmetry in a pennate diatom. A, apical axis; B, pervalvar axis; C, transapical axis; a, valvar plane; b, apical plane; c, transapical plane. (After Hendey.)

enhanced our appreciation of the more detailed aspects of frustules.[9] Such a complex wall structure has generated an abundance of terms in the description of the elements of the wall, and a revised terminology has been formulated (Anon., 1975) in the hope that such a standarization of terms will result in greater clarity of information. The present treatment follows this recently proposed terminology.

The vegetative cell of most members of this class can be regarded as having either bilateral or radial symmetry. A cell with bilateral symmetry can be positioned along three axes (Fig. 7.37): the apical axis (A), the pervalvar axis (B), and the transapical axis (C). Three planes correspond to these three axes: the valvar plane (a), the apical plane (b), and the transapical plane (c). In a cell without bilateral symmetry only the pervalvar axis and the valvar plane are present. The frustule can be regarded as being composed of two overlapping halves of a close-fitting container, the two opposing, distal surfaces being the **valves,** one larger (**epivalve**) and the other slightly smaller (**hypovalve**). Located between the valves is the **girdle,** which is subdivided into two overlapping portions: the **epicingulum** and the **hypocingulum** (Fig. 7.47*b*). The epivalve and epicingulum comprise the **epitheca,** and the hypovalve and hypocingulum comprise the **hypotheca.**

A simple girdle consists of only two wall pieces, the two cingula, whereas a compound girdle has additional pieces, called **intercalary bands,** which are elements of wall material located proximal to the valve. Some diatoms are able to grow considerably in length (along the pervalvar axis) by the addition of such intercalary bands. The shape of these bands may be a complete loop, an open loop, or smaller imbricate scales, such as in *Rhizosolenia* (Fig. 7.38).

The components of the frustule are further divided into a variety of structures, which can be visualized with the light microscope. An **elevation** is a raised area of the valve wall, which does not project laterally outside of the margin of the valve. Eleva-

[9] For recent examples of fine-structural studies of diatoms, the reader is referred to the following accounts: Gerloff, 1970; Wornardt, 1971; Hendey, 1971; Ross and Sims, 1972; Round, 1973B; Venkateswarlu and Round, 1973; Gerloff and Helmcke, 1975A, 1975B; Fryxell, 1975; Paasche, et al., 1975; Housley, et al., 1975; Brooks, 1975; and Hasle and Evensen, 1976.

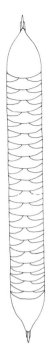

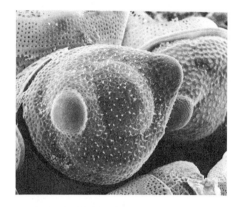

Fig. 7.38 *Rhizosolenia*. Girdle view. The elongated pervalvar axis is due to the many intercalary bands. × 165. (After Hendey.)

Fig. 7.39 *Biddulphia biddulphiana*. A scanning electron micrograph of valve showing elevations. × 742. (Courtesy of Michael A. Hoban.)

tions are common in the genus *Biddulphia* (Fig. 7.39). If the outgrowth of the valve projects beyond the valve margin, it is termed a **seta,** and the structure is different from that of the valve. Setae are prominent in the large genus *Chaetoceros* (Fig. 7.40) and are significant in increasing the surface area, providing a greater surface: volume ratio, which, in turn, enhances the ability of this planktonic alga to stay afloat (Smayda, 1970). A marginal ridge of continuous or interrupted wall material may occur at

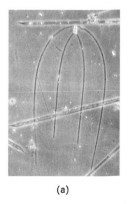

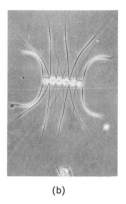

(a)   (b)

Fig. 7.40 *Chaetoceros*. (*a*) *C. peruvianum* Brightwell, a solitary species. (*b*) *C. affine* Lauder, a filamentous species. (*a*) × 53; (*b*) × 80.

Fig. 7.41 *Skeletonema tropicum.* Scanning electron micrograph of two half-cells joined by marginal spines. × 2400. (Courtesy of Dr. Greta Fryxell.)

the periphery of the valve. In *Skeletonema* the marginal ridge consists of long, straight spines, which make contact between adjacent cells (Fig. 7.41) and thus unite the cells into filaments.

Several types of markings on the valve surfaces may be present: **puncta,** which may be irregularly arranged or arranged in regular lines called **striae; areolae** (Fig. 7.42*a*), which correspond to pores or chambers within the valve wall; and **costae** (Fig. 7.42*b*), which are elongate thickenings of the valve due to heavy silica deposition and which sever as strengthening ribs. Valves in some diatoms consist of only one layer of silica, but valves containing **locules** can be considered as consisting of two

Fig. 7.42 (*a*) Areolae of *Stictodiscus johnsonianus.* (*b*) Costae of *Diatoma vulgare.* (*a*) × 427; (*b*) × 1080. [(*a*) after Bold, *Morphology of Plants*, 3rd ed., Harper & Row, Publ.; (*b*) courtesy of Dr. Kent McDonald.]

(a)

(b)

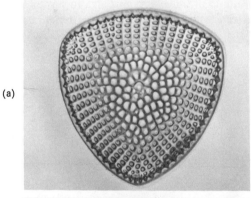

layers of silica separated by vertical walls and having a honeycomb appearance (Fig. 7.43). The latter type of wall is referred to as loculate. The individual locules are usually constricted on one side, where there is an opening termed the **foramen,** and covered by a thin, perforated layer of silica, the **velum,** on the opposite side. A velum that is regularly pitted with small holes is called a **cribrum.** Depending on the species, the foramen of a locule may be to the outside and the velum on the inside, or the reverse situation may occur. Areas of valves devoid of such markings (puncta, areolae, or costae) are referred to as **hyaline fields,** and these areas may run along the apical axis in diatoms lacking raphes (see below) or be elsewhere on the valve.

The valves of some diatoms may have an opening or fissure running along the apical axis (Fig. 7.47*a*). This is the **raphe.** If present, the raphe is usually on both epi-

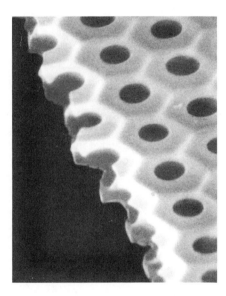

Fig. 7.43 *Odontella chinensis.* Scanning electron micrograph of a frustule showing the honeycomb nature of the loculate wall. × 15,180. (Courtesy of Michael Hoban.)

valve and hypovalve, but in some genera (e.g., *Cocconeis* and *Achnanthes*) a raphe is present on only one of the valves. The raphe is not continuous but is interrupted in the central area by solid wall, this region being called the **central nodule. Polar nodules** lie at the two opposite ends of the raphe. Diatoms with a raphe are capable of a gliding movement (p. 413).

The frustules of some diatoms bear special processes or projections that are noteworthy. One such structure is the **labiate process** (Fig. 7.44), which is an opening through the valve. It projects inward, terminating in a flattened tube with a longitudinal slit surrounded by two liplike edges, and may also extend outward from the valve surface as a tube or lack any external portion. Labiate processes are detectable with the light microscope but are more readily discernible with the scanning electron microscope. A valve may bear a single, two, a few, or numerous labiate processes. In some centric diatoms, labiate processes may bear precise spatial relationships with

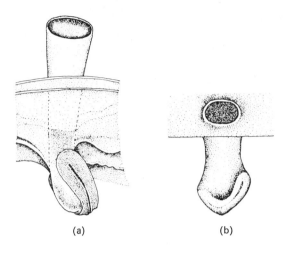

(a)   (b)

Fig. 7.44 Types of labiate processes in *Thalassiosira* based upon scanning electron micrographs. (*a*) *T. tumida.* (*b*) *T. bioculata.* (After Hasle.)

openings in the valves or with setae arising from the valves. The setae in species of *Odontella* (Fig. 7.45) represent projections from labiate processes, the base of a seta having a typical labiate structure on the inside of the valve. Likewise, the central spine in *Ditylum brightwellii* (T. West) Grun. ex Van Heurek represents a labiate process. The hyaline rays, or slits in the valve, in *Asteromphalus* (Fig. 7.46*a*) each have a labiate process situated at the outer end of the slit near the valve margin (Hasle, 1972).

The function of labiate processes in centric diatoms is not understood, but it has been speculated (Simonson, 1970; Hasle, 1974) that the labiate process was a predecessor of the raphe system. Labiate processes occur in both centric and pennate diatoms (Hasle, 1972). Among the pennate diatoms, however, a labiate process is absent in those taxa with well-developed raphes (the Biraphidae and Monoraphidae) but present in the Raphidioideae, a group in which the raphe supposedly first appeared (Hasle, 1974).

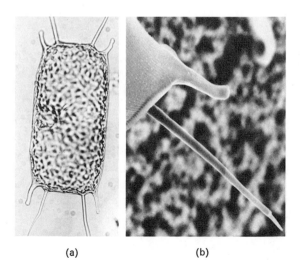

(a)   (b)

Fig. 7.45 *Odontella regia.* (*a*) Light micrograph of cell in girdle view. Valves bear both elevations (shorter) and setae (longer). (*b*) Scanning electron micrograph of elevation and seta, the latter representing the external portion of a labiate process. (*a*) × 300; (*b*) × 1406. (Courtesy of Michael Hoban.)

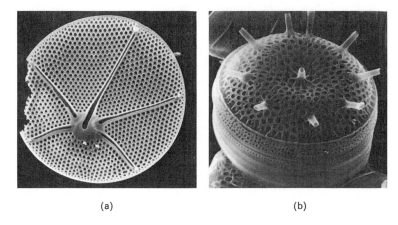

(a)   (b)

Fig. 7.46 (*a*) *Asteromphalus arachna*. Inside view of the valve showing rays extending inward from labiate processes located near the valve margin. (*b*) *Thalassiosira nordenskioeldii*. Scanning electron micrograph of valve with strutted processes. (*a*) × 1700; (*b*) × 6375. (Courtesy of Dr. Greta Fryxell.)

Another type of structure in the frustule, seen in *Thalassiosira*, is the strutted process (Fig. 7.46*b*), which is a tubule extending outward from the valve. Threads of chitan are extruded from strutted processes (McLachlan, et al., 1965), the threads from the central strutted processes being involved in chain formation (Fryxell and Hasle, 1972). In some species of *Thalassiosira* a single delicate thread holds cells together in chains; in other species more than one thread is involved.

Reproduction

Vegetative cell division is the ordinary method of reproduction in diatoms. During the processes of mitosis and cytokinesis, the two valves of the parental cell move somewhat apart, and the division of the protoplast occurs centripetally in a plane parallel to the valves (Fig. 7.47*b*, *c*, *d*). The events of nuclear division have been inves-

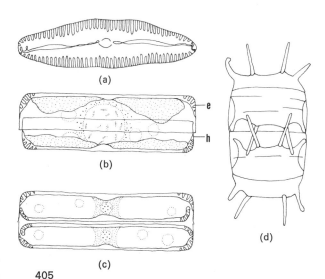

(a)

(b)

(c)

(d)

Fig. 7.47 (*a*)–(*c*) *Pinnularia*. (*d*) *Odontella*. (*a*) Valve view, with costae and raphe along apical axis indicated. (*b*) Girdle view. *e*, epicingulum; *h*, hypocingulum; (*c*) Girdle view of recently divided cell. (*d*) Recently divided cell, the two products still attached. [(*a*)–(*c*) after Bold.]

tigated with the electron microscope in some detail in *Lithodesmium* (Manton, et al., 1969A, 1969B, 1970A, 1970B), *Diatoma* (Pickett-Heaps, et al., 1975), and *Melosira* (Tippit, et al., 1975). The formation of the new valves inside the cleavage furrow has also been reported on (Dawson, 1973A). The valves of the parent both serve as epivalves in the two products of division, which results in one of the two cells being slightly smaller than the parent. Thus in a given diatom population a progressive diminution of cell size occurs with continued division (Rao and Desikachary, 1970).

Although this reduction in size is experienced in most species, some species manage to maintain a constant size through an indefinite period of vegetative division (Wiedling, 1948), which might be explained by the plasticity of their cell wall (Hendey, et al., 1954) or by the open nature of their girdle elements. Interestingly, some species maintain a constant size under favorable growth conditions but undergo diminution of cell size in unfavorable growth conditions. Enlargement of small cells to the maximum size possible in that species can occur vegetatively (Stosch, 1965B) by the partial or complete extrusion of the protoplast from the small frustule and the regeneration of a new frustule. This phenomenon is known to occur both in nature and in culture. Under *in vitro* conditions nutritional manipulation may be used to induce the cells to undergo the enlargement process.

The more prevalent method of attaining maximum size is by the sexual process. Sexuality in diatoms seems to be closely associated with cell size such that in some species only cells of a size less than a critical level can undergo sexuality (Geitler, 1935; Drebes, 1966). Sexual reproduction is oogamous in centric diatoms and isogamous in pennate diatoms. The product of sexual fusion is the **auxospore** (zygote), which is not a resting stage but characteristically increases in volume immediately after the gametes have fused. The actual process of gametogenesis, although invariably associated with meiosis, demonstrates many variations among the diatoms. In *Stephanopyxis palmeriana* (Grev.) Grunow, which produces a single egg per cell, the fertile female protoplast (Fig. 7.48*d, e*) swells and stretches, causing the intercalary bands to be separated and the protoplast to be exposed, permitting entry of sperm (Drebes, 1966). Following fertilization, the zygote of this species greatly expands, forming a spherical auxospore surrounded by a fertilization membrane termed the **perizonium**[10] (Fig. 7.48*f*).

The protoplast of male cells (Fig. 7.48*a*) undergoes divisions to form spherical, chloroplast-containing bodies called **microspores**. In *Stephanopyxis palmeriana*, 4, 6, or 8 microspores are produced (Fig. 7.48*b*); in other species as many as 128 microspores might be formed. The microspores are essentially spermatogonia, each undergoing meiosis. Cytokinesis does not occur; instead, the four haploid nuclei are simply

---

[10]The term **perizonium** was redefined by von Stosch (1962) to restrict it to pennate diatoms. The auxospore membrane is the outermost boundary of the zygote, whereas the perizonium, according to von Stosch, is an inner silicified membrane within which the initial cell is formed. In the great majority of centric diatoms only one membrane is present, the fertilization membrane (also referred to in the literature as a "perizonium"), which is silicified, and the initial cell is developed within it. In the centric diatom *Bacteriastrum hyalinum* Lauder, however, the thin, evanescent auxospore membrane is replaced by a more robust and silicified inner membrane (Drebes, 1972), which thus resembles the situation in pennate diatoms.

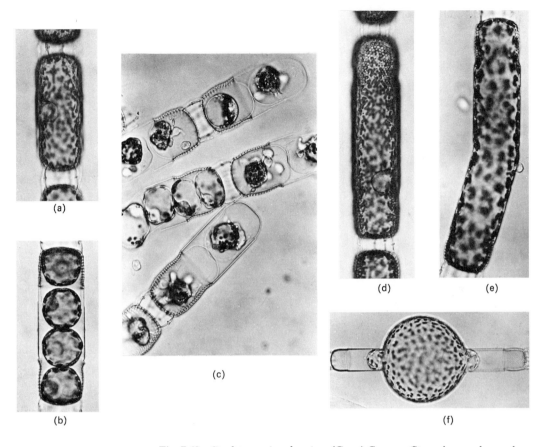

(a)

(b)

(c)

(d)

(e)

(f)

Fig. 7.48 *Stephanopyxis palmeriana* (Grev.) Grunow. Stages in sexual reproduction. (*a*) A vegetative cell prior to microspore formation. (*b*) A cell containing four microspores. (*c*) Following meiosis, four colorless sperm are cut off from each spermatogonium, or microspore. (*d*) Oogonium containing an enlarged nucleus undergoing meiosis. (*e*) Oogonium, with sperm attached. (*f*) Enlarged auxospore withdrawing from the oogonial cell. (*a*)–(*e*) × 360; (*f*) × 173. (After Drebes; permission of Helgoländer Wiss. Meeresunters.)

detached, each with a small amount of cytoplasm (Fig. 7.48*c*). Each generates a flagellum in the process and swims away as a uniflagellate, colorless sperm. The vestigial mass of cytoplasm of each microspore, containing the chloroplasts, degenerates.

Pennate diatoms also undergo meiosis during gametogenesis, but only one or two nuclei of the meiotic tetrad of a sexual cell are functional, the remaining nuclei degenerating. The gametes are amoeboid and morphologically isogamous, although behavioral anisogamy, in which one gamete is active and the other passive, also is known. More frequently both gametes are active and approach each other. Associated with their emergence from their respective frustules is the production of copious amounts of mucilage, which embeds the copulating cells (Fig. 7.49). In some diatoms, such as *Eunotia*, copulation tubes are also produced at a particular site (Geitler, 1969),

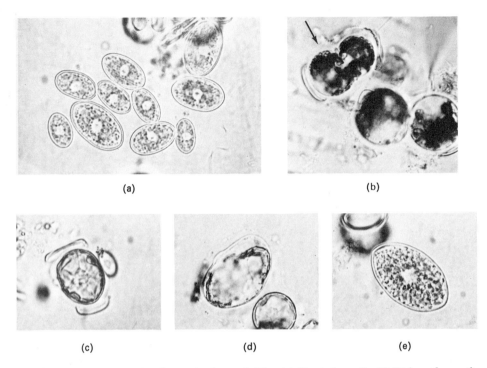

(a)

(b)

(c)   (d)   (e)

Fig. 7.49 *Cocconeis placentula* Ehr. (*a*) Vegetative cells. (*b*) Union of gametic protoplasts (arrow) of adjacent cells. (*c*), (*d*) Auxospore formation. (*e*) Regenerated cell. Note its size relative to those in (*a*). × 540. (After Bold, *Morphology of Plants*, 3rd ed., 1973, Harper & Row, Publ.)

and contact is brought about by chemotropic growth. Two main patterns of reproduction occur in pennate diatoms: Each parental cell produces a single gamete, a single auxospore resulting; or each parental cell produces two gametes, two auxospores resulting. The superfluous nuclei of the meiotic tetrad degenerate.

A third variation of auxospore formation, which is less common, is **autogamous** reproduction, in which the two haploid nuclei from a single cell fuse to form the auxospore. Auxospores may also arise parthenogenetically by a process of **apogamy,** in which the nucleus of the parental cell undergoes divisions not involving a reduction in chromosome number.

A new cell, invariably much larger than the cell or cells giving rise to the auxospore, is developed within the perizonium of the auxospore. The position of the auxospore relative to the parental cell(s) is typically constant for any species and therefore is taxonomically useful. Auxospores may be free, not having any direct contact with the parent cell, or terminal, borne at the end of the parent cell. Other patterns include a lateral position (Fig. 7.50*a*), in which the auxospore is located in the girdle region of the parent cell with its pervalvar axis transverse in relation to that of the parent cell, or an intercalary position (Fig. 7.50*b*), in which the auxospore is attached to the valves of parent cells or their remains, the cingula being discarded during the enlargement of the auxospore.

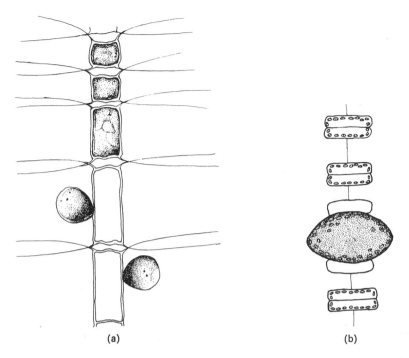

Fig. 7.50 Auxospore formation. (*a*) Lateral auxospores in *Chaetoceros*. (*b*) Intercalary auxospores in *Thalassiosira*. (*a*) × 354; (*b*) × 438. (After Drebes.)

The causes of sexual induction in diatoms have generated research interest, and various factors have been determined to influence auxosporogenesis, such as temperature and light conditions (Stosch and Drebes, 1964; Drebes, 1966; Holmes, 1966) and nutrition. Sexualization was brought about in the marine centric diatom *Lithodesmium undulatum* Ehr. by transferring the material into fresh culture medium at a higher temperature (24°C instead of 15°C) and with much stronger light (Manton, et al., 1968A). A comparable increase in temperature and light intensity induces spermatogenesis in the centric diatom *Coscinodiscus asteromphalus* Ehr. (Werner, 1971), the spermatogonia appearing as soon as 24 hours after the start of induction. If cells of the appropriate size range in *Stephanopyxis palmeriana* (previously maintained at 15°C and 400 Lux) are exposed to a temperature of 21°C and light intensity in the range of 3000–5000 Lux, gamete production is induced. In the closely related species *Stephanopyxis turris* (Grev. et Arn.) Ralfs, which occurs in colder waters, an increase in light intensity alone can induce sexuality (Stosch and Drebes, 1964). The formation of auxospores in the marine pennate diatom *Navicula ostrearia* is induced (Neuville and Daste, 1975) by subjecting a culture to a photoperiod of only 6 hours of light per day, but only cells less than a critical size will undergo sexualization.

An example of a nutritional connection to gamete formation was observed (Steele, 1965) in *Ditylum brightwellii*, in which the formation of eggs and sperm was correlated

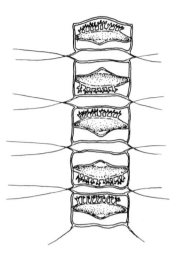

Fig. 7.51 *Chaetoceros.* A chain of cells containing resting spores. × 186. (After Drebes.)

with the absence of manganese in the medium. The conversion of vegetative cells of *Cyclotella meneghiniana* Kuetz. into male gametes and auxospores was caused by an increase in the sodium concentration, whereas cells of the related species *C. cryptica* Reimann, Lewin, et Guillard undergo spermatogenesis when subjected to the same relatively high levels of sodium (Schultz and Trainor, 1968).

Not to be confused with auxospores, which are never resting stages, are **resting spores,** which are formed by most benthic and neritic centric diatoms and constitute the principal method for these diatoms to persist during unfavorable conditions of growth. The protoplasm of the parental frustule darkens and is contracted, and the resting spore develops a very heavy, siliceous wall of its own and usually sinks to the bottom. Resting spores may be formed singly, in pairs, or in series of four (Anon., 1975). Typically, they consist of only two valves and lack a girdle. If the resting spore is completely enclosed within the parent cell, it is endogenous (Fig. 7.51). If it lies partly within the parent cell and the other half free, it is semiendogenous. If the finished resting spore is not enclosed at all by the parent cell, it is exogenous. Resting spores may have spherical to ovoid shapes and may be variously sculptured (Heimdal, 1974; Hargraves, 1976). Their morphology is always different from that of the vegetative cell (Hendey, 1964). In some neritic plankton species, such as in *Chaetoceros* and *Thalassiosira*, the valves of the resting spore are quite dissimilar from each other, one being much more convex in girdle view than the other (Fig. 7.51).

The formation of resting spores has been studied in *Stephanopyxis palmeriana*, and the "trigger" has been determined (Drebes, 1966) to be phosphate deficiency in the medium. A temperature of 12°C was most effective in the induction of the resting spores, but they also appeared at temperatures up to 24°C. Germination of resting spores is brought about by transferring them into fresh, complete medium. The resting spores elongates and divides, the valves of the resting spore being reused in some species but discarded in other species. It is interesting to note that the growth conditions inducing the formation of resting spores differ from those inducing auxospore formation, but in both processes cells must be of a particular size class to undergo these changes. In reference to the formation of resting spores in *Stephanopyxis turris* vegetative cells must have a diameter larger than a critical size to form spores (Stosch and Drebes, 1964), but in *S. palmeriana* the diameter of vegetative cells must be less than a critical value and can be as small as the lower limit of vegetative existence (Drebes, 1966).

Silicification

Silicon metabolism in diatoms has been the focus of much research interest in recent years (Busby and Lewin, 1967; Coombs and Volcani, 1968; Lewin and Reimann, 1969; Darley, 1974). In general, diatoms need their cell wall to live,[11] and silicon has been shown to be an absolute requirement for mitosis and frustule formation (Darley, 1969; Darley and Volcani, 1969). Silicon is present in the frustule as hydrated amorphous silica, $SiO_2 \cdot nH_2O$, and its morphogenesis has been followed at the electron-microscope level (Stoermer, et al., 1965; Reimann, et al., 1965, 1966). Following the division of the parent protoplast into two products such that each is bounded by its own plasma membrane, the formation of new wall components is indicated by the appearance of vesicles in the region of the new frustule beneath the plasma membrane, the vesicles being derived from Golgi bodies (Coombs, et al., 1968; Dawson, 1973B). These vesicles fuse laterally, and their common membrane is referred to as the **silicalemma** (Reimann, et al., 1966). A rapid deposition of silica takes place, the silica becoming tightly bound to the silicalemma and somehow assuming the precise shape characteristic of the particular species. The valve is formed first, the central or raphe region being filled in prior to the deposition of silica at the mantle edge, or peripheral region. After the valve is mature, the first girdle band is developed by the aggregation of Golgi vesicles and their fusion into the silicalemma of this piece of frustule. Once frustule formation is completed, a new plasma membrane appears beneath the wall, and the outerlying membranes are lost (Dawson, 1973B).

In addition to the silica component, the frustule contains a fraction of organic material. The proteinaceous compounds seem to be derived from the original silicalemma, which adheres to the frustule, whereas carbohydrates seem to be continually added to the wall (Coombs and Volcani, 1968). Silicic acid is the form of silicon taken up by diatoms from their environment. The uptake and concentration of silica within the vesicle of the silicalemma require active transport (Lewin and Chen, 1968), and a process of polymerization and solidification of the silica takes place (Darley, 1974).

The silica of living diatoms does not leach out into the medium, but that of dead cells does, suggesting that the silicalemma might function to retard silica dissolution. Nonetheless, diatom frustules are essentially quite resistant to natural degradation, and their accumulation over geologic periods has resulted in significant deposits in

---

[11] An exception is *Phaeodactylum tricornutum* Bohlin, a much studied, weakly silicified diatom, which is unusual in that it is polymorphic. Its "fusiform" condition does not have an organized siliceous wall but contains as much silica as the partially silicified cells of the "oval" form (Lewin, et al., 1958).

Fig. 7.52 Deposit of diatomaceous earth, or diatomite, being mined at Lompoc, California. (Courtesy of Johns-Manville Products Corp.)

various places in the world, one of the most well known in the United States being in the region of Lompoc, California. Termed **diatomaceous earth**, or diatomite, this siliceous material is mined (Fig. 7.52) and used for a variety of commerical purposes (Conger, 1936; Patrick and Reimer, 1966).

The periodicity and the succession of various species of freshwater planktonic diatoms have been related to concentrations of silica (Kilham, 1971). When diatom species have differing abilities to utilize silicate or other nutrients (Tilman and Kilham, 1976), then competition for those nutrients along nutrient ratio gradients determines species composition and succession (Titman, 1976). The kinetics of growth in marine diatoms can also be correlated with silicate levels. In *Thalassiosira pseudonana* Hasle et Heimdal, a clone from the Sargasso Sea was more efficient at taking up silicate at low levels, whereas an estuarine clone was less efficient at low levels but at the same time had a higher maximal growth rate, when the silicate level was not limiting (Guillard, et al., 1973). Physiological races thus can occur within species of phytoplankton.

The organic component of the frustule in *Melosira nummuloides* (Dillw.) C.Ag. has been shown (Crawford, 1973) to be several layers thick, probably polysaccharide in composition, and located inside the siliceous wall. It is laid down after the frustule has been formed. Such a finding differs from observations on *Cylindrotheca fusiformis* Reimann et Lewin (Reimann, et al., 1965) and *Navicula pelliculosa* (Breb.) Hilse (Reimann, et al., 1966), in which the organic layer lies outside of the siliceous wall. The wall of the auxospore in *Melosira nummuloides* has been shown also to consist of two layers, but unlike the vegetative cell in this species the organic layer overlies the siliceous layer and may function as a protective layer during expansion of the developing auxospore (Crawford, 1974). The silica layer consists of scales, which are overlapping, perforate plates.

Movement in Diatoms

The gliding motility in diatoms that have a raphe system has long been noticed, but the mechanism has been variously interpreted. A relationship to the flow of cytoplasm in the raphe has been suggested, and the electron microscope has revealed the presence of a system of fibrils in the region of the raphe as well as crystalloid bodies (Drum and Hopkins, 1966). It has been recognized that diatom locomotion is dependent on the adhesion of the cells to a substratum, the adhesion being brought about by material secreted through the raphe system. From the crystalloid bodies is generated mucous material that facilitates locomotion, and inhibition of motility can be caused by the addition of mucous-dispersing agents. A mucilaginous trail deposited behind moving diatoms can be detected by special stains. Also, if the protoplast is plasmolyzed away from the cell apices, locomotion is prevented. Streaming of the protoplast within the active ventral raphe system is consistently in the direction opposite to that of the diatom motion.

Paths of moving diatoms are variable, depending primarily on the shape of the raphe (Nultsch, 1956). *Navicula* thus has a straight movement; *Amphora* has a curved movement; and *Nitzschia* has two differing curved movements with varying radii. Positive or negative phototaxis is demonstrated in these gliding diatoms, a positive reaction being expressed in white light by most diatoms (Halldal, 1962).

One of the most spectacular patterns of motility is seen in *Bacillaria paradoxa* Mueller, a colony of pennate cells that continually line up and then synchronously slide apart with precise rhythmic patterns. If the colonies are quite large, the movements are no longer synchronous (Patrick and Reimer, 1966). Adjacent cells in a colony manage to interlock by siliceous extensions from the edges of the frustule (Drum and Pankratz. 1966).

A different aspect of motility concerns the migration within mud sediments of some diatoms. The endogenous rhythmic upward movement of cells of *Hantzschia* during daytime low tides and the downward migration during high tides or at night were described by Palmer and Round (1967). The mechanism postulated a dual clock, one with a periodicity of 24.8 hours (a lunar day) and the other with a periodicity of 24.0 hours (a solar day).

Habitats of Diatoms

The two major modes of existence of diatoms are benthic and planktonic. Benthic forms may live upon substrata, such as rock, sand, or mud, or be epiphytic on plant life or epizoic on animals (Russell and Norris, 1971). An endozoic existence has also been recorded (Apelt, 1969) for the pennate diatom *Licmophora*, in which naked cells live symbiotically within the flatworm *Convoluta convoluta*. Epiphytic forms may be attached by a stalklike secretion of mucilage, such as in *Cymbella* and *Gomphonema*

(Dawson, 1973C) or with the entire valve attached by a coating of mucilage, such as in *Cocconeis*, or they may continually glide over the surface of the plant, much as those living on inanimate substrata. A surprisingly rich flora can occur in what casually might appear to be a uniform substratum. A recent report (Lee, et al., 1975) of the diatoms epiphytic on the green alga *Enteromorpha* (p. 171) growing in a Long Island (New York) salt marsh listed a total of 218 species or varieties of diatoms on this particular host; however, the summer epiphytic flora was dominated by only 6 of these diatom species. Some colonial diatoms secrete a common tubular envelope, and the individual motile cells can glide inside the hollow tube. These colonial genera, such as in *Schizonema* and *Berkeleya*, reach macroscopic sizes and can confuse a collector by their resemblance to brown algae of the family Ectocarpaceae (p. 275). The individual cells within a colony of *Schizonema* resemble *Navicula*, causing some phycologists not to recognize the colony as a distinct taxon. Likewise, the individual cells in the colony of *Berkeleya* resemble *Amphipleura*, but reasons have been offered (Cox, 1975A, B) for the retention of *Berkeleya* as a separate genus.

Planktonic diatoms occur in both freshwater and marine bodies of water, although holoplanktonic types (those that do not require substratum for completion of their life cycles but are completely at the mercy of water current and the winds) are restricted to the marine environment (Hendey, 1964). Freshwater and brackish water diatoms collected as plankton are invariably neritic as are marine species in inshore or littoral waters. Neritic species occur along shorelines. Three categories of neritic marine diatoms were recognized by Hendey (1964):

1. **holoplanktonic**—species associated with a coastline but living an oceanic existence in the sense that they are not dependent on the bottom to complete their life cycle.
2. **meroplanktonic**—species that are pelagic for only a portion of their life cycle, spending the remainder of their existence on the bottom.
3. **tychopelagic**—species that actually spend the major portion of their life cycle attached to a fixed substratum but enter the surface layers of the sea when forcibly torn from their usual habitat.

The adaptation to an oceanic existence has been approached by both structural and physiological modifications of the cells (Fig. 7.53). To maintain themselves in the **euphotic zone,** these cells often have very elaborate horns, setae, or other projections that result in a greater surface area relative to the volume of the cell. Their sinking is thus slowed down. The very large, flattened, discoid cells of some species of *Coscinodiscus* and *Plantoniella sol* (Wallich) Schütt likewise are more buoyant in the planktonic habitat. The asymmetric shape of the valves of the elongate cells in *Rhizosolenia* (Fig. 7.38) effectively forces the cell to curve into a horizontal position rather than to sink vertically. The selective absorption of monovalent ions over the relatively heavier divalent ions in the cell sap of some species is a physiological adaptation, again resulting in a less dense and more buoyant cell.

The formation of filaments effectively increases the friction against the medium and accentuates the activity of cells by microcurrents (Smayda, 1970). Filaments of *Chaetoceros* and *Bacteriastrum* bear elaborate setae and spines, whereas those of *Eucampia*, *Steptotheca*, and some species of *Fragilaria* form flattened ribbons.

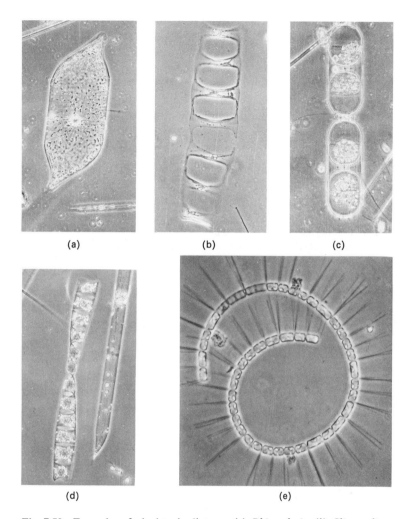

Fig. 7.53 Examples of planktonic diatoms. (*a*) *Rhizosolenia.* (*b*) *Climacodium.* (*c*) *Stephanopyxis.* (*d*) *Hemiaulus* (to left) and *Rhizosolenia* (to right). (*e*) *Chaeto-ceros.* (*a*) × 84; (*b*) × 119; (*c*), (*d*) × 140; (*e*) × 119. [(*e*) courtesy of Joseph Morgan.]

A recent comparison of some freshwater and brackish-water diatom populations demonstrated that the length of valve processes can be influenced by salinity. A fresh-water form with short processes and a brackish-water form with long processes were shown (Hasle and Evensen, 1976) to be conspecific and correctly identified as *Skele-tonema potamos* (Weber) Hasle. Cultures of these populations had extremely short processes at a salinity of 0‰ but much longer processes at salinities of 2‰ or more.

Some filamentous species also produce mucous investments in which the cells are embedded, such as in *Chaetoceros armatum* T. West, *C. sociale* Lauder, *Thalassiosira subtilis* (Ostenfeld) Gran, and *Asterionella socialis* Lewin et Norris. Some of these diatoms occur in the surf zone, and their blooms may result in extensive accumulations of brown organic material washed up on the beach (Lewin and Norris, 1970).

Classification

The present treatment follows the thinking of Simonsen (1972) in retaining the diatoms as a class within the division Chrysophycophyta and in recognizing two major groups, which are given ordinal status: the Centrales and the Pennales. These two groups were originally regarded by Schutt in 1896 as two major subdivisions in the family Bacillariaceae, which included all the diatoms. They were later afforded ordinal status (Karsten, 1928), and although the terms Centrales and Pennales are not based upon any generic names, they are still legitimately used (Simonsen, 1972).

Other interpretations include those by Hendey (1937), who included all diatoms in one class with the single order Bacillariales; Silva (1962B), who set up the division Bacillariophyta with the two classes Centrobacillariophyceae and Pennatibacillariophyceae; and Patrick and Reimer (1966, 1975), who included all diatoms in the single class Bacillariophyceae, comprised of nine orders. By recognizing the diatoms as a class within the division Chrysophycophyta, the present treatment emphasizes the many similarities, such as nature of the photosynthetic reserves, pigment composition, and flagellar structure, shared by the diatoms with other members of this algal division. A close phylogenetic relationship appears to exist between the Bacillariophyceae and the other classes of the Chrysophycophyta.

The following list represents one scheme of classification for the diatoms as two orders comprised of suborders:

Order Centrales: structure of the valve is arranged in reference to a central point on the valve (centric or radial valve) or in reference to two, three, or more points (gonioid valve) such that a biangular, triangular, or polygonal valve is evident.

Suborder Coscinodiscineae: valves disciform, flat or convex (possibly concave), without prominent processes or intercalary bands; diameter of frustules greater than thickness.

Suborder Rhizosoleniineae: frustules long, cylindrical; girdle view usually evident; many intercalary bands present.

Suborder Biddulphiineae: valve bipolar or multipolar; angles of valves provided with spines, elevations, ocelli, or other projections.

Order Pennales: structure of the valve is arranged in reference to a central line (pennate) or in reference to a point not on the valve surface (trellisoid); raphe or a hyaline field in the axial area present.

Suborder Araphidineae: hyaline field present in the axial area of the valves.

Suborder Raphidioidineae: rudimentary raphe present at ends of the valves.

Suborder Monoraphidineae: raphe present on one valve; hyaline field present on the araphid valve.

Suborder Biraphidineae: raphe present on both valves.

8

Division Pyrrhophycophyta

Introduction

The division Pyrrhophycophyta is comprised of the dinoflagellates, a diverse assemblage of biflagellated unicellular organisms, which constitute an important component of marine, brackish, and fresh bodies of water. In addition to the flagellated form, which is the most prevalent expression, nonmotile forms occur in diverse adaptations to their habitats. These forms include coccoid, filamentous, palmelloid, and amoeboid members. This variation in morphological types is paralleled by a nutritional diversity. Besides photosynthesis, heterotrophic nutrition is well developed in this group, saprophytic, parasitic, symbiotic, and holozoic patterns all being represented. Many of the photoautotrophic species, particularly those that are marine, are **auxotrophic** for various vitamins (Hutner and Provasoli, 1964). Some of the parasitic types are structurally very specialized, and it is only by the release of their morphologically distinctive motile cells that their affinities with the Pyrrhophycophyta can be recognized.

The division is characterized by the presence of chlorophyll a and c; β carotene; and the unique xanthophylls[1] peridinin, neoperidinin, dinoxanthin, and neodinoxanthin. The food reserve is starch, and the wall, or **theca,** when present, is composed primarily of cellulose. The dinoflagellate nucleus is another distinctive trait in that the chromosomes do not go through a coiling and uncoiling cycle but remain permanently condensed and thus visible at all stages of the nuclear cycle. This behavior contrasts sharply with that of the chromosomes of the eukaryotic cells, and the term **mesokaryotic** has been introduced (Dodge, 1966) to distinguish dinoflagellate cells as different from prokaryotic and eukaryotic cells.

The flagella of motile cells in this division are also distinctive. The typical pattern is a pair of unequal, heterodynamic flagella, which have independent beating patterns

[1]Fucoxanthin is present in some dinoflagellates lacking peridinin, the significance of which is discussed on p. 438.

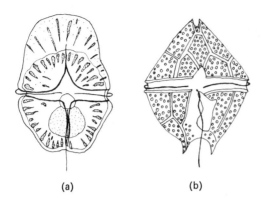

(a)   (b)

Fig. 8.1 Examples of dinoflagellates. (*a*) *Gymnodinium mirabile* Penard, a naked type. (*b*) *Peridinium conicoides* Paulsen, an armored type. (*a*) × 429; (*b*) × 317. [(*a*) after Kofoid and Swezy; (*b*) after Kamaji.]

(Jahn, et al., 1963). In the class Dinophyceae the two flagella are situated in grooves or depressions (Fig. 8.1). An acronematic, posteriorly directed flagellum is located in a longitudinally oriented groove, the **sulcus**, and a flattened or ribbonlike flagellum is located in a transverse groove, the **cingulum,** which encircles the cell in the equatorial region or closer to one or the other pole. The transverse flagellum coils around the cell, and its beat causes the cell both to turn and to be propelled in an anterior direction. This flagellum (Fig. 8.2) has an outer axoneme and an inner portion that has the appearance of a striated strand (Leadbeater and Dodge, 1967B). The inner margin of the flagellum is anchored to the surface of the cingulum by delicate threads, and the resultant beat of the flagellum is hemihelical (Taylor, 1975).

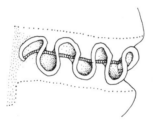

Fig. 8.2 Diagrammatic representation of proximal portion of transverse flagellum deduced from SEM micrographs. (After Taylor.)

The Pyrrhophycophyta has been traditionally subdivided into two major classes, the Desmophyceae and the Dinophyceae, although some current treatments (Dodge, 1975B) prefer to recognize a single class because of basic ultrastructural similarities among the members. In schemes that recognize two classes, the Desmophyceae includes cells that have flagella originating from the anterior end and that can be divided into equivalent right and left halves (Fig. 8.3). The Dinophyceae, on the other hand, includes cells that have flagella originating from a ventral position, one flagellum located in the sulcus and the other flagellum located in the cinglum, which may have a median, or equatorial, position (Fig. 8.1), as in *Gymnodinium* (Fig. 8.1*a*), which has more or less equivalent anterior and posterior halves, the **epicone** and **hypocone,** respectively. Or the cingulum may be located more proximately to the apical pole, as in *Amphidinium* (Fig. 8.4*a*), with a reduced epicone, or more proximately to the antapical pole, as in *Katodinium* (Fig. 8.4*b*), which has a reduced hypocone.

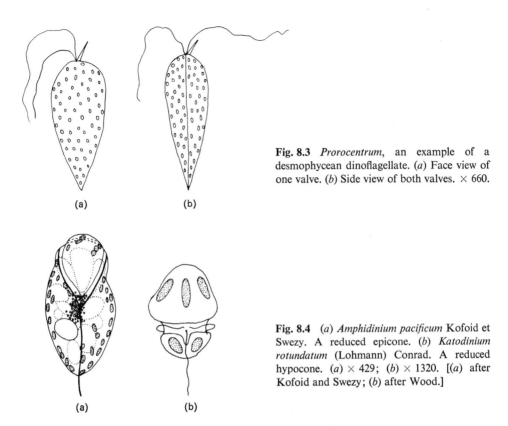

Fig. 8.3 *Prorocentrum*, an example of a desmophycean dinoflagellate. (*a*) Face view of one valve. (*b*) Side view of both valves. × 660.

Fig. 8.4 (*a*) *Amphidinium pacificum* Kofoid et Swezy. A reduced epicone. (*b*) *Katodinium rotundatum* (Lohmann) Conrad. A reduced hypocone. (*a*) × 429; (*b*) × 1320. [(*a*) after Kofoid and Swezy; (*b*) after Wood.]

The Cell Covering

Another means of subdividing the dinoflagellates is on the basis of the cell covering of the vegetative cell, which includes any phase of the life cycle of the dinoflagellate other than the resting or encysted stage. This cell covering is termed the **amphiesma** (Loeblich, 1970), and cells may be unarmored (naked) or armored (thecate). It has been pointed out that transitional types of cell coverings occur, thin plates being present in some apparently naked cells (Dodge and Crawford, 1969A). Despite the naked or armored nature of dinoflagellates, the amphiesma is basically the same (Loeblich, 1970), consisting of several layers of membranes, which may or may not contain fibrillar material. The exact location of the plasma membrane has been debated. The outermost continuous membrane has been regarded (Dodge and Crawford, 1970A) as the plasma membrane. This attitude was supported by subsequent ultrastructural studies by Wetherbee (1975A, 1975B), who pointed out that this outermost membrane is the only membrane to persist and surround the cell during division of the cell. The divergent viewpoint (Loeblich, 1970; Kubai and Ris, 1969; Kalley and Bisalputra, 1971; Sweeney, 1976A) is that the plasma membrane lies as the innermost membrane layer, the rest, the amphiesma, lying external to it. Most investigators agree that the cell covering consists of several layers of membranes. Four distinct membranes have been observed in *Ceratium* (Wetherbee, 1975B); these include (centrifugally) an inner plate membrane, a thecal membrane, an outer plate membrane, and the cell membrane.

The armored dinoflagellates are distinguished by the formation of **thecal plates,** which has been shown (Wetherbee, 1975C) to result from two mechanisms of deposition. One method involves the deposition of material by elongate vesicles at the sutures of adjacent plates; the other method takes place by the flattening out and fusing of vesicles containing precursor material, lying just beneath the plates. The theca may consist of only two plates, as in *Prorocentrum* (Fig. 8.5), or as many as 100. The margins of these plates are slightly beveled and overlap (Cox and Arnott, 1971), the margins expanding to facilitate an increase in the size of the cell. A pellicle layer, lying internal to the theca, has been noted (Loeblich, 1970) in some armored dinoflagellates. This **pellicle** is thinner than the theca and has a fibrous organization but lacks any sign of a plate arrangement.

The number and arrangement of plates in the theca are one of the most useful criteria in the systematics of armored dinoflagellates. Formulas, or tabulations, for characterizing genera have been devised and are especially useful in the Peridiniales. The median girdle divides the cell into two halves, the **epitheca** and the **hypotheca,** comparable to the epicone and hypocone of unarmored dinoflagellates. The epitheca is divided into two complete transverse series: apical (') and precingular ("), both of which are counted from the ventral side in a clockwise sequence when viewing the

(a)

(b)

Fig. 8.5 *Prorocentrum mariae-lebouriae* (Parke et Ballantine) Faust. (*a*) The almost spherical outline of the cells is evident in this scanning electron micrograph. (*b*) The surface of the valve has small spines, the amorphous material being detritus. (*a*) × 2088; (*b*) × 6380. (After Dr. Maria A. Faust; permission of Journal of Phycology.)

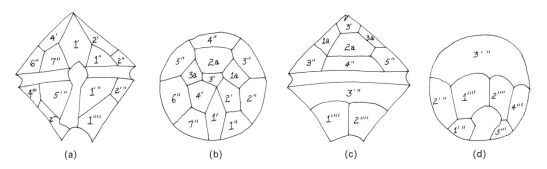

Fig. 8.6 *Peridinium*. Tabulation of the apical, precingular, postcingular, and antapical series of plates. (*a*) Ventral view. (*b*) Apical view, (*c*) Dorsal view. (*d*) Antapical view.

anterior pole. An incomplete series of plates on the dorsal surface of the epitheca occurs in some genera; these plates, numbering from one to three, are called anterior intercalary plates (a). The hypotheca is also divided into two transverse series: post-cingular ($'''$) and antapical ($''''$). The hypotheca of some genera includes an incomplete series, the posterior intercalary plates (p).

The following formula represents the plate pattern for a typical species of *Peridinium*: 4′ 3a 7″ 5‴ 2⁗, and Fig. 8.6 corresponds to such a tabulation. Some genera, including *Peridinium*, are large, and the species demonstrate a degree of variation in their plate patterns. Accordingly, the plate patterns possible in the genus *Peridinium* as a whole might be expressed by the following formula: 2–5′ 0–8a 6–7″ 5–6‴ Op 2⁗. Thecal plates are also present in the cingulum and the sulcus, but these regions are more difficult to analyze, careful microdissection being required. Although earlier workers often omitted designating the number of plates in the cingulum and in the sulcus, an attempt has been made to include them in more recent descriptions of Peridiniales, despite the tedious effort needed. Most genera have six plates in the cingulum, which is designated by 6C. *Ceratium* and *Peridinium* may have as few as four cingular plates. The number of cingular and sulcal plates is still unknown for many genera, and accordingly these data are not always included in the tabulations. Two plates are particularly useful in the identification of species within the genus *Peridinium*, namely, the first apical plate and the second apical intercalary plate. These plates may be four-, five-, or six-sided and are termed *ortho*, *meta*, and *para*, respectively.

The Dinophysiales is an order of armored dinoflagellates with a plate morphology and arrangement entirely different from that of the Peridiniales. All genera in this order have a total number of 18 plates (Norris and Berner, 1970) and have a relatively large hypotheca and a relatively small epitheca. The hypotheca is comprised of 4 plates, 2 of which are very large. Although members of the Peridiniales tend to be flattened dorsiventrally, members of the Dinophysiales tend to be flattened laterally. Conspicuous projections, developed into wings or "lists," may be developed from the cingular regions, and in some genera, such as *Ornithocercus* and *Histioneis* (Fig. 8.7),

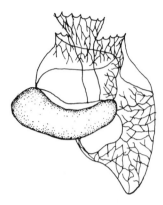

Fig. 8.7 *Histioneis pietschmannii* Bohm. The cingular list is inflated into a chamber. × 581. (After Wood.)

the cingular list forms a large, hollow chamber in which symbiotic blue-green algal cells may be harbored (Norris, 1967B). Since these dinoflagellate genera are often distributed in tropical oceanic waters in which the nitrate concentration may be a limiting factor, it can be speculated that these **consortiums** with blue-green algae, which possibly fix nitrogen into usable form, provide ammonia to the dinoflagellate host.

The Prorocentrales represent a third order of armored dinoflagellates, much simpler in their thecal structure than the Peridiniales and the Dinophysiales. Instead of the complicated series of articulated, overlapping plates present in these latter two orders, the theca in the Prorocentrales may be comprised of only two large plates, or valves. Pores for the flagella may be located on one of the valves, or the flagella may emerge between the anterior margins of the two valves (Dodge, 1965). An alternative arrangement is that a separate, small plate (or plates) may be present in addition to the large valves, and the flagellar pores may occur on this third plate. Both types of thecal layers occur within the genus *Prorocentrum* (p. 449).

In the various orders of armored dinoflagellates, Prorocentrales, Peridiniales, and Dinophysiales, a diversity of ornamentation in the thecal plates exists (Dodge and Crawford, 1970A; Dodge, 1973). Thecal plates may bear spines (Figs. 8.5*b* and 8.8*a*), ridges, or reticulations (Fig. 8.8*c*) or they may be perforated with openings (Fig. 8.8*b*) for the discharge of **trichocysts** (p. 429), or various combinations of these patterns may occur.

Reproduction

The most common mode of reproduction in dinoflagellates is simply by cell division, which may involve a longitudinal, transverse, or oblique bipartitioning of the parental cell. Three different methods of division are known. Naked dinoflagellates such as *Gymnodinium* pinch apart by a process of constriction, the new amphiesma being formed around each product during the course of their separation. For armored dinoflagellates two alternative modes exist. Prior to division the parental theca may be shed in a process called **ecdysis,** necessitating the formation of a new theca for each

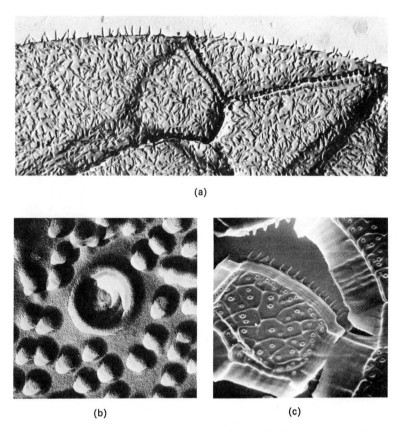

(a)

(b)   (c)

Fig. 8.8   (*a*), (*b*) *Ensiculifera loeblichii* Cox et Arnott. (*a*) Carbon replica of theca showing four plates prior to their expansion, (*b*) A single trichocyst pore surrounded by spines, (*c*) *Peridinium cinctum* (O. F. Müll.) Ehr. Single plate showing reticulations and trichocyst pores and regions of plate expansion (megacytic zones). (*a*) × 13,566; (*b*) × 108,500; (*c*) × 4650. [(*a*), (*b*) after Cox and Arnott; (*c*) courtesy of Dr. E. R. Cox.]

product of division. A variation of this mode is the discarding of the parental theca following cytokinesis. The other method is the splitting of the parental theca into two portions, each product of division retaining one-half of the parental theca and synthesizing the missing portion anew as in *Ceratium* (Fig. 8.9*a*). In *Prorocentrum* (p. 449) and *Dinophysis* (Fib. 8.9*b*) the longitudinal division separates the two large valves of the parental cell, each product of division retaining one valve and forming the second valve. In genera with transverse or oblique divisions, the plane of division passes through the region of the cell from which the flagella emerge.

Filamentous organizations are possible in a few genera of this division, as in *Dinothrix*, which consists of small, branched, attached filaments of both freshwater and marine distribution. *Dinothrix* is classified in the Dinotrichales, an order of only two photosynthetic genera, which release zoospores resembling *Gymnodinium* (p. 443).

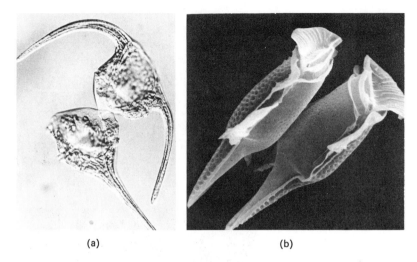

(a)   (b)

Fig. 8.9 Examples of division in dinoflagellates. (*a*) *Ceratium tripos* (O. F. Müll.) Nitzsch. Two products of a recent division remain attached, each new cell retaining one half of the parental theca and about to regenerate the other half, (*b*) *Dinophysis* sp. Two products of division each retain one parental valve and produce a second valve, the new valves being the inner pair. (*a*) × 2808; (*b*) × 382. [(*a*) after Wetherbee; permission of Journal of Ultrastructure Research; (*b*) courtesy of Elaine Stamman.]

Upon division, cells do not separate, and a filamentous habit results. Other examples of filamentous genera include *Haplozoon* (p. 438), a member of the parasitic order Blastodiniales. In some motile dinoflagellates the cells are held together in loose chains, as in some species of *Ceratium* and *Gonyaulax* (Fig. 8.13*a*). These formations are not truly filamentous and easily become separated.

Sexual reproduction has been described in a number of dinoflagellates. Both isogamous and anisogamous conditions occur, although the former is more common. Homothallic and heterothallic examples have been reported. Several species of *Ceratium*, both marine and freshwater types, have been investigated by von Stosch (1964, 1965A, 1972, 1973) in regard to their nuclear cycles. For *C. tripos* (O. F. Müller) Nitzsch several small forms had been described and regarded as juvenile or accessory stages, but they proved to be in reality male gametes. In *C. horridum* (Cleve) Gran such small cells were seen to attach themselves to female cells (Fig. 8.10), which were similar in appearance to the vegetative cells. The male gamete was resorbed into the female cell, the zygote remaining motile. The zygote in *C. cornutum* (Ehr.) Clap. and Lachm. is a resting cyst, whereas in other species it is a nonmotile stage lacking a wall (Stosch, 1972). All species of *Ceratium* examined were haploid and haplobiontic, meiosis occurring in marine species at variable times after a protracted growth phase of the nonresting **planozygote.** In freshwater species a resting cyst is formed, and meiosis occurs either in the cyst envelope, from which two swarmers are released, or later within the single swarmer that is released from the cyst.

A nitrogen-deficient medium was observed (Pfiester, 1975) to induce sexuality in a freshwater *Peridinium*. Thecate cells released small, naked cells, which acted as gametes. Pairs of gametes settled next to each other to initiate fusion, but the partially fused gametes resumed their motility, the fusion being completed while the gametes are in motion. A theca was produced by the zygote within about 24 hours after the completion of fusion, and the thick-walled zygote remained motile for the next 12–13 days, although the gametic nuclei did not fuse. The zygote then entered a resting phase lasting about 2 months. At germination, meiosis of the zygote nucleus occurred, and a single vegetative cell emerged from the very thick-walled hypnozygote. The reproductive behavior in a second species of *Peridinium* was observed (Pfiester, 1976) to conform to this pattern just described, except for the planozygote being uninucleate from the start.

The fact that meiosis is taking place can be detected by the occurrence of a "nuclear cyclosis," in which the zygotic nucleus reaches a maximal volume and the nuclear contents can be seen to rotate. This process is reported to be correlated with the process of chromosomal pairing. Not only has it been described in several species of *Ceratium* (von Stosch, 1972) but also in freshwater species of *Gymnodinium* and *Woloszynskia* (von Stosch, 1973). The species of *Gymnodinium* and *Woloszynskia* studied were isogamous and produced planozygotes, which eventually formed **hypnozygotes,** with spiny to warty ornamentation. In the species of *Gymnodinium* the hypnozygote encysted, releasing a single biflagellate swarmer, which functioned as a **meiocyte,** first undergoing "nuclear cyclosis" followed by two divisional sequences. In the species of *Woloszynskia*, nuclear cyclosis (evidence of meiosis) was observed

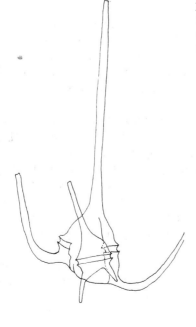

Fig. 8.10 *Ceratium horridum* (Cleve) Gran. Process of conjugation. The small male cell is attached to the ventral region of the female cell. (After von Stosch.)

to occur within the hypnozygote, and four swarmers eventually were released upon excystment.

In contrast to the haploid condition of *Ceratium*, *Gymnodinium*, and *Woloszynskia*, the large, colorless genus *Noctiluca* (p. 444) is apparently diploid. Cells of *N. miliaris* Suriray were described (Zingmark, 1970A) to undergo a process of differentiation, the original nucleus first dividing seemingly by meiosis into four nuclei. These nuceli continue division, four distinct clusters of nuclei being at least temporarily discernible. More than 1000 nuclei can eventually result from these divisions, and uniflagellated, naked gametes are differentiated, each incorporating one of the nuclei. Fusion and zygote formation were witnessed, and the development of a typical vacuolate vegetative cell from a zygote was followed.

Unequivocal evidence for the occurrence of genetic recombination in dinoflagellates has been recently furnished from two sources. Using chemically induced carotene-deficient strains of *Crypthecodinium cohnii* (Seligo) Chatton in Grasse, Tuttle and Loeblich (1974) mixed together combinations of mutant strains. This wild type of nonphotosynthetic alga has carotenes only. Various types of albino mutants were placed in a medium containing low concentrations of nitrogen and phosphorus, since it had been recognized that pairs of motile cells sometimes fuse in such a deficient medium. After incubation, yellow cells (i.e., wild type) did not appear in cultures of the individual mutant strains, but small percentages of yellow cells appeared in cultures of some of the combinations of mutant strains. These results are evidence of genetic recombination by independent assortment. The second source of evidence for genetic recombination involved tetrad analysis using flagellar mutants (Beam and Himes, 1974).

Encysted Stages

Many dinoflagellates are capable of forming encysted stages, which are resting cells formed in response to unfavorable conditions (Loeblich and Loeblich, 1966; Wall and Dale, 1968; Reid, 1972). The wall of the cyst is much more resistant to acid and alkali treatment, and many fossil forms are thought to represent the encysted stages because their thecae are not affected by such harsh treatment, which causes the plates of extant dinoflagellates to separate. Cysts fall into two categories: those that resemble the shape of the vegetative cells and those that are very different, appearing spherical and covered with tubular projections. The former type (Fig. 8.11a) is formed internally within a vegetative cell, and ridges between adjacent plates correspond to the suture lines in the vegetative cell. The latter type includes **hystrichospheres** (Fig. 8.11b) and are abundant in the fossil record (Sarjeant, 1974), although the thecae of vegetative cells have not been fossilized. Relatively recently hystrichospheres were recognized (Evitt, 1963B) to represent the encysted stages of dinoflagellates, and it was realized that the tubular projections extended outward to the inside of the thecal plates.

When cysts rupture, usually a definite opening, called the **archeopyle,** is formed and its position is of taxonomic utility. The archeopyle may result from the loss of

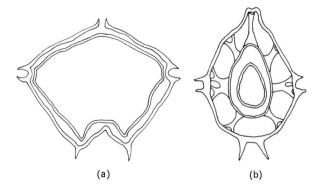

Fig. 8.11 Cyst formation (diagrammatic representations). (*a*) *Peridinium leonis* Pavillard, with a close correspondence between cyst and vegetative theca. (*b*) *Gonyaulax digitale* (Pouchet) Kofoid, the cyst being a typical hystrichosphere. (After Evitt and Davidson.)

(a)   (b)

the apical series of plates, or of a dorsal apical intercalary plate, or of a middorsal precingular plate, the type of archeopyle being constant per species. Several genera of extant thecate dinoflagellates are known to produce encysted stages, including *Gonyaulax*, *Peridinium*, and *Pyrodinium* (Evitt and Davidson, 1964; Wall, et al., 1967; Wall and Dale, 1969). These distinctive cysts are especially significant to paleontologists because of the presence of homologous spores in the fossil record (Sarjeant, 1965; Schrader, 1966). One interesting example is the discovery (Wall and Dale, 1969) that the cyst stage of the extant bioluminescent dinoflagellate *Pyrodinium bahamense* Plate had been earlier known as a fossil form, *Hemicystodinium zoharyi* (Rossingnol) Wall. The morphological characteristics of the cyst stages are proving to be valuable in better understanding the phylogenetic relationships among the Peridiniales (Wall and Dale, 1968).

Very rarely an internal skeleton may be present in dinoflagellates. Five-rayed stars composed of silica and termed *pentasters* are present in a relatively few extant species of dinoflagellates, as in the genus *Actiniscus*, which was first described as a fossil by Ehrenberg. A recent description (Bursa, 1969) of one of these relict forms occurring in Canadian Arctic lakes referred to the extremely delicate nature of their pellicle, the peripheral or perinuclear arrangement of the pentasters, which may number as many as 14 per cell, and the presence of a central capsule containing the nucleus. These cells are capable of holozoic nutrition. The resemblance to silicoflagellates (p. 372) and coccolithophorids (p. 381) of the Chrysophycophyta can also be noted. In the genus *Plectodinium*, which resembles *Gymnodinium* (p. 443), distinct bundles of siliceous rods occur within the cytoplasm (Biecheler, 1934).

The Nucleus

The nucleus in dinoflagellates has long been the focus of attention because of its many unusual properties, including the persistence of the chromosomes in a condensed configuration during interphase (Fig. 8.12) and the absence of centromeres or a spindle (Dodge, 1963A). These unusual features led to the suggestion (Dodge, 1966) of the term **mesokaryotic** to emphasize the special attributes of the typical dinoflagellate nucleus. The nucleus in dinoflagellates is relatively large, often occupying

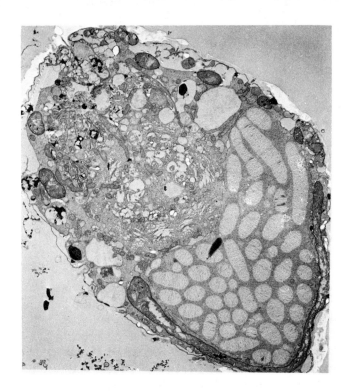

Fig. 8.12 *Prorocentrum micans.* Interphase nucleus, showing condensed chromosomes as is typical of mesokaryotic nuclei. × 3604. (Courtesy of Richard Zingmark.)

about one-half the volume of the cell. Nuclear shapes are variable, ranging from spherioids to U-, V-, or Y-shaped configurations. Chromosome counts range from 12 to around 400 (Loeblich, 1976), and a low count of 4 was attributed to *Syndinium* (Ris and Kubai, 1974). The number is difficult to determine (Dodge, 1963B, 1966), some evidence suggesting that counts may be variable within a species.

At the time of division, chromosomes divide longitudinally, the split starting at one end of the chromosome and moving along the entire length. The nucleus is invaded by cytoplasmic channels that pass from one pole to the other. Groups of microtubules have been observed (Leadbeater and Dodge, 1967A) to be present in these channels. The nuclear envelope and the nucleolus are persistent throughout nuclear division. The chromosomes upon dividing assume a V or Y configuration, and the apices of such chromosomes have been observed (Kubai and Ris, 1969) to be closely associated with the nuclear envelope surrounding a cytoplasmic channel. This association suggests that the cytoplasmic channels serve as a mechanism for the movement of the chromosomes. Some similarity to the segregation of genetic material seen in bacteria, which involves an attachment or association with membranes, has been noted (Kubai and Ris, 1969). This similarity is supported by biochemical properties such as the lack of basic proteins (histones) in dinoflagellate chromosomes and in the nucleoplasm of prokaryotic cells and the lack of coiling in both. The occurrence of acid-soluble proteins in the chromosomes of free-living dinoflagellates has been reported (Rizzo and Noodén, 1972, 1974A, B).

The account of nuclear division above, based upon studies of free-living dino-flagellates, is sharply different, however, from the observations (Ris and Kubai, 1974) of mitosis in a parasitic dinoflagellate, *Syndinium.* Four V-shaped chromosomes are permanently attached to a specific area of the nuclear envelope in that organism, and the chemistry and behavior of the chromosomes are not typical of dinoflagellates, making it doubtful whether some of the parasitic genera are appropriately placed with the dinoflagellates. These differences motivated Loeblich (1976) to recognize *Syndinium* as belonging to a new class, the Syndiniophyceae (p. 440).

Distinctive Organelles

Many unusual cytoplasmic structures are present among dinoflagellates. In certain marine and freshwater dinoflagellates a saclike structure, termed a **pusule,** is located near the flagellar insertion, with a small opening to the outside. In a survey of 40 dinoflagellate species the pusule was observed (Dodge, 1972) to have a basic structure of a system of vesicles that are lined by two appressed membranes. The inner membrane is the invaginated plasma membrane from the flagellar canal. The pusule may consist of a simple vesicle, or it may be more elaborate, made up of vesicles leading into a collecting chamber that in turn is connected to the flagellar canal. The pusule has been regarded as a possible flotation device (Norris, 1966) or to have an osmoregulatory role (Dodge, 1972).

The Pyrrhophycophyta is distinctive among algal divisions in the occurrence of a variety of eyespots or light-sensitive organelles (Dodge, 1969B, 1971A), ranging from simple collections of carotenoid-containing lipid globules without any relationship to membranes to more complex associations of layers of lipid droplets surrounded by membrane envelopes. In *Glenodinium foliaceum* Stein the eyespot is associated with an unusual organelle composed of a stack of up to 50 flattened vesicles (Dodge and Crawford, 1969B). The most elaborate of any light-sensitive organelle occurring in the algae is the **ocellus,** present in a few rather rare genera of marine dinoflagellates, including *Warnowia.* Ocelli are conspicuous, complex structures in the cell, consisting of a flattened or cup-shaped, red- or black-pigmented portion appressed against a larger refractive portion, the lens. The ultrastructure of the ocellus has been described in *Nematodinium* (Mornin and Francis, 1967) and in two other genera (Greuet, 1968), and experimental evidence (Francis, 1967) has demonstrated that light from outside the cell can be brought into focus on a layer lining the cup called the *retinoid.*

Trichocysts, which may be present in the hundreds per cell, occur in a wide variety of dinoflagellates, both naked and armored. These trichocysts are discharged into the medium apparently by a rapid hydration process, the discharged rodlike, proteinaceous structures measuring up to 200 μm in length, which is much longer than the cells from which they were discharged. The trichocysts arise in vesicles from dictyosomes (Bouck and Sweeney, 1966), first appearing as elongate crystals within a single membrane. They migrate to the cell surface and, if the cell is armored, situate themselves beneath the narrow trichocyst pores that perforate the thecal plates (Fig. 8.8*b*). Trichocysts are similar to those of the Chloromonadophyceae (p. 396) as well as those of the ciliate *Paramecium.*

Nematocysts, or **cnidocysts,** bearing a remarkable resemblance to comparable structures occurring in the Coelenterata, are elaborate ejectile organelles present in only two dinoflagellate genera, *Nematodinium* and *Polykrikos*. Originally it was thought that they might have been "captured" from coelenterates, but some structural differences exist (Mornin and Francis, 1967; Greuet, 1971). Usually only about 8 to 10 nematocysts are present per cell.

Muciferous bodies are located just beneath the cell membrane and are merely vesicles of mucilaginous material. They often are aggregated in the region of the sulcus, and their release in some species is correlated with the **psammophilous** existence of these species, facilitating the attachment of the cells to sand grains along the seashore.

Bioluminescence and Circadian Rhythms

The phenomenon of **bioluminescence** is well-known in several marine dinoflagellates, although its function is not understood. First recorded in *Noctiluca* (p. 444), bioluminescence has also been recorded in species of *Gonyaulax*, *Pyrocystis*, and *Pyrodinium*. Small crystalline inclusions called "scintillons" were initially regarded (DeSa, et al., 1963) as the source of the photoemission. These particles were later shown to be composed of guanine, however, and biochemical studies (Fogel, et al., 1972) indicated that they were not directly involved in the particulate bioluminescence. Similar crystals occur in both luminescent and nonluminescent dinoflagellates. A structural basis for bioluminescence has been searched for. Stacks of flattened lamellae associated with large mitochondria at the surface of the cell have been observed (Dodge, 1971A) in a luminescent *Peridinium*. In *Gonyaulax polyedra* Stein polyvesicular bodies have been reported (Schmitter, 1971) to be present in the peripheral cytoplasm, and these bodies do occur in the purified preparations from extracts of *Gonyaulax polyedra* that exhibit luminescence. Particles and crystalline inclusions have also been described (Soyer, 1968; Fuller, et al., 1972) to be present in some luminescent dinoflagellates.

In vivo bioluminescence emission spectra with a peak of intensity around 480 nm have been reported (Soli, 1966; Taylor, et al., 1966; Swift and Taylor, 1967; Biggley, et al., 1969) for various species, which suggests that the substrate-enzyme complex of luciferin-luciferase, which characterizes all bioluminescent organisms, is the same in the various luminescent dinoflagellates examined thus far. The complex is also dinoflagellate-specific. Two types of bioluminescence have been described (Sweeney and Hastings, 1962) in *Gonyaulax polyedra*. When the cells are vigorously agitated, they give off a flash of light. The amount of this stimulated luminescence can be measured by a photomultiplier photometer. When grown under natural illumination or under a light-dark (LD) cycle, the amount of light emitted under agitated conditions (for maximum stimulation) is observed to be low during the period of illumination and to reach a peak during the middle of the dark period (Fig. 8.13*a*). By transferring such a culture from a LD cycle to a DD cycle (i.e., continued darkness), it can be observed (Sweeney and Hastings, 1957; Sweeney, 1963) that an **endogenous circadian rhythm** is indeed present, the cycle of peaks of bioluminescence continuing for several days and the amplitude gradually decreasing (Fig. 8.13*b*). A long, persisting periodicity in stimulated luminescence can also be demonstrated in this same species

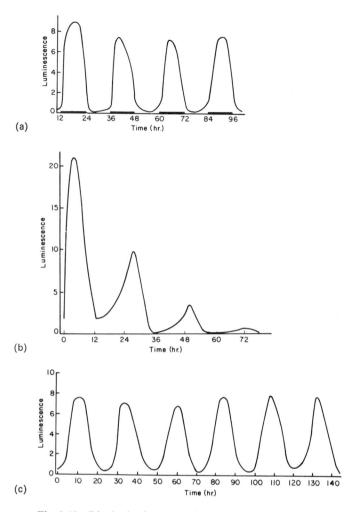

Fig. 8.13 Rhythmic phenomena in *Gonyaulax polyedra*. (*a*) Cyclic expression of stimulated luminescence, with peaks in the middle of the dark period. (*b*) Demonstration of the endogenous nature of luminescence; peaks continue to occur for a few cycles even under continuous illumination, (*c*) the rhythm persists indefinitely under conditions of low intensity, although the emission of this stimulated luminescence is not as great as that under a light:dark regime. (After Sweeney and Hastings and Sweeney.)

by placing the cells in LL conditions of low-intensity light (1000 lux). Although the peak of light emission is not so great as that reached under LD conditions, the circadian rhythm continues indefinitely (Fig. 8.13c). The second type of luminescence occurring in this species is a glowing reaction or spontaneous luminescence (McMurray and Hastings, 1972), under nonagitated conditions rather than the flashing reaction under agitated conditions. When cells are placed near a photometer in darkness at the beginning of the dark period, cells give off light, even though not agitated. The amount of light produced gradually increases, reaching a peak at midnight, or the midpoint of the dark phase. This "glow rhythm" is different from the photoemission of stimulated cells (Sweeney, 1969). Therefore, two components of bioluminescence in *Gonyaulax polyedra* display a circadian rhythm, one related to the luminescence capacity, which is the amount of light released in a sudden flash, and the other related to luminescence intensity or "glow", which is a continuous, low-intensity emission. The periodicity in luminescence has been correlated with nightly increases in the luciferin (substrate) and luciferase (enzyme) levels.

Gonyaulax polyedra is recognized to have two other rhythmic phenomena (Bode and Sweeney, 1963; Sweeney, 1969; McMurray and Hastings, 1972), namely, photosynthetic capacity and cell division. Whether measured by oxygen production or carbon dioxide fixation, photosynthesis demonstrates a circadian periodicity, the peak occurring in the middle of the day and the rhythm continuing under constant light conditions (Sweeney, 1960). Assays over time for the activity of ribulose diphosphate carboxylase, the enzyme that initiates the series of dark reactions, showed a periodicity sufficient to account for the circadian rhythm. Cell divisions in this species are concentrated into a 30-minute span during any 24-day period when grown under a LD cycle (Sweeney and Hastings, 1958). When the LD cycle is 12: 12, the time of maximal cell division is at "dawn." This periodicity in cell division can be detected under LL (i.e., constant light) conditions, indicating that it is endogenous. These diverse circadian rhythms of stimulated and spontaneous bioluminescence, photosynthetic capacity, and cell division in *G. polyedra* seem to be coordinated within one circadian system (McMurray and Hastings, 1972). A membrane model for the generation of circadian oscillations has been proposed (Njus, et al., 1974). At the ultrastructural level, events have been correlated with these cyclic phenomena both in regard to the chloroplasts (Herman and Sweeney, 1975) and to the number and size of particles on the extracellular face of the peripheral vesicles of the cytoplasmic membrane (Sweeney, 1976A).

A different type of circadian rhythm involving phototactic behavior has been noted (Forward, 1970; Forward and Davenport, 1970; Hand and Schmidt, 1975) in *Gyrodinium dorsum* Kofoid.

Red Tides and Toxins

Dinoflagellates are widely recognized to produce "blooms" or "red tides," in which the concentration of cells may be so great as to color the ocean, locally, red, reddish brown, or yellow (Prakash and Taylor, 1966; Holmes, et al., 1967; Sweeney, 1976B). Patches up to several square kilometers in extent may be discolored, the most intense blooms usually occurring in areas protected from the wind. Surface waters of these blooms often contain 1 to 20 million cells per liter. Species of *Prorocentrum*, *Gymnodinium*, *Gonyaulax*, *Ceratium*, and *Cochlodinium* have been reported to produce blooms. Some of these blooms are associated with the production of toxins (Torpey and Ingle, 1966; Schantz, 1967, 1971; Ray and Aldrich, 1967; Sasner, et al., 1972; Steidinger and Joyce, 1973), resulting in fish kills and mortality of other marine organisms; yet some dinoflagellate blooms produce no toxic effects (Dragovich, et al., 1965; Wilton and Barham, 1968).

The toxic blooms of dinoflagellates fall into three categories (Steidinger, 1973): (1) blooms that kill fish but few invertebrates (*Gymnodinium breve* Davis, the Florida red tide organism, is an example); (2) blooms that kill primarily invertebrates (several species of *Gonyaulax* are of this type); (3) blooms that kill few marine organisms but the toxins are concentrated within the siphons or digestive glands of filter-feeding bivalve molluscans (clams, mussels, oysters, scallops, etc.) causing **paralytic shellfish poisoning** (= **PSP**). The most notorious PSP-causing dinoflagellate on the Pacific coast is chain-forming *Gonyaulax catenella* Whedon et Kofoid, its poison being called (Evans, 1971) **saxitoxin,** a neurotoxin 100,000 times more potent than cocaine (Steidinger and Joyce, 1973). The toxin has been isolated from Alaska butter clams and California mussels and has been chemically characterized (Wong, et al., 1971) as well as produced under *in vitro* conditions by cultivation of *G. catenella* (Proctor, et al., 1975). Saxitoxin acts to prevent normal transmission across neuromuscular synapses by interfering with the movement of sodium ions through excitable membranes (Kao, 1972). Interestingly, mussels may become too toxic for human consumption when concentrations of *G. catenella* reach only 100–200 cells per milliliter, but concentrations of at least 20,000–30,000 cells per milliliter must be reached before a bloom is apparent (Schantz, et al., 1966). Normally, the toxicity in the mussels disappears within 2–3 weeks after a bloom, but much longer retentions have also been reported.

Another chain-forming species is *Gonyaulax monilata* Howell (Fig. 8.14), which has been associated (Williams and Ingle, 1972) with fish kills in offshore waters off the west coast of Florida. It is not so serious a problem as *Gymnodinium breve*,[2] and interestingly a bloom of *G. breve* was reported to occur following an outbreak of *Gonyaulax monilata*. The toxin of *G. monilata* has been obtained (Aldrich, et al., 1967) under *in vitro* conditions.

The pharmacognosy of *Gymnodinium breve* toxin has been analyzed (Sievers, 1969; Abbott and Paster, 1970; Steidinger, et al., 1973), and the toxin was found to be milder than saxitoxin. *Gonyaulax excavata* (Braarud) Balech[3] has caused PSP on the Atlantic coast, this toxin being more toxic than saxitoxin. More than 1600 cases of PSP have been reported worldwide, but statistically PSP does not constitute a major public health problem. In lethal concentrations in humans, death results from respiratory and cardiovascular arrest within 12 hours after consumption of toxic bivalves (Steidinger, 1973).

Researchers have long asked the question of what triggers such a massive bloom of dinoflagellates, and various answers have emerged. Red tides have occurred essen-

[2]Blooms of *Gymnodinium breve* are almost entirely restricted to the west coast of Florida but rarely can occur on the east coast, if unusual current configurations take place and carry "seed" populations from the Gulf out through the Florida Keys (Murphey, et al., 1975).

[3]The species regarded as causing toxic blooms in waters had been identified as *G. tamarensis* Lebour, but evidence has been presented (Loeblich and Loeblich, 1975) that *G. tamarensis*, described from England, is not luminescent or toxic and lacks a ventral pore, whereas the New England red tide organism is luminescent and toxic and has a ventral pore.

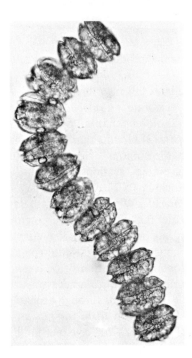

Fig. 8.14 *Gonyaulax monilata* Howell, a chain-forming dinoflagellate associated with toxin production. × 240. (Courtesy of Florida Department of Natural Resources Marine Research Laboratory.)

tially on all shores of North America, from Florida to New England, the Gulf of Mexico, and the Pacific coast. A survey of localities of the occurrence of red tides suggests certain factors for their expression. One correlation is an association with the surfacing of nutrients by upwelling of ocean currents or by tidal turbulence as in the Bay of Fundy (Hutner and McLaughlin, 1958). Red tide dinoflagellates may have a competitive advantage over coastal diatoms during times of upwelling because of their ability to take up and assimilate nitrate in the dark, plus other properties (Harrison, 1976).

Blooms in other areas, as the Florida coast, seem to be set off by heavy rains on the land, the runoff washing phosphates into the sea and also lowering the salinity, which is associated with a promotion of dinoflagellate growth. It is also known that vitamin B_{12}, which is required by most marine dinoflagellates, may also be washed into the sea from the soil and marsh areas, where it is being produced by both bacteria and blue-green algae. However, no apparent correlation could be established (Stewart, et al., 1966) between levels of cells of *Gymnodinium breve* and distribution of B_{12}. The observation by Wilson (1966) that chelated iron promoted the growth of *G. breve* led some researchers to suggest that humic substances, which are naturally occurring chelators in the soil, are leached into the ocean from runoff and contribute to blooms of *G. breve* in the coastal waters of the Gulf of Mexico. Humic substances in small amounts have been shown (Prakash and Rashid, 1968) to exert a stimulatory effect on the growth of dinoflagellates. It has also been recognized (Wilson, 1967; Reid, 1972) that species involved in toxic blooms are cyst producers, the cyst stages occurring in sediments, and that the cycle of encystment-excystment is a regular occurrence for many estuarine and **neritic** species causing the blooms (Steidinger and Ingle, 1972; Steidinger, 1973).

Specialized Modes of Existence: Symbiosis and Parasitism

A wide variety of marine invertebrates, including sponges, jellyfish, sea anemones, corals, gastropods, and turbellarians, and some protistans, including ciliates, radiolarians, and foraminiferans, harbor within them golden spherical cells termed **zooxanthellae** (Fig. 8.15). A thin periplast surrounds these coccoid cells; older cells become more thick-walled and cystic (Kevin, et al., 1969). Healthy, vegetative cells routinely undergo division into two equal products (Freudenthal, 1962). Cells can also transform their contents into a single, naked zoospore, which resembles the genus *Gymnodinium* (p. 443), thus revealing their relationship to the Pyrrhophycophyta.

Cultures of zooxanthellae have been isolated and grown from a wide variety of hosts and diverse geographical localities, and the suggestion has been made (McLaughlin and Zahl, 1966) that a single pandemic species, *Symbiodinium microadriaticum* Freudenthal, may be involved. Size differences exist in zooxanthellae from different hosts examined *in vivo*, but the size differential disappears when the zooxanthellae are grown in culture (Taylor, 1969C). A significant distinction lies, however, in the type of motile cell released by zooxanthellae. Whereas gymnodinioid-type swarmers are formed by the great majority of zooxanthellae from various hosts (these zooxanthellae being assigned to *Symbiodinium*), the swarmers released by the symbiont of the colonial hydrozoan *Velella velella* (this symbiont being identified as the dinoflagellate *Endodinium*) resemble *Amphidinium*, a naked dinoflagellate with a reduced epicone (D. Taylor, 1971). Taylor (1971) concluded that special generic recognition should not be given to zooxanthellae, and accordingly he transferred *Symbiodinium microadriaticum* to *Gymnodinium* on the basis of its motile cell and similarly *Endodinium chattonii* Hovasse to *Amphidinium*. However, objection has been raised (Sournia, et al., 1975) of the transfer of *Endodinium* to *Amphidinium* because of ultrastructural differences.

The translocation of photosynthate from zooxanthellae to the host organism has been substantiated by numerous investigations (Lenhoff, et al., 1968; von Holt and von Holt, 1968; D. Taylor, 1969A, 1969B, 1973, 1974; Smith, et al., 1969; Trench, 1971; Pearse and Muscatine, 1971). Glycerol has been shown (Muscatine, 1967) to be the

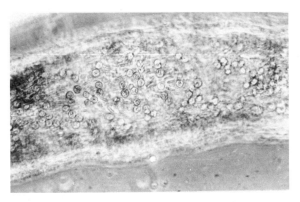

Fig. 8.15 Zooxanthellae present in the tentacle of a sea anemone. × 174.

principal product excreted by zooxanthellae of corals, and, interestingly, high levels of excretion under *in vitro* conditions occur only in the presence of host tissue. A stimulatory effect on the zooxanthellae is evident. Values of more than 60% of the carbon fixed in photosynthesis being transferred from the alga to the host have been measured (von Holt and von Holt, 1968; Taylor, 1969B); lesser values have also been obtained (Trench, 1971). The endocytosis of zooxanthellae by host amoebocytes in the giant clam *Tridacna* has been reported (Fankboner, 1971).

The degree of dependency of the host on the symbiotic algae is variable. A predatory host such as the sea anemone *Anemonia* is dependent on exogenous food and can survive indefinitely without the contribution of its symbiont. If deprived of its exogenous food supply and forced to rely only on the excreted photosynthate, death results in most cases (D. Taylor, 1969A). The metabolic requirements of the coelenterate *Zoanthus*, however, are satisfied by the photosynthetic activity of the zooxanthellae (von Holt and von Holt, 1968). It has been maintained (Johannes and Coles, 1969) that hermatypic (i.e., reef-building) corals obtain the bulk of their energy requirements from their zooxanthellae and are unable to utilize zooplankton alone. This hypothesis, however, has been questioned (Goreau, et al., 1971), on account of the superbly efficient and voraciously carnivorous heterotrophy of the corals.

It is also known that the rate of calcification is at least $10\times$ greater in the light than in the dark (Goreau, 1961; Freudenthal, et al., 1966) and that hermatypic corals lacking zooxanthellae have a uniformly low rate of calcification. The growing tips of the hermatypic coral *Acropora* were found (Pearse and Muscatine, 1971) to contain lower concentrations of zooxanthellae than regions farther away from the tip, and yet light enhancement of calcification rates is greatest in the tips. These data proved to be consistent with the concept that photosynthetic products from the algae accelerate calcification in the tips by the experimental demonstration that ^{14}C-labeled photosynthates are translocated from the algae-rich regions of the coral to the algae-poor tips. On the basis of the stable carbon isotopic composition, it has been determined (Land, et al., 1975) that most of the organic carbon in the tissues of the hermatypic corals living in the shallow waters off Jamaica had passed through the zooxanthellae.

Symbiosis remains one of the more fascinating phenomena in biology, and a large literature exists (Droop, 1963; Henry, 1966; Trager, 1970) on the topic. In addition to dinoflagellate zooxanthellae, algae classified in other divisions have also been demonstrated to be symbiotic. The term **zooxanthella** may be broadly defined as to include golden, yellowish, brownish, and even reddish symbiotic algal cells assignable to the Pyrrhophycophyta, the Bacillariophyceae of the Chrysophycophyta (Ax and Apelt, 1965), and the Cryptophycophyta (Smith, 1950; Barber, et al., 1969; Taylor, et al., 1971). The term **zoochlorella** is applied to symbiotic green cells assignable to the Chlorophycophyta and include the genera *Platymonas*, *Pyramimonas*, *Chlorella*, and *Oöcystis* among others. These symbioses occur both in freshwater and marine habitats. Blue-green algae are also present in a variety of symbiotic relationships (Norris, 1967B).

What were originally considered to be dinophycean zooxanthellae living symbiotically in some marine slugs (sacoglossans) were recognized (Taylor, 1968; Trench,

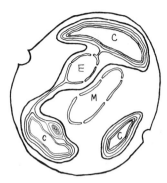

Fig. 8.16 Schematic representation of *Peridinium balticum* (Levander) Lemmerm., in which a chrysophycean endosymbiont with a eukaryotic nucleus (E) and chloroplasts (C) is contained within the protoplast of the colorless host with its own mesokaryotic nucleus (M). (After Tomas and Cox.)

1969) on closer inspection to be isolated chloroplasts from the green alga *Codium* (p. 192). In this symbiotic relationship the sacoglossan uses its stylet to consume the protoplasm of the *Codium*, the algal chloroplasts becoming incorporated into the epithelial cells lining the digestive canal of the slug. The chloroplasts remain intact and have been demonstrated to function as endosymbiotic organelles in the host tissue (Trench, et al., 1969; Greene, 1970; Trench, et al., 1973A, B). The slugs use the photosynthetic products from the chloroplasts to synthesize mucopolysaccharides, which are excreted as mucus from the pedal gland (Trench, et al., 1972; Trench, 1973).

A different type of symbiotic relationship has recently been discovered (Tomas and Cox, 1973A) in the Pyrrhophycophyta. The presence of two nuclei in *Glenodinium foliaceum*,[4] one of which was typically eukaryotic and the other of which was mesokaryotic, had earlier been reported (Dodge, 1971B). This unusual occurrence of two types of nuclei was also observed (Tomas, et al., 1973) in *Peridinium balticum* (Levander) Lemmermann (Fig. 8.16). A more careful scrutiny of the membrane relationships within the cell of *Peridinium balticum* led to the appreciation (Tomas and Cox, 1973A) that the eukaryotic nucleus belonged to a chrysophycean endosymbiont within the cytoplasm of a colorless dinoflagellate host, the two protoplasts being separated by their adjacent plasma membranes (Fig. 8.16). The evidence for this interpretation included the presence of fucoxanthin (Tomas and Cox, 1973B)[5] and the absence of peridinin, the former pigment being characteristic of most Chrysophycophyta and the latter indicative of the Pyrrhophycophyta (see Table 1.1). In addition, studies on the ultrastructure of the chloroplasts in *P. balticum* revealed (Tomas and Cox, 1973A; Dodge, 1975A) the presence of "girdle lamellae," which are membrane configurations lying just within the chloroplast envelope and encircling the stacks of photosynthetic lamellae within the chloroplast; such an arrangement is again typical of the Chrysophycophyta rather than the Pyrrhophycophyta (Dodge and Crawford, 1971). Furthermore, an internal pyrenoid within the chloroplast is present, whereas pyrenoids are generally stalked and projecting from the chloroplasts in those dinoflagellates that have pyrenoids.

[4]Also regarded as a *Peridinium* (Jeffrey, et al., 1975).
[5]Other reports of fucoxanthin in dinoflagellates had been published earlier (Riley and Wilson, 1967; Mandelli, 1968).

More recent surveys of both chloroplast ultrastructure (Dodge, 1975A) and pigment composition (Jeffrey, et al., 1975) in this division have revealed interesting variations. Fucoxanthin has been found to be the predominant carotenoid in other dinoflagellates, as in species of *Woloszynskia* (Whittle and Casselton, 1968), *Gymnodinium* (Riley and Wilson, 1967), and *Prorocentrum* (Jeffrey, et al., 1975, as *Exuviaella*). The species of *Gymnodinium* examined (Dodge, 1975A) also have internal pyrenoids but, interestingly, have a single nucleus, the mesokaryotic type. Dodge (1975A) has speculated that it is possible to regard the lack of chloroplasts as the "primitive," or original, condition, from which endosymbiosis with autotrophic chrysophycean algae might have been established, as in *Peridinium balticum* and *Glenodinium foliaceum*. The next step could have conceivably involved the loss of the eukaryotic nucleus of the endosymbiont but the retention of the internal pyrenoid and fucoxanthin. Finally, the presence of peridinin, stalked pyrenoids, and only one type of nucleus (mesokaryotic) would represent the "advanced" or acquired condition. It is true that heterotrophic nutrition is still widespread in this division.

Parasitism is a well-developed mode of nutrition in the Pyrrhophycophyta. Parasitic species are often exceedingly specialized in their structure and reveal their correct placement in this division either by their gymnodinioid motile stages or by their persistent chromosomes. Although many of these genera, such as *Myxodinium* (Cachon, et al., 1969), *Blastodinium* (Soyer, 1969, 1971), and *Syndinium* (Ris and Kubai, 1974), have mesokaryotic nuclei typical for this division, other genera, such as *Paradinium* (Cachon, et al., 1968) and *Amoebophyra* (Cachon and Cachon, 1970), lack such a nucleus and may be wrongly classified here.

Both **endoparasites** and **ectoparasites** are known. *Haplozoon* includes about a dozen parasites in the intestines of annelids. A recently described example is *H. axiothellae* Siebert (Siebert, 1973), which has a colonial organization (Fig. 8.17*a*) and is

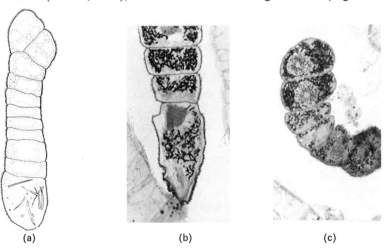

(a)   (b)   (c)

Fig. 8.17 *Haplozoon axiothellae* Siebert. (*a*) Camera lucida drawing of colony composed of trophocyte (T), row of gonocytes (G), and a pair of quadrinucleate sporocytes (S), (*b*) Attachment of colony to gut lining of host. (*c*) Longitudinal section of gonocytes. (*a*) × 510; (*b*), (*c*) × 826. (After Siebert; permission of Journal of Phycology.)

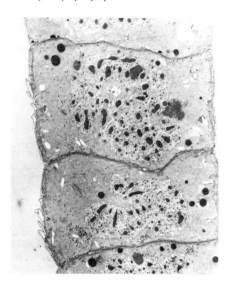

Fig. 8.18 *Haplozoon axiothellae.* Electron micrograph of cellular details, including thecal covering and typical mesokaryotic nucleus. × 3360. (After Siebert and West; permission of Protoplasma.)

composed of three types of cells: **trophocyte,** the anterior-most cell that attaches the parasite to the gut lining (Fig. 8.17*b*); **gonocyte,** an intercalary cell in the zone of cell division (Fig. 8.17*c*); and **sporocyte,** the reproductive cell type, which here becomes quadrinucleate and is apparently shed. The cells are covered with a theca (Fig. 8.18), and the chromosomes remain condensed during interphase (Siebert and West, 1974).

Although *Haplozoon* is a completely heterotrophic genus, most species of *Blastodinium* are facultative heterotrophs, living in the intestines of marine copepods. Individuals of *Blastodinium* are two-celled, the trophocyte remaining quiescent and the other cell undergoing numerous divisions into many sporocytes, which are released by the rupture of the envelope. The sporocytes give rise to gymnodinioid swarmers, and the trophocyte divides, cutting off a new reproductive cell. A list of these parasitic genera has been recently published (Sournia, et al., 1975). A compilation of their hosts and of the mode of attachment has also appeared (Chatton, 1952).

Endocytosis, or **phagotrophy,** which is the holozoic engulfment of particulate food, is of sporadic occurrence among dinoflagellates (Biecheler, 1952), including *Gymnodinium* (p. 443), *Gyrodinium*, *Peridinium* (p. 446), *Crypthecodinium*, and *Ceratium* (p. 446). In a marine species and a freshwater species of *Ceratium*, endocytosis seems to accompany photosynthesis (Norris, 1969B; Dodge and Crawford, 1970B). The ventral region of *Ceratium hirundinella* (O. F. Müller) Schrank, a heavily armored freshwater species, is not thecate, and food is engulfed through a conspicuous sulcal aperture.

Classification

The division Pyrrhophycophyta is recognized to contain five classes, two of which, the Ebriophyceae and the Ellobiophyceae, are relatively small and poorly known. Two of the other classes, the Dinophyceae and the Desmophyceae, have been merged by some workers (Dodge, 1975B; Loeblich, 1976) into a single class, the

Dinophyceae. The following outline is based upon the treatments of Loeblich (1970) and Bourrelly (1970), except for the exclusion of the Cryptophyceae (p. 566) included by Bourrelly in this division. A new class, the Syndiniophyceae, was recently established (Loeblich, 1976) to include intracellular parasites with anomalous nuclear characteristics, based upon the work of Ris and Kubai (1974). A brief characterization of each order is also presented.

Class Ebriophyceae
   Order Ebriales: cells colorless, biflagellate, lacking resistant outer covering; an internal siliceous skeleton[6] present.

Class Ellobiophyceae
   Order Thalassomycetales: cells parasitic, attached, lacking a theca but with a complex pellicle: motile stages of the gymnodinioid type (Galt and Whisler, 1970).

Class Syndiniophyceae
   Order Syndiniales: cells parasitic, without a cellular covering of plates; nuclei with low chromosome numbers; chromosomes V-shaped and with detectable amounts of basic protein; mitosis associated with centrioles.

Class Dinophyceae: motile cells biflagellate, one flagellum located in a transversely aligned groove and with a vibrating beat, the other flagellum beating in a longitudinally aligned groove and extending in a posterior direction from the cell.
   Order Blastodiniales: attached parasitic forms; amoeboid, coccoid, or multicellular colonies; cell covering nonthecate or possibly with a layer of thin plates (not completely understood); mostly marine.
   Order Coccidiniales: parasitic forms, marine; cell covering not understood.
   Order Dinamoebales: free-living amoeboid cell, solitary, nonthecate; cyst stage with a wall, releasing gymnodinioid zoospores upon germination.
   Order Dinophysiales: cells motile, solitary, free-living (autotrophic or heterotrophic), with a theca of 18 (rarely 19) plates, including 2 that are much larger than the others.
   Order Dinotrichales: immobile, filamentous forms, branched, uniseriate, with a cell wall; gymnodinioid motile stages.
   Order Gloeodiniales: palmelloid vegetative stage of cells aggregated in mucilaginous aggregations; motile cells *Hemidinium*-like, with a layer of thin thecal plates and an abbreviated cingulum.
   Order Gymnodiniales: cells motile, solitary, free-living; cell covering without thecal plates (but with peripheral vesicles that may have contents homologous to thecal plates).
   Order Noctilucales: cells solitary, free-living, colorless, holozoic, lacking a theca; meiosis at gametogenesis (in *Noctiluca*).

---

[6]Others (e.g., Fritsch, 1935) include this small group in the Silicoflagellata (p. 372) of the Chrysophycophyta.

Order Peridiniales: cells free-living, motile, with a theca composed of many plates; photosynthetic or colorless, sometimes holozoic.

Order Phytodiniales: cells immobile, coccoid, free-living, attached or unattached, with a firm wall; reproduction by autospores or gymnodinioid motile cells.

Order Pyrocystales: cells immobile, usually solitary, free-living, pigmented, coccoid to crescent-shaped, unarmored but with a firm wall; reproduction by gymnodinioid motile cells or aplanospore formation.

Order Zooxanthellales:[7] cells coccoid, photosynthetic, living symbiotically in various Protozoans and Metazoans, with a cell covering of membranes, some of which enclose thick amorphous material.

Class Desmophyceae: motile cells with two apically or subapically inserted flagella, one directed longitudinally forward and the other beating in a plane perpendicular to the first and encircling the first.

Order Desmocapsales: vegetative cells immobile, grouped in a gelatinous, palmelloid mass; pigmented.

Order Prorocentrales: cells motile, solitary, free-living, with a theca basically of two major parts, each concave; flagella apically inserted.

Order Protaspidales: cells colorless, with two subapically inserted flagella; nonthecate cell cover.

REPRESENTATIVE GENERA

Order 1. Dinophysiales

DINOPHYSIS Ehrenberg   Photosynthetic and widespread in tropical and subtropical waters, the armored cells of *Dinophysis* (Gr. *dinein*, to whorl + Gr. *physis*, nature) are laterally compressed and have a cingulum close to the anterior end, resulting in a shortened epitheca. The cellulosic theca extends out from the cingulum and/or the sulcus as membranous structures termed **lists**. The lists may be supported by ribs and in some species are developed into prominent winglike or flange-like expanses (Fig. 8.19), which seem to aid in flotation. As for other members of this order, the theca is composed of 18 plates.[8] Two of the four hypothecal plates are very much enlarged.

The epitheca in *Dinophysis* is usually difficult to detect because of the anterior cingular list projecting beyond it. Closely related *Phalacroma* is distinguished (Wood, 1954) by the fact that its epitheca extends beyond the anterior cingular list. Some workers (Abé, 1967B; Balech, 1967B) have suggested the merger of *Phalacroma* into *Dinophysis*, but their distinction is maintained by others (e.g., Steidinger and Williams, 1970).

[7]The status of this order is questionable, following the proposal (D. Taylor, 1971) to include dinoflagellate zooxanthellae within *Gymnodinium* and *Amphidinium* of the Gymnodiniales.

[8]Tai and Skogsberg (1934) first realized that the theca was made up of 17 plates, but Norris and Berner (1970) counted an additional small plate. The only apparent exception is *Latifascia*, which has 19 plates (Balech, 1967B).

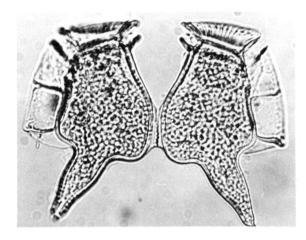

Fig. 8.19   *Dinophysis* sp., a species from the Gulf of Mexico. Pair of cells following division. × 718.

ORNITHOCERUS Stein   Distinguished from *Dinophysis* by the much greater development of the anterior cingular list, which at times is up to one and one-half times the width of the cell, *Ornithocerus* (Gr. *ornitho*, bird + Gr. *kerkos*, tail) includes some spectacularly sculptured cells (Fig. 8.20), with a funnel-shaped anterior cingular list and rib-supported sulcal lists. They have provided rich material for investigations with the scanning electron microscope (F. Taylor, 1971).

Cell division also has been examined (F. Taylor, 1973) with the SEM. Binary fission is longitudinal, and a region called a **megacytic zone**[9] is developed, which is a region of

[9]This term is generally applied to the zone of growth in all armored dinoflagellates and refers to the region between adjacent plates where new thecal material is added to allow for enlargement of the cell. New material is usually unevenly added to individual thecal plates (Cox and Arnott, 1971).

Fig. 8.20   *Ornithocercus magnificus* (Stein) Kofoid et Skogsberg. Scanning electron micrograph of a specimen collected in the Gulf of Mexico.   × 705. (Courtesy of Dr. Greta Fryxell.)

lateral thecal growth, encircling the body of the theca. It is more noticeable in the dorsal region of the dividing cell. The megacytic zone enables both halves of the parent cell to contribute to the formation of new complementary halves. When the two products of division are securely enclosed by a combination of the old and the new theca, the megacytic zone becomes resorbed. The megacytic zone is typical for cell division in the order Dinophysiales. Chloroplasts are absent in the species of this genus, which are more commonly distributed in tropical seas (Abé, 1967C). Six species have been reported (Norris, 1969A) from the Gulf of Mexico.

Order 2. Gymnodiniales

GYMNODINIUM Stein   With a temperate to tropical distribution in fresh, brackish, and marine bodies of water, *Gymnodinium* (Gr. *gymnos*, naked + Gr. *dinein*, to whorl) has the cingulum in a median position. The cingulum may form a complete equatorial circle by the ends meeting on the ventral surface (Fig. 8.1*a*), or the ends may be slightly offset. If the ends of the cingulum are conspicuously displaced (i.e., they are separated by about one-fifth the length of the cell), the cells are referred to *Gyrodinium* rather than *Gymnodinium*. There is evidence, however, that this distinction is artificial (Norris, 1966), particularly since some cultured species exhibit a degree of **pleiomorphism** in respect to the degree of spiral of the cingulum (Kimball and Wood, 1965). If the degree of torsion of the cingulum is such as to make one and one-half or more turns as it passes down the long axis of the cell, the cells are assigned to the genus *Cochlodinium*. It has been suggested (Norris, 1966) that this structural streamlining and elongation of the dinoflagellate cell afford an apparent advantage in natural selection by the reduction of friction in the spirally swimming cells.

As the name indicates, this genus is comprised essentially of naked forms. It is well-known that these organisms are so delicate as not to be preserved in plankton samples fixed with formaldehyde. Their ability to tolerate being examined under a light microscope is minimal due to their extreme sensitivity to heat and light intensity. The amphiesma (p. 419) of most Gymnodiniums consists simply of three layers of membranes, made up of flattened, appressed vesicles lying beneath the plasma membrane (Dodge and Crawford, 1969A; Dodge, 1974B). A layer of numerous microtubules lies beneath the innermost membrane. Some species have vesicles containing dark-staining ingredients (Dodge, 1970) or a very thin, platelike structure homologous with the plates of armored dinoflagellates. Delicate striations may also be present, their presence or absence being useful in delimiting subgenera. The closely related *Woloszynskia* is distinguished by the presence of thin plates of uniform thickness within its vesicles, the plates being thick enough to be detected by the light microscope.

In their classic monograph of unarmored dinoflagellates, Kofoid and Swezy (1921) presented colored plates of a variety of Gymnodiniums, which beautifully depicted the full range of colors in the spectrum possible in this group. Many species contain chloroplasts; heterotrophic nutrition is also prevalent, some species being holozoic. The color of photosynthetic members ranges from yellow to shades of gold and brown. The cytoplasm may be variously hued in shades of purple, blue, red, pink, and green, however, especially due to the pigmentation of the pusules, which are common in this genus (Dodge, 1974B). Trichocysts are present in some species but absent in the type species *G. fuscum* (Ehr.) Stein (Dodge and Crawford, 1969A). The phototactic response of the spectacular *G. splendens* Lebour has been investigated (Forward, 1974).

AMPHIDINIUM Claparède et Lachmann   Distinguished from *Gymnodinium* by its relatively small epicone at the anterior end (Fig. 8.4*a*), *Amphidinium* (Gr. *amphi*, around + *dinein*, to whorl) is perhaps the simplest member of the Gymnodiniales, the cell body lacking any torsion and the cingulum located at the far anterior end, at least on the dorsal side. A displacement of the ends of the cingulum usually does not occur. The great majority of species are photosynthetic, inhabiting freshwater, brackish water, and marine water, both warm and cold temperate. The largest group of species shows a strong dorsiventral flattening of the body, which is rare in *Gymnodinium*.

The nucleus is centrally or posteriorly located. Pusules are generally common. The cell covering may be smooth, finely striated, or furrowed. It has been shown (Dodge and Crawford, 1968) to be composed of three membranes with an underlying system of microtubules.

The reproductive cycle of *Amphidinium carteri* Hulburt has been investigated (Vien, 1967A, 1968). Asexual multiplication involves binary fission or a series of vegetative divisions during which the cells remain interconnected. The zygote resulting from sexual reproduction is naked but nonmotile; it undergoes meiosis to produce four meiospores, which recycle the vegetative phase. Thus, the pattern expressed is H, h. Motility in this same species has also been reported (Gittleson, et al., 1974).

Order 3. Noctilucales

NOCTILUCA Suriray in Lamarck   One of the most familiar algae collected in neritic phytoplankton because of its large size (up to 2 mm in diameter) and its luminescent capabilities, *Noctiluca* (L. *nox*, night + L. *lucere* to shine) is a naked cell (Fig. 8.21), spherical to somewhat reniform, colorless, and with a characteristic tentacle. The cell is highly vacuolated, which probably affords a degree of buoyancy to the cell, and strands of cytoplasm extend out from the conspicuous central nuclear mass. The cingulum is scarcely distinguishable, and the transverse flagellum is reduced to a toothlike structure. Although the longitudinal flagellum is small, the sulcus is deep and elaborately developed into an oral groove and a **cytostome,** the latter functioning as the locus for food engulfment. Beneath the cytostome emerges the tentacle, which as a length about the same as that of the cell. Although fragile and easily shed, it is quite thick, transversely striated, and beats with broad undulations (Chatton, 1952).

Nutrition is holozoic; *Noctiluca* is known to devour voraciously blooms of other dinoflagellates. It was the first dinoflagellate in which bioluminescence was reported. *Noctiluca miliaris* Sur., the only species in the genus, and *Gonyaulax polyedra* are the

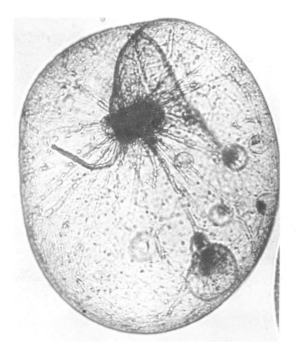

Fig. 8.21 *Noctiluca*, bearing a whiplike tentacle. × 270. (Courtesy of Dr. Richard Zingmark.)

two primary dinoflagellates responsible for bioluminescence in neritic waters, other dinoflagellates being the cause in **pelagic** waters.

Cells of *Noctiluca* are able to swim directly at the surface of the water, their buoyancy being enhanced by such physiological properties as the selective exclusion of the heavier divalent ions (calcium and sulfate) and the relatively high concentrations of the lighter ions (sodium and ammonium) in the cytoplasm (Kesseler, 1966).

Asexual reproduction occurs by longitudinal binary fission, preceded by a resorption of the differentiated cytoplasm. A second type of reproduction, which is reportedly sexual (Zingmark, 1970A), involves successive divisions of the nucleus into a large number of nuclei (at times in excess of 1000). This sequence commences apparently with meiosis, and at least until the 256-nucleate stage the nuclei are maintained in four distinct groups (Zingmark, 1970A). Motile cells with a single longitudinal flagellum are then differentiated from these nuclei. They may be clustered to one side of the parent cell (Chatton, 1952; Drebes, 1974), forming a discoid or dome-like aggregation. The isogametes copulate, fusion being possible from the products of a single fertile cell (i.e., a monoecious condition). The zygote enlarges by growth and vacuolation into the typical vegetative cell.

The nucleus of the vegetative cell is atypical in that the nucleoplasm is homogeneous and chromosomes are not distinct (Zingmark, 1970B). During the process of division into gametes, the chromosomes do not become condensed and discernible until toward the 32-nucleate stage (Soyer, 1972). Typical mesokaryotic nuclei persist for the remainder of the process of gametogenesis.

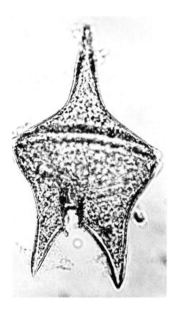

Fig. 8.22 *Peridinium.* × 379. (Courtesy of Michael A. Hoban.)

Order 4. Peridiniales

PERIDINIUM Ehrenberg   Perhaps the most common marine dinoflagellate and also of frequent freshwater occurrence, *Peridinium* (Gr. *peri*, around + Gr. *dinein*, to whorl) is primarily a genus of heterotrophic species, despite its usual automatic inclusion as a primary producer when preserved samples are counted in productivity studies. No other genus of dinoflagellate has been so intensively investigated in reference to its thecal arrangement, the papers of Wood (1954) and Balech (1967A) being examples. The body shape (Fig. 8.22) is variable, but often there is a dorsiventral flattening, a tendency prevalent in the Peridiniales and in contrast to the lateral flattening observed in the Dinophysiales. Horns are common in many species, both apical and antapical in position. The cingulum is generally median, but it is displaced in either an ascending or a descending manner in some species.

The thecal plates may bear pits, knobs, spines, or ridges. The association of the plates with the membranes of the amphiesma has been investigated (Messer and Ben-Shaul, 1969; Kalley and Bisalputra, 1970, 1971). The thecal tabulation of type species of *Peridinium cinctum* (Müller) Ehr., a freshwater species, is 4′ 7″ 5‴ 2‴′ (Bourrelly, 1968B), with five plates in the cingulum. The shapes of the first apical plate and the second apical intercalary plate are useful in placing a species within one of the many groups within the genus.

CERATIUM Schrank   A heavily armored genus in freshwater, brackish, and marine systems, *Ceratium* (Gr. *kerastes*, horned) is quite variable in regard to its appearance (Fig. 8.23). The epitheca is usually drawn out into a single horn, and the hypotheca is developed into two or three horn-like processes. Cingular lists may be

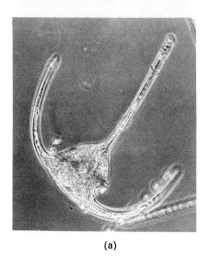

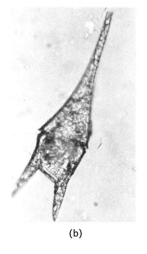

(a)   (b)

Fig. 8.23 *Ceratium.* (*a*) × 250; (*b*) × 211.

present. The thecal plates have the following tabulation: 4′ 5″ 5‴ 2‴′. In addition, the ventral region of the cingulum has a well-developed plate structure, from two to five plates being reported (Loeblich, 1970).

    Four basic cell shapes are presented in Fig. 8.24, corresponding to four subgenera (Sournia, 1967). A distinct apical horn may be absent, and the epitheca may instead be inflated in subgenus Archaeceratium (Fig. 8.24*a*). The cell may be elongated along the longitudinal axis, the length being more than 10× the width, in subgenus Amphiceratium (Fig. 8.24*b*). The antapical horns may be directed in a posterior direction, one being smaller than the other, as in the subgenus Ceratium (= Biceratium) (Fig. 8.24*c*). The antapical horns may be similar in size but diverge right and left, as in the subgenus Orthoceratium (= Euceratium) (Fig. 8.24*d*).

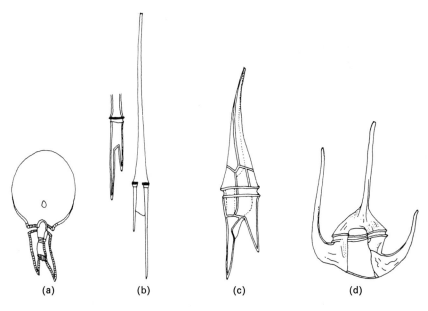

(a)   (b)   (c)   (d)

Fig. 8.24 Subgenera of *Ceratium*. (*a*) Archaeceratium. (*b*) Amphiceratium. (*c*) Ceratium. (*d*) Orthoceratium. (After Yamaji.)

Cell division is the type (p. 423) in which the two products of division of each retain one-half of the original theca (Fig. 8.9*a*) and synthesize the new portion (Wetherbee, 1975B). The cells eventually disassociate. In some species, however, such as *C. lunula* Schimper and *C. vultur* Cleve, loose filamentous colonies are formed.

Most species are photosynthetic, but the holozoic uptake of particulate food (p. 439) may supplement autotrophic nutrition, such as in *C. hirundinella*.

Order 5. Pyrocystales

PYROCYSTIS Thomson in J. Murray   Cell shape is variable in this rather small genus of bioluminescent, nonmotile dinoflagellates. Depending on the species, cells may be spherical, such as in *Pyrocystis* (Gr. *pyr*, fire + Gr. *kystis*, sac) *pseudonoctiluca* Thomson ex Murray; spindle-shaped (Fig. 8.25*a*), such as in *P. fusiformis* Thomson ex Murray; or lunate (Fig. 8.25*b*), such as in *P. lunula* (Schütt) Schütt. The cell wall consists of a firm but thin cellulosic layer, chemically similar to the wall material of dinoflagellate cysts (Swift and Remsen, 1970). These coccoid stages are consequently regarded as cysts. Cells are photosynthetic, the chloroplasts migrating to the corners or periphery of the cell away from the nucleus when light emission is strongest (Swift and Taylor, 1967).

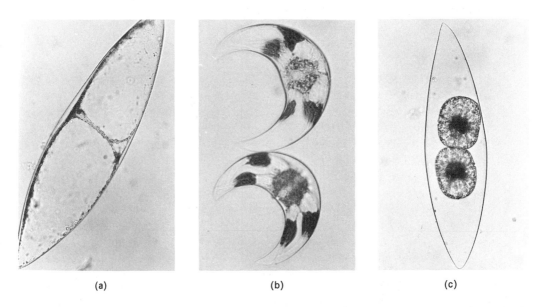

(a)   (b)   (c)

Fig. **8.25** *Pyrocystis.* (*a*) *P. fusiformis* Thomson ex Murray. (*b*) *P. lunula* (Schütt) Schütt. (*c*) *P. fusiformis*, reproductive stage. (*a*) × 171; (*b*) × 325; (*c*) × 208. [(*a*) and (*c*) after Swift and Durbin; (*b*) after Swift and Taylor; permission of Journal of Phycology.]

In a study of three different species, an asexual reproductive process was described (Swift and Durbin, 1971) to involve the division of the crescent-shaped cells into either aplanospores or gymnodinioid zoospores with a single longitudinally directed flagellum. This process began with the retraction of the peripheral protoplasm from the wall and its aggregation in the center of the cell, although it remained attached to the cell wall by cytoplasmic strands. Transverse division of the protoplasm then occurred, the products becoming detached from the parental walls (Fig. 8.25c). In the case of aplanospore formation, the initially small products of division quickly swell up to a size and shape equivalent to that of the parental cell, bursting through the original wall if it still remained. In the case of zoospore formation, either one or two cells are formed per parental cell; the zoospores also have the ability to increase in volume at a rapid rate, for example, as much as a twenty-twofold increase in the case of *P. fusiformis*. This is a noteworthy feature of this genus (Swift and Durbin, 1971).

Similarities to the lunate colorless genus *Sporodinium* (= *Dissodinium*), which is an ectoparasite on copepods eggs (Drebes, 1974), deserve mention. Some of the stages in the life cycle of the parasite and *Pyrocystis* are equivalent, and it had been thought (Sournia, 1973) that these two genera are congeneric. Recent work (Loeblich, 1974) has presented evidence that they are distinct. *Pyrocystis lunula* is photosynthetic, luminescent, has a horseshoe-shaped nucleus and has a single nonmotile stage, which produces two uniflagellate zoospores. *Sporodinium pseudocalanii* Gönnert is weakly pigmented (probably parasitic), has a spherical nucleus, and has two nonmotile stages (a spherical stage and a lunate stage). The lunate stage forms 4-8 biflagellated zoospores; the spherical nonmotile stage cleaves up into the lunate cells.

Order 6. Prorocentrales

PROROCENTRUM Ehrenberg   The theca of this armored dinoflagellate consists basically of two valves, which are usually laterally compressed. Cells of *Prorocentrum* (L. *prora*, prow + L. *centrum*, center) are spherical, ovate (Fig. 8.5a), or teardrop-shaped in face view (Fig. 8.3a) and compressed and saucer-shaped in side view. Two heterodynamic flagella emerge from the apical end of the cell, one of which is directed in the long axis of the cell and the other encircles the first and is perpendicular to it in its orientation. A spinelike process may also arise at the anterior end of the cell. The genus *Prorocentrum* was formerly regarded as bearing such a spine in contrast to *Exuviaella*, which lacked a spine. But this trait was shown to be an unreliable criterion, and the merger of *Exuviaella* into *Prorocentrum* was suggested (Abé, 1967A).

A different basis for distinguishing these two genera was later proposed (Loeblich, 1970), namely, the presence of pores from which the flagella arise in *Prorocentrum* and the absence of flagellar pores in *Exuviaella*, the flagella simply emerging between the two valves. The flagellar pores may be located on one of the large valves or on a special pore plate. In *P. mariae-lebouriae* (Parke et Ballentine) Faust the pores were observed (Faust, 1974) to be surrounded by a total of eight small, thick plates, which are located in a V-shaped depression on one of the valves. Other workers (Dodge and

Bibby, 1973; Dodge, 1975B) continue to support the suggested merger of these two genera, *Prorocentrum* and *Exuviaella*.

The number of species of *Prorocentrum* (including *Exuviaella*) was reduced from 64 to 21 (Dodge, 1975B), the species falling into five fairly distinct groups. The general form and structure of the two main thecal plates, or valves, proved to be the most useful taxonomic characteristic. Most keys (e.g., Steidinger and Williams, 1970) make an attempt to permit the identification of species at the light-microscope level, while recognizing the finer details discernible with the electron microscope (Dodge, 1965; Faust, 1974). Valves of some species lack obvious pores or ornamentation; those of others may have only trichocyst pores. The valves of still other species may be provided with depressions, pores, or small spines (Fig. 8.5*b*).

Most species of this genus are photosynthetic, with one or several chloroplasts present in a cell. The genus is widely distributed in fresh, brackish, and marine bodies of water. Reproduction is by longitudinal division of a cell into two products, each retaining one of the parental valves. A special mode of vegetative multiplication resulting in the formation of four products of division has also been described (Vien, 1967B).

9

Division Rhodophycophyta

General Features

The division Rhodophycophyta, comprising the red algae, is readily distinguishable from other groups of eukaryotic algae by the following combination of characteristics: the complete absence of any flagellate stages; the presence of accessory photosynthetic pigments called **phycobilins** (phycoerythrin and phycocyanin); the occurrence of nonaggregated photosynthetic lamellae, or thylakoids, within the chloroplasts (Fig. 9.1); so-called **Floridean starch** as the food reserve; and the existence of oogamous sexual reproduction involving specialized female cells termed **carpogonia** and male elements termed **spermatia**. However, sexuality is apparently lacking in some members.

Phycoerythrin, which is typically the predominant accessory pigment, is responsible for imparting the red coloration to this assemblage of plants, often completely masking the presence of chlorophyll *a*. Although in recent times regarded (O'hEocha, 1971; Dixon, 1973) to be either an artifact of extraction or a slightly modified form of chlorophyll *a*, chlorophyll *d* is most recently (Meeks, 1974) accepted as a genuine pigment and restricted to the red algae. Yet it is a minor component, is not present in all species, and has not been detected in the Bangiophycidae. Brilliant red pigmentation is particularly well expressed in sublittoral members or those occurring in shaded situations. Due to photodestruction of the phycoerythrin many red algae do not appear reddish at all, and a full range of pigments is exhibited, especially in those of intertidal distribution, including violet, purple, brownish, black, yellow, and greenish forms, a phenomenon that is usually confusing to the inexperienced collector. Members of the same species in a given population may have a variable pigmentation as one observes them over their vertical distribution. This capability of the algae in altering their proportion of pigments in response to differing qualities of incident light is referred to as chromatic adaptation (Crosett, et al., 1965).

The phycobilins (red-colored phycoerythrin and the blue-colored phycocyanin) differ from chlorophylls and carotenoids in being water-soluble pigments, and they

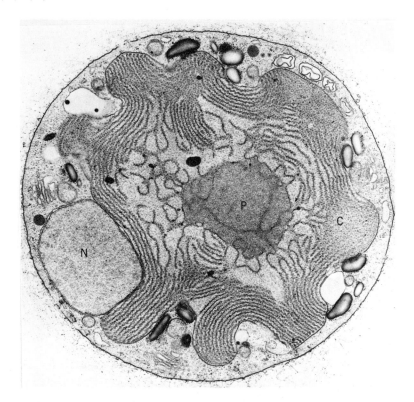

Fig. 9.1 *Porphyridium purpureum*. Cellular details including the massive, stellate chloroplast (C) with a central pyrenoid region (P), nucleus (N), and peripheral starch granules. × 13,110. (After Gantt and Conti; permission of Rockefeller University Press for Journal of Cell Biology.)

are fairly tightly joined to a protein, the combination being referred to as a **biliprotein.** The phycobilins consist of a tetrapyrrolic structure, which shows some resemblance to the bile pigments of animals. These phycobilins are observed elsewhere only in the Cyanochloronta and the Cryptophycophyta. Some variations in the absorption spectra are displayed for both phycoerythrin and phycocyanin. The designations R-, B-, and C- have been used as prefixes to both phycoerythrin and phycocyanin, originally referring to the taxonomic groupings: the class Rhodophyceae (excepting the Bangiales), the Bangiales, and the Cyanophyceae (of the division Cyanochloronta), respectively, from which the pigments were obtained and reflecting differences in the peaks of light absorption. It is now recognized (O'hEocha, 1962, 1971; Goodwin, 1974), however, that a strict correlation does not exist between the type of biliprotein and the taxonomic category. For example, in surveys of many species of red algae and blue-green algae (Hirose and Kumano, 1966; Hirose, et al., 1969) five different types of phycoerythrin were detected on the basis of absorption spectra, with some types being characteristic of both red algae as well as blue-green algae; so now the prefixes simply denote the different absorption spectra.

Aggregates of phycobiliproteins have been visualized with the electron microscope to be located on the outer surfaces (i.e., the stroma side) of the photosynthetic lamellae (the thylakoids) and are termed **phycobilisomes** (Fig. 9.2). These granules were first noted (Gantt and Conti, 1965) in *Porphyridium purpureum*[1] (Bory) Drew et Ross, and they appear as spherical particles about 35 nm in diameter. Subsequent investigations (Gantt and Conti, 1966; Gantt, 1975) have resulted in models attempting to elucidate the mode of attachment of the phycobilisomes to the thylakoid surface and the sequence of energy transfer to chlorophyll *a*. Freeze-etch investigations (Neushul, 1970) have revealed the orderly arrangement of these particles on the membrane surfaces; they are usually in rows forming a two-dimensional crystalline array. In those species in which phycocyanin predominates over phycoerythrin, resulting in blue-green-hued red algae, such as *Porphyridium aerugineum* Geitler (Gantt, et al., 1968), *Rhodella violaceum* (Kornm.) Wehrm. (Wehrmeyer, 1971), and *Batrachospermum virgatum* Sirod. (Lichtle and Giraud, 1970), the phycobilisomes appear discoid or cylindrical rather than spherical. However, this correlation of shape of phycobilisome and ratio of phycobilin pigment may not be valid (Wildman and Bowen, 1974). The apparent absence of phycobilisomes in some red algal species (Brown and Weier, 1970; Scott and Dixon, 1973A) may be the result of fixation procedures (Dodge, 1973). It is obvious that additional species need examination.

The food reserve of the red algae is a starch, essentially similar to the branched type, or amylopectin, of higher plants (Meeuse, 1962). Treatment with iodine results in a yellow or brown staining reaction, and the name **Floridean starch** is usually ap-

[1]The correct name for this species is *P. purpureum* (Drew and Ross, 1964), although *P. cruentum* (S. F. Gray) Naeg. commonly appears in the literature.

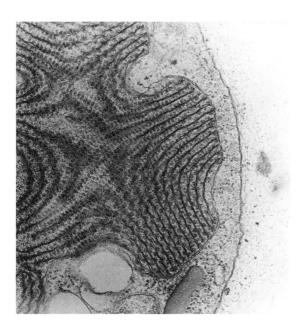

Fig. 9.2 *Porphyridium purpureum.* Portion of chloroplast with particulate phycobilisomes covering the outer surface of the thylakoids. × 38,480. (After Gantt and Conti; permission of Brookhaven Symposia in Biology.)

plied to distinguish this glucan from that of higher plants. Other reserves have been recorded in the red algae, including sugars and glycosides such as trehalose, floridoside, isofloridoside, maltose, and sucrose (Craigie, 1974). The Floridean starch granules are always located scattered in the cytoplasm, never within the chloroplast proper. Pyrenoids are present in the chloroplasts of some red algae, particularly in members of the Bangiophycidae and the order Nemalionales of the Florideophycidae. Starch may or may not be associated with the pyrenoids. Three different types of pyrenoids have been described (Dodge, 1973) in the red algae: a simple-internal type; a compound-internal type, in which the pyrenoid is traversed by the thylakoids; and a multiple-stalked type, in which the pyrenoid is attached to several separate portions of the chloroplast.

The cell walls in the Rhodophycophyta consist of an inner rigid component made up of microfibrils and an outer more amorphous component consisting of mucilage or slime. An outermost cuticle has been detected (Hanic and Craigie, 1969) and can be peeled away from the underlying layer; in *Porphyra* this cuticle was shown to be primarily protein. The fibrillar component of most red algae is composed of cellulose, which occurs in randomly arranged microfibrils. Cellulose has long been recognized to be absent in Bangiophycidae, such as *Porphyra* and *Bangia*. Microfibrils *are* detectable and have been shown (Frei and Preston, 1964) to be xylan as the main portion of the wall. The outer, nonfibrillar part of the wall of *Porphyra* is mannan.

The mucilaginous, amorphous matrix of the wall in red algae is usually a sulfated galactan (Percival and McDowell, 1967; Mackie and Preston, 1974), of which agar, porphyran, furcelleran, and carrageenan are examples. This water-soluble fraction of the cell wall may amount to 70% (Dixon, 1973), and the commercial value of the extracted product has been discussed in Chapter 1. Analogous to the physical properties of alginate, these sulfated galactans are capable of forming gels under the appropriate conditions.

The capacity to deposit calcium carbonate in the cell wall is expressed in scattered taxonomic groupings of red algae. This is true of the family Corallinaceae; however, when members are grown under laboratory conditions, they are often much less calcified. Two tropical genera of the order Nemalionales, *Galaxaura* and *Liagora*, also have walls becoming impregnated with calcium carbonate, giving them a stony or chalky texture. The calcification in *Liagora* (p. 490) may range from barely present to a heavy encrustation. Two crystalline forms of anhydrous calcium carbonate occur in the red algae, calcite and aragonite; these crystals differ from each other both in their shape and in other physical properties.

A distinctive feature of the cross wall separating cells of many red algae is the **pit connection** (Fig. 9.3), which is neither a "pit" nor an intercellular connection but a distinct lens-shaped plug (Ramus, 1971) held within a septal aperture because of its equatorial groove. The plugs can be cleanly isolated by treating the material with cellulase. Chemical analyses (Ramus, 1971) of the plugs of a species of *Griffithsia* (p. 547) showed the composition was an acid polysaccharide-protein complex.

The ultrastructure has also been studied by a number of investigators (Bouck, 1962; Ramus, 1969B, C; Sommerfeld and Leeper, 1970; Lee, 1971; Hawkins, 1972; Evans, 1974). Its development has been followed (Ramus, 1969B) in the genus

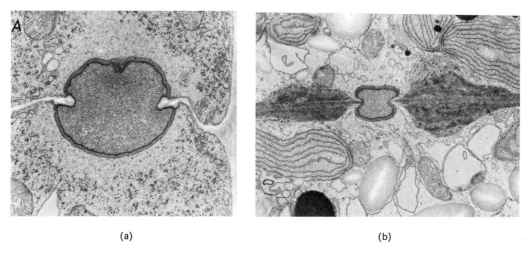

(a)   (b)

Fig. 9.3 *Pseudogloiophloea confusa.* Pit connections. (*a*) Simple septum surrounding central aperture from the fleshy gametophyte phase. (*b*) Septum greatly
thickened around the central aperture in the filamentous phase of the gametophyte. (*a*) × 11,180; (*b*) × 13,340. [(*a*) after Ramus; permission of Journal of
Phycology; (*b*) after Ramus; permission of Rockefeller University Press for
Journal of Cell Biology.]

Pseudogloiophloea (p. 492). An annular ingrowth of wall material at first separates the
two products of cell division, leaving an aperture connecting the cytoplasm of the two
cells. Vesicles then deposit material in this aperture, and the condensation of this
material brings about the plug. The final form of the plug consists of a circular septal
aperture and a lens-shaped plug fitted into the aperture. On each side of the plug is a
membrane-bounded plug cap (Lee, 1971). Although the plasmalemma appears to be
continuous from cell to cell, there seems to be no cytoplasmic continuity or transfer
of material between adjacent cells.

It is noteworthy that within the same species of red algae it is possible to observe
two types of pit connections (Ramus, 1969C). A simple septum with a central, circular aperture characterizes the pits (Fig. 9.3*a*) occurring in the fleshy phase of *Pseudogloiophloea confusa* (Setch.) Levr., the type usually seen in red algae; however, in the
filamentous phase of this species the septa are greatly thickened around the aperture
of the pits (Fig. 9.3*b*), forming a distinct ringlike structure.

Some phylogenetic significance was originally attached to the presence or absence
of pit connections, the Bangiophycidae seeming to lack them. Structures resembling
the pit connections characteristic of members of the Florideophycidae have been
detected (Magne, 1960) with the light microscope for *Rhodochaete* and with the electron microscope (Lee and Fultz, 1970; Bourne, et al., 1970; Lee, 1971) in the *Conchocelis* phase of *Porphyra* but not in the thallus stage. The systematic value of this
trait is thus lessened (Dixon, 1963).

Two types of pit connections can be distinguished on the basis of their ontogeny.
A **primary pit connection** results from the more or less equal division of a cell into two

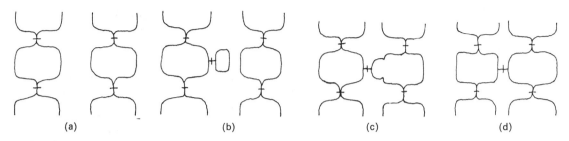

(a)   (b)   (c)   (d)

Fig. 9.4 Outline of events in the formation of a secondary pit connection, involving the cutting off of a small cell from one of two adjacent cells and the fusion of this small cell with the adjacent cell, resulting in a secondary pit connection.

progeny cells, the plug remaining as an indication of their relatedness. A **secondary pit connection,** on the other hand, is a mechanism in which neighboring filaments or cells may establish mutual contact secondarily, creating linkups within the thallus. Such a connection is facilitated by the cutting off of a small cell by an unequal division and the fusion of the small cell with a neighboring cell or filament (Fig. 9.4). Whereas primary pit connections are commonly produced throughout the Florideophycidae, secondary pit connections are not produced in the Nemalionales, some Gigartinales, some Cryptonemaiales, nor in most of the family Ceramiaceae. Therefore, secondary pit connections represent a derived feature within the Florideophycidae.

Secondary pit connections may be established between red algal host and red algal parasite (Pocock, 1956; Goff, 1976B). The phenomenon of a relatively high incidence of parasitism (see p. 516) in the red algae, often involving closely related genera, may be correlated with this ability to form secondary pit connections to facilitate translocation of materials. As previously pointed out, however, the mechanism for such a flow is not understood in terms of the ultrastructural observations of no communication between adjacent cells.

The absence of flagellate stages remains one of the most distinctive characteristics of the Rhodophycophyta.[2] It is recognized that both a change in shape and amoeboid movement for a period of time after release are exhibited in, for example, monospores of *Erythrocladia* (p. 471) (Nichols and Lissant, 1967) and *Acrochaetium* (p. 471) (White and Boney, 1969). With the absence of any flagellated stages, it can be expected that centrioles would also be lacking, which is true. A structure superficially resembling a centriole has been described (McDonald, 1972) and termed a **polar ring**. It was observed to be located at either pole only during prophase of mitosis in *Membranoptera* (p. 555) and is simply a short, hollow cylinder. It was suggested that it may serve as a "microtubule-organizing center" (Pickett-Heaps, 1969 1972D) in conjunction with nuclear division.

A few genera of unicellular red algae are known as well as simple filamentous and colonial forms. The great majority of Rhodophyceae are filamentous, foliose, or

[2]The published figures of Simon-Bichard-Breaud (1971, 1972) of what are alleged to be intracellular flagella in spermatangia of *Bonnemaisonia* (p. 496) do not present convincing ultrastructural details.

more massive forms. Despite this diversity of appearance the final shape is almost always the result of the aggregation of filaments into pseudoparenchymatous constructions. The usual interpretation (Dixon, 1973) is that the thallus of red algae is the result of filamentous growth, which may demonstrate a variable degree of consolidation. Some forms are held merely by the production of copious mucilage that serves to hold the filaments into a loosely organized whole. The largest sizes encountered among the red algae never approach those attained among the brown algae. Some species of *Gigartina* (p. 523) on the California coast reach lengths of almost 1 m (Smith, 1944), whereas specimens of the foliose alga *Schizymenia borealis* Abbott from Puget Sound, Washington, attain widths of 2 m (Abbott, 1967). Although filamentous or pseudoparenchymatous modes of development are almost the rule in the red algae, a few examples of the formation of a parenchyma are also known, as in *Bangia* and *Porphyra* (p. 477) and in certain Delesseriaceae (p. 555).

The manner of growth seen in most Rhodophyceae is by an apical cell or apical cells. This apical cell may be a single conspicuous dome-shaped structure (Fig. 9.5*a*), cutting off segments distally; this pattern would constitute a uniaxial system of growth. Or there may be a number of filaments each terminated by an apical initial, each cutting off segments; this pattern would constitute a **multiaxial** or **fountain-type** of growth (Fig. 9.5*b*). In the case of the foliose alga *Weeksia fryeana* Setch. a careful ontogenetic examination (Norris, 1971) revealed that the apical cell in the initially uniaxial bladelet ceased dividing and lateral initials continue growth, transforming the growth into the multiaxial type.

The degree of elongation and enlargement that segment cells may undergo can be very great, and this aspect of the development of thalli in the red algae has been analyzed in some detail (Dixon, 1971B, 1973). Not only are axial cells produced in a regular sequence, but the lateral filaments arising from the axial cells are also produced in a highly ordered sequence, this arrangement corresponding to the age of the laterals. This organization is under a fairly rigid control, and this control also applies to the patterns of cell enlargement. Cells in angiosperms are recognized to enlarge in volume

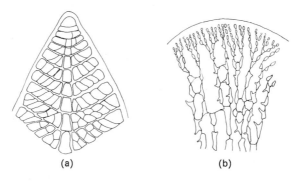

(a)   (b)

Fig. 9.5 Patterns of construction in the red algae. (*a*) Uniaxial development in *Delesseria*. (*b*) Multiaxial development in *Chondrus*. [(*a*) after Rosenvinge; (*b*) after Kylin.]

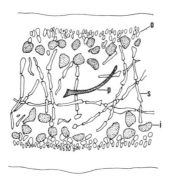

Fig. 9.6 *Weeksia fryeana* Setchell. Transection of mature blade revealing cells of outer cortex (o), inner cortex (i), primary axial system (p), and secondary medulla (s). × 125. (After Norris.)

sometimes 200 times over that of the original volume; the magnitude of enlargement in red algal cells is several times greater even than this amount (Dixon, 1971B).

The degree of cellular specialization that is achieved in the Phaeophycophyta is not expressed in members of the Rhodophycophyta; yet some cellular differentiation does exist in red algae. The usual distinction between small cortical cells with many chloroplasts and larger medullary cells with fewer or no chloroplasts is discernible in most larger forms. Even in such relatively simple forms as *Ceramium* (p. 550) the larger, less pigmented axial cells are separated by bands of small, heavily pigmented cells forming distinctive nodes at the junctures of the axial cells (Fig. 9.82). In some foliose forms such as *Weeksia* a transection through the blade (Fig. 9.6) reveals essentially four different cell types: outer cortical cells, inner cortical cells, primary axial cells, and cells of the secondary medulla, which is initiated from the inner cortex.

The anatomy of the medulla is variable among the red algae. It may be composed of uniformly filamentous cells, as in *Iridaea* and *Girgartina* (p. 523). It may consist of enlarged, more or less isodiametric cells, as in *Gracilaria* (p. 534). The medulla of *Callophyllis* is a mixture of large and small cells, the small cells forming filaments among the large cells. *Gelidium* (p. 498) and related genera are easily recognized by the presence of thick-walled, longitudinally coursing hyphae, usually positioned between the subcortex and the outer medulla. Transections through the axis of *Neoagardhiella* (p. 530) reveal an innermost zone of slender, longitudinally directed filaments surrounded by large, rounded medullary cells.

Vesicle, or secretory, cells are of sporadic distribution in all orders of Florideophycidae. Their nature and composition are variable, some being characterized by clear, highly refractive contents (Fig. 9.7), as those in some species of *Antithamnion* (p. 544); at maturity they lack a nucleus and chloroplasts. They may contain an elongate crystal, or raphide, its presence or absence apparently being species-specific. Vesicle cells in the family Bonnemaisoniaceae (p. 495) are especially prominent and usually contain high concentrations of bromine and/or iodine. By means of electron microprobe spectroscopy bromine and iodine were demonstrated (Wolk, 1968) to be much more heavily concentrated in vesicle cells of *Bonnemaisonia* (p. 496) than in neighboring cells. It was also recognized that the refractive inclusions so characteristic of these vesicle cells would not form if bromide were deleted from the culture medium.

Nevertheless, halogen metabolism is not restricted to these specialized secretory cells. Many genera in which halogen metabolism is well developed, as *Laurencia* (p.

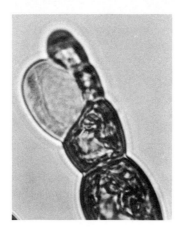

Fig. 9.7 Gland cell of *Antithamnion defectum* Kylin is distinctive because of its highly refractive contents. × 1020. (Courtesy of David N. Young.)

564), do not have secretory cells. Halogen metabolism in the red algae has been recently reviewed (Fenical, 1975). Although some uncertainty still exists as to the complete function of the halogen-containing metabolites so prevalent in red algae, evidence exists that the compounds show antibacterial properties and that halogen-containing ketones from *Asparagopsis* and *Bonnemaisonia* are toxic, thus possibly discouraging invertebrate predators. It is evident that allellopathic interactions between red algae and their environment are operating.

Attachment to the substrate for red algae is provided by a variety of devices. Single-celled rhizoids are produced in many members, the tip of the rhizoid flaring out into a broad discoid attachment or branching into an elaborate unicellular holdfast (Fig. 9.8*a*). Multicellular rhizoidal attachments (Fig. 9.8*b*) are also produced, as, for example, the holdfast of *Porphyra* (p. 477), *Nemalion* (p. 489), and *Membranoptera* (p. 555).

Two other morphological features deserve mention: hairs and tendrils. Colorless **hairs** are of common occurrence in many Florideophycidae, typically emerging from cortical cells. The novice observer can be easily confused by these hairs when searching for a carpogonium, as in *Nemalion*. The function of these hairs, which are often

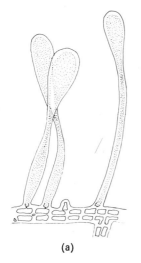
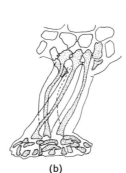

Fig. 9.8 Examples of holdfasts. (*a*) Unicellular rhizoids of *Polysiphonia rubrorhiza* Hollenb. (*b*) Multicellular holdfast of *Platysiphonia decumbens* Wynne. (*a*) × 50; (*b*) 123. [(*a*) after Hollenberg; (*b*) after Wynne.]

(a)   (b)

deciduous, is essentially unknown (West, 1972B; Dixon, 1973), although they may serve to protect the alga from excessive illumination, as is apparently the situation in some brown algae, or act as a nutrient-absorbing surface, as in *Acetabularia* (p. 219) and *Sphacelaria* (p. 298) (Hoek and Flinterman, 1968).

Tendrils and other hooklike appendages are specialized structures for attachment usually to other algae. Crozier-hooked laterals are produced by *Hypnea musciformis* (Wulfen) Lamour., *Bonnemaisonia hamifera* Hariot (Fig. 9.8*b*) and *Cystoclonium purpureum* (Huds.) Batt. var. *cirrhosum* Harv., invariably effecting an entangled habit for these algae.

Reproduction and Life Histories

It has previously been stressed that the lack of flagellate cells is one of the most distinctive traits in the division Rhodophycophyta. In all the red algae in which sexual reproduction is known to occur, it is invariably oogamous, involving nonmotile male elements termed **spermatia** and specialized female cells termed **carpogonia** in the Florideophycidae. Although sexual reproduction is absent in most members of the Bangiophycidae, the evidence for its occurrence in some species seems reasonably valid, even with conflicting facts still needing resolution (Dixon, 1973, p. 174).

The most prevalent pattern of alleged sexuality in the Bangiophycidae, occurring in *Bangia* and *Porphyra*, is the fusion of a small colorless spermatium with a slightly differentiated thallus cell, which has an emergent portion, the **prototrichogyne**, apparently providing a receptive surface. Other observations have reported a narrow tube or canal, which facilitates migration of the alleged sperm nucleus toward the undifferentiated thallus cell. The possibility has also been raised that this canal may be the result of a fungal infestation, since fungi are often observed to infect *Porphyra* (Kazama and Fuller, 1970).

The zygote in *Bangia* and *Porphyra* immediately undergoes successive divisions, producing a number of **carposporangia**, each of which forms a **carpospore**. An alternate, dissimilar phase in the life history of *Bangia* and *Porphyra* is produced by the germination of the carpospores. This entangled mass of branching, uniseriate filaments is termed the "*Conchocelis*" phase, since it was originally regarded as an independent alga. This reportedly diploid phase[3] in the life history typically produces either **monospores**, asexually recycling the "*Conchocelis*" phase, or larger "**conchospores**," which are released from so-called fertile cell rows. Evidence has been presented (Giraud and Magne, 1968) that these fertile cell rows are the site of meiosis in *Porphyra*.

A more complicated but less controversial pattern of sexual reproduction is encountered in members of Florideophycidae. The male gametes, called **spermatia**, are produced in reproductive structures called **spermatangia**, from which they are released by a rupturing mechanism of the wall due to the secretion of material at the base of the spermatangium (Scott and Dixon, 1973B). The liberated spermatium has

---

[3]Many conflicting reports exist concerning the nature of the life history in *Porphyra*, and these statements and variations in the pattern of reproduction are discussed in greater detail on p. 475.

a thin layer of mucilage but no rigid wall. Regeneration of spermatangia occurs in some red algae, as evidenced by the accumulated wall layers left from secondary spermatangia. Some old reports have indicated that the entire spermatangium is dehisced and functions as the male gamete. The spermatangia are always produced in abundance from special cortical cells (Fig. 9.28*d*) or from special branchlets, as in *Polysiphonia* (Fig. 9.94*a*).

The **carpogonium** typically is an elongate cell, relatively enlarged at the base and extending out distally into a process called a **trichogyne**, to which spermatia may become attached (Fig. 9.28*a*). Upon fertilization of the carpogonium by the spermatium, the carpogonium may either directly or indirectly produce a phase called the **carposporophyte**. If directly, the carpogonium proceeds to divide, and a mass of cells referred to as the carposporophyte results. The term **gonimoblast** is essentially synonymous with carposporophyte. The term **cystocarp** has a different connotation, since it usually includes female gametophytic tissue surrounding the carposporophyte.

In the Florideophycidae the carposporophyte may be entirely converted into fertile cells, the **carpospores**. Or the carposporophyte may consist of sterile filaments bearing carpospores; this situation may be referred to as gonimoblast filaments bearing carpospores. A variation that is also recognized is that the carpospores may be produced by a process involving meiosis, with the sporangium being termed a **carpotetrasporangium.** Such carpotetrasporangia are known for some members of the Helminthocladiaceae (p. 490) and Phyllophoraceae (p. 527).

Most commonly, rather than the fertilized carpogonium directly undergoing division to give rise to the carposporophyte, the initiation of the carposporophyte is indirectly brought about. The diploid nucleus resulting from the fusion of spermatium and carpogonium is transferred to another cell, from which the carposporophyte is generated. This other cell is termed the **auxiliary cell**, or generative auxiliary cell.

The transfer process may be a direct fusion between the carpogonium and the auxiliary cell by their physical proximity and contact; or it may be facilitated by the cutting off of a small cell from the carpogonium, which acts as an intermediary between carpogonium and auxiliary cell; or a long **connecting filament** may issue from the fertilized carpogonium, which effects the transfer of the diploid nucleus to the auxiliary cell. This filament, which is also termed an **ooblast**, may first make contact with a cell in the carpogonial branch before it proceeds out to locate the auxiliary cell or auxiliary cells. In this situation the length of filament running between carpogonium and first-contacted cell in the carpogonial branch is termed a **primary ooblast**; the length of filament running between that point and an auxiliary cell is termed a **secondary ooblast**.

The spatial relationship between the carpogonium and the auxiliary cell happens to be constant for a given species of Florideophycidae, and this fact serves as perhaps the most fundamental criterion in categorizing these algae into discrete taxonomic assemblages. This is discussed further on p. 465.

In the typical life-history pattern in the Florideophycidae the carpospore is released and germinates into a diploid phase termed the **tetrasporophyte**, which eventually becomes reproductively mature and produces **tetrasporangia**. These tetrasporan-

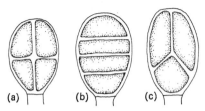

Fig. 9.9 Patterns of tetrasporangial division. (*a*) Cruciate or cruciform. (*b*) Zonate. (*c*) Tetrahedral.

gia constitute the site of meiosis, the four haploid nuclei being incorporated into four products termed **tetraspores**. The pattern of division of the tetrasporangium is fairly useful in taxonomic relationships of these algae, and three basic types of division are recognized: cruciate, or cruciform (Fig. 9.9*a*); zonate (Fig. 9.9*b*); and tetrahedral (Fig. 9.9*c*). In general, a given alga has only one pattern of division, but in some species, especially among members of the Nemalionales, as *Acrochaetium* (p. 481) and *Pseudogloiophloea* (p. 492), the pattern is quite flexible and thus not reliable as a taxonomic trait.

From the preceding discussion of the basic events in the typical Florideophycean alga, it is next possible to formulate a life-history pattern (Fig. 9.10). It is readily apparent that this pattern is distinctive among the algae in having three different phases: a gametophyte, a carposporophyte, and a tetrasporophyte. This circumstance has led some workers to refer to members of the subclass Florideophycidae as having triphasic life histories. Two of the phases, the gametophyte and the tetrasporophyte, are independent and free-living, whereas the carposporophyte is essentially parasitic in that it is attached and dependent on the gametophyte, from which it arises by sexual reproduction.

Alternative patterns are superimposed upon the basic scheme presented in Fig. 9.10. The basic distinction is whether the gametophyte and the tetrasporophyte are similar in appearance, that is, **isomorphic**, or dissimilar in appearance, that is, **heteromorphic**. Examples of the former include *Polysiphonia* (p. 561), *Antithamnion* (p. 544), *Callophyllis* (p. 507), and *Chondrus* (p. 525). Heteromorphic patterns, involving a very different morphological appearance between the gametophyte and sporophyte, are present in such genera as *Nemalion* (p. 489), *Pseudogloiophloea* (p. 492), *Liagora* (p. 490), *Gloiosiphonia* (p. 503), and *Gigartina*, namely, in *G. agardhii* S. et G. and *G. papillata* (C. Ag.) J. Ag. (p. 524).

Numerous variations exist in reference to the basic triphasic pattern. Although the tetrasporangium is the typical site of meiosis in the red algae (Austin, 1960;

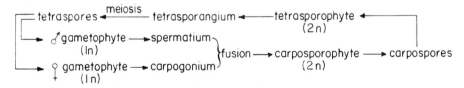

Fig. 9.10 The typical triphasic life history of most Florideophycidae, involving gametophyte(s), carposporophyte, and tetrasporophyte.

Tözün, 1974) and synaptonemal complexes have been detected (Kugrens and West, 1972A, 1972B) with the electron microscope, evidence indicates that in some species the tetrasporangia are apomeiotic or at least not regularly meiotic. Asexual cycles, consisting of tetrasporic phases being recycled through such apomeiotic tetrasporangia, are known for certain species of *Rhodochorton* (West, 1970A), *Antithamnion* (Sundene, 1962), and *Heterosiphonia* (West, 1970B) and the genera *Palmaria* and *Halosaccion* (Sparling, 1961; Guiry, 1974). An example of irregular or facultative meiosis is known for a species of *Lithophyllum* in the Corallinaceae (p. 508). Three different types of sporangia were noted (Suneson, 1950) in the material: quadrinucleate tetrasporangia, quadrinucleate **bisporangia**, and binucleate bisporangia. Evidence indicated that meiosis occurred in the first two types but not in the third type of sporangium. Reports of tetrasporangia and gametangia (carpogonia and spermatangia) borne on the same thallus are not uncommon (Dixon, 1961; West and

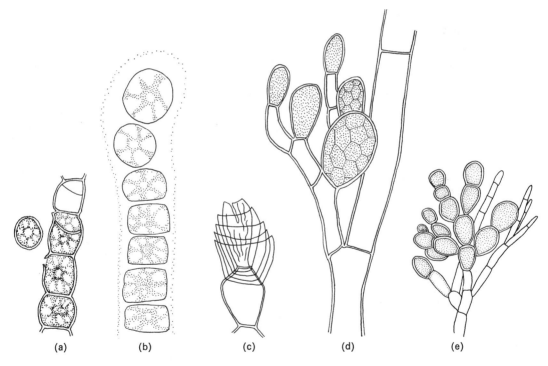

(a)   (b)   (c)   (d)   (e)

Fig. 9.11 Patterns of spore formation. (*a*) *Erythrotrichia carnea* (Dillw.) J. Ag. A differentiated cell, or monosporangium, undergoes an unequal cell division, the smaller cell being released as the monospore. (*b*) *Goniotrichum elegans* (Chauvin) Zanard. Monospores are formed by a direct transformation of vegetative cells. (*c*) *Acrochaetium virgatulum* (Harv.) J. Ag. Evidence of successive proliferation of sporangia by the remaining walls. (*d*) *Tiffaniella snyderae* (Farlow) Abbott. Polysporangia. (*e*) *Seirospora seirosperma* (Harv.) Dixon. Seirosporangia. (*a*) × 442; (*b*) × 911; (*c*) × 878; (*d*) × 230; (*e*) × 174. [(*a*) and (*e*) after Taylor; (*b*) and (*d*) after Smith; (*c*) after Boney.]

Norris, 1966; Edelstein and McLachlan, 1967B). Such anomalous situations are of sporadic occurrence throughout the orders of Florideophycidae.

A variety of other reproductive structures occurs throughout the Rhodophycophyta, one of the most widespread being monospores. A **monospore** is an asexual reproductive cell that is released as a single cell from a differentiated cell, or **monosporangium** (Fig. 9.11*a*), or from an undifferentiated cell (Fig. 9.11*b*). Monospore reproduction of both types is the sole method of reproduction for many members of the Bangiophycidae, and the pattern of spore formation is accepted as an important taxonomic trait (p. 469). Monosporangia are met with in many genera of the Nemalionales and may be the only method of reproduction for some species of *Acrochaetium* or serve as an accessory means of reproduction (as in *Pseudogloiophloea*, *Pikea*, and *Ceramium*). Monospores are always pigmented and larger than spermatia. Within the original wall of a monosporangium that has released its contents, further production of additional monosporangia may occur; this proliferation is evident (Fig. 9.11*c*) from the remains of the walls of the successive monosporangia (Boney, 1967A).

Bisporangia are of somewhat more restricted distribution but are in many Corallinaceae. They may be the only type of sporangium or be present intermingled with tetrasporangia. In most cases they are regarded as homologous to tetrasporangia, this fact being supported by the presence of two nuclei in each cell of the bisporangium. Uninucleate bisporangia are more apt to be explained by an apomeiotic division.

The application of the terms *polysporangia* and *parasporangia* has been somewhat confused. Both sporangial types contain more than four spores. The term **polysporangium** is restricted (Dixon, 1973) for sporangia homologous to tetrasporangia, and therefore they typically replace tetrasporangia on a given plant. The term **parasporangium** is retained for sporangia not homologous with tetrasporangia, and, therefore, meiosis does not occur within them. Yet in most cases the cytological data have not yet been obtained, which precludes a definite understanding at this time. Polysporangia (Fig. 9.11*d*) are known in *Pleonosporium* and *Spermothamnion* of the Ceramiaceae (p. 544) and in certain members of the Champiaceae (p. 537). Parasporangia occur in certain species of *Ceramium* and in *Plumaria elegans* (Bonnem.) Schm., both in the Ceramiaceae.

A rather special type of reproduction, the formation of catenate series of spores, or **seirospores**, is observed (Fig. 9.11*e*) in *Seirospora* of the Ceramiaceae. A direct transformation of vegetative cells in terminal series brings about the seirosporangia, which individually release their contents as seirospores (Dixon, 1971A). One other special type of reproduction is the formation of multicellular asexual structures termed **gemmae**, as happens in the freshwater red alga *Hildenbrandia rivularis* (Liebm.) J. Ag. (Nichols, 1965).

Classification

The division Rhodophycophyta is generally regarded to contain the single class Rhodophyceae, which in turn is divided into two subclasses, the Bangiophycidae and the Florideophycidae. The former is the smaller category, generally accepted as consisting of less advanced forms, and is comprised of a greater proportion of freshwater members than contained in the latter subclass. The freshwater members of the Florideophycidae are almost entirely restricted to the order Nemalionales, which is usually conceded to be the simplest order of Florideophycidae.

The two subclasses can be distinguished by the following characteristics:

Bangiophycidae—
1. Cells are always uninucleate.
2. A single stellate chloroplast in an axile position is the typical arrangement (a few exceptions are known, such as *Goniotrichopsis* and *Rhodospora* with many small chloroplasts per cell).
3. Intercalary cell divisions are generally present (*Rhodochaete* is an exception, having apical growth).
4. Unicellular and multicellular forms occur.
5. Pit connections are almost entirely absent (the "*Conchocelis*" phase of *Porphyra* is an exception).
6. Sexual reproduction is usually absent; it is probably present in some species of *Porphyra* and possibly in *Bangia* but remains controversial (see p. 475).

Florideophycidae—
1. Cells are multinucleate in most species, except for apical or reproductive cells.
2. Several to many small, discoid chloroplasts per cell, peripherally located, is the most common pattern; a single stellate chloroplast does occur in some Nemalionales. In some situations the initially single chloroplast becomes fragmented as the cell enlarges and ages.
3. Apical cell divisions are most common (intercalary divisions may be encountered in Corallinaceae, and some Delesseriaceae).
4. Only multicellular forms occur.
5. Prominent pit connections are present.
6. Sexual reproduction is widespread, although accessory asexual reproductive patterns are also known.

The basis of the currently accepted system of classification of the red algae, particularly the Florideophycidae, is credited to Schmitz (1883) and his earlier important contribution of the discovery of the auxiliary cell (p. 461). Kylin (1932) subsequently refined the system, and even though critical new information on life histories has necessitated fresh interpretations, the taxonomic scheme as presented in Kylin's monumental treatise (1956) still stands as the standard method of delineating the orders. Basically, the generative auxiliary cell is employed as the primary criterion to distinguish the orders of Florideophycidae (Papenfuss, 1957): its presence or absence, the manner in which it is formed and its location, and the time of its formation (before or after fertilization of the carpogonium). The following key serves to separate the orders of the Rhodophycophyta:

1. Pit connections usually absent; cell divisions intercalary (except for *Rhodochaete*); sexual reproduction mostly absent but when present, involving spermatia and little modified thallus cells, which undergo division directly to form spores 2

1. Pit connections prominent; cell divisions apical; sexual reproduction usually well developed, involving spermatia and specialized female cells termed carpogonia 6
  2. Unicellular forms, sometimes forming loose aggregations or colonies ....Porphyridiales (p. 467)
  2. Multicellular forms (filamentous, discoid, tubular, or foliose) 3
3. Thalli filamentous, uniseriate, branched, with apical growth ....Rhodochaetales
3. Thalli filamentous or otherwise, with growth by means of intercalary divisions 4
  4. Thallus filamentous, becoming parenchymatous (larger central cells surrounded by one or more layers of small cortical cells) ....Compsopogonales (p. 470)
  4. Thalli filamentous or otherwise; if becoming parenchymatous, not as above 5
5. Thalli filamentous, branched or unbranched; asexual reproduction only, by monospores, not involving special cell division; sexual reproduction absent ....Goniotrichales (p. 469)
5. Thalli filamentous, discoid, foliose, saccate, or parenchymatous; asexual reproduction by monospores produced from narrow, branching filaments, derived from vegetative cells, which form a reticulate pattern in the intercellular matrix around the vegetative cells (Boldiaceae) or from entire thallus cells (Bangiaceae) or by unequal cleavage of a thallus cell (Erythropeltidaceae); sexual reproduction present in some genera.................................Bangiales (p. 470)
  6. Carposporophyte developed directly from the fertilized carpogonium or from the hypogynous cell (the cell subtending the carpogonium) or indirectly from the cell resulting from the fusion of the carpogonium and its hypogynous cell ....Nemalionales (p. 480)
  6. Carpospophyte not developed from carpogonium but from a generative auxiliary cell (not the hypogynous cell)............................7
7. Generative auxiliary cell produced only *after* fertilization; produced from supporting cell ....Ceramiales (p. 542)
7. Generative auxiliary cell present (though not necessarily recognizable) prior to fertilization 8
  8. Generative auxiliary cell an intercalary vegetative cell, part of an ordinary vegetative filament ....Gigartinales (p. 520)
  8. Generative auxiliary cell not an intercalary vegetative cell 9
9. Generative auxiliary cell occurring in an accessory (i.e., nonvegetative) filament, either borne on the supporting cell (procarpial) of the carpogonial branch or remote from the carpogonial branch (nonprocarpial) ....Cryptonemiales (p. 499)
9. Generative auxiliary cell the terminal cell of a two-celled filament borne on the supporting cell of the carpogonial branch (always procarpial) ....Rhodymeniales (p. 536)

It is immediately obvious from the distinctions above that these criteria are difficult to use by anyone other than a specialist. Even experts may disagree upon the interpretation of what is happening in a given postfertilization stage being observed under the microscope. Often one is faced with the situation that a particular collection of an alga being examined will be lacking female plants entirely or it will have female plants lacking the appropriate stages for the determination of the critical stages. For many red algae auxiliary cells have not yet been reported for various reasons (Dawson, 1966), and thus it is by inference or by the employment of alternative criteria that one is able to place the alga into an appropriate genus, family, or order. Since the auxiliary cell, upon receipt of the zygote nucleus, initiates the carposporophytic generation in the triphasic life history (p. 462) of red algae (Papenfuss, 1957), there seems to exist reasonable justification to continue to attach much weight to its nature in the broad categorization of the Florideophycidae. Selected representatives of some of the orders of Bangiophycidae and Florideophycidae are presented in the following account.

SUBCLASS BANGIOPHYCIDAE

Order 1. Porphyridiales

This order of unicellular red algae contains about eight genera, which may be found as solitary cells or massed together into irregular colonies held in mucilage. One of the generic distinctions (Kylin, 1956) is the shape of the chloroplasts, stellate configurations (Fig. 9.1) being the most common type (*Porphyridium*, *Chroothece*, and *Rhodella*) but other configurations being possible: numerous discoid chloroplasts in *Rhodospora* (Geitler, 1955) and a cup-shaped chloroplast with three to seven lobes encircling a large pyrenoid in *Rhodosorus* (Fig. 9.12) (Giraud, 1962). The nature of the gelatinous layer is also useful, an irregular, watery envelope being present in *Porphyridium* and a firmer, stratified layer surrounding the cells of *Chroothece*. The

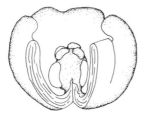

Fig. 9.12 *Rhodosorus marinus* Geitler. Representation of lobed chloroplast and emergent central pyrenoid. × 8500. (After Giraud.)

recently established genus *Rhodella* (Evans, 1970) is closely related to *Porphyridium* in many respects but has a much branched chloroplast, the arms of which may appear as unconnected lobes upon sectioning of a cell. Also, the pyrenoid is connected to the arms of the chloroplast by narrow isthmuses of the chloroplast rather than occupying a central position as in *Porphyridium*.

Motility of a gliding or amoeboid pattern is possible in these unicellular algae, as in *Petrovanella* and *Porphyridium* (Pringsheim, 1968B; Lin, et al., 1975). The cells may have a pronounced amoeboid shape and be larger than other cells in the population (Fig. 9.13*b*). Several cells may be advanced together in a positive phototactic response (Fig. 9.13*c*). The ultrastructure of such motile cells has been shown (Lin, et al., 1975) to be morphologically polarized, a large fibrillar vesicle being located at the posterior end and associated with the production of a mucilage trail. *Porphyridium* is the most

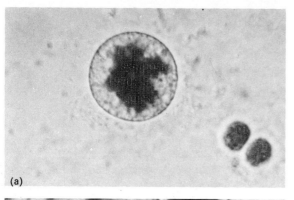

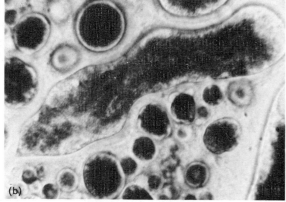

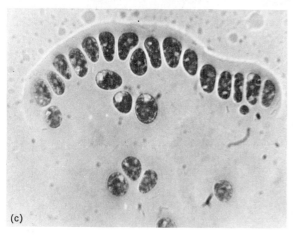

Fig. 9.13 *Porphyridium*. (*a*) *P. sordium* Geitler. The large axial, stellate chloroplast fills most of the protoplast. (*b*) *P. purpureum* (Bory) Drew and Ross. Large amoeboid cell, surrounded by spherical vegetative cells of typical size. (*c*) *P. purpureum*. A group of phototactic wandering cells showing a strongly phototactic response. (*a*) × 1189; (*b*) × 663; (*c*) × 615. (Courtesy of Dr. Franklyn Ott.)

commonly collected member. Other genera, such as *Rhodosorus* (Ott, 1967; West, 1969B) and *Chroothece* (F. Ott, personal communication), are of rare occurrence in the United States.

The proposal has been made (Feldmann, 1955; Dixon, 1973) to merge this order into the filamentous order Goniotrichales on the basis of the resemblance of the unicellular genus *Chroothece* to stages of the filamentous genus *Asterocytis* (p. 470). The latter genus becomes disorganized into a unicellular condition when grown under low salinity (Lewin and Robertson, 1971). The present treatment continues to recognize unicellular Bangiophycidae as a distinct order. Only the genus *Porphyridium* is discussed in detail in this account.

PORPHYRIDIUM Näg.   A diverse spectrum of habitats is occupied by the genus *Porphyridium* (Gr. *porphyra*, purple, + *idion*, similarity): freshwater, brackish water, and marine water as well as the surface of moist soils, or pots in greenhouses, upon which growths may form noticeable reddish coatings. A single prominent stellate chloroplast is present in each cell (Fig. 9.13), with a central pyrenoid region within the chloroplast (Fig. 9.1). The ultrastructure of *Porphyridium* has been reported in great detail (Gantt and Conti, 1965, 1966; Gantt, et al., 1968; Chapman and Lang, 1973), and it has served as a useful tool in investigations of phycobilisomes (p. 453) (Gantt and Lipschultz, 1972; Gantt, 1975) and of polysaccharide production (Ramus, 1972B; Ramus and Groves, 1972).

Cells of *Porphyridium* are essentially without walls, in that a skeletal or microfibrillar component is lacking. Amorphous mucilaginous material is constantly being excreted from the cell, forming an encapsulating halo around the cell. This material is a water-soluble, polyanionic sulfated polysaccharide (Ramus, 1972), which is produced in copious amounts during stationary phase of growth. It is solubilized as it moves away from the cell into the surrounding medium, and it is constantly being replenished at the cell surface by dictyosome-mediated processes (Ramus and Groves, 1972). The growth of *Porphyridium* has also been investigated (Pringsheim and Pringsheim, 1949; Jones, et al., 1963; Pringsheim, 1968B).

The taxonomy and nomenclature of *Porphyridium* have been the subject of several recent papers (Drew and Ross, 1964; Sommerfeld and Nichols, 1970; Ott, 1972). According to Ott (1972), color is the only valid criterion for specific separation in this genus. Environmental changes may cause slight color variations. But the significant color variations hold up even when the various species are grown under monochromatic illumination. The correct name of the type species is *P. purpureum* (Bory) Drew et Ross, *P. cruentum* (S. F. Gray) Näg. being a synonym.

Order 2. Goniotrichales

The members of this order of Bangiophycidae tend to be minute epiphytes, usually on larger seaweeds and escaping notice. They often appear as contaminants in cultures when one is attempting to isolate the host alga. Marine and ubiquitous in occurrence, they are characterized chiefly by their mode of asexual reproduction. Monospores are formed from entire vegetative cells and released singly by the dissolution or opening of the gelatinous matrix around the parental cell. This has been referred to as Type 2 spore formation (Fig. 9.11*b*) by Drew (1956) and is the only pattern present in this order. It is distinct from Type 1 spore formation (Fig. 9.11*a*), occurring in the Erythropeltidaceae (p. 471), which involves the release of monospores from differentiated monosporangia. It is also different from the formation of many spores by successive divisions of the original cell, the pattern characteristic of the Bangiaceae (p. 474) and referred to as Type 3 spore formation. Drew's designation

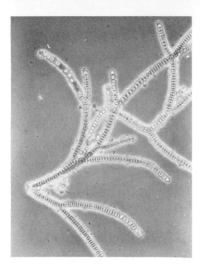

Fig. 9.14 *Goniotrichum elegans.* Habit. × 280.

of patterns of spore formation has been subjected to some modification (Conway, 1964; Richardson, 1972).

GONIOTRICHUM Kütz. Thalli of *Goniotrichum* (Gr. *gonia*, an angle, + Gr. *trichion*, a hair) are small, irregularly branched filaments (Fig. 9.14), the cells embedded in a well-developed mucilaginous matrix. A single stellate chloroplast is located in an axile position within each cell and is responsible for the reddish-purple color of the tufts. The filaments are rarely mutiseriate at their ends. Monospores are formed by the direct transformation of vegetative cells and are often sloughed off from the tips of the filaments.

ASTEROCYTIS Gobi Morphologically, the small thalli of *Asterocytis* (Gr. *aster*, a star, + *kutis*, a hollow) closely resemble those of *Goniotrichum*. The most conspicuous difference is the blue-green pigmentation of *Asterocytis*, phycocyanin being the dominant pigment. The alteration of this genus from the typical branched, filamentous habit to a unicellular condition due to reduced salinity or other unfavorable environmental factors has been noted (Lewin and Robertson, 1971).

Order 3. Compsopogonales

The freshwater genus *Compsopogon* (Gr. *kompsos*, elegant, + Gr. *pogon*, beard), occurring in tropical and subtropical regions (Krishnamurthy, 1962), is assigned to its own order, largely on the basis of its distinctive development. Uniseriate filaments are developed by diffuse growth, they may branch, and eventually the uniseriate filaments become corticated with small cells surrounding the central axial row. This mature structure (Fig. 9.15) is not met with in other Bangiophycidae.

Young cells contain a single parietal chloroplast, which later becomes fragmented into many small, discoid chloroplasts. Although Type 1 spore formation, involving the unequal division of the parental wall by an oblique or curving wall, is the usual method of monospore production, some variations have been noted (Nichols, 1964A) in cultured material.

470

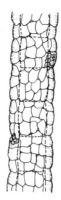

Fig. 9.15 *Compsopogon coeruleus* (Balbis) Mont. Corticated axis. × 100. (After Nichols.)

Order 4. Bangiales

This group comprises the largest order of the subclass Bangiophycidae and contains three families, which are distinguished as follows:

Erythropeltidaceae: monospores formed by a curving wall laid down to cut off a single small cell from the larger parental cell; holdfast usually a small pad or basal cushion, lacking rhizoids.

Boldiaceae: monospores formed from uniseriate, branching filaments that form a network in the intercellular matrix external to and among groups of vegetative cells, from which the filaments are derived.

Bangiaceae: monospores formed from entire vegetative cells, never by the unequal division of the original cell; holdfast rhizoidal.

Family 1. Erythropeltidaceae

The habit of genera contained in this family is quite variable: creeping filaments, becoming discoid or a polystromatic cushion (*Erythrocladia*); erect filaments (branched or unbranched, uniseriate or multiseriate), attached either by a single cell or by a multicellular disc (*Erythrotrichia*); or blades arising from a cushion-like base (*Smithora*). Another blade-forming genus, *Porphyropsis*, which will not be discussed below, has an unusual ontogeny, recalling that of some species of *Monostroma* (p. 169). A prostrate cushion forms a blister, which opens up into usually several blades (Murray, et al., 1972). Three representative genera are discussed below.

ERYTHROCLADIA Rosenv.   The small epiphytes of *Erythrocladia* (Gr. *erythros*, red, + Gr. *klados*, a branch) are common on larger seaweeds but inconspicuous. The thallus consists of creeping filaments, radiating outward from the point of original spore germination and forming either a closed disc (Fig. 9.16) or a more diffuse network. Monospores are formed by the older cells rather than by marginal cells, where growth is occurring (Heerebout, 1968). Attention was focused upon the extent of

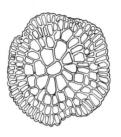

Fig. 9.16 *Erythrocladia subintegra* Rosenv. Surface view of disc. × 325. (After G. M. Smith).

variation possible in one species, *E. subintegra* Rosenv., in culturing studies by Nichols and Lissant (1967). They also pointed out the ability of monospores to go through a phase of amoeboid movement before settling down.

ERYTHROTRICHIA Aresch.   The many species of *Erythrotrichia* (Gr. *erythros*, red, + *trichion*, a hair) show considerable variation from one to the next, some being uniseriate and others being multiseriate, some being simple and others branched. Usually it is an erect system, arising from a basal disc or single cell of attachment. Chloroplasts are central and stellate in contrast to the parietal, plate-shaped ones in *Erythrocladia*. Monospores are cut off by unequal divisions of the parental cell (Fig. 9.11*a*). Sexual reproduction by fusion of two adjacent cells in the thallus has been reported (Heerebout, 1968); however, Dixon (1973) expressed the attitude that the evidence was not very convincing. Heerebout (1968) also reported the spontaneous occurrence of a "*Conchocelis*" phase in his cultures of *Erythrotrichia*, resembling that of *Bangia* and *Porphyra* (p. 477).

SMITHORA Hollenb.   The genus *Smithora* (named for *Gilbert Morgan Smith*, American phycologist) was established (Hollenberg, 1959) to receive the single species *Porphyra naiadum* Anderson, primarily on the basis of the manner of blade production. Clusters of blades are produced from a common cushion-like base (Fig. 9.17*a*). Differences in pigmentation and in methods of reproduction have also been referred to. So-called deciduous sori are produced along the margins. These are differentiated areas of the blade, from which masses of monospores are released. The differentiation and release of these spores have been investigated with the electron microscope (McBride and Cole, 1971). The sori easily detach and float away. Of less common occurrence is the production of apparent spermatangia, which are cut off singly from larger cells in portions of the blade that have become distromatic. Sexual reproduction, however, remains doubtful, although a "*Conchocelis*" phase, similar to that occurring in the Bangiaceae (p. 474), has been observed in culture (Richardson and Dixon, 1969).

An usual feature of the chloroplast in *Smithora* has been the observation (McBride and Cole, 1969) that the photosynthetic lamellae are locally associated into stacks, which is not known for other red algal chloroplasts. This association is still primitive and not extensive, most thylakoids remaining unassociated.

Smithora is fairly common on the Pacific coast of North America, ranging from Baja California, Mexico, to Alaska. In the field the purplish-red blades are observed invariably to be epiphytes on the seagrasses *Zostera* and *Phyllospadix*. Blades on *Zostera*, growing in sheltered bays, may be 6 cm wide or more, whereas those on surf-tolerant *Phyllospadix* are usually 1–2 cm wide. Bidirectional transfer of labeled products between *Smithora* and its seagrass hosts has been demonstrated (Harlin, 1973B). The lack of a penetrating holdfast indicates that such a structure is not a requisite for exchange of materials. Although blades of *Smithora* would appear to be obligate epiphytes, their growth on artificial substrate under field conditions has also been reported (Harlin, 1973A).

Heerebout (1968) has excluded this genus from the Erythropeltidaceae because it is so different from the other genera in the family; yet he did not suggest where it should be placed. Its pad-like base, which is persistent and able to regenerate new crops of blades, lacks rhizoids and shows greater affinity with the Erythropeltidaceae than the Bangiaceae (Hollenberg, 1959).

Family 2. Boldiaceae

This monotypic family was established (Herndon, 1964) to include the single freshwater genus *Boldia*, which is distinctive both for its saccate (Fig. 9.17*b*) or tubular shape and also for the manner in which monospores are produced. There is no evidence of sexual reproduction or meiosis.

BOLDIA Herndon   The ontogenic development of *Boldia* (named for *Harold C. Bold*, American phycologist-morphologist) is unusual in several respects. A monospore germinates to give rise to a prostrate, monostromatic, discoid growth,

(a)   (b)

Fig. 9.17   (*a*) *Smithora naiadum* (Anderson) Hollenb. Blades epiphytic on seagrass. (*b*) *Boldia erythrosiphon* Herndon. (*a*) × 0.14; (*b*) × 0.6. (*a*) courtesy of Dr. Marilyn M. Harlin.

resembling *Erythrocladia* (p. 471). By a series of anticlinal and periclinal divisions a cushion-like mound of cells is developed (Nichols, 1964B). The outer layer of cells undergoes active division, giving rise to a hollow saccate portion at the distal end. The basal portion functions as the region of attachment by rhizoid formation of the cells in this portion. The thalli may reach 20 cm in height.

Other distinctive features include the manner in which monospores are produced. They are cut off superficially from uniseriate branching filaments that course through the gelationous matrix of the sac, having developed from vegetative cells composing the monostromatic sacs (Herndon, 1964). Irregularly lobed, ribbonlike chloroplasts, lacking pyrenoids, are located in the periphery of each cell.

These brownish-red to olivaceous thalli, occurring in streams, have been recorded from scattered stations in the eastern half of the United States (Whitford and Schumacher, 1969). A second species *B. angustata* Deason et Nichols, has been described (Deason and Nichols, 1970), in which the erect saccate thalli are developed from apically growing, uniseriate filaments derived from a prostrate disc.

Family 3. Bangiaceae

This family, containing only three genera (*Bangia*, *Porphyra*, and *Porphyrella*), is distinguished from the Erythropeltidaceae in that spore formation involves successive walls being laid down, more or less at right angles to those of the preceding division. The original cell that undergoes this process does not increase significantly in size. This has been designed Type 3 spore formation (Drew, 1956) and results in spores of two size categories. One size category consists of relatively large, pigmented spores, either 4, 8, or 16 per packet (Fig. 9.20*d*, *f*), depending on the species, and the other consists of relatively small, colorless spores, 16, 32, 64, 128, or 256 per packet (Fig. 9.20*e*, *f*), again depending on the species. The term *carpospore* has been applied to the former, the implication being that they are the products of division of the zygote, whereas the latter have been referred to as spermatia, since they have been considered to be the male gametes. Some authors (Kunieda, 1939; Kornmann, 1961C) have reported that a sexual process does occur, involving the fusion of the spermatia with a cell of the thallus, which has produced a small protuberance, or **prototrichogyne**[4] (Tseng and Chang, 1955). Other workers (Krishnamurthy, 1959; Richardson and Dixon, 1968) have seen no evidence of sexual reproduction in this family, which has led Conway (1964) to suggest the terms α *spores* for the larger "carpospores" and β *spores* for the smaller "spermatia," since these terms would not connote sexual reproduction.

---

[4]An unusual process of reproduction, which superficially resembles sexual reproduction because of the presence of very conspicuous trichogyne-like processes, was described (Conway and Cole, 1973) in a species of *Porphyra*. Thallus cells divided longitudinally into four carpogonia-like cells with projections extending out from the matrix. These cells were shed and divided into a group of small spores, which next produce a "*Conchocelis*" stage with conchosporangia. The various stages were observed to have $n = 3$, with thus no evidence of sexuality.

An important breakthrough in our understanding of the life history of this family was the recognition by the English phycologist Kathleen Drew-Baker (Drew, 1949) that the small filamentous alga *Conchocelis rosea* Batt., which usually occurs as pinkish growths on shells, is an alternate stage in the life history of *Porphyra*. This alternation between large blades of *Porphyra* and its inconspicuous "*Conchocelis*" phase (Fig. 9.19*b*) thus is a life-history pattern involving dissimilar phases. The complete cycle in *Porphyra* was later reported (Hollenberg, 1958), and a similar pattern in *Bangia* has also been described (Richardson and Dixon, 1968).

Controversy has long existed concerning the relationship between *Bangia* and *Porphyra* and their "*Conchocelis*" stages. It is now accepted that for a number of species of *Porphyra* a difference in the ploidy level exists between the blade and the "*Conchocelis*" stage, the blade being haploid and the carpospores and "*Conchocelis*" stage being diploid (Yabu, 1969A, 1972; Kito, et al., 1971). A fusion of the male and female nucleus within the carpogonium and the production of carpospores without meiosis has been observed (Yabu, 1969B) in *P. tenera* Kjellm., the species most commonly used for nori (p. 28). Meiosis has been shown (Giraud and Magne, 1968) to occur in the production of so-called fertile cell rows, or **conchosporangia**, in the "*Conchocelis*" stage (Fig. 9.19*b*), which means that the large conchospores released from these fertile cell rows are haploid, germinating into the blade stage; smaller monospores produced by the filaments are, on the other hand, diploid, recycling the "*Conchocelis*" stage.

Yet some species apparently lack sexuality, even though their cycles have both blade and "*Conchocelis*" stages. Of eight species examined cytologically from British Colombia, Cole and Conway (1974) determined that six have a haploid "*Conchocelis*" stage and only two have a diploid "*Conchocelis*" stage. In addition, not all species of *Porphyra* have this filamentous alternate stage; of a total of 15 species examined by Cole and Conway (1974), 13 produced "*Conchocelis*" stages, the other 2 species producing directly by monospores. **Monospores** of Drew's Type 2[5] (p. 469), which are essentially vegetative cells that become disassociated from the thallus, are an accessory means of recycling the phase, being produced under certain environmental conditions, for example in *Bangia* (Dixon and Richardson, 1970), when α spores and β spores are not being produced. It has also been recognized (Dixon and Richardson, 1969) that the basal parts of the blade may perennate, generating a new blade, or may form monospores. One other possibility is the direct production of blades from the "*Conchocelis*" stage (Miura, 1961).

Although the alternation between the blade and the "*Conchocelis*" stage may or may not be linked to a cytological alternation, it has been recognized that for many

---

[5]A modification of the definition of Drew's Type II spores in the Bangiaceae has been proposed (Richardson, 1972). The system was based on the classification on cell fate and cell function. Germination is either in the unipolar or bipolar manner. With photoperiods of greater than 12 hours of light, carpospores from the macroscopic thallus and monospores from the *Conchocelis* phase both give rise to the *Conchocelis* phase. With photoperiods of less than 12 hours of light, monospores and carpospores from the macroscopic thallus and conchospores from the *Conchocelis* phase give rise to the macroscopic thallus.

species of *Porphyra* and for *Bangia* a photoperiodic control is in operation, regulating the type of spores produced and the phase of the life history brought about. In general, the fertile cell rows (Fig. 9.19*b*) of the *"Conchocelis"* stage are induced by short days, and the conchospores that are released germinate into the macroscopic thallus of *Porphyra* and *Bangia*. This photoinduction of conchosporangial production in *Porphyra*, which was first noted by Kurogi (1959), was later shown (Dring, 1967; Rentschler, 1967) to be a phytochrome-mediated process. The participation of this pigment system, which has been known to occur in a variety of plants in which green pigments (chlorophyll *a* and *b*) predominate, was supported by light-break experiments with red and far-red light (Rentschler, 1967). The critical daylength for the production of the conchospores in *Porphyra* was 10 hours; longer periods of daylight inhibited their formation.   WHICH SPECIES WINTER TOO?!.

A similar phenomenon of photoperiodic control was demonstrated in *Bangia* (Richardson and Dixon, 1968; Richardson, 1970; Dixon and Richardson, 1970). The reciprocal photoperiodic conversion of the *Porphyra* or *Bangia* stage to the filamentous stage is triggered by relatively long daylengths. Typically, photoregimes of more than 12 hours of continuous light per 24 hours will cause the carpospores to be released and germinate into the *"Conchocelis"* stage. A different effect of daylength was observed (Suto, 1972) on the growth of *Porphyra umbilicalis* (L.) J. Ag. With a 13 : $\overline{11}$ photoregime dioecious blades were produced; with an 11 : $\overline{13}$ photoregime monoecious blades, with small male patches, were produced; with a 9 : $\overline{15}$ photoregime monoecious blades were produced, with half of the blade male and the other half female.

An interpretation other than a photoperiodic mechanism was given for *Porphyra miniata* (C. Ag.) C. Ag. It was observed (Chen, et al., 1970) that in this species carpospores germinate into the filamentous stage, which is apparently perennial and can propagate itself indefinitely by monospores. Fertile cell rows were formed at relatively higher temperatures and low light intensities *and under long photoperiods* (i.e., 10–16 hours of light). The critical event was the release of the conchospores, which required a low temperature of 5°C.

BANGIA Lyngb.   Thalli of *Bangia* (named after *N. H. Bang*, a Danish botanist) are filiform, unbranched cylinders with the cells embedded in a firm gelatinous sheath. The initially uniseriate filament becomes multiseriate with age (Fig. 9.18) and thus parenchymatous in construction. The cells each contain a large stellate chloroplast with a central pyrenoid. The masses of thalli often occur in dense purplish-black tufts on rock or wood substrates, and, in reference to the marine species, *B. fuscopurpurea* (Dillw.) Lyngb., may become dried out during periods of low tide on account of its typically upper littoral habitat.

A second species, *B. atropurpurea* (Roth) C. Ag., occurring in freshwater habitats (Belcher, 1960), may not be distinct from the marine species. Geesink (1973) was able to adapt gradually a marine isolate to an almost freshwater medium and likewise adapt a freshwater isolate to seawater, this adaptation using monospores. He concluded that *B. fuscopurpurea* and *B. atropurpurea* were conspecific, the latter name

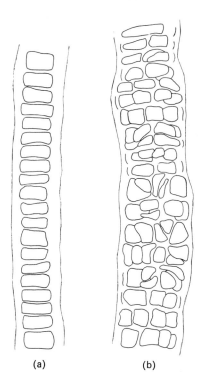

Fig. 9.18 *Bangia   fuscopurpurea*   (Dillw.) Lyngb. (*a*) Uniseriate axis; (*b*) Multiseriate axis. × 330.

(a)   (b)

having priority. His freshwater isolate did not produce a "*Conchoselis*" stage, whereas the marine isolate did.

Although photoperiod has been demonstrated (Richardson, 1970) to regulate the conversion of *Bangia* to the "*Conchocelis*" stage (p. 476), other evidence (Sommerfeld and Nichols, 1973) showed that the type of spore produced by the *Bangia* phase was primarily dependent on temperature. At higher temperatures only monospores were produced, whereas at lower temperatures both monospores and carpospores were produced. At this lower temperature (9°C) photoperiod played a role in sporogenesis. Long daylengths induced carpospores (which developed into the "*Conchocelis*" stage) and short daylengths induced monospores (which developed into the *Bangia* phase).

PORPHYRA C. Ag.   Blades of *Porphyra* (Gr. *porphyra*, purple) are frequently collected on rocky shores from polar to tropical seas. Most species are of intertidal occurrence, but some inhabit the sublittoral zone. Blades (Fig. 9.19*a*) may reach 75 cm in some species, and both **epiphytic** and **saxicolous** types are known. The base of the blade is attached to the substratum by numerous rhizoidal cells.

A large number of species have been described in this genus, the North Pacific seemingly being the region of their greatest abundance and diversity (Fukuhara, 1968; Krishnamurthy, 1972; Kurogi, 1972; Conway, et al., 1976). The genus is sub-

(a)   (b)

Fig. 9.19 *Porphyra kanakaensis* Mumford. (*a*) Foliose thallus of a monoecious species. (*b*) "*Conchocelis*" stage bearing conchosporangial branches. (*a*) × 0.18; (*b*) × 120. [(*a*) after Mumford; permission of Syesis; (*b*) courtesy of Dr. Thomas F. Mumford, Jr.]

divided into three groups, based upon blade thickness and chloroplast number: species with monostromatic blades containing a single chloroplast per cell (Fig. 9.20*a*), species with monostromatic blades containing two chloroplasts per cell (Fig. 9.20*b*) and species with distromatic blades containing a single chloroplast per cell (Fig. 9.20*c*). Other taxonomic criteria include the shape of the blade and the distribution of reproductive cells: monoecious, dioecious, or a combination of monoecious and dioecious blades in the population. Some species are protandrous, the male regions being early sloughed away, leaving the blade exclusively female.

One of the more reliable characteristics, but a somewhat difficult one to use, is the pattern of division of the sexual reproductive cells (Kurogi, 1972). The original vegetative cell that undergoes successive division to form spermatangia and the thallus cell that acts as a carpogonium and subsequently divides into carpospores can be thought of as a cube that becomes divided along its three axes. By a careful examination of male and female regions in surface view (Fig. 9.20*f*) and cross-sectional view (Fig. 9.20*d, e*), one can formulate the pattern of divisions, this pattern generally being a species-specific trait.

The growing season of any particular species is apt to be restricted. A summer group of species and a winter group of species have been recognized (Kurogi, 1972), the summer group being more common in boreal regions. The blade may have an early period of monospore production, which is followed by sexual reproduction. The "*Conchocelis*" stage (Fig. 9.19*b*) represents the phase of the alga present when the

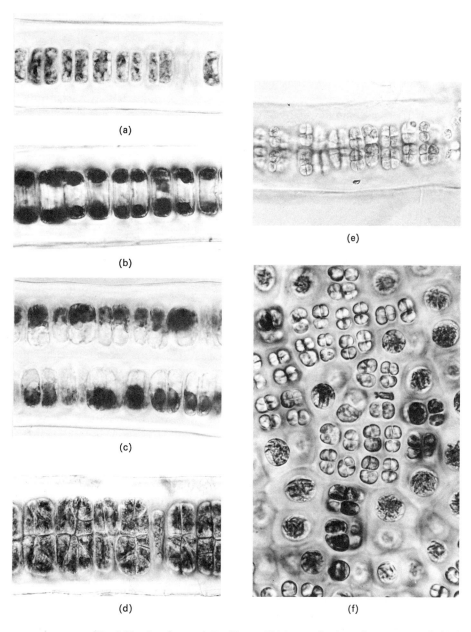

Fig. 9.20 *Porphyra.* (*a*) *P. abbottae* Krishnam. Section of monostromatic blade with one chloroplast per cell. (*b*) *P. kanakaensis.* Section of monostromatic blade with two chloroplasts per cell. (*c*) *P. schizophylla* Hollenb. Section of distromatic blade. (*d*) *P. brumalis* Mumford. Section of blade bearing packets of carpospores. (*e*) *P. brumalis.* Section of blade bearing packets of spermatangia. (*f*) *P. miniata* (C. Ag.) C. Ag. Surface view of blade bearing packets of carpospores and spermatangia mixed with vegetative cells. (*a*) × 461; (*b*) × 324; (*c*) × 378; (*d*) × 394; (*e*) × 332; (*f*) × 560. [(*a*) courtesy of Dr. T. F. Mumford, Jr.; (*b*), (*d*), and (*e*) after Mumford; permission of Syesis; (*c*) and (*f*) after Conway, Mumford, and Scagel; permission of Syesis.]

blade phase is absent. *"Conchocelis"* stages of different species are recognized to be distinctive in regard to color, rhythm of spore liberation, and response to environmental factors. The variation possible in single species of *Porphyra* grown in the laboratory may be extensive (Suto, 1972). A final point is that hybridization has been carried out among different species of *Porphyra* (Suto, 1963).

SUBCLASS FLORIDEOPHYCIDAE

Order 5. Nemalionales

Perhaps no other order of red algae has seen such dramatic changes made in our understanding of its life histories as the Nemalionales. Life history investigations on many of the genera have caused the classical interpretation of a haplobiontic life history as typical to be discarded. It is now recognized that no single life history can be applied in this order and that a broad spectrum of life histories is encountered. Some of these will be presented in the discussion of the individual genera.

The rather conservative approach is taken in the present account of including in this order several families that other contemporary workers would separate as distinct orders. Bonnemaisoniaceae was separated as an order by Feldmann (1952), on the basis of the heteromorphic life history (Feldmann and Feldmann, 1942), the first group of red algae in which such a pattern was discovered. The subsequent discovery of such a pattern in many other genera of not only the order Nemalionales but other orders has lessened the significance of the life-history pattern as an ordinal trait. Nonetheless, additional supporting evidence has recently been offered (Chihara and Yoshizaki, 1972B) for retaining the Bonnemaisoniales as a distinct group. Similarly, the Gelidiaceae has been distinguished as a separate order, but the merits of such a move are dubious (Dixon, 1973).

The general consensus is that the individual families within the order Nemalionales are fairly clearly delineated, whereas the entire assemblage is obviously heterogeneous. The one shared trait is the lack of an auxiliary cell (p. 461). Following fertilization, the carposporophyte is developed either directly from the carpogonium, as in *Acrochaetium* (p. 481), *Nemalion* (p. 489), and *Bonnemaisonia* (p. 496), or from the cell immediately below the carpogonium, as in *Pseudogloiophloea* (p. 492), or from some other cell, as in *Gelidium* (p. 498). A diversity of forms, ranging from small uniseriate, branched filaments to more massive, multiaxial forms, is expressed by the members of this order. Representative genera of seven families are presented below.

1. Uniaxial in development 2
1. Multiaxial in development 5
    2. Typically uniseriate branched thalli, lacking cortication ....Acrochaetiaceae
    2. Axes covered by loose or firm cortication 3
3. Cortication of loose consistency; freshwater ....Batrachospermaceae
3. Cortication of firm to solid consistency; marine 4

4. Cystocarps surrounded by definite pericarp; life history
usually comprised of dissimilar phases . . . . .Bonnemaisoniaceae
4. Cystocarps not surrounded by a definite pericarp but embedded
within the axis; life history comprised of similar phases   . . . . . . .Gelidiaceae
5. Cystocarp surrounded by a definite pericarp with an ostiole. . . .Chaetangiaceae
5. Cystocarp not enclosed by a pericarp but occurring among
the cortical filaments 6
6. Carpogonial branch with a variable number of cells, straight,
clearly modified from a cortical filament . . . . .Nemalionaceae
6. Carpogonial branch with a definite number of cells, curved,
accessory in nature. . . . . .Helminthocladiaceae

Family 1. Acrochaetiaceae

Algae of this family are usually small, branched filaments with apical growth and uncorticated, uniaxial construction. The organization tends to be heterotrichous, the prostrate system or the erect system possibly being more well developed. Attachment may be by means of a single cell or a consolidated disc. Epiphytic, endophytic, epizoic, endozoic, and saxicolous types are all known. The most common pattern of reproduction is simply by monospores, many species apparently having only this means of reproduction. Some species produce bisporangia, and others, such as *Rhodochorton concrescens* Drew, are known to produce only tetrasporangia, which seem to be apomeiotic (West, 1970A). Sexual reproduction is also known, and this topic will be discussed in greater detail in reference to particular genera below. It is primarily a family of marine distribution, but some freshwater species also occur.

Unfortunately, the status of the genera and criteria used in separating them in this family are at present not agreed upon. A more traditional attitude (Papenfuss, 1945) employs the number and shape of the chloroplasts as the primary means of distinguishing the genera. Cells contain a few to many small, discoid chloroplasts in *Rhodochorton*; cells contain one parietal or laminate chloroplast in *Acrochaetium*, and cells contain one or more spiral chromatophores in *Audouinella*. A different view (Smith, 1944) was to restrict species with monosporangia to *Acrochaetium* and species with tetrasporangia to *Rhodochorton*; some species produce both. A recent treatment (Woelkerling, 1971, 1973) places all sexually reproducing species of *Rhodochorton*, *Acrochaetium*, and other related genera into one large genus, *Audouinella*; nonsexually reproducing species were assigned to *Colaconema*, which is equivalent to a form genus of Fungi Imperfecti. The shape of the chloroplast was regarded as too variable to be used reliably in specific and/or generic separation, as could be shown in cultures of a single species (Fig. 9.21). Number of pyrenoids, sporangial dimensions, and cell dimensions were regarded as the most reliable criteria. Despite these reservations, the present authors have elected to retain *Acrochaetium* and *Rhodochorton* as distinct from *Audouinella* on the basis of chloroplast morphology.

ACROCHAETIUM Naeg. *Acrochaetium* (Gr. *akros*, topmost + *chaite*, bristle) has a widespread distribution but is usually not noticed due to the small size of the

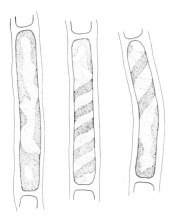

Fig. 9.21 *Acrochaetium pectinatum* (Kylin) Hamel. Variations in chloroplast shape in vegetative cells. × 1400. (After Woelkerling, as *Audouinella pectinatum.*)

tufts of filaments or their endophytic or endozoic nature. Cells usually have a single parietal, platelike chloroplast with a pyrenoid (Fig. 9.22*a*); monosporangia are randomly or regularly arranged on the filaments. Hairs (Fig. 9.22*b*) may also be present in some species.

The life history of *Acrochaetium pectinatum* (Kylin) Hamel has been investigated (West, 1968) in culture. With a photoregime of 12 hours or more of light per 24 hours, there was an exclusive production of monosporangia, which recycled the phase. With a shorter than 12-hour daylength, tetrasporangia were produced as well as monosporangia. The tetraspores developed into sexual plants, which had a much smaller stature than the tetrasporic plants. These plants were initially unisexual, but as female

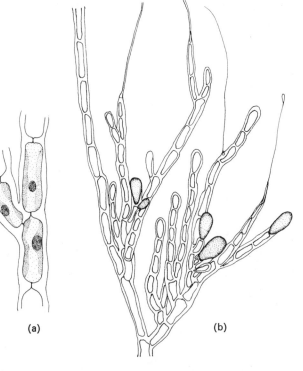

(a)   (b)

Fig. 9.22 (*a*) *Acrochaetium dasyae* Collins. Chloroplast shape in vegetative cells. (*b*) *A. pectinatum*. Upper portion of thallus bearing monosporangia and hairs. (*a*) × 540; (*b*) × 525. [(*a*) after Woelkerling, as *Audouinella dasyae*; (*b*) after Woelkerling, as *Audouinella pectinatum.*]

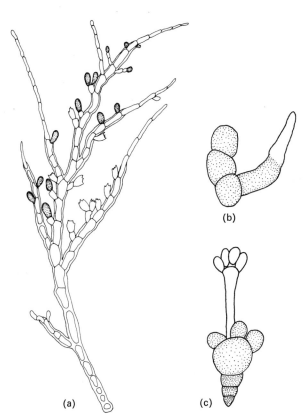

(b)

(c)

(a)

Fig. 9.23 *Kylinia.* (*a*) *K. alariae* (Jonsson) Kyl. (*a*) Habit of thallus with monosporangia and single cell of attachment. (*b*) *K. rosulata* Rosenv. Plant with carpogonium. (*c*) Plant with androphore bearing spermatangia. (*a*) × 320; (*b*), (*c*) × 1800. [(*a*) after Woelkerling; (*b*) and (*c*) after Boney and White.]

plants grew larger, they also produced spermatangia. No postfertilization stages were obtained in the cultures. (However, see below under *Kylinia.*) The induction of tetrasporangia was not regarded as a genuine photoperiodic response in that a light break in the middle of the dark period during short day length regime did not inhibit tetrasporangial production.

KYLINIA Rosenv. The circumscription of this genus (named for the Swedish phycologist, *Harald Kylin*) has changed over the years (Woekerling, 1971). The original diagnostic feature was the production of spermatangia on special colorless stalk cells (called androphores). This description was later changed to restrict *Kylinia* to marine species with unicellular holdfasts, as in *K. alariae* (Jonsson) Kyl. (Fig. 9.23a). As is typical for the entire family, carpogonia are borne as sessile structures (Fig. 9.23b); spermatangia are produced often in clusters from the tips of androphores (Fig. 9.23c) (Boney and White, 1967). Culturing studies of sexual stages of *K. rosulata* Rosenv. have demonstrated (Boillot and Magne, 1973) the production of a larger tetrasporophyte apparently identical to *Acrochaetium pectinatum* (see above).

RHODOCHORTON Naeg. The small tufted filaments of *Rhodochorton* (Gr. *rhodon*, a rose + *chorton*, grass) have cells containing several small, discoid chloroplasts (Fig. 9.24a) with several pyrenoids. Tetrasporangia are frequently borne in clusters usually at the tips of the filaments (Fig. 9.24b). Some species are less than 1 mm in height; others may be 1–2 cm in height.

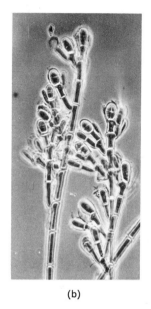

(a)   (b)

Fig. 9.24 *Rhodochorton.* (*a*) *R. floridulum*
(Dillw.) Naeg. Vegetative cells with many small
chloroplasts. (*b*) *R. purpureum.* Tetrasporangial
production. (*a*) × 500; (*b*) × 175. [(*a*) after
Woelkerling, as *Audouinella floridulum*.]

One of the most widely distributed species is *R. purpureum* (Lightf.) Rosenv.,
which typically occurs in shaded areas of the upper littoral zone. Throughout its
range only tetrasporic plants had been observed. Field and culture studies of populations from Alaska, Washington, and California by West (1969A) revealed that tetrasporangia were induced under short photoregimes (8–12 hours of light per 24 hours).
Although the tetraspore germlings from the Alaska and Washington clones did not
progress beyond a few cells, those of the California clone developed into microscopic
unisexual gametophytes. The male plants bore terminal clusters of spermatangia, and
female plants bore terminal sessile carpogonia, singly or in clusters. The postfertilization development was most unusual. A lateral bulge developed from the fertilized
carpogonium, and a club-shaped primary **gonimoblast** emerged from this bulge. The
carpogonium underwent a transverse division, and from the lower cell secondary
gonimoblast cells were formed. These clustered gonimoblast cells all enlarged and
extended out across the substratum, becoming attached by rhizoids after the female
gametophyte died. These filaments (= gonimoblasts) continued to grow apically,
became branched, and eventually—under a short photoregime—produced tetrasporangia. One interpretation of this development in *R. purpureum* is that the carposporophytic and tetrasporophytic phases have become combined into a single phase.
West (1969A) has presented some interesting phylogenetic implications concerning
this pattern.

Sporulation in *Rhodochorton purpureum* has been subjected to more detailed
analysis (West, 1972A). Clones from Alaska, Washington, and California sporulated
only when grown in short daylengths, whereas a clone from Chile was not influenced
by daylength. The relationship among temperature, daylength, and sporulation in a
natural population is presented in Fig. 9.25.

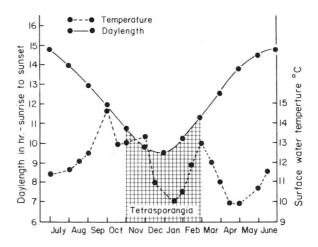

Fig. 9.25 Representation of tetrasporangial production in *Rhodochorton pur-pureum*. Daylength appears to be the primary control of induction. (Based on West.)

Family 2. Batrachospermaceae

This family includes several genera of freshwater red algae, characterized by uniaxial construction with cortical filaments arising from the central filament. The cortical filaments are abundantly branched but loosely grouped. Carpospores are borne terminally on the gonimoblast filaments. Only the genus *Batrachospermum* is treated here.

BATRACHOSPERMUM Roth   Plants of *Batrachospermum* (Gr. *batrachos*, a frog + Gr. *sperma*, seed) are soft, gelatinous, branched axes with a beaded appearance (Fig. 9.26*a*) and in varying shades of blue-green, olive, violet, or gray. They are strictly freshwater in occurrence, most often inhabiting cold running streams or cold spring-fed ponds and lakes throughout the world (Israelson, 1942; Mori, 1975).

A series of segments is cut off from the apical cell (Fig. 9.27*a*), and each segment produces four lateral projections, which are cut off as **pericentral cells.** The axial cells do not divide again but continue to increase in length and breadth farther away from the apex. Each pericentral cell produces lateral branches, which repeatedly branch but are of limited growth. The production of these lateral whorls of determinate branches at the nodes of the axial cells results in the characteristic beaded appearance of the plant (Fig. 9.26*a*). In addition to the laterally projecting filaments, rhizoidal filaments are developed from the lowermost cells of the nodal branches. These rhizoidal filaments grow downward and ensheath the enlarged axial cells, giving the central axis of the plant a multicellular appearance (Smith, 1950). This characteristic of a closely appressed cortex around the central axis is used (Bourrelly, 1970) to distin-

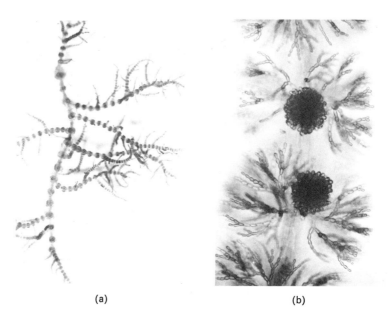

(a)   (b)

Fig. 9.26 *Batrachospermum*. (*a*) Habit of vegetative thallus. (*b*) Compact carposporophytes borne among the nodal tufts. (*a*) × 5.6; (*b*) × 104.

guish this family from two other smaller families of freshwater Nemalionales, the
Lemaneaceae and Thoreaceae, which are not treated in this account.

Carpogonial branches are of two types in the genus (Dixon, 1958). In some species
the carpogonial branch arises from a cell of a lateral branch of limited growth, whereas
in other species the carpogonial branch is equivalent to a modified lateral branch of
unlimited growth. In this latter type the carpogonial branch may be quite long (Fig.
9.27*b*, *c*), sometimes more than 10 cells in length and may bear lateral branches due
to the dichotomization of the apical cell (Dixon, 1958). After fertilization the zygote
nucleus migrates into a protrusion at the base of the carpogonium (Fig. 9.27*d*) and
cells are cut off in this basal portion (Fig. 9.27*e*) that initiate the carposporophyte
(Fig. 9.26*b*). Various patterns of carposporophyte development have been used (Kumano, 1970) to illustrate phylogenetic relationships within the genus. Carpospores
germinate into a microscopic filamentous system resembling *Audouinella* of the
Acrochaetiaceae (p. 481). During a part of the year *Batrachospermum* persists as this
microthallus; when conditions favor the growth of the macrothallus, apical cells from
the filamentous phase give rise to the characteristic adult plant.

The seasonal periodicity of *Batrachospermum macrosporum* Mont. and *Audouinella violacea* (Kuetz.) Hamel, which were regarded as distinct species, was followed
(Dillard, 1966). *Batrachospermum* was dominant in the summer, and *Audouinella* was
dominant in the winter. An artificial stream apparatus demonstrated that water
temperature was of primary importance in the initiation of seasonal dominance but
that light could also become dominant during the summer or the winter. *Audouinella*

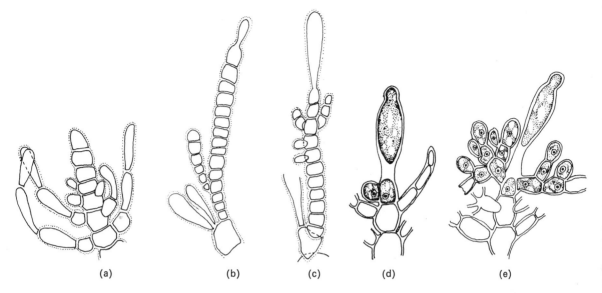

Fig. 9.27 *Bactrachospermum.* (*a*) Diagram of apex. (*b*) Young carpogonial branch with a five-celled lateral. (*c*) Mature carpogonial branch with laterals. (*d*) Early stage in development of gonimoblast from base of carpogonium. (*e*) Portion of mature carposporophyte. (*a*)–(*c*) × 396; (*d*) × 792; (*e*) × 660. [(*a*) from Smith; (*b*) and (*c*) from Dixon; (*d*) and (*e*) from Kylin.]

required low light and low water temperature; *Batrachospermum* required low light and high water temperature. The critical temperature bringing about the change in dominance by one or the other was 15°C.

An investigation (Rider and Wagner, 1972) of two species of *Batrachospermum* at two sites showed that *B. moniliforme* Roth is better able to tolerate high light intensity than *B. vagum* (Roth) C. Ag. The latter species grew year-round in a dark-water stream protected from high-intensity light by the filtering effect of the bog water, and it was also influenced by water velocity. A different study (Woelkerling, 1975) of *Batrachospermum* reported that the seasonal appearance did not seem to be correlated with light intensity or temperature.

Family 3. Nemalionaceae

Multiaxial growth occurs in this family of a few marine genera. The term *fountain type* of growth has been used to describe the appearance of many longitudinally coursing filaments at the apex (Fig. 9.5*b*), with enlarged colorless medullary cells in the central portion and smaller pigmented cortical cells toward the thallus surface. The mechanism that enables a thallus with such a pattern of development to produce branches is not understood.

Formerly, the members of this family and those of the following family, Helminthocladiaceae, were placed in a single grouping, but the present treatment follows other recent authors (Abbott and Doty, 1960; Chihara and Yoshizaki, 1972A), who

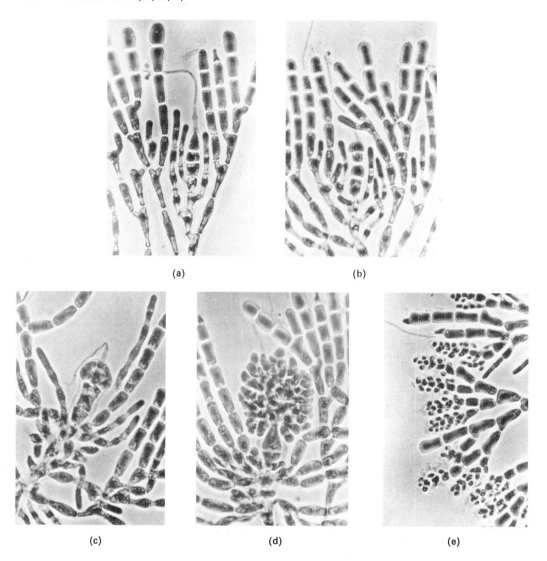

Fig. 9.28 *Nemalion helminthoides* (Velley) Batters. (*a*) Carpogonial branch. (*b*) First division of fertilized carpogonium. (*c*) Postfertilization development. (*d*) Carposporophyte. (*e*) Spermatangial production. × 792.

have segregated them into two families on the basis of the structure of the carpogonial branch. In the Nemalionaceae, carpogonial branches (Fig. 9.28*a*) resemble cortical filaments and consist of cells in a straight row. In the Helminthocladiaceae, however, the carpogonial branches (Fig. 9.31*a*) are curved and borne laterally from the cortical filaments. They are accessory rather than being modified cortical filaments.

In addition to the genus *Nemalion*, which is discussed in greater detail below, this family contains *Yamadaella*, recently established by Abbott (1970). This genus, which was based upon a species formerly placed in *Liagora* (p. 490), is distinctive in that the fertilized carpogonium produces a diffuse system of branches that terminate in tetrasporangia, thus, a **carpotetrasporophyte.** Such a postfertilization development is, therefore, intermediate between that described for *Rhodochorton purpureum* (p. 484) and that occurring in some species of *Liagora*, as *L. tetrasporifera* Børg. (p. 492).

NEMALION Duby   The elastic, gelatinous thalli of *Nemalion* (Gr. *nema*, thread) are most likely to be collected in surf-washed mid-littoral rocky areas. *Nemalion* has also been described (Söderström, 1970), however, as thriving in sheltered and somewhat brackish habitats. Opinion differs whether a single polymorphic species, containing simple to much branched forms, should be included within *N. helminthoides* (Velley) Batt., the type species. Some authors (Womersley, 1965; Hollenberg and Abbott, 1966) have expressed the opinion that intergrading forms occur even in the same population in regard to branching, inner vegetative construction, and the reproductive structures, thus concluding that *N. helminthoides* is highly variable. Another interpretation (Söderström, 1970) is that *N. helminthoides* is an unbranched or rarely branched alga (Fig. 9.29*a*) with blunt apices, a somewhat cartilaginous consistency, and is typically unisexual, whereas *N. multifidum* (Weber et Mohr) J. Ag. is usually

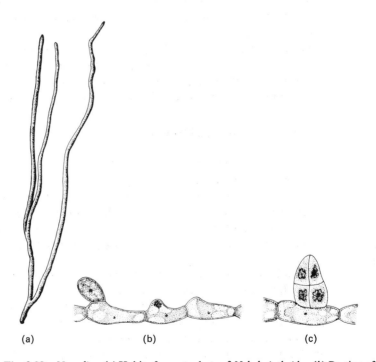

(a)   (b)   (c)

Fig. 9.29   *Nemalion.* (*a*) Habit of gametophyte of *N. helminthoides.* (*b*) Portion of filamentous phase of *N. vermiculare* Suringar with young tetrasporangium. (*c*) Tetrasporangia in *N. vermiculare.* (*a*) × 0.35; (*b*), (*c*) × 290. [(*b*) and (*c*) from Umezaki.]

much branched with tapering apices, a loose mucilaginous consistency, and is unisexual or bisexual. Other species are also known, as *N. vermiculare* Sur. from Japan.

Until relatively recently only sexual plants were known in this genus. Because of this fact and of some evidence purporting that meiosis took place in the fertilized carpogonium, it was long held that *Nemalion* was a haplobiontic alga, the entire life history being haploid except for the fertilized carpogonium. The discovery of an alternation in dissimilar phases in related genera motivated workers to look more closely at *Nemalion*, especially to follow the development of the germinating carpospores. Such culturing studies (Umezaki, 1967, 1972; Fries, 1967, 1969) have revealed that a microscopic filamentous stage, resembling *Acrochaetium* (p. 481), alternates as the tetrasporophyte, producing tetrasporangia (Fig. 9.29*b*, *c*) and monosporangia. The tetraspores return the large sexual plant, and the monosporangia recycle the sporophyte.

The sexual plant in *Nemalion* may be bisexual, the spermatangia often appearing first (= *protandrous*) and the carpogonia appearing later. The spermatangia (Fig. 9.28*d*) are borne in clusters from cortical filaments. The small colorless spermatia are released from the spermatangia. Carpogonial branches occur among cortical filaments and consist of usually three to five cells, the carpogonium proper being terminated by an elongate trichogyne, to which spermatia become attached. Following fertilization and the migration of the male nucleus to and union with the female nucleus, the carpogonium undergoes a transverse division. The lower cell does not further divide, but the upper cell divides several times longitudinally, producing a compact knot of cells (Fig. 9.28*b*). These cells continue to divide, and this rounded compact mass of cells constitutes the carposporophyte, which eventually fragments releasing the carpospores. The entire carposporophyte (Fig. 9.28*c*) is apparently fertile. There is no envelope or pericarp surrounding the carposporophyte.

Family 4. Helminthocladiaceae

As discussed above, this family differs from the preceding group in the presence of laterally borne, accessory carpogonial branches (Fig. 9.31*a*), being dissimilar in appearance to the cortical filaments upon which they are borne. About six marine genera are contained in this group, including *Cumagloia* on the west coast of North America, *Helminthora* and *Dermonema* of tropical and subtropical seas, and the large tropical genus *Liagora*, which is discussed in greater detail below.

LIAGORA Lamour.   Plants of *Liagora* (L. *Liagora*, a nereid, marine worm) may be barely to heavily impregnated with calcium carbonate; some species do not become calcified at all. It is a large pantropical genus with many species occurring in the waters of Hawaii and Florida (Abbott, 1945; Taylor, 1960). Plants are usually terete, very soft and lubricous in texture, or stiff if becoming heavily calcified, and may reach heights of 20 cm. Branching is alternate, dichotomous, or irregular. As is characteristic of the family, the internal thallus structure is a multiaxial construction, with medullary and cortical filaments (Fig. 9.30).

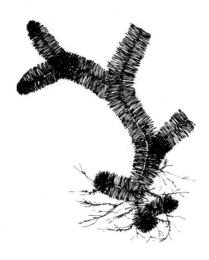

Fig. 9.30 *Liagora farinosa* Lamour. Habit of cultured gametophyte. × 5. (From von Stosch.)

Species of *Liagora* are either unisexual or bisexual (Desikachary, 1956; Chihara and Yoshizaki, 1972A). Carpogonial branches are present in abundance in the upper portions of the plant. The carpogonial branch (Fig. 9.31*a*) is usually recurved proximally, and the trichogyne extends beyond the periphery of the thallus. Following fertilization, the carpogonium divides transversely, the lower cell being the **stalk cell** (Desikachary, 1956) and the upper cell being the **gonimoblast initial**, from which the much branched gonimoblast, or carposporophyte, is developed (Fig. 9.31*b*). One

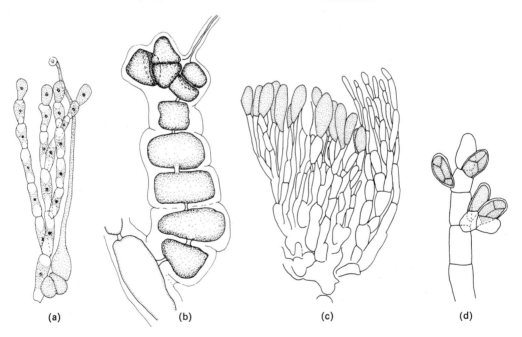

(a)   (b)   (c)   (d)

Fig. 9.31 *Liagora*. (*a*) Carpogonial branch of *L. maxima* Batters. (*b*) Early and (*c*) later stages of postfertilization development in *L. japonica* Yamada; *gf*, gonimoblast filament; *tr*, trichogyne. (*d*) Tetrasporangial production on microscopic tetrasporic phase. (*a*) × 510; (*b*) × 955; (*c*) × 315; (*d*) × 214. [(*a*) after Desikachary; (*b*) and (*c*) after Chihara and Yoshizaki; (*d*) after von Stosch.]

difference from the previous family is that involucral filaments are formed by vegetative cells and ultimately almost completely surround the carposporophyte (Fig. 9.31c).

It was stated above that most members of the Nemalionales had originally been regarded as being haplobiontic. Mullahy (1952) and Magne (1961, 1964) presented serious questions concerning the validity of this concept with their careful cytological studies of stages before and after fertilization of representative members of the Nemalionales. Magne observed that the zygotic nucleus underwent mitotic divisions rather than meiotic divisions and was forced to conclude that the life history was probably not haplobiontic. The first authenticated verification of Magne's hypothesis was von Stosch's (1965C) culturing study of a species of *Liagora* from the Mediterranean. Von Stosch cultured carpospores and found that they gave rise to creeping *Acrochaetium*-like filaments, which produced monospores and tetraspores (Fig. 9.31d). The tetraspores were demonstrated to be the products of meiosis and returned the larger thalli of *Liagora*, by first producing filamentous protonemal growths, from which buds arose that developed into the macroscopic plants.

Some species of *Liagora* are known to have compact carposporophytes that produce tetrasporangia rather than carposporangia, as *L. tetrasporifera* Børg. and *L. papenfussii* Abbott. Evidence has been presented (Couté, 1971) that meiotic divisions occur during the formation of the tetrad of spores; therefore, these species would be haplobiontic, lacking a free-living tetrasporic stage.

Family 5. Chaetangiaceae

This family includes about seven genera of marine algae, with a multiaxial construction. It is distinguished from the Nemalionaceae and Helminthocladiaceae by the presence of an enclosing **pericarp** of slender filaments around the carposporophyte, with an **ostiole** for the release of the carpospores. The plants of this family are of moderate size, erect, and with a bushy habit. Some genera are soft and gelatinous, such as *Pseudogloiophloea* and *Scinaia*; others are lightly to heavily calcified, as *Galaxaura*. The outer cells of the thallus may be enlarged into utricles and contiguous, forming a continuous epidermal layer. The family is also recognized by the more complicated postfertilization stages, which are discussed below.

PSEUDOGLOIOPHLOEA   The dichotomously branched fronds (Fig. 9.32) of *Pseudogloiophloea* (Gr. *pseudes*, false + Gr. *gloios*, viscid + Gr. *phloios*, bark) are gelatinous, collapsing when exposed by the outgoing tide. These fleshy gametophytes are bisexual. Spermatangial filaments (Fig. 9.33a) arise from small assimilatory filaments between large colorless utricles. The carpogonial branches are three-celled. Sterile filaments are produced from the lowermost cell of the carpogonial branch; the **hypogynous cell** cuts off two cells, one of which divides again, such that the mature carpogonial branch has four cells located beneath the carpogonium (Fig. 9.33b). After fertilization, a single gonimoblast initial is produced directly from the carpogonial base (Ramus, 1969A), unlike certain other species of this genus in which the

Fig. 9.32 *Pseudogloiophloea confusa* (Setch.) Levr. Habit of fleshy phase, the gametophyte. × 0.48. (After Ramus; published in 1969 by The Regents of the University of California; reprinted by permission of the University of California Press.)

5 cm

gonimoblast is initiated from an hypogynous cell. The maturing carposporophyte (Fig. 9.33c) is surrounded by a pericarp, composed of colorless filaments produced from the basal cell of the carpogonial branch.

Only the sexual phase had been recognized in *Pseudogloiophloea confusa* (Setch.) Levr. until the culturing investigation by Ramus (1969A). He determined that the carpospores produced *Acrochaetium*-like plants. These filaments asexually recycled themselves by monospores, but when placed in seawater medium lacking the usual nitrate additive the terminal cells in some of the filaments produced tetrasporangia

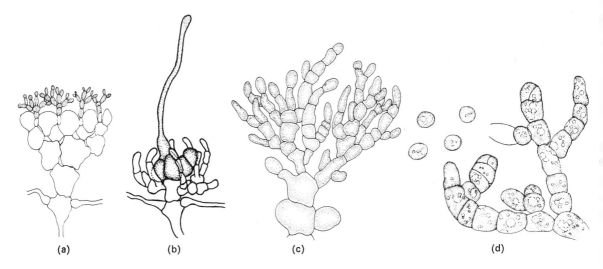

(a)   (b)   (c)   (d)

Fig. 9.33 *Pseudogloiophloea confusa.* (a) Production of spermatangial filaments from gaps between large, colorless utricles. (b) Carpogonial branch. (c) Portion of carposporophyte. (d) Tetrasporangial production on microscopic diploid phase. (a) × 680; (b) × 480; (c) × 450; (d) × 650. (After Ramus.)

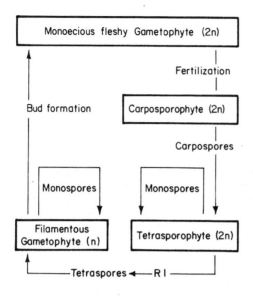

Fig. 9.34 Schematic representation of the life history of *Pseudogloiophloea confusa*. "R!" indicates the presumed point of meiosis. (After Ramus; published in 1969 by The Regents of the University of California; reprinted by permission of the University of California Press.)

(Fig. 9.33*d*), which showed zonate, cruciate, and tetrahedral cleavage patterns. These tetraspores gave rise to *Acrochaetium*-like plants, resembling the tetrasporic filaments. These filaments similarly recycled themselves by monospores. When placed in relatively bright light (400 ft-c), however, and at 17°C the filaments formed buds, which developed into the characteristic, dichotomously branched fleshy plants. Thus, the life history of this species proved to consist of two free-living generations: a gametophyte ($n = 4$ or 5) and a strikingly dissimilar tetrasporophyte ($2n = 8$ or 10). Figure 9.34 outlines this sequence, which is under the regulation of nutrients (i.e., nitrate) as well as intensity of illumination.

GALAXAURA   Occupying a somewhat anomalous position, *Galaxaura* (Gr. *gala*, milk + Gr. *auros*, gold) (Fig. 9.35) is unusual in that the life history consists of

Fig. 9.35 *Galaxaura.* Habit. × 0.53.

two free-living macroscopic stages, gametophyte and tetrasporophyte, that may exhibit strong dimorphism (Svedelius, 1944). This seems to be the only genus to show such dimorphism, which involves not only general morphology but anatomical details such as the construction of the cortex. Many species have been described (Papenfuss and Chiang, 1969), and much work remains to match up the "pairs of species" that in reality represent phases of the same alga.

Family 6. Bonnemaisoniaceae

This family has generated much excitement ever since the discovery of the existence of heteromorphic life histories among some of its members. The original work (Feldmann and Feldmann, 1942) linked together *Bonnemaisonia hamifera* Hariot with *Trailliella intricata* Batt., a monotypic genus that had earlier been placed in the Ceramiaceae (p. 544) and previously known only as a tetrasporic plant. Similarly, *Asparagopsis armata* Harv. and *Falkenbergia ruflanosa* (Harv.) Schmitz were recognized to represent gametophytic and tetrasporic phases, respectively, in the life history. *Falkenbergia* had erroneously been classified as a genus of Rhodomelaceae (p. 559), in which only tetrasporangia had been observed. Finally, *Hymenoclonium serpens* (Crouan fr.) Batt. was connected as a protonemal phase in the life history of *Bonnemaisonia asparagoides* (Woodw.) C. Ag., with uncertainty expressed (Feldmann, 1966) regarding if and when meiosis might occur.

In addition to the haplobiontic type, as in *Bonnemaisonia asparagoides*, and a diplobiontic type involving heteromorphic phases, as in *Bonnemaisonia hamifera* and *Asparagopsis armata*, a diplobiontic pattern involving isomorphic phases occurs in species of *Leptophyllis* and *Delisea* from Australia (Levring, 1953). It is evident that flexibility exists in the characterization of the pattern of life history in this family (Chihara, 1961, 1962).

Plants are generally of moderate size and of uniaxial construction. Growth is apical; the axial row of cells is surrounded by short, branched cell rows constituting a pseudoparenchymatous cortex with a continuous surface. Vesicle cells (p. 458) seem to be present in all species.

In respect to reproduction, three-celled carpogonial branches occur throughout the family, but two different patterns of postfertilization have been reported (Chihara and Yoshizaki, 1972B). The carposporophyte develops either directly from the fertilized carpogonium, as in *Bonnemaisonia* (Svedelius, 1933; Hudson and Wynne, 1969) or from the hypogynous cell, as in *Asparagopsis* (Svedelius, 1933). Carpospores are relatively large and borne singly on the gonimoblast filaments. The carposporophyte is surrounded by a well-developed **pericarp**, which is a protective urn-shaped structure developed from female tissue and projecting beyond the surface of the thallus.

The proposal by Feldmann (1952), influenced by the heteromorphic life history of this group, to recognize the order Bonnemaisoniales has been recently offered support (Chihara and Yoshizaki, 1972B). Similarities to the Ceramiales include the uniaxial construction and the presence of a pericarp. Along with members of the Chaetangiaceae, members of the Bonnemaisoniaceae have many discoid chloroplasts lacking pyrenoids, a feature shared by orders of Florideophycidae other than most Nemaionales (Hara, 1972).

The phenomenon of perennation and vegetative propagation has been studied in reference to *Asparagopsis armata* in British waters (Dixon, 1965). Evidence suggests that this species was introduced into Europe from Australia. In the northern limits of distribution, tetrasporangia are not produced in the *Falkenbergia* phase, and carposporangia are infrequent in the *Asparagopsis* phase. Dixon (1965) offered the idea that these two phases can vegetatively multiply themselves independently, and therefore the asexual cycles prevail rather than the theoretical life history. Similarly, in *Bonnemaisonia hamifera*, which is an import into the Atlantic from the Pacific Ocean, female plants occur but no male plants in many areas (Feldmann and Feldmann, 1942; Taylor, 1962A; Chen, et al., 1969). If conditions are optimal, including frequently renewed clear water, however, the male plants are also present, and the sexual cycle is completed (Floc'h, 1969).

BONNEMAISONIA C. Ag.   The much branched bright red thalli of *Bonnemaisonia* (named for the French naturalist *Theophile Bonnemaison*) have cylindrical to compressed axes. Bilateral branching occurs in most species, with branches arranged in an opposite or alternate arrangement. Opposite branches are dissimilar, one being short and determinate and the other being potentially long and indeterminate. Branching tends to be radial or irregular in *B. hamifera* (Fig. 9.36*a*), which occurs on the

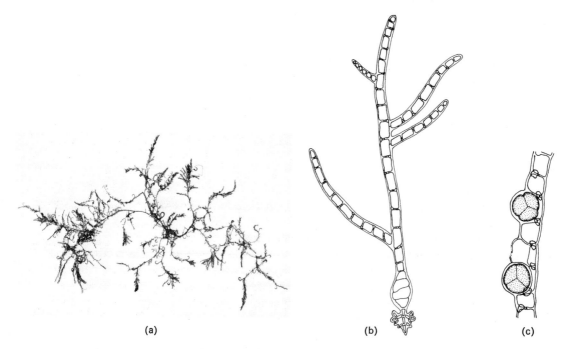

(a)   (b)   (c)

Fig. 9.36 *Bonnemaisonia hamifera* Hariot. (*a*) Habit. (*b*) "*Trailliella*" stage of *Bonnemaisonia hamifera*. (*c*) Production of tetrasporangia by the "*Trailliella*" stage. (*a*) × 0.52; (*b*) × 65; (*c*) × 133. [(*b*) and (*c*) from Chihara.]

Atlantic coast. Enlarged hook-shaped branches (Fig. 9.8*c*) occur in addition to the usual small branchlets. Carpospores of this species were cultured (Feldmann and Feldmann, 1942), and uniseriate branched filaments bearing small gland cells (Fig. 9.36*b*) were obtained. This morphologically dissimilar phase, identifiable as the genus *Trailliella*, produces tetrahedrally divided tetrasporangia, and the life history was completed (Harder and Koch, 1949) by the germination of tetraspores and their subsequent development into the macroscopic sexual phase. *Bonnemaisonia nootkana* (Esper) Silva, which has a Pacific North American distribution, has been demonstrated (Chihara, 1965) also to have a "*Trailliella*" stage as its sporophyte.

Family 7. Gelidiaceae

This family is characterized by apical growth and a diplobiontic life history. It has been regarded as a distinct order (Kylin 1923; Fan, 1961A; Papenfuss, 1966) from the Nemalionales; yet as in other Nemalionales generative auxiliary cells (p. 461) are not present (Dixon, 1970). Carpogonia occur in sessile positions on intercalary cells of filaments of the third order. Sessile carpogonia also occur in the Acrochaetiaceae. Following fertilization, the carpogonium fuses completely with the **supporting cell,** and from this inflated cell gonimoblast filaments are developed that branch and produce terminal carpospores. Other variations in the way in which the carposporophyte develops have been described (Dixon, 1959, 1973), but it is generally agreed that auxiliary cells are absent. In the vicinity of the carposporophyte chains of nutritive cells are produced by the vegetative tissue, and the diffusely developed carposporophyte apparently derives some nourishment from these nutritive cells.

Tetrasporophytes bear cruciately divided tetrasporangia. Spermatangia are produced in sori at the apices of the male plants, a small clear space below the apex indicating the presence of this sorus. A vegetative characteristic that unites all the genera (excepting *Gelidiella*) is the presence of rhizoidal filaments (Fig. 9.37*a*) densely

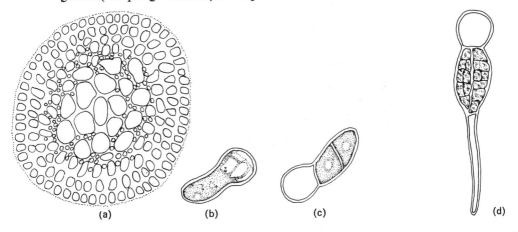

(a)   (b)   (c)   (d)

Fig. 9.37 (*a*) *Gelidium*. Presence of narrow, rhizoidal filaments in the medulla. (*b*)–(*d*) *Gelidiella*. Stages in tetraspore germination, with evacuation of the original spore. (*a*) × 165; (*b*)–(*d*) × 396. [(*b*)–(*d*) after Chihara and Kimura.]

coursing among the larger medullary cells or located as a layer between the cortex and medulla. Tetraspores of *Gelidiella*, a genus known so far only from tetrasporic specimens, exhibit the characteristic pattern of germination of the order (Fig. 9.37*b–d*), thus showing their affinity (Chihara and Kamura, 1963).

About a dozen genera of marine distribution are categorized in this order, most of which are small and monotypic. Two genera, *Gelidium* and *Pterocladia*, are large and taxonomically perplexing (Stewart, 1968). The primary distinction between these two genera is that the former has cystocarps (p. 461) with two ostioles, or openings, and the latter has cystocarps with a single ostiole. This trait does not easily solve the problem of identification since female plants are often rare, a disproportionate number of specimens in a population being tetrasporic. Only the genus *Gelidium* is discussed in further detail.

GELIDIUM Lamour.   Thalli range in shape from terete to broadly flattened axes and in size from about 1 cm tall, as in *G. pusillum* (Stackh.) LeJolis, to 50 cm tall, as *G. robustum* (Gardn.) Hollenb. et Abbott, both on the Pacific coast. Plants of *Gelidium* (Gr. *gelidus*, congealed, which refers to their being boiled down to produce gels) are always stiff and cartilaginous. Branching is bilateral and compoundly pinnate (Fig. 9.38), and the axes are perennial.

The chief source of **agar** (p. 28) is *Gelidium*, agar having been manufactured in Japan since around 1760 (Tilden, 1937). Its chief use in America and Europe is in the preparation of microbiological media. When the supply of Japanese agar was cut off during World War II, *Gelidium nudifrons* Gardner was harvested from Mexico, and a production plant was operating in San Diego, California, to provide the critically needed agar. An excellent literature survey of the Gelidiaceae, with special emphasis

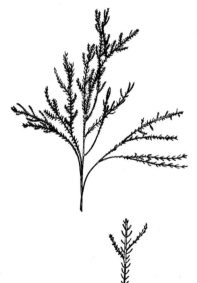

Fig. 9.38 *Gelidium* sp. Habit. × 0.6 (× 2.5 for detail of apex).

on *Gelidium*, has been recently published (Santelices, 1974), with such topics covered as taxonomy, ecology, and economic utility.

Order 6. Cryptonemiales

This fairly large order of marine red algae, containing about 10 families, is not always easily distinguishable from the Gigartinales, which is to follow in this book, and much recent work (Papenfuss, 1951B; Norris, 1957; Balakrishnan, 1960; Dixon, 1973) has directed attention to the problem of overlapping examples and better distinguishing these groups. The characteristic still used as a means of separating the two orders is that in the Cryptonemiales, the auxiliary cell, the cell that typically initiates the carposporophyte, is produced in an accessory or specialized branch of the female thallus, whereas in the Gigartinales the auxiliary cell is produced in a vegetative filament.

Vegetatively, plant bodies in this order may be filamentous or pseudoparenchymatous (blades, terete axes, or crusts). Soft, lubricous forms, as *Gloiosiphonia*, occur, as well as calcified forms as in the Corallinaceae (p. 508). Life histories are typically diplobiontic and isomorphic, but culturing studies of some members demonstrated that heteromorphic patterns are also present, as in *Pikea* and *Gloiosiphonia*.[6] Families are distinguished on such bases as the pattern of division of the tetrasporangia (zonate or cruciate), the type of development (uniaxial or multiaxial), and relationship of carpogonial branch to auxiliary cell (procarpial or nonprocarpial) (p. 466). In regard to this last trait, however, it is now recognized that this condition is variable within one family (see the Kallymeniaceae, p. 505). The six representative families treated in the present account are distinguished by the following key.

1. Thalli calcified, crustose or erect with segmented organization; reproductive structures produced in conceptacles ....Corallinaceae
1. Thalli noncalcified 2
   2. Thalli parasitic (on members of the Rhodomelaceae, p. 559) ....Choreocolaceae
   2. Not as above 3
3. Carpogonial branches and auxiliary cell branches scattered in the thallus (although in some Dumontiaceae the auxiliary cells are nonfunctional) 4
3. Carpogonia and auxiliary cells in one and the same branch system 5
   4. A connecting filament from the fertilized carpogonium making contact with another cell in the carpogonial branch prior to development of carposporophyte ....Dumontiaceae
   4. A connecting filament not making contact with any cell in the carpogonial branch but directly connecting with an auxiliary cell in a remote auxiliary cell branch ....Cryptonemiaceae

---

[6]The tetrasporophytes of *Pikea* (p. 501) and *Gloiosiphonia* (p. 503) are both small discs or crusts, composed of two to several cell layers, but that of *Acrosymphyton* (Dumontiaceae) is unusual in that it is a small compact phase resembling *Hymenoclonium serpens*, similar to the phase in *Bonnemaisonia asparagoides* (p. 496). This phase in *Acrosymphton* has been observed (Cortel-Breeman, 1975) to form tetrasporangia under short-day conditions.

5. Carpogonial branch and auxiliary cell branch borne on a common supporting cell, which is not involved in the initiation of the carposporophyte ....Gloiosiphoniaceae

5. Supporting cell of carpogonial branch(es) involved in the formation of a fusion cell following fertilization of the carpogonium and functioning either as the auxiliary cell or the site of origin of a connecting filament to a remote auxiliary cell ....Kallymeniaceae

Family 1. Dumontiaceae

Carpogonial branches and auxiliary cell branches are borne remotely from one another in this family, and the auxiliary cell branch is sparingly branched. The carpogonial branches (Fig. 9.39*a*) is some genera may be very long, for example, up to 15 cells long in *Cryptosiphonia* and 12 to 20 cells in *Pikea*. Typical post fertilization events might involve the carpogonium sending out a connecting filament, which fuses with a cell in the carpogonial branch (Fig. 9.39*b*). From this fusion one or more connecting filaments proceed out and more or less "seek out" auxiliary cells, with which they make contact and fuse (Fig. 9.39*c*). From this auxiliary cell a carposporophyte is then produced, this condition being nonprocarpial in that the carpogonial branch and auxiliary cells are spatially separated.

In an investigation of some genera, including *Weeksia*, which are members of the Dumontiaceae, Abbott (1968) recognized that the carposporophytes were produced from a cell in the carpogonial branch following fusion of the carpogonium with this

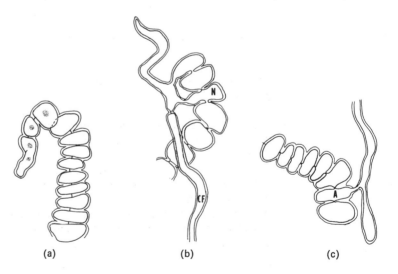

(a)   (b)   (c)

Fig. 9.39 Reproduction in the Cryptonemiales (*Farlowia*). (*a*) Carpogonial branch. (*b*) Connection of primary ooblast filament with a nutritive cell in the carpogonial branch (N) and the continuation of the connecting filament (CF) toward an auxiliary cell. (*c*) Contact of the connecting filament with an auxiliary cell (A). (*a*)–(*c*) × 340. (After Abbott.)

cell, with no participation by auxiliary cells, which nonetheless are present on specialized auxiliary cell branches. She interpreted this anomalous pattern as procarpial, since the carposporophyte is initiated from a cell in a proximate position to the carpogonium, and removed the genera having this pattern into the family Weeksiaceae. The presence of both procarpial and nonprocarpial types in another family of this order (the Kallymeniaceae), which demonstarate obvious relationships and means of derivation, would argue that there is no justification to recognize the Weeksiaceae, when the means of deriving the procarpial types from the nonprocarpial types are evident.

Tetrasporangia in most genera of this family are cruciately divided, but zonately divided tetrasporangia are produced in certain genera, as the tropical genus *Dudresnaya* and the North Pacific genus *Constantinea*. Both uniaxial and multiaxial patterns are present in this family, the former type being present in *Cryptosiphonia* and *Pikea* and the latter type in *Weeksia*, *Dilsea*, and *Constantinea*. All of these genera mentioned are present on the Pacific coast of North America. In an ontogenetic study of the developing blades of *Weeksia*, it has been demonstrated (Norris, 1971) that a uniaxial pattern present in the youngest stages is replaced by a multiaxial pattern in later stages by the primary axial filament ceasing its growth and the initials of secondary filaments taking over.

PIKEA Harvey   Thalli of *Pikea* (named for Capt. *Nicholas Pike* of Brooklyn, original collector) consist of tufts of much branched, narrow, compressed axes (Fig. 9.40), with pinnate branching and apical growth. Two species are recognized (Abbott,

Fig. 9.40   *Pikea robusta* Abbott. Habit. × 0.5.

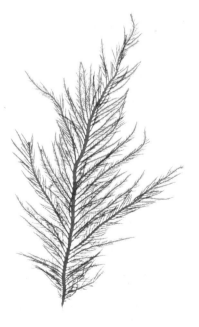

1968), *P. californica* Harvey being the type. Fertile material is rarely collected, about 1 in 50 specimens being cystocarpic. Cystocarps are present in enlarged tips of branches. Tetrasporic plants were unknown until the discovery (Scott and Dixon, 1971) that carpospores develop into a small crustose tetrasporangial phase, which produces monosporangia and cruciately divided tetrasporangia. Monospores recycled the microscopic crusts; tetraspores returned the large sexual plants.

Family 2. Gloiosiphoniaceae

Uniaxial construction is the rule in this family. The main axis bears transverse whorls of short, simple branches, the basal cells of which are large and branch out in both vertical and horizontal directions (Fig. 9.41b). Secondary branches give rise to progressively smaller-celled branches, which ultimately unite with branches of adjacent lateral filaments to form a cortex of continuous small cells. The thallus surface may have a banded aspect from the junctures of adjacent branch systems of successive branch whorls.

Carpogonia and auxiliary cells are produced on the same branch systems; thus, **procarps** are present (Fig. 9.41c). The three-celled carpogonial branch, including a large **hypogynous cell**, is borne on a **supporting cell** that also bears an auxiliary cell

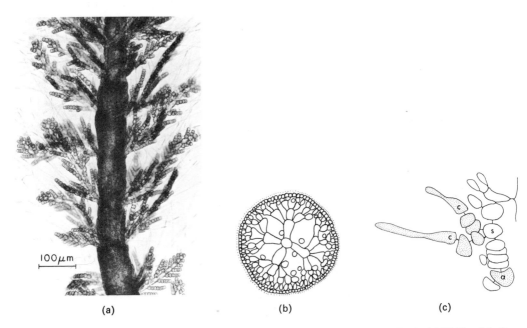

100 μm

(a)   (b)   (c)

Fig. 9.41 *Gloiosiphonia capillaris* (Hudson) Carm. ex Berk. (*a*) Habit of thallus grown in culture. (*b*) Diagrammatic representation of transection of thallus. (*c*) Carpogonial branch system; *a*, auxiliary cell; *c*, carpogonium; *s*, supporting cell. [(*a*) after Edelstein; permission of Phycologia; (*b*) and (*c*) after Kylin.]

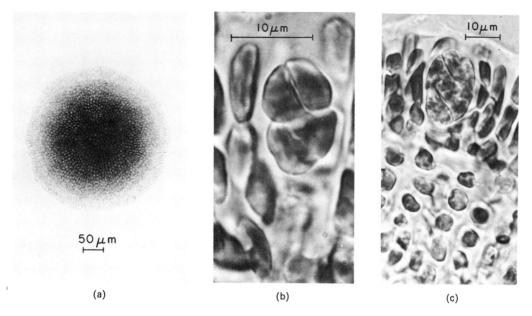

Fig. 9.42 *Gloiosiphonia capillaris.* (*a*) Tetrasporic disc. (*b*) and (*c*) Tetrasporangia. (After Edelstein; permission of Phycologia.)

branch of several cells. In the genus *Schimmelmannia*, occurring on the Pacific coast, the terminal cell of the auxiliary cell branch acts as the auxiliary cell (Abbott, 1961); however, in *Gloiosiphonia* (see below) the **auxiliary cell** is one of the intercalary cells of a six- to eight-celled auxiliary cell branch. The zygote nucleus is transferred to the auxiliary cell either by a small **connecting cell** (in *Schimmelmannia*) or by a **connecting filament** (in *Gloiosiphonia*).

GLOIOSIPHONIA Carm. ex Berk.   The bright red, gelatinous, cylindrical thalli of *Gloiosiphonia* (Gr. *gloios*, viscid + Gr. *siphon*, a tube) may be variously branched. Whorls of short branches characterize *G. verticillaris* Farlow (Fig. 9.41*a*), a Pacific coast species. Profuse, irregular branching characterizes *G. capillaris* (Hudson) Carm. ex Berk., a species occurring on both the Atlantic and Pacific coasts (Edelstein, 1972). Usually several thalli arise from the same base, and they may be hollow, particularly at their bases.

The life history of *Gloiosiphonia capillaris* has been investigated (Edelstein, 1970; Edelstein and McLachlan, 1971). Carpospores were observed to give rise to small compact discs (Fig. 9.42*a*), which were composed or erect filaments embedded in a mucilaginous matrix but growing with a monostromatic margin. Tetrasporangia (Fig. 9.42*b*), demonstrating all three patterns of division (cruciate, zonate, and tetrahedral) were developed in the crust, and the released tetraspores germinated into a new generation of gametophytes, which proved to be bisexual. The basal system of the large gametophyte, which resembled the small crutose sporophyte, persisted in

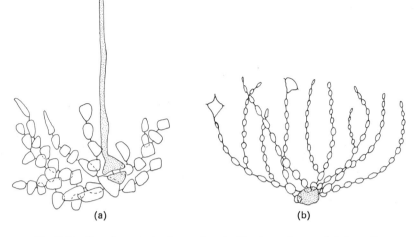

Fig. 9.43 Female reproductive system in Cryptonemiaceae. (a) Ampulla containing a carpogonial branch (shaded). (b) Ampulla containing an auxiliary cell (shaded). (a) × 520; (b) × 253. (After Y.-M. Chiang.)

culture for more than 18 months (Edelstein and McLachlan, 1971), and it would appear that this basal portion of the gametophyte and the crustose sporophyte could both serve to perennate the alga during the season when the conspicuous gametophytic phase was not present.

Family 3. Cryptonemiaceae

This family is characterized as being nonprocarpial, the carpogonia and auxiliary cells being formed in separate accessory branch systems. These systems, which originate usually at the boundary between the cortex and the medulla, are much branched and have a compact knotted appearance; they are called **ampullae**. The two-celled carpogonial branch may have a sterile lateral of several cells arising from the lower cell, and the supporting cell of this carpogonial branch is usually an intercalary cell in the ampulla (Fig. 9.43a). The auxiliary cell is likewise an intercalary cell in an ampulla, generally in the lower portion and with a cluster of filaments arising from it (Fig. 9.43b). Following fertilization, one or more connecting filaments arise from the carpogonium and make direct contact with the auxiliary cell(s), from which the carposporophytes are produced toward the thallus surface. The connecting filament may continue on to make contact with auxiliary cells in other ampullae. A pericarp is usually present surrounding the cystocarp (Chiang, 1970).

The life history of this family is diplobiontic with isomorphic tetrasporophytes and gametophytes.[7] Gametophytes are separately sexed. Tetrasporangia are cruciately

[7]An anomalous pattern has been reported (van den Hoek and Cortel-Breeman, 1970) for *Halymenia floresia* (Clem.) C. Ag. in the Mediterranean, in which carpospores grew into *Acrochaetium*-like plants, which formed monospores. Some of the *Acrochaetium*-like plants differentiated into *Halymenia* sporophytes, which produced monospores in monosporangial sori but no tetrasporangia. These monospores proceeded to germinate into a new crop of *Halymenia* gametophytes. Yet isomorphic tetrasporophytes of this species are known in the field.

Fig. 9.44 *Cryptonemia borealis* Kylin. Habit. × 0.23.

divided. This family includes several genera of so-called "flat reds" (Fig. 9.44), which may be difficult to identify without reproductive material. These include *Halymenia*, *Cryptonemia*, and *Grateloupia*. Anatomical features of the cortex and medulla have been used to distinguish the genera (Abbott, 1968). Studies on reproductive stages (Balakrishnan, 1961A, 1961B; Chiang, 1970) have been helpful in better understanding boundaries.

Family 4. Kallymeniaceae

Members of this family have multiaxial construction and shapes ranging from blades, as in *Kallymenia* and *Pugetia*, to much branched, flattened axes of narrow to broad width, as in *Callophyllis*. Most of the genera lack procarps (*Pugetia* and *Kallymenia*), but procarps are present in a few (*Callophyllis*). A transition proceeding from a nonprocarpial system with many carpogonial branches per supporting cell to a procarpial system with only one carpogonial branch per supporting cell has been formulated by Norris (1957). The obvious interrelationships that exist going from one stage to the next in this scheme (Fig. 9.45) would preclude an arbitrary division of this family into subdivisions based upon the usual distinction of procarpial versus a nonprocarpial condition.

Pugetia occupies a seemingly "primitive" position in the family in reference to its female reproductive system (Fig. 9.45*a*). Many carpogonial branches are borne per supporting cell, and these supporting cells may function as auxiliary cells, if contacted by a connecting filament issuing from a carpogonium of a separate supporting cell. This feature suggests the homology of auxiliary cell systems and carpogonial branch systems. From this nonprocarpial, polycarpogonial condition, two trends are evident, one proceeding to a procarpial condition and the other proceeding to a monocarpogonial system. In *Kallymenia*, which represents an intermediate level of advancement, many carpogonial branches are borne per supporting cell, and the auxiliary cell is a

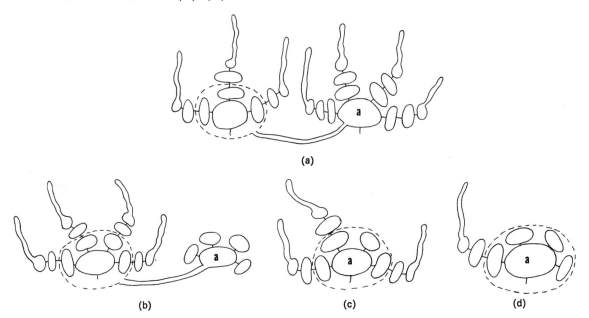

Fig. 9.45 Examples of female reproductive systems in the Kallymeniaceae. (*a*) *Pugetia*. (*b*) *Kallymenia*. (*c*) *Callophyllis*, a species with many carpogonial branches per supporting cell. (*d*) *Callophyllis*, a species with a single carpogonial branch per supporting cell. *a*, auxiliary cell. Dotted line represents fusion cell. (After Norris.)

separate system not bearing carpogonial branches (Fig. 9.45*b*). Following fertilization in *Kallymenia*, a large **fusion cell** is formed involving the supporting cell and the lower cells of the other carpogonial branches. Many long, nonseptate connecting filaments arise from this fusion cell and seek out remote auxiliary cells (Norris, 1957, 1964; Hommersand and Ott, 1970). Finally, in *Callophyllis* the supporting cell functions as the auxiliary cell. It bears either a single carpogonial branch (Fig. 9.45*d*) or many carpogonial branches (Fig. 9.45*c*), depending on the species. A fusion cell, involving the auxiliary cell and the lower cells of the carpogonial branches or any sterile filaments, in the case of the monocarpogonial condition, is formed, and the carposporophyte is produced.

The three important American representatives of this family are *Kallymenia*, *Callophyllis*, and *Pugetia*. *Pugetia* and *Callophyllis* have a somewhat similar internal vegetative structure (Fig. 9.46*a*) of large medullary cells interspersed with small photosynthetic cells. The medulla of *Kallymenia* is composed of rhizoidal filaments intermixed with enlarged **stellate cells** (Fig. 9.46*b*), containing dense, refractive contents. These stellate cells make contact via secondary pit connections with one another and with the rhizoidal cells (Codomier, 1971), this resulting in an interconnected internal tissue that probably can better withstand wave action (Norris, 1957).

Probably more than any other family of red algae, the Kallymeniaceae presents one of the most difficult female reproductive systems to interpret. This complexity is

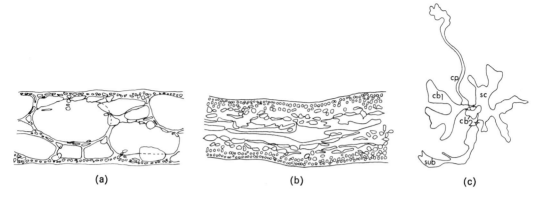

(a)   (b)   (c)

Fig. 9.46 (*a*) *Callophyllis.* Internal anatomy. (*b*) *Kallymenia.* Internal anatomy. (*c*) *Callophyllis firma* (Kyl.) Norris. Carpogonial branch. *cb₁*, first cell of carpogonial branch; *cb₂*, second cell of carpogonial branch; *cp*, carpogonium; *sc*, supporting cell; *sub*, subsidiary cell. (*a*) × 140; (*b*) × 205; (*c*) × 330. (After Norris.)

caused by the irregularly shaped carpogonial branches (Fig. 9.46*c*) and the complicated and elaborate series of cell fusions that occur following fertilization. The study of *Kallymenia reniformis* (Turner) J. Ag. by Hommersand and Ott (1970) reflects the diligence required in unraveling the reproductive story.

Life histories in this family are diplobiontic, involving isomorphic sexual and tetrasporangial phases (Murray and Dixon, 1972). Gametophytes are unisexual. Tetrasporophytes bear cruciately divided tetrasporangia.

CALLOPHYLLIS Kützing   The west coast of North America seems to be the center of distribution for *Callophyllis* (Gr. *kallos*, beauty + Gr. *phyllon*, a leaf), about 10 species currently being recognized (Abbott and Norris, 1965). The primary criterion used in distinguishing the species concerns the female reproductive system; a monocarpogonial series and a polycarpogonial series are maintained. Several pairs of species are known (Abbott and Norris, 1965) that are vegetatively similar but can be distinguished on this basis, as the monocarpogonial *C. obtusifolia* J. Ag. and the polycarpogonial *C. pinnata* Setch. et Sweezy. In some species the female plants bear scattered cystocarps (Fig. 9.47); in other series the cystocarps are restricted to the margins. Degree of branching and the type of branching are also useful, but the production of lateral proliferations with age may cause some confusion. Species of this genus are generally of sublittoral occurrence, the attractive, bright red thalli often being washed ashore for collectors.

Two species that had been placed in the genus *Euthora*, *E. cristata* (C. Ag.) J. Ag., occurring on the Atlantic coast, and *E. fruticulosa* (Rupr.) J. Ag., occurring in the North Pacific, have been shown (Hooper and South, 1974) to represent a single species, exhibiting a range from very narrow forms to more broadly branched forms. Their carpogonial branch system agrees with that of *Callophyllis*, and, therefore, these forms should be recognized as the single species, *C. cristata* (C. Ag.) Kütz.

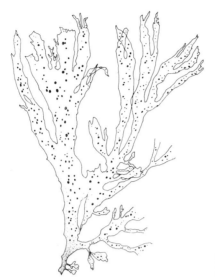

Fig. 9.47 *Callophyllis pinnata* Setch. et Swezy. Habit. × 0.23.

Family 5. Corallinaceae

This group of calcareous forms is the largest family in the order and one of the largest and most important families of red algae. Extending from pole to pole, they occur in tropical (Taylor, 1950; Littler, 1973A, B), temperate (Cabioch, 1971A, 1972A; Masaki, 1968; Johansen, 1969), and boreal and subarctic seas (Adey, 1966, 1970A; Lebednik, 1977A). They present an exciting diversity of adaptations to the marine environment. Their members include colorless parasites on other coralline algae, such as *Kvaleya* (Adey and Sperapani, 1971); some that are parasites only during their juvenile stages on other coralline algae, such as species of *Amphiroa* and *Pseudolithophyllum* (Cabioch, 1969); some species that grow almost exclusively on other coralline algae (Adey and Johansen, 1972); and *Schmitziella*, the only noncalcified genus in the family, which grows endophytically under the cuticle of a *Cladophora* (p. 183). Their members also include massive forms, such as species of *Clathromorphum* in the far North Atlantic and North Pacific.

In general, the coralline algae seem to flourish best in cold, agitated waters, but they also seem to tolerate extremes of temperature, particularly the crustose forms (Littler, 1972). In addition, they are known to inhabit great depths of the ocean due to their capacity to tolerate low light intensity, again the crustose types penetrating more deeply than articulated types. It has been demonstrated (Adey, 1970B) that the depth and geographical distributional patterns of boreal-subarctic crustose coralline-algae are controlled by temperature and light, the light dependence decreasing when the temperature dropped below 4–6°C, the winter temperatures. Some crustose coralline algae are restricted to small rocks, and others occur only on ledges. Such a parameter takes on significance in the distribution of these species (Adey, 1970C, 1971).

The early history of the classification of coralline algae is noteworthy and has been recently reviewed (Lebednik, 1977A). The two types of coralline algae, **articulated** and **nonarticulated** (see below), have somewhat different backgrounds in regard to how they were classified. It was not until the mid-nineteenth century that they were finally categorized together in a single plant family. The earliest concept of the articulated members, in the early eighteenth century, was that they were plants on the basis of their attached, arborescent habit. By the time of Linnaeus, however, during the mid-eighteenth century, the articulated coralline algae, along with assorted other articulated calcareous Rhodophyceae and Chlorophyceae as well as corals and other animals, were regarded as colonial animals consisting of numerous polyps and termed "zoophytes." In the early nineteenth century the term **nullipore** was introduced by Lamarck to distinguish the coralline algae, both articulated and nonarticulated types, from the porous corals, the nullipores lacking obvious pores. Shortly thereafter careful observations of *Corallina* and other articulated forms led Schweigger to reassert the plantlike nature of the articulated coralline algae. This move left the nonarticulated types somewhat in an uncertain systematic position until 1837, at which time their plant nature was recognized by Phillippi.

Taxonomically, the Corallinaceae are distinguishable as a family on several bases. One unifying trait is their calcareous nature (except for *Schmitziella* and one species of *Melobesia*). The reproductive organs are sunken into cavities called **conceptacles**. The spermatangial (Fig. 9.48*a*) and the carpogonial conceptacles (Fig. 9.48*b*) have a single opening, or ostiole; tetrasporangial conceptacles (Fig. 9.48*d*) have either a single ostiole or many, depending on the genus. Only *Sporolithon* (= *Archaeolithothamnium*), which interestingly was first known only as a fossil genus and later recognized to be extant as well, has tetrasporangia in sori.

Two types of growth occur in members of this family, most members having both types: apical growth and intercalary growth. A radial, longitudinal section through the growing margin (Fig. 9.49*a*) of a crustose representative is used as an example to explain how the intercalary meristem is derived from the apical meristem. Figure 9.50 is a schematic representation of the process. All meristematic cells are shown in black. Several terms are used to designate the various zones of tissues. The apical cells at the margin (left in Fig. 9.50) undergo anticlinal divisions and give rise to cells of the **hypothallus** (stippled). By periclinal or oblique divisions another zone of meristematic cells is derived on the upper surface. These latter cells become the intercalary meristem, cutting off cells both outwardly (**epithallus**) and inwardly (**perithallus**). The epithallial cells usually have a different morphology from either hypothallial or perithallial cells and differ in position. The small outermost cells of the epithallus are called **cover cells**.

The hypothallus of most coralline algae is, therefore, the result of apical growth; in some species the marginal meristem may be covered by an epithallial layer, making the meristematic cells intercalary in position. The medullary tissue of erect axes is homologous to the hypothallus and thus the product of apical growth. The vertical filaments of crusts (Fig. 9.50) and the lateral (i.e., cortical) filaments of erect axes typically have meristematic cells below the surface (Fig. 9.49*b*) and thus are the result

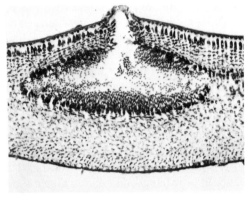

(a)

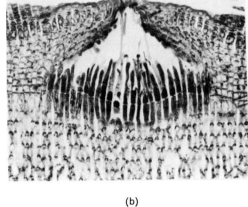

(b)

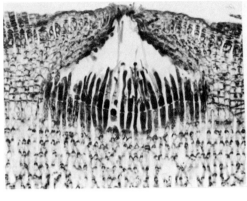

(c)

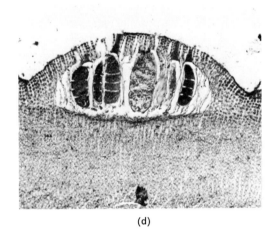

(d)

(e)

Fig. 9.48 Conceptacles in the Corallinaceae.
(a) Spermatangial conceptacle of *Clathromorphum reclinatum* (Fosl.) Adey. (b) Conceptacle containing carpogonial branches of *Clathromorphum loculosum* (Kjellm) Fosl. (c) Conceptacle containing a mature carposporophyte of *Bossiella californica* ssp. *schmittii* (Manza) Johansen. (d) Tetrasporangial conceptacle of *Mesophyllum lamellatum* (Setch. and Fosl.) Adey. (e) Bisporangial conceptacle of *Lithothamnium pacificum* (Fosl.) Fosl. (a) × 128; (b) × 221; (c) × 113; (d) × 143; (e) × 204. [(a), (b), (d), and (e) courtesy of Philip Lebednik; (c) courtesy of H. W. Johansen.]

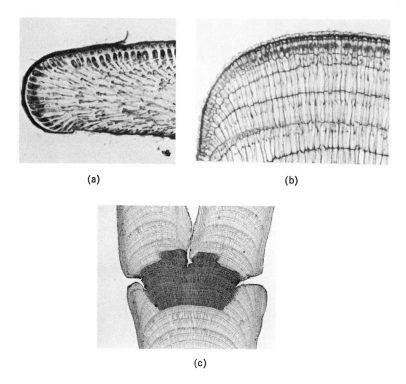

(a)   (b)

(c)

Fig. 9.49 Anatomy in the Corallinaceae. (*a*) Longitudinal section through margin of *Clathromorphum reclinatum.* (*b*) Longitudinal section through apex of *Amphiroa ephedraea* (Lamarck) Decn. (*c*) Longitudinal section through geniculum of *Amphiroa capensis.* (*a*) × 135; (*b*) × 108; (*c*) × 26. [(*a*) after Lebednik; permission of Syesis; (*b*) and (*c*) after Johansen; permission of Journal of Phycology.]

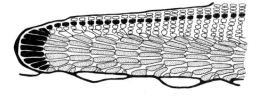

Fig. 9.50 Diagrammatic representation of vegetative development in *Mesophyllum*, as seen in vertical radial section. Meristematic cells are shown in black; hypothallial cells are shaded. The apical meristematic cells at the margin (left) give rise directly to the hypothallial cells by means of divisions anticlinal to the filament axis and indirectly to other meristematic cells on the upper surface by means of branching (periclinal) divisions. These derived meristematic cells give rise to the cells of the perithallus (proximally) and the epithallus (distally). (Courtesy of Dr. Philip A. Lebednik.)

of intercalary growth. The transformation form apical growth to intercalary growth occurs near the thallus margin in the zone where the filaments bend upward to a vertical orientation.

Another important term is **geniculum**, meaning "knee" or "joint." Certain coralline algae have uncalcified portions (Fig. 9.49c) alternating with calcified portions of the thallus, the latter being referred to as **intergenicula** or segments. Dating back to the mid-nineteenth century, two major subdivisions of the family had been recognized: those with genicula and thus having an **articulated** structure (Fig. 9.51a, b) and those lacking such genicula and thus being **nonarticulated** or crustose (Fig. 9.51c, d). Some

(a)   (b)

(c)   (d)

Fig. 9.51 Examples of coralline algae. (a) *Corallina mediterranea* Aresch. (b) *Bossiella californica* (Decn.) Silva. (c) *Clathromorphum loculosum.* (d) *Lithothamnium pacificum* Fosl. (a) × 2.9; (b) × 1.8; (c) × 0.27; (d) × 3.9. [(a) courtesy of Dr. H. W. Johansen; (b) after Johansen, permission of Phycologia; (c) and (d) courtesy of P. Lebednik.]

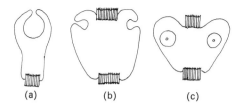

Fig. 9.52   Positions of conceptacles. (a) Axial. (b) Marginal. (c) Lateral. (After Johansen.)

of the species in this latter category do have branching systems, but they are never articulated. The great majority of the nonarticulated coralline algae have prostrate crustose habits.

More recently the family has been subdivided into a total of seven subfamilies[8] by Johansen (1969), four including nonarticulated types and three including articulated types. Characteristics deemed significant included the form of the tetrasporangial (or bisporangial) conceptacle. In the subfamily Melobesioideae each tetrasporangium has its own overlying pore. Two variations exist within this subfamily; the tissue between the sporangia may break down, leaving the mature sporangia in a common chamber with a multiporate roof, as in *Lithothaminium*, or the intersporangial tissue may persist, resulting in only a single tetrasporangium in each chamber, as in *Sporolithon*.[8] The other six subfamilies have uniporate tetrasporangial conceptacles; i.e., all tetrasporangia emerge through the same single ostiole.

Other traits that are used (Johansen, 1969; Adey and Johansen, 1972) to separate subfamilies include presence or absence of genicula, presence or absence of secondary pit connections, and the nature of the genicula, whether unizonal or multizonal. *Schmitziella* is retained in its own subfamily on the basis of its rudimentary conceptacles. The subfamily Corallinoideae contains genera with unizonal genicula and lacking secondary pit connections and genicula, which may be unizonal (*Lithothrix*) or multizonal (*Amphiroa*) (Fig. 9.49c).

Within the subfamily Corallinoideae, which contains the major group of articulated genera, the place of origin and development of the conceptacles is useful in distinguishing genera. Three alternatives exist: axial, in which the conceptacles are produced in a terminal position on the axis (Fig. 9.52a); marginal, in which the conceptacles are produced on the edges of the flattened segments (Fig. 9.52b); and lateral, in which conceptacles arise on the flattened surfaces of the segments (Fig. 9.52c). In the large genus *Bossiella* (Johansen, 1971), occurring in the Pacific Ocean, conceptacles are exclusively lateral. In the cosmopolitan genus *Corallina*, conceptacles have an axial position. In the large Pacific genus *Calliarthron* (Johansen, 1969), conceptacles are both marginal and lateral. On the other hand, in *Serraticardia*, which is a smaller genus also with a Pacific Ocean distribution, conceptacles may be both axial and lateral.

In a study of an articulated coralline alga the earliest indication of the initiation of a male or female conceptacle was observed (Johansen, 1973) to be the secretion of

[8]Johansen's tribe Sporolitheae has subsequently been elevated (Cabioch, 1971B) to the rank of subfamily, the Sporolithoideae. Disagreement exists whether the correct generic name should be *Sporolithon* or *Archaeolithothamnium*.

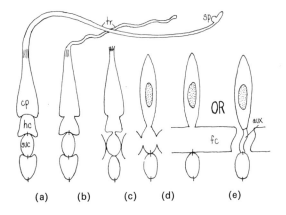

Fig. 9.53 Diagrammatic representations of fertilization and postfertilization in the Corallinaceae (*Calliarthron tuberculosum*). *aux*, auxiliary cell; *cp*, carpogonium; *fc*, fusion cell; *hc*, hypogynous cell; *sp*, spermatium; *suc*, supporting cell; *tr*, trichogyne. (After Johansen; published in 1969 by The Regents of the University of California; reprinted by permission of the University of California Press.)

a cap of deeply staining material below the epithallus. The cuticle and cover cells above the cap are gradually sloughed away, and the upper portions of the cortical cells below the cap also disintegrate. The lower nucleate portions of the cortical cells below the cap recover and become the initials of the reproductive cells.

The initiation of conceptacles in crustose forms involves the elongation of intercalary meristematic cells, accompanied by a simultaneous secretion of mucilage. This discoid area of elongate cells undergoes other changes to be transformed into the conceptacle (Lebednik, 1977A).

Carpogonial branches arise from the floor of a conceptacle and are always two-celled, arising singly from a supporting cell (Fig. 9.53a). Following fertilization, the carpogonium fuses with the hypogynous cell (Fig. 9.53b), and next there is a fusion with the supporting cell (Fig. 9.53c). In many species there is a general fusion of the supporting cells from the nonfertilized carpogonial branches with the earlier-mentioned fusion cell (Fig. 9.53d, e), and this resultant fusion cell, the **placental cell**, proceeds to bud off large carpospores from its margins. The supporting cell of the carpogonial branch thus functions as the auxiliary cell and is a cell in a specialized or accessory system, conforming to the order Cryptonemiales. It is thus a procarpial condition (however, see next paragraph for a different interpretation).

New information on the subfamily Melobesioideae has provided a different interpretation of postfertilization at least for that subfamily. Lebednik (1975) reported that within the female conceptacle an inner disc of carpogonial branches is surrounded by an outer ring of auxiliary cell branches (Fig. 9.54a). After fertilization a fusion cell is formed (Fig. 9.54b) within the carpogonial complex. From the edge of the fusion cell, connecting filaments frow outward, fusing with the auxiliary cell branch systems (Fig. 9.54c). Carpospores are then produced from the apices of the connecting filaments (Fig. 9.54d). He concluded that although the auxiliary cell branches would be accessory, conforming to the order, the separation of the carpogonia from the auxiliary cell branches constitutes a nonprocarpial condition, again, in this particular subfamily, the Melobesioideae.

Life histories in this family are generally diplobiontic with isomorphic gametophytic and tetrasporophytic stages. The first species to have its cycle completed in

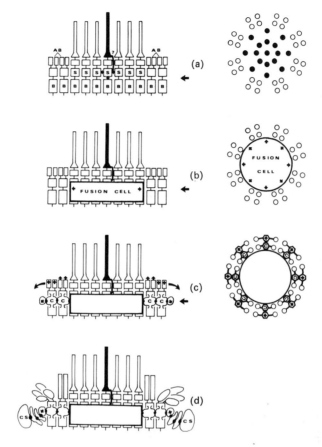

Fig. 9.54 Schematic representation of postfertilization development in *Meso-phyllum*; vertical median (left) and horizontal (right) sections of cystocarpic conceptacles. AB, auxiliary cell branch system; S, supporting cell; B, basal cell; C, connecting filament; CS, carpospore; M, modified cell. (*a*) Outgrowth from carpogonial base to supporting cell (or another cell, indicated by "?"); in horizontal view carpogonial branch systems are shown in black and auxiliary cell branch systems in white. (*b*) Formation of fusion cell from supporting and basal cells. (*c*) Growth of connecting filament (C) from fusion cell to periphery and modification of upper cells of the auxiliary cell branch systems (indicated by arrows). (*d*) Formation of carpospores and enlargement of upper cells of auxiliary cell branches. (After Lebednik.)

culture was *Corallina officinalis* L. (Stosch, 1969). Tetrasporangia are always zonately divided (Fig. 9.48*d*). Bisporangia (Fig. 9.48*e*) are produced in several species. When they are binucleate it is evident they are homologous to tetrasporangia and products of meiosis. Uninucleate bisporangia have been shown (Suneson, 1950), on the other hand, to represent the products of a mitotic division.

The coralline algae have been investigated in regard to their yearly reproductive cycles (Adey, 1964, 1965, 1966), spore dimensions, and spore germination characteristics, and it has been determined (Cabioch, 1972A; Chihara, 1973, 1974) that they can be divided into distinct categories. For the nonarticulated coralline algae, two groups were recognized (Chihara, 1974); the *Lithophyllum* group, which reproduces in the summer, has relatively small spores and has a distinctive pattern of spore cleavage, and the *Lithothamnium* group, which reproduces throughout the year or in seasons other than summer, has relatively large spores and has a different sequence of divisions by the germinating spores. For the articulated coralline algae there were two groups paralleling the nonarticulated groups in reference to pattern of spore germination: The *Amphiroa* group was similar to the *Lithophyllum* group, whereas the *Corallina* group was similar to the *Lithothamnium* group (Chihara, 1973). In addition to these major types of spore germination, Cabioch (1972A) recognized a third pattern, her *Neogoniolithon* mode, in certain nonarticulated members, and Chihara (1974) noted several other minor variations in the *Fosliella-Heteroderma* complex of nonarticulated coralline algae. The conclusion drawn from these investigations is that the earlier subdivision of the family into two subfamilies based upon an articulated versus nonarticulated distinction is not natural, since some nonarticulated types are more closely related to articulated types than to other nonarticulated types, and *vice versa*.

The diagnostic value of using reproductive structures to distinguish genera of articulated coralline algae has been discussed (Johansen, 1970). Although *Jania* and *Corallina* both have conceptacles in axial positions and appear superficially similar, significant differences exist in regard to the shape of their spermatangial conceptacles and the fusion cell in their carposporophytic conceptacles, their being thin and broad in *Corallina* but thick and narrow in *Jania*. The phenotypic variability in articulated coralline algae, such as differences in intergenicular size and distances between successive branches, can be significant in populations of a single species, reflecting environmental differences as depth (Johansen and Colthart, 1975). Striking differences in the development of male conceptacles were noted (Lebednik, 1977B) between the crustose genera *Clathromorphum* and *Mesophyllum*.

This family has been the subject of extensive reviews in recent years, the focus being upon both nonarticulated members (Littler, 1972; Adey and MacIntyre, 1973) and articulated members (Johansen, 1974).

Family 6. Choreocolaceae

This family contains extremely reduced parasites, which form whitish or almost colorless, erumpent cushions on their red algal hosts. Two of these genera, *Choreocolax* and *Harveyella*, which have been recorded from both the Atlantic and Pacific coasts, are discussed in greater detail below. A third genus, *Holmsella*, occurs in Europe and is parasitic on *Gracilaria* (p. 534). The major product of photosynthesis in *Gracilaria* was demonstrated (Evans, et al., 1973) to be floridoside. Autoradiographic analyses indicated that ^{14}C was transferred to *Holmsella* and accumulated there as floridoside, mannitol, and starch. It was postulated that the transfer was effected through the endophytic filaments arising at the base of the parasite and penetrating between the host cells. Although secondary pit connections were not detected between *Holmsella* and its host, they have been observed in other parasitic relationships (see below).

CHOREOCOLAX Reinsch   The genus *Choreocolax* (Gr. *chorein*, to penetrate + Gr. *kolaks*, a parasite) includes *C. polysiphoniae* Reinsch, which forms small, whitish cushions on various species of *Polysiphonia* (p. 561) and occasionally on the related genus *Pterosiphonia*. In an ultrastructural examination (Kugrens and West, 1973B), its cells were observed to be thick-walled and to lack starch in cortical cells but to have abundant starch in medullary cells. Apparent chloroplasts, lacking any internal lamellar system, seemed to represent reduced organelles that were probably nonfunctional. Many secondary pit connections were present between host and parasite, although whether a transfer of food reserves from host to parasite actually occurs was not investigated. According to Kugrens and West (1973B) such pit connections are common between host and **adelphoparasite** (p. 535) but apparently rare for host and **alloparasite**.

HARVEYELLA Schmitz et Reinke   The small, bulbose pustules (Fig. 9.55) of *Harveyella* (named for the Irish phycologist, *W. H. Harvey*) occur on the host genera *Odonthalia* and *Rhodomela* of the Rhodomelaceae (p. 559). *Harveyella*, therefore, is an alloparasite (p. 536), not closely related to its host(s). Careful field and laboratory studies (Goff, 1976B; Goff and Cole, 1976A, B) have demonstrated that the initial spore germination of *Harveyella* occurs in host wounds caused primarily by grazing isopods and amphipods. Rhizoidal cells penetrate between the host cells and establish

Fig. 9.55   *Harveyella mirabilis.* Bumpy cystocarpic pustules and smooth male pustules on the host *Odonthalia floccosa.* (After Goff and Cole; permission of Phycologia.)

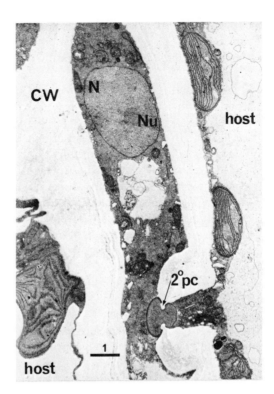

Fig. 9.56 *Harveyella mirabilis.* A rhizoidal cell of the parasite is attached to an adjacent host cell by means of a secondary pit connection (*2°pc*). (After Goff in Protoplasma.)

secondary pit connections (Goff, 1976A) with the host cells (Fig. 9.56). A rapid growth of the rhizoidal cells of *Harveyella* within the host leads to the rupturing of the host's outer wall and the development of the emergent colorless parasite. Morphological and cytological studies (Goff and Cole, 1973, 1976B) have shown that the life history (Fig. 9.57) of *Harveyella mirabilis* (Reinsch) Schmitz et Reinke on its host *Odonthalia floccosa* (Esper) Falkenberg consists of an alternation of isomorphic phases, with unisexual gametophytes. The cystocarpic pustules usually have a bumpy appearance; the male pustules have a smooth appearance (Fig. 9.55).

Sections through the parasite and host reveal three zones (Fig. 9.58). Zone 1 is the cortex of the parasite, where tetraspores, spermatia, and procarps of the three different reproductive plants are borne. Zone II is the medulla of the host. Some host cells (stippled in Fig. 9.58) are scattered in both zones I and II, this arrangement seemingly being nutritionally significant (Goff, 1976B). Zone III is the interdigitation zone in which the rhizoidal cells of the parasite are in close association with the host cells. Some damage to the host cells was observed in this region, but no such damage was noted in zones I and II.

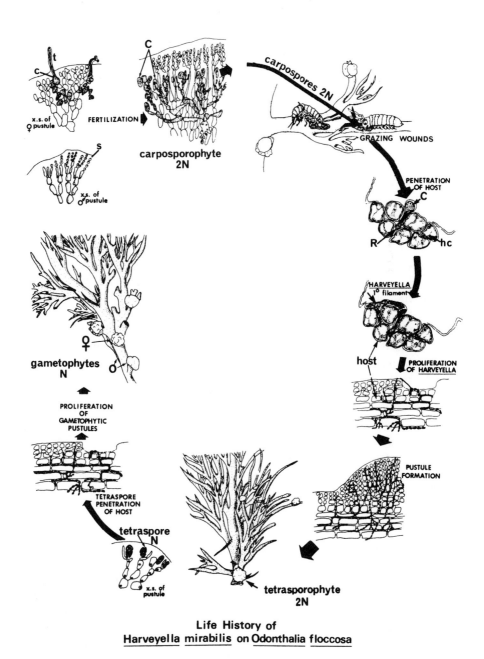

Life History of
Harveyella mirabilis on Odonthalia floccosa

Fig. 9.57 Life history of *Harveyella mirabilis* on *Odonthalia floccosa*. (After Goff and Cole; reproduced by permission of the National Research Council of Canada from the Canadian Journal of Botany, vol. 54, pp. 281–292, 1976.)

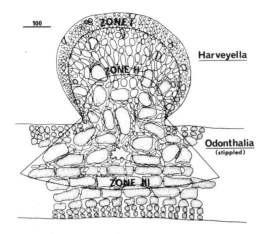

Harveyella

Odonthalia
(stippled)

Fig. 9.58 Section of parasite (*Harveyella mirabilis*) in host (*Odonthalia floccosa*). Interdigitation zone is zone III, although host cells (stippled) also occur in zones I and II. (After Goff; permission of Journal of Phycology.)

Carposporophyte development has been studied (Goff and Cole, 1975) and a similarity to certain families in the orders Cryptonemiales and Gigartinales was noted as well as the lack of a similarity to the Ceramiales, confirming *Harveyella's* reputation as an alloparasite. A four-celled carpogonial branch is borne on a large intercalary supporting cell, which acts as the auxiliary cell. A short connecting filament transfers the zygote nucleus to the auxiliary cell, and postfertilization events lead to the development of a carposporophyte, composed of filaments bearing terminally borne carpospores. As the gonimoblast develops, medullary cells of the female gametophyte in the vicinity of the gonimoblast initial elongate, forming long columnar cells with chains of small cortical cells at their ends. The elongation of these columnar cells creates an open chamber in which the gonimblast filaments are both horizontally and vertically branched. Carpospores are released through an ostiole.

Autoradiographic studies indicated that $H^{14}CO_3^-$ is photosynthetically assimilated by the host and subsequently transferred to *H. mirabilis*. This transfer took place primarily from host medullary cells to adjacent rhizoidal cells of the parasite. The translocated compound was similar to glucuronic and galacturonic acids on the basis of chromatographic separations.

Order 7. Gigartinales

In terms of number of families, this order is the largest in the Rhodophycophyta, now containing about two dozen families (Kylin, 1956; Papenfuss, 1966; Searles, 1968; Womersley, 1971; Kraft, 1973; Min-Thien and Womersley, 1976). It is a large, heterogeneous, and admittedly ill-defined assemblage. The primary distinguishing trait of the order, as defined by Kylin (1932), is that the auxiliary cell is not borne on a special, or accessory, filament but is an intercalary cell of an ordinary vegetative filament of the thallus. The overlapping nature of some genera with both the Gigartinales and the Cryptonemiales, such as the recently described Australian genus *Adelophyton* (Kraft, 1975), emphasizes the blurred nature of the distinction between these orders. Although theoretically the difference might appear to be a natural means to separate these two orders, in practice it is difficult and sometimes almost impossible to determine if an auxiliary cell is borne on a filament of vegetative or accessory (= specialized) origin (Searles, 1968; Goff and Cole, 1975).

The direction of development of the young carposporophyte, toward the thallus interior or the exterior, has been employed as a major taxonomic criterion at the family level, but this distinction is perhaps not so useful as formerly believed, since the carposporophyte may be developed laterally (Kraft and Abbott, 1971) or other vegetative and reproductive features may indicate more compelling justification for placing in the same family genera with both inwardly and outwardly developing carposporophytes (Searles, 1968).

Additional characters of significance in separating families include construction (uniaxial versus multiaxial), division pattern of tetrasporangia (cruciate versus zonate), and the spatial relationship of carpogonium to auxiliary cell (procarpial versus nonprocarpial). These distinguishing characteristics parallel the separation of families in the Cryptonemiales.

Most members of this order are moderate-sized to large fleshy algae, but crusts (the family Cruoriaceae), softly gelatinous types (e.g., *Calosiphonia*), and even one moderately calcified genus (*Titanophora*) are also members of this large order. Representatives of only five Gigartinalean families are treated in the present account. A key to their separation follows.

1. The gonimoblast developing from the auxiliary cell, following contact by the connecting filament 2
1. The gonimoblast developing from the connecting filament, following contact of the connecting filament with the auxiliary cell . . . . Cruoriaceae
   2. The gonimoblast developing toward the interior of the thallus 3
   2. The gonimoblast developing toward the exterior of the thallus . . . . Gracilariaceae
   3. Tetrasporangia zonately divided . . . . Solieriaceae
   3. Tetrasporangia cruciately divided 4
      4. Medulla composed of large cells, pseudoparenchymatous in appearance . . . . Phyllophoraceae
      4. Medulla composed of narrow filaments, nonparenchymatous in appearance . . . . Gigartinaceae

Family 1. Gigartinaceae

This family contains mostly cartilaginous forms, often flattened but also terete or compressed, simple or branched, and of multiaxial construction. Thalli consist of a basal expanded crust that grows radially over the substratum and erect axes. The crust is perennial, giving rise to crops of annual or more persistent upright portions, in which the reproductive structures are produced. The activity of the apical meristem of erect axes either slows down or ceases when these axes are only a few millimeters high, and a diffuse secondary meristem is developed in an intercalary position (Norris and Kim, 1972). New medullary filaments grow out, and much of the

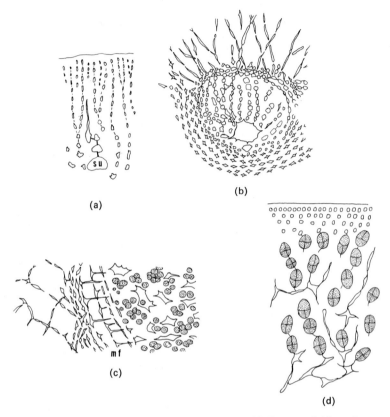

Fig. 9.59 Reproductive structures in *Gigartina*. (*a*) Carpogonial branch system on a supporting cell (*su*). (*b*) Early development of gonimoblasts and surrounding medullary filaments. (*c*) Portion of an older cystocarp surrounded by a special investment of medullary filaments (*mf*). (*d*) Tetrasporangia formed from series of cells in the inner cortex. (*a*) × 250; (*b*), (*c*) × 135; (*d*) × 175. [(*a*)–(*c*) after Mikami; (*d*) after Kylin.]

expansion is caused by the elongation of these new medullary filaments. The resultant structure is composed of a loosely filamentous medulla and a cortex of small, pigmented cells, usually arranged in densely crowded, branched, anticlinal rows.

The genera included in the Gigartinaceae demonstrate a rather monotonous uniformity in regard to their sexual development (Mikami, 1965). Carpogonial branches are three-celled, a single carpogonial branch occurring per supporting cell (Fig. 9.59*a*). This supporting cell, which is in a vegetative filament of the inner cortex, is connected to the surrounding cells by secondary pit connections and may bear sterile filaments toward the outer cortex. Following fertilization, the carpogonium transfers the zygote nucleus to the supporting cell by a broad connection, the supporting cell thus functioning as the auxiliary cell. The family is, therefore, procarpial.

Many gonimoblast filaments are developed from the auxiliary cell toward the interior of the thallus (Fig. 9.59*b*). The carposporophyte is more or less intermingled with the remains of the medulla. A special rhizoidal envelope, consisting of cytoplasmic-rich, nutritive filaments, is developed to surround the carposporophyte (Fig. 9.59c) in the genera *Gigartina, Iridaea,* and *Rhodoglossum.*[9] *Chondrus* lacks such a special filamentous investment around the carposporophyte. Carpospores are usually disseminated by the eventual disintegration of the thallus, since they are deeply embedded in the mature tissue. In *Gigartina,* however, the carposporophytes are emergent and provided with ostioles for spore release.

Tetrasporangia are always cruciately divided. They may be differentiated from subcortical cells, as in *Rhodoglossum,* or from special medullary filaments (Fig. 9.59d), as in *Iridaea* and *Chondrus.* The tetrasporangia are typically produced in large masses.

The life history of this family is typically diplobiontic and isomorphic, involving unisexual gametophytes. But an interesting exception has been discovered in some species of *Gigartina,* discussed below.

GIGARTINA Stackhouse   Close to 100 species have been described and assigned to *Gigartina* (Gr. *gigarton,* a grape stone) in both northern and southern hemispheres. The genus seems to be most diversified in the North Pacific, and species on the California coast, as *G. exasperata* Harv. et Bail., *G. corymbifera* (Kütz.) J. Ag., and *G. harveyana* (Kütz.) S. et G., are perhaps the most massive of all red algae. Shapes (Fig. 9.60) range from simple to much branched, dichotomously or pinnately.

[9]The generic separation of these three taxa has recently been questioned by Kim (1976), who concluded that *Iridaea* and *Rhodoglossum* should be merged with *Gigartina.* He also suggested that those species of *Gigartina* in subgenus Mastocarpus with an alternation of heteromorphic phases (p. 524) be removed (as the genus *Mastocarpus* Kuetz.) from the Gigartinaceae, a transfer supported by additional reproductive differences.

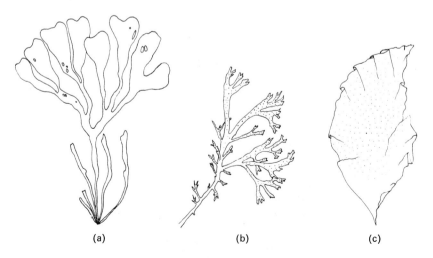

| (a) | (b) | (c) |

Fig. 9.60 *Gigartina.* (*a*) *G. volans* (C. Ag.) J. Ag. (*b*) *G. agardhii* S. et G. (*c*) *G. corymbifera* (Kuetz.) J. Ag. (*a*) × 0.3; (*b*) × 0.6; (*c*) × 0.13.

This genus is usually separated from the other genera by the numerous short papillae or projections that cover the surfaces of the thalli. The carpogonial branches are produced in these papillae in the female plants, but male plants may lack them. Tetrasporangia may be located in the papillae or scattered around their bases.

For most Gigartinas an alternation of isomorphic phases has been recognized, even though the frequency of the unisexual plants, particularly male individuals, may be much less than that of tetrasporic plants. In some species tetrasporophytes were seemingly absent in the life history, until culturing studies (West, 1972C) demonstrated that the tetrasporophyte may be a small crustose phase and the life history accordingly an alternation of hetermorphic phases. Tetraspores of the crustose alga *Petrocelis franciscana* S. et G., an apparent member of the Cruoriaceae (p. 529), were released, and the germlings developed into crustose discs with marginal meristems. From these discs were produced erect multiaxial blades, which were morphologically similar to *Gigartina agardhii* S. et G. To verify the relationship, cultures were established (Polanshek, 1974) from carpospores of *G. agardhii* from the field, and they produced crustose algae resembling *Petrocelis* (Fig. 9.61), which, however, did not become fertile. Reciprocal crosses were established between field-collected gametophytes of *G. agardhii* and gametophytes derived from tetraspores of *Petrocelis*, and the various crosses all resulted in carposporophyte development, indicating that *Gigartina agardhii* and *Petrocelis franciscana* are involved in the same life history. Subsequent culturing work involving *P. middendorffii* (Rupr.) Kjellm., a species of the North Pacific, demonstrated (Polanshek and West, 1975) interfertility of the gametophytes with the gametophytes derived from the more southerly occurring *P. franciscana*, causing these workers to reduce *P. franciscana* to synonymy with *P. middendorffii*.

Fig. 9.61 *Petrocelis middendorfii* (Rupr.) Kjellm. Intercalary production of tetrasporangia. × 169. (Courtesy of Dr. Alan Polanshek.)

Another species of the subgenus Mastocarpus of *Gigartina* in which tetrasporic plants seemed to be lacking is *G. stellata* (Stackh.) Batt. of the North Atlantic. An exclusively apomictic life history was reported (Chen, et al., 1974; Edelstein, et al., 1974) for this species from Nova Scotia. But a variable situation was also reported (West and Polanshek, 1975) in other regions of its range. In strains from Wales, Scotland, and Maine a direct development of the multiaxial, erect, dichotomously forked blades from carpospores of similar plants occurred in some strains, and a production of crusts resembling *Petrocelis* from carpospores occurred in other strains. The latter alternation of dissimilar phases was comparable to that demonstrated in *G. agardhii* and seemingly involved sexual reproduction.

In an ecological study of *Gigartina stellata*, Burns and Mathieson (1972B) observed that salinity was an important factor in limiting the distribution of this species. It is a perennial plant, largely restricted to the low littoral zone. These workers also determined that summer harvesting of *G. stellata* allowed maximum regrowth, as long as the holdfast is not damaged.

CHONDRUS Stackh.   Thalli of *Chondrus* (Gr. *chondros*, cartilage) tend to occur in bushy clumps and usually have branched, flattened blades terminating a narrow, cartilaginous stalk. Although about seven species of variable shape, size, and branching pattern are recognized (Mikami, 1965) from the North Pacific, a single species, *C. crispus* Stackh. (Fig. 9.62a), occurs on both sides of the North Atlantic. This dichotomously branched species, familiarly known as "Irish moss," is well-known because of its importance in the phycocolloid industry and is one of the primary sources of carrageenan (p. 28). A recent publication (Harvey and McLachlan, 1972) has been

Fig. 9.62 (*a*) *Chondrus crispus* Stackh. Habit. (*b*) *Iridaea cordata* var. *splendens* (S. et G.) Abbott. Habit. (*a*) × 0.49; (*b*) × 0.18.

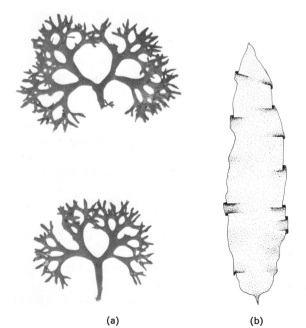

(a)   (b)

devoted to reviewing the status of our knowledge of the taxonomy, life history, ecology, and biochemistry of this species.

This genus is distinguished (Kim, 1976) from the other members of the family by the fact that the carposporophyte is usually not surrounded by special medullary filaments, the carposporophytes being developed in the blade, not in papillae. The life history of *Chondrus crispus* consists of an alternation of isomorphic phases. Gametophytes are separately sexed. The complete cycle, which required 20 months in culture (Chen and McLachlan, 1972), was followed. Starting with tetraspores, gametophytes were obtained, which produced carpospores; the carpospores developed into tetrasporophytes, which produced tetraspores; the tetraspores gave rise to a second generation of gametophytes. Male plants, which are generally shorter and less branched than female plants, bear whitish patches on the branch tips. The branch tips of female plants bear procarps, consisting of a three-celled carpogonial branch on an enlarged supporting cell, which functions as the auxiliary cell. After the direct fusion of carpogonium with auxiliary cell, numerous colorless, branched gonimoblast filaments arise from the auxiliary cell (Prince and Kingsbury, 1973A) and course somewhat loosely through the medulla, forming short chains of carpospores.

The ecology of *Chondrus crispus* has been intensively studied and reported (Prince and Kingsbury, 1973B, 1973C; Mathieson and Prince, 1973; Mathieson and Burns, 1975). A tolerance to a broad range of temperatures, light conditions, and salinities has been shown. The well-recognized extreme morphological variability in this species is perhaps tied in with this ecological tolerance to varying conditions of growth. Maximum growth occurs in spring and summer. The production of carpospores begins in the summer; tetrasporic plants show a minor peak of spore release in late spring and a major peak in the summer (Mathieson and Prince, 1973).

An interesting physiological phenomenon has been recognized in *Chondrus* as well as in other genera of this family, in which a differing distribution of galactans, i.e., carrageenans, occurs in gametophytic and tetrasporophytic plants. In *Chondrus crispus* (McCandless, et al., 1973) sexual plants produce κ-carrageenan, which gels, or precipitates, in the presence of potassium ion; μ-carrageenan, which is the presumed precursor of κ-carrageenan; and a third galactan, which is not λ-carrageenan, which does not gel in the presence of potassium ion. By means of a fluorescent antibody staining technique, using antibodies specific to either κ-carrageenan or λ-carrageenan, it has been possible to demonstrate (Gordon-Mills and McCandless, 1975; Hosford and McCandless, 1975) on the basis of the staining reaction that κ-carrageenan was present in the vegetative tissue of the female gametophyte as well as the cross walls of tetraspores (the first cells of the gametophytic phase) and that λ-carrageenan was present in the vegetative tissue of the tetrasporophyte as well as the cell walls of the carposporophyte (derived from the zygote nucleus). Thus, the differentiation of the haploid and diploid phases is detectable from the earliest stages. This fractionation of types of carrageenans is paralleled in *Gigartina* (Chen, et al., 1973; Pickmere, et el., 1973) and *Iridaea* (McCandless, et al., 1975).

Rates of photosynthesis and respiration in *Chondrus crispus* are adversely affected by a high degree of dehydration, more so than in *Gigartina stellata* (Mathieson and Burns, 1971). Spores of *C. crispus* germinated and grew over a broad range of salinities

(Burns and Mathieson, 1972A). A correlation between the ratio of the carrageenan fractions and plant age, reproduction, and habitat has been noted (Fuller and Mathieson, 1972; Mathieson and Tveter, 1975).

IRIDAEA Bory   One of the commonest red algae on the west coast of North America, *Iridaea* (L. *iris*, stem: *irid-*, rainbow) contains simple foliose forms as well as forked or lobed forms of littoral and sublittoral habitats. One of the most prevalent species on the California coast is *Iridaea cordata* (Turner) Bory var. *splendens* (S. et G.) Abbott, in which the usually simple (Fig. 9.62*b*) but occasionally deeply cleft blades may reach lengths of more than 100 cm and give off a brilliant iridescence when viewed underwater. The genus has an almost exclusively eastern Pacific distribution, only two species extending westward to Japan (Abbott, 1971). Its counterpart genus in the western Pacific is *Chondrus*, which has about the same number of species.

Although as many as 18 species of *Iridaea* have been described from the eastern and north Pacific, this number was reduced to 7 species, one including two forms, in a monographic treatment by Abbott (1971). She also evaluated (1972B) the taxonomic criteria that have been employed in separating the species and stressed the fact that the great variation in external form in this genus cannot be appreciated from examining herbarium specimens on account of extreme shrinkage. Field studies revealed that certain blade shapes were correlated with particular tidal levels.

Tetrasporangia in *Iridaea* are cut off from special medullary filaments, as in *Chondrus*, but a distinction from *Chondrus* can be made by an inspection of carposporophytic specimens. A surrounding network of special sterile filaments is present around the carposporophytes in *Iridaea*, but such is lacking in *Chondrus*.

Mariculture of *Iridaea*, which along with *Gigartina* is a commercially important seaweed because of the high carrageenan content, in Puget Sound has been investigated (Jamison and Beswick, 1972; Waaland, 1973). These genera are dominant only in areas where there is an outcropping of hard clay. The limiting factor is suitable hard substrate. In southern Puget Sound 99% of the available substrate is loose sand or mud interspersed with gravel, which is unsuitable for *Iridaea* and *Gigartina*.

Family 2. Phyllophoraceae

This family resembles the Gigartinaceae in several respects, such as the multiaxial construction, the inwardly directed development of the carposporophyte, and the multiple number of gonimoblast initials cut off by the auxiliary cell. It can be distinguished by the nature of the medulla, which is large-celled rather than filamentous, giving a pseudoparenchymatous aspect. Plus, the tetrasporangia occur in **nemathecia,** which are raised areas on the surface, rather than immersed in the thallus. Thalli may be flattened or terete. A dichotomous branching pattern is prevalent among members. The reproductive development is basically similar to that of the Gigartinaceae (Mikami, 1965).

Several different life-history patterns are known to occur in the Phyllophoraceae, sometimes variations existing within the same genus, as in *Phyllophora*. Examples of these variations are discussed below.

PHYLLOPHORA Grev.   The erect thalli of *Phyllophora* (Gr. *phyllon*, a leaf +
Gr. *phoreo*, I bear) are either simple or dichotomously branched, the axes arising
from a discoid base and flattened, linear or wedgeshaped, terminal portions borne on
cylindrical stipes (Fig. 9.63*a*). Small proliferous blades often appear on the margins
of the older blades. Many of the species in this genus are sublittoral in their occurrence.

The taxonomy and nomenclature of *Phyllophora* have been treated (Newroth and
Taylor, 1971), and keys have been contributed (Newroth, 1971B) for the identification
of the species. In *P. pseudoceranoides* (Gmelin) Newroth et Taylor an alternation of
isomorphic phases has been recognized (Newroth, 1972), with unisexual gametophytes.
In *P. truncata* (Pallas) Zinova (Fig. 9.63*g*), however, culturing and cytological studies
(Newroth, 1971A) demonstrated an abbreviated life history, with a diploid tetrasporangial phase on a haploid bisexual gametophyte, seemingly as a result of fertilization. Meiosis occurred in the tetrasporangia, the tetraspores developing into
apparent gametophytes.

Several vegetatively similar genera have been segregated out of *Phyllophora*.
These include *Petroglossum* (Hollenberg, 1943), which was distinguished by the production of spermatangial sori continuous with the thallus surface rather than in pits
as in *Phyllophora; Ozophora*, an old genus that was reestablished by Abbott (1969) on
the basis of special spermatangial leaflets; and *Schottera* (Guiry and Hollenberg,
1975) on the basis of its unique manner of carposporophyte development.

GYMNOGONGRUS Martius   The dichotomously branched axes of *Gymnogongrus* (Gr. *gymnos*, naked + Gr. *gongros*, an excrescence) may be terete or slightly
compressed, as in *G. linearis* (Turner) J. Ag., the most commonly occurring species of

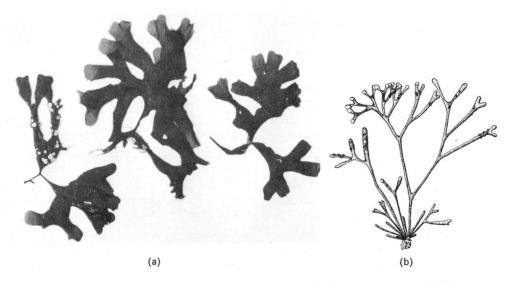

(a)   (b)

Fig. 9.63   (*a*) *Phyllphora truncata* (Pallas) Zinova. Habit. (*b*) *Gymnogongrus
linearis* (Turner) J. Ag. Habit. (*a*) × 0.63; (*b*) × 0.25.

this genus on the west coast. It grows from central California to southern British Columbia (Markham and Newroth, 1972). The axes may be markedly flattened and thin, as in *G. platyphyllus* Gardner, with the same latitudinal range. Two different types of life history are known in this genus. In *G. linearis* only spermatangial or cystocarpic plants are known (Newroth and Markham, 1972), the carpospores seemingly developing parthenogenetically into a new crop of female plants. Cystocarps persist for several years before spores are shed. In *G. platyphyllus* only one type of plant occurs, and it bears nemathecia of tetrasporangia and occasional spermatangial areas. These wartlike nemathecia, which are conspicuously raised areas of the thallus containing reproductive structures, were formerly regarded as a parasitic genus (*Actinococcus*).

Thalli of *G. linearis* (Fig. 9.63*b*) typically occur in dense clumps of perhaps 50–100 individuals arising from a common discoid base. It is encountered in areas periodically buried under sand (Markham and Newroth, 1972), which happens during the summer months in populations along the coasts of Oregon and British Columbia.

Family 3. Cruoriaceae

This family contains crustose forms that adhere tightly to the substratum. The thallus is composed of sparingly branched, parallel, vertical filaments, which may be embedded in mucilage and thus only loosely united, as in *Cruoria* (L. *cruor*, blood), or firmly united to one another and lacking a mucilaginous matrix, as in *Haematocelis* (Gr. *haima*, blood + Gr. *kelis*, a spot).

The auxiliary cell, which is an ordinary vegetative cell remote from the carpogonium, is contacted by the connecting filament issuing from the fertilized carpogonium, but the gonimoblast is developed from the connecting filament rather than from the auxiliary cell. The gonimoblast usually consists of just a few cells, all of which are released as carpospores. Tetrasporangia, which are produced on isomorphic tetrasporophytes, are zonately divided in *Cruoria* (Fig. 9.64) and laterally inserted on the vertical filaments. *Cruoriopsis* has cruciately or irregularly divided tetrasporangia.

Fig. 9.64 *Cruoria pellita* (Lyngb.) Fries. Habit of tetrasporic thallus. × 116. (From Newton.)

The crustose genus *Petrocelis* (Gr. *petros*, a rock + Gr. *kelis*, a spot or stain), forming almost tar-like, adherent, purplish patches in the mid-littoral zones on both Atlantic and Pacific coasts and previously placed in this family, has been linked (West, 1972C) to the life history of certain species of *Gigartina* (p. 523). Some populations of certain Gigartinas have a dissimilar crustose stage bearing tetrasporangia (Fig. 9.61), which has passed as the genus *Petrocelis*.

Family 4. Solieriaceae

Multiaxial construction characterizes the members of this family as do the lack of procarps, zonately divided tetrasporangia, and the formation of only one initial gonimoblast cell from the auxiliary cell in an inward direction (Min-Thien and Womersley, 1976). A diverse spectrum of shapes is encountered among the genera of this family: terete, branched axes in *Neoagardhiella* and *Solieria*; flattened blades bearing proliferations in *Opuntiella*; flattened, strap-shaped, branched axes in *Sarcodiotheca;* much branched, fleshy, cartilaginous axes in *Eucheuma*; and the small globose genus *Gardneriella*, which is parasitic on *Neoagardhiella*.

The carposporophytes are either deeply immersed in the thallus or may be projecting. The carposporophyte may have a small-celled, inner sterile region, as in *Neoagardhiella* and *Sarcodiotheca*, or have a central spherical fusion cell, as in *Solieria* and *Eucheuma*. A pericarp is usually present, provided with an ostiole. The zonately divided tetrasporangia are scattered over the thallus surface among the cortical cells. Only three representative genera are discussed below.

NEOAGARDHIELLA Wynne et Taylor   This genus has a temperate to subtropical distribution, occurring on both the Atlantic and Pacific coasts as well as in the Gulf of Mexico. Thalli of *Neoagardhiella* (Gr. *neos*, new; named for the Swedish phycologist *J. G. Agardh*) are much branched, cylindrical axes (Fig. 9.65*a*), with a somewhat crisp texture, arising from a discoid holdfast. Branches are usually somewhat spindle-shaped, tapering toward their bases. A resemblance to *Gracilaria* (p. 534) is evident, but a transection of an axis of *Neoagardhiella* reveals a distinctive anatomical organization of the medulla: a central core of longitudinally directed, narrow filaments surrounded by large, rounded medullary cells and an outermost cortical region of small, pigmented cells. Another difference is that the tetrasporangia are zonately divided in this genus but cruciately divided in *Gracilaria*.

The type species of this genus, *N. baileyi* (Harv. ex Kütz.) Wynne et Taylor, has been examined by many workers in regard to reproduction and postfertilization development. Carpogonial branches are three-celled and occur near the very tip of the branches. Following fertilization, a connecting filament arises from the carpogonium and makes contact with a remote auxiliary cell; the auxiliary cell is an ordinary vegetative cell in an intercalary position and distinguishable prior to fertilization (Smith, 1955). The carposporophyte is then produced toward the interior of the thallus. Terminal carpospores are produced, and the very central portion of the carposporophyte consists of small, sterile cells.

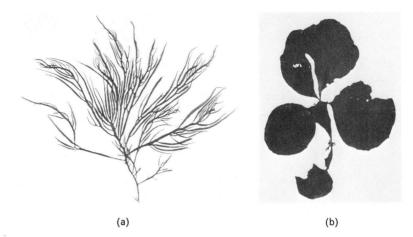

(a)   (b)

Fig. 9.65 (*a*) *Neoagardhiella baileyi* (Harv. ex Kütz.) Wynne and Taylor. Habit. (*b*) *Opuntiella californica* (Farl.) Kyl. Habit. (*a*) × 0.31; (*b*) × 0.19.

Two distinctive entities were found (Taylor and Rhyne, 1970) to be passing as one species, *Agardhiella tenera*. A closer examination of reproductive material led Wynne and Taylor (1973) to recognize one entity, with a more southerly range, as belonging to the related genus *Solieria* and the second entity, with a more northerly range, as conforming to a distinct genus, *Neoagardhiella*. A second species, *N. ramosissima* (Harv.) Wynne et Taylor, occurs in Florida and North Carolina.

OPUNTIELLA Kylin   The undivided, orbicular blades of *Opuntiella* (similar in outline to the cactus genus *Opuntia*), a monotypic Pacific genus with a range from southern California to Alaska, present a distinctive appearance because of the production of proliferous secondary blades from the thallus margins (Fig. 9.65*b*). *Opuntiella californica* (Farlow) Kylin tends to be a sublittoral alga, but it may extend into the lower littoral zone. Very large cylindrical vesicle, or gland, cells are present in the cortex close to the blade surface. The carposporophytes of this genus have a large fusion cell in the central region; a surrounding filamentous involucre is absent.

EUCHEUMA J. Ag.   Economically valuable because of its high carrageenan content, *Eucheuma* (Gr. *eu*, true + Gr. *cheuma*, basin) is a large tropical genus, exceedingly difficult taxonomically and exhibiting great morphological diversity. Species may be much branched (radially or bilaterally), terete or flattened, and even foliose, as *E. procrusteanum* Kraft from the Philippines (Kraft, 1969). Thalli are often very fleshy and cartilaginous, forming clumps or mounds of congested axes. Generic traits include a large fusion cell Fig. 9.67*b*) located in the center of the carposporophyte, the fusion cell resulting from a fusion of the auxiliary cell with sterile gonimoblast filaments. The carposporophytes are usually in papillae projecting from the thallus surface; these papillae may be stalked (Fig. 9.66*c*). The medulla is very densely compact (Fig. 9.66*b*).

(a)   (b)   (c)

Fig. 9.66 *Eucheuma isiforme* (C. Ag.) J. Ag. (*a*) Habit. (*b*) Portion of transection showing dense central medulla. (*c*) Stalked cystocarps. (*b*) × 53; (*c*) × 0.8. (Courtesy of Dr. Donald Cheney.)

In a biosystematic study of *Eucheuma* in Florida, two groups of species were differentiated (Cheney and Dawes, 1974), one conforming to the circumscription of *Eucheuma* and the other differing in many respects. The former group, including *E. isiforme* (C. Ag.) J. Ag. (Fig. 9.66*a*), has thalli with terete axes, a large fusion cell, and a compact central portion in the medulla. The latter group, including several species formerly placed in *Eucheuma*, has thalli with flattened axes, a small-celled central carposporophyte (Fig. 9.68) and a loose medullary construction, which evidence suggests the removal of this complex from *Eucheuma*.

A typical diplobiontic life history seems to be the rule in *Eucheuma*. Male plants have spermatangial sori, the spermatia being budded from the tips of elongate cells (Fig. 9.67*c*). Female plants have three-celled carpogonial branches (Fig. 9.67*a*), produced just inside the outer cortex. Following fetilization, connecting filaments make contact with remote auxiliary cells, from which the carposporophyte is developed. The conspicuous fusion cell (Fig. 9.67*b*) is surrounded by gonimoblast filaments bearing carpospores, and the entire carposporophyte is surrounded by a dense filamentous nutritive tissue. Tetrasporic plants bear zonately divided tetrasporangia (Fig. 9.67*d*) in the outer cortex. Other data derived from culturing and transplant studies as well as electrophoretic analyses supported this separation.

A series of ecological papers has appeared (Dawes, et al., 1974; Dawes, et al., 1974; Mathieson and Dawes, 1974), focusing on the seasonal growth, phenology, and variation of carrageenan and total carbohydrate content over the year.

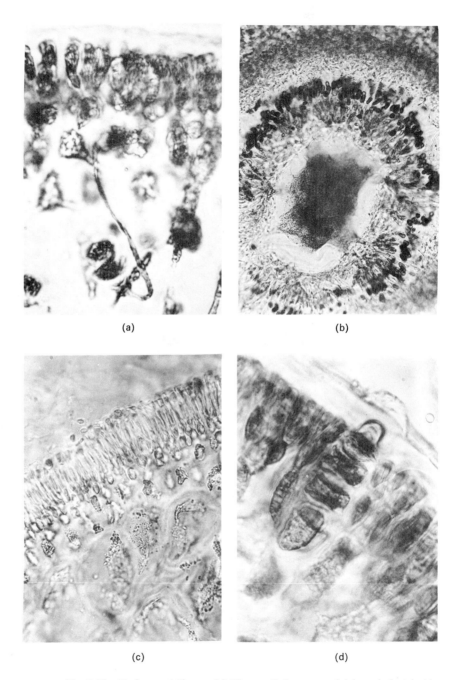

(a)

(b)

(c)

(d)

Fig. 9.67 *Eucheuma isiforme.* (*a*) Three-celled carpogonial branch just inside outer cortex. (*b*) Mature cystocarp with its characteristic central fusion cell. (*c*) Spermatia production from a spermatangial sorus. (*d*) Zonately divided tetrasporangia. (*a*) × 900; (*b*) × 98; (*c*) × 675; (*d*) × 1800. (Courtesy of D. Cheney.)

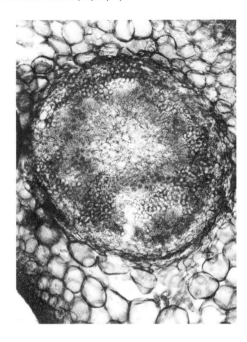

Fig. 9.68 *Eucheuma gelidium* (J. Ag.) J. Ag. Sectional view of mature cystocarp, showing the small-celled construction of the central region in this particular species. × 109. (Courtesy of D. Cheney.)

Family 5. Gracilariaceae

Although about eight genera are contained in this family, only *Gracilaria* is significant in terms of number of species and distribution throughout the world. The members of this family share the characteristics of a carposporophyte developed toward the outside of the thallus, cruciately divided tetrasporangia, and a pseudoparenchymatous construction such that the cells of the medulla are more or less isodiametric rather than filamentous.

GRACILARIA Grev.   *Gracilaria* (L. *gracilis*, slender) is a commercially valuable agarophyte (Levring, et al., 1969; Kim, 1970), and the presence of its many species (about 100) distributed throughout temperate and tropical seas (Michanek, 1971) causes it to be ranked as one of the more valuable red algae. Thalli are usually highly branched, terete to flattened forms, with a cartilaginous texture. Clumps of more or less dichotomously branched *G. verrucosa* (Hudson) Papenf. (Fig. 9.69) may be 30–40 cm across. This species extends from Canada to Florida and the Gulf of Mexico on the east coast and from southern British Columbia to Baja California and the Gulf of California on the west coast of North America. This species, which is the only source of agar in some countries such as Italy (Simonetti, et al., 1970), thrives in turbid, shallow lagoon bays, particularly near freshwater inflow containing large amounts of nutrients. The life history of *G. verrucosa*, the type species, was shown (Ogata, et al, 1972) to consist of an alternation of isomorphic phases, with unisexual gametophytes. An anomalous pattern, however, was also reported (Cabioch, 1972B) in the same species.

Fig. 9.69 *Gracilaria foliifera* (Forssk.) Boerg.
Habit. × 0.31.

Gracilaria has a uniaxial construction, a single apical cell terminating each branch tip, but the central axial filament is soon unrecognizable. Spermatia are produced in cavities or shallow depressions on the male plants. Carpogonial branches are two-celled and difficult to detect. Cystocarps are usually prominent, hemispherical structures, projecting from the thallus surface; a pericarp is developed over the gonimoblast filaments, with an ostiole through which the carpospores are released. A section through the cystocarp usually reveals nutritive filaments connecting the carposporo-phyte to the overlying pericarp and the presence of a relatively large-celled central region. *Gracilariopsis* was segregated from *Gracilaria* (Dawson, 1949) on the basis of the absence of such nutritive filaments and the small-celled central region of the car-sporophyte. This difference, however, was later pointed out to be unreliable (Papen-fuss, 1967).

GRACILARIOPHILA Setch. et Wilson   This parasitic genus contains minute, white to yellowish plants that are barely detectable to the eye. Thalli of *Gracilariophila* (Gr. *philos*, loving) are only 1–2 mm in diameter and form an erumpent, globose pustule on the host *Gracilaria*. The epithet of one species, *G. oryzoides* Setch. et Wilson, alludes to its resemblance to a grain of rice (the genus *Oryza*). Interestingly, the parasite is often on portions of its host, such as *G. sjoestedtii* Kylin, that become buried under sand. The two species of *Gracilariophila* occur on the west coast of North America.

This genus is an example of the category of red algal parasites termed **adel-phoparasites**, which include closely related parasite and host pairs (Feldmann and Feldmann, 1958). Of the approximately 40 parasitic genera, about 90% have this close systematic relationship (Dawson, 1966), the remainder essentially being unrelated and termed **alloparasites.** The theories that have been proposed to explain this phenomenon of a high incidence of parasitism among the red algae have been reviewed (Fan, 1961B). According to one hypothesis, originally proposed by Setchell (1918), parasites might have arisen from spores that germinated *in situ* and mutated to become parasitic and dependent on the host. The *in situ* germination both of tetraspores and carpospores is not an uncommon event among the red algae, and this interpretation for the possible derivation of adelphoparasites is an attractive explanation.

A contrasting hypothesis for the origin of parasites was offered by Sturch (1926), who attempted to explain the existence of alloparasites, such as *Harveyella* (p. 517). The parasite might have been initially merely an epiphyte on a given host; epiphytism of algae on other algae is of frequent occurrence. If **haustoria** or other means of attachment from the epiphyte were able to absorb food material, the epiphyte might have undergone a transition to a parasitic mode of existence. The ability of red algae to form secondary pit connections (p. 456) might have been a significant factor in this establishment of a physiological communication between host and epiphyte.

Order 8. Rhodymeniales

This order, containing about three dozen genera placed among three families, is characterized by multiaxial growth and a procarp (Fig. 9.70a) consisting of a three- or four-celled carpogonial branch and an auxiliary cell, which usually terminates a two-celled filament also arising from the supporting cell. Postfertilization events

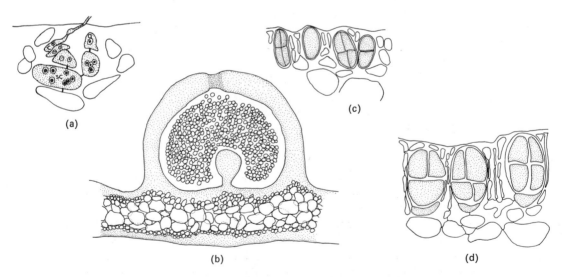

Fig. 9.70 (*a*) Procarp in *Lomentaria articulata* (Huds.) Lyngb.; *c*, carpogonium; *a*, auxiliary cell; *sc*, supporting cell. (*b*) Sectioned cystocarp in *Rhodymenia pseudopalmata* (Lamour.) Silva. (*c*) Tetrasporangia in *Rhodymenia pseudopalmata*. (*d*) Tetrasporangia in *Palmaria palmata* (Linn.) Stackh. (*a*) × 550; (*b*) × 90; (*c*), (*d*) × 280. [(*a*) after Bliding; (*b*) after Sparling; (*c*) and (*d*) after Guiry.]

involve the rapid fusion of the cells of the carpogonial branch, the cutting off of a connecting cell from the fused carpogonial branch cells, and the contact of this connecting cell with the nearby auxiliary cell. Typically, a large fusion cell is evident at the base of the carposporophyte (Fig. 9.70b). The gonimoblast is always developed outwardly, and a domed pericarp with an ostiole is characteristic for the order. Tetrasporangia are cruciately or tetrahedrally divided; polysporangia are produced in some genera. The life history is generally diplobiontic, but anomalous life histories are also known (in the Palmariaceae). Vesicle cells, which are cut off from the inner cortex, are of frequent occurrence in the members of this order.

The Rhodymeniales has been traditionally divided into two families (Sparling, 1957), but a third family, the Palmariaceae, has been recently recognized (Guiry, 1974). The following key separates these three families:

1. Tetrasporangia tetrahedrally divided (excepting *Coeloseira*, which has polysporangia); hollow portions always present and provided with longitudinal filaments around the periphery of the lumen ....Champiaceae
1. Tetrasporangia cruciately divided (excepting *Hymenocladia*, which has tetrahedrally divided tetrasporangia); thallus solid or hollow; hollow portions lacking longitudinal filaments 2
   2. A carposporophytic phase included in the life history; tetrasporangia not associated with a stalk cell (Fig. 9.70c) ....Rhodymeniaceae
   2. A carposporophytic phase not included in the life history; tetrasporangia associated with a stalk cell (Fig. 9.70d)........Palmariaceae

Family 1. Champiaceae

All members of this family have at least portions of their thalli hollow, and a longitudinal section through an axis reveals elongate filaments bordering this cavity. The terete axes of *Lomentaria* are entirely hollow, whereas those of *Champia* are transected by multicellular diaphragms, the axes being constricted at these diaphragms. It is noteworthy that *Lomentaria baileyeana* (Harv.) Farl. and *Champia parvula* (C. Ag.) Harv. have distributional ranges from northern Massachusetts to tropical waters (Taylor, 1962; Humm, 1969; Edwards, 1970A). In addition to the differences mentioned in the key (above), another distinction of Champiaceae from Rhodymeniaceae is that only the terminal cells of the gonimoblast filaments develop into carpospores, whereas all cells of the carposporophyte in the Rhodymeniaceae become carpospores. The exception to this generalization is *Lomentaria*, which agrees with the Rhodymeniaceae.

CHAMPIA Lamour. Thalli of *Champia* (named after the French naturalist *L. A. Deschamps*) are soft, pinkish, much branched axes (Fig. 9.71), with a nodulose aspect due to the regular pattern of constrictions. The cystocarps are external, the prominent hemispherical pericarp having a distinct ostiole. The tetrahedrally divided tetrasporangia are scattered in the cortex of the diploid plants. *Champia parvula* (C. Ag.) Harv. occurs in the Atlantic Ocean and the Gulf of Mexico.

Fig. 9.71 *Champia parvula* (C. Ag.) Harv. Habit. × 1.2.

GASTROCLONIUM Kütz. Plants of *Gastroclonium* (Gr. *gastro*, belly + Gr. *klon*, branch) are coarser and more massive than those of *Champia*, the thallus consisting of solid, terete axes in the lower portions and ultimate branches, which are hollow and septate, as in *Champia*. The carpospores are borne directly on a large fusion cell, and the surrounding pericarp lacks an ostiole. Diploid plants have both tetrasporangia and occasional polysporangia (p. 464). The only species of this genus from North America is *G. coulteri* (Harv.) Kylin, which is common from southern British Columbia to Mexico (Scagel, 1957). Some species have iridescent properties (Feldmann, 1970).

Family 2. Rhodymeniaceae

Members of this family may be hollow, as *Chrysymenia*; solid, as *Rhodymenia*; or solid below and hollow terminally, as *Botryocladia*. The flattened blades of *Fryeella*, a genus recorded (Hollenberg and Abbott, 1966) from Canada to Baja California superficially resemble *Rhodymenia*, but the thalli are hollow and transected by multicellular diaphragms. The life history of this family is diplobiontic, with isomorphic tetrasporic and unisexual phases. Two representative genera are elaborated upon below.

RHODYMENIA Grev. *Rhodymenia* (Gr. *rhodon*, a rose + Gr. *hymen*, a membrane), a genus with more than 40 species, is well represented in the northern hemisphere, with a great variation in the form of the thallus. The type species, *R. pseudopalmata* (Lamour.) Silva, occurring in the Gulf of Mexico (Edwards and Kapraun, 1973) and the tropical Atlantic northward to North Carolina, consists of flattened, dichotomously branched axes (Fig. 9.72*a*) up to 10 cm high and attached by a basal disc. On the other hand, the spectacular perforate blades (Fig. 9.72*b*) of *R. pertusa* (P. et R.) J. Ag., distributed from Oregon to Alaska, may be up to 1 m in length (Scagel, 1967). Some of these species, such as *R. pacifica* Kyl. and *R. californica* Kyl., have cylindrical, horizontal rhizomes, or runners, which may contact the substrate to propagate new plants. In a monograph of *Rhodymenia*, Dawson (1941) employed criteria observed in tetrasporic plants to distinguish sections within the genus.

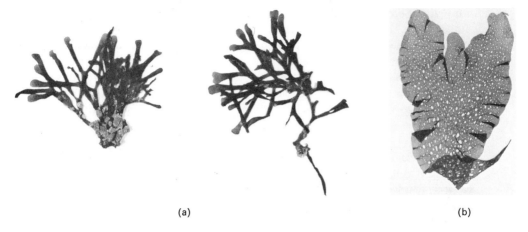

(a)   (b)

Fig. 9.72   *Rhodymenia*. (*a*) *R. pseudopalmata* (Lamour.) Silva. Habit. (*b*) *R. pertusa* (P. and R.) J. Ag. Habit. (*a*) × 0.5; (*b*) × 0.11.

BOTRYOCLADIA Kylin   The lower portion of the thallus *Botryocladia* (Gr. *botrys*, a cluster of grapes + *klados*, a branch) is solid and cylindrical, each branch of the axis being terminated in a swollen hollow portion. The west coast species, *B. pseudodichotoma* (Farl.) Kyl., consists of a well-developed system of alternately branched axes with very much enlarged saccate tips (Fig. 9.73) filled with mucilage. Thalli of *B. occidentalis* (Børg.) Kyl., a species distributed from Florida (Dawes, 1974) to North Carolina, bear numerous small bladders, or sacs, each about 5–10 mm long. Other species of this genus have been recently described from various parts of the world (Feldmann, 1945; Feldmann and Bodard, 1965; Ganesan and Lemus, 1972), including *B. spinulifera* Taylor et Abbott (1973) from the Virgin Islands, which has vesicles less than 3 mm long.

Family 3. Palmariaceae

Some species of *Rhodymenia*, including the widespread *R. palmata* (L.) Grev., and the genera *Halosaccion* and *Leptosarcus* have long been a vexing problem in that female plants are unknown. Only tetrasporangial and somewhat scarcer male plants

Fig. 9.73   *Botryocladia pseudodichotoma* (Farl.) Kylin. Habit. × 0.66.

are found. Some of these algae were cultured (Sparling, 1961) to determine if a dissimilar phase might be involved in their life history, but tetraspore germlings developed into plants similar in form to the original sporophytes, making it unlikely that heteromorphic generations were involved. The anomalous nature of the tetrasporangia in *Rhodymenia palmata* and in *Halosaccion* was noted by Guiry (1974). Tetrasporangia in these plants occur in sori distributed over nearly the entire thallus surface. Cortical cells undergo an unequal periclinal division, the inner cell functioning as a "stalk cell" and the outer cell functioning as the tetrasporangial cell. The developing tetrasporangium (Fig. 9.70*d*) causes surrounding cortical cells to be stretched and modified, and after release of the tetraspores the stalk cell may divide to form a secondary tetrasporangial cell. This situation is contrasted with that in Rhodymenias with diplobiontic life histories. The tetrasporangial sori are restricted to the tips of the branches, and the cortical cells enlarge into tetrasporangia, causing little modification of surrounding cortical tissue. The cell undergoes an equal, periclinal division, followed by two anticlinal divisions to produce the cruciately divided tetrasporangium (Fig. 9.70*c*). There is no stalk cell, and secondary tetrasporangia are not formed.

The conclusion from the evidence above by Guiry (1974) is that *Rhodymenia palmata* is not congeneric with *R. pseudopalmata*, the type of the genus, but has greater affinities with *Halosaccion*. There are sufficient differences, however, to retain *R. palmata* as a genus distinct from *Halosaccion*, and the old name *Palmaria* was reinstated. For the same reasons cited above, the family Palmariaceae was established.

PALMARIA Stackhouse   Blades of *Palmaria* (L. *palma*, palm of a hand) *palmata* (L.) Stackh., one of the most ubiquitous red algae on temperate shores of North America, are usually deeply divided (Fig. 9.74) but may be simple. Proliferous branching from the margins, especially near the base, is frequent. The thallus is

Fig. 9.74   *Palmaria palmata*. Habit. × 0.19.

Fig. 9.75 *Halosaccion glandiforme* (Gmelin) Rupr. Mid-littoral rocks covered with specimens of this alga.

gradually attenuated toward the base rather than stipitate, and attachment to the substratum is by means of a disc. The structure of the blade in sectional view is pseudoparchymatous, with larger vacuolate cells in the medulla and small, pigmented cells in the cortex.

Dried blades of *P. palmata* are called "dulse," and it is commonly eaten, or simply chewed like tobacco, in certain regions, such as Maine and the maritime provinces of Canada. For poor peasants on the west coast of Ireland, dulse was the only addition to potatoes in many of their meals (Newton, 1931), and it was regarded as having medicinal properties.

HALOSACCION Kütz.   The hollow thalli of *Halosaccion* (Gr. *hals*, the sea + Gr. *sakkos*, sack) may be simple, as in *H. glandiforme* (Gmel.) Rupr., or much branched, as in *H. ramentaceum* (L.) J. Ag. There are no diaphrams present in the hollow interior. The range of *H. glandiforme* extends from the Bering Sea south to central California, and it is a prevalent alga of the mid-littoral zone, often forming dense stands (Fig. 9.75) that give a yellowish-green aspect to this zone. Its abundance can be just as great at central California sites as at localities in the Aleutian Islands. The saccate thalli, which arise in clusters from a common holdfast, may be 25 cm long. Branching occasionally occurs, especially upon injury. When thalli out of water are tightly squeezed, fine jets of water are expelled through minute pores.

The narrow, irregularly and proliferously branched thalli of *H. ramentaceum* have a firm to fleshy consistency and may reach 40 cm in height. Most often inhabiting tidal pools, this species is a cold-water alga, ranging from northern Massachusetts to above the Arctic Circle. It also occurs on the Alaskan coast.

The reproductive phenology of species of *Halosaccion*, as well as that of *Palmaria palmata*, was followed (Lee and Kurogi, 1972) in northern Japan, and it was reported that the production of tetrasporangia and spermatangia, the only reproductive stages, lasted for about 8 months of the year, the peak occurring in winter. This schedule is contrasted with that of most red algae having the more typical diplobiontic life histories (i.e., production of tetrasporangia, carposporangia, and spermatangia by three phases, p. 462), in which reproduction is often restricted to just the three to four summer months. The distinctiveness of the Palmariaceae is thus supported by these data.

Order 9. Ceramiales

The largest order of red algae in terms of number of genera, the Ceramiales is also the most clearly defined. The distinctive ordinal characteristic is that the auxiliary cell is formed after fertilization, invariably from the supporting cell. The most striking characteristic of the order is the uniformity in the female reproductive system. Carpogonial branches are always four-celled (Fig. 9.76*a*). Usually a single carpogonial branch is produced per supporting cell, but a pair of carpogonial branches is produced

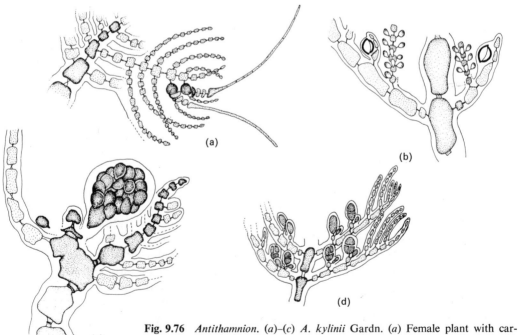

Fig. 9.76 *Antithamnion.* (*a*)–(*c*) *A. kylinii* Gardn. (*a*) Female plant with carpogonial branches borne on swollen, basal cell of branchlet. (*b*) Male plant with spermatangia. (*c*) Carposporophyte. (*d*) *A. defectum* Kyl. Tetrasporangia production. (*a*), (*b*) × 330; (*c*) × 250; (*d*) × 75. (After Wollaston.)

per supporting cell in a few Ceramiaceae and Delesseriaceae (Hommersand, 1963). The supporting cell is ordinarily a **pericentral cell**, which is a cell that is cut off from an axial cell. Following fertilization, the supporting cell cuts off the auxiliary cell,[10] the entire order thus being procarpial.

Transfer of the diploid nucleus is accomplished by two different mechanisms: (1) The carpogonium may cut off a connecting cell, which then fuses with the auxiliary cell, which takes place in the majority of Ceramiaceae and Dasyaceae. (2) The carpogonium fuses directly with the auxiliary cell, which takes place in nearly all Delesseriaceae and Rhodomelaceae. The carposporophyte is then developed from the auxiliary cell by its initial division into a gonimoblast initial and a foot cell. The gonimoblast initial in the Ceramiaceae cuts off from one to five gonimolobes. A phylogenetic sequence is evident (Hommersand, 1963) in which one type of gonimoblast is entirely converted into carpospores, an intermediate situation being the transformation of about one-half of the cells of the goninoblast into carpospores, and the third situation consisting of the gonimoblast having only terminal carpospores.

Life histories in this order are basically diplobiontic, with isomorphic gametophytic and sporophytic generations. The gametophytes are typically unisexual. Many culturing studies (West and Norris, 1966; Edwards, 1968, 1969B, 1970B, 1973; Waaland and Kemp, 1972) have confirmed this pattern, often referred to as the "*Polysiphonia*-type life history" (Kylin, 1938). Tetrasporangia may be cruciately divided, which is true for many Ceramiaceae, as *Antithamnion* (Fig. 9.77c), or tetrahedrally divided, which is true for most members of the other three families of this order. Polysporangia are also known, for example, in *Spermothamnion* and *Pleonosporium*, both genera of the Ceramiaceae.

Four families are generally recognized as belonging to this order, and the distinctions are reasonably clear-cut, except for some intermediate types linking together the Delesseriaceae and Rhodomelaceae (Wynne, 1969B). The following key is used to separate these families:

> 1. Carposporophyte exposed or with a loose investment of involucral filaments; vegetative axial cells lacking pericentral cells or corticated by cells not equal in length to the axial cells ....Ceramiaceae
> 1. Carposporophyte enclosed by firmly constructed, ostiolate pericarp; pericentral cells present, equal in length (at least initially) to the axial cells....2
>   2. Axes sympodially developed; tetrasporangia borne in **stichidia**; pericentral cells cut off in a spiral, clockwise manner[11] ....Dasyaceae
>   2. Axes monopodially developed, with rare exception; tetrasporangia usually not borne in stichidia; pericental cells not cut off in a spiral, clockwise manner 3

[10]Some rare exceptions to this generalization have been noted (Dixon, 1964), in which the supporting cell acts directly as the auxiliary cell, but arguments have been made (Papenfuss, 1966) that the integrity of the order Ceramiales nonetheless remains intact.

[11]Exceptions to this pattern have been reported (Parsons, 1975) in *Heterosiphonia* (p. 552) and *Thuretia* both in vegetative axes and in female fertile segments. The alternating sequence observed in these genera parallels that of the Rhodomelaceae, and Parsons (1975) expressed the idea that this criterion is therefore significant at the generic level rather than at the family level.

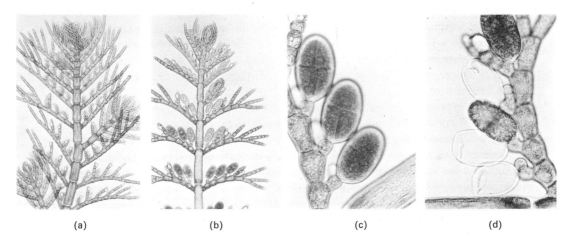

(a)   (b)   (c)   (d)

Fig. 9.77   *Antithamnion*. (*a*) Male plant with gland cells and spermatangia. (*b*) Tetrasporic plant. (*c*) Cluster of cruciately divided tetrasporangia. (*d*) Discharged tetrasporangia. (*a*), (*b*) × 56; (*c*), (*d*) × 400. (Courtesy of Dr. J. A. West.)

3. Flattened, foliose forms common; never more than four pericentral cells formed; lateral pericentral cells cut off first, followed by abaxial and adaxial pericentral cells (except *Sarcomenia* group); tetrasporangial initials cut off prior to their being covered by "cover cells" . . . . .Delesseriaceae
3. Flattened, foliose forms rare (*Amplisiphonia*, for example); four or more pericentral cells, sequence of their appearance such that the last formed is opposite the first; "cover cells" present prior to formation of tetrasporangial initial . . . . .Rhodomelaceae

Selected representatives from these four families are discussed in the account that follows.

Family 1. Ceramiaceae

This large family includes some very small, delicate forms, such as *Callithamnion* and *Antithamnion*, but also some more massive, robust forms such as *Ptilota* and *Microcladia*. The axial cells are completely naked in genera such as *Antithamnion* and *Griffithsia*, or they may be covered by partial cortication (such as in *Ceramium*, which may have cortication restricted to the junctures of adjacent cells) or complete, extensive cortication, such as in *Ptilota*. One other feature of this family is the absence of a definite pericarp over the gonimoblast; it is either naked or loosely covered by an involucre of sterile filaments.

ANTITHAMNION Nägeli   With a wide-ranging distribution in temperate seas of the world, *Antithamnion* (Gr. *anti*, opposite + Gr. *thamnion*, a small bush) is a genus of small, delicate plants, consisting of uniseriate axes without cortication. Each axial cell bears a pair of determinate branchlets (Fig. 9.77a), which are equal in length and simple or distichously branched. The basal cell of each determinate branch is small and may produce a lateral branch initial, rhizoids, or reproductive organs. Gland cells are of frequent occurrence in this genus, usually occurring on short determinate branches arising on the upper (adaxial) side of the branchlet. Cells in this genus are uninucleate.

Female plants bear carpogonial branches on the small basal cell of a determinate branch (Fig. 9.76a). The auxiliary cell is cut off from this basal cell after fertilization, and following the transfer of the diploid nucleus, the carposporophyte is initiated. The mature carposporophyte (Fig. 9.76c) is naked. The typically ephemeral male plants produce whorls of spermatangia (Fig. 9.76b) from the cells of special branches arising adaxially from the lateral branchelts (Fig. 9.77a). The cruciately divided tetra-sporangia (Fig. 9.77c) arise often in clusters (Fig. 9.76d) or singly (Fig. 9.77b) on pedicels from the upper surfaces of the lateral branchlets. The wall of the empty tetrasporangium remains (Fig. 9.77d) after the spores are released.

Antithamnion has been the subject of several recent monographic treatments (Wollaston, 1968, 1971; L'Hardy-Halos, 1968, 1970). The circumscription of *Anti-thamnion* has been refined by the detailed studies by Wollaston (1971) on west coast species of the genus, and she has transferred some species that had been classified in *Antithamnion* to *Antithamnionella* as well as to two newly created genera, *Scagelia* and *Hollenbergia*. She has also referred (1972A) to certain differences, such as in the position of the gland cells and the position of carpogonial branches, shared by the Antithamnions of Pacific North America, as contrasted with Antithamnions of Europe and Australia, suggesting that the two groups might possibly have developed independently.

Culturing experiments with strains from different localities have been useful in demonstrating the interfertility or intersterility of such strains (Sundene, 1959, 1975). Genetic studies (Rueness and Rueness, 1975) have enabled us to appreciate that a single gene may be involved in the expression of distinct varieties, as in *Antithamnion plumula* (Ellis) Thur., in which the var. *plumula* is under the control of the dominant allele and var. *bebbii* is under control of the recessive allele.

A typical alternation of isomorphic phases has been observed (Sundene, 1964B) in some species of *Antithamnion*. For *A. boreale* (Gobi) Kjellm., only tetrasporangial plants are known to occur in nature, and culturing studies (Sundene, 1962) with this species resulted in three successive generations of tetrasporangial plants. Anomalous conditions have been observed in other species, such as *A. occidentale* Kylin,[12] in which male plants eventually started to produce tetrasporangia (West and Norris, 1966). The spores germinated into male plants lacking tetrasporangia. The occurrence of tetrasporangia and sexual structures on the same plants has been recorded (Sundene, 1964B; Rueness and Rueness, 1973) for *A. tenuissimum* (Hauck) Schiffner. Culturing and cytological studies have shown (Rueness and Rueness, 1973) that male plants bearing tetrasporangia are haploid, the tetrasporangia being apomeiotic. Such tetraspores recycle the sexual plants, the life history of *A. tenuissimum* being of the "*Polysiphonia-type*."

[12]Later made the type species of *Scagelia* by Wollaston (1971).

A closely related genus, occurring on the west coast of North America, is *Platythaminon*, which is distinguished (Kylin, 1925; Wollaston, 1972B) in having four determinate branchlets arising from each axial cell, one pair of which is long and the other pair is short. The five species placed in this genus were discussed by Wollaston (1972B). Gland cells are produced in abundance laterally on cells of the branchlets.

CALLITHAMNION Lyngbye   The delicate, filamentous thalli of *Callithamnion* (Gr. *kallos*, beauty + Gr. *thamnion*, a small bush) are usually small tufts, rarely reaching more than about 10 cm in height. The great majority of species, which have a fairly cosmopolitan distribution in temperate seas of the world, have uncorticated axes or rhizoidal cortication near the base, but the heavily corticated *C. pikeanum* Harvey of Pacific North America is a noteworthy exception. The pattern of branching is alternate (Fig. 9.78a), distinguishing it from *Antithamnion* with its opposite branching pattern. Contributions to the systematics of this genus have been made, including those of Boddeke (1958) and Harris (1962).

The life history of *Callithamnion byssoides* Arnott ex Harv. in Hook; has been shown (Edwards, 1969B) to consist of isomorphic gametophytic and sporophytic phases, as is characteristic for the order. Some cases of plants bearing both sexual organs and tetrasporangia on the same plants have been observed (Hassinger-Huizinga, 1952; West and Norris, 1966). Spermatangia (Fig. 9.78a) are borne directly on cells of determinate branches. Carpogonial branches develop from special cells occurring laterally in the upper parts of the indeterminate axes (Fig. 9.78b). The growth and maturation of the carposporophyte (Fig. 9.78c) required only 5 days, the entire life cycle (from tetraspore through production of carpogonia, spermatia, and carpospores back to tetraspores) requiring in culture only about a month (Edwards, 1969B). The naked carposporophytes are usually lobed. Tetrasporangia (Fig. 9.78d) are tetrahedrally divided.

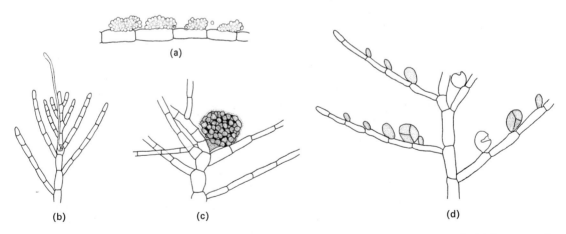

Fig. 9.78 *Callithamnion byssoides* Arnott ex Harv. in Hook. (*a*) Spermatangia production. (*b*) Carpogonial branch. (*c*) Carposporophyte. (*d*) Tetrasporangial production. (*a*) × 150; (*b*) × 80; (*c*), (*d*) × 67.

Cells of *Callithamnion* are multinucleate, nuclear divisions normally being synchronous. Development is monopodial (L'Hardy-Halos, 1971), although it may appear sympodial or even dichotomous in some species on account of segment cells below the apical cell rapidly giving rise to secondary initials that falsely give the impression of replacing the primary axis.

GRIFFITHSIA C. Ag.   A significant fact concerning *Griffithsia* (named for *Amelia W. Griffiths*, British phycologist) is the very large size of the vegetative cells. In many of the species, such as *G. globulifera* Harvey, which ranges from the tropics of the Atlantic northward to Massachusetts (Taylor, 1962A), and *G. pacifica* Kylin, which ranges from the Gulf of California and Baja California northward to southern British Columbia (Dawson, 1961), the individual cells of the uniseriate uncorticated axes are visible to the unaided eye, and at maturity they contain thousands of nuclei. Certain species, such as *G. tenuis* C. Ag., with a tropical and subtropical distribution, have smaller sized vegetative cells. Short, branched filaments of small cells may occur on younger plants; they elongate into hairlike structures in older portions of the plant and are ultimately shed.

The large vegetative cells of female plants cut off small, three-celled branches from their distal ends. The intermediate cell of these branches cuts off three pericentral cells, one of which acts as the supporting cell of a single carpogonial branch and a single sterile cell. The procarp (Fig. 9.79) is enveloped in mucilage. Following fertilization, the supporting cell cuts off the auxiliary cell (however, see footnote 10 on

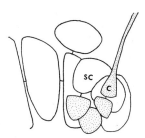

Fig. 9.79 *Griffithsia corallina* (Lightf.) C. Ag. Procarp. *c*, carpogonium; *sc*, supporting cell; cells of carpogonial branch are stippled. (After Kylin.)

p. 543, which refers to rare instances of the supporting cells acting directly as the auxiliary cell, as has been reported in *G. globulifera*). A general fusion of cells occurs, and the carposporophyte is developed, most of the cells of which produce carpospores. Usually the lowermost cell of the original three-celled fertile branch remains distinct, and enlarged involucral cells are budded off from it and loosely surround the carposporophyte (Fig. 9.80*b*).

Distribution of spermatangia varies from species to species. Male plants of *G. globulifera* can be easily spotted in the field because of the very swollen terminal cells of the axes. Much branched spermatangial branches form a cap over the distal end of these enlarged terminal cells. Tetrasporic plants of *G. globulifera* can also be distinguished in the field because the tetrahedrally divided tetrasporangia form obvious rings around the junctures of the large vegetative cells. In *G. tenuis* C. Ag. tetraspo-

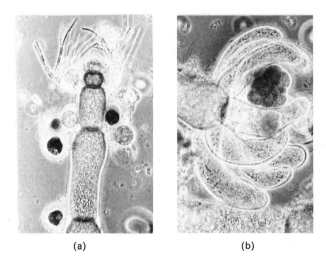

(a)   (b)

Fig. 9.80 *Griffithsia tenuis* C. Ag. (*a*) Tetrasporangia borne at nodal regions; (*b*) Carposporophyte surrounded by involucral cells. × 99.

rangia are also borne at the junctures of axial cells but are fewer in number (Fig. 9.80*a*).

Griffithsia pacifica Kylin has served as a useful tool in analyzing processes such as regeneration of plants from single cells and the phenomenon of cell repair following injury. Whole new plants may be regenerated from single cells (Duffield, et al., 1972). Cells may be regarded as either shoot cells or rhizoidal cells. Shoot cells grow by apical division at the rate of one to two cells per day. Branches are formed by the budding of subapical cells. Unlike *Callithamnion* (p. 546), in which every shoot cell (i.e., cell of an indeterminate axis) branches, cells of the shoots in *Griffithsia* may or may not branch; those cells that branch are termed **nodal cells**, and those that do not branch are termed **internodal cells**. Both light intensity and photoperiod have been shown (Waaland and Cleland, 1972) to affect morphogenesis. Grown under a 16 : 8 photoregime, plants show a **diurnal rhythm** in regard to both cell division and cell elongation, this rhythm persisting for at least seven cycles under conditions of continuous light. The rates of cell division of apical cells and cell elongation are relatively insensitive to light intensity and photoperiod. High light intensities and long photoperiods promote branching, however, by increasing the number of nodal cells over the number of internodal cells. Specifically, plants are highly branched when grown at 300 ft-c light intensity but remain unbranched at 50 ft-c (Waaland and Cleland, 1972).

Intercalary cells seem to be prevented from dividing by the presence of the adjacent cells. The removal of an adjacent cell (by rupturing it) results, however, in the formation of either a new apical cell of a shoot or a rhizoid, depending on whether the ruptured cell was above or below the regenerating cell. A rhizoid is regenerated from the superjacent cell (Fig. 9.81*a*) next to a dead cell, whereas a shoot is regenerated

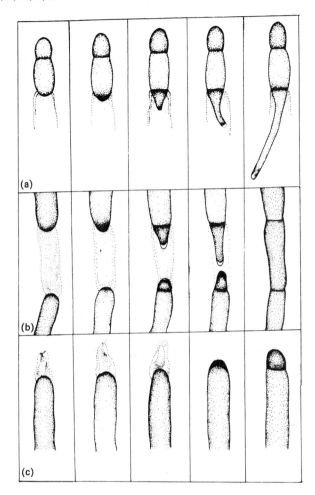

Fig. 9.81 Regeneration in *Griffithsia pacifica* Kylin. (*a*) Process of filament regeneration by rhizoid formation from the upper half of a severed shoot filament. (*b*) Process of cell repair by formation of a repair rhizoid from the superjacent cell and a repair shoot from the subjacent cell and their fusion in the space where an internodal cell had been punctured. (*c*) Process of filament regeneration by shoot formation from the lower half of a severed filament. × 53. (After Waaland and Cleland; permission of Protoplasma.)

from the subjacent cell (Fig. 9.81*c*). Thus, two filaments are regenerated from the original filament that was cut into two. This phenomenon differs from the process of cell repair (Fig. 9.81*b*), which involves the superjacent cell and the subjacent cell forming a repair rhizoid and a repair shoot, respectively, the growth of these two processes toward each through the lumen of the dead cell (Fig. 9.81*b*), and their subsequent fusion and expansion in volume to fill the cavity of the dead cell. Cell repair occurs when the adjacent cells are still near each other because of the retention of the wall of the original ruptured cell (Waaland and Cleland, 1974).

The process of cell elongation in the long, cylindrical cells of *Griffithsia pacifica* has also been investigated (Waaland, et al., 1972). Intercalary shoot cells have been shown to have a unique process termed *bipolar band growth*, which involves wall extension being localized to two narrow bands, one at each end of the cell. Rhizoidal cells, on the other hand, have tip growth.

CERAMIUM Roth   The central axial filaments of *Ceramium* (Gr. *keramium*, a vessel) are relatively large-celled with bands of small corticating cells developed at the junctures of these axial cells. The extent of this nodal cortication varies. In some species the bands of corticating cells are restricted to the nodes (Fig. 9.82*a*). In other species, such as *C. rubrum* (Huds.) J. Ag., which is widespread in the North Atlantic, and *C. pacificum* (Coll.) Kyl., which has an extensive range on the Pacific coast, the axes become completely corticated by the upward and/or downward growth of the corticating filaments.

Branching in many species is dichotomous, and tips of thalli often are forcipate, or pincerlike. Spermatangia are borne on the cortical bands. One or two carpogonial branches are produced from a first-formed pericentral cell at a node (Dixon, 1960). If fertilization occurs, a carposporophyte is developed, encircled by a loose involucre of several adventitious branches (Hommersand, 1963). Tetrasporangia are borne at the nodes and, depending on the species, may be protruding (Fig. 9.82*b*) or partially or completely embedded in the nodal cortication.

Several species of *Ceramium* have been observed in culture, and a diplobiontic life history, typical of the order, has been demonstrated (Edwards, 1973; Rueness, 1973). Edwards (1973) observed that in the northern part of the range of certain

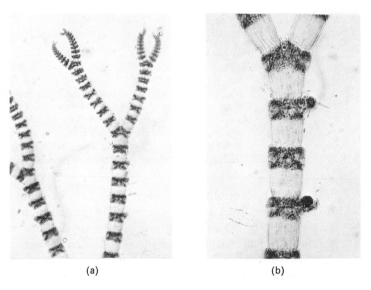

(a)   (b)

Fig. 9.82 *Ceramium.* (*a*) Axis with discrete bands of nodal cortication. (*b*) Tetrasporangia production. (*a*) × 131; (*b*) × 125.

Fig. 9.83 *Neoptilota hypnoides* (Harv.) Kyl. Habit. × 0.35.

species the gametophytes do not become fertile probably due to unfavorable environmental conditions. In culture, however, the gametophytes derived from tetraspore germlings did become sexual, indicating that they have the reproductive potential if conditions are favorable. In addition to the normal life history, some plants of *C. strictum* Harv. were observed in culture (Rueness, 1973) to produce **parasporangia**, which are reproductive structures forming spores that have the same nuclear phase as the parent plant. These parasporangia, which superficially resemble gonimoblasts but lack the subtending adventitious lateral branches, gave rise to new generations of parasporangium-bearing plants.

PTILOTA C. Ag.   With an apparent preference for colder seas, *Ptilota* (Gr. *ptilotos*, winged) includes several species occurring on the more northerly shores of the Atlantic and the Pacific Ocean. On the east coast *Ptilota plumosa* (Huds.) C. Ag. and *P. serrata* Kuetz. occur in the maritime provinces of Canada (Mathieson, et al., 1969; South and Cardinal, 1970), the latter species also rarely extending its range south of Cape Cod (Taylor, 1962A). *Ptilota filicina* J. Ag. is a rather common alga on the west coast, bushy, much branched, and often 30–35 cm tall; it ranges from central California to the Aleutians.

The main axes in *Ptilota* are prominent, heavily corticated, and compressed, bearing bilaterally arranged ultimate branches. The final branches often are dissimilar in appearance, but both are capable of indeterminate growth. This latter trait distinguishes *Ptilota* from *Neoptilota* (Abbott, 1972A; Widdowson, 1974), to which genus were assigned (Kylin, 1956) some of the west coast species of *Ptilota*. The closely related *Neoptilota*, which includes *N. californica* (Harv.) Kyl. and *N. densa* (C. Ag.) Kyl., has opposite pairs of ultimate branches that are very dissimilar in appearance, one member of the pair being short and potentially indeterminate and the other being larger, expanded, and determinate. The ultrastructure of tetrasporogenesis and spermatogenesis has been reported (Scott and Dixon, 1973A, B) in *Neoptilota hypnoides* (Harv.) Kyl. (Fig. 9.83).

Family 2.   Dasyaceae

All members of this family have **sympodial development**, which involves the apical cell continually being replaced by a lateral that at least temporarily assumes the role of the primary axis. A continual shifting of the growing point characterizes sympodial development and is contrasted with **monopodial development**, in which the primary axis remains as the main agent of growth, cutting off laterals. Pericentral cells are cut off in a clockwise spiral. Five pericentral cells are commonly present, but four or six also occur.

Spermatangia are produced on uncorticated uniseriate filaments of male plants. Female plants bear carpogonial branches enclosed in a pericarp, always with one carpogonial branch per supporting cell. Two groups of sterile cells are present in each procarp. The carposporophyte bears carpospores in chains. Tetrasporangia are borne in **stichidia** (Fig. 9.84) on the sporophytes, which are enlarged, specialized branches.

The family Dasyaceae is the smallest family in the order. Two genera are discussed in the following account.

HETEROSIPHONIA Mont.   Several species of this genus are recorded from the temperate and tropical waters of the Atlantic and Pacific. *Heterosiphonia* (Gr. *heteros*, different + Gr. *siphon*, a tube) is easily studied because of its relatively simple construction and lack of extensive cortication (Fig. 9.85*a*). The ultimate branches are

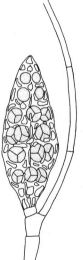

Fig. 9.84 *Dasya baillouviana* (Gmel.) Mont. Stichidia of tetrasporangia. × 98.

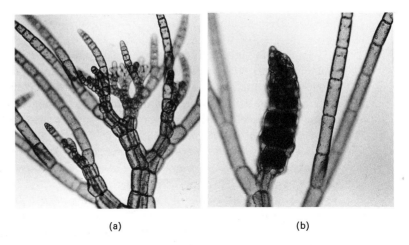

(a)   (b)

Fig. 9.85 *Heterosiphonia japonica* Yendo. (*a*) Apex of cultured plant; sympodial branching of main polysiphonous axis is evident. (*b*) Pedicellate stichidium of tetrasporangia. (After J. A. West; permission of Madroño.)

monosiphonous. Five pericentral cells are formed at some distance from the apex, and slight cortication is usually developed.

A succession of tetrasporic plants (Fig. 9.85*b*) was obtained in culture of *H. japonica* Yendo (West, 1970B). This is correlated with the almost exclusive occurrence of sporophytes in the field, but sexual plants have been occasionally collected.

DASYA C. Ag.   One of the most aesthetically pleasing red algae, *Dasya* (Gr. *dasus*, thick) has heavily corticated, terete axes covered with fine hairs. *Dasya baillouviana* (Gmelin) Mont. is widespread on the Atlantic coast, ranging from the tropical Atlantic northward to northern Massachusetts, where it is a summer annual. The delicate, showy thalli (Fig. 9.86) may reach 50 cm in height and are freely branched. The distinctive stichidia (Fig. 9.84) bear several tetrahedrally divided tetrasporangia at each level.

Fig. 9.86 *Dasya baillouviana*. Habit. × 0.25.

Family 3. Delesseriaceae

Some of the most spectacularly attractive red algae are members of this family, containing nearly 100 genera in tropical, temperate, and polar seas. Most genera have a foliose form, the blade being constructed of laterally coherent filaments. Two basic patterns of growth are apparent in this family: (1) a single, enlarged apical cell, which undergoes unequal divisions, cutting off segments proximally (Fig. 9.87a); (2) a marginal meristem, which consists of initials undergoing periclinal and anticlinal divisions, resulting in the formation of a parenchyma. It is obvious that the second type is derived from the first; juvenile stages of the second type start as uniaxial systems, gradually converting over to the marginal pattern.

Four pericentral cells are present in forms with a uniaxial system. The two lateral pericentral cells are usually first to be cut off (except for the *Sarcomenia* group, such as *Platysiphonia*, p. 555). These lateral pericentral cells continue to divide, assuming the role of secondary initials. Laterals produced from these secondary initials may similarly assume the function of initials (tertiary initials), continuing growth also in the same plane and resulting in "wings" or membranous extensions from the primary axial row of cells. The apical cells of the secondary axial rows always reach the margin of the blade; the tertiary apical cells reach the margin in some genera (such as *Hypoglossum* and *Branchioglossum*) but not in others (such as *Membranoptera* and *Delesseria*).

Two subfamilies have been recognized, but at present only one criterion (Mikami, 1971) serves to distinguish the genera in these two groupings. The procarps are borne along the primary axial row in the Delesserioideae, whereas the procarps are randomly

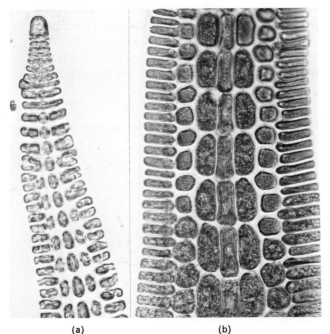

(a)   (b)

Fig. 9.87 *Platysiphonia decumbens* Wynne. (a) Apical region. (b) Blade of maximum width development. (a) × 440; (b) × 484. (After Wynne.)

distributed over the blade surface in the Nitophylloideae. Another trait that is useful is that intercalary cell divisions rarely occur in the primary axial row in members of the Delesserioideae, but they regularly occur (Fig. 9.91) in the primary axial row of those members of the Nitophylloideae that have the uniaxial pattern of growth. All members of the Delesserioideae have a single apical cell, but members of the Nitophylloideae have either a single apical cell or marginal initials.

Branching in this family may occur marginally due to the outgrowth of a secondary initial that essentially becomes a primary apical cell; this type of branching occurs in *Membranoptera* (below) and *Phycodrys*. (p. 557). Branching may occur along the primary axis, or midrib. This type of branching, which occurs in *Delesseria* (p. 556) and *Platysiphonia* (below), is termed *endogenous* and involves the cutting off of a primary initial after the pericentral cells have been formed. It is contrasted with exogenous branching, in which a branch initial is cut off prior to the formation of pericentral cells; this latter type of branch formation is more typical of the Rhodomelaceae (p. 559).

PLATYSIPHONIA Børg.   The delicate, narrow axes of *Platysiphonia* (Gr. *platy*, flat + Gr. *siphon*, a tube) are branched from the primary axial row. Three species, *P. clevelandii* (Farl.) Papenf., *P. parva* Silva and Cleary, and *P. decumbens* Wynne, occur on the west coast of North America. In the first two species, two flanking cells are cut off from the lateral pericentral cells, these two flanking cells each being one-half of the length of the pericentral cells. The entire thallus width is accordingly only five cells wide, and in the field these plants might pass as specimens of *Polysiphonia* (p. 561) because of the very limited development of their **alae**. In *P. decumbens* an additional pair of cells called "flanking cell derivatives" is cut off by the flanking cells, resulting in a blade seven cells wide (Fig. 9.87*b*). The narrow blades of *P. decumbens* become attached by the formation of complex multicellular holdfasts (Fig. 9.8*b*), which arise from several marginal cells (Wynne, 1969B). In this genus, along with other members of the *Sarcomenia* group of the Delesseriaceae, the sequence of the cutting off of pericentral cells is unusual for the family in that the abaxial pericentral cell is cut off prior to the cutting off of the lateral pericentral cells (Silva and Cleary, 1954).

MEMBRANOPTERA   Species of *Membranoptera* (Gr. *membrana*, parchment + Gr. *pteron*, wing) are usually small, delicate blades (Fig. 9.88) marginally branched and with faint venation. Intercalary cell divisions do not occur in the primary axial row, as is typical of the subfamily Delesserioideae, nor in the secondary axial row. Tetrasporangia are produced in sori usually on both sides along the midrib. Cystocarps are developed along the primary axial rows (Fig. 9.88), including those of marginal branches in which the initially secondary initials have assumed the role of primary initials. This pattern may give the false impression of cystocarps being scattered over the blade surface.

An alternation of isomorphic phases and dioecious gametophytes was demonstrated (West and Norris, 1966) in *Membranoptera platyphylla* (S. and G.) Kylin, which constitutes one of the first reports of such a life history based on a unialgal

Fig. 9.88 *Membranoptera multiramosa* Gardner. Cystocarpic specimen, bearing cystocarps along primary axial rows. × 11.

culture. A different species that also occurs on the west coast of North America, *M. multiramosa* Gardner, was shown to have monoecious gametophytes in culture (Waaland and Kemp, 1972).

DELESSERIA Lamour. *Delesseria* (named for the French naturalist, *Baron Delessert*) includes bright red, membranous thalli, up to 30 cm tall and attached by a cartilaginous stipe. The common west coast species, *D. decipiens* J. Ag. (Fig. 9.89), has branches arising from the conspicuous midrib, the branching extending out to four or five orders. The wings, or **alae**, are monostromatic except for lateral veins. The primary axial row becomes well developed and cylindrical and bears special fertile leaflets. One anatomical distinction from *Membranoptera* is that in *Delesseria* the cells of the secondary axial row can undergo intercalary divisions, but such divisions do not occur in *Membranoptera*.

Fig. 9.89 *Delesseria decipiens* J. Ag. Habit. × 0.23.

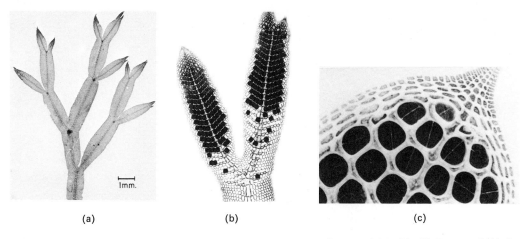

(a)   (b)   (c)

Fig. 9.90 *Caloglossa leprieurii* (Mont.) J. Ag. (*a*) Habit. (*b*) Tetrasporic blade. (*c*) Tetrasporangia with cover cells. (*b*) × 48; (*c*) × 348. [(*a*) after Papenfuss; (*b*) and (*c*) courtesy of Charles Yarish and Kwok Lee.]

CALOGLOSSA (Harv.) J. Agardh   Perhaps the most widely distributed genus of the family, *Caloglossa* (Gr. *kallos*, beauty + Gr. *glossa*, tongue) occurs from tropical to temperate regions, reaching the shores of Maryland, New Jersey, and Connecticut on the Atlantic coast (Taylor, 1962A). Similar to several species of *Bostrychia* of the Rhodomelaceae, *Caloglossa leprieurii* (Mont.) J. Ag. is an alga inhabiting protected, brackish water habitats and is commonly present on the roots of mangroves. The narrow, forking blades are branched from the margins (Fig. 9.90*a*) and are usually somewhat prostrate on the substratum (mud, wood, rocks, etc.), attached by clusters of rhizoids arising from the ventral surface.

The vegetative structure and reproductive development have been investigated by Papenfuss (1961), and the development of both **exogenous** and **endogenous** branches was described. The primary branches of the thallus are exogenous in origin, being initiated from a segment cell of the apical cell and later occupying a marginal position. A small number of endogenous branches arise by the production of dorsal laterals from the midrib of nodal regions of the thalli. The monostromatic **alae** are built of series of cells produced from the lateral pericentral cells. Tetrasporangia are formed in acropetal succession (Fig. 9.90*b*), by any cells of the alae, except for a few rows of marginal cells. They are protected by cover cells (Fig. 9.90*c*).

PHYCODRYS Kütz.   The handsome, leaf-like thalli of *Phycodrys* (Gr. *phykos*, a seaweed + *drus*, an oak tree) have pronounced lateral venation and may have entire or serrate margins. Blades are monostromatic except for the midrib and nerves. The commonly occurring Atlantic species, *P. rubens* (L.) Batt. (Fig. 9.91), is branched from the margins, as is typical for the genus. *Phycodrys setchellii* Skottsberg occurs in California (Smith, 1944). Plants often appear oppositely branched because of the gradual sloughing away of the blades, leaving the persistent midrib and opposite nerves with their secondary blades.

Growth occurs by means of a single apical cell. Intercalary divisions take place in cells of the primary axial row. A characteristic indicating its position in the sub-

Fig. 9.91 *Phycodrys rubens* (L.) Batters. Habit. × 0.6.

family Nitophylloideae is the scattered distribution of the procarps over the blade surface. Unlike their restriction to the primary axial row in the Delesserioideae, the procarps (= carpogonial branches) may arise at random. Mature cystocarps have an ostiole and carpospores produced in catenate series.

POLYNEURA Kylin   This genus is placed in the subfamily Nitophylloideae, as evidenced by the scattered arrangement of the cystocarps over the blade surface and the occurrence of intercalary divisions in the primary axial row (Fig. 9.92). The fan-shaped blades (Fig. 9.93) of *Polyneura* (Gr. *polys*, many + Gr. *neuron*, a nerve) become deeply lobed or lacerated with age. Although the marginal areas of the blades are monostromatic, they become polystromatic with age, and an interconnected network of veins is expressed in the mature blades. Discrete tetrasporangial sori are randomly distributed over both blade surfaces.

Polyneura latissima (Harv.) Kyl. is a common seaweed on the Pacific coast, extending from Baja California northward to British Columbia. Blades, which are attached by a nonstipitate holdfast, may become reproductive when less than 1 cm tall or at their maximum height of 20 cm. This species tends to occur in somewhat sheltered habitats, from the lower littoral to sublittoral zones.

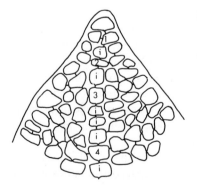

Fig. 9.92 *Polyneura latissima* (Harv.) Kyl. Representation of blade apex, with segment cells (1, 2, 3, 4) and cells derived from intercalary divisions (i) in primary row indicated. × 560. (After Kylin.)

Fig. 9.93 *Polyneura latissima.* Habit. × 0.25.

A less frequent alga that superficially resembles *Polyneura* with its fan-shaped blades and well-developed, anastomosing venation is *Polyneuropsis*, which has an overlapping distribution in the northern range of the former genus. *Polyneuropsis* can be distinguished by the stipitate nature of the blades and the nature of the holdfast, which acts as a stolon, establishing new plants. The reproductive stages of *Polyneuropsis* are most distinctive in that male, female, and tetrasporic plants bear their reproductive cells on special bladelets rather than directly on the thallus surface (Wynne, et al., 1973).

Family 4. Rhodomelaceae

In terms of number of genera, the Rhodomelaceae is the largest family of red algae, including more than 100 genera. Five seems to be the basic number of **pericentral cells**, in that the two basal cells of **trichoblasts** (which are simple or branched filaments cut off from the apex of the plant) always have 5 pericentral cells, even though the vegetative axes may have from 4 to 24 pericentral cells. The pericentral cells are cut off in an alternating sequence such that the last one formed is located opposite the first formed.

The sexual organs are typically borne on the trichoblasts, which are **exogenous** branches in that they arise from a subapical cell before pericentral cells are cut off from the axial cell. Spermatangia originate in various ways, depending on the particular genus. They may develop on unspecialized branches, but more often they will be borne on trichoblasts (Fig. 9.94*a*), the spermatangia forming a concentrated, cylindrical cluster. These male trichoblasts in *Laurencia* (p. 564) are located in depressions near the thallus apex.

A pericarp is present prior to fertilization (Fig. 9.94*b*), making it somewhat difficult to study the carpogonial branch. The four-celled carpogonial branch is produced by a pericentral cell on the second segment from the base of the trichoblast, a single procarp being present per trichoblast (Fig. 9.97*a*). Following fertilization, the auxiliary cell is cut off from the supporting cell (Fig. 9.97*b*), and this auxiliary

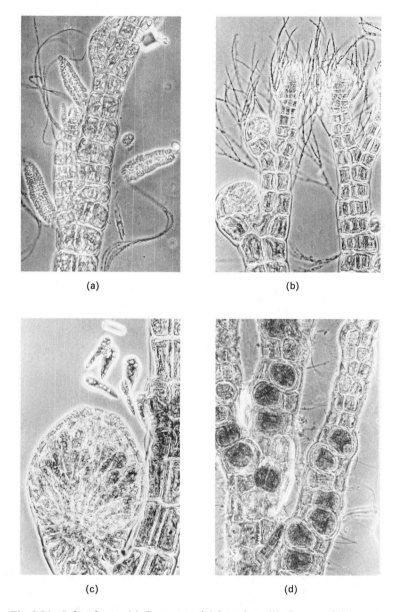

(a)

(b)

(c)

(d)

Fig. 9.94 *Polysiphonia.* (*a*) Spermatangial branches. (*b*) Carpogonial branch surrounded by pericarp. (*c*) Mature cystocarp. (*d*) Tetrasporangia. (*a*) × 111; (*b*) × 176; (*c*) × 174; (*d*) × 101.

cell functions directly as the gonimoblast initial (Hommersand, 1963), rather than dividing transversely into a foot cell and a gonimoblast initial (as in the Ceramiaceae) or becoming septated into a series of gonimoblast initials (as in the Dasyaceae and Delesseriaceae). Nearly all the gonimblast cells that are not transformed into carpospores are eventually incorporated into a single, large fusion cell.

The prefertilization pericarp, for example, in *Pterosiphonia* (p. 563), consists of cortical filaments lacking pericentral cells. If fertilization occurs, the terminal cells of these cortical filaments function as apical initials and divide, their segments each cutting off two pericentral cells to the outside (Hommersand, 1963). These filaments converge distally, leaving an opening, the ostiole of the mature pericarp (Fig. 9.94c).

Tetrasporangia are produced by a pericentral cell. The pericentral cell first divides longitudinally, cutting off two cover cells, and then it cuts off the tetrasporangium distally, leaving only the stalk cell. Tetrasporangia are always tetrahedrally divided (Fig. 9.94d).

The ultrastructure of the formation of spermatia, carpospores, and tetrasporangia in this family has been investigated in some depth (Tripodi, 1971; Kugrens and West, 1972B, C, and 1973A, 1974). Some unusual processes in carpospore formation have also been described (Wetherbee and Wynne, 1973), in which a variety of cytoplasmic inclusions and reserves was seen to be associated closely with a highly convoluted nuclear envelope.

In such a large family that has an almost monotonous sameness in regard to the postfertilization development, it might be anticipated that a tremendous diversity as to vegetative development has been evolved. Perhaps the most significant separation in reference to vegetative development concerns whether the growing tips exhibit radial construction or dorsiventral construction. In the first category are included *Polysiphonia* (below) with radial branching as well as *Pterosiphonia* (p. 563) with distichous branching (Fig. 9.96a), whereas the second category includes *Bostrychia*, with incurved tips and bilateral branching and two genera with creeping main axes, *Lophosiphonia* with mainly endogenous branching (Hollenberg, 1968A; Rueness, 1971) and *Herposiphonia* with exogenous branching (Hollenberg, 1968C). This dorsiventral series culminates in flattened, foliose forms, such as *Amplisiphonia* (Fig. 9.96b), which superficially resembles members of the Delesseriaceae (p. 554).

In most genera of the Rhodomelaceae the pericentral cells remain the length of the axial cell that originally formed them. In some genera, however, such as *Rhodomela*, *Odonthalia*, and *Bostrychia*, the pericentral cells divide transversely. Some genera become so massively corticated that the pericentral cells are not readily detectable, except upon sectioning near the apex. These fleshy or even cartilaginous types include *Laurencia* (p. 564), *Chondria*, and *Acanthophora.* CARAGEENOPHYTE

A significant number of rhodomelaceous genera are parasites, including *Levringiella*, a parasite on *Pterosiphonia*, and *Janczewskia*, a parasite on *Laurencia Erythrocystis*, which is also parasitic on *Laurencia*, still retains some photosynthetic pigments and may be several centimeters tall; thus, it is not so reduced as the other parasites mentioned. A few representative genera of the Rhodomelaceae are treated in greater depth below.

POLYSIPHONIA Grev.   A genus with more than 150 species and of widespread distribution throughout the seas of the world, *Polysiphonia* (Gr. *polys*, many + Gr. *siphon*, a tube) is a radially constructed, usually freely branched alga (Fig. 9.95) with apical growth and a **polysiphonous** construction. The apical cell cuts off segments proximally, the segments elongating and cutting off pericentral cells of the same length

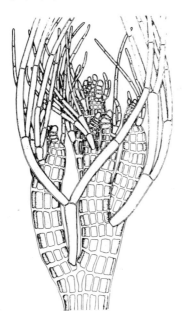

Fig. 9.95 *Polysiphonia hendryi* var. *gardneri* (Kyl.) Hollenb. Polysiphonous construction of the uniseriate axes is evident. × 230. (After Hollenberg; permission of American Journal of Botany.)

as the central siphon, or axial cell. Depending on the species, the number of pericentral cells may be as few as 4 (the subgenus *Oligosiphonia*) or up to 24, the subgenus *Polysiphonia* including species with more than 4 pericentral cells. The pericentral cells are connected to the central cell by primary pit connections and become connected to each other peripherally as well as to cells immediately above and below them by secondary pit connections (p. 456). The "siphons" may be aligned parallel to the long axis of the plant or be somewhat spirally placed. Cortication may be added, sometimes making the plant very coarse, such as in *P. elongata* (Huds.) Harv. and *P. nigrescens* (Huds.) Grev., both Atlantic species, and *P. brodiaei* (Dillw.) Grev. from the Atlantic and the Pacific coasts (Lauret, 1971).

Branching may be of various origins, and these have been discussed by Hollenberg (1942). **Exogenous** branches arise from segments immediately below the apical cell and are cut off before pericentral cells are formed. **Trichoblasts**, which are simple or branched colorless, hairlike laterals in *Polysiphonia*, arise exogenously. Trichoblasts are often dediduous, leaving a scar cell, which is the persistent basal cell of the trichoblast. New branches may arise from these scar cells, these branches being termed **cicatrigenous branches** and indirectly of exogenous origin. Branches that arise from the central cell after the pericentral cells have been cut off are of **endogenous** origin, and they are less common in *Polysiphonia*. Unlike endogenous branches, exogenous branches typically have a reduced number of pericentral cells in their basal segments. Whether the branching is endogenous or exogenous is a useful trait to distinguish species.

The life history in *Polysiphonia* involves an alternation of free-living unisexual gametophytes with an isomorphic tetrasporophytic phase, which has been confirmed by culturing studies (Edwards, 1968, 1970B). Male plants bear spermatangia on special branchlets (Fig. 9.94a), which are equivalent to trichoblasts; the pericentral cells of these laterals cut off spermatangial initials, from which the spermatia are budded off in profusion. The female plants also bear special short laterals (Fig. 9.94b), in which sterile filaments form an urn-shaped envelope, the pericarp, over the carpogonial branch, the latter being cut off from a fertile pericentral cell at the base of the short lateral. Other details of this development have been previously described (p. 559). The carposporophyte is formed with a large, basal fusion cell from which are budded the carposopores (Fig. 9.94c). In *Polysiphonia* the tetrasporangia are borne singly (Fig. 9.94d) per fertile tier on the tetrasporic plants.

Although a notoriously difficult group taxonomically, several regional mono-graphs of *Polysiphonia* have appeared, including those by Hollenberg (1942, 1944, 1968A, B), Segi (1951), and Lauret (1967, 1970). Relatively little utilization of hybridization studies has been applied toward better understanding the taxonomy of red algal genera. One interesting example, however, obtained in *Polysiphonia* is the investigation of Rueness (1973). By carrying out reciprocal crosses between sexual plants of a population of *P. hemisphaerica* Aresch. from Scandinavia and those of a population of *P. boldii* Wynne et Edwards from the Gulf of Mexico, Rueness demonstrated that these widely disjunct populations were interfertile. The culturing of carpospores produced viable tetrasporophytes, which proceeded to form tetrasporangia. The tetraspores of these hybrid plants had negligible viability, however, suggesting that a partial isolating mechanism was present. He concluded his study by regarding them as varieties within *P. hemisphaerica*.

PTEROSIPHONIA Falkenb.   Unlike the radial pattern of branching seen in *Polysiphonia*, a distichous pattern of branching is evident in *Pterosiphonia* (Gr. *pteron*, a wing + Gr. *siphon*, a tube). Thalli arise from a creeping portion, the erect axes tending to be **percurrent** with alternately pinnate branches (Fig. 9.96a). Trichoblasts are absent, unlike the apices of *Polysiphonia*, where they are typically present. But like

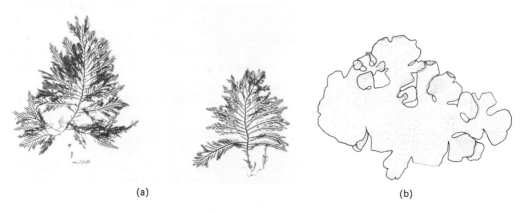

(a)   (b)

Fig. 9.96   (a) *Pterosiphonia*. Habit. (b) *Amplisiphonia pacifica* Hollenb. Habit. (a) × 0.46; (b) × 2.9.

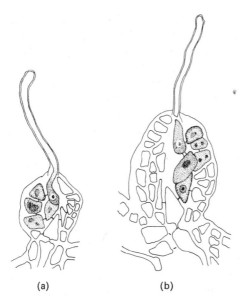

Fig. 9.97 *Laurencia venusta* Yamada. (*a*) Carpogonial branch is not covered by pericarp prior to fertilization. (*b*) Auxiliary cell is cut off from the supporting cell after fertilization. × 700. (After Saito.)

(a)   (b)

Polysiphonia, tetrasporic plants bear a single tetrasporangium per segment in fertile regions. The number of pericentral cells ranges from 5 to 12, depending on the species. Some species, such as *P. bipinnata* (P. et R.) Falkenb. on the Pacific coast, are ecorticate; whereas others, such as *P. baileyi* (Harv.) Falkenb. also on the Pacific coast, are heavily corticated. Contributions concerning the structure of *Pterosiphonia* and the relationship with *Polysiphonia* have been made (Ardré, 1967A, B).

AMPLISIPHONIA Hollenberg   The flattened blades of *Amplisiphonia* (L. *amplus*, ample, full + Gr. *siphon*, a tube) are prostrate on the substratum, attached by many unicellular rhizoids. Growth occurs at the margin by means of numerous apical cells. Segments from the apical cells cut off five pericentral cells, three cells being dorsal and two being ventral around the axial cell. These polysiphonous, forking filaments are laterally united to produce the foliose, dorsiventrally organized thallus (Fig. 9.96*b*). In the original description of this genus (Hollenberg, 1939), tetrasporangia were described as being produced in somewhat modified flattened lobes of the thallus. More recently, the sexual stages have been described (Hollenberg and Wynne, 1970). Male plants bear special spermatangial branches along the thallus margins, whereas female plants also are fringed by procarps within pericarps arising from the margins. The single species, *A. pacifica* Hollenb., occurs on the Pacific coast of North America.

LAURENCIA Lamour.   The heavily corticated thalli of *Laurencia* (named after the French naturalist, M. *de la Laurencie*) consist of terete or compressed axes (Fig. 9.98), typically with pinnate branching. The apical cell is usually sunken into a depression, and the pericentral cells are noticeable only near the apex because of the abundant cortication that is quickly developed. Sexual plants are isomorphic with the

Fig. 9.98 *Laurencia spectabilis* P. et R. Habit. × 2.

tetrasporic plants. Male plants produce spermatangia within shallow to deep cavities. According to Saito (1964, 1965) the carpogonial branch (Fig. 9.97*a*) is not initially covered by a pericarp and is still exposed when the auxiliary cell (Fig. 9.97*b*) is cut off from the supporting cell.

Tetresporangia are borne in the cortical tissue, and the tetrasporangia have been shown (Saito, 1969A) to be cut off from their parent cell adaxially in several California species examined, whereas they are cut off abaxially from parent cells in Japanese species. Excellent monographic treatments (Saito, 1969B; Saito and Womersley, 1974) have been provided for this difficult genus, which is widespread both in tropical and temperate seas of the world.

10

Division Cryptophycophyta

The division Cryptophycophyta contains a relatively small group of biflagellate organisms, the cryptomonads, whose asymmetric cells are flattened dorsiventrally and bounded by a periplast. The **periplast** consists of a plasmalemma with granular or fibrillar material to the outside and layered material to the inside, such that a somewhat rectangular or hexagonal surface pattern is detectable (Hibberd, et al., 1971; Gantt, 1971). Cells exhibit a degree of firmness coupled with some flexibility. The two flagella invariably arise ventrally from within a depression or furrow, the opening of which is close to the anterior end of the cell. The flagella may be equal or subequal in length, homodynamic or heterodynamic in behavior, and bear hairs of the same thickness and stiffness as those present on flagella of the Chrysophycophyta and Phaeophycophyta. A significant difference, however, is that hairs occur on both flagella rather than only on one of the two flagella in at least two genera of this division. In *Cryptomonas* (Fig. 10.1) and *Hemiselmis* the longer of the two flagella bears two opposite rows of hairs, and the shorter flagellum bears only a single row of hairs (Hibberd, et al., 1971). The difference in length between the hairs on the longer flagellum and those on the shorter flagellum is a further unique trait. The occurrence of two pleuronematic flagella is a distinctive feature for this division.[1]

A broad range in pigmentation is evident in this group of organisms,[2] and some colorless genera are also known. Cells may be red, blue, olive-yellow, brown, or green. The treatment by Butcher (1967) of representatives of this division from the British coastal waters includes colored plates that aptly depict variations in pigmentation. The potential change in pigmentation with age was stressed by Butcher's culturing

---

[1]The shorter flagellum in *Chroomonas* is apparently devoid of hairs (Hibberd, et al., 1971), but the investigators admitted that it was uncertain whether this apparent absence was genuine or due to techniques of preparation.

[2]Facultative heterotrophy is also present in these algae. Cannibalistic ingestion of other cryptomonad cells has been recorded (Wawrik, 1970A) in *Cryptomonas*.

work, which demonstrated that the red appearance of younger cultures of *Chroomonas salina* (Wislouch) Butch. is converted into a greenish hue in older cultures of this species. The instability of color as a taxonomic trait, particularly at the generic level, is obvious, causing Butcher (1967) to merge *Rhodomonas* within *Cryptomonas* and *Cryptochrysis* within *Chroomonas,* since differences in pigmentation had been the only means of formerly separating these pairs of genera.

Usually only one or a pair of chloroplasts is present in a cell; the only exception is *Cyanomonas* with numerous small chloroplasts. The chloroplast is surrounded by four membranes, the outer two being the chloroplast endoplasmic reticulum (Gibbs, 1962C; Dodge, 1973). An unusual feature is the somewhat loosely paired arrangement of the thylakoids within the chloroplasts (Wehrmeyer, 1970B), a pattern that can be considered (Lucas, 1970A) to be intermediate between the singly occurring thylakoids in red algal chloroplasts and the common arrangement of thylakoids in groups of three prevalent in many algal classes. Another distinctive characteristic is the apparent presence of the phycobiliproteins within the intrathylakoidal spaces rather than on the stromal side of these lamellae. These pigments do not occur as discrete phycobilisome-type aggregations (Gantt, et al., 1971). Some variations in the typical paired arrangement of thylakoids have been noted (Taylor and Lee, 1971; Dodge, 1973), whereby thylakoids might occur singly possibly due to culturing conditions or in larger aggregations. Starch, which is the major photosynthate, is present as a single large granule or many smaller granules, usually in close proximity to a pyrenoid (Lucas, 1970B) and lying within the perichloroplastic matrix, which is the space between the chloroplast envelope proper and the outerlying endoplasmic reticulum (Bisalputra, 1974). Eyespots occur in some cryptomonads. Situated within the chloroplast, they consist

Fig. 10.1 Flagella of *Cryptomonas ovata* Ehr. Two pleuronematic flagella arise from a subapical position. (After Hibberd, Greenwood, and Griffiths.)

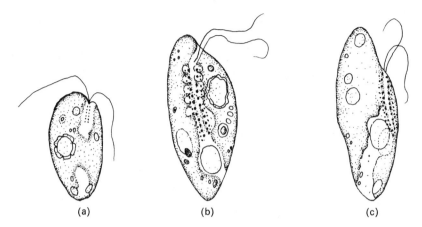

Fig. 10.2 (*a*) *Chroomonas salina* (Wislouch) Butcher. (*b*), (*c*) *Cryptomonas major* Butcher, front (*b*) and side (*c*) views. (*a*) × 2362; (*b*), (*c*) × 2554. (After Butcher.)

of a single layer of granules. The eyespot usually occupies a median position within the cell, close to the nucleus and at the periphery of the chloroplast.

Photosynthetic pigments include chlorophylls *a* and *c*, α and β carotene (the former being predominant over the latter), and some distinctive xanthophylls (Haxo and Fork, 1959; Goodwin, 1974). Biliproteins are also present (O'hEocha and Raferty, 1959; O'hEocha, 1971) and responsible for the reddish to bluish hues observed in many cryptomonads. In the colorless species *Chilomonas paramecium* Ehr., a leuco-plast is present, which is an extensive organelle containing starch granules and sur-rounded by a chloroplast endoplasmic reticulum (Sepsenwol, 1973).

With the single exception of the marine genus *Hillea*, all cryptomonads contain **ejectosomes**, which appear in the light microscope as small, refractive dots. They are especially evident lining the sides of the gullet, arranged in two rows in *Chroomonas* (Fig. 10.2*a*) or in several rows in *Cryptomonas* (Fig. 10.2*b*, *c*). A second, smaller cate-gory of peripherally located ejectosomes may also be present, such as in *Cryptomonas* (Lucas, 1970A) and *Chroomonas* (Dodge, 1969A). These ejectosomes are analogous to the trichocysts of dinoflagellates (p. 429) and ciliates, but because of their very different structure it has been suggested (Dodge, 1973) that the term *ejectosome* be applied to this organelle in crytomonads to distinguish it from the trichocysts of dinoflagellates.

The undischarged ejectosome is a cylinder of membranous material (Anderson, 1962; Schuster, 1970), made up of a central reel and a larger outer coil. Upon ejection, the central reel is seemingly shot out, pulling the larger coil along with it (Mignot, et al., 1968). As cells dry out or experience irritation, the ejectosomes are spewed out and unreeled as long threads. The ejectosomes lining the gullet are released into the gullet. The formation of ejectosomes is associated with the Golgi body (Wehrmeyer, 1970A).

Members of this division occur in both freshwater (Huber-Pestalozzi, 1950; Pringsheim, 1968A) and marine (Butcher, 1967) habitats, some species in the latter category, such as *Chroomonas salina*, being euryhaline and tolerating salt-marsh pools as well as the open waters of estuaries. Most forms are photosynthetic, but saprophytic nutrition also occurs. Some of the autotrophic species have been demonstrated to be auxotrophic for various vitamins. Although the free-swimming condition is the usual mode of existence, palmelloid phases can also be formed, the nonmotile cells even retaining their flagella tightly coiled around the cells within the mucilaginous colonies. Some members of this division are known to be **zooxanthellae** (p. 435), residing in the tissues of host invertebrates or within certain marine ciliates, blooms of which bring about "red tides" (Barber, et al., 1969).

The primary method of reproduction is simply by longitudinal cell division, the cell dividing either in a free-swimming or nonmotile condition. Sexual reproduction is doubtful.

Appendix

Cultivation of Algae in the Laboratory

Introduction

Advances in the techniques of growing algae, both freshwater and marine, in the laboratory have had a tremendous impact on phycology. These improvements in algal cultivation have made possible great advances in our knowledge of algal life histories, physiology, taxonomy, genetics, biochemistry, and ultrastructure (Bold, 1974). Most relevant in the present connection is that these techniques have resulted in great improvement in phycological laboratory instruction. It is now possible, thanks to many past investigators who deposited their cultures in such great collections of living algae as those at the University of Texas at Austin, Austin, Texas, and at Cambridge, England, among others, to obtain and to maintain for study in the living condition many algae that formerly were available to students only as bleached specimens stored in formaldehyde.

The purpose of the following paragraphs is not to provide a comprehensive manual on the techniques of algal cultivation. They have been included, rather, to be immediately available and useful to instructors and students interested in growing algae. For more comprehensive treatments of the topic, the publications of Bold (1942), Pringsheim (1946A), Lewin (1959), Starr (1964, 1966, 1971B), Stein (1973), O'Kelley (1974), and James (1974), among others may be referenced.

Culture Media

Depending on one's objectives, one may use culture media completely defined chemically, or those that contain such undefined ingredients as soil, soil extract, pea cotyledons, or rice grains, for example. Obviously, for precise physiological experiments, defined media would be required.

Some Media for Freshwater Algae[1]

BOLD'S BASAL MEDIUM (Bischoff and Bold, 1963)

Six stock solutions (in distilled or deionized water) 400 ml in volume should be prepared, each containing one of the following salts in the concentration listed:

Salt	Grams
$NaNO_3$	10.0 g
$CaCl_2 \cdot 2H_2O$	1.0 g
$MgSO_4 \cdot 7H_2O$	3.0 g
K_2HPO_4	3.0 g
KH_2PO_4	7.0 g
NaCl	1.0 g

To 940 ml distilled water, add 10 ml of each stock solution and 1.0 ml of each of the stock trace-element solutions prepared as follows:

1. 50 g EDTA and 31 g KOH dissolved in 1 liter distilled H_2O (or 50 g Na_2EDTA).
2. 4.98 g $FeSO_4 \cdot 7H_2O$ dissolved in 1 liter of acidified water (acidified H_2O: 1.0 ml H_2SO_4 dissolved in 1 liter distilled H_2O).
3. 11.42 g H_3BO_3 dissolved in 1 liter distilled H_2O.
4. The following, in amounts indicated, all dissolved in 1 liter distilled water: $ZnSO_4 \cdot 7H_2O$, 8.82 g; $MnCl_2 \cdot 4H_2O$, 1.44 g; MoO_3, 0.71 g; $CuSO_4 \cdot 5H_2O$, 1.57 g; $Co(NO_3)_2 \cdot 6H_2O$, 0.49 g.

This may be enriched by substituting 30 ml of stock $NaNO_3$ per liter to the definitive solution (3 × N BBM). Alternately, many algae thrive when urea is substituted as the nitrogen source; it may be provided at the level of 3× or 6× the level of nitrogen in BBM.

Vitamins, most frequently B_1, B_6, and B_{12}, may enhance the growth of algae in BBM. These may be added to a liter of BBM as 5 ml of Eagle's mixture[2] and B_{12} (cyanocobalamine) at concentrations of 0.1 ml of a 1.0 mg/ml solution (equivalent to 100 μg/liter).

BBM is useful in cultivating a variety of green algae (as well as moss protonema, fern gametophytes, etc.).

CHU 10 MEDIUM (CHU, 1942)

Salt	Grams/Liter
$Ca(NO_3)_2$	0.04
K_2HPO_4	0.01 or 0.005
$MgSO_4 \cdot 7H_2O$	0.025
Na_2CO_3	0.02
Na_2SiO_3	0.025
$FeCl_3$	0.0008

[1]See also Nichols (1973).
[2]"TC-Vitamins Minimal Eagle, 100 ×" (Difco Laboratories, Detroit, Mich.).

Chu's (1942) medium and that of Hughes, et al. (1958) (original and as variously modified) have proved useful in growing blue-green algae.

MODIFIED CHU NO. 10 SOLUTION
(Modified by Wright and Guillard; and by Van Dover)

1. Make stock solutions by dissolving the salts listed in the amounts indicated (in grams) each in 100 ml of distilled or deionized water; autoclave them and keep sterile.

Salt	Grams
$CaCl_2 \cdot 2H_2O$	3.67
$MgSO_4 \cdot 7H_2O$	3.69
$NaHCO_3$	1.26
K_2HPO_4	0.87
$NaNO_3$	8.5
$Na_2SiO_3 \cdot 9H_2O$ (metasilicate)	2.84

2. Prepare an iron solution by dissolving 3.35 g citric acid ($C_6H_8O_7 \cdot H_2O$) in 100 ml distilled water; then add 3.35 g ferric citrate ($FeC_6H_5O_3 \cdot 5H_2O$), autoclave to dissolve, dispense in sterile tubes and keep sterile, refrigerated in darkness (wrapped in aluminum foil).
3. Prepare a trace-elements solution by dissolving the salts in the amounts (milligrams) indicated together in 1 liter of distilled water. Autoclave and keep sterile.

Salt	Milligrams
$CuSO_4 \cdot 5H_2O$	19.6
$ZnSO_4 \cdot 7H_2O$	44.0
$CoCl_2 \cdot 6H_2O$	20.0
$MnCl_2 \cdot 4H_2O$	36.0
$Na_2MoO_4 \cdot 2H_2O$	12.6
H_3BO_3	618.4

To prepare the definitive solution, add aseptically 1 ml of each of the six stock solutions in step 1 to a liter of sterile double distilled or deionized water. Then add aseptically 1 ml of stock solution in step 2 and 1 ml of the trace elements solution in step 3. Dispense aseptically into sterile containers.

HUGHES, GORHAM, AND ZEHNDER'S (1958) MEDIUM

Salt	Grams/Liter
NaNO$_3$	0.496
K$_2$HPO$_4$	0.0399
MgSO$_4$·7H$_2$O	0.075
CuCl$_2$·2H$_2$O	0.036
Na$_2$CO$_3$	0.020
Na$_2$SiO$_3$·9H$_2$O	0.058
Ferric citrate	0.006
Citric acid	0.006
EDTA	0.001
Gaffron's trace-element solution	0.08 ml
Distilled or deionized water	to 1 liter

GAFFRON'S TRACE-ELEMENT SOLUTION

Salt	Grams/Liter
H$_3$BO$_3$	3.100
MnSO$_4$·4H$_2$O	2.230
ZnSO$_4$·7H$_2$O	0.287
(NH$_4$)$_6$Mo$_7$O$_{24}$·4H$_2$O	0.088
CO(NO$_3$)$_2$·4H$_2$O	0.146
Na$_2$WO$_4$·2H$_2$O	0.033
KBr	0.119
KI	0.083
Cd(NO$_3$)$_2$·4H$_2$O	0.154
NiSO$_4$(NH$_4$)$_2$SO$_4$·6H$_2$O	0.198
VOSO$_4$·2H$_2$O	0.020
Al$_2$(SO$_4$)$_3$·K$_2$SO$_4$·24H$_2$O	0.474

ALLEN'S (1968) MODIFICATION OF HUGHES, GORHAM, AND ZEHNDER'S (1958) MEDIUM

Salt	Grams/Liter
NaNO$_3$	1.5
K$_2$HPO$_4$	0.039
MgSO$_4$·7H$_2$O	0.075
CaCl$_2$·2H$_2$O	0.027
Na$_2$CO$_3$	0.020
Na$_2$SiO$_3$·9H$_2$O	0.058
Ferric citrate[3]	0.006
Citric acid	0.006
EDTA	0.001
Allen's trace-element solution	1.0 ml
Distilled or deionized water	to 1 liter
pH of medium	7.8

[3]Autoclave separately; add aseptically after medium has stood for 24 hours.

To prepare solid media, equal volumes of double-strength mineral base and double-strength agar are separately sterilized and combined after cooling to 48 °C.

ALLEN'S (1968) TRACE-ELEMENT SOLUTION (Modified)

Dissolve in 1 liter of distilled water:

Salt	Grams
H_3BO_3	2.86
$MnCl_2 \cdot 4H_2O$	1.81
$ZnSO_4 \cdot 7H_2O$	0.222
$Na_2MoO_4 \cdot 2H_2O$	0.391
$CuSO_4 \cdot 5H_2O$	0.079
$Co(NO_3)_2 \cdot 6H_2O$	0.0494

All the media above may be solidified by adding 15 g/liter of agar.

PRINGSHEIM'S BIPHASIC SOIL-WATER METHOD (PRINGSHEIM, 1946A, B; STARR, 1964)

Select a soil with some, but not too much, humus content. Introduce a pinch of $CaCO_3$ into the bottom of a culture tube or other glass container and cover to a depth of $\frac{1}{4}-\frac{1}{2}$ in. with garden soil. Then add distilled or deionized water until the container is three-quarters full. Plug the container with cotton or cover loosely and *steam*[4] it for 1 hour on two successive days. When cool, inoculate. The writers have found that addition of one-third BBM and distilled water often prolongs the life of the culture. Furthermore, the soil-water containers may be sterilized by autoclaving once for 1 hour. This is an especially useful medium for maintaining cultures for teaching.

Some Media for Marine Algae

OTT'S (1965) ARTIFICIAL SEAWATER

Distilled Water.............1000 ml

Salt	Grams/Liter
NaCl	21
$MgSO_4 \cdot 7H_2O$	6
$MgCl_2 \cdot 6H_2O$	5
$CaCl_2 \cdot 2H_2O$	1
KCl	0.8
NaBr	0.1
$NaNO_3$	0.2
$NaHCO_3$	0.2
H_3BO_3	0.06
$Na_2SiO_3 \cdot 9H_2O$	0.01
$Sr(NO_3)_2$	0.03
Na_2HPO_4	0.02

[4]The writers have found autoclaving to be satisfactory.

To the above, add 1 ml each of the micronutrients listed under the formula for BBM on p. 572.

This artificial seawater may be used in preparing Erdschreiber or von Stosch's enrichment media, etc.

In addition to this synthetic seawater, various "mixes" of sea salts are available commercially. These include "Instant Ocean" (Aquarium Systems, Inc., 33208 Lakeland Blvd., Eastlake, Ohio.), Utility Marine Mix (Utility Chemical Co., 6th Ave. & Wait St., Paterson, N.J.), Dayno (Dayno Sales Co., 678 Washington St., Lynn, Mass.), and "Rila Marine Mixture" (Rila Products, Teaneck, N.J.)

ERDSCHREIBER MEDIUM (Starr, 1964, p. 1038)

1000 ml filtered seawater
50 ml soil-water supernatant[5]
0.2 g $NaNO_3$
0.03 g $Na_2HPO_4 \cdot 12H_2O$

First day: Filter seawater through No. 1 filter paper and then heat to 73°C.
Second day: 1. Again heat seawater to 73°C.
                      2. Autoclave salt solutions (made up in distilled water so that 1 ml of each solution gives required amount for 1 liter of culture medium).
Third day: Add cold salt solutions to cold soil-water supernatant; then add these to cold seawater. Dispense in sterile tubes, flasks, or petri dishes.

VON STOSCH'S ENRICHMENT MEDIUM
(Stosch, 1964B)

Salt

$NaNO_3$ . . . . .500 μmol (= 115 mg N)
NaH_2PO_4. . . . . 30 μmol (= 0.93 mg P)
$FeSO_4$ . . . . 1 μmol (= 55.8 μg Fe)
$MnCl_2$ . . . . 0.1 μmol (= 5.9 μg Mn)
Ethylene diaminetetraacetic acid,
Disodium salt $\cdot 2H_2O$ . . . . . . 10 μmol (= 3.7 mg)
Seawater 1020 ml

[5] Supernatant from Pringsheim's soil-water medium.

PROVASOLI'S ENRICHED SEAWATER (PES)
(Provasoli, 1963, 1968; McLachlan, 1973;
West, Personal Communication)

Stock solutions (Each in 100 ml water)[6]		Milliliters of stock solutions to be added
$NaNO_3$	35 g	10
Na_2 glycerophosphate[7]	5 g	10
Vitamin B_{12}	1 mg	10
Thiamine	50 mg	10
Biotin	0.5 mg	10
Tris buffer[8]		—
Fe (as EDTA 1:1 molar)		250
$Fe(NH_4)_2(SO_4) \cdot 6H_2O$, 351 mg		
+ Na_2EDTA, 300 mg/500 ml		
P II trace metals		250
		Bring total volume to 1250 ml with distilled or deionized water.

Add 20 ml of the above stock solution mixture to 1000 ml of filtered sea water to prepare full-strength medium.

P II Trace Metals Solution

Dissolve the following in 1000 ml H_2O:

Salt	
H_3BO_3	1.14 g
$FeCl_3 \cdot 6H_2O$	49 mg
$MnSO_4 \cdot 4H_2O$...........	164 mg
$ZnSO_4 \cdot 7H_2O$	22 mg
$CoSO_4 \cdot 7H_2O$	4.8 mg
Na_2EDTA	1 g

In addition to these marine media, a number of others have proved useful (see McLachlan, 1973 and Page, 1973).

Conditions of Cultivation

Algal cultures should not be placed in direct sunlight but grown, rather, in a north window or under fluorescent illumination. For active growth, an intensity of 300–500 ft-c at the level of the cultures is recommended. For maintaining the cultures

[6]Distilled or deionized H_2O.

[7]Sigma Chemical Co., DL-glycerophosphate.

[8]Consult Sigma Technical Bulletin No. 106B Revised March 1972 for Trizma mixing table, which presents amounts of HCl and base to be mixed for obtaining a desired pH. (Example: For a pH of 7.8 at 25°C, mix 5.32 g Trizma HCl and 1.97 g Trizma base.)

in a less active state, they may be stored farther from the light source at lower intensities (ca. 75 ft-c). Most blue-green and red algae grow better at low light intensities. Photoperiods of 12 hours of light, 12 hours of dark and 18 hours of light and 6 hours of dark have proved satisfactory. Temperatures of 20–21°C are favorable for the growth of many algae. Those from colder ocean waters require lower temperatures.

Sources of Algal Cultures

Cultures for phycological instruction may be obtained in the United States from the Culture Collection of Algae, Department of Botany, University of Texas, Austin, Texas; from the Carolina Biological Supply Company, Burlington, N.C.; from the General Biological Supply House, 8200 S. Hoyne Avenue, Chicago, Illinois; or from Ward's Natural Science Establishment, P.O. Box 1712, Rochester, New York.[9]

Of course, cultures may also be started from fresh collections in the field by isolating single cells, colonies, filaments, or fragments and introducing them into suitable culture media. Microalgae may be isolated also by the techniques of streaking on appropriate culture media solidified with 1.5% agar. In this method an inoculating loopful of liquid containing the microalgae is spread with the loop across the surface of the agarized culture medium in petri dishes in a zigzag pattern. The petri dishes are stored, bottom side up under bright light. When colonies have grown (3–4 days to 2 weeks), those with differing configurations (an indication that they are different taxonomically), as viewed under stereoscopic binocular microscope at 14 × may be isolated with drawn-out disposable Pasteur pipettes and inoculated into tubes of liquid culture media.

Hoshaw (1961) and Starr (1964) have summarized methods for inducing sexual reproduction in a number of algae, and in this connection, the reference volume by Stein (1973) should be consulted.

[9] Also P.O. Box 1749, Monterey, California.

Glossary

Acronematic: referring to a flagellum that is smooth and terminated with a fine fibril, thus tapered at its distal end (Fig. 3.2*j*).

Adelophycean: referring to the inconspicuous phase in the life history of an alga, alternating with the macroscopic (delophycean) phase.

Adelphoparasite: a parasite that is closely related to its host (e.g., both placed in the same family).

Agar: a phycocolloid characteristic of the Rhodophyceae; it is a sulfated polysaccharide (a galactan) occurring in the cell walls and intercellular spaces, extracted primarily from *Gelidium* and *Gracilaria* and used commercially for its gelling properties.

Akinete: a thick-walled nonmotile spore derived from a vegetative cell that thickens its wall (Fig. 2.8*a*).

Ala (Alae): wing, such as the membranous expanses from the central midrib in foliose members of the Delesseriaceae (Fig. 9.87).

Algin: the soluble sodium salt of alginic acid.

Alginate: salt form of alginic acid, such as with calcium or barium ions.

Alginic acid: a phycocolloid produced in the walls of Phaeophyceae, consisting of a polysaccharide made up of β-1,4-linked D-mannuronic acid and 1,4-linked L-guluronic acid units, the ratios of these two acids being variable.

Alloparasite: a parasite that is not closely related taxonomically to its host; placed in remote taxa, such as in different orders.

Alternation of generations: a sequence in a life cycle in which a haploid, gamete-producing phase is followed by a diploid, spore-producing phase; the spores of the latter reinitiate the haploid phase (Fig. 1.4).

Amoeboid: having the changeable shape of an amoeba.

Amphiesma: the cellular covering of a dinoflagellate vegetative cell, comprised of only membranes in naked species or thecal plates enclosed in membranes in armored species.

Ampulla (Ampullae):   accessory branch systems, usually congested in appearance and occurring in the inner cortex, producing either carpogonial branches or auxiliary cells (i.e., in separate ampullae in Cryptonemiaceae) or both carpogonial branches and an auxiliary cell (such as in Endocladiaceae).

Androphore: a branch bearing antheridia.

Androsporangium (Androsporangia):   a cell or sporangium producing androspores (Fig. 3.97c).

Androspore: a zoospore that grows into a dwarf male filament in the green algal order Oedogoniales.

Anisogamous: referring to gametes that consistently differ in size, the larger designated "female" and the smaller "male" (Fig. 1.3m).

Anterior invagination: a depression at or near the apex of Euglenoids (Fig. 5.2).

Antheridium (Antheridia):   the gametangium that produces sperm in oogamous sexual reproduction (Fig. 1.1b).

Apical cell: a prominent meristematic cell at the tip of a plant, e.g., *Dictyota* (Fig. 6.36a).

Apical growth: Growth at the apex of a plant.

Aplanosporangium (Aplanosporangia):   a cell or sporangium that contains aplanospores (Fig. 1.3f).

Aplanospore: an ontogentically potential zoospore that has omitted the motile period (Fig. 1.3f).

Apogamy: formation of an organism without involving the fusion of gametes.

Apomeiosis: nuclear division without meiosis.

Archeopyle: a pore in the cyst wall through which the protoplast escapes or is liberated at germination.

Areola (Areolae):   the regularly repeated perforation through the siliceous layer of a frustule, usually covered on one side by a velum.

Articulated: segmented or jointed in appearance, as thalli of *Halimeda* (Fig. 3.123) and some Corallinaceae (Fig. 9.51a, b).

Assimilatory filament: photosynthetic filament.

Autocolony: a colony formed in the asexual reproduction of coenobia that is a miniature of the parental colony (Fig. 1.3j).

Autogamous: refers to sexual reproduction in which the zygote is formed by the fusion of two haploid nuclei from one individual.

Autosporangium: a cell or sporangium that contains autospores (Fig. 1.3h).

Autospore: a nonmotile spore that is the miniature of the cell from which it is derived; not potentially flagellate, ontogenetically. (Fig. 1.3g, h).

Auxiliary cell (or generative auxiliary cell):   a cell from which the carposporophyte is produced following transfer to it of the zygote nucleus or one of its diploid progeny (in Florideophycidae of the Rhodophycophyta).

Auxospore: the zygote of diatoms originally so-called because of its rapid increase in size (Fig. 7.48f).

Auxotrophic: requiring for growth an exogenous source of one or more vitamins.

Axenic culture: a population of individuals of one strain or species free from other strains or species (usually bacteria and fungi, as used in phycology).

Axoneme (also axial complex):   the $9 + 2$ complex of microtubules contained within the flagellum and enclosed by a surrounding membrane.

Benthos, benthic: bottom-dwelling; nonplanktonic; attached to or resting on the substrate.

Biliprotein: a chromoprotein in which the prosthetic group is a pigment tightly bound by covalent linkage(s) to its apoprotein.

Binary fission: division into two products.

Bioluminescence: the emission of visible light by living organisms such as certain dinoflagellates.

Bipartition: division into two products.

Bisexual: referring to an individual organism or clone whose gametes unite sexually.

Bisporangium (Bisporangia):   a sporangium whose contents divide into two spores.

Blade: the broad, membranous distal portions of kelp plants (Fig. 6.50).

Brittlewort: a name applied to calcareous members of the Charophyta.

Canal: a tubelike area connection the anterior invagination or reservoir to the surrounding medium in Euglenoids (Fig. 5.2).

Capitulum: a cell in the antheridium of Charophyta from which the antheridial filaments arise (Fig. 4.4c).

Carpogonium (Carpogonia):   the oogonium, usually with a trichogyne, in Rhodophycophyta (Fig. 9.28a).

Carpospore: red algal spore that arises as a result of fertilization.

Carposporophyte: a phase produced following fertilization in the Florideophycidae, comprised of gonimoblast filaments bearing carpospores.

Carpotetrasporangium: the reproductive structure produced by a carpotetrasporophyte.

Carpotetrasporophyte: a special type of carposporophyte, resulting from fertilization in Florideophycidae, in which the gonimoblast filaments produce carpotetrasporangia rather than carpospores.

Carrageenan: a phycocolloid characteristic of some Rhodophyceae (such as Gigartinaceae, Solieriaceae, Phyllophoraceae, and Hypneaceae); a sulfated galactan located extracellularly; various fractions are recognized based on solubility differences.

Caulerpicin: a toxin produced by *Caulerpa*.

Cauloid: stemlike organ.

Cell plate: a thickened structure arising in the cytoplasm between telophasic nuclei.

Central nodule: a thickened region in the center of the valves of motile pennate diatoms, separating the fissures of the raphe.

Chasmolithic: living in rock fissures.

Chemotaxis: movement of an organism toward or away from a chemical stimulus.

Chloroplast: a double-membrane-bounded organelle containing elaborated membranous sacs known as thylakoids; the membranes contain chlorophyll *a* and other components of the photosynthetic light reactions.

Chloroplast endoplasmic reticulum: an extension of the endoplasmic reticulum to enclose the chloroplast, this membrane system in some cases also interconnecting with the outer nuclear envelope.

Choanoflagellate: a flagellated organism in which a ringlike extension of rhizopodia forms a collar around the flagellum (Fig. 7.13).

Cicatrigenous branch: a branch arising from a scar cell, which is a basal cell of a shed trichoblast, such as in *Polysiphonia* (p. 561); thus, it is indirectly exogenously developed.

Cingulum: (*a*) in Pyrrhophycophyta a constriction with a transverse orientation, the girdle region of the cell; (*b*) in Bacillariophyceae the girdle or region of the frustule connecting the two distal valves.

Clonal culture: a population descended from one individual.

Cnidocyst: a complex ejectile organelle (also termed *nematocyst*) present in a few dinoflagellate genera, such as *Nematodinium*.

Coccolith: minute, calcareous, ornamented scale on cell surface of a coccolithophorid (Fig. 7.22).

Coccolithophorid: a unicellular organism classified within the order Prymnesiales of the Prymnesiophyceae, which bears variously sculptured calcareous scales (coccoliths) upon its cell surface.

Coccosphere: a cell covering of coccoliths in which the coccoliths hold together to form an intact shell of scales (Fig. 7.21).

Coenobium: a colony in which the number of cells is fixed at its origin and not augmented subsequently (Fig. 1.2*d*).

Coenocyst: a multinucleate, thick-walled cyst (Fig. 3.52-9).

Coenocytic: multinucleate and without transverse walls.

Conceptacle: a near-spherical invagination or cavity containing reproductive structures, such as in Fucales (Fig. 6.76) and Corallinaceae (Fig. 9.48).

Conchosporangium: a type of enlarged sporangium usually produced in series ("fertile cell rows") by the *Conchocelis* phase of several members of the Bangiales.

Conchospore: a spore produced and released singly by a conchosporangium.

Conjugation: connection of cells by papillae and tubes through which amoeboid gametes move or in which they conjugate.

Conjugation canal or tube: the product of union of two conjugation papillae (in the Zygnemataceae and Desmidiaceae) in which the gametes migrate (Fig. 3.168*c*).

Conjugation papilla: protuberances emerging from contiguous sexually mature cells in conjugate green algae (Fig. 3.165*c*).

Connecting cell: a cell that effects the transfer of the zygote nucleus from the fertilized carpogonium to the auxiliary cell.

Connecting filament: a filament that effects the transfer of the zygote nucleus from the fertilized carpogonium to the auxiliary cell, either directly or indirectly; if the latter, the filament may fuse with a cell in the carpogonial branch before proceeding out to contact auxiliary cells.

Consortium: an association, or "living together," of organisms, ranging from epiphytism without obvious interdependence to symbiosis with definite mutual dependencies between the partners.

Contractile vacuole: a vacuole that by contraction forces water from a cell (Fig. 3.7*b*).

Convergent evolution: the independent development of characteristics in remotely related organisms, the end products of this evolution looking very similar.

Corona: the 5 or 10 cells at the apices of the oogonia of Charophyta (Fig. 4.4).

Cortex: the outermost layer of cells or tissue in a thallus.

Corticolous: inhabiting bark.

Costa (Costae): an elongated, solid thickening of the valve of a diatom frustule.

Cover cells: (in Ceramiales of Rhodophycophyta) cells that are cut off in association with tetrasporangia, serving as superficial, protective covers; or (in Corallinaceae of the Rhodophycophyta) the outermost cells of the epithallus of the vegetative plant body.

Cribrum (Cribra): a velum that has many regularly arranged pores in a diatom frustule.

Crown cells: see corona.

Cryptostoma (Cryptostomata): small invaginations in the cortex of Fucales, in which sterile hairs are produced, the hairs often extending out through the opening of the cryptostoma.

Cyst: a (usually) thick-walled dormant cell.

Cystocarp: the carposporophyte and any immediately surrounding envelope or pericarp provided by the gametophyte.

Cytostome: a special mouth-like opening (in *Noctiluca*) into which food particles may be drawn into the cell.

Defined medium: one in which the ingredients and their concentration are known.

Delophycean: the conspicuous or macroscopic phase in the life history of an alga.

Desmid: unicellular or filamentous green algae that have conjugation of nonflagellate, amoeboid gametes.

Desmoschisis: sometimes called "vegetative cell division" (*sensu* Fritsch); division of cells, in which the parental wall forms part of the wall of the cellular progeny (p. 127) (Fig. 3.61).

Diatom: member of the class Bacillariophyceae.

Diatomaceous earth: siliceous deposits made up of the sedimentary buildup of diatom frustules.

Diffuse growth: generalized, not localized, growth.

Dioecious: having male and female gametes or gametes of different mating type produced by different individuals or clones.

Diplobiontic: having two free-living phases in the life history of an organism.

Diploid: having twice the basic (haploid) number of chromosomes.

Diurnal rhythm: a cycle based upon a daily periodicity; see also endogenous circadian rhythm.

Ecdysis: a process in which the thecal wall layer of a dinoflagellate cell is shed.

Ectocarpin: an erotactin (sex hormone) produced by female gametes of the brown alga *Ectocarpus*, causing cluster formation by the male gametes and their accumulation at the source of the hormone.

Ectoparasite: a parasite living on the outside of its host.

Edaphic: of the soil.

Ejectosome: a type of ejectile organelle occurring in most cryptomonads, which are structurally different from trichocysts but analogous to them in function.

Eleutheroschisis: cellular division in which the walls of the cellular products are entirely new and free from the parental walls (Fig. 1.3*d, e*).

Elevation: raised region of a valve wall of a diatom frustule, not projecting laterally beyond the valve margin and with much the same structure as the valve.

Endedaphic: living in soil.

Endochite: the innermost layer of the Fucalean oogonium.

Endocytic: living within a cell.

Endocytosis: phagotrophic engulfment of particulate matter.

Endogenous branch: a branch arising from a central cell after formation of pericentral cells, such as in *Polysiphonia*.

Endogenous circadian rhythm: the occurrence of a natural phenomenon, such as bioluminescence or cell division, with a periodicity of 24 hours, the peaks of activity continuing at least for a few cycles even if external variables such as photoperiod are absent.

Endolithic: rock-penetrating.

Endoparasite: a parasite living within the tissue of its host.

Endophytic: living within the tissue of the host plant (Fig. 2.2).

Endospore: spores of blue-green algae formed by internal divisions of the protoplast of a parental cell (Fig. 2.8*b*).

Endozoic: living within an animal.

Enrichment culture: a culture to which nutrients have been added to encourage the growth and reproduction of one or more species.

Epicingulum: that portion of a girdle of a frustule adjacent to an epivalve.

Epicone: the portion of a dinoflagellate cell anterior to the cingulum.

Epidaphic: living on the soil surface.

Epilithic: living upon stones.

Epipelic: attached to mud or sand.

Epiphytic: living upon a plant.

Epithallus (also epithallium): that portion of the vegetative thallus (in Corallinaceae) in which the cells or filaments are developed outwardly from an intercalary meristem.

Epitheca: the epivalve and epicingulum of a diatom frustule.

Epivalve: the flattened or convex plate of a frustule opposite to the hypovalve and formed prior to the hypovalve and thus usually slightly larger.

Epizoic: living upon an animal.

Eukaryotic: having membrane-bounded nuclei.

Euphotic zone: that upper layer of water receiving sufficient light for photosynthesis to occr.

Euryhaline: having a broad tolerance to varying salinity.

Eutrophic: rich in nutrients.

Exochite: the outermost layer of the oogonium in the Fucales.

Exogenous branch: a branch arising directly from a branch primordium cut off from a subapical cell.

Exospore: spores of blue-green algae abstricted from the protoplast of the parental cell (Fig. 2.8*c*).

Facultative gametes: zoospores that may function as gametes.

Facultative heterotrophy: the ability of photosynthetic organisms to carry on heterotrophic nutrition also if the circumstances present themselves.

Fertile sheet: the cellular layer or layers lining the inside of the conceptable from which the reproductive structures are produced, such as the oogonia and antheridia in the Fucales (Fig. 6.76).

Fertilization tubule: a minute tubule that connects uniting gametes of *Chlamydomonas reinhardtii* and perhaps other species.

Flagellar swelling: a thickening near the flagellar base of Euglenoids (Fig. 5.2).

Floridean starch: the storage product of Rhodophycophyta, which is an amylopectin (α-1, 4-linked glucan with some α-1, 6-linkages, resulting in a branched polymer). It is present as granules outside the chloroplasts.

Foramen (Foramina): a small aperture or opening through the constriction of an areola of a diatom frustule.

Fountain-type growth: multiaxial growth, typically comprised of many filaments in which cell divisions are occurring.

Fragmentation: formation of new individuals from segments arising by the breakup of parental ones.

Free-nuclear division: repeated mitosis without intervening cytokineses.

Frustule: the siliceous wall of a diatom.

Fucoidan: water-soluble, sulfated polysacchraides in the cell walls of brown algae, characterized by the presence of sulfate esters and L-fucose.

Fucosan versicles: small, refractive vesicles present in the cells of brown algae containing fucosan, which demonstrates tanninlike properties; also called physodes.

Fucoserraten: an erotactin produced by eggs of *Fucus* and attracting sperm.

Furcellaran: a sulfated polysaccharide wall component with mucilaginous properties produced by the red alga *Furcellaria*; chemically similar to carrageenan.

Fusion cell: a cell produced by the union of the protoplasts of two or more cells.

Gametangium: a container in which gametes are produced (Fig. 1.1*b, e*).

Gamete: a sexually active cell capable of uniting with a compatible cell to form a zygote (Fig. 1.1*a*).

Gametic meiosis: meiosis occurring during the formation of gametes.

Gametophore: a branch bearing a gametangium or gametangia.

Gametophyte: the sexual, gamete-producing phase in the life history of a plant.

Gamone: agglutinating substance that functions in gametic union.

Gemma (Gemmae): a multicellular, asexual reproductive structure, including buds or fragments of a plant.

Generative auxiliary cell: see *auxiliary cell.*

Geniculum (Genicula): the axial portion of a thallus that is flexible, such as the noncalcified zones between the calcified segments (intergenicula) of articulated corallines (Fig. 9.51*a, b*) and the green alga *Halimeda* (Fig. 3.123).

Genophore: a gene-containing body, such as the discrete ringlike structure occurring in some algal chloroplasts.

Girdle: that portion of a diatom frustule between the valves; also called *cingulum.*

Girdle lamella (-e): a lamella, or membranous sac, encircling the periphery of a chloroplast, lying just within the chloroplast envelope.

Gland cell: a special, usually highly refractive cell occurring in the thalli of some red algae, which may function in secretion or storage of material.

Gliding movement: movement of organisms without flagella or pseudopodia when in contact with a substrate.

Gonidium: an asexual, nonmotile reproductive cell, as in *Volvox.*

Gonimoblast: a filament bearing a carpospore or carpospores or the entire collection of these filaments comprising the carposporophyte.

Gonimoblast initial: the primordium arising from a fertilized carpogonium or from a "diploidized" auxiliary cell that develops the gonimoblast or carposporophyte.

Gonocyte: a type of cell in multicellular parasitic dinoflagellates, which undergo cell division; they are located in the middle of the colony with a trophocyte at the anterior end and sporocytes to the posterior end (Fig. 8.17*a*).

Granum (Grana): broadly defined as a stack(s) of thylakoids within a chloroplast, such that the membranes of adjacent thylakoids are fused.

Growth: increase in size and volume, often, but not always, involving differentiation.

Growth, apical: growth at the tip of a plant or its branches.

Growth, basal: growth at the base of a plant or its branches.

Growth, generalized: diffuse or nonlocalized growth.

Growth, intercalary: growth localized at one or more loci between the apex and base.

Growth, localized: nondiffuse or nongeneralized growth.

Gynandrosporous: an Oedogonialean species in which the androspore develops on the female filament (Fig. 3.97*a*).

Haematochrome: astaxanthin or 3: 3′ diketo 4: 4′ dihydroxy-β-carotene.

Hairs: colorless, typically elongate, unicellular or multicellular structures.

Haplobiontic: with a single free-living phase in the life history of an organism.

Haploid: with a single basic set of chromosomes characteristic of the species.

Haplostichous: type of construction (in certain brown algal orders) in which the thallus is composed of free or consolidated filaments but lacking a true parenchymatous organization.

Hapteron (-a): multicellular attaching organs of the kelps (Fig. 6.70).

Haptonema: a short to long cellular extension with a distinctive fine structure (Fig. 7.2*c*), characteristic of the class Prymnesiophyceae.

Haustorium: that portion of a parasite that penetrates host cells and absorbs nutrients.

Hemiblastic: pattern of branching (in *Sphacelaria*) in which the branch primordium occupies the height of a superior segment, i.e., the upper half of the length of a primary segment cut off from the apical cell (Fig. 6.32*b*).

Hermaphroditic: producing both male and female gametes on the same individual (Fig. 3.93*a*).

Heterococcolith: type of coccolith having calcite (or aragonite) crystals of more than one shape and size.

Heterocyst: a thick-walled, usually translucent cell of certain Cyanochloronta thought to be the site of nitrogen fixation (Fig. 2.8*a*).

Heterodynamic flagella: flagella that have independent patterns of beating, not correlated with each other.

Heterogamous: referring to sexual reproduction in which the gametes are not morphologically identical.

Heteromorphic: morphologically different.

Heteroplastidy: having two kinds of plastids (chloroplasts and starch-storing leucoplasts).

Heterotrichy: a filamentous system composed of both an erect and prostrate portion (Fig. 3.83*b*).

Heterotrophic: nonautotraphic (nutrition).

Holdfast: an attaching cell or cells (Fig. 3.75*b*).

Holoblastic: referring to branching pattern (in *Sphacelaria*) in which the branch primordium originates from the entire segment of an apical cell (as contrasted with hemiblastic) (Fig. 6.32*c*).

Holocarpic: refers to the conversion of the entire thallus into a reproductive structure.

Holococcolith: type of coccolith having a single type of calcite (or aragonite) crystal.

Holoplanktonic: refers to organisms that spend their entire existence as free-floating individuals.

Hormogonium: a (usually) motile segment of a Cyanochlorontan filament capable of growing into another filament (Fig. 2.7).

Hyaline field: region on the silica layer of a frustule not penetrated by areolae or puncta.

Hypha (Hyphae):   the elongate cells in the medulla of kelps and fucoids.

Hypnospore: a thick-walled aplanospore.

Hypnozygote: a thick-walled zygote.

Hypocingulum: that portion of the girdle of a frustule adjacent to the hypovalve.

Hypocone: the portion of a dinoflagellate cell posterior to the cingulum.

Hypogynous cell: the cell subtending (directly beneath) the carpogonium.

Hypolithic: living on the lower surface of stones.

Hypothallus (also hypothallium):   that portion of the vegetative thallus (in Corallinaceae) derived from anticlinal divisions of marginal apical cells and which eventually is located in the lower portion of the crustose alga, covered by the perithallus and epithallus (Fig. 9.50).

Hypotheca: that portion of a diatom frustule consisting of hypovalve and hypocingulum.

Hypovalve: the flattened or convex plate of a frustule opposite to the epivalve.

Hystrichosphere: cysts, or resting stages, of fossil and some recent dinoflagellates, which bear characteristic projections and markings and often an apparent excystment aperture (the archeopyle).

Idioandrosporous: species of Oedogonialean algae in which the androspore is produced by a special filament (other than the female) (Fig. 3.97*b*, *c*).

Inducing factors: a chemical produced in certain algae (e.g., *Volvox*) (and ferns) that effects the sexual maturation of immature individuals.

Intercalary band: an element of frustule in the girdle region, located proximal to the valves.

Intergeniculum (Intergenicula):   a nonflexible region of the axis of segmented thalli, such as in the articulated corallines and the green alga *Halimeda*, located between the genicula.

Internode: that portion of an axis intermediate between two nodes.

Intertidal: occurring between the levels of low and high tides; thus exposed at low tide; also referred to as eulittoral.

Involucre: a sterile group of cells or filaments forming an envelope around a reproductive structure(s).

Isogamous: gametes which are morphologically indistinguishable (Fig. 1.1*a*).

Isomorphic: morphologically similar.

Isthmus: the equatorial region connecting the semicells of desmids.

Karyogamy: union of nuclei.

Kelp: member of the brown algal order Laminariales.

Kinetochore: a region connecting the two chromatids of a chromosome in some algae at the time of nuclear division, which has a layered appearance at the electron microscope level; microtubules make contact with the chromosome at the kinetochore, the kinetochore splitting and migrating to opposite poles at anaphase.

Labiate process: a tube or an opening located on the valve of a frustule, with a characteristic liplike structure surrounding a narrow slit at the end of the internal portion (Fig. 7.44).

Lamina: blade.

Laminaran: a storage product of brown algae; a polysaccharide made up of β-1, 3-linked glucans, with some β-1, 6-linkages providing a branching polymer; it is soluble in the cell.

Leucoplast: a colorless, often starch-storing, plastid.

List: a cellulosic extension of the cell wall in some armored dinoflagellates, usually extending out from the cingulum and/or sulcus.

Littoral: near the shore (see *intertidal*).

Locule: a chamber within the frustule of a diatom, having a constricted opening on one side and a velum on the opposite side; or, a compartment of reproductive organ.

Lorica: a nonliving investment of protoplasts, sometimes remote from them (difficult to distinguish from cell walls) (Fig. 5.6).

Macrandrous: species of Oedogonialean algae in which the male and female filaments are approximately similar in diameter (Fig. 3.95*a*).

Macrogametangium: a gametangium containing relatively large locules, thus producing the larger of two different size categories of gametes.

Macrothallus: the large, conspicuous phase in the life history of an organism, as contrasted with the microthallus.

Maintenance culture: a natural collection of algae retained in the laboratory.

Manubrium: a columnar cell that connects the pedicel to the shield cell in Charophytan antheridia.

Mastigoneme: a fine, hairlike appendage of flagella (Fig. 3.2*o*).

Medulla: central region of a thallus.

Megacytic zone: a region of growth or expansion between adjacent plates of armored dinoflagellates where new thecal material is added to allow enlargement of the cell or following cell division (Fig. 8.9*b*).

Meiocyte: a cell that undergoes meiosis.

Meiosis: nuclear division in which the chromosome number is reduced by one-half and genetic segregation occurs.

Meiosporangium (Meiosporangia):   a sporangium in which meiosis occurs; or, (a special meaning in reference to *Feldmannia*, p. 280) a sporangium with chambers of intermediate size.

Meriblastic: pattern of branching (in *Sphacelaria*) in which the branch primordium originates from only a part of a secondary segment.

Meristoderm: a superficial layer of embryonic (dividing) cells covering the thalli of Laminariales.

Meroplanktonic: refers to organisms that spend a part of their life cycle as planktonic and a part as benthic (as in a resting stage), which restricts such organisms largely to neritic conditions.

Merotomy: the cutting up or division of cells into several portions, with or without nuclei, e.g., the experimental grafting with *Acetabularia*.

Mesochite: the middle layer in the oogonial wall of Fucalean algae.

Mesokaryotic: refers to the dinoflagellate nucleus, in which the chromosomes persist in a condensed, discrete condition at all times.

Microgametangium: type of gametangium with relatively small locules, which produce gametes of a smaller size category compared with those from a macrogametangium.

Microsporangium: a sporangium producing microspores.

Microspores (in diatoms):   the products of division of a diatom cell prior to spermatogenesis; the spherical microspores undergo meiosis, each giving rise to four sperm (Fig. 7.48*b*).

Microthallus: the small, inconspicuous phase in the life history of an organism, alternating with the macrothallus.

Microzoospore: a zoospore of a small size relative to others produced by the organism.

Mid-littoral: inhabiting the central region of the intertidal zone.

Mitosporangium: a sporangium in which the products (spores) are produced by mitotic divisions.

Monad: a single organism, usually implying a free-living, unicellular, flagellate stage.

Monoecious: producing male and female gametes on the same individual (Fig. 3.93*a*).

Monopodial: method of development in which the primary axis is maintained as the main line of growth and secondary laterals or offshoots are produced from it.

Monosporangium: a sporangium that produces a single spore.

Monospore: a nonflagellate spore produced singly from a sporangium (Fig. 9.11*a*).

Monostromatic: composed of a single layer of cells.

Muciferous bodies: organelles containing mucilage, which can be ejected, of common occurrence in the euglenoids and dinoflagellates; lacking the complex ultrastructure of trichocysts.

Multiaxial: an axis composed of many elongate filaments (Fig. 9.5*b*).

Nannandrous: species of Oedogonialean algae in which the male filament is a dwarf epiphyte on the female (Fig. 3.97*b*).

Nanoplankton: those organisms from 2 to 20 μm in diameter, ultrananoplankton being less than 2 μm (Dussart, 1965); the organisms would pass through the mesh of the plankton net.

Necridium: a dead cell between two hormogonia in filamentous blue-green algae (Fig. 2.7).

Nematocyst: a structurally complex ejectile organelle produced in a few dinoflagellate genera; also called *cnidocyst*.

Neritic: living in coastal waters.

Neuston: community of organisms living at the water-atmosphere interface.

Neustonic: living at the interface of water and the atmosphere.

Nitrogen fixation: incorporation of atmospheric nitrogen as a nitrogen source into the cells of certain blue-green algae and bacteria.

Node: the site on an axis from which leaves and/or branches arise.

Nonarticulated: having a construction lacking segmentation or separation into genicula and intergenicula.

Nullipore: obsolete term formerly applied to the nonarticulated coralline algae that seemingly lacked pores, to contrast them with the true corals.

Oceanic: living out at sea, away from and independent of coastal influences.

Ocellus: complex light-perceiving organelle present in a few dinoflagellate genera, consisting of large refractive lens and a pigment-containing cup.

Oligotrophic: waters having less than 100 ppm of solutes.

Ooblast: a connecting filament that transfers the zygote nucleus from a carpogonium to an auxiliary cell.

Oogamous: sexual reproduction in which a small motile sperm or a nonmotile spermatium unites with a nonmotile egg (Fig. 1.3*n*).

Oogonium (Oogonia): the female gametangium which produces an egg or eggs (Fig. 1.1*b*).

Ostiole: an opening as in the conceptacle of fucalean algae.

Palmella stage: nonmotile cells embedded in an amorphous gelatinous matrix.

Pantacronematic: type of flagellum with two rows of mastigonemes, flimmer, or fibrils and a terminal fibril (Fig. 3.2*m*).

Pantonematic: type of flagellum with two rows of mastigonemes, flimmer, or fibrils but no terminal fibril (Fig. 3.2*k*).

Paraflagellar rod: a zone of material at one side of the $9 + 2$ flagellar fibrils running the length of the flagellar axis in Euglenoids.

Paramylon: a β-1, 3-glucose polymer, a food reserve of Euglenophycophyta.

Paraphysis (Paraphyses): a sterile structure among sporangia or gametangia (Fig. 6.45*b*).

Parasporangium (Parasporangia): a sporangium producing many spores and which is not homologous with a tetrasporangium.

Paraspore: a spore produced within a parasporangium.

Parenchyma: tissue composed of living thin-walled cells, most often functioning in photosynthesis or storage.

Parthenogenetic: production of a new individual from a single, unfertilized gamete, often the egg.

Peduncle: a cytoplasmic extension used as a stalk or means of affixing the cell.

Pelagic: living in the open oceans rather than in coastal or inland waters; contrasted with neritic.

Pellicle (in dinoflagellates): that portion of the cell covering of all armored dinoflagellates that surrounds the cell after the theca is shed by ecdysis.

Pellicle (in Euglenoids): a proteinaceous surface layer, composed of overlapping strips, below the plasma membrane (Fig. 5.4*b*).

Percurrent: extending from base to apex, such as a well-developed primary axis.

Pericarp: a sterile covering around a carposporophyte (in the Florideophycidae).

Pericentral cell: one of a ring of cells cut off by a central or axial cell.

Periplast: a cell covering (in Cryptophycophyta) consisting of a cell membrane with an underlying layer of plates or membranes and an overlying layer of granular material; it has greater resiliency than a simple plasma membrane but is not as elaborate as a pellicle.

Perithallus (also perithallium):   that portion of the vegetative thallus (in Corallinaceae) in which the cells or filaments are developed inwardly from the intercalary meristem.

Perizonium: an outerlying membrane surrounding the auxospore and derived from the fertilization membrane.

Peroxisome: probable site of catalase activity in cells.

Phagotrophic: holozoic, endocytic, phagocytotic, or engulfing food particles in nutrition.

Phagotrophy: a type of nutrition involving the engulfment of particulate food (also endocytosis).

Phialopore: an intercellular space in the autocolonies of certain coenobic Volvocalean algae, through which the colony everts.

Photoautotrophic or Phototrophic: nutrition in which, with the aid of light energy, inorganic compounds are used in the synthesis of protoplasm.

Photoauxotrophic: photosynthetic but requiring vitamins.

Phragmoplast: microfibrils parallel to the spindle axis at telophase across which a cell plate is deposited (Fig. 3.1d).

Phycobilin: biliprotein pigments of blue-green and red algae.

Phycobilisome: the cellular organelle on the surface of thylakoids in which the biliprotein pigments are present in blue-green and red algae (Fig. 2.5).

Phycocyanin: blue biliprotein pigment of blue-green and red algae.

Phycoerythrin: red biliprotein pigment of blue-green and red algae.

Phycoplast: an assemblage of microtubules perpendicular to the spindle and at the equator of the cell at telophase (Fig. 3.1a, b).

Phycovirus: virus growing in algal cells.

Physiological anisogamy: isogamy with physiological differences between the uniting isogametes.

Pit connection: a discrete lens-shaped plug held within an aperture of the cross wall between two adjacent cells in red algae.

Placental cell: a cell that usually results from the fusion of auxiliary cell and nearby cells and that proceeds to give rise to carpospores.

Placoderm desmid: members of the Desmidiaceae in which the cell walls of the semicells are of different age (Fig. 3.176).

Plankton: the community of minute organisms suspended in water.

Planozygote: a motile zygote.

Plasmodesma (Plasmodesmata): delicate protoplasmic connections between cells arising by incomplete cytokinesis.

Plasmodial: having a naked, multinucleate mass of protoplasm.

Plasmogamy: union of gametic cells [Fig. 3.9(15)].

Pleiomorphism (= polymorphism): the occurrence of many and variable expressions in a genetically uniform group of organisms.

Plethysmothallus: in some brown algae, a diploid, microscopic phase, usually having an appearance resembling *Ectocarpus* or *Streblonema*, capable of asexually multiplying itself by zoospores.

Pleuronematic: type of flagellum with one or more rows of hairlike processes (Fig. 3.2*k*, *l*, *m*); subdivided into stichonematic, pantonematic, or panacronematic.

Plurilocular: having many small chambers or locules (Fig. 6.4*a*).

Plurispores: the motile cells produced by a plurilocular organ.

Pneumatocyst: an air bladder.

Polar nodule (in Bacillariophyceae): thickened regions at opposite ends of the valve of motile pennate forms.

Polar nodule (in Cyanochloronta): thickenings in the heterocyst walls at the points of attachment of heterocysts to adjacent cells (Figs. 2.8*a*, 2.10).

Polar ring: a short hollow cylinder located at each pole of the spindle of a dividing nucleus (in Florideophycidae).

Polyeder or polyhedral cell: the angular cell produced by the zoospores that arise in zygote germination of *Hydrodictyon* and *Pediastrum* (Fig. 3.58*e*).

Polyglucan granules: polymers of glucose somewhat like animal glycogen.

Polysiphonous: having many siphons, or pericentral cells (as in *Polysiphonia*) (Fig. 9.95).

Polysporangium (Polysporangia): a sporangium producing many spores and which is homologous with a tetrasporangium.

Polystichous: type of construction (in certain brown algal orders) in which the thallus is composed of a true parenchymatous organization.

Polystromatic: composed of many cellular layers.

Porphyran: a sulfated polysaccharide composed of galactose units present in *Porphyra*.

Primary filament: type of filament comprising (along with secondarily formed fibers) the medulla of thalli of Fucalean algae.

Primary nucleus: the first nucleus in certain siphonous green algae, such as *Acetabularia*, which becomes relative enlarged (Fig. 3.160).

Primary ooblast: the portion of a connecting filament linking the fertilized carpogonium to a cell in the carpoginial branch.

Primary pit connection: a pit connection developed between the two products of a cell division.

Primary spermatogenous cell: a cell that functions as an initial to form a male gametangium.

Procarp: the close spatial association of a carpogonium and an auxiliary cell, occurring in the same branch system.

Progressive cleavage: gradual cytokinesis of multinucleate protoplasm ultimately into uninucleate segments.

Prokaryotic: lacking membrane-bounded DNA (and Golgi apparatus, mitochondria, and plastids).

Propagule: a multicellular structure functioning for asexual reproduction.

Protonema (Protonemata), primary: the product of the germination of the zygote in Charophyta.

Protosphere: a globose cell developed from a germinating zygote, which next passes into a juvenile siphonous stage.

Prototrichogyne: the emergent portion of a cell in a thallus of Bangiophycidae that functions as a female gamete, this exposed portion being receptive to contact by spermatia.

Psammophilous: "sand-loving;" living in a sandy habitat.

Pseudoflagellum: an immobile flagellum (Fig. 3.41c).

Pseudoparenchymatous: parenchyma-like because of interweaving in the growth of contiguous filaments.

Punctum (Puncta): a marking on a diatom frustule.

Pure culture: see axenic culture.

Pusule: an osmoregulatory organelle in both freshwater and marine dinoflagellates, of variable form, consisting of two closely appressed membranes that bound a vesicle open to the outside.

Pyrenoid: a differentiated region of a plastid, a center for starch formation in green algae (Fig. 3.7).

Raphe: an elongate fissure, or opening, or pair of such fissures in the valve(s) of certain pennate diatoms.

Receptacle: a fertile area on which gametangia or sporangia arise.

Reproduction, asexual: increase in number of individuals not involving gametic union.

Reproduction, sexual: increase in the number of individuals involving usually union of gametes (plasmogamy), of their nuclei (karyogamy), association of their chromosomes and meiosis.

Reservoir: the base of the flask-like invagination in Euglenoids (Fig. 5.2).

Resting spore: a cell in the life history of an organism, usually asexually formed, which is typically dormant and capable of tolerating unfavorable conditions.

Rhizoid: an anchoring and/or absorptive organ lacking vascular tissue and a root cap.

Rhizoplast: a rootlike striated band arising from the basal body of a flagellum and usually directed to the nuclear envelope.

Saxicolous: growing on rocks or rocky substrata.

Saxitoxin: a neurotoxin produced by *Gonyaulax catenella*.

Scalariform conjugation: sexual reproduction in Zygnematacean algae in which the connected filaments and their conjugation canals give a ladderlike appearance (Fig. 3.165*c*, *d*).

Secondary fibers: a type of filament produced secondarily in the thalli of Fucales, which contribute to the formation of the medulla.

Secondary ooblast: the portion of a connecting filament linking a cell in the carpogonial branch other than the carpogonium with an auxiliary cell.

Secondary pit connection: a pit connection developed between two adjacent cells by the cutting off of a small cell from one of the pair of adjacent cells and the fusion of that small cell with the other member of the pair (Fig. 9.4).

Segregative cell division: type of cell division characteristic of the green algal order Siphonocladales in which the protoplast is cleaved into spherical masses, which then proceed to expand and function as independent entities (Fig. 3.147).

Seirosporangium (Seirosporangia): a type of sporangium in which branched or unbranched series of sporangia are terminally produced.

Seirospore: a spore produced by a seirosporangium.

Semicell: one-half of a Desmidiacean cell (Fig. 3.177).

Separation disc: see *necridium*.

Seta (Setae): a stiff hair, bristle, or bristle-like process; in diatoms, a hollow projection of the frustule extending beyond the valve margin.

Sheath: an (often pectinaceous) investment outside the cell wall of certain algal cells (Fig. 3.35).

Shield cell: a wall cell of the Charophytan antheridium (Fig. 4.4*c*).

Sieve area: a field of pores lined by plasmalemma and through which photoassimilates are translocated; the pores may be numerous and small, as in *Laminaria* (Fig. 6.54) or few and large, as in *Macrocystis*.

Sieve element: a cell having sieve areas across which photoassimilates are translocated from one cell to another. These sieve elements may be randomly oriented in relation to each other or may be superimposed in longitudinal series, constituting sieve tubes.

Sieve tube: a longitudinal series or file of sieve elements, present in *Nereocystis* and *Macrocystis* of the brown algal order Laminariales.

Silicalemma: the membrane of the vesicle in which silica depostion occurs in frustule formation of diatoms; the silica adheres tightly to this membrane, which presumably has a role in the very specific silicification process; this membrane is situated in the central region between the two products of cell division just beneath inside the plasmalemma.

Silicoflagellate: an organism of the Chrysophycean order Dictyochales, with a siliceous skeleton (Fig. 7.15).

Sinus: the invaginated region at the isthmus of certain Desmideaceae (Fig. 3.177*a*).

Siphon: designation for the large, multinucleate cells of certain green and red algae.

Siphonaxanthin: a carotenoid pigment, showing some similarities to siphonein, occurring in the chloroplasts of members of the Caulerpales and some members of the Siphonocladales and rarely in other green algae.

Siphonein: a carotenoid pigment occurring in the chloroplasts of members of the Caulerpales and of very rare occurrence in other green algae (*cf.* Goodwin, 1974).

Sorus (i):   a group or cluster of reproductive organs (Fig. 6.65).

Spermatangium (a):   male reproductive structure in red algae, which produces a spermatium.

Spermatium (a):   male gamete in red algae, nonmotile and colorless; released from spermatangium.

Sporangium: a spore-producing structure.

Spore: a cellular agent of asexual reproduction.

Sporic meiosis: meiosis that occurs during sporogenesis.

Sporocyte: a diploid cell that undergoes meiosis to form spores.

Sporophyll: a fertile, spore-producing leaf.

Sporophyte: the diploid (usually) spore-producing plant or phase in a life cycle.

Sporopollenin: a carotenoid polymer that is a constituent of the outer wall layer of *Chlorella*.

Stalk cell: a cell to which is attached a reproductive structure.

Statospore: a resting spore of some Chrysophycean and Xanthophycean algae, consisting of two pieces.

Stellate cell: a type of medullary cell in some red algae, with a stellate shape.

Stephanokont: having a ring or crown of flagella.

Stichidium (Stichidia):   a specialized branch bearing tetrasporangia (in Florideophycidae), usually enlarged over the vegetative axes.

Stichonematic: type of flagellum with a single row of mastigonemes, or hairs (Fig. 3.2*l*).

Stigma (Stigmata):   the red eyespot or pigmented area in certain motile algae involved directly or indirectly in light perception (Fig. 3.7).

Stipe: stalk, as in kelps (Fig. 6.50).

Stonewort: a calcified member of the *Charophyta*.

Stria (Striae):   a row of puncta, areolae, or an elongate chamber in the frustule of a diatom.

Stromatolite: a fossilized, calcareous aggregate of blue-green algae.

Subaerial: exposed to the air or atmosphere; not submerged.

Sublittoral: at a depth below the lowest level of low tide.

Sulcus: a longitudinal furrow or depression on the ventral side of a dinoflagellate cell, with a flagellum located in this region.

Supporting cell: (*a*) in Oedogoniales the cell below the oogonium with which the latter arises by division of an oogonial mother cell; (*b*) in Florideophycidae the cell bearing a carpogonial branch.

Supralittoral: above the uppermost limit of high tide; usually the spray zone.

Swarmer: see *zoospore*.

Sympodial: method of development in which the primary axis is continually being replaced by lateral axes which become temporarily dominant but also soon are replaced by their own laterals.

Tendril: a twining or clasping organ.

Terminal conjugation: in most texts, the so-called "lateral conjugation" of Zygnematacean algae of contiguous cells of one filament by formation of conjugation canals near the terminal walls (Fig. 3.166*a*).

Terminal nodule: site of the terminal pore of a raphe on the value of a motile pennate diatom.

Tetrasporangium (Tetrasporangia): a cell in which a diploid nucleus undergoes meiosis and four haploid spores (tetraspores) are produced (in the Florideophycidae).

Tetraspore: a spore produced within a tetrasporangium (in the Rhodophycophyta) (Fig. 9.9).

Tetrasporophyte: a diploid phase producing tetrasporangia.

Thallus (Thalli): a plant body not differentiated into vascularized roots, stems, and leaves.

Theca: a cellulosic wall covering in some dinoflagellates, sometimes arranged in separate plates.

Thecal plate: a unit or portion of a theca, composed of cellulose.

Thylakoid: a photosynthetic lamella or sac (Figs. 2.5, 3.7).

Trabeculum (Trabecula): an extension of the wall into the central lumen of the cell (characteristic of *Caulerpa*) (Fig. 3.132).

Transition zone: the region between stipe and blade where active cell division is localized in Laminarialean thalli; thus an intercalary meristem.

Trichoblast: a simple or branched filament, pigmented or colorless, arising exogenously at the apices of thalli in the Rhodomelaceae.

Trichocyst: an ejectile organelle of complex internal structure, membrane-bounded, which, when stimulated, is discharged as an elongate thread or fibril.

Trichocyst pore: an opening or aperture in the thecal plate of armored dinoflagellates through which trichocysts may be discharged (Fig. 8.8*b*).

Trichogyne: a receptive protuberance or elongation of a female gametangium (oogonium or carpogonium) to which male gametes become attached (Fig. 9.28*a*).

Trichothallic growth: a method of growth (in the Phaeophycophyta) in which the site of active cell division is at the base of a filament or group of filaments.

Trophocyte: a type of cell in multicellular parasitic dinoflagellates that is the single cell attaching the colony to the host (Fig. 8.16*a*).

Tube cell: the helical cells enveloping Charophytan oogonia (Fig. 4.6).

Tychopelagic: living a benthic, or attached, existence in littoral habitats but entering the plankton when dislodged from the usual fixed condition.

Unialgal culture: a culture containing only one strain or species of alga.

Uniaxial: having an axis composed of a single filament (Fig. 9.5*a*).

Unilocular zoosporangium: a zoosporangium in which all the zoospores are produced in a single cavity (or cell) (Fig. 6.4*b*).

Unispores: the motile spores produced by a unilocular organ.

Utricle: the dilated, or swollen, terminal portion of a filament or tube, as in *Codium* (Fig. 3.122).

Valves: the opposite faces, or distal plates, of a diatom frustule, typically flattened or somewhat convex.

Vegetative cell division: see *desmoschisis.*

Velum (Vela): (*a*) in Dasycladales a protective covering over emergent lateral branches (Fig. 3.159); (*b*) in Bacillariophyceae a thin, perforated layer of silica over an areola.

Vesicle: membrane-bound organelle.

Water bloom: a concentrated population of planktonic algae macroscopically apparent.

Wrack: a tangled mass of Fucalean seaweeds on the seashore.

Zoochlorella: a symbiotic green-pigmented algal cell living in the tissue of an animal host.

Zoosporangium (Zoosporangia): a sporangium that produces zoospores (Fig. 1.3*d, e*).

Zoospore: a flagellated, asexual reproductive cell (Fig. 1.3*e*).

Zooxanthella: a symbiotic golden-pigmented algal cell living in the tissue of an animal host: usually a Pyrrhophycophytan alga but can also belong to other algal classes.

Zygote: the cellular product of gametic union [Fig. 3.9(17)].

Zygotic meiosis: meiosis that occurs during zygote maturation or germination.

References

General References

ABBOTT, I. A., AND G. J. HOLLENBERG. 1976. Marine Algae of California. Stanford University Press, Stanford, California. xii + 827 pp.

BONEY, A.D. 1966. A Biology of Marine Algae. Hutchinson Educational Ltd., London. 216 pp.
_____. 1975. Phytoplankton. Studies in Biology Ser., 52. Crane, Russak & Co., Inc., New York. vii + 116 pp.
BOURRELLY, P. 1966. Les Algues d'eau douce. Initiation à la systématique. I. Les algues vertes. N. Boubée & Cie., Paris. 511 pp.

_____. 1968. Les Algues d'eau douce. Initiation à la systématique. II. Les algues jaunes et brunes, Chrysophycées, Phéophycées, Xanthophycées et Diatomées. Ed. N. Boubée & Cie., Paris. 438 pp.

_____. 1970. Les Algues d'eau douce. Initiation à la systématique. III. Les algues bleues et rouges, les Eugléniens, Peridiniens et Cryptomonadines. Ed. N. Boubée et Cie., Paris. 512 pp.

_____. 1972. Les Algues d'eau douce. Initiation à la systématique. I. Les algues vertes. Revised ed. N. Boubée & Cie., Paris. 569 pp.

CHAPMAN, F. J. 1970. Seaweeds and their Uses. 2nd ed. Methuen and Co., London. 304 pp.
_____, and D. J. CHAPMAN. 1973. The Algae. 2nd ed. Macmillan and Co., New York. 497 pp.

DAWSON, E. Y. 1956. How to Know the Seaweeds. W. C. Brown Co., Dubuque, Iowa. 197 pp.
_____. 1966. Marine Botany, an Introduction. Holt, Rinehart and Winston, New York. xii + 371 pp.
DIXON, P. S. 1973. Biology of the Rhodophyta. Hefner Press, New York. xiii + 285 pp.
DODGE, J. D. 1973. The Fine Structure of Algal Cells. Academic Press, London and New York, x + 261 pp.

FLÜGEL, E. (Ed.). 1977. Fossil Algae, Recent Results and Developments. Springer-Verlag, Berlin, New York. xiii + 375 pp.
FOTT, B. 1971. Algenkunde. 2nd ed. Gustav Fischer, Jena. 581 pp.
FRITSCH, F. E. 1935. The Structure and Reproduction of Algae. Vol. I. University Press, Cambridge. 791 pp.

_____. 1945. The Structure and Reproduction of Algae. Vol. II. University Press, Cambridge. 939 pp.

GEORGE, E. A. 1976. A guide to algal keys. Br. Phycol. J. 11: 49–55.

GODWARD, M. B. E. (Ed.). 1966. The Chromosomes of the Algae. St. Martin's Press, New York. 212 pp.

JACKSON, D. F. (Ed.). 1964. Algae and Man. Plenum Press, New York. 434 pp.

_____. 1968. Algae, Man and the Environment. University Press, Syracuse. 554 pp.

KYLIN, H. 1956. Die Gattungen der Rhodophyceen. C. W. K. Gleerups, Lund. 673 pp.

LEWIN, R. A. (Ed.). 1976. The Genetics of Algae. Bot. Monogr. 12. Blackwell Sci. Publications, Oxford. 360 pp.

MORRIS, I. 1967. An Introduction to the Algae. Hutchinson University Library, London. 189 pp.

OLTMANNS, F. 1922–23. Morphologie und Biologie der Algen. 2nd ed. Vol. 1–III Gustav Fischer, Jena. Vol. I, vi + 459 pp.; vol. II, iv + 439 pp.; vol. III, vii + 558 pp.

PAPENFUSS, G. F. 1955. Classification of the algae. *In* A Century of Progress in the Natural Sciences, 1853–1953. Calif. Acad. of Sciences, San Francisco, pp. 115–224.

PRESCOTT, G. W. 1962. Algae of the Western Great Lakes Area. W. C. Brown Co., Dubuque, Iowa. 977 pp.

_____. 1968. The Algae: a Review. Houghton-Mifflin Co., Boston. 436 pp.

_____. 1970. How to Know the Fresh-water Algae. W. C. Brown Co., Dubuque, Iowa. 348 pp.

ROSOWSKI, J. R., AND B. C. PARKER (Ed.). 1971. Selected Papers in Phycology. University of Nebraska, Lincoln. 876 pp.

ROUND, F. E. 1965. The Biology of the Algae. Edward Arnold, London. 269 pp.

SCHWIMMER, M., AND D. SCHWIMMER. 1955. The Role of Algae and Plankton in Medicine. Grune and Stratten, Inc., New York. 85 pp.

SMITH, G. M. 1950. The Fresh-water Algae of the United States. 2nd ed. McGraw-Hill Book Co., New York. 719 pp.

_____. (Ed.). 1951. Manual of Phycology. Chronica Botanica, Waltham, Mass. 375 pp.

_____. 1955. Cryptogamic Botany. Vol. I. McGraw-Hill, New York. 546 pp.

_____, G. J. HOLLENBERG, AND I. A. ABBOTT. 1969. Marine Algae of the Monterey Peninsula, California. 2nd ed., incorporating the 1966 Supplement. Stanford University Press, Stanford, California. xi + 752 pp.

STEIN, J. (Ed.). 1973. Handbook of Phycological Methods. I. Isolation and Purification. University Press, Cambridge. xii + 448 pp.

STEWART, W. D. P. (Ed.). 1974. Algal Physiology and Biochemistry. Univ. Calif. Press, Berkeley and Los Angeles. 989 pp.

TAFT, C. E. 1965. Water and Algae—World Problems. Educational Pub., Inc., Chicago. 236 pp.

TAYLOR, F. J. R. 1976. Flagellate phylogeny: a study in conflicts. J. Protozool. 23: 28–40.

TAYLOR, W. R. 1957. Marine Algae of the Northeastern Coast of North America. Revised ed. University of Michigan Press, Ann Arbor. 509 pp.

_____. 1960. Marine Algae of the Eastern Tropical and Subtropical Coasts of the Americas. University of Michigan Press, Ann Arbor. 870 pp.

TIFFANY, L. H. 1958. Algae, the Grass of Many Waters. 2nd ed. Chas. C. Thomas, Springfield, Illinois. 199 pp.

TILDEN, J. E. 1935. The Algae and their Life Relations; Fundamentals of Phycology. University of Minnesota Press, Minneapolis. 550 pp.

WERNER, D. (Ed.). 1977. The biology of diatoms. Bot. Monogr. 13. Blackwell Scientific Publications, Oxford. 520 pp.

Special References

AARONSON, S. 1973A. Digestion in phytoflagellates. *In* Lysosomes in Biology and Pathology (Ed., J. T. Dingle), vol. 3, pp. 18–37, North-Holland Pub., Amsterdam.

————. 1973B. Particle aggregation and photoautotrophy by *Ochromonas*. Arch. Mikrobiol. 92: 39–44.

————, B. DE ANGELIS, O. FRANK, AND H. BAKER. 1971. Secretion of vitamins and amino acids into the environment by *Ochromonas danica*. J. Phycol. 7: 215–218.

ABBOTT, B. C., AND Z. PASTER. 1970. Action of toxins from *Gymnodinium breve*. Toxicon 8: 120. (Abstr.).

ABBOTT, I. A., 1945. The genus *Liagora* (Rhodophyceae) in Hawaii. Occas. Papers of Bernice P. Bishop Museum, Honolulu, 18: 145–169.

————. 1961. On *Schimmelmannia* from California and Japan. Pacific Naturalist 2: 379–386.

————. 1967. Studies in the foliose red algae of the Pacific coast II. *Schizymenia*. Bull. So. Calif. Acad. Sci. 66: 161–174.

————. 1968. Studies in some foliose red algae of the Pacific coast. III. Dumontiaceae, Weeksiaceae, Kallymeniaceae. J. Phycol. 4: 180–198.

————. 1969. Some new species, new combinations, and new records of red algae from the Pacific Coast. Madroño 20: 42–53.

————. 1970. *Yamadaella*, a new genus in the Nemaliales (Rhodophyta). Phycologia 9:115–123.

————. 1971. On the species of *Iridaea* (Rhodophyta) from the Pacific coast of North America. Syesis 4: 51–72.

————. 1972A. Taxonomic and nomenclatural notes on North Pacific marine algae. Phycologia 11: 259–265.

————. 1972B. Field studies which evaluate criteria used in separating species of *Iridaea* (Rhodophyta). *In* Contributions to the Systematics of Benthic Marine Algae of the North Pacific (Ed., I. A. Abbott and M. Kurogi), pp. 253–264. Jap. Soc. Phycology, Kobe.

————, AND M. S. DOTY. 1960. Studies in the Helminthocladiaceae. II. *Trichogloeopsis*. Am. J. Bot. 47: 632–640.

————, AND R. E. NORRIS. 1965. Studies on *Callophyllis* (Rhodophyceae) from the Pacific Coast of North America. Nova Hedwigia 10: 67–84.

ABE, K. 1935. Zur Kenntniss der Entwicklungsgeschichte von *Heterochordaria*, *Scytosiphon*, und *Sorocarpus*. Sci. Rep. Tohoku Imp. Univ., Ser. 4, Biol. 9: 329–337.

————. 1936. Kernphasenwechsel von *Heterochordaria abietina*. Science Reports of the Tohoku Imp. Univ.; 4th series, Biology. Vol. XI. No. 2, Sendai, Japan. pp. 239–241.

ABÉ, T. H. 1967A. The armoured Dinoflagellata. II. Prorocentridae and Dinophysidae(A). Pub. Seto Mar. Biol. Lab. 14: 369–389.

————. 1967B. The armoured Dinoflagellata: II. Prorocentridae and Dinophysidae (B). *Dinophysis* and its allied genera. Pub. Seto. Marine Biol. Lab. (Kyoto Univ.) 15: 37–75.

————. 1967C. The armoured Dinoflagellata: II. Prorocentridae and Dinophysidae (C). *Ornithocercus*, *Histioneis*, *Amphisolenia* and others. Pub. Seto Marine Biol. Lab. (Kyoto Univ.) 15 (2): 79–116.

ADEY, W. H. 1964. The genus *Phymatolithon* in the Gulf of Maine. Hydrobiologia 24: 377–420.

————. 1965. The genus *Clathromorphum* (Corallinaceae) in the Gulf of Maine. Hydrobiologia 26: 539–573.

————. 1966. The genera *Lithothamnium*, *Leptophytum* (nov. gen.) and *Phymatolithon* in the Gulf of Maine. Hydrobiologia 28: 321–370.

————. 1970A. The crustose corallines of the Northwestern North Atlantic, including *Lithothamnium lemoineae* n. sp. J. Phycol. 6: 225–229.

————. 1970B. The effects of light and temperature on growth rates in boreal-subarctic crustose corallines. J. Phycol. 6: 269–276.

————. 1970C. Some relationships between crustose corallines and their substrate. Soc. Scientia Islandica (1970): 21–25.

_____. 1971. The sublittoral distribution of crustose corallines on the Norwegian coast. Sarsia 46: 41–58.

_____, AND H. W. JOHANSEN. 1972. Morphology and taxonomy of Corallinaceae with special reference to *Clathromorphum*, *Mesophyllum*, and *Neopolyporolithon* gen. nov. (Rhodophyceae, Cryptonemiales). Phycologia 11: 159–180.

_____, AND I. G. MacINTYRE. 1973. Crustose coralline algae: a re-evaluation in the geological sciences. Geol. Soc. Amer. Bull. 84: 883–904.

_____, AND C. P. SPERAPANI. 1971. The biology of *Kvaleya epilaeve*, a new parasitic genus and species of Corallinaceae. Phycologia 10: 29–42.

AFZELIUS, B. A. 1961. Flimmer flagellum of the sponge. Nature 191: 1318–1319.

AHMADJIAN, V. 1967. A guide to the algae occurring as lichen symbionts: isolation, culture, cultural physiology, and identification. Phycologia 6: 127–160.

ALDRICH, D. V., S. M. RAY, AND W. B. WILSON. 1967. *Gonyoulax monilata*: population growth and development of toxicity in cultures. J. Protozool. 14: 636–639.

ALLEN, D. M., AND D. H. NORTHCOTE. 1975. The scales of *Chrysochromulina chiton*. Protoplasma 83: 389–412.

ALLEN, M. A. 1958. The biology of a species complex in *Spirogyra*. Ph. D. Thesis. Indiana University (Libr. Congr. Card No. Nme.: 58–7901), Univ. Microfilms, Ann Arbor, Mich.

ALLEN, M. B., C. S. FRENCH, AND J. S. BROWN. 1960. Chlorophylls in algal groups. *In* Comparative Biochemistry of Photoreactive Systems (Ed., M. B. Allen), pp. 33–52, Academic Press, New York.

ALLEN, M. M. 1968. Simple conditions for growth of unicellular blue-green algae on plates. J. Phycol. 4: 1–4.

ALSTON, R. E. 1958. An investigation of the purple vacuolar pigment of *Zygogonium ericetorum* and the status of "algal anthocyanins" and "phycoporphyrins." Am. J. Bot. 45: 688–692.

ANDERSON, E. 1962. A cytological study of *Chilomonas paramecium* with particular reference to the so-called trichocysts. J. Protozool. 9: 380–395.

ANDERSON, E. K., AND W. J. NORTH. 1967. Zoospore release rates in giant kelp *Macrocystis*. Bull. So. Calif. Acad. Sci. 66: 223–232.

ANDREIS, C. 1975. A survey of different types of vacuoles in the Cyanophyceae. Giornale Bot. Ital. 109: 193–203.

ANON. 1975. Proposals for a standardization of diatom terminology and diagnoses. Beihefte zur Nova Hedw. 53: 323–354.

APELT, G. 1969. Die Symbiose zwischen dem acoelen Turbellar *Convoluta convoluta* und Diatomeen der Gattung *Licmophora*. Mar. Biol. 3: 165–187.

ARCE, G., AND H. C. BOLD. 1958. Some Chlorophyceae from Cuban soils. Am. J. Bot. 45: 492–503.

ARCHIBALD, P. A. 1972. The genus *Nautococcus* Korschikov (Chlorophyceae, Chlorococcales). Phycologia 11: 207–212.

_____. 1973. The genus *Neochloris* Starr. Phycologia 12: 187–193.

_____. 1975. *Trebouxia* de Puymaly (Chlorophyceae, Chlorococcales) and *Pseudotrebouxia* gen. nov. (Chlorophyceae, Chlorosarcinales). Phycologia 14: 125–137.

_____, AND H. C. BOLD. 1970. Phycological Studies XI. The genus *Chlorococcum* Meneghini. The University of Texas Publication 7015. 115 pp.

ARDRÉ, F. 1967A. Remarques sur la structure des *Pterosiphonia*, (Rhodomélacées, Céramiales) et leurs rapports systématiques avec les *Polysiphonia*. Rev. Algol. 9: 37–77.

_____. 1967B. Nouvelles remarques sur la structure des *Pterosiphonia* (Rhodomélacées, Céramiales), et leurs rapports systématiques avec les *Polysiphonia*. C. R. Acad. Sc. (Paris). 264: 2192–2195.

ARNESON, R. 1973. *Pseudotetracystis*, a new Chlorosarcinacean alga. J. Phycol. 9: 10–14.

ARNOLD, C. A. 1952. A specimen of *Prototaxites* from the Kettle Point Black Shale of Ontario. Palaeontographica 93(B): 45–56.

ARNOLD, C. G., O. SCHIMMER, F. SCHOTZ, AND H. BATHELT. 1972. Die Mitochondrien von *Chlamydomonas reinhardii*. Arch. Mikrobiol 81: 50–67.

ARNOTT, H. J., AND R. M. BROWN, JR. 1967. Ultrastructure of the eyespot and its possible significance in phototaxis of *Tetracystis excentrica*. J. Protozool. 14: 529–539.

———, AND P. L. WALNE. 1967. Observations of the fine structure of the pellicle pores of *Euglena granulata*. Protoplasma 64: 330–344.

ASHFORD, B. K., R. CIFERRI, AND L. M. DALMAN. 1930. A new species of *Prototheca* and a variety of the same isolated from the human intestine. Arch. Protistenk. 70: 619–623.

ASHKENAZY, I. 1975. Manna from the sea. Oceans 8(3): 8–17.

ASMUND, B. 1959. Electron microscope observations on *Mallomonas* species and remarks on their occurrence in some Danish ponds and lakes. III. Dansk Bot. Arkiv 18(3): 1–50.

ATKINSON, A. W., JR., B. E. S. GUNNING, AND P. C. C. JOHN. 1972. Sporopollenin in the cell wall of *Chlorella* and other algae: Ultrastructure, chemistry and incorporation of ^{14}C acetate, studied in synchronous cultures. Planta 107: 1–32.

ATKINSON, K. M. 1972. Birds as transporters of algae. Br. Phycol. J. 7: 319–321.

AUSTIN, A. P. 1960. Life history and reproduction of *Furcellaria fastigiata* (L.) Lam. 2. The tetrasporophyte and reduction division in the tetrasporangium. Ann. Bot., N. S. 23: 296–310.

AX, P., AND G. APELT. 1966. Die "Zooxanthellen" von *Convoluta convoluta* (Turbellaria Acoela) entstehen aus Diatomeen. Erster Nachweis einer Endosymbiose zwischen Tieren und Kieselalgen. Naturwissenschaften 52(15): 444–446.

BAILEY, D., A. P. MAZURAK, AND J. R. ROSOWSKI. 1973. Aggregation of soil particles by algae. J. Phycol. 9: 99–101.

BAILEY, G. P., R. REZAK, AND E. R. COX. 1976. A revision of generic concepts of living members in the subfamily Acetabularieae (Dasycladaceae, Dasycladales) based on scanning electron microscopy. Phycologia 15: 7–18.

BAKER, A. F., AND H. C. BOLD. 1970. Phycological Studies X. Taxonomic studies in the Oscillatoriaceae. The University of Texas Pub. 7004. 105 pp.

BAKER, J. R. J., AND L. V. EVANS. 1971. A myrionematoid variant of *Ectocarpus fasciculatus* Harv. Br. Phycol. J. 6: 73–80.

———, AND ———. 1973A. The ship-fouling alga *Ectocarpus* I. Ultrastructure and cytochemistry of plurilocular reproductive stages. Protoplasma 77: 1–13.

———, AND ———. 1973B. The ship-fouling alga *Ectocarpus* II. Ultrastructure of the unilocular reproductive stages. Protoplasma 77: 181–189.

BALAKRISHNAN, M. S. 1960. Reproduction in some Indian red algae and their taxonomy. *In* Proceedings Symposium Algology New Delhi (Ed., P. Kachroo), pp. 85–98.

———. 1961A. Studies on Indian Cryptonemiales. I. *Grateloupia* C. A. Ag. J. Madras Univ. 31B: 11–35.

———. 1961B. Studies on Indian Cryptonemiales. III. *Halymenia* C. A. Ag. J. Madras Univ. 31B: 183–217.

BALECH, E. 1967A. Dinoflagellates and tintinnids in the northeastern Gulf of Mexico. Bull. Marine Sci. 17: 280–298.

———. 1967B. Dinoflagelados nuevos o interesantes del Golfo de Mexico y Caribe. Revista Mus. Argent. Cienc. Nat. 2(3): 77–126.

BARBER, R. T., A. W. WHITE, AND H. W. SIEGELMAN, 1969. Evidence for a Cryptomonad symbiont in the ciliate, *Cyclotrichium meunieri*. J. Phycol. 5: 86–88.

BARGHOORN, E. S., AND S. A. TYLER. 1965. Microorganisms from the Gunflint chert. Science 147: 563–577.

BARTLETT, C., P. L. WALNE, O. J. SCHWARTZ, AND D. H. BROWN. 1972. Large scale isolation and purification of eyespot granules from *Euglena gracilis* var. *bacillaris*. Pl. Physiol 49: 881–885.

BARTLETT, R. B., AND G. R. SOUTH. 1973. Observations of life-history of *Bryopsis hypnoides* Lamour. from Newfoundland: a new variation in culture. Acta Bot. Neerlandia 22: 1–5.

BASCHNAGEL, R. A. 1966. New fossil algae from the Middle Devonian of New York. Trans. Amer. Micrbiol. Soc. 85: 297–302.

BASTIA, D., K. CHIANG, AND H. SWIFT. 1969. Chloroplast dedifferentiation and redifferentiation during zygote maturation and germination in *Chlamydomonas reinhardii*. J. Cell Biol. 43: 11A.

BATRA, P. P., AND G. TOLIN. 1964. Phototaxis in *Euglena*. I. Isolation of the eyespot granules and identification of the eye-spot pigments. Biochim. Biophys. Acta 79: 371–378.

BAULD, J., AND T. D. BROCK. 1974. Algal excretion and bacterial assimilation in hot spring algal mats. J. Phycol. 10: 101–106.

BAZIN, M. J. 1968. Sexuality in a blue-green alga: genetic recombination in *Anacystis nidulans*. Nature 28: 282–283.

BEAM, C. A., AND M. HIMES. 1974. Evidence for sexual fusion and recombination in the dinoflagellate *Crypthecodinium* (*Gyrodinium*) *connis*. Nature 250: 435–36.

BEANEY, W. D., AND L. R. HOFFMAN. 1968. Two new species of *Oedocladium*. J. Phycol. 4: 221–229.

BECH-HANSEN, C. W., AND L. C. FOWKE. 1972. Mitosis in *Mougeotia* sp. Canad. J. Bot. 50: 1811–1816.

BELCHER, J. H. 1960. Culture studies of *Bangia atropurpurea* (Roth) Ag. New Phytol. 59: 367–373.

_____. 1968. A study of *Pyramimonas reticulata* Korschikov (Prasinophyceae) in culture. Nova Hedw. 15: 179–190.

_____. 1969A. Some remarks upon *Mallomonas papillosa* Harris and Bradley and *M. calceolus* Bradley. Nova Hedw. 18: 257–270.

_____. 1969B. A re-examination of *Phaeaster pascheri* Scherffel in culture. Br. Phycol. J. 4: 191–197.

_____. 1969C. Further observations on the type species of *Pyramimonas* (*P. tetrarhynchus* Schmarda) (Prasinophyceae): an examination by light microscopy together with notes on its taxonomy. Bot. J. Linn. Soc. 62: 241–253.

_____, AND E. M. F. SWALE. 1967A. Observations of *Pteromonas tenuis* sp. nov. and *P. angulosa* (Carter) Lemmerman (Chlorophyceae, Volvocales) by light and electron microscopy. Nova Hedw. 13: 353–359.

_____, AND _____. 1967B. *Chromulina placentula* sp. nov. (Chrysophyceae), a freshwater nannoplankton flagellate Br. Phycol. Bull. 3: 257–267.

BELLY, R. T., M. R. TANSEY, AND T. D. BROCK. 1973. Algal excretion of ^{14}C-labeled compounds and microbial interactions in *Cyanidium caldarium* mats. J. Phycol. 9: 123–127.

BENDIX, S. W. 1960. Phototaxis. Bot. Rev. 26: 145–208.

BENEDETTI, P. A., AND A. CHECCURRI. 1975. Paraflagellar body (PFB) pigments studied by fluorescence microscopy in *Euglena gracilis*. Plant Science Letters 4: 47–51.

BENSON, A. A., AND L. MUSCATINE. 1974. Wax in coral mucus: energy transfer from corals to reef fishes. Limnol. and Oceanog. 19: 810–814.

BERGER, S., W. HERTH, W. W. FRANKE, H. FALK, AND H. SPRING. 1975. Morphology of nucleocytoplasmic interactions during the development of *Acetabularia* cells. II. The generative phase. Protoplasma 84: 223–256.

BERGMAN, K., U. W. GOODENOUGH, D. A. GOODENOUGH, J. JAURTZ, AND H. MARTIN. 1975. Gametic differentiation in *Chlamydomonas reinhardtii*. II. Flagellar membranes and the agglutination reaction. J. Cell Biol. 67: 606–622.

BERKALOFF, C. 1962. L'ultrastructure des globules iridescents de *Dictyota dichotoma*. J. Microscopie 1: 313–316.

_____. 1967. Modifications ultrastructurales du plaste et de divers autres organites cellulaires au cours du développement et de l'enkystement du *Protosiphon botryoides* (Chlorophycées). J. Microscopie 6: 839–852.

BERLAND, B. R., D. J. BONIN, A. L. CORNU, S. Y. MAESTRINI, AND J. MARINO. 1972. The antibacterial substances of the marine alga *Stichochrysis immobilis* (Chrysophyta). J. Phycol. 8: 383–392.

BERNSTEIN, E., AND T. L. JOHN. 1955. Certain aspects of sexuality of two species of *Chlamydomonas*. J. Protozool. 2: 81–85.

BERNSTEIN, I. L., AND R. S. SAFFERMAN. 1970. Viable algae in house dust. Nature 227: 851–852.

_____, AND _____. 1973. The biomedical impact of algae. Carolina Tips 36: 45–48.

BIEBEL, P. 1964. The sexual cycle of *Netrium digitus*. Am. J. Bot. 51: 697–704.

_____. 1973. Morphology and life cycles of saccoderm desmids in culture. Beih. Nova Hedw. 42: 39–47.

_____, AND R. REID. 1965. Inheritance of mating types and zygospore morphology in *Netrium digitus* var. *lamellosum*. Proc. Penn. Acad. Sci. 39: 134–137.

BIECHELER, B. 1934. Sur un dinoflagellée à capsule périnculéaire *Plectodinium* n. gen., *nucleovalvatum* n. sp., et sur relations du Péridiniens avec les Radiolaires. C. R. Acad. Sci. (Paris) 198: 601–603.

_____. 1952. Recherches sur les Péridiniens. Bull. Biol. Fr. Belg. suppl. 36: 1–149.

BIGGLEY, W. H., E. SWIFT, R. J. BUCHANAN, AND H. H. SELIGER. 1969. Stimulable and spontaneous bioluminescence in the marine dinoflagellates, *Pyrodinium bahamense*, *Gonyaulax polyedra*, and *Pyrocystis lunula*. J. Gen. Physiol. 54: 96–122.

BILLARD, C., AND P. GAYRAL. 1972. Two new species of *Isochrysis* with remarks on the genus *Ruttnera*. Br. Phycol. J. 7: 289–297.

BIRKBECK, T. E., K. D. STEWART, AND K. R. MATTOX. 1974. The cytology and classification of *Schizomeris leibleinii* (Chlorophyceae). II. The structure of the quadriflagellate zoospores. Phycologia 13: 71–79.

BIRKENES, E., AND T. BRAARUD. 1952. Phytoplankton in the Oslo fjord during a "*Coccolithus huxleyi*-summer." Av. norske Vidensk. Akad. Oslo (Mat.—Nat. Kl.) 2: 1–23.

BISALPUTRA, T. 1974. Plastids. Chapter. 4 in Algal Physiology and Biochemistry (Ed., W. D. P. Stewart), Univ. of Calif. Press, Berkeley and Los Angeles, XI + 998 pp.

_____, F. M. ASHTON, AND T. E. WEIER. 1966. Role of dictyosomes in wall formation during cell division of *Chlorella vulgaris*. Am. J. Bot. 53: 213–216.

_____, AND A. A. BISALPUTRA. 1969. The ultrastructure of chloroplast of a brown alga *Sphacelaria* sp. I. Plastid DNA configuration—The chloroplast genophore. J. Ultrastr. Res. 29: 151–170.

_____, AND H. BURTON. 1970. On the chloroplast DNA-membrane complex in *Sphacelaria* sp. J. Microscopie 9: 661–666.

_____, B. R. OAKLEY, D. C. WALKER, AND C. M. SHIELDS. 1975. Microtubular complexes in blue-green algae. Protoplasma 86: 19–28.

_____, C. M. SHIELDS, AND J. W. MARKHAM. 1971. *In situ* observations of the fine structure of *Laminaria* gametophytes and embryos in culture. I. Methods and the ultrastructure of the zygote. J. Microscopie 10: 83–98.

_____, AND J. R. STEIN. 1966. The development of cytoplasmic bridges in *Volvox aureus*. Can. J. Bot. 44: 1697–1702.

_____, AND T. E. WEIER. 1963. The cell wall of *Scenedesmus quadricauda*. Am. J. Bot. 50: 1011–1019.

_____, AND _____. 1964. The pyrenoid of *Scenedesmus quadricauda*. Am. J. Bot. 51: 881–892.

_____, _____, E. B. RISLEY, AND A. H. P. ENGELBRECHT. 1964. The pectic layer of the cell wall of *Scenedesmus quadricauda*. Am. J. Bot. 51: 548–551.

BISCHOFF, H. W. 1959. Some observations on *Chlamydomonas microhalophila* sp. nov. Biol. Bull. 117: 54–62.

_____, AND H. C. BOLD. 1963. Phycological Studies. IV. Some algae from Enchanted Rock and related algal species. The Univ. of Texas Pub. No. 6318. 95 pp.

BLACK, M. 1963. The fine structure of the mineral parts of the Coccolithophoridae. Proc. Linn. Soc. London 174: 41–46.

_____. 1965. Coccoliths. Endeavour 24: 131–137.

_____. 1971. The systematics of coccoliths in relation to the palaeontological record. pp. 611–624. *In* "The Micropalaeontology of Oceans" (Ed. B. M. Funnell and W. R. Riedel), Cambridge University Press.

BLACK, R. 1974. Some biological interactions affecting intertidal populations of the kelp *Egregia laevigata*. Mar. Biol. 28: 189–98.

BLACKLER, H., AND A. KATPITIA. 1963. Observations on the life-history and cytology of *Elachista fucicola*. Trans. Bot. Soc. Edinb. 39: 392–395.

BLACKMAN, F. F. 1900. The primitive algae and the flagellata. An account of modern work bearing on the evolution of the algae. Ann. Bot. 14: 647–688.

_____, AND A. G. TANSLEY. 1903. A revision of the classification of the green algae. Reprinted, with some rearrangements, from the New Phytologist, Vol. I, 1902. 64 pp.

BLANKLEY, W. F. 1969. Heterotrophic growth and calcification in coccolithophorids. XI Intern. Botan. Congr. Abstracts, p. 16. Seattle.

BLIDING, C. 1957. Studies in *Rhizoclonium* I. Life history of two species. Bot. Notiser 110: 271–275.

_____. 1963. A critical survey of European taxa in Ulvales I. Opera Botanica 8(3): 1–160.

_____. 1968. II. Bot. Notiser 121: 535–629.

BLINKS, L. R. 1951. Physiology and biochemistry of algae. *In* Manual of Phycology (Ed., G. M. Smith). pp. 263–284, Chronica Botanica, Waltham, Mass.

_____. 1955. Some electrical properties of large plant cells. *In* Electrochemistry in Biology and Medicine (Ed., T. Shedlovsky). pp. 187–212. Wiley, New York.

BLUM, J. J., J. R. SOMMER, AND V. KAHN. 1965. Some biochemical, cytological, and morphogenetic comparisons between *Astasia longa* and a bleached *Euglena gracilis*. J. Protozool. 12: 202–209.

BLUM, J. L. 1972. Vaucheriaceae. *In* North American Flora. Ser. II, Part 8, 64 pp. The New York Botanical Garden.

BOALCH, G. T. 1961A. Studies on *Ectocarpus* in culture. I. Introduction and methods of obtaining uni-algal and bacteria-free cultures. J. Mar. Biol. Assn. U. K. 41: 279–286.

_____. 1961B. Studies on *Ectocarpus* in culture. II. Growth and nutrition of a bacteria-free culture. *Ibid.* 41: 287–304.

BOASSON, R., AND S. P. GIBBS. 1973. Chloroplast replication in synchronously dividing *Euglena gracilis*. Planta 115: 125–134.

BODDEKE, R. 1958. The genus *Callithamnion* Lyngb. in the Netherlands. A taxonomic and oecological study. Acta Bot. Neerlandica 7: 589–604.

BODE, V. C., AND B. M. SWEENEY. 1963. Daily rhythm of luciferin activity in *Gonyaulax polyedra*. Science 141: 913–915.

BOILLOT, A., AND F. MAGNE. 1973. Le cycle biologique de *Kylinia rosulata* Rosenvinge (Rhodophycées, Acrochaetiales). Soc. Phycol. Fr. Bull. 18: 47–53.

BOLD, H. C. 1942. The cultivation of algae. Bot. Rev. 8: 69–138.

_____. 1949. The morphology of *Chlamydomonas chlamydogama* sp. nov. Bull. Torrey Bot. Club 76: 101–108.

_____. 1951. Cytology of algae. Chapter 11. *In* Manual of Phycology (Ed., G. M. Smith), pp. 203–227. Chronica Botanica Co., Waltham, Mass.

_____. 1958. Three new chlorophycean algae. Am. J. Bot. 45: 737–743.

_____. 1970. Some aspects of the taxonomy of soil algae. Ann. N.Y. Acad. Sci. 175: 601–616.

_____. 1973. Morphology of Plants, 3rd ed. Harper and Row. New York. 668 pp.

_____. 1974. Phycology, 1947–1972. Ann. Missouri Bot. Gard. 61: 14–44.

_____, AND F. J. MacENTEE, S. J. 1973. Phycological notes. II. *Euglena myxocylindracea* sp. nov. J. Phycol. 9: 152–156.

_____, AND R. C. STARR. 1953. A new member of the Phacotaceae. Bull. Torrey Bot. Club 80: 178–186.

BONEY, A. D. 1965. Aspects of the biology of the seaweeds of economic importance. Adv. Marine Biol. 3: 105–253.

_____. 1966. A Biology of Marine Algae. Hutchinson Educational, Ltd. London. 216 pp.

_____. 1967A. Spore emission, sporangium proliferation and spore germination *in situ* in monosporangia of *Acrochaetium virgatulum*. Br. Phycol. Bull. 3: 317–326.

_____. 1967B. Experimental studies on the benthic phases of Haptophyceae. II. Studies on "aged" benthic phases of *Pleurochrysis scherffellii*. E. G. Pringsh. J. Exp. Mar. Biol. Ecol. 1: 7–33.

_____. 1970. Scale-bearing phytoflagellates. Oceanogr. Marine Biol. Ann. Rev. 8: 251–305.

_____, AND A. BURROWS. 1966. Experimental studies on the benthic phases of Haptophyceae. I. Effects of some experimental conditions on the release of coccolithophorids. J. Mar. Biol. Assn. U. K. 46: 295–319.

_____, AND E. B. WHITE. 1967. Observations on *Kylinia rosulata* from S. W. England. J. Mar. Biol. Assn. U. K. 47: 591–596.

BOOTH, W. E. 1941. Algae as pioneers in plant succession and their importance in erosion control. Ecology 22: 38–46.

BORDEN, C. A., AND J. R. STEIN. 1969. Reproduction and early development in *Codium fragile* (Suringar) Hariot: Chlorophyceae. Phycologia 8: 91–99.

BOUCK, G. B. 1962. Chromatophore development, pits and other fine structure in the red alga *Lomentaria baileyana* (Harv.) Farlow. J. Cell Biol. 12: 553–569.

_____. 1965. Fine structure and organelle associations in brown algae. J. Cell Biol. 26: 523–537.

_____. 1970. The development and postfertilization fate of the eyespot and the apparent photoreceptor in *Fucus* sperm. Ann. N.Y. Acad. Sci. 175: 673–685.

_____. 1971. The structure, origin, isolation, and composition of the tubular mastigonemes of the *Ochromonas* flagellum. J. Cell Biol. 50: 362–384.

_____. 1972. Architecture and assembly of mastigonemes. *In* Advances in Cell and Molecular Biology (E. J. Du Praw, Ed.) pp. 237–271.

_____, AND D. L. BROWN. 1973. Microtubule biogenesis and cell shape in *Ochromonas*. I. The distribution of cytoplasmic and mitotic microtubules. J. Cell Biol. 56: 340–359.

_____, AND BEATRICE M. SWEENEY. 1966. The fine structure and ontogeny of trichocysts in marine dinoflagellates. Protoplasma 61: 205–223.

BOURNE, V. L., E. CONWAY, AND K. COLE. 1970. On the ultrastructure of pit connections in the conchocelis phase of the red alga *Porphya perforata* J. Ag. Phycologia 9: 79–81.

BOURRELLY, P. 1957. Recherches sur les Chrysophycées. Morphologie, phylogénie, systématique. Rev. Algol., Mém. Hors-Sér. No. 1, 412 pp.

_____. 1962A. *Chlamydobotrys, Pyrobotrys* ou *Uva*. Rev. Algol. 2: 126–128.

_____. 1962B. Chrysophycées et Phylogénie. Vorträgen Gesamtgebiet der Botanik, Deutsch. Botan. Gesells. Neue Folge No. 1, pp. 32–36.

_____. 1963. Loricae and cysts in the Chrysophyceae. Ann. N.Y. Acad. Sci. 108: 421–429.

_____. 1966. Les Algues d'eau douce. Initiation à la systématique. I. Les algues vertes. N. Boubée and Cie. Paris. 511 pp.

_____. 1968A. Les Algues d'eau douce. Initiation à la systématique. II: Les algues jaunes et brunes, Chrysophycées, Phéophycées, Xanthophycées et Diatomées. Ed. N. Boubée and Cie, Paris. 438 pp.

_____. 1968B. Note sur les Péridiens d'eau douce. Protistologica 4: 5–14.

_____. 1970. Les Algues d'eau douce. Initiation à la systématique. III. Les algues bleues et rouges, les Eugléniens, Peridiniens et Cryptomonadines. Ed. N. Boubée et Cie, Paris. 512 pp.

_____. 1972. Les Algues d'eau douce. Initiation à la systématique. I. Les algues vertes. Revised ed. N. Boubée and Cie. Paris. 569 pp.

BOVEE, E. C. 1960. *Stephanosphaera pluvialis*, alga of mountain pools. Turtox News 38: 99–101.

BOWEN, W. R. 1967. Ultrastructural aspects of the cell boundary of *Haematococcus pluvialis*. Trans. Am. Microscop. Soc. 86: 36–43.

BRAARUD, T., G. DEFLANDRE, P. HALLDAL, AND E. KAMPTNER. 1955. Terminology, nomenclature, and systematics of the Coccolithophoridae. Micropaleontology 1: 157–159.

BRACHET, J. L. A. 1965. *Acetabularia*. Endeavour 24: 155–161.

_____, AND S. BONOTTO (Ed.). 1970. Biology of *Acetabularia*. Academic Press, N.Y. 300 pp.

BRADLEY, W. H. 1970. Eocene algae and plant hairs from the Green River formation of Wyoming. Am. J. Bot. 57: 782–785.

BRADLEY, S., AND N. S. CARR. 1976. Heterocyst and nitrogenase development in *Anabaena cylindrica*. J. Gen. Microbiol 96: 175–184.

BRAMLETTE, M. N. 1958. Significance of coccolithophorids in calcium-carbonate deposition. Bull. Geol. Soc. Am. 69: 121–126.

BRANDHAM, P. E. 1965. Polyploidy in desmids. Can. J. Bot. 43: 405–417.

BRANNING, T. G. 1976. Giant kelp: its comeback against urchins, sewage. Smithsonian 7(6): 102–109.

BRÅTEN, T. 1971. The ultrastructure of fertilization and zygote formation in the green alga *Ulva mutabilis* Føyn. J. Cell Sci. 9: 621–635.

_____. 1973. Autoradiographic evidence for the rapid disintegration of one chloroplast in the zygote of the green alga *Ulva mutabilis*. J. Cell Sci. 12: 385–389.

_____. 1975A. Observations on mechanisms of attachment in the green alga *Ulva mutabilis* Føyn.

An ultrastructural and light microscopical study of zygotes and rhizoids. Protoplasma 84: 161–173.

————. 1975B. Ultrastructural localization of phosphohydrolases in gametes, zygotes and zoospores of *Ulva mutabilis* Føyn. J. Cell Sci. 17: 647–653.

————, AND A. Løvlie. 1968. On the ultrastructure of vegetative and sporulating cells of the multicellular green alga *Ulva mutabilis* Føyn. Nytt Mag. Bot. 15: 209–219.

————, AND Ø. Nordby. 1973. Ultrastructure of meiosis and centriole behaviour in *Ulva mutabilis* Føyn. J. Cell Sci. 13: 69–81.

BRAVO, L. M. 1965. Studies on the life history of *Prasiola meridionalis*. Phycologia 4: 177–194.

BRAY, D. F., K. NAKAMURA, J. W. COSTERTON AND E. B. WAGENAAR. 1974. Ultrastructure of *Chlamydomonas eugametos* as revealed by freezeetching: cell wall, plasmalemma and chloroplast membrane. J. Ultrastr. Res. 47: 125–141.

BRILL, B. 1973.Untersuchungen zur Ultrastruktur der Choanocyte von *Ephydatia fluviatilis* L. Z. Zellforsch. Mikrosk. Anat. 144: 231–245.

BROCK, T. D. 1967. Life at high temperatures. Science 158: 1012–1019.

————. 1973. Lower pH limit for the existence of blue-green algae: Evolutionary and ecological implications. Science 179: 480–483.

BROOK, A. J. 1968. The discoloration of roofs in the United States and Canada by algae. J. Phycol. 4: 250.

BROOKS, A. E. 1966. The sexual cycle and intercrossing in the genus *Astrephomene*. J. Protozool. 13: 367–375.

BROOKS, M. 1975. Studies on the genus *Coscinodiscus*. I. Light, transmission and scanning electron microscopy of *C. concinnus* Wm. Smith. Bot. Mar. 18: 1–13.

BROWN, D. L., AND T. E. WEIER. 1970. Ultrastructure of the freshwater algae *Batrachospermum*. I. Thin-section and freeze-etch analysis of juvenile and photosynthetic filament vegetative cells. Phycologia 9: 217–235.

BROWN, R. M., JR., 1969. Observations on the relationship of the Golgi apparatus to wall formation in the marine chrysophycean alga, *Pleurochrysis scherffelii*. J. Cell Biol. 41: 109–123.

————. 1971. Studies of Hawaiian freshwater and soil algae. I. The atmospheric dispersal of algae and fern spores across the island of Oahu, Hawaii. *In* Parker, B. C. and R. M. Brown, Jr. (Ed.). Contributions in Phycology pp. 175–188. Allen Press, Lawrence, Kansas.

————. 1972. Algal viruses. *In* Advances in Virus Research. 17: 243–277.

————, AND H. J. ARNOTT. 1970. Structure and function of the algal pyrenoid. I. Ultrastructure and cytochemistry during zoosporogenesis in *Tetracystis excentrica*. J. Phycol. 6: 4–22.

————, AND H. C. BOLD. 1964. Phycological Studies V. Comparative studies of the algal genera *Tetracystis* and *Chlorococcum*. The Univ. Texas Pub. 6417. 213 pp.

————, W. W. FRANKE, H. KLEINIG, H. FALK, AND P. SITTE. 1969. Cellulosic wall component produced by the Golgi apparatus of *Pleurochrysis scherffelii*. Science 166: 894–896.

————, ————, ————, ————, AND ————. 1970. Scale formation in chrysophycean algae. I. Cellulosic and noncellulosic wall components made by the Golgi apparatus. J. Cell Biol. 45: 246–271.

————, W. HERTH, W. W. FRANKE, AND D. ROMANOVICZ. 1973. The role of the Golgi apparatus in the biosynthesis and secretion of a cellulosic glycoprotein in *Pleurochrysis*: a model system for the synthesis of structural polysaccharides. *In* Biogenesis of Plant Cell Wall Polysaccharides (Ed., F. Loewus) pp. 207–257. Academic Press, New York.

————, SISTER C. JOHNSON, O. P., AND H. C. BOLD. 1968. Electron and phase-contrast microscopy of sexual reproduction in *Chlamydomonas moewusii*. J. Phycol. 4: 100–120.

————, AND R. J. McLEAN. 1969. New taxonomic criteria in the classification of *Chlorococcum* species. II. Pyrenoid fine structure. J. Phycol. 5: 114–118.

BUETOW, D. E. (Ed.). 1968. The biology of *Euglena*. Academic Press. New York. Vol. 1, 361 pp. Vol. 2, 417 pp.

BUFFALOE, N. 1958. A comparative cytological study of four species of *Chlamydomonas*. Bull. Torrey Bot. Club 85: 157–178.

BURKHOLDER, P., L. M. BURKHOLDER, AND L. ALMODOVAR. 1960. Antibiotic activity of some marine algae of Puerto Rico. Botanica Marina 2: 149–156.

BURNS, R. L., AND A. C. MATHIESON. 1972A. Ecological studies on economic red algae. II. Culture studies of *Chondrus crispus* Stackhouse and *Gigartina stellata* (Stackhouse) Batters. J. Exp. Mar. Biol. Ecol. 8: 1–6.

————, AND ————. 1972B. Ecological studies of economic red algae. III. Growth and reproduction of natural and harvested populations of *Gigartina stellata* (Stackhouse) Batters in New Hampshire. J. Exp. Mar. Biol. Ecol. 9: 77–95.

BURR, F. A., AND R. F. EVERT. 1972. A cytochemical study of the wound-healing protein in *Bryopsis hypnoides*. Cytobios 6: 199–215.

————, AND M. D. MCCRACKEN. 1973. Existence of a surface layer on the sheath of *Volvox*. J. Phycol. 9: 345–346.

————, AND J. A. WEST. 1970. Light and electron microscope observations on the vegetative and reproductive structures of *Bryopsis hypnoides*. Phycologia 9: 17–37.

————, AND ————. 1971A. Comparative ultrastructure of the primary nucleus in *Bryopsis* and *Acetabularia*. J. Phycol. 7: 108–113.

————, AND ————. 1971B. Protein bodies in *Bryopsis hypnoides*: their relationship to wound-healing and branch septum development. J. Ultrastr. Res. 35: 476–498.

BURROWS, E. M. 1964. An experimental assessment of some of the characters used for specific delimitation in the genus *Laminaria*. J. Mar. Biol. Assn. U. K. 44: 137–143.

————, AND S. M. LODGE. 1951. Autecology and the species problem in *Fucus*. J. Mar. Biol. Assn. U. K. 30: 161–176.

————, AND ————. 1953. Culture of *Fucus* hybrids. Nature 172: 1009–1010.

BURSA, A. S. 1969. *Actiniscus canadensis* n. sp., *A. pentasterias* Ehrenberg v. *arcticus* n. var., *Pseudoactiniscus apentasterias* n. gen., n. sp., marine relicts in Canadian Arctic lakes. J. Protozool. 16: 411–418.

BUSBY, W. F., AND J. C. LEWIN. 1967. Silica uptake and silica shell formation by synchronously dividing cells of the diatom *Navicula pelliculosa* (Breb.) Hilse. J. Phycol. 3: 127–131.

BUTCHER, R. W. 1967. An Introductory Account of the Smaller Algae of British Coastal Waters. Part IV: Cryptophyceae. Ministry of Agric., Fish., and Food, Fish. Invest., Ser. IV, vi + 54 pp., 20 pls.

CABIOCH, J. 1969. Sur le mode de développement de quelques *Amphiroa* (Rhodophycées, Corallinacées). C. R. Acad. Sci. (Paris). 269D: 2338–2340.

————. 1971A. Étude sur les Corallinacées. 1. Caractères généraux de la cytologie. Cahiers Biol. Mar. 12:121–186.

————. 1971B. Essai d'une nouvelle classification des Corallinacées actuelles. C. R. Acad. Sc. (Paris). 272D: 1616–1619.

————. 1972A. Étude sur les Corallinacées. II. La morphogenèse: conséquences systématiques et phylogénétiques. Cahiers Biol. Mar. 13: 137–287.

————. 1972B. Un nouveau cas d'anomalie du cycle des Gigartinales (Algues Floridées). C. R. Acad. Sci. (Paris). 275D: 1979–1981.

CABRERA, S. M. 1970. Sobre el ciclo biologico de *Giffordia mitchellae* (Harvey) Hamel (Phaeophyta, Ectocarpaceae). Boletin Sociedad Argentina Botanica 13: 31–41.

CACHON, J., AND M. CACHON. 1970. Ultrastructure des *Amoebophyridae* (peridiens Duboscquodinida). II. Systemes atractophoriens et microtubulaires; leur intervention dans la mitose. Protistologica 6: 57–70.

————, ————, AND F. BOUQUAHEUX. 1969. *Myxodinium pipiens* gen. nov., sp. nov., peridinien parasite d'*Halosphaera*. Phycologia 8: 157–165.

————, ————, AND C. K. PYNE. 1968. Structures et ultrastructure de *Paradinium poucheti* Chatton 1910, et position systematique des Paradinides. Protistologica 4: 303–311.

CAIN, J. 1965. Nitrogen utilization in 35 freshwater chlamydomonad algae. Can. J. Bot. 43: 1367–1378.

CAIN, J. R., K. R. MATTOX, AND K. D. STEWART. 1973. The cytology of zoosporogenesis in the filamentous green algal genus *Klebsormidium*. Trans. Am. Micro. Soc. 92: 398–404.

————, ———— AND ————. 1974. Conditions of illumination and zoosporogenesis in *Klebsormidium flaccidum*. J. Phycol. 10: 134–136.

CALVERT, H. E., C. J. DAWES, AND M. A. BOROWITZKA. 1976. Phylogenetic relationships of *Caulerpa* (Chlorophyta) based on comparative chloroplast ultrastructure. J. Phycol. 12: 149–162.

CAMERON, R. E. 1962. Species of *Nostoc* Vaucher occurring in the Sonoran desert in Arizona. Trans. Am. Micr. Soc. 81: 379–384.

CAPLIN, S. M. 1968. *Fucus* life cycle: time to correct the text books. BioScience 18: 193.

CARAM, B. 1961. Sur l'alternance de générations et de phases cytologique chez le *Sauvageaugloia griffithsiana* (Grev.) Hamel. C. R. Acad. Sc. (Paris). 252: 594–596.

————. 1965. Recherches sur la reproduction et le cycle sexué de quelques Phéophycées. Vie et Milieu 16: 21–226.

————. 1972. Le cycle de reproduction des Phéophycées-Phéosporées et ses modifications. Soc. Bot. Fr., Mém. 1972: 151–160.

CARDINAL, A. 1964. Étude sur les Ectocarpacées de la Manche. Beihefte z. Nova Hedw. 15.86 pp.

CAREFOOT, J. R. 1966. Sexual reproduction and intercrossing in *Volvulina steinii*. J. Phycol. 2: 150–156.

————. 1967. Nutrition of *Volvulina* Playfair. J. Protozool. 14: 15–18.

CARMICHAEL, W. W., D. F. BIGGS, AND P. R. GORHAM. 1975. Toxicology and pharmacological action of *Anabaena flos-aquae* toxin. Science 187: 542–544.

CARR, D. J., AND M. M. ROSS. 1963. Studies on the morphology and physiology of germination of *Chara gymnopitys* A. Br. II. Factors in germination. Port. Acta. Biol. 8 (A.1–3): 41–56.

CARR, N. G., AND B. A. WHITTON (Ed.). 1973. The Biology of Blue-Green Algae. Univ. Calif. Press. Berkeley and Los Angeles. 676 pp.

CASSIE, V. 1969. A free-floating *Pseudobryopsis* (Chlorophyceae) from New Zealand. Phycologia 8: 71–76.

CASTENHOLZ, R. W. 1968. The behavior of *Oscillatoria terebriformis* in hot springs. J. Phycol. 4: 132–139.

————. 1969. Thermophilic blue-green algae and the thermal environment. Bact. Rev. 33: 476–504.

————. 1973. Movements (of blue-green algae). *In* Carr, N. G., and B. A. Whitton (Ed.). The Biology of the Blue-Green Algae. Univ. Calif. Press. Berkeley and Los Angeles. pp. 320–329.

CATT, J. W., G. J. HILLS, AND K. ROBERTS. 1976. A structural glycoprotein containing hydroxyproline isolated from the cell wall of *Chlamydomonas reinhardii*. Planta 131: 165–171.

CAVALIER-SMITH, T. 1975. Electron and light microscopy of gametogenesis and gamete fusion in *Chlamydomonas reinhardii*. Protoplasma 86: 1–18.

————. 1976. Electron microscopy of zygospore formation in *Chlamydomanas reinhardii*. Protoplasma 87: 297–315

CHAN, K. 1973. A review of the genus *Coelastrum* (Chlorophyceae). J. Chinese University of Hong Kong 1: 275–281.

————. 1974. Ultrastructure of pyrenoid division in *Coelastrum* sp. Cytologia 39: 531–536.

CHAN, K-Y., AND S. L. LING WONG. 1975. Ultrastructural observations on *Coelastrum reticulatum*. Cytologia 41: 663–675.

CHANTANACHAT, S., AND H. C. BOLD. 1962. Phycological Studies. II. Some algae from arid soils. The University of Texas Pub. No. 6218. Austin, Texas. 75 pp.

CHAPMAN, A. R. O. 1972. Morphological variation and its taxonomic implications in the ligulate members of the genus *Desmarestia* occurring on the west coast of North America. Syesis 5: 1–20.

————. 1973. A critique of prevailing attitudes toward control of seaweed zonation on the seashore. Bot. Mar. 16: 80–82.

————. 1975. Inheritance of mucilage canals in *Laminaria* (section Simplices) in eastern Canada. Br. Phycol. J. 10: 219–223.

————, AND E. M. BURROWS. 1970. Experimental investigations into the controlling effects of light conditions on the development and growth of *Desmarestia aculeata* (L.) Lamour. Phycologia 9 :103–108.

_____, AND _____. 1971. Field and culture studies of *Desmarestia aculeata* (L.) Lamour. Phycologia 10: 63–76.

CHAPMAN, R. L., AND N. J. LANG. 1973. Virus-like particles and nuclear inclusions in the red alga *Porphyridum purpureum* (Bory) Drew et Ross. J. Phycol. 9: 117–122.

CHAPMAN, V. J. 1954. The Siphonocladales. Bull. Torr. Bot. Club. 81: 76–82.

_____. 1962. A contribution to the ecology of *Egregia laevigata* Setchell. I. Taxonomic status and morphology. II. Desiccation and growth. III. Photosynthesis and respiration; Conclusions. Bot. Mar. 3: 33–45; 46–55; 101–122.

_____. 1970. Seaweeds and their Uses. 2nd ed. Methuen. London. 304 pp.

_____, AND D. J. CHAPMAN. 1973. The Algae. 2nd ed. Macmillan and Co., New York. 497 pp.

CHARDARD, M. R. 1975. Origin of the cell-wall microfibrils of the green alga *Closterium acerosum* Ehrenb. and the role of the Golgi apparatus. C. R. Acad. Sci. (Paris) 280D: 25–28.

CHARTERS, A. C., M. NEUSHUL, AND C. BARILOTTI. 1969. The functional morphology of *Eisenia arborea*. Proc. Intern. Seaweed Symp. 6: 89–105.

CHASEY, D. 1974. The three-dimensional arrangement of radial spokes in the flagella of *Chlamydomonas reinhardii*. Exp. Cell Res. 84: 374–378.

CHATTON, E. 1952. Classe des Dinoflagellés ou Péridiniens. *In* Traité de Zoologie. (Ed., P. P. Grassé). Vol. 1. pp. 309–390. Masson et Cie, Paris.

CHEIGNON, M. 1964. Ultrastructure du gamète mâle d' *Ascophyllum nodosum*. C. R. Acad. Sci. (Paris) 258: 676–678.

CHEN, L. C.-M., T. EDELSTEIN, AND J. MCLACHLAN. 1969. *Bonnemaisonia hamifera* Hariot in nature and in culture. J. Phycol. 5: 211–220.

_____, _____, AND _____. 1974. The life history of *Gigartina stellata* (Stackh.) Batt. (Rhodophyceae, Gigartinales) in culture. Phycologia 13: 287–294.

_____, _____, E. OGATA, AND J. MCLACHLAN. 1970. The life history of *Porphyra miniata*. Can. J. Bot. 48: 385–389.

_____, AND J. MCLACHLAN. 1972. The life history of *Chondrus crispus* in culture. Can. J. Bot. 50: 1055–1060.

_____, _____, A. C. NEISH, AND P. F. SHACKLOCK. 1973. The ratio of kappa- to lambda-carrageenan in nuclear phases of the rhodophycean algae, *Chondrus crispus* and *Gigartina stellata*. J. Mar. Biol. Assn. U. K. 53: 11–16.

CHENEY, D. P., AND C. J. DAWES. 1974. Biosystematic studies of the genus *Eucheuma* (Rhodophyta, Solieriacear). J. Phycol. 10 (Suppl.): 4 (Abstr.)

CHENG, T.-H. 1969. Production of kelp—a major aspect of China's exploitation of the sea. Econ. Bot. 23: 215–236.

CHI, E. Y. 1971. Brown algal pyrenoids. Protoplasma 72: 101–104.

CHIANG, Y.-M. 1970. Morphological studies of red algae of the family Cryptonemiaceae. Univ. Calif. Pub. Bot. 58. vi + 95 pp.

CHIHARA, M. 1961. Life cycle of the Bonnemaisoniaceous algae in Japan (1). Sc. Rep. Tokyo Kyoiku Daigaku, sect. B, 10: 121–153.

_____. 1962. Life cycle of the Bonnemaisoniaceous algae in Japan (2). Tokyo Univ. of Education, Science Reports. Sect. B. 11: 27–54.

_____. 1963. The life history of *Prasinocladus ascus* as found in Japan, with special reference to the systematic position of the genus. Phycologia 3: 19–28.

_____. 1965. Germination of the carpospores of *Bonnemaisonia nootkana*, with special reference to the life cycle. Phycologia 5: 71–79.

_____. 1973. The significance of reproductive and spore germination characteristics to the systematics of the Corallinaceae: articulated coralline algae. Jap. J. Bot. 20: 369–379.

_____. 1974. The significance of reproductive and spore germination characteristics to the systematics of the Corallinaceae: nonarticulated coralline algae. J. Phycol. 10: 266–274.

_____, AND T. HORI. 1972. The fine structure of *Prasinocladus ascus* and *Platymonas* species found in Japan with special reference to their taxonomy. Proc. Intern. Seaweed Symp. 7: 188–191.

_____, AND S. KAMURA. 1963. On the germination of tetraspores of *Gelidiella acerosa*. Phycologia 2: 69–74.

_____, AND M. YOSHIZAKI. 1972A. Reproductive system of *Liagora japonica* (Nemaliales, Rhodophyta). Bull. Nat'l. Sci. Mus. 15: 395–401.

_____, AND _____. 1972B. Bonnemaisoniaceae: their gonimoblast development, life history and systematics. *In* Contributions to the Systematics of Benthic Marine Algae of the North Pacific (Ed. I. A. Abbott and M. Kurogi), Jap. Soc. Phycology, Kobe. pp. 243–251.

CHRETIENNOT, M.-J. 1973. The fine structure and taxonomy of *Platychrysis pigra* Geitler (Haptophyceae). J. Mar. Biol. Assn. U. K. 53: 905–914.

CHRISTENSEN, T. 1956. Studies on the genus *Vaucheria* III. Remarks on some species from brackish water. Bot. Notis. 109: 275–280.

_____. 1962. Alger. *In* Böcher, T. W., Lange, M., and Sørensen, T. (Ed.). Systematisk Botanik. Vol. 2, No. 2. Munksgaard, Copenhagen. 178 pp.

_____. 1964. The gross classification of algae. *In* Jackson, D. F. (Ed.). Algae and Man. Plenum Press, New York. pp. 59–64.

_____. 1969. *Vaucheria* collections from Vaucher's region. K. Danske Vidensk. Selsk. Biol. Skrifter 16(4). 36 pp.

_____. 1971. The gross classification of algae. *In* Rosowski, J. R. and B. C. Parker (Ed.). Selected Papers in Phycology. University of Nebraska. Lincoln.

CHRISTIE, A. O., AND M. SHAW. 1968. Settlement experiments with zoospores of *Enteromorpha intestinalis* (L.) Link. Br. Phycol. Bull. 3: 529–534.

CHU, S. P. 1942. The influence of the mineral composition of the medium on the growth of planktonic algae. I. Methods and culture media. J. Ecol. 30: 284–325.

CHURCHILL, A. C., AND H. W. MOELLER. 1972. Seasonal patterns of reproduction in New York populations of *Codium fragile* (Sur.) Hariot subsp. *tomentosoides* (van Goor) Silva. J. Phycol. 8: 147–152.

CLAASEN, M. I. 1973. Freshwater algae of South Africa. I. Notes on *Gloeotrichia ghosei* R. N. Singh. Br. Phycol. J. 8: 325–331.

CLARK, R. L., AND T. E. JENSEN. 1969. Ultrastructure of akinete development in *Cylindrosperum* sp. Cytologia 34: 439–448.

CLAYTON, M. N. 1972. The occurrence of variant forms in cultures of species of *Ectocarpus* and *Giffordia*. Br. Phycol. J. 7: 101–108.

_____. 1974. Studies in the development, life history and taxonomy of the Ectocarpales in southern Australia. Austral. J. Bot. 22: 743–813.

_____, AND S. C. DUCKER. 1970. The life history of *Punctaria latifolia* Greville (Phaeophyta) in southern Australia. Austral. J. Bot. 18: 293–300.

CLEARE, M., AND E. PERCIVAL. 1972. Carbohydrates of the freshwater alga *Tribonema aequale*. I. Low molecular weight and polysaccharides. Br. Phycol. J. 7: 185–193.

_____, AND _____. 1973. Carbohydrates of the freshwater alga *Tribonema aequale*. II. Preliminary photosynthetic studies with ^{14}C. Br. Phycol. J. 8: 181–184.

CLENDENNING, K. 1964. Photosynthesis and growth in *Macrocystis pyrifera*. Proc. Intern. Seaweed Symp. 41: 55–65.

_____. 1971. Organic productivity in kelp areas. *In* The Biology of Giant Kelp Beds (Ed., W. J. North), Beihefte z. Nova Hedw. 32: 257–263.

CLOUD, P. E., JR., G. R. LICORI, L. R. WRIGHT, AND B. W. TROXEL. 1969. Proterozoic eucaryotes from Eastern California. Proc. Nat. Acad. Sci. 63: 623–630.

_____, M. MOORMAN, AND D. PIERCE. 1975. Sporulation and ultrastructure in a late proterozoic cyanophyte: some implications for taxonomy and plant phylogeny. Quart. Rev. Biol 50: 131–150.

COBB, H. D. 1963. An *in vivo* absorption spectrum of the eyespot of *Euglena mesnili*. Texas J. Science 15: 231–235.

CODOMIER, L. 1971. Recherches sur les *Kallymenia* (Cryptonemiales, Kallymeniacées). Vie et Milieu, Ser. A: Biol. Mar., 22: 1–54.

COESEL, P. F. M., AND R. M. V. TEXEIRA. 1974. Notes on sexual reproduction in desmids. I. Zygospore formation in nature (with special reference to some unusual records of zygotes). Acta Bot.

Neerland. 23: 361–368. II. Experiences with conjugation experiments in uni-algal cultures. Acta Bot. Neerland. 23: 603–64.

COLE, G. T., AND M. J. WYNNE. 1974. Endocytosis of *Microcystis aeruginosa* by *Ochromonas danica*. J. Phycol. 10: 397–410.

COLE, K., AND E. CONWAY. 1974. Cytological considerations of life histories in the genus *Porphyra*. Proc. Inter. Seaweed Symp. 8, p. A60. (Abstr.).

COLEMAN, A. W. 1959. Sexual isolation in *Pandorina*. J. Protozool. 6: 249–264.

_____. 1962. Sexuality. Chapter 48. *In* (R. A. Lewin, Ed.) Physiology and Biochemistry of Algae. Academic Press, New York. 929 pp.

_____. 1963. Immobilization, agglutination and agar precipitin effects of antibodies to flagella of *Pandorina* mating types. J. Protozool. 10: 141–148.

_____. 1975. Long-term maintenance of fertile algal clones: experience with *Pandorina* (Chlorophyceae). J. Phycol. 11: 282–286.

_____, AND J. ZOLLNER. 1977. Cytogenetic polymorphism within the species *Pandorina morum* (Volvocaceae). Arch. Protistenk. 119:

COLIJN, F., AND C. VAN DEN HOEK. 1971. The life-history of *Sphacelaria furcigera* Kütz. (Phaeophyceae) II. The influence of daylength and temperature on sexual and vegetative reproduction. Nova Hedw. 21: 901–922.

COLINVAUX, L. H., AND E. A. GRAHAM. 1964. A new species of *Halimeda*. Nova Hedw. 7: 5–10.

_____, K. M. WILBUR, AND N. WATABE. 1965. Tropical marine algae: growth in laboratory culture. J. Phycol. 1: 69–78.

CONGER, P. S. 1936. Significance of shell formation in diatoms. pp. 325–344. Smithsonian Report, Smithsonian Inst., Washington, D.C.

CONRAD, W. 1931. Recherches sur les Flagellates de Belgique. I. Mem. Mus. Roy. Hist. Nat. Belgique 8, No. 47. 65 pp.

CONWAY, E. 1964. Autecological studies of the genus *Porphyra*. I. The species found in Britain. Br. Phycol. Bull. 2: 342–348.

_____, AND K. COLE. 1973. Observations on an unusual form of reproduction in *Porphyra* (Rhodophyceae, Bangiales). Phycologia 12: 213–225.

_____, T. F. MUMFORD, JR., AND R. F. SCAGEL. 1976. The genus *Porphyra* in British Columbia and Washington. Syesis 8: 185–244.

COOK, A. H., AND J. A. ELVIDGE. 1951. Fertilization in the Fucaceae. Investigations on the nature of the chemotactic substance produced by eggs of *Fucus serratus* and *F. vesiculosus*. Proc. R. Soc. London B 138: 97–114.

COOK, P. W. 1962. Growth and reproduction of *Bulbochaete hiloensis* in unialgal culture. Trans. Am. Microsc. Soc. 81: 384–395.

_____. 1963. Variation in vegetative and sexual morphology among some small curved species of *Closterium*. Phycologia 3: 1–18.

COOKE, W. B. 1968A. Studies in the genus *Prototheca*. I. Literature review. J. Elisha Mitchell Sci. Soc. 84: 213–216.

_____. 1968B. Studies in the genus *Prototheca*. II. Taxonomy. J. Elisha Mitchell Sci. Soc. 84: 217–220.

COOKE, W. J. 1975. The occurrence of an endozoic green alga in the marine mollusc, *Clinocardium nuttallii* (Conrad, 1837). Phycologia 14: 35–39.

COOMBS, J., AND B. E. VOLCANI. 1968. Studies on the biochemistry and fine structure of silica shell formation in diatoms. Chemical changes in the wall of *Navicula pelliculosa* during its formation. Planta 82: 280–292.

_____, J. A. LAURITUS, W. M. DARLEY, AND B. E. VOLCANI. 1968. Studies on the biochemistry and fine structure of silica shell formation in diatoms. V. Effects of colchicine on wall formation in *Navicula pelliculosa* (Breb) Hilse. Z. Pflanzenphys. 59: 124–152.

CORTEL-BREEMAN, A. M. 1975. The life history of *Acrosymphyton purpuriferum* (J. Ag.) Sjöst. (Rhodophyceae, Cryptonemiales). Isolation of tetrasporophytes. Acta Bot. Neerl. 24: 111–127.

COSS, R. A. 1974. Mitosis in *Chlamydomonas reinhardtii*: basal bodies and the mitotic apparatus. J. Cell Biol. 63: 325–329.

_____, AND J. D. PICKETT-HEAPS. 1973. Gametogenesis in the green alga *Oedogonium cardiacum*. I. The cell divisions leading to the formation of spermatids and oogonia. Protoplasma 78: 21–39.

_____, AND _____. 1974A. Gametogenesis in the green alga *Oedogonium cardiacum*. II. Spermiogenesis. Protoplasma 81: 297–311.

_____, AND _____. 1974B. The effects of isopropyl *N*-phenyl carbonate on the green alga *Oedogonium cardiacum*. I. Cell division. J. Cell Biol. 63: 84–98.

COUTÉ, A. 1971. Sur le cycle morphologique du *Liagora tetrasporifera* comparé à celui du *Liagora distenta* (Rhodophycées, Némalionales, Helinthocladiacées). C. R. Acad. Sci. (Paris). 273D: 626–629.

COX, EDMOND R., AND T. R. DEASON. 1968. *Axilosphaera* and *Heterotetracystis*, new chlorosphaeracean genera from Tennessee soil. J. Phycol. 4: 240–249.

_____, AND J. HIGHTOWER. 1972. Some corticolous algae of McMinn County, Tennessee, U.S.A. J. Phycol. 8: 203–205.

COX, EILEEN J. 1975A. A reappraisal of the diatom genus *Amphipleura* KÜtz. light and electron microscopy. Br. Phycol. J. 10: 1–12.

_____. 1975B. Further studies on the genus *Berkeleys* Grev. Br. Phycol. J. 10: 205–17.

COX, ELENOR R., AND H. J. ARNOTT. 1971. The ultrastructure of the theca of the marine dinoflagellate, *Ensiculifera loeblichii* sp. nov. *In* Contributions in Phycology (Ed., B. C. Parker and R. M. Brown, Jr.). Allen Press, Lawrence, Kansas. pp. 121–136.

_____, AND H. C. BOLD. 1966. Taxonomic investigations of *Stigeoclonium*. The Univ. Texas Pub. 6612. 167 pp.

CRAIGIE, J. S. 1974. Storage products, Chapter 7 in Stewart, W. D. P. (Ed.). Algal Physiology and Biochemistry. Univ. Calif. Press. Berkeley and Los Angeles, 989 pp.

CRAWFORD, R. M. 1973. The organic component of the cell wall of the marine diatom *Melosira nummuloides* (Dillw.) C. Ag. Br. Phycol. J. 8: 257–266.

_____. 1974. The auxospore wall of the marine diatom *Melosira nummuloides* (Dillw.) C. Ag. and related species. Br. Phycol. J. 9: 9–20.

CRAWLEY, J. C. W. 1963. The fine structure of *Acetabularia mediterranea*. Exp. Cell Res. 32: 368–378.

_____. 1966. Some observations on the fine structure of the gametes and zygotes of *Acetabularia*. Planta 69: 365–376.

CROFT, W. N., AND E. A. GEORGE. 1959. Blue-green algae from the Middle Devonian of Rhynie, Aberdeenshire. Bull. Br. Museum (N. History) Geology 3: 341–353.

CROSSETT, R. N., E. A. DREW, AND A. W. D. LARKUM. 1965. Chromatic adaptation in benthic marine algae. Nature 207: 547–548.

DAHL, A. L. 1971. Development, form and environment in the brown alga *Zonaria farlowii* (Dictyotales). Bot. Mar. 14: 76–112.

DAILY, F. K. 1975. A note concerning calcium carbonate deposits in charophytes. Phycologia 14: 331–332.

DALES, R. P. 1960. On the pigments of the Chrysophyceae. J. Mar. Biol. Assn. U. K. 39: 693–699.

DAMMANN, H. 1930. Entwicklungsgeschichtliche und zytologische Untersuchungen an Helgoländer Meeresalgen. Wiss. Meeresuntersuch. Abt. Helgoland, N. F. 18(4). 36 pp.

DANGEARD, P. 1963. Sur le développement de *Punctaria latifolia* Grev. récolté dans le Bassin d'Arachon. Botaniste 46: 205–222.

_____. 1965. Recherches sur le cycle evolutif de *Leathesia difformis* (L.) Areschoug. Botaniste 48: 5–43.

_____. 1966. Sur un *Myriotrichia* Harvey récolté à Saint-Vaast la Hougue (Cotentin). Botaniste 49: 79–98.

_____. 1969A. A propos des travaux récents sur le cycle évolutif de quelques Phéophycées, Phéosporées. Botaniste 52: 59–102.

_____. 1969B. Sur le développement du *Stilophora rhizodes* (Ehr.) J. Agardh. Botaniste 51: 95–116.

DARDEN, W. H., JR. 1966. Sexual reproduction in *Volvox aureus*. J. Protozool. 13: 239–255.

_____. 1970. Hormonal control of sexuality in the genus *Volvox*. Ann. N.Y. Acad. Sci. 175: 757–763.

_____. 1971. A new system of male induction in *Volvox aureus* M 5. Biochem. Biophys. Res. Comm. 45: 1205–1211.

_____. 1973. Formation and assay of a *Volvox* factor-histone complex. Microbios 8: 167–174.

_____. 1974. Hormonal control of sexuality in algae. *In* Hormonal Control of Development (Ed., LeBue) Academic Press, New York.

DARLEY, W. M. 1969. Silicon and the division cycle of the diatoms *Navicula pelliculosa* and *Cylindrotheca fusiformis*. Proc. N. Am. Paleontological Convention, Pt. G, pp. 994–1009.

_____. 1974. Silicification and calcification. Chapter 24 in Algal Physiology and Biochemistry (Ed., W. D. P. Stewart), Univ. of Calif. Press, Berkeley and Los Angeles, xi + 989 pp.

_____, AND B. E. VOLCANI. 1969. A silicon requirement for deoxyribonucleic acid synthesis in the diatom *Cylindrotheca fusiformis* Reimann and Lewin. Exptl. Cell Res. 58: 335–342.

DA SILVA, E. J., AND H. G. GYLLENBERG. 1973. A taxonomic treatment of the genus *Chlorella* by the technique of continuous classification. Arch. Mikrobiol 87: 99–117.

DAVIES, J. M., N. C. FERRIER, AND C. S. JOHNSTON. 1973. The ultrastructure of the meristoderm cells of the hapteron of *Laminaria*. J. Mar. Biol. Assn. U. K. 53: 237–246.

DAVIS, J. S. 1964. Colony form in *Pediastrum*. Bot. Gaz. 25: 129–131.

_____. 1967. The life cycle of *Pediastrum simplex*. J. Phycol. 3: 95–103.

DAWES, C. J. 1965. An ultrastructure study of *Spirogyra*. J. Phycol. 1: 121–127.

_____. 1971. Indole-3-acetic acid in the green algal coenocyte *Caulerpa prolifera* (Chlorophyceae, Siphonales). Phycologia 10: 375–379.

_____. 1974. Marine Algae of the West Coast of Florida. Univ. Miami Press, Coral Gables. xvi + 201 pp.

_____, J. M. LAWRENCE, D. P. CHENEY, AND A. C. MATHIESON. 1974. Ecological studies of Floridian *Eucheuma* (Rhodophyta, Gigartinales). III. Seasonal variation of carrageenan, total carbohydrates, protein, and lipid. Bull. Mar. Sci. 24: 186–199.

_____, A. C. MATHIESON, AND D. P. CHENEY. 1974. Ecological studies of Floridian *Eucheuma* (Rhodophyta, Gigartinales). I. Seasonal growth and reproduction. Bull. Mar. Sci. 24: 235–273.

_____, AND E. RHAMSTINE. 1967. An ultrastructural study of the giant green algal coenocyte *Caulerpa prolifera*. Am. J. Bot. 56: 8–16.

DAWSON, E. Y. 1941. A review of the genus *Rhodymenia* with descriptions of new species. Allan Hancock Pacific Exped. 3: 123–181.

_____. 1944. The marine algae of the Gulf of California. Allan Hancock Pac. Exped. 3: 189–453.

_____. 1945. Marine algae associated with upwelling along the northwestern coast of Baja California, Mexico. Bull. S. Calif. Acad. Sci. 44: 57–71.

_____. 1949. Studies of northeast Pacific Gracilariaceae. Allan Hancock Found. Pub., Occas. Pap. No. 7. 105 pp.

_____. 1951. A further study of upwelling and associated vegetation along Pacific Baja California, Mexico. J. Mar. Res. 10: 39–58.

_____. 1961. Literature of benthic algae from the eastern Pacific. Pacific Science 15: 370–461.

_____. 1966. Marine Botany, an Introduction. Holt, Rinehart and Winston, New York. xii + 371 pp.

_____, M. NEUSHUL, AND R. D. WILDMAN. 1960. Seaweeds associated with kelp beds along southern California and northwestern Mexico. Pac. Naturalist 1(14), 81 pp.

DAWSON, P. A. 1973A. Observations on the structure of some forms of *Gomphonema parvulum* Kütz. II. The internal organization. J. Phycol. 9: 165–175.

_____. 1973B. Observations on the structure of some forms of *Gomphonema parvulum* Kütz. III. Frustule formation. J. Phycol. 9: 353–365.

_____. 1973C. Observations on some species of the diatom genus *Gomphonema* C. A. Agardh. Br. Phycol. J. 8: 413–423.

DEASON, T. R. 1965. Some observations on the fine structure of vegetative and dividing cells of *Chlorococcum echinozygotum* Starr. J. Phycol. 1: 97–102.

_____. 1967. *Pulchrasphaera*, a new Chlorococcalean genus. J. Phycol. 3: 19–21.

_____. 1971. The genera *Spongiococcum* and *Neospongiococcum*. I. The genus *Spongiococcum* and the multinucleate species of the genus *Neospongiococcum*. Phycologia 10: 17–27.

_____, AND E. R. COX. The genera *Spongiococcum* and *Neospongiococcum*. II. Species of *Neospongiococcum* with labile walls. Phycologia 10: 255–262.

_____, W. H. DARDEN, AND S. ELY. 1969. The development of sperm packets of the MS strain of *Volvox aureus*. J. Ultrastr. Res. 26: 85–94.

_____, AND H. W. NICHOLS. 1970. A new Bangiophycidean alga from Alabama. J. Phycol. 6: 39–43.

DEFLANDRE, G. 1950. Contribution a l'étude des silicoflagellidés actuels et fossiles. Microscopie 2: 72–108; 117–142; 191–210.

_____. 1952. Classe des silicoflagellidés. *In* Traité de Zoologie (Ed., P.-P. Grassé). Masson et Cie, Paris. pp. 425–438.

DEGREEF, J. A., AND R. CAUBERGS. 1970. Chlorophyll *c* in *Vaucheria*. Naturwissensch. 57: 673–674.

✓ DEIG, F. E., D. W. EHRESMANN, M. T. HATCH, AND D. J. RIEDLINGER. 1974. Inhibition of herpesvirus replication by marine algae extracts. Antimicrobial Agents and Chemotherapy 6: 524–525.

DESA, R., J. W. HASTINGS, AND A. E. VATTER. 1963. Luminescent 'crystalline' particles: an organized subcellular bioluminescent system. Science 141: 1269–1970.

DESIKACHARY, T. V. 1956. Observations on two species of *Liagora* (Rhodophyta). Pac. Sci. 10: 423–430.

_____. 1959. Cyanophyta. Indian Council Agr. Res. New Delhi. 686 pp.

_____. 1970. Taxonomy of blue-green algae—problems and prospects. Schweiz. Z. Hydrol. 32: 490–494.

_____ (Ed.) 1972. Taxonomy and Biology of Blue-green Algae. Univ. Madras. Centre for Advanced Study in Botany. 591 pp.

DIAMOND, J., AND J. S. SCHIFF. 1974. Isolation and characterization of mutants of *Euglena* resistant to streptomycin. Pl. Science Letters 3: 259–295.

DIAZ-PIFERRER, M. 1965. A new species of *Pseudobryopsis* from Puerto Rico. Bull. Mar. Sci. 15: 463–474.

_____. 1967. Algas de importancia economica. El Farol. 24: 18–22.

_____. 1969. Distribution of the marine benthic flora of the Caribbean Sea. Carib. J. Sci. 9: 151–178.

_____, AND E. M. BURROWS. 1974. The life history of *Bryopsis hypnoides* Lamour. from Anglesey, North Wales and from the Caribbean. J. Mar. Biol. Assn. U. K. 54: 529–38.

DIEHN, B. 1969A. Two perpendicularly oriented pigment systems involved in phototaxis in *Euglena*. Nature 221: 366–367.

_____. 1969B. Action spectra of phototactic responses in *Euglena*. Biochimica and Biophysica Acta 177: 136–143.

_____. 1973. Phototaxis and sensory transduction in *Euglena*. Science 181: 1009–1015.

DILLARD, G. E. 1966. Seasonal periodicity of *Batrachospermum macrosporum* Mont. and *Audouinella violacea* (Kuetz). Ham. in Turkey Creek, Moore County, North Carolina. J. Elisha Mitchell Sci. Soc. 82: 204–207.

_____, AND D. DAPRA. 1976. Observations on the genus *Uva* Playfair. Am. Midl. Nat. 86: 208–212.

DIXON, P. S. 1958. The development of carpogonial branches and lateral branches of unlimited growth in *Batrachospermum vagum*. Bot. Notiser 111: 645–649.

_____. 1959. The structure and development of the reproductive organs and carposporophyte in two British species of *Gelidium*. Ann. Bot., N. S. 23: 397–407.

_____. 1960. Studies on marine algae of the British Isles: the genus *Ceramium*. J. Mar. Biol. Assn. U. K. 39: 331–374.

_____. 1961. The occurrence of tetrasporangia and carposporophytes on the same thallus in *Euthora cristata* (L. ex Turn.) J. Ag. Can. J. Bot. 39: 541–543.

_____. 1964. Auxiliary cells in the Ceramiales. Nature 201: 519–520.

_____. 1965. Perennation, vegetative propagation and algal life histories, with special reference to *Asparagopsis* and other Rhodophyta. Botanica Gothoburg. 3: 67–74.

_____. 1970. The Rhodophyta: some aspects of their biology, II. Oceanogr. Mar. Biol. Ann. Rev. 8: 307–352.

_____. 1971A. Studies on the genus *Seirospora*. Botaniste 44: 35–48.

_____. 1971B. Cell enlargement in relation to the development of thallus form in Florideophycidae. Br. Phycol. J. 6: 195–205.

_____. 1973. Biology of the Rhodophyta. Hafner Press, New York. xiii + 285 pp.

_____, AND W. N. RICHARDSON. 1969. The life histories of *Bangia* and *Porphyra* and the photoperiodic control of spore production. Proc. Intern. Seaweed Symp. 6: 133–139.

_____, AND _____. 1970. Growth and reproduction in red algae in relation to light and dark cycles. Ann. N.Y. Acad. Sci. 175: 764–777.

DOBELL, C. 1932. Antony van Leewenhoek and his "Little Animals." Staples Press, London.

DODGE, J. D. 1963A. The nucleus and nuclear division in the Dinophyceae. Arch. Protistenk. 106: 442–452.

_____. 1963B. Chromosome numbers in some marine dinoflagellates. Bot. Mar. 5: 121–127.

_____. 1965. Thecal fine-structure in the dinoflagellate genera *Prorocentrum* and *Exuviaella*. J. Mar. Biol. Assn. U. K. 451: 607–614.

_____. 1966. The Dinophyceae. *In* The Chromosomes of the Algae (Ed., M. B. E. Godward). Arnold, London. pp. 96–115.

_____. 1969A. The ultrastructure of *Chroomonas mesostigmatica* Butcher (Cryptophyceae). Arch. Mikrobiol 69: 266–280.

_____. 1969B. A review of the fine structure of algal eyespots. Br. Phycol. J. 4: 199–210.

_____. 1970. A survey of thecal fine structure in the Dinophyceae. Bot. J. Linn. Soc. 63: 53–67.

_____. 1971A. Fine structure of the Pyrrophyta. Bot. Rev. 37: 481–508.

_____. 1971B. A dinoflagellate with both a mesocaryotic and a eucaryotic nucleus. I. Fine structure of the nuclei. Protoplasma 73: 145–157.

_____. 1972. The ultrastructure of the dinoflagellate pusule: a unique osmo-regulatory organelle. Protoplasma 75: 285–302.

_____. 1973. The Fine Structure of Algal Cells. Academic Press, London and New York, x + 261 pp.

_____. 1974A. Fine structure and phylogeny in the algae. Sci. Prog. (Oxford) 61: 257–274.

_____. 1974B. A redescription of the dinoflagellate *Gymnodinium simplex* with the aid of electron microscopy. J. Mar. Bio. Assn. U. K. 54: 171–177.

_____. 1975A. A survey of chloroplast ultrastructure in the Dinophyceae. Phycologia 14: 253–263.

_____. 1975B. The Prorocentrales (Dinophyceae). II. Revision of the taxonomy within the genus *Prorocentrum*. Bot. J. Linn. Soc. 71: 103–125.

_____. 1975C. The fine structure of *Trachelomonas* (Euglenophyceae). Arch. Protistenk. 117: 65–77.

_____, AND B. T. BIBBY. 1973. The Prorocentrales (Dinophyceae). I. A comparative account of fine structure in the genera *Prorocentrum* and *Exuviaella*. Bot. J. Linn. Soc. 67: 175–187.

_____, AND R. M. CRAWFORD. 1968. Fine structure of the dinoflagellate *Amphidinium carteri* Hulbert. Protistologica 4: 231–242.

_____, AND _____. 1969A. The fine structure of *Gymnodinium fuscum* (Dinophyceae). New Phytologist 68: 613–618.

_____, AND _____. 1969B. Observations on the fine structure of the eyespot and associated organelles in the dinoflagellate *Glenodinium foliaceum*. J. Cell Sci. 5: 479–493.

_____, AND _____. 1970A. A survey of thecal fine structure in the Dinophyceae. Bot. J. Linn. Soc. 63: 53–67.

_____, AND _____. 1970B. The morphology and fine structure of *Ceratium hirundinella* (Dinophyceae). J. Phycol. 6: 137–149.

_____, AND _____. 1971. A fine-structural survey of dinoflagellate pyrenoids and food-reserves. Bot. J. Linn. Soc. 64: 105–115.

DOTY, M. S. 1946. Critical tide factors that are correlated with vertical distribution of marine algae and other organisms along the Pacific Coast. Ecology 27: 315–328.

_____, AND G. ANGUILAR-SANTOS. 1966. Caulerpicin, a toxic constituent of *Caulerpa*. Nature 211: 990.

_____, AND _____. 1970. Transfer of toxic algal substances in marine food chains. Pac. Sci. 24: 351–355.

DRAGOVICH, A., J. A. KELLY, JR., AND R. D. KELLY. 1965. Red water bloom of a dinoflagellate *Ceratium furca* Ehr. in Hillsborough Bay, Florida. Nature 207: 1209–1210.

DREBES, G. 1966. On the life history of the marine plankton diatom *Stephanopyxis palmeriana*. Helgoländ. Wissen. Meeresunters. 13: 101–114.

_____. 1972. The life history of the centric diatom *Bacteriastrum hyalinum* Lauder. Beihefte Nova Hedw. 39: 95–110.

_____. 1974. Marines Phytoplankton. Eine Auswahl der Helgoländer Planktonalgen (Diatomeen, Peridineen). Georg Thieme Verlag, Stuttgart. vi + 186 pp.

DREW, K. M. 1949. *Conchocelis*-phase in the life-history of *Porphyra umbilicas* (L.) Kütz. Nature 164: 748.

_____. 1956. Reproduction in the Bangiophycidae. The Botanical Review 22: 553–610.

_____, AND R. Ross. 1964. Some generic names in the Bangiophycidae. Taxon 14: 93–99.

DRING, M. J. 1967. Phytochrome in red alga, *Porphyra tenera*. Nature 215: 1411–1412.

_____, AND K. Lüning. 1975A. Induction of two-dimensional growth and hair formation by blue light in the brown alga *Scytosiphon lomentaria*. Z. Pflanzenphys. 75: 107–117.

_____, AND _____. 1975B. A photoperiodic response mediated by blue-light in the brown alga *Scytosiphon lomentaria*. Planta 125: 25–32.

DROOP, M. R. 1956A. *Haematococcus pluvialis* and its allies. I. The Sphaerellaceae. Rev. Algol., N. S. 2: 53–71.

_____. 1956B. *Haematococcus pluvialis* and its allies. II. Nomenclature in *Haematococcus*. Rev. Algol., N. S. 3: 182–192.

_____. 1961. *Haematococcus pluvialis* and its allies. III. Organic nutrition. Rev. Algol., N. S. 4: 247–258.

_____. 1963. Algae and invertebrates in symbiosis. *In* (P. S. Nutman and B. Mosse, Ed.) Symbiotic Associations. Symposium Society General Microbiol. 13: 171–199. Cambridge Univ. Press. London.

DROUET, F. 1968. Revision of the classification of Oscillatoriaceae. Monogr. Acad. Nat. Sci. Philad. 15.370 pp.

_____. 1973. Revision of the Nostocaceae with cylindrical trichomes. Hafner Press. New York. 292 pp.

_____, AND W. A. DAILY. 1956. Revision of the coccoid Myxophyceae. Butler Univ. Botanical Studies 12: 1–218.

DRUEHL, L. D. 1967. Distribution of two species of *Laminaria* as related to some environmental factors. J. Phycol. 3: 103–108.

_____. 1968. Taxonomy and distribution of northeast Pacific species of *Laminaria*. Can. J. Bot. 46: 539–548.

_____. 1970. The pattern of Laminariales distribution in the northeast Pacific. Phycologia 9: 237–247.

_____, AND S. I. C. HSIAO. 1969. Axenic culture of Laminariales in defined media. Phycologia. 8: 47–49.

DRUM, R. W., AND J. T. HOPKINS. 1966. Diatom locomotion: an explanation. Protoplasma 62: 1–33.

_____, AND H. S. PANKRATZ. 1966. Locomotion and raphe structure of the diatom *Bacillaria*. Nova Hedw. 10: 315–317.

DUBE, M. A. 1967. On the life history of *Monostroma fuscum* (Postels et Ruprecht) Wittrock. J. Phycol. 3: 64–73.

DUBOIS-TYLSKI, T. 1973. Étude ultrastructurale de la conjugaison et de la jenne zygospore chez *Closterium moniliferum* (Bory) Ehrbg. Ann. des Sci. Nat. Bot. et Biol. Végét. 14: 41–52.

_____. 1975. Aspects ultrastructuraux de l'induction sexuelle chez *Closterium moniliferum* (Bory) Ehrb. Beiheifte Nova Hedw. 42: 91–101.

DUBOURSKY, N. 1974. Selectivity of ingestion and digestion in the chrysomonad flagellate *Ochromonas malhamensis*. J. Protozool. 21: 295–298.

DUCKER, S. C. 1965. The structure and reproduction of the green alga *Chlorodesmis bulbosa*. Phycologia 4: 149–162.

―――――. 1967. The genus *Chlorodesmis* (Chlorophyta) in the Indo-Pacific region. Nova Hedw. 13: 145–182.

―――――. 1969. Additions to the genus *Chlorodesmis* (Chlorophyta). Phycologia 8: 17–20.

―――――, W. T. WILLIAMS, AND G. N. LANCE. 1965. Numerical taxonomy of the Pacific forms of *Chlorodesmis* (Chlorophyta). Austral. J. Bot. 13: 489–499.

DUCKETT, J. G., R. TOTH, AND S. L. SONI. 1975. An ultrastructural study of the *Azolla, Anabaena azollae* relationship. New Phytol. 75: 111–118.

DUFFIELD, E. C. S., S. D. WAALAND, AND R. CLELAND. 1972. Morphogenesis in the red alga, *Griffithsia pacifica*: regeneration from single cells. Planta 105: 185–195.

DUNN, J. H., R. D. SIMON, AND C. P. WOLK. 1971. Incorporation of amino sugars in walls during heterocyst differentiation. Develop. Biol. 26: 159–164.

―――――, AND C. P. WOLK, 1970. Composition of the cellular envelopes of *Anabaena cylindrica*. J. Bact. 103: 153–158.

DURANT, J. P., L. SPRATTLING, AND J. C. O'KELLEY. 1968. A study of the effect of light intensity, periodicity, and wavelength on zoospore production by *Protosiphon botryoides* Klebs. J. Phycol. 4: 356–362.

DUSSART, B. H. 1965. Les différentes catégories de plancton. Hydrobiologia 26: 72–74.

DYKSTRA, R. 1971. *Borodinellopsis texensis* gen. et. sp. nov. a new alga from the Texas Gulf Coast. *In* Contributions in Phycology (Ed., B. C. Parker and R. M. Brown, Jr.). Allen Press, Lawrence, Kansas. pp. 1–8.

DYNESIUS, R. A., AND P. L. WALNE. 1975. Ultrastructure of the reservoir and flagella in *Phacus pleuronectes* (Euglenophyceae). J. Phycol. 11: 125–130.

EARLE, S. A. 1969. Phaeophyta of the eastern Gulf of Mexico. Phycologia 7: 71–254.

EATON, J. W., J. G. BROWN, AND F. E. ROUND. 1966. Some observations on polarity and regeneration in *Enteromorpha*. Br. Phycol. Bull. 3: 53–62.

EDELSTEIN, T. 1970. The life history of *Gloiosiphonia capillaris* (Hudson) Carmichael. Phycologia 9: 55–59.

―――――. 1972. On the taxonomic status of *Gloiosiphonia californica* (Farlow) J. Agardh (Cryptonemiales, Gloiosiphoniaceae). Syesis 5: 227–234.

―――――, L. CHEN, AND J. MCLACHLAN. 1968. Sporangia of *Ralfsia fungiformis* (Gunn.) Setchell et Gardner. J. Phycol. 4: 157–160.

―――――, ―――――, AND ―――――. 1970. The life cycle of *Ralfsia clavata* and *R. borneti*. Can. J. Bot. 48: 527–531.

―――――, ―――――, AND ―――――. 1971. On the life histories of some brown algae from eastern Canada. Can. J. Bot. 49: 1247–1251.

―――――, ―――――, AND ―――――. 1974. The reproductive structures of *Gigartina stellata* (Stackh.) Batt. (Gigartinales, Rhodophyceae) in nature and culture. Phycologia 13: 99–108.

―――――, AND J. MCLACHLAN. 1967A. Investigations of the marine algae of Nova Scotia. III. Species of Phaeophyceae new or rare to Nova Scotia. Can. J. Bot. 45: 203–210.

―――――, AND ―――――. 1967B. Cystocarps and tetrasporangia on the same thallus in *Membranoptera alata* and *Polysiphonia urceolata*. Br. Phycol. Bull. 3: 185–7.

―――――, AND ―――――. 1971. Further observations on *Gloiosiphonia capillaris* (Hudson) Carmichael in culture. Phycologia 10: 215–219.

EDWARDS, L. K. 1968. Some non-aquatic epiphytic and lithophilous algae. M. A. Thesis. The University of Texas at Austin.

EDWARDS, P. 1968. The life history of *Polysiphonia denudata* (Dillwyn) Kützing in culture. J. Phycol. 4: 35–37.

―――――. 1969A. Field and cultural studies on the seasonal periodicity of growth and reproduction of selected Texas benthic marine algae. Contrib. Mar. Sci., The University of Texas at Austin, 14: 59–114.

―――――. 1969B. The life history of *Callithamnion byssoides* in culture. J. Phycol. 5: 266–268.

————. 1970A. Illustrated Guide to the Seaweeds and Sea Grasses in the vicinity of Port Aransas, Texas. Contrib. Mar. Sci., The University of Texas at Austin, 15 (Suppl.), 128 pp.

————. 1970B. Field and cultural observations on the growth and reproduction of *Polysiphonia denudata* from Texas. Br. Phycol. J. 5: 145–153.

————. 1973. Life history studies of selected British *Ceramium* species. J. Phycol. 9: 181–184.

————, AND D. F. KAPRAUN. 1973. Benthic marine algal ecology in the Port Aransas, Texas area. Contrib. Mar. Sci. 17: 15–52.

EGEROD, L. E. 1952. An analysis of the siphonous Chlorophycophyta. Univ. of Calif. Pub. Bot. 25: 325–454.

EHARA, T., I. SHIHIRA-ISHIKAWA, T. OSAFUNE, E. HASE, AND I. OHKURO. 1975. Some structural characteristics of chloroplast degeneration in cells of *Euglena gracilis* z during their heterotrophic growth in darkness. J. Electron Microscopy 24: 253–261.

EL-ANI, A. S. 1967. Life cycle and variation of *Prototheca wickershamii*. Science 156: 1501–1503.

ELLIS, D. V., AND R. T. WILCE. 1961. Arctic and subarctic examples of intertidal zonation. Arctic 14: 224–235.

ELLIS, R. J., AND L. MACHLIS. 1968A. Nutrition of the green alga *Golenkinia*. Am. J. Bot. 55: 590–599.

————, AND ————. 1968B. Control of sexuality in *Golenkinia*. Am. J. Bot. 55: 600–610.

ENGELMANN, T. 1882. Über Licht- und Farbenperzeption niederster Organismen. Pflügers Arch. f. d. ges. Physiol. d. Menchen u. d. Tiere. 29: 387–400.

EPPLEY, R. W., AND C. R. BOVELL. 1958. Sulfuric acid in *Desmarestia*. Biol. Bull. 115: 101–106.

ETTL, H. 1956. Ein Beitrag zur Systematik der Heterokonten. Bot. Notiser 109: 411–445.

————. 1964. Über eine besondere Form von *Asterococcus superbus* und deren systematische Stellung. Österr. Bot. Zeit. 4: 354–365.

————. 1966. *Pedinomonadineae*, eine Gruppe kleiner, asymmetrischer Flagellaten der Chlorophyceen. Österr. Bot. Zeit. 3: 511–528.

————. 1967. Die Gattung *Pedinomonas* Korschikoff. Arch. Protistenk. 110: 1–11.

————. 1972. *Pedinomonas minor* Korschikoff, ein einfacher Modellorganismus aus der Bereiche der kleinsten autotrophen Flagellaten. Arch. Hydrobiol. Suppl. 41: 48–56.

————. 1976. Die Gattung *Chlamydomonas* Ehrenberg. Beihefte Nova Hedw. 49: 1–1122.

————, AND I. MANTON. 1964. Die feinere Struktur von *Pedinomonas minor* Korschikoff. Nova Hedw. 8: 421–451.

————, D. G. MÜLLER, K. NEUMANN, H. A. VON STOSCH, AND W. WEBER. 1967. Vegetative Fortpflanzung, Parthenogenese und Apogamie bei Algen. *In* Handbuch Pflanzenphysiologie (W. Ruhland, Ed.). 18: 597–776. Springer. Heidelberg.

EVANS, E. H., I. FOULDS, AND N. G. CARR. 1976. Environmental conditions and morphological variation in the blue-green alga *Chlorogloea fritschii*. J. Gen. Microbiol. 92: 147–155.

EVANS, L. V. 1965. Cytological studies in the Laminariales. Ann. Bot., N. S. 29: 541–562.

————. 1966. Distribution of phyrenoids among some brown algae. J. Cell Sci. 1: 449–454.

————. 1968. Chloroplast morphology and fine structure in British Fucoids. New Phytol. 67: 173–178.

————. 1970. Electron microscopical observations on a new red algal unicell, *Rhodella maculata* gen nov., sp. nov. Br. Phycol. J. 5: 1–13.

————. 1974. Cytoplasmic organelles. Chapter 3 *in* Algal Physiology and Biochemistry (Ed., W. D. P. Stewart). Univ. of California Press, Berkeley. 989 pp.

————, AND M. S. HOLLIGAN. 1972. Correlated light and electron microscope studies on brown algae. I. Localization of alginic acid and sulphated polysaccharides in *Dictyota*. New Phytol. 71: 1161–1172.

————, M. SIMPSON, AND M. E. CALLOW. 1973. Sulphated polysaccharide synthesis in brown algae. Planta 110: 237–252.

EVANS, M. H. 1971. A comparison of the biological effects of paralytic shellfish poisons from clam, mussel and dinoflagellate. Toxicon 9: 139.

EVITT, W. R. 1963A. Occurrence of the freshwater alga *Pediastrum* in Cretaceous marine sediments. Am. J. Sci. 261: 890–893.

_____. 1963B. A discussion and proposals concerning fossil dinoflagellates, hystrichospheres and acritarchs. Proc. Natl. Acad. Sci. (U.S.) 49: 158–164; 298–302.

_____, AND S. E. DAVIDSON. 1964. Dinoflagellate studies. I. Dinoflagellate cysts and thecae. Stanford Univ. Pub. Univ. Ser. Geol. Sci. 10, 12 pp.

FAGERBERG, W. R., AND C. J. DAWES. 1973. An electron microscopic study of the sporophytic and gametophytic plants of *Padina vickersiae* Hoyt. J. Phycol. 9: 199–204.

FALK, H., AND H. KLEINIG. 1968. Feinbau und Carotenoide von *Tribonema* (Xanthophyceae). Arch. Mikrobiol. 61: 347–362.

FAN, K.-C. 1959. Studies on the life histories of marine algae. I. *Codiolum petrocelidis* and *Spongomorpha coalita*. Bull. Torrey Bot. Club 86: 1–12.

_____. 1961A. Morphological studies of the Gelidiales. Univ. Calif. Pub. Bot. 32: 315–368.

_____. 1961B. Studies on *Hypneocolax*, with a discussion on the origin of parasitic red algae. Nova Hedw. 3: 119–128.

FANKBONER, P. V. 1971. Intracellular digestion of symbiotic Zooxanthellae by host amoebocytes in giant clams (Bivalvia: Tridacnidae), with a note on the nutritional role of the hypertrophied siphonal epidermis. Biol. Bull. 141: 222–234.

FARNHAM, W. F., R. L. FLETCHER, AND L. M. IRVINE. 1973. Attached *Sargassum* found in Britain. Nature 243: 231–232.

FAURE-FRÉMIET, E., AND C. ROUILLER. 1957. Le flagelle interne d'une Chrysomonadale: *Chromulina psammobia*. C. R. Acad. Sci. (Paris). 244: 2655–2657.

FAUST, M. A. 1974. Micromorphology of a small dinoflagellate *Prorocentrum mariae-lebouriae* (Parke and Ballantine) comb. nov. J. Phycol. 10: 315–322.

FEIGE, W. 1969. Die Feinstruktur der Epithelien von *Ephydatia fluviatilis*. Zool. JB. (Anat.) 86: 177–237.

FELDMANN, G. 1945. Révision du genre *Botryocladia* Kylin (Rhodophycées-Rhodyméniacées). Bull. Soc. Hist. Nat. l'Afrique du Nord 35: 49–61.

_____. 1966. Sur le cycle haplobiontique du *Bonnemaisonia asparagoides* (Woodw.) Ag. C. R. Acad. Sci. (Paris) 262D: 1695–1698.

_____. 1970. Sur l'ultrastructure de l'appareil irisant du *Gastroclonium clavatum* (Roth) Ardissone (Rhodophycées). C. R. Acad. Sci. (Paris). 270D: 1244–1246.

_____, AND M. BODARD. 1965. Une nouvelle espèce de *Botryocladia* des cotes du Sénégal. Bull. Inst. Oceanogr. 65 (1342). 14 pp.

FELDMANN, J. 1949. L'ordre des Scytosiphonales. Trav. bot. déd. à R. Maire, Alger, pp. 103–115.

_____. 1950. Sur l'existence d'une alterance de générations entre l'*Halicystis parvula* Schmitz et le *Derbesia tenuissima* (De Not.) Crn. C. R. Acad. Sci. (Paris). 230: 322–323.

_____. 1952. Les cycles de reproduction des algues et leur rapports avec la phylogénie. Rev. Cytol. Biol. Vég. 13: 1–49.

_____. 1954. Sur la classification des Chlorophycées siphonées. VIII Congrès Intern. Bot. Rapp. Comm. Sect. 17, pp. 96–98.

_____. 1955. Un nouveau genre de Protofloridée: *Colacodictyon*, nov. gen. Bull. Soc. Bot. Fr. 102: 23–28.

_____. 1969. *Pseudobryopsis myura* and its reproduction. Am. J. Bot. 56: 691–695.

_____, AND G. FELDMANN. 1942. Recherches sur les Bonnemaisoniacées et leur alternance de générations. Ann. Sci. Nat. Bot., Sér. 11, 3: 75–175.

_____, AND _____. 1958. Recherches sur quelques Floridées parasites. Rev. Gén. de Bot. 65: 49–128.

FENICAL, W. 1975. Halogenation in the Rhodophyta—a review. J. Phycol. 11: 245–259.

FISCHER-ARNOLD, G. 1963. Untersuchungen über die Chloroplastenbewegung bei *Vaucheria sessilis*. Protoplasma 56: 495–520.

FISHER, K. A., AND N. J. LANG. 1971A. Ultrastructure of the pyrenoid of *Trebouxia* in *Ramalina menziesii* Tuck. J. Phycol. 7: 25–37.

_____, AND _____. 1971B. Comparative ultrastructure of cultured species of *Trebouxia*. J. Phycol. 7: 155–165.

FITZGERALD, G. P. 1969. Some factors in the competition or antagonism among bacteria, algae, and aquatic weeds. J. Phycol. 5: 351–359.

_____. 1971. The biotic relationships within water blooms. *In* J. R. Rosowski and B. C. Parker (Ed.). Selected papers in Phycology. pp. 26–32.

FJERDINGSTAD, E. J. 1961. Ultrastructure of the collar of the choanoflagellate *Codonosiga botrytis* (Ehrenb.). Z. Zellforsch. Mikrosk. Anat. 54: 499–510.

_____, K. KEMP, E. FJERDINGSTAD, AND L. VANGAARD. 1974. Chemical analyses of red "snow" from East-Greenland with remarks on *Chlamydomonas nivalis* (Bau.) Wille. Arch. Hydrobiol. 73: 70–83.

FLEMING, H., AND R. HASELKORN. 1973. Differentiation in *Nostoc muscorum*. Nitrogenase is synthesized in heterocysts. Proc. Natl. Acad. Sci. 70: 2727–2731.

FLETCHER, R. L., AND S. M. FLETCHER. 1975A. Studies on the recently introduced brown alga *Sargassum muticum*(Yendo) Fensholt. I. Ecology and reproduction. Bot. Mar. 18: 149–156.

_____, AND _____. 1975B. Studies on the recently introduced brown alga *Sargassum muticum* (Yendo) Fensholt. II. Regenerative ability. Bot. Mar. 18: 157–162.

FLOC'H, J. 1969. On the ecology of *Bonnemaisonia hamifera* in its preferred habitats on the western coast of Brittany (France). Br. Phycol. J. 4: 91–95.

FLOYD, G. L., K. D. STEWART, AND K. R. MATTOX. 1971. Cytokinesis and plasmodesmata in *Ulothrix*. J. Phycol. 7: 306–309.

_____, _____, AND _____. 1972A. Cellular organization, mitosis and cytokinesis in the Ulotrichalean alga *Klebsormidium*. J. Phycol. 8: 176–184.

_____, _____, AND _____. 1972B. Comparative cytology of *Ulothrix* and *Stigeoclonium*. J. Phycol. 8: 68–81.

FOGEL, M., R. E. SCHMITTER, AND J. W. HASTINGS. 1972. On the physical identity of scintillons: bioluminescent particles in *Gonyaulax polyedra*. J. Cell Sci. 11: 305–317.

FOGG, G. E. 1971. Extracellular products of algae in freshwater. Arch. Hydrobiol. Beih. Ergegn. Limnol. 5: 1–25.

_____. 1974. Nitrogen fixation. Chapter 20 *in* Stewart, W. D. P. (Ed.) Algal Physiology and Biochemistry. Univ. Calif. Press. Berkeley and Los Angeles. 989 pp.

_____, W. D. P. STEWART, P. FAY, AND A. E. WALSBY. 1973. The blue-green algae. Academic Press. London and New York. 459 pp.

FOREMAN, R. E. 1976. Physiological aspects of carbon monoxide production by the brown alga *Nereocystis luetkeana*. Can. J. Bot. 54: 352–360.

FORK, D. C., AND J. S. BROWN. 1975. A comparison of light-induced shifts in carotenoid absorption in representatives of different algal groups. Carnegie Institution Year Book 74: 776–779.

FORSBERG, C. 1963. Sterile germination of oospores of *Chara* and seed of *Najas marina*. Physiol. Plant. 18: 128–137.

_____. 1965. Nutritional studies of *Chara* in axenic culture. Physiol. Plant. 18: 275–290.

FÖRSTER, H., AND L. WIESE. 1954. Gamonwirkungen bei *Chlamydomonas eugametos*. Z. Naturforsch. 9b: 548–550.

_____, AND _____. 1955. Gamonwirkungen bei *Chlamydomonas reinhardtii*. Z. Naturforsch. 10b: 91–92.

_____, _____, AND G. BRAUNITZER. 1956. Über das agglutinierend wirkende Gynogamon von *Chlamydomonas eugametos*. Z. Naturforsch. 11b: 315–317.

FORWARD, R. B., JR. 1970. Change in the photoresponse action spectrum of the dinoflagellate *Gyrodinium dorsum* Kofoid by red and far-red light. Planta 92: 248–258.

_____. 1974. Phototaxis by the dinoflagellate *Gymnodinium splendens* Labour. J. Protozool. 21: 312–315.

_____, AND D. DAVENPORT. 1970. The circadian rhythm of a behavioral photoresponse in the dinoflagellate *Gyrodinium dorsum*. Planta 92: 259–266.

FOSTER, M. S. 1975A. Algal succession in a *Macrocystis pyrifera* forest. Mar. Biol. 32: 313–329.

_____. 1975B. Regulation of algal community development in a *Macrocystis pyrifera* forest. Mar. Biol. 32: 331–342.

_____, M. Neushul, and E. Y. Chi. 1972. Growth and reproduction of *Dictyota binghamiae* J. G. Agardh. Bot. Mar. 15: 96–101.

Fott, B. 1962. Taxonomy of *Mallomonas* based on electron micrographs of scales. Preslia 34: 69–84.

_____. 1964. Hologamic and agamic cyst formation in loricate Chrysomonads. Phykos 3: 15–18.

_____. 1968. VIII. Klasse Chloromonadophyceae *In* Das Phytoplankton des Süsswassers. (Ed., G. Huber-Pestalozzi). Stuttgart. pp. 79–93.

_____. 1969. Studies in Phycology. Academia. Prague. 304 pp.

_____. 1971. Algenkunde. 2nd ed. Gustav Fischer. Jena. 581 pp.

_____. 1972. Chlorophyceae (Tetrasporales). Das Phytoplankton des Süsswassers. Part 6 of Volume 16 of Die Binnengëwasser. E. Schweizerbart. Stuttgart. 116 pp. + 47 plates.

_____. 1974. The phylogeny of eukaryotic algae. Taxon 23: 446–461.

_____, and T. Kalina. 1962. Über die Gattung *Eremosphaera* DeBary und deren taxonomische Gliederung. Preslia 34: 348–358.

_____, and M. Nováková. 1969. A monograph of the genus *Chlorella*. The fresh water species. pp. 10–74 in Fott, B, (Ed.). Studies in Phycology. Academia. Prague.

_____, and _____. 1971. Taxonomy of the palmelloid genera *Gloeocystis* Nägeli and *Palmogloea* Kützing (Chlorophyceae). Arch. Protistenk 113: 322–333.

Fowke, L. C., and J. D. Pickett-Heaps. 1969. Cell division in *Spirogyra*. 1969A. I. Mitosis. J. Phycol. 5: 240–259. 1969B. II. Cytokinesis. J. Phycol. 5: 273–281.

_____, and _____. 1971. Conjugation in *Spirogyra*. J. Phycol. 7: 285–294.

Fox, J. E. 1958. Meiosis in *Closterium*. News Bull. Phycol. Soc. Am. 11: 63.

Föyn, B. 1943. Lebenzyklus, Cytologie und Sexualität der Chlorophycee *Cladophora suhriana* Kützing. Arch. Protistenk. 83: 154–177.

_____. 1958. Über die Sexualität und der Generationswechsel von *Ulva mutabilis* (N. S). Arch. Protistenk. 102: 473–480.

Frame, P. W., and T. Sawa. 1975. Comparative anatomy of Charophyta: II. The axial nodal complex—an approach to the taxonomy of *Lamprothamnium*. J. Phycol. 11: 202–205.

Francis, D. 1967. On the eyespot of the dinoflagellate *Nematodinium*. J. Exp. Biol. 47: 495–502.

Franke, W. W., S. Berger, H. Falk, H. Spring, U. Scheer, W. Herth, M. F. Trendelenburg, and H. G. Schweiger. 1974. Morphology of the nucleo-cytoplasmic interactions during the development of *Acetabularia* cells. I. The vegetative phase. Protoplasma 82: 249–282.

_____, and W. Herth. 1973. Cell and lorica fine structure of the Chrysomonad alga, *Dinobryon sertularia* Ehr. (Chrysophyceae). Arch. Mikrobiol. 91: 323–344.

_____, H. Spring, U. Scheer, and H. Zerban. 1975. Growth of the nuclear envelope in the vegetative phase of the green alga *Acetabularia*. J. Cell Biol. 66: 681–689.

Fraser, T. W. 1975. Involvement of the Golgi apparatus and microtubules in the formation and positioning of the phycoplast of *Bulbochaete hiloensis* (Nordst.) Tiffany. Protoplasma 83: 103–110.

_____, and B. E. S. Gunning. 1969. The ultrastructure of the plasmodesmata in the filamentous green alga, *Bulbochaete hiloensis* (Nordst.) Tiffany. Planta 88: 244–254.

_____, and _____. 1973. Ultrastructure of the hairs of the filamentous green alga *Bulbochaete hiloensis* (Nordst.) Tiffany: an apoplastidic plant cell with a well developed Golgi apparatus. Planta 113: 1–19.

Frederick, S. E., P. J. Gruber, and N. E. Tolbert. 1973. The occurrence of glycolate dehydrogenase and glycolate oxidase in green plants: an evolutionary survey. Pl. Physiol. 52: 318–323.

Frei, E., and R. D. Preston. 1964. Non-cellulosic structural polysaccharides in algal cell walls. II. Association of xylan mannan in *Porphyra umbilicalis*. Proc. Roy. Soc. London, Ser, B, 160: 314–327.

Freudenthal, H. 1962. *Symbiodinium* gen. nov. and *Symbiodinium microadriaticum* sp. nov., a zooxanthella: taxonomy, life cycle, and morphology. J. Protozool. 9: 45–52.

_____, J. Lee, and J. McLaughlin. 1966. Symbionts of the sea. Nat. Hist. 75(9): 46–50.

Friedmann, I. 1959. Structure, life-history and sex determination of *Prasiola stipitata* Suhr. Ann. Bot., N. S. 23: 571–594.

_____. 1971. Light and scanning electron microscopy of the endolithic desert algal habitat. Phycologia 10: 411–428.

_____, A. L. COLWIN, AND L. H. COLWIN. 1968. Fine structural aspects of fertilization in *Chlamydomonas reinhardii*. J. Cell Sci. 3: 115–128.

_____, AND M. GALUN. 1972. Desert algae, lichens and fungi. *In* Brown, G. W. JR., (Ed.) Desert Biology II. Academic Press. New York.

_____, Y. LIPKIN, AND R. OCAMPO-PAUS. 1967. Desert Algae of the Negev (Israel). Phycologia 6: 185–200.

_____, AND I. MANTON. 1960. Gametes, fertilization and zygote development in *Prasiola stipitata* Suhr. I. Light microscopy. Nova Hedw. 1: 333–344. II. Electron microscopy. Nova Hedw. 1: 443–462.

_____, AND R. OCAMPO. 1976. Endolithic blue-green algae in the dry valleys: primary producers in the Antarctic desert ecosystem. Science 193: 1247–1249.

FRIES, L. 1967. The sporophyte of *Nemalion multifidum* (Weber et Mohr) J. Ag. Sv. Bot. Tidskr. 61: 457–462.

_____. 1969. The sporophyte of *Nemalion multifidum* (Weber et Mohr) J. Ag. found on the Swedish west coast. Sv. Bot. Tidskr. 63: 139–141.

FRITSCH, F. E. 1935. The Structure and Reproduction of the Algae. Vol. I. University Press, Cambridge. 791 pp.

_____. 1942. The interrelations and classification of Myxophyceae (Cyanophyceae). New Phytol. 4: 134–148.

_____. 1945. The Structure and Reproduction of the Algae. Vol. II. University Press, Cambridge. 939 pp.

_____. 1947. The status of the Siphonocladales. J. Indian Bot. Soc., 1946: 29–48.

FRY, W. L., AND H. P. BANKS. 1955. Three new genera of algae from the upper Devonian of New York. J. Paleontol. 2a: 37–44.

FRYXELL, G. A. 1975. Three new species of *Thalassiosira* with observations on the occluded process, a newly observed structure of diatom valves. Beihefte Nova Hedw. 53: 57–75.

_____, AND G. R. HASLE. 1972. *Thalassiosira eccentrica* (Ehrenb.) Cleve, *T. symmetrica* sp. nov., and some related centric diatoms. J. Phycol. 8: 297–317.

FUKUHARA, E. 1968. Studies on the taxonomy and ecology of *Porphyra* of Hokkaido and its adjacent waters. Bull. Hokkaido Regional Fisheries Res. Lab. (Fisheries Agency) 34: 40–99.

FULCHER, R. G., AND M. E. McCULLY. 1969A. Histological studies on the genus *Fucus*. IV. Regeneration and adventive embryony. Can. J. Bot. 47: 1643–1649.

_____, AND _____. 1969B. Laboratory culture of the intertidal brown alga *Fucus vesiculosus*. Can. J. Bot. 47: 219–222.

FULLER, C. W., P. KREISS, AND H. H. SELIGER. 1972. Particulate bioluminescence in dinoflagellates: dissociation and partial reconstitution. Science 177: 884–885.

FULLER, S. W., AND A. C. MATHIESON. 1972. Ecological studies of economic red algae. IV. Variations of carrageenan concentration and properties in *Chondrus crispus* Stackhouse. J. Exp. Mar. Biol. Ecol. 10: 49–58.

GAARDER, K. R. 1971. Comments on the distribution of coccolithophorids in the oceans. *In* The Micropalaeontology of Oceans (Ed., B. M. Funnell and W. R. Riedel). University Press, Cambridge. pp. 97–103.

_____, AND G. R. HASLE. 1971. Coccolithophorids of the Gulf of Mexico. Bull. Mar. Sci. 21: 519–544.

GAILLARD, J. 1972. Quelques remarques sur le cycle reproducteur des Dictyotales et sur ses variations. Soc. Bot. Fr. Mém. 1972: 145–150.

GALLOIS, R. W. 1976. Coccolith blooms in the Kimmeridge Clay and origin of North Sea oil. Nature 259: 473–475.

GALT, J. H., AND H. C. WHISLER. 1970. Differentiation of flagellated spores in *Thalassomyces* ellobiopsid parasite of marine Crustacea. Arch. Mikrobiol. 71: 295–303.

GANESAN, E. K., AND A. J. LEMUS. 1972. Studies on the marine algal flora of Venezuela, IV. *Botryocladia papenfussii* sp. nov. (Rhodophyceae-Rhodymeniales). Phycologia 11: 25–31.

GANTT, E. 1971. Micromorphology of the periplast of *Chroomonas* sp. (Cryptophyceae). J. Phycol. 7: 177–184.

————. 1975. Phycobilisomes: light-harvesting pigment complexes. BioScience 25: 781–788.

————, AND S. F. CONTI. 1965. The ultrastructure of *Poryphyridium cruentum*. J. Cell Biol. 26: 365–381.

————, AND ————. 1966. Granules associated with the chloroplast lamellae of *Porphyridium cruentum*. J. Cell Biol. 29: 423–434.

————, M. R. EDWARDS, AND S. F. CONTI. 1968. Ultrastructure of *Porphyridium aerugineum* a blue-green colored Rhodophytan. J. Phycol. 4: 65–71.

————, ————, AND L. PROVASOLI. 1971. Chloroplast structure in the Cryptophyceae. Evidence for phycobilisomes within intrathylakoidal spaces. J. Cell Biol. 48: 280–290.

————, AND C. A. LIPSCHULTZ. 1972. Phycobilisomes of *Porphyridium cruentum*. I. Isolation. J. Cell Biol. 54: 313–324.

GARDNER, K. H., AND J. BLACKWELL. 1971. The substructure of the cellulose microfibrils from the cell walls of the algae *Valonia ventricosa*. J. Ultrastr. Res. 36: 725–731.

GAUTHIER-LIÈVRE, L. 1963–64. Oedogoniacées africaines. J. Cramer, Weinheim. 558 pp. 104 plates.

————. 1965. Zygnémacées africaines. Beihefte Nova Hedw. 20: 1–210.

GAWLIK, S. R., AND W. F. MILLINGTON. 1969. Pattern formation and the fine structure of the developing cell wall in colonies of *Pediastrum boryanum*. Am. J. Bot. 56:1084–1093.

GAYRAL, P. 1965. *Monostroma* Thuret, *Ulvaria* Rupr. emend. Gayral, *Ulvopsis* Gayral (Chlorophycées, Ulotrichales): structure, reproduction, cycles, position systématique. Rev. Gén. Bot. 72: 627–638.

————, AND J. FRESNEL-MORANGE. 1971. Résultats préliminaires sur la structure et la biologie de la Coccolithacée *Ochrosphaera neapolitana* Schussnig. C. R. Acad. Sci. (Paris). 273D: 1683–1686.

————, AND C. HAAS. 1969. Étude comparée des genres *Chrysomeris* Carter et *Giraudyopsis* P. Dang. position systématique des Chrysomeridaceae (Chrysophyceae). Rev. Gén. Bot. 76: 659–666.

————, ————, AND H. LEPAILLEUR. 1972. Alternance morphologique de générations et alternance de phases chez les Chrysophycées. Soc. Bot. Fr., Mém. 1972: 215–230.

————, AND H. LEPAILLEUR. 1971. Étude de deux Chrysophycées filamenteuses: *Nematochrysopsis roscoffensis* Chadefaud, *Nematochrysis hieroglyphica* Waern. Rev. Gén. Bot. 78: 61–74.

GEESINK, R. 1973. Experimental investigations on marine and freshwater *Bangia* (Rhodophyta) from The Netherlands. J. Exp. Mar. Biol. and Ecol. 11: 239–247.

GEITLER, L. 1931. Untersuchungen über das sexuelle Verhalten von *Tetraspora lubrica*. Biol. Zent. 51: 173–187.

————. 1932. Cyanophyceae. *In* Kryptogamenflora von Deutschland, Österreich und der Schweiz (L. Rabenhorst, Ed.). Vol. 14. Akademische Verlags Gesellschaft. Leipzig. pp. 673–1056.

————. 1935. Reproduction and life history in diatoms. Bot. Rev. 1(5): 149–161.

————. 1955. Die atmophytische Bangioidee *Rhodospora*. Österr. Bot. Zeitschrift 102: 25–29.

————. 1960. Schizophyzeen. *In* Handbuch der Pflanzenanatome (Zimmerman, W., and P. Ozenda, Ed.). Borntraegar. Berlin. 131 pp.

————. 1967. *Gloeochrysis pyrenigera* n. sp. und der Chromatophor von *Phaeodermatium*. Österr. Bot. Zeitschr. 114: 115–118.

————. 1969. Comparative studies on the behavior of allogamous pennate diatoms in auxospore formation. Am. J. Bot. 56: 718–722.

————, AND H. SCHIMAN-CZEIKA. 1970. Über das sogenannte Palmellastadium von *Phaeothamnion confervicola*. Österr. Bot. Zeitsch. 118: 293–296.

GEMEINHARDT, K. 1930. Silicoflagellatae. *In* Kryptogamen-Flora von Deutschland, Österreich und der Schweiz. (Ed., L. Rabenhorst). Vol. 10(2). Akademische Verlags Gesellschaft. v. Leipzig. 87 pp.

GERISCH, G. 1959. Die Zellendifferenzierung bei *Pleodorina californica* Shaw und die Organization der Phytomonadinenkolonien. Arch. Protistenk. 104: 292–358.

GERLOFF, J. 1967. Eine neue Phaeophycee aus dem Süsswasser: *Pseudobodanella peterfii* nov. gen. et nov. spec. Rev. Roum. Biol.-Bot. 12: 27–35.

_____. 1970. Elektronenmikroskopische Untersuchungen an Diatomeenschalen VII. Der Bau der Schale von *Planktoniella sol* (Wallich) Schütt. Beihefte Nova Hedw. 31: 203–234.

_____, AND J.-G. HELMCKE. 1975A. Der Feinbau der Schalen von *Diploneis papula* (A. S.) Cleve, *Diploneis smithii* (Breb.) Cleve und *Diploneis parca* (A. S.) Boyer. Willdenowia 7: 539–563.

_____, AND _____. 1975B. Vol. 10, *In* Diatomeenschalen im Elektronenmikroskopischen Bild, (Ed., J.-G. Helmcke, W. Krieger, and J. Gerloff), J. Cramer, Germany.

GERRATH, J. F., AND K. H. NICHOLLS, 1974. A red snow in Ontario caused by the dinoflagellate, *Gymnodinium pascheri*. Can. J. Bot. 52: 683–685.

GIBB, D. C. 1957. The free-living forms of *Ascophyllum nodosum* (L.) LeJol. J. Ecol. 45: 49–83.

GIBBS, S. P. 1962A. The ultrastructure of the pyrenoids of algae, exclusive of green algae. J. Ultrastr. Res. 7: 247–261.

_____. 1962B. The ultrastructure of the chloroplasts of algae. J. Ultrastr. Res. 7: 418–435.

_____. 1962C. Nuclear envelope chloroplast relationships in algae. J. Cell Biol. 14: 433–444.

_____. 1962D. Chloroplast development in *Ochromonas danica*. J. Cell Biol. 15: 343-361.

_____. 1970. The comparative ultrastructure of the algal chloroplast. Ann. N.Y. Acad. Sci. 175: 454–473.

_____, D. CHENG, AND T. SLANKIS. 1974. The chloroplast nucleoid in *Ochromonas danica*. I. Three-dimensional morphology in light- and dark-grown cells. J. Cell Sci. 16: 557–577.

_____, R. A. LEWIN, AND D. E. PHILPOTT. 1958. The fine structure of the flagellar apparatus of *Chlamydomonas moewusii*. Exptl. Cell Res. 15: 619–622.

GIBOR, A. 1966. *Acetabularia*: a useful giant cell. Sci. Am. 215(5): 118–124.

_____. 1973A. Observations on the sterile whorls of *Acetabularia*. Protoplasma 78: 195–202.

_____. 1973B. *Acetabularia*. Physiological role of their deciduous organelles. Protoplasma 78: 461–465.

GILES, K. L., AND V. SARAFIS. 1972. Chloroplast survival and division *in vitro*. Nature 236: 56–57.

GILLHAM, N. W. 1969. Uniparental inheritance in *Chlamydomonas reinhardii*. Am. Naturalist 103: 355–388.

GINSBURG, R., AND N. LAZAROFF. 1973. Ultrastructural development of *Nostoc muscorum* A. J. Gen. Microbiol 75: 1–9.

GIRAUD, A., AND F. MAGNE. 1968. La place de la meiose dans le cycle de developpement de *Porphyra umbilicalis*. C. R. Acad. Sci. (Paris). 267: 586–588.

GIRAUD, G. 1962. Les infrastructures de quelques algues et leur physiologie. J. Microscopie 1: 251–274.

GITTLESON, S. M., S. K. HOTCHKISS, AND F. G. VALENCIA. 1974. Locomotion in the marine dinoflagellate *Amphidinium carterae* (Hulburt). Trans. Am. Microsc. Soc. 93: 101–105.

GLEZER, Z. I. 1970. Silicoflagellatophyceae. *In* Cryptogamic Plants of the U.S.S.R. Vol. 7. Komarov Inst. Bot., Acad. Sci., U.S.S.R. Israel Program for Sci. Translations, Jerusalem. iv + 361 pp.

GODWARD, M. B. E. 1961. Meiosis in *Spirogyra crassa*. Heredity 16: 53–62.

_____. 1966. The Chromosomes of the Algae. Edward Arnold (Publishers). Ltd. London. 212 pp.

GOFF, L. J. 1976A. Solitary bodies (S-bodies) in the parasitic red alga *Harveyella mirabilis* (Choreocolaceae, Cryptonemiales) Protoplasma 89: 189–195.

_____. 1976B. The biology of *Harveyella mirabilis* (Cryptonemiales, Rhodophyceae). V. Host response to parasite infection. J. Phycol. 12: 313–328.

_____, AND K. COLE. 1973. The biology of *Harveyella mirabilis* (Cryptonemiales, Rhodophyceae). I. Cytological investigations of *Harveyella mirabilis* and its host, *Odonthalia floccosa*. Phycologia 12: 237–245.

_____, AND _____. 1975. The biology of *Harveyella mirabilis* (Cryptonemiales, Rhodophyceae). II. Carposporophyte development as related to the taxonomic affiliation of the parasitic alga, *Harveyella mirabilis*. Phycologia 14: 227–238.

_____, AND _____. 1976A. The biology of *Harveyella mirabilis* (Cryptonemiales, Rhodophyceae). III. Spore germination and subsequent development. Can. J. Bot. 54: 268–280.

_____, AND _____. 1976B. The biology of *Harveyella mirabilis* (Cryptonemiales, Rhodophyceae). IV. Life history and phenology. Can. J. Bot. 54: 181–292.

GOJDICS, M. 1953. The genus *Euglena*. Univ. Wis. Press. Madison. 268 pp.

GOLDSTEIN, M. 1964. Speciation and mating behavior in *Eudorina*. J. Protozool. 11: 317–344.

_____. 1967. Colony differentiation in *Eudorina*. Can. J. Bot. 45: 1591–1596.

_____, AND S. MORRALL. 1970. Gametogenesis and fertilization in *Caulerpa*. *In* Phylogenesis and Morphogenesis in the Algae. (Ed., J. F. Fredrick and R. M. Klein.). Ann. N.Y. Acad. Sci. 175: 660–672.

GOODBAND, S. J. 1971. The taxonomy of *Sphacelaria cirrosa* (Roth) Ag., *Sphacelaria fusca* (Huds.) Ag., and *Sphacelaria furcigera* (Kütz.) Sauv. A simple statistical approach. Ann. Bot. 35: 957–980.

_____. 1973. Observations on the development of endophytic filaments of *Sphacelaria bipinnata* (Kütz.) Sauv. on *Halidrys siliquosa* (L.) Lyngb. Br. Phycol. J. 8: 175–179.

GOODENOUGH, U. 1970. Chloroplast division and pyrenoid formation in *Chlamydomonas reinhardii*. J. Phycol. 6: 1–6.

GOODENOUGH, U. W., AND R. L. WEISS. 1975. Gametic differentiation in *Chlamydomonas reinhardtii*. III. Cell wall lysis and microfilament-associated mating structure in wild-type and mutant strains. J. Cell Biol. 67: 623–631.

GOODWIN, T. W. 1974. Carotenoids and biliproteins. Chapter 6 in Stewart, W. D. P. (Ed.). Algal Physiology and Biochemistry. Univ. Calif. Press. Berkeley and Los Angeles. 989 pp.

GORDON-MILLS, E. M., AND E. L. McCANDLESS. 1975. Carrageenans in the cell walls of *Chondrus crispus* Stack. (Rhodophyceae, Gigartinales). I. Localization with fluorescent antibody. Phycologia 14: 275–281.

GOREAU, T. F. 1961. Problems of growth and calcium deposition in reef corals. Endeavor 20: 32–39.

_____, N. I. GOREAU, AND C. M. YONGE. 1971. Reef corals: autotrophs or heterotrophs? Biol. Bull. 141: 247–260.

_____, AND E. A. GRAHAM. 1967. A new species of *Halimeda* from Jamaica. Bull. Mar. Sci. 17: 432–441.

GORHAM, P. R. 1964. Toxic algae. *In* D. F. Jackson (Ed.) Algae and Man. Plenum Press. New York. pp. 307–336.

GOTELLI, I. B., AND R. CLELAND. 1968. Differences in the occurrence and distribution of hydroxyproline-proteins among the algae. Am. J. Bot. 55: 907–914.

GOWANS, C. S. 1960. Some genetic investigations on *Chlamydomonas eugametos*. Zeit. Vererbungsl. 91: 63–73.

GRAHAM, E. A. 1975. Fruiting in *Halimeda* (Order Siphonales) 1. *Halimeda cryptica* Colinvaux and Graham. Bull. Mar. Sci. 25: 130–133.

GRAMBAST, L. J. 1974. Phylogeny of the *Charophyta*. Taxon 23: 463–481.

GRANT, M. C., AND V. M. PROCTOR. 1972. *Chara vulgaris* and *C. contraria*: Patterns of reproductive isolation for two cosmopolitan species complexes. Evolution 26: 267–281.

GRAY, B. H., C. A. LIPSCHULTZ, AND E. GANTT. 1973. Phycobilisomes from a blue-green alga, *Nostoc* species. J. Bact. 116: 471–475.

GRAY, J. 1960. Fossil Chlorophycean algae from the Miocene of Oregon. J. Paleont. 34: 453–463.

GREEN, B. R. 1973. Evidence for the occurrence of meiosis before cyst formation in *Acetabularia mediterranea* (Chlorophyceae, Siphonales). Phycologia 12: 233–235.

GREEN, J. C., AND B. S. C. LEADBEATER. 1972. *Chrysochromulina parkeae* sp. nov. (Haptophyceae) a new species recorded from S. W. England and Norway. J. Mar. Biol. Assn. U. K. 52: 469–474.

_____, AND M. PARKE. 1974. A reinvestigation by light and electron microscopy of *Ruttnera spectabilis* Geitler (Haptophyceae), with special reference to the fine structure of the zoids. J. Mar. Biol. Assn. U. K. 54: 539–550.

_____, AND _____. 1975. New observations upon members of the genus *Chrysotila* Anand, with remarks upon their relationships within the Haptophyceae. J. Mar. Biol. Assn. U. K. 55: 109–121.

GREENE, R. W. 1970. Symbiosis in sacoglossan opisthobranchs: functional capacity of symbiotic chloroplasts. Mar. Biol. 7: 138–142.

GREUET, C. 1968. Organisation ultrastructurale de l'ocelle de deux Péridiniens Warnowiidae, *Erythropsis pavillardi* Kofoid et Swezy et *Warnowia pulchra* Schiller. Protistologica 4: 209–230.

——. 1971. Étude ultrastructurale et évolution des cnidocystes de *Nematodinium*, Peridinien Warnowiidae Lindemann. Protistologica 7: 345–355.

GRIFFIN, D. G. III, AND V. W. PROCTOR. 1964. A population study of *Chara zeylanica* in Texas, Oklahoma and New Mexico. Am. J. Bot. 51: 120–124.

GRIFFITHS, D. A., AND D. J. GRIFFITHS. 1969. The fine structure of autotrophic and heterotrophic cells of *Chlorella vulgaris* (Emerson strain). Pl. Cell Physiol. 10: 11–19.

GRIFFITHS, D. W. 1970. The pyrenoid. Bot. Rev. 36: 29–58.

GRILLI, C. M. 1974. A light and electron microscopic study of the blue-green algae living either in the coralloid roots of *Macrozamia communis* or isolated in culture. Giornale Bot. Ital. 108: 161–173.

GRILLI-CAOLA, M. 1975. A light and electron microscope study of blue-green algae growing in the coralloid-roots of *Encephalartos altensteinii* and in culture. Phycologia 14: 25–33.

GROBE, B., AND C-G. ARNOLD. 1975. Evidence of a large, ramified mitochondrium in *Chlamydomonas reinhardii*. Protoplasma 86: 291–294.

GROOVER, R. D., AND H. C. BOLD. 1968. Phycological notes. I. *Oocystis polymorpha* sp. nov. Southwestern Naturalist 13: 129–135.

——, AND ——. 1969. Phycological Studies. VIII. The taxonomy and comparative physiology of the Chlorosarchinales and certain other edaphic algae. Univ. Texas Pub., No. 6907, Austin, 165 pp.

——, AND SR. A. M. HOFSTETTER., O. P. 1969. *Planophila terrestris*, a new green alga from Tennessee soil. Tulane Studies in Zoology and Botany 15: 75–80.

GROSS, I. 1931. Entwicklungsgeschichte, Phasenwechsel, und Sexualität bei der Gattung *Ulothrix*. Arch. Protistenk. 73: 206–234.

GROTE, M., AND M. PFAUITSCH. 1977. Licht- und rasterelektronenmikroskopische Beobachtungen zum Konjugationsprozess bei *Spirogyra majuscula* (Jachalzen). Cytobiologie 14: 222–228.

GRUBER, H. E., AND B. ROSARIO. 1974. Variation in eyespot ultrastructure in *Chlamydomonas reinhardii* (*ac*-31). J. Cell Sci. 15: 451–494.

GUILLARD, R. R. L., P. KILHAM, AND T. A. JACKSON. 1973. Kinetics of silicon-limited growth in the marine diatom *Thalassiosira pseudonana* Hasle and Heimdal (= *Cyclotella nana* Hustedt). J. Phycol. 9: 233–237.

——, AND C. J. LORENZEN. 1972. Yellow-green algae with chlorophyllide c. J. Phycol 8: 10–14.

GUIRY, M. D. 1974. A preliminary consideration of the taxonomic position of *Palmaria palmata* (Linnaeus) Stackhouse = *Rhodymenia palmata* (Linnaeus) Greville. J. Mar. Biol. Assn. U. K. 54: 509–528.

——, AND G. J. HOLLENBERG. 1975. *Schottera* gen. nov. and *Schottera nicaeensis* (Lamour. ex Duby) comb. nov. (= *Petroglossum nicaeense* (Lamour. ex Duby) Schotter) in the British Isles. Br. Phycol. J. 10: 149–164.

GUTTMAN, H. N., AND H. ZIEGLER. 1974. Clarification of structures related to function in *Euglena gracilis*. Cytobiologie 9: 10–22.

HAAS-NIEKERK, T., DE. 1965. The genus *Sphacelaria* Lyngbye (Phaeophyceae) in the Netherlands. Blumea 13: 145–161.

HALFEN, L. N. 1973. Gliding motility of *Oscillatoria*: ultrastructural and chemcial characterization of the fibrillar layer. J. Phycol. 9: 248–253.

——, AND R. W. CASTENHOLZ. 1971A. Gliding motility in the blue-green alga *Oscillatoria princeps*. J. Phycol. 7: 133–145.

——, AND ——. 1971B. Energy expenditure for gliding motility in a blue-green alga. J. Phycol. 7: 258–260.

HALLDAL, P. 1962. Taxes. *In* Physiology and Biochemistry of Algae (Ed., R. A. Lewin). Academic Press, N.Y. pp. 583–593.

_____. 1964. Phototaxis in Protozoa. *In* Biochemistry and Physiology of Protozoa. (Ed., S. H. Hutner). Academic Press, New York. pp. 277–296.

HAMEL, G. 1939. Sur la classification des Ectocarpales. Bot. Notiser 1939: 65–70.

HÄMMERLING, J. 1963. Nucleo-cytoplasmic interactions in *Acetabularia* and other cells. Ann. Rev. Plant Phys. 14: 65–92.

_____: 1964. Gibt es bei Dasycladaceen Zoosporen? Ann. Biol., Ser. 4, 3: 33–36.

HAND, W. G., AND J. A. SCHMIDT. 1975. Phototactic orientation by the marine dinoflagellate *Gyrodinium dorsum* Kofoid. II. Flagellar activity and overall response mechanism. J. Protozool. 22: 494–498.

HANIC, L. A. 1965. Life history studies on *Urospora* and *Codiolum* from southern British Columbia. Ph. D. Thesis. University of British Columbia, Vancouver.

_____, AND J. S. CRAIGIE. 1969. Studies on the algal cuticle. J. Phycol. 5: 89–102.

HARA, Y. 1972. An electron microscopic study on the chloroplasts of the Rhodophyta. Proc. Intern. Seaweed Symp. 7: 153–158.

HARDER, R., AND W. KOCH. 1949. Life-history of *Bonnemaisonia hamifera* (*Trailliella intricata*). Nature 163: 106.

HARGRAVES, P. E. 1976. Studies on marine plankton diatoms. II. Resting spore morphology. J. Phycol. 12: 118–128.

HARLIN, M. M. 1973A. "Obligate" algal epiphyte: *Smithora naiadum* grows on a synthetic substrate. J. Phycol. 9: 230–232.

_____. 1973B. Transfer of products between epiphytic marine algae and host plants. J. Phycol. 9: 243–248.

HARRIS, D. O. 1969. Nutrition of *Platydorina caudata* Kofoid. J. Phycol. 5: 205–210.

_____. 1970. An autoinhibitory substance produced by *Platydorina caudata* Kofoid. Plant Phys. 45: 210–214.

_____. 1971. Growth inhibitors produced by the green algae (Volvocaceae). Arch. Mikrobiol. 76: 47–50.

_____. 1972. Life history and growth inhibition studies in *Platydorina caudata* (Volvocaceae). Bull. Soc. Bot. France, Mémoires 1972: 161–172.

_____, AND C. CALDWELL. 1974. Possible mode of action of a photosynthetic inhibitor produced by *Pandorina morum*. Arch. Microbiol. 95: 193–204.

_____, AND D. E. JAMES. 1974. Toxic Algae. Carolina Tips 37: 13–14.

_____, AND M. C. PAREKH. 1974. Further observations on an algicide produced by *Pandorina morum*, a colonial green flagellate. Microbios 9: 259–265.

_____, AND R. C. STARR. 1969. Life history and physiology of reproduction of *Platydorina caudata* Kofoid. Arch. Protistenk. 111: 138–155.

HARRIS, K., AND D. E. BRADLEY. 1960. A taxonomic study of *Mallomonas*. J. Gen. Microbiol. 22: 750–777.

HARRIS, R. E. 1962. Contribution to the taxonomy of *Callithamnion* Lyngbye emend. Naegeli. Bot. Notiser 115: 18–28.

HARRISON, W. G. 1976. Nitrate metabolism of the red tide dinoflagellate *Gonyaulax polyedra* Stein. J. Exp. Mar. Biol. Ecol. 21: 199–209.

HARTSHORNE, J. N. 1953. The function of the eyespot in *Chlamydomonas*. New Phytol. 52: 292–297.

HARVEY, M. J., AND J. McLACHLAN. 1973. *Chondrus crispus*. Proc. Nova Scotian Inst. Sci. 27 (Suppl.) xii + 155 pp.

HASEGAWA. Y. 1972. Forced cultivation of *Laminaria*. Proc. Intern. Seaweed Symp. 7: 391–394.

HASLE, G. R. 1972. Two types of valve processes in centric diatoms. Beihefte Nova Hedw. 39: 55–78.

_____. 1974. The "mucilage pore" of pennate diatoms. Beihefte Nova Hedw. 45: 167–186.

_____, AND D. L. EVENSEN. 1976. Brackish water and freshwater species of the diatom genus *Skeletonema*. II. *Skeletonema potamos* comb. nov. J. Phycol. 12: 73–82.

HASSINGER-HUIZINGA, H. 1952. Generationswechsel und Geschlechtsbestimmung bei *Callithamnion corymbosum* (Sm.) Lyngb. Arch. Protistenk. 98: 91–124.

HASTINGS, P. J., E. E. LEVINE, E. COSBEY, M. O. HUDDOCK, N. W. GILLHAM, S. J. SUZYCKI, R. LOPPES, AND R. P. LEVINE. 1965. The linkage groups of *Chlamydomonas reinhardtii*. Microbiol. Gen. Bull. 23: 17–19.

HAUG, A., B. LARSEN, AND E. BAARDSETH. 1969. Comparison of the constitution of alginates from different sources. Proc. Intern. Seaweed Symp. 6: 443–451.

HAUPT, W. 1963. Photoreceptorprobleme der Chloroplastenbewegung. Ber. Deutsch. Bot. Gesel. 76: 313–322.

HAWKINS, A. F., AND G. F. LEEDALE. 1971. Zoospore structure and colony formation in *Pediastrum* ssp. and *Hydrodictyon reticulatum* (L.) Lagerh. Ann. Bot. 35: 201–211.

HAWKINS, E. K. 1972. Observations on the developmental morphology and fine structure of pit connections in red algae. Cytologia 37: 759–768.

HAXO, F. T., AND L. R. BLINKS. 1950. Photosynthetic action spectra of marine algae. J. Gen. Physiol. 33: 389–422.

_____, AND D. C. FORK. 1959. Photosynthetically active accessory pigments of cryptomonads. Nature 184: 1051–1052.

HAYWARD, J. 1974. Studies on the growth of *Stichococcus bacillaris* Naeg. in culture. J. Mar. Biol. Assn. U. K. 54: 261–265.

HEATH, I. B., AND W. M. DARLEY. 1972. Observations on the ultrastructure of the male gametes of *Biddulphia laevis* Ehr. J. Phycol. 8: 51–59.

HECKY, R. E., K. MOPPER, P. KILHAM, AND E. T. DEGENS. 1973. The amino acid and sugar composition of diatom cell-walls. Mar. Biol. 19: 323–331.

HEEREBOUT, G. R. 1968. Studies on the Erythropeltidaceae (Rhodophyceae-Bangiophycidae). Blumea 16: 139–157.

HEIMDAL, B. R. 1974. Further observations on the resting spores of *Thalassiosira constricta* (Bacillariophyceae). Norw. J. Bot. 21: 303–307.

HELLEBUST, J. A. 1965. Excretion of some organic compounds by marine phytoplankton. Limnol. and Oceanogr. 10: 192–206.

_____. 1974. Extracellular products. Chapter 30 *In* Stewart, W. D. P. (Ed.) Algae Physiology and Biochemistry. Univ. of Calif. Press. Berkeley and Los Angeles. 989 pp.

_____, AND A. HAUG. 1972A. Photosynthesis, translocation, and alginic acid synthesis in *Laminaria digitata* and *Laminaria hyperborea*. Can. J. Bot. 50: 169–176.

_____, AND _____. 1972B. *In situ* studies on alginic acid synthesis and other aspects of the metabolism of *Laminaria digitata*. Can. J. Bot. 50: 177–184.

HENDEY, N. I. 1937. The plankton diatoms of the Southern Seas. Discovery Reports 16: 151–364.

_____. 1964. An Introductory Account of the Smaller Algae of British Coastal Waters. Part V. Bacillariophyceae (Diatoms). Ministry of Agric., Fisheries and Food, Fishery Investigation Ser. IV. xxii + 317 pp.

_____. 1971. Electronmicroscope studies and the classification of diatoms. *In* The Micropalaeontology of Oceans" (Ed., B. M. Funnell and W. R. Riedel). Cambridge Univ. Press. pp. 625–631.

_____, D. H. CUSHING, AND G. W. RIPLEY. 1954. Electron microscope studies of diatoms. J. Roy. Microsc. Soc. 74: 22–34.

HENRY, M. S. (Ed.). 1966. Symbiosis. Vol. 1. Academic Press, New York. 478 pp.

HERMAN, E. M., AND B. M. SWEENEY. 1975. Circadian rhythm of chloroplast ultrastructure in *Gonyaulax polyedra*, concentric organization around a central cluster of ribosomes. J. Ultrastr. Res. 50: 347–354.

HERNDON, W. R. 1958. Studies on Chlorosphaeracean algae from soil. Am. J. Bot. 45: 298–308.

_____. 1964. *Boldia*: a new Rhodophycean genus. Am. J. Bot. 51: 575–581.

HERTH, W., A. KUPPEL, AND W. W. FRANKE. 1975. Cellulose in *Acetabularia* cyst walls. J. Ultrastr. Res. 50: 289–292.

HEYWOOD, P. 1972. Structure and origin of flagellar hairs in *Vacuolaria virescens*. J. Ultrastr. Res. 39: 608–623.

_____. 1973. Nutritional studies on the Chloromonadophyceae: *Vacuolaria virescens* and *Gonyostomum semen*. J. Phycol. 9: 156–159.

_____, AND M. B. E. GODWARD. 1972. Centrometric organization in the Chloromonadophycean alga *Vacuolaria virescens*. Chromosoma 39: 333–339.

_____, AND P. J. MAGEE. 1976. Meiosis in Protists. Some structural and physiological aspects of meiosis in algae, fungi and Protozoa. Bact. Rev. 40: 192–240.

HIBBERD, D. J. 1971. Observations on the cytology and ultrastructure of *Chrysamoeba radians* Klebs (Chrysophyceae). Br. Phycol. J. 6: 207–223.

_____. 1973. Observations on the ultrastructure of flagellar scales in the genus *Synura* (Chrysophyceae). Arch. Mikrobiol. 89: 291–304.

_____. 1975. Observations on the ultrastructure of the choanoflagellate *Codosiga botrytis* (Ehr.) Saville-Kent with special reference to the flagellar apparatus. J. Cell Sci. 17: 191–219.

_____. 1976. The ultrastructure and taxonomy of the Chrysophyceae and Prymnesiophyceae (Haptophyceae): a survey with some new observations of the ultrastructure of the Chrysophyceae. Bot. J. Linn. Soc. 72: 55–80.

_____, A. D. GREENWOOD, AND H. B. GRIFFITHS. 1971. Observations on the ultrastructure of the flagella and periplast in Cryptophyceae. Brit. Phycol. J. 6: 61–72.

_____, AND G. F. LEEDALE. 1970. Eustigmatophyceae—a new algal class with unique organization of the motile cell. Nature 225: 759–760.

_____, AND _____. 1971A. Cytology and ultrastructure of the Xanthophyceae. II. The zoospore and vegetative cell of coccoid forms, with special reference to *Ophiocytium majus* Naegeli. Br. Phycol. J. 6: 1–23.

_____, AND _____. 1971B. A new algal class—the Eustigmatophyceae. Taxon 20: 523–525.

_____, AND _____. 1972. Observations on the cytology and ultrastructure of the new algal class, Eustigmatophyceae. Ann. Bot. 36: 49–71.

HILENSKI, L. L., P. L. WALNE, AND F. SNYDER. 1976. Aliphatic chains of esterified lipids in isolated eyespots of *Euglena gracilis* var. *bacillaris*. Plant Physiol. 57: 645–646.

HILL, G. J. C., AND L. MACHLIS. 1968. An ultrastructural study of vegetative cell division in *Oedogonium borisianum*. J. Phycol. 4: 261–271.

_____, AND _____. 1970. Defined media for growth and gamete production by the green alga *Oedogonium cardiacum*. Plant Physiol. 46: 224–226.

HILLIARD, D. K. 1971. Observations on the lorica structure of some *Dinobryon* species (Crysophyceae), with comments on related genera. Österr. Bot. Z. 119: 25–40.

HILLIS, L. W. 1959. A revision of the genus *Halimeda* (order Siphonales). Pub. Inst. Mar. Sci., University of Tex. 6: 321–403.

HILLS, G. J. 1973. Cell wall assembly *in vitro* from *Chlamydomonas reinhardii*. Planta 115: 17–23.

_____, M. GURNEY-SMITH, AND K. ROBERTS. 1973. Structure, composition and morphogenesis of the cell wall of *Chlamydomonas reinhardii*. II. Electron microscopy and optical diffraction analysis. J. Ultrastruct. Res. 43: 179–192.

HIROSE, H., AND S. KUMANO. 1966. Spectroscopic studies on the phycoerythrins from Rhodophycean algae with special reference to their phylogenetic relations. Bot. Mag., Tokyo 79: 105–113.

_____, _____, AND K. MADONO. 1969. Spectroscopic studies on phycoerythrins from Cyanophycean and Rhodophycean algae with special reference to their phylogenetical relations. Bot. Mag., Tokyo 82: 197–203.

_____, AND K. YOSHIDA. 1964. A review of the life history of the genus *Monstroma*. Bull. Jap. Soc. of Phycol. 12: 19–31.

HOBBS, M. J. 1971. The fine structure of *Eudorina illinoiensis* (Kofoid) Pascher. Br. Phyc. J. 6: 81–103.

_____. 1972. Eyespot fine structure in *Eudorina illinoiensis*. Br. Phycol. J. 7: 347–355.

HOEK, C. VAN DEN. 1963. Revision of the European species of *Cladophora* E. J. Brill. Leiden. 248 pp, 55 plates.

_____, AND A. M. CORTEL-BREEMAN. 1970. Life-history studies on Rhodophyceae II. *Halymenia floresia* (Clem.) Ag. Acta Bot. Neerl. 19: 341–362.

_____, AND _____, H. RIETEMA, AND J. B. W. WANDERS. 1972. L'interprétation de données obtenues, par des cultures unialgales, sur les cycles évolutifs des algues. Quelques exemples tirés des recherches conduites au laboratoire de Groningue. Soc. Bot. Fr. Mém. 1972: 45–66.

_____, AND A. FLINTERMAN. 1968. The life-history of *Sphacelaria furcigera* Kütz. (Phaeophyceae). Blumea 16: 193–242.

HOFFMAN, L. R. 1960. Chemotaxis of *Oedogonium* sperms. Southwestern Naturalist 5: 111–116.

_____. 1965. Cytological studies of *Oedogonium* I. Oospore germination in *O. foveolatum*. Am. J. Bot. 52: 173–181.

_____. 1967. Observations on the fine structure of *Oedogonium*. III. Microtubular elements in the chloroplasts of *Oe. cardiacum*. J. Phycol. 3: 212–221.

_____. 1968. Observations in the fine structure of *Oedogonium*. V. Evidence for the *de novo* formation of pyrenoids in the zoospores of *Oe. cardiacum*. J. Phycol. 4: 212–218.

_____. 1970. Observations on the fine structure of *Oedogonium*. VI. The striated component of the compound flagellar "roots" of *O. cardiacum*. Can. J. Bot. 48: 189–196.

_____. 1971. Observations on the fine structure of *Oedogonium*. VII. The oogonium prior to fertilization. *In* Contributions in Phycology. (Ed., B. C. Parker and R. M. Brown, Jr.) Allen Press, Lawrence, Kansas. pp. 93–106.

_____. 1973A. Fertilization in *Oedogonium* I. Plasmogamy. J. Phycol. 9: 62–84.

_____. 1973B. II. Polyspermy. J. Phycol. 9: 296–301.

_____. 1974. III. Karyogamy. Am. J. Bot. 61: 1076–1090.

_____. 1976. Fine structure of *Cylindrocapsa* zoospores. Protoplasma 87: 191–219.

_____, AND C. S. HOFMANN. 1975. Zoospore formation in *Cylindrocapsa*. Can. J. Bot. 53: 439–451.

_____, AND I. MANTON. 1962. Observations on the fine structure of the zoospore of *Oedogonium cardiacum* with special reference to the flagellar apparatus. J. Exptl. Bot. 13: 443–449.

_____, AND _____. 1963. Observations on the fine structure of *Oedogonium*. II. The spermatozoid of *O. cardiacum* Amer. J. Bot. 50: 455–463.

HOHAM, R. W. 1973. Pleiomorphism in the snow alga, *Raphidonema nivale* Lagerh. (Chlorophyta), and a revision of the genus *Raphidonema* Lagerh. Syesis 6: 243–253.

_____. 1974A. New findings in the life history of the snow alga, *Chlainomonas rubra* (Stein et Brooke) *comb. nov.* (Chlorophyta, Volvocales). Syesis 7: 239–247.

_____. 1974B. *Chlainomonas kolii* (Hardy et Curl.) *comb. nov.* (Chlorophyta, Volvocales), a revision of the snow alga, *Trachelomonas kolii* Hardy et Curl (Euglenophyta, Euglenales). J. Phycol. 10: 392–396.

_____. 1975A. The life history and ecology of the snow alga *Chloromonas pichinchae* (Chlorophyta, Volvocales). Phycologia 14: 213–226.

_____. 1975B. Optimum temperatures and temperature ranges for growth of snow algae. Arctic and Alpine Research 7: 13–24.

_____. 1976. The effect of coniferous litter and different snow meltwaters upon the growth of two species of snow algae in axenic culture. Arctic and Alpine Research 8: 377–386.

_____, AND J. E. MULLET. 1977. The life history and ecology of the snow alga *Chloromonas cryophila* sp. nov. (Chlorophyta, Volvocales) Phycologia 16: 53–68.

HOLDSWORTH, R. H. 1968. The presence of a crystalline matrix in pyrenoids of the diatom, *Achnanthes brevipes*. J. Cell Biol. 37: 831–837.

HOLLENBERG, G. J. 1935. A study of *Halicystis ovalis*. I. Morphology and reproduction. Am. J. Bot. 22: 782–812.

_____. 1939. A morphological study of *Amplisiphonia*, a new member of the Rhodomelaceae. Bot. Gaz. 101: 380–390.

_____. 1942. An account of the species of *Polysiphonia* on the Pacific coast of North America. I. Oligosiphonia. Am. J. Bot. 29: 772–785.

_____. 1943. New marine algae from southern California. II. Am. J. Bot. 30: 571–579.

_____. 1944. An account of the species of *Polysiphonia* on the Pacific coast of North America. II. Polysiphonia. Am. J. Bot. 31: 474–483.

_____. 1958. Culture studies of marine algae. III. *Porphyra perforata*. Am. J. Bot. 45: 653–656.

_____. 1959. *Smithora*, an interesting new algal genus in the Erythropeltidaceae. Pacif. Nat. 1(8): 3–11.

_____. 1968A. An account of the species of *Polysiphonia* of the Central and Western tropical Pacific Ocean. I. Oligosiphonia. Pacif. Sci. 22: 56–98.

_____. 1968B. An account of the species of the red alga *Polysiphonia* of the central and western tropical Pacific Ocean II. Polysiphonia. Pac. Sci. 22: 198–207.

_____. 1968C. An account of the species of the red alga *Herposiphonia* occurring in the Central and Western Tropical Pacific Ocean. Pacif. Sci. 22: 536–559.

_____. 1969. An account of the Ralfsiaceae (Phaeophyta) of California. J. Phycol. 5: 290–301.

_____, AND I. A. ABBOTT. 1966. Supplement to Smith's Marine Algae of the Monterey Peninsula. Stanford University Press, Stanford, Calif. xi + 130 pp.

_____, AND M. J. Wynne. 1970. Sexual plants of *Amplisiphonia pacifica* (Rhodophyta). Phycologia 9: 175–178.

HOLM-HANSEN, O. 1968. Ecology, physiology and biochemistry of blue-green algae. Ann. Rev. Microbiol. 22: 47–70.

HOLMES, R. W. 1966. Short-term temperature and light conditions associated with auxospore formation in the marine centric diatom *Coscinodiscus concinnus* W. Smith. Nature 209: 217–218.

_____. 1977. *Lauderia annulata*—a marine centric diatom with an elongate bilobed nucleus. J. Phycol. 13: 180–183.

_____, P. M. WILLIAMS, AND R. W. EPPLEY. 1967. Red water in La Jolla Bay, 1964–1966. Limnol. Oceanogr. 12: 503–512.

HOLMGREN, P. R., H. P. HOSTETTER, AND V. E. SCHOLES. 1971. Ultrastructural observation of cross-walls in the blue-green alga *Spirulina major*. J. Phycol. 7: 309–311.

HOLT, C. VON, AND M. VON HOLT. 1968. Transfer of photosynthetic products from zooxanthellae to coelenterate hosts. Comp. Biochem and Physiol. 24: 73–81.

HOMMERSAND, M. H. 1963. The morphology and classification of some Ceramiaceae and Rhodomelaceae. Univ. Calif. Pub. Bot. 35: 165–366.

_____, AND D. W. OTT. 1970. Development of the carposporophyte of *Kallymenia reniformis* (Turner) J. Agardh. J. Phycol. 6: 322–331.

HONIGBERG, B. M., *et al.* 1964. A revised classification of the phylum Protozoa. J. Protozool. 11: 7–20.

HOOPER, R., AND G. R. SOUTH. 1974. A taxonomic appraisal of *Callophyllis* and *Euthora* (Rhodophyta). Br. Phycol. J. 9: 423–428.

HOPKINS, A. W., AND G. E. MCBRIDE, 1976. The life history of *Coleochaete scutata* (Chlorophyceae) studied by a Feulgen microspectrophotometric analysis of the DNA cycle. J. Phycol. 12: 29–35.

HORI, T. 1971. Survey of pyrenoid distribution in brown algae. Bot. Mag., Tokyo 84: 231–242.

_____. 1972A. Further survey of the pyrenoid distribution in Japanese brown algae. Bot. Mag. Tokyo 85: 125–134.

_____. 1972B. Ultrastructure of the pyrenoid of *Monostroma* (Chlorophyceae) and related genera. *In* Contributions to the Systematics of Benthic Marine Algae of the North Pacific (Ed., I. A. Abbott and M. Kurogi). Japanese Society of Phycology. pp. 17–32.

_____. 1973. Comparative studies of pyrenoid ultrastructure in algae of the *Monostroma* complex. J. Phycol. 9: 190–199.

_____, AND M. CHIHARA. 1974A. Light and electron microscope observations on the developmental sequence of *Prasinocladus marinus*. Sci. Reports of the Tokyo Kyoiku Daigaku. Sec. B. 15: 265–271.

_____, AND _____. 1974B. Studies on the fine structure of *Prasinocladus ascus* (Prasinophyceae). Phycologia 13: 307–315.

_____, AND R. UEDA. 1967. Electron microscope studies on the fine structure of plastids in siphonous green algae with special reference to their phylogenetic relationships. Science Report Tokyo Kyoiku Daigaku, Sec. 12: 225–244.

HOSFORD, S. P. C., AND E. L. MCCANDLESS, 1975. Immunochemistry of carrageenans from gametophytes and sporophytes of certain red algae. Can. J. Bot. 53: 2835–2841.

HOSHAW, R. W. 1961. Sexual cycles of three green algae for laboratory study. Am. Biol. Teacher 23: 489–499.

_____. 1965. A cultural study of sexuality in *Sirogonium melanosporum*. J. Phycol. 1: 134–138.

_____. 1968. Biology of the filamentous conjugating algae. *In* D. F. Jackson (Ed.) Algae, Man and the Environment. Syracuse Univ. Press. Syracuse, New York. 554 pp.

_____, AND R. L. HILTON, JR. 1966. Observations on the sexual cycle of the saccoderm desmid *Spirotaenia condensata*. J. Arizona Acad. Sci. 4: 88–92.

HOUSLEY, H. L., R. W. SCHEETZ, AND G. F. PESSONEY. 1975. Filament formation in the diatom *Skeletonema costatum*. Protoplasma 86: 363–369.

HOYT, W. D. 1927. The periodic fruiting of *Dictyota* and its relation to the environment. Am. J. Bot. 14: 592–619.

HSIAO, S. I. C. 1969. Life history and iodine nutrition of the marine brown alga, *Petalonia fascia* (O.F. Müll.) Kuntze. Can. J. Bot. 47: 1611–1616.

_____, AND L. D. DRUEHL. 1971. Environmental control of gametogenesis in *Laminaria saccharina*. I. The effects of light and culture media. Can. J. Bot. 49: 1503–1508.

_____, AND _____. 1973A. Environmental control of gametogenesis in *Laminaria saccharina*. II. Correlation of nitrate and phosphate concentrations with gametogenesis and selected metabolites. Can. J. Bot. 51: 829–839.

_____, AND _____. 1973B. Environmental control of gametogenesis in *Laminaria saccharina*. IV. *In situ* development of gametophytes and young sporophytes. J. Phycol. 9: 160–164.

HUBER-PESTALOZZI, G. 1941. Das Phytoplankton des Süsswassers. Systematik und Biologie. Teil 2(1). Chrysophyceen, farblose Flagellaten, Heterokonten. *In* Die Binnengewässer (Ed., A. Thienemann), Stuttgart. 365 pp.

_____. 1950. Das Phytoplankton des Süsswassers. Teil 3. Cryptophyceen, Chloromonadinen, Peridineen *In* Die Binnengewässer (Ed., A. Thienemann). Stuttgart. 322 pp.

HUDSON, P. R., AND J. R. WAALAND. 1974. Ultrastructure of mitosis and cytokinesis in the multinucleate green alga *Acrosiphonia*. J. Cell Biol. 62: 274–294.

_____, AND M. J. WYNNE. 1969. Sexual plants of *Bonnemaisonia geniculata* (Nemaliales). Phycologia 8: 207–213.

HUGHES, E. O., P. R. GORHAM, AND A. ZEHNDER. 1958. Toxicity of a unialgal culture of *Microcystis aeruginosa*. Can. J. Microbiol. 4: 225–236.

HUIZING, H. J., AND H. RIETEMA. 1975. Xylan and mannan as cell wall constituents of different stages in the life-histories of some siphoneous green algae. Br. Phycol. J. 10: 13–16.

HUMM, H. J. 1956. Rediscovery of *Anadyomene menziesii*, a deep-water alga from the Gulf of Mexico. Bull. Marine Sci. (Miami) 6: 346–348.

_____. 1969. Distribution of marine algae along the Atlantic coast of North America. Phycologia 7: 43–53.

HUNTSMAN, S. 1972. Organic excretion by *Dunaliella tertiolecta*. J. Phycol. 8: 59–63.

HUSTEDE, H. 1964. Entwicklungsphysiologische Untersuchungen über den Generationswechsel zwischen *Derbesia neglecta* Berth. und *Bryopsis Halymeniae* Berth. Bot. Mar. 6: 134–142.

HUTH, K. 1970. Bewegung und Orientierung bei *Volvox aureus* Ehrb. I. Mechanismus der phototaktischen Reaktion. Z. Pflanzenphysiol. 62: 436–450.

HUTNER, S. H., AND J. J. A. MCLAUGHLIN. 1958. Poisonous tides. Sci. Am. 199: 92–98.

_____, AND L. PROVOSOLI. 1964. Nutrition of algae. Ann. Rev. Plant Physiol. 15: 37–56.

HYAMS, J., AND D. CHASEY. 1974. Aspects of the flagella apparatus and associated microtubules in a marine alga. Exp. Cell Res. 84: 381–387.

_____, AND D. R. DAVIS. 1972. Induction and characterization of cell wall mutants of *Chlamydomonas reinhardi*. Mutation Res. 14: 381–389.

ICHIMURA, T. 1972. Sexual cell division and conjugation-papilla formation in sexual reproduction of *Closterium strigosum*. Proc. Intern. Seaweed Symp. 7: 208–214.

_____, AND M. M. WATANABE. 1976. Biosystematic studies of the *Closterium peracerosum— strigosum—littorale* complex. I. Morphological variation among the inbreeding populations and an experimental demonstration for source of the cell size variation. Bot. Mag. Tokyo 89: 123–140.

IKUSHIMA, N., AND S. MARUYAMA. 1968. The protoplasmic connection in *Volvox*. J. Protozool. 15: 136–141.

IRVINE, D. E. 1956. Notes on British species of the genus *Sphacelaria* Lyngb. Trans. Bot. Soc. Edinb. 37: 24–45.

ISHIURA, M., AND K. IWASA. 1973A. Gametogenesis in *Chlamydomonas*. I. Effect of light on the induction of sexuality. Plant and Cell Phys. 14: 911–921.

_____, AND _____. 1973B. Gametogenesis in *Chlamydomonas*. II. Effect of cyclohexamide on the induction of sexuality. Plant and Cell Physiol. 14: 923–933.

_____, AND _____. 1973C. Gametogenesis in *Chlamydomonas*. III. Daily fluctuation of sex competence. Plant and Cell Physiol. 14: 935–939.

ISLAM, A. K. M. N. 1963. Revision of the genus *Stigeoclonium*. Beih. Nova Hedw. 10. J. Cramer. Weinheim. 164 pp.

ISRAELSON, G. 1942. The freshwater Florideae of Sweden. Studies on their taxonomy, ecology, and distribution. Symbolae Bot. Upsal. 6(1). 134 pp.

IYENGAR, M. O. P., AND K. R. RAMANATHAN. 1940. On sexual reproduction in a *Dictyosphaerium*. J. Indian Bot. Soc. 18: 195–200.

JAENICKE, L. 1977. Sex hormones of brown algae. Naturwissensch. 64: 69–75.

JAFFE, L. F. 1968. Localization in the developing *Fucus* egg and the general role of localizing currents. Advan. Morphog. 7: 295–328.

JAHN, T. L., AND E. C. BOVEE. 1964. Protoplasmic movements and locomotion of protozoa. *In* Hutner, S. H. (Ed.) Biochemistry and Physiology of Protozoa. Vol. 3. Academic Press, New York. pp. 61–129.

_____, W. M. HARMON, AND M. LANDMAN. 1963. Mechanisms of locomotion in flagellates. I. *Ceratium*. J. Protozool. 10: 358–363.

_____, M. D. LANDMAN, AND J. R. FONSECA. 1964. The mechanism of locomotion of the flagellates. II. Function of the mastigonemes of *Ochromonas*. J. Protozool. 11: 291–296.

JAMES, D. E. 1974. Culturing algae. Carolina Biological Supply Co. 22 pp.

JAMISON, D. W., AND R. A. BESWICK. 1972. The future of seaweed culture in the State of Washington. Proc. Intern. Seaweed Symp. 7: 346–350.

JAROSCH, R. 1970. On the flagellar waves of *Synura bioreti* and the mechanics of the uniplanar waves. Protoplasma 69: 201–214.

JAVORNICKY, P. 1967. Some interesting algal flagellates. Folia Gebot. Phytotaxon. 2: 43–67.

JEFFREY, S. W. 1968. Pigment composition of siphonales algae in the brain coral *Favia*. Biol. Bull. 135: 141–148.

_____, M. SIELICKI, AND F. T. HAXO. 1975. Chloroplast pigment patterns in dinoflagellates. J. Phycol. 11: 374–384.

JENSEN, J. B. 1974. Morphological studies in Cystoseiraceae and Sargassaceae (Phaeophyceae) with special reference to apical organization. Univ. Calif. Pub. Bot. 68, vi + 61 pp.

JENSEN, T. E., AND L. M. SICKO. 1972. The fine structure of the cell wall of *Gloeocapsa alpicola*, a blue-green alga. Cytobiol. 6: 439–446.

JOHANNES, R. E., AND S. L. COLES. 1969. The role of zooplankton in nutrition of scleractinian corals. Symp. Corals, Coral Reefs, Jan. 1969, Mar. Biol. Assn. India, Mandapan Camp, 1969: 8.

JOHANSEN, H. W. 1969. Morphology and systematics of coralline algae with special reference to *Calliarthron*. Univ. Calif. Pub. Bot. 49. vii + 98 pp.

_____. 1970. The diagnostic value of reproductive organs in some genera of articulated coralline red algae. Br. Phycol. Bull. 5: 79–86.

_____. 1971. *Bossiella*, a genus of articulated corallines (Rhodophyceae, Cryptonemiales) in the eastern Pacific. Phycologia 10: 381–396.

_____. 1973. Ontogeny of sexual conceptacles in a species of *Bossiella* (Corallinaceae). J. Phycol. 9: 141–148.

_____. 1974. Articulated coralline algae. Oceanogr. Mar. Biol. Ann. Rev. 12: 77–127.

_____, AND B. J. COLTHART. 1975. Variability in articulated coralline algae (Rhodophyta). Nova Hedw. 26: 135–149.

JOHNSON, J. H., AND O. A. HØEG. 1961. Studies of Ordovician algae. Colorado School of Mines Quarterly 56: V–120.

————, AND K. KONISHI. 1956. Studies of Mississippian algae. Colorado School of Mines Quarterly 51: V–131.

————, AND ————. 1958. Studies of Devonian algae. Colorado School of Mines 53: V–114.

————, ————, AND R. REZAK. 1959. Studies of Silurian (Gotlandian) algae. Colorado School of Mines 54: 1–173.

JOHNSON, U. G., AND K. R. PORTER. 1968. Fine structure of cell division in *Chlamydomonas reinhardii*. J. Cell Biol. 38: 403–425.

JOHNSTON, H. W. 1965. The biological and economic importance of algae, Part I. Tuatara 13: 90–104.

————. 1966. The biological and economic importance of algae, Part 2. Tuatara 14: 30–63.

————. 1970. The biological and economic importance of algae, Part 3. Edible algae of fresh and brackish waters. Tuatara 18: 19–35.

————. 1976. The biological and economic importance of algae, Part 4. Industrial culturing of algae. Tuatara 22: 1–114.

JOLY, A. B., AND E. C. DE OLIVEIRA FILHO. 1967. Two Brazilian Laminarias. Pub. Inst. Pesquisas da Marinha, Minist. da Marinha, Rio de Janeiro, No. 4, 13 pp.

————, AND ————. 1968. Notes on Brazilian algae II. A new *Anadyomene* of the deep water flora. Phykos 7: 27–31.

JONES, G., AND W. FARNHAM. 1973. Japweed: new threat to British coasts. New Scientist 60: 394–395.

JONES, R. F., J. R. KATER, AND S. J. KELLER. 1968. Protein turnover and macromolecular synthesis during growth and gametic differentiation in *Chlamydomonas reinhardtii*. Biochim. Biophys. Acta 157: 589–598.

————, AND R. A. LEWIN. 1960. The chemical nature of the flagella of *Chlamydomonas moewusii*, Expt. Cell Res. 19: 408–410.

————, H. L. SPEER, AND W. KURY. 1963. Studies on the growth of the red alga *Porphyridium cruentum*. Physiologia Plantarum 16: 636–643.

————, AND L. WIESE. 1962. Studies on the mating reaction in *Chlamydomonas*. J. Gen. Physiol. 46: 358A (Abstr.).

JONES, W. E., AND M. S. BABB. 1968. The motile period of swarmers of *Enteromorpha intestinalis* (L.) Link. Br. Phycol. Bull. 525–528.

JÓNSSON, S. 1962A. Sur la reproduction de l'*Anadyomene stellata* (Wulf.) Ag. de la Méditerranée. C. R. Acad. Sci. (Paris). 255: 1983–1985.

————. 1962B. Recherches sur des Cladophoracées marines (structure, reproduction, cycles comparés, conséquences systématiques). Theses presentees a la Faculté des Sci. de Univ. Paris. Masson & Cie, Edit. Paris. pp. 25–230.

————. 1965. La validité et le délimitation de l'ordre des Siphonocladales. Travaux dédies a Lucien Plantefol. Masson and Cie, Paris. pp. 391–406.

JORDAN, A. J., AND R. L. VADAS. 1972. Influence of environmental parameters on intraspecific variation in *Fucus vesiculosus*. Mar. Biol. 14: 248–252.

JORDE, I., AND N. KLAVESTAD. 1959. Observations on *Ectocarpus*, *Feldmannia*, *Pylaiella*, and *Stictyosiphon* in Hordangerfjord, West Norway. Nytt Mag. Bot. 7: 145–156.

KADIS, S., A. CIEGLER, AND S. J. AJL (Ed.). 1971. Microbiol Toxins. VII. Algal and fungal toxins. Academic Press. New York. xvi + 401 pp.

KAIN, J. M. 1964. Aspects on the biology of *Laminaria hyperborea* III. Survival and growth of gametophytes. J. Mar. Biol. Assn. U. K. 44: 415–433.

————. 1967. Populations of *Laminaria hyperborea* at various latitudes. Helgoländer Wissen. Meeresunters. 15: 489–499.

————. 1969. The biology of *Laminaria hyperborea*. V. Comparison with early stages of competitors. J. Mar. Biol. Assn. U. K. 49: 455–473.

————. 1975. The biology of *Laminaria hyperborea* VII. Reproduction of the sporophyte. J. Mar. Biol. Assn. U. K. 55: 567–582.

KALINA, T. 1969. *Gloeochrysis montana* n. sp. und *Poteriochromonas stipitata* Scherffel (Chrysophyceae) aus Krkonose (Riesengebirge). Österr. Bot. Z. 117: 139–145.

KALINSKY, R. G. 1971. *Pedinomonas minor* Korschikoff, a rare alga in Ohio. J. Phycol. 7: 82–83.

KALLEY, J. P., AND T. BISALPUTRA. 1970. *Peridinium trochoideum*: the fine structure of the theca as shown by freeze-etching. J. Ultrastr. Res. 31:95–108.

_____, AND _____. 1971. *Peridinium trochoideum*: the fine structure of the thecal plates and associated membranes. J. Ultrastr. Res. 37: 521–531.

KANN, E. 1972. Zur Systematik und Ökologie der Gattung *Chamaesiphon* (Cyanophyceae). 1. Systematik. Arch. Hydrobiol. 41 (suppl.) 117–171.

KANTZ, T., AND H. C. BOLD. 1969. Phycological Studies. IX. Morphological and taxonomic investigations of *Nostoc* and *Anabaena* in culture. Univ. Texas Pub. 6924. Austin 67 pp.

KAO, C. Y. 1972. Pharmacology of tetrodotoxin and saxitoxin. Fed. Proc. 31: 1117.

KAPRAUN, D. F. 1970. Field and cultural studies of *Ulva* and *Enteromorpha* in the vicinity of Port Aransas, Texas. Contributions in Marine Science 15: 205–283.

_____, AND E. H. FLYNN. 1973. Culture studies of *Enteromorpha linza* (L.) J. Ag. and *Ulvaria oxysperma* (Kützing) Bliding (Chlorophyceae, Ulvales) from Central America. Phycologia 12: 145–152.

KARAKASHIAN, S. J., AND M. W. KARAKASHIAN. 1965. Evolution and symbiosis in the genus *Chlorella* and related algae. Evolution 19: 368–377.

_____, _____, AND M. A. RUDZINKA. 1968. Electron microscopic observations on the symbiosis of *Paramecium bursaria* and its intracellular algae. J. Protozool. 15: 113–128.

KARN, R. C., R. C. STARR, AND G. A. HUDOCK. 1974. Sexual and asexual differentiation in *Volvox obversus* (Shaw) Printz, Strains WD3 and WD7. Arch. Protistenk. 116: 142–148.

KARSTEN, G. 1928. Bacillariophyta (Diatomeae). *In* Engler and Prantl, (Ed.), Die Natürlichen Pflanzenfamilien: Peridineae, Diatomae, Myxomycetes, 2nd. ed., Vol. 2. W. Engelmann Leipzig. pp. 105–303.

KASAHARA, K. 1973. The development of the mucilage gland of two Japanese species of *Alaria*. Bot. Mag., Tokyo 86: 169–181.

KATES, J. R., K. S. CHIANG, AND R. F. JONES. 1968. Studies on DNA replication during synchronized vegetative growth and gametic differentiation in *Chlamydomonas reinhardtii*. Exp. Cell. Res. 49: 121–135.

_____, AND R. F. JONES. 1964. The control of gametic differentiation in liquid cultures of *Chlamydomonas*. J. Cell. Comp. Phys. 63: 157–164.

_____, AND _____. 1966. Pattern of CO_2 fixation during vegetative development and gametic differentiation in *Chlamydomonas reinhardtii*. J. Cell. Physiol. 67: 101–106.

KAZAMA, F., AND M. S. FULLER. 1970. Ultrastructure of *Porphyra perforata* infected with *Pythium marinum*, a marine fungus. Can. J. Bot. 48: 2103–2107.

KENYON, C. N. 1972. The fatty acid composition of unicellular strains of blue-green algae. J. Bact. 109: 827–834.

KESLING, R. V., AND A. GRAHAM. 1962. *Ischadites* is a dasycladacean alga. J. Paleont. 36: 943–952.

KESSELER, H. 1966. Beitrag zur Kenntnis der chemischen und physikalischen Eigenschaften des Zellsaftes von *Noctiluca miliaris*. Veröff. Inst. Meeresf. Bremerhaven Sb. II: 357–368.

KESSLER, E. 1972. Physiologische und biochemische Beiträge zur Taxonomie der Gattung *Chlorella*. VII. Die Thermophile von *Chlorella vulgaris* f. *tertia* Fott et Nováková. Arch. Mikrobiol. 87: 243–248.

_____. 1974. Physiologische und biochemische Beiträge zur Taxonomie der Gattung *Chlorella*. IX. Salzresistenz als taxonomische Merkmal. Arch. Mikrobiol. 100: 51–56.

_____. 1976. Comparative physiology, biochemistry and the taxonomy of *Chlorella* (Chlorophyceae). Plant Systematics and Evolution 125: 125–138.

_____, F. CZYGAN, B. FOTT, AND M. NOVÁKOVÁ. 1968. Über *Halochlorella rubescens* Dangeard. Arch. Protistenk. 110: 462–467.

KEVIN, M. J., W. T. HALL, J. J. A. MCLAUGHLIN, AND P. A. ZAHL. 1969. *Symbiodinium microadriaticum* Freudenthal, a revised taxonomic description, ultrastructure. J. Phycol. 5: 341–350.

KHALEAFA, A. F., M. A. M. KHARBOUSH, A. METWALLI, A. F. MOHREN, AND A. SERUR. 1975. Antibiotic (fungicidal) action from extracts of some seaweeds. Bot. Mar. 18: 163–165.

KIES, L. 1964. Über die experimentelle Auslösung von FortpflanzungsVorgängen und die Zygotenkeimung bei *Closterium acerosum* (Schrank) Ehrenbg. Arch. Protistenk. 107: 331–350.

———. 1967. Oogamie bei *Eremosphaera viridis* De Bary. Flora, Abt. B. 157: 1–12.

———. 1968. Über die Zygotenbildung bei *Micrasterias papillifera* Bréb. Flora, Abt. B, 157: 301–313 .

———. 1970A. Elektronmikroskopische Untersuchungen über Bildung und Struktur der Zygo-
tenwand bei *Micrasterias papillifera* (Desmidiaceae). I. Das Exospor. Protoplasma 70: 21–47.

———. 1970B. II. Die Struktur von Mesospor und Endospor. Protoplasma 71: 139–146.

———. 1975. Elektromikroskopische Untersuchungen über die Konjugation bei *Micrasterias papillifera*. Beih. z. Nova Hedw. 42: 139–154.

KILHAM, P. 1971. A hypothesis concerning silica and the freshwater planktonic diatoms. Limnol. Oceanogr. 16: 10–18.

KIM, D. H. 1970. Economically important seaweeds, in Chile. I. *Gracilaria*. Bot. Mar. 13: 140–162.

———. 1976. A study of the development of cystocarps and tetrasporangial sori in Gigartinaceae (Rhodophyta, Gigartinales). Nova Hedw. 27: 1–146.

KIMBALL, J. F., JR., AND E. J. F. WOOD. 1965. A dinoflagellate with characters of *Gymnodinium* and *Gyrodinium*. J. Protozool. 12: 577–580.

KING, J. M. 1971. Comparative studies of some palmelloid green algae. Ph. D. Dissertation. The University of Texas at Austin, Texas. 254 pp.

———. 1973. *Gloeococcus minutissimus* sp. nov. isolated from soil. J. Phycol. 9: 349–352.

KITO, H., E. OGATA, AND J. MCLACHLAN. 1971. Cytological observations on three species of *Por-phyra* from the Atlantic. Bot. Mag. Tokyo 84: 141–148.

KIVIE, P. A., AND M. VESK. 1974. Pinocytotic uptake of protein from the reservoir of *Euglena*. Arch. Mikrobiol. 96: 155–159.

KLAVENESS, D. 1972A. *Coccolithus huxleyi* (Lohmann) Kamptner. I. Morphological investigations on the vegetative cell and the process of coccolith formation. Protistologica 8: 335–346.

———. 1972B. *Coccolithus huxleyi* (Lohm.) Kamptn. II. The flagellate cell, aberrant cell types, vegetative propagation and life-cycles. Br. Phycol. J. 7: 309–318.

KLEIN, K. M., AND A. CRONQUIST. 1967. A consideration of the evolutionary and taxonomic signifi-
cance of some of some biochemical, micromorphological and physiological characters in the Thallophytes. Quart. Rev. Biol. 42: 105–296.

KLYVER, F. D. 1929. Notes on the life history of *Tetraspora gelatinosa* (Vauch.) Desr. Arch. Protistenk. 66: 290–296.

KNIGHT, M. 1929. Studies in the Ectocarpaceae. II. The life-history and cytology of *Ectocarpus siliculosus*, Dillw. Trans. Roy. Soc. Edin. 56: 307–332.

———. 1931. Nuclear phases and alternation in algae. Phaeophyceae. Beih. Bot. bl., Dresden, 48: 15–37.

———, AND M. PARKE. 1950. A biological study of *Fucus vesiculosus* L. and *F. serratus* L. J. Mar. Biol. Assn. U. K. 29: 439–514.

KNOEPFFLER-PÉGUY, M. 1970. Quelques *Feldmannia* Hamel, 1939 (Phaeophyceae Ectocarpales) des côtes d'Europe. Vie et Milieu, sér. A, 21: 137–188.

———. 1974. Le genre *Acinetospora* Bornet 1891. Vie et Milieu, Sér. A: Biol. Mar. 24: 43–72.

KOCHERT, G. 1968. Differentiation of reproductive cells in *Volvox carteri*. J. Protozool. 15: 438–452.

———, AND I. YATES. 1974. Purification and partial characterization of a glycoprotein sexual inducer from *Volvox carteri*. Proc. Nat. Acad. Sci. USA 71: 1211–1214.

KOFOID, C. A., AND O. SWEZY. 1921. The free-living unarmored Dinoflagellata. Mem. Univ. Calif. 5: 1–562.

KÖHLER, K. 1956. Entwicklungsgeschichte, Geschlechtsbestimmung und Befruchtung bei *Chaeto-morpha*. Arch. Protistenk. 101: 223–268.

KOL, E. 1968. Kryobiologie. Biologie und Limnologie des Schnees und Eices I. Kryovegetation. *In* Thienemann, A. Die Binnengewässer Vol. 24. E. Schweizerbart'sche Verlagsbuchhandlung. Stuttgart. 216 pp.

KOMAREK, J. 1970. Generic identity of the "*Anacystis nidulans*" *Synechococcus leopoliensis* (Racib.) Komarek. strain Kratz-Allen/Bloom. 629 with *Synechococcus* Nag. 1849. Arch. Protistenk. 112: 343–364.

————, AND J. LUDVIK. 1971. Die Zellwandultrastruktur als taxonomisches Merkmal in der Gattung *Scenedesmus*. 1. Die Ultrastukturelemente. Arch. Hydrobiol./Suppl. 39: 301–333.

————, AND ————. 1972. Die Zellwandultrastruktur als taxonomisches Merkmal in der Gattung *Scenedesmus*. 2. Taxonomische Answertung der untersuchten Arten. Arch. Hydrobiol. Suppl. 41 (Algological Studies 6): 11–47.

KOMARKOVA-LEGENEROVÁ, J. 1969. The systematics and ontogenesis of the genera *Ankistrodesmus* Cord and *Monorophidium* gen. nov. *In* Fott, B. (Ed.) Phycological Studies. Academia. Prague. pp. 75–144.

KOOP, H.-U. 1975A. Germination of cysts in *Acetabularia mediterranea*. Protoplasma 84: 137–46.

————. 1975B. Über den Ort der Meiose bei *Acetabularia mediterranea*. Protoplasma 85: 109–114.

KORNMANN, P. 1938. Zur Entwicklungsgeschichte von *Derbesia* und *Halicystis*. Planta 28: 464–470.

————. 1953. Der Formenkreis von *Acinetospora crinita* (Carm.) nov. comb. Helgoländ. Wissen. Meersunt. 4: 205–224.

————. 1954. *Giffordia fuscata* (Zan.) Kuck. nov. comb., eine Ectocarpaceae mit heteromorphen monophasischen Generationen. Helgoländ. Wissen. Meeresunt. 5: 51–52.

————. 1955. Beobachtungen an *Phaeocystis*-Kulturen. Helgoländ. Wissen. Meeresunt. 5: 218–233.

————. 1956A. Über die Entwicklung einer *Ectocarpus confervoides* Form. Pubbl. Zool. Napoli 28: 32–43.

————. 1956B. Zur Morphologie und Entwicklung von *Percursaria percursa*. Helgoländ. Wissen. Meeresunter. 5: 259–272.

————. 1961A. Über *Spongomorpha lanosa* und ihre Sporophytenformen. Helgoländ. Wissen. Meeresunter. 7: 195–205.

————. 1961B. Über *Codiolum* und *Urospora*. Helgoländ. Wissen. Meeresunter. 8: 42–57.

————. 1961C. Die Entwicklung von *Porphyra leucosticta* im Kulturversuch. Helgoländ. Wissen. Meeresunter. 8: 167–175.

————. 1962A. Zur Entwicklung von *Monostroma grevillei* und zur systematische Stellung von *Gomontia polyrhiza*. Ber. Deutsch. Bot. Gesell. Neue Folge 1: 37–39 (Algen Symposium Göttingen).

————. 1962B. Die Entwicklung von *Monostroma grevillei*. Helgoländ. Wiss. Meeresunter. 8: 195–202.

————. 1962C. Die Entwicklung von *Chordaria flagelliformis*. Helgoländ. Wissen. Meeresunter. 8: 265–279.

————. 1962D. Der Lebenzyklus von *Desmarestia viridis*. Helgoländ Wissen. Meeresunter. 8: 287–292.

————. 1963A. Die Lebenszyklus einer marinen *Ulothrix*-Art. Helgoländ. Wissen. Meeresunter. 8: 357–360.

————. 1963B. Die Ulotrichales, neu geordnet auf der Grundlage entwicklungsgeschlichtler Befunde. Phycologia 3: 60–68.

————. 1964A. Über *Monstroma bullosum* (Roth) Thuret und *M. oxyspermum* (Kütz.) Doty. Helgoländ. Wissen. Meeresunter. 11: 13–21.

————. 1964B. Die *Ulothrix*—Arten von Helgoländ. I. Helgoländ. Wissen. Meeresunter. 11: 27–38.

————. 1964C. Der Lebenzyklus von *Acrosiphonia arcta*. Helgoländ Wissen. Meeresunter. 11: 110–117.

————. 1964D. Zur Biologie von *Spongomorpha aeruginosa* (L.) van den Hoek. Helgoländ. Wissen Meeresunter. 11: 200–208.

————. 1965A. Was ist *Acrosiphonia arcta*? Helgoländ. Wissen. Meeresunter. 12: 40–51.

————. 1965B. Zur Analyse des Wachstums und des Aufbau von *Acrosiphonia*. Helgoländ. Wissen. Meeresunter. 12: 219–238.

————. 1966. Wachstum und Zellteilung bei *Urospora*. Helgoländ. Wissen. Meeresunter. 13: 73–83.

_____. 1967. Wachstum und Aufau von *Spongomorpha aeruginosa* (Chlorophyta) Acrosiphoniales. Blumea 15: 9–16.

_____. 1968. Das Wachstum einer *Chaetomorpha*—Art von List/Sylt. Helgoländ. Wissen. Meeresunter. 18: 194–207.

_____. 1969. Gesetzmassigkeiten des Wachstums und der Entwicklung von *Chaetomorpha darwinii* (Chlorophyta, Cladophorales). Helgoländ. Wissen. Meeresunter. 19: 335–354.

_____. 1970A. Der Lebenszyklus von *Acrosiphonia grandis* (Acrosiphoniales; Chlorophyta) Mar. Biol. 7: 324–331.

_____. 1970B. Phylogenetische Beziehungen in der Grünalgengattung *Acrosiphonia*. Helgoländ. Wissen. Meeresunter. 21: 292–304.

_____. 1972A. Ein Beitrag zur Taxonomie der Gattung *Chaetomorpha* (Cladophorales, Chlorophyta). Helgoländ. Wissen. Meeresunter. 23: 1–31.

_____. 1972B. Les sporophytes vivant en endophyte de quelques Acrosiphoniacées et leurs rapports biologiques et taxonomiques. Soc. Bot. Fr. Mém. 1972: 75–86.

_____. 1973. Codiolophyceae, a new class of Chlorophyta. Helgoländ. Wissen. Meeresunter. 25: 1–13.

_____, AND P. SAHLING. 1962. Zur Taxonomie und Entwicklung der *Monostroma*—Arten von Helgoländ. Helgoländ. Wissen. Meeresunter. 8: 302–320.

_____, AND _____. 1974. Prasiolales (Chlorophyta) von Helgoländ. Helgoland. Wissen. Meeresunter. 26: 99–133.

KORSCHIKOFF, A. 1953. Protococcineae. Kiev, AN URSR. 439 pp. (In Ukranian.)

KRAFT, G. T. 1969. *Eucheuma procrusteanum*, a new red algal species from the Philippines. Phycologia 8: 215–219.

_____. 1973. The morphology of *Cubiculosporum koronicarpus* gen. et sp. nov., representing a new family in the Gigartinales (Rhodophyta). Am. J. Bot. 60: 872–882.

_____. 1975. Consideration of the order Cryptonemiales and the families Nemastomataceae and Furcellariaceae (Gigartinales, Rhodophyta) in light of the morphology of *Adelophyton corneum* (J. Agardh.) gen. et comb. nov. from southern Australia. Br. Phycol. J. 10: 279–290.

_____, AND I. A. ABBOTT. 1971. *Predaea weldii*, a new species of Rhodophyta from Hawaii, with an evaluation of the genus. J. Phycol. 7: 194–202.

KRATZ, W. A., AND J. MYERS. 1955. Nutrition and growth of several blue-green algae. Am. J. Bot. 42: 282–287.

KRAUSS, R. W. 1962. Mass culture of algae for food and other organic compounds. Am. J. Bot. 49: 425–435.

KREGER, D. R. 1962. Cell walls. Chapter 19 (pp. 315–335) *In* Lewin, R. A. (Ed.). Physiology and Biochemistry of Algae. Academic Press., New York. 929 pp.

KRICHENBAUER, H. 1937. Beitrag zur Kenntnis der Morphologie und Entwicklungsgeshichte der Gattungen *Euglena* and *Phacus*. Arch. Protistenk. 90: 88–122.

KRISHNAMURTHY, V. 1959. Cytological investigations on *Porphyra umbilicalis* (L.) Kütz. var. *laciniata* (Lightf.) J. Ag. Ann. Bot., N. S. 23: 147–176.

_____. 1962. The morphology and taxonomy of the genus *Compsopogon* Montagne. J. Linn. Soc. (Bot.) 58: 207–222.

_____. 1972. A revision of the species of the algal genus *Porphyra* occurring on the Pacific Coast of North America. Pac. Sci. 26: 24–49.

KRISTIANSEN, J. 1960. Some cases of sexuality in *Kephyriopsis* (Chrysophyceae). Bot. Tids. 56: 128–131.

_____. 1961. Sexual reproduction in *Mallomonas caudata*. Bot. Tids. 57: 306–309.

_____. 1963A. Sexual and asexual reproduction in *Kephyrion* and *Stenocalyx* (Chrysophyceae). Bot. Tids. 59: 244–254.

_____. 1963B. Observations on the structure and ecology of *Synura splendida*. Bot. Tids. 59: 281–289.

_____, AND P. L. WALNE. 1976. Structural connections between flagellar base and stigma in *Dinobryon*. Protoplasma 89: 371–374.

KROES, H. W. 1971. Growth interactions between *Chlamydomonas globosa* Snow and *Chlorococcum ellipsoideum* Deason and Bold under different experimental conditions, with special attention to the role of pH. Limnol. Oceanogr. 16: 869–879.

KRONSTEDT, E., AND B. WALLES. 1975. On the presence of plastids and the eyespot apparatus in a porfiromycin-bleached strain of *Euglena gracilis*. Protoplasma 84: 75–82.

KUBAI, D. F., AND H. RIS. 1969. Division in the dinoflagellate *Gyrodinium cohnii* (Schiller). A new type of nuclear reproduction. J. Cell Biol. 40: 508–528.

KUCKUCK, P. 1912. Neue Untersuchungen über *Nemoderma* Schousboe. Wissen. Meeresunter., Abt. Helgoländ, N. F. 5: 117–152.

———. 1954. Ectocarpaceen-Studien II. *Streblonema*. Helgoländ. Wissen. Meeresunter. 5: 103–117. ("herausgegeben von P. Kornmann").

———. 1955. Ectocarpaceen-Studien III. *Protectocarpus* nov. gen. Helgoländ. Wissen. Meeresunter. 5: 119–140. ("herausgegeben von P. Kornmann").

———. 1956. Ectocarpaceen-Studien IV. *Herponema*, *Kützingiella* nov. gen., *Farlowiella* nov. gen. Helgoländ. Wissen. Meeresunter. 5: 292–325. ("herausgegeben von P. Kornmann").

KUGRENS, P., AND J. A. WEST. 1972A. Synaptonemal complexes in red algae. J. Phycol. 8: 187–191.

———, AND ———. 1972B. Ultrastructure of tetrasporogenesis in the parasitic red alga *Levringiella gardneri* (Setchell) Kylin. J. Phycol. 8: 370–383.

———, AND ———. 1972C. Ultrastructure of spermatial development in the parasitic red algae *Levringiella gardneri* and *Erythrocystis saccata*. J. Phycol. 8: 331–343.

———, AND ———. 1973A. The ultrastructure of carpospore differentiation in the parasitic red alga *Levringiella gardneri* (Setch.) Kylin. Phycologia 12: 163–173.

———, AND ———. 1973B. The ultrastructure of an alloparasitic red alga *Choreocolax polysiphoniae*. Phycologia 12: 175–186.

———, AND ———. 1974. The ultrastructure of carposporogenesis in the marine red alga *Erythrocystis saccata*. J. Phycol. 10: 139–147.

KUMANO, S. 1970. On the development of the carposporophytes in several species of the Batrachospermaceae with special reference to their phylogenetical relations. Bull. Jap. Soc. Phycol. 28: 116–120.

KUNEIDA, H. 1939. On the life history of *Porphyra tenera* Kjellman. J. Coll. Agric. Univ. Tokyo 14: 377–405.

KUNISAWA, R., AND G. COHEN-BAZIRE. 1970. Mutations of *Anacystis nidulans* that affect cell division. Arch. Mikrobiol. 71: 49–59.

KUROGI, M. 1959. Influences of light on the growth and maturation of *Conchocelis*-thallus of *Porphyra*. I. Effect of photoperiod on the formation of monosporangia and liberation of monospores (1). Bull. Tohoku Reg. Fish. Res. Lab. 15: 33–42.

———. 1972. Systematics of *Porphyra* in Japan. *In* Contributions to Systematics of Benthic Marine Algae of the North Pacific (Ed., I. A. Abbott and M. Kurogi). Japan. Soc. Phycology, Kobe. pp. 167–191.

KYLIN, H. 1923. Studien über die Entwicklungsgeschichte der Florideen. K. Sv. Vet. Akad. Handl. 63(11). 139 pp.

———. 1925. The marine red algae in the vicinity of the Biological Station at Friday Harbor, Wash. Lunds Univ. Årsskr., N. F., Avd. 2, 21(9). 87 pp.

———. 1932. Die Florideenordnung Gigartinales. Lunds Univ. Årsskr., N. F., Avd. 2, 26(6). 104 pp.

———. 1937. Bemerkungen über die Entwicklungsgeschichte einiger Phaeophyceen. Lunds Univ. Arsskr., N. F., Avd. 2, 33(1). 33 pp.

———. 1938. Beziehungen zwischen Generationswechsel und Phylogenie. Arch. Protistenk. 90: 432–447.

———. 1956. Die Gattungen der Rhodophyceen. Gleerup, Lund. xv + 673 pp.

LACALLI, T. C. 1973. Cytokineses in *Micrasterias rotata*. Problems of directed primary wall deposition. Protoplasma 78: 433–442.

LACKEY, J. B. 1939. Notes on plankton flagellates from the Scioto River. Lloydia 2: 128–143.

LAMONT, H. C. 1969. Sacrificial cell death and trichome breakage in an Oscillatoriacean blue-green alga: the role of murein. Arch. Mikrobiol. 68: 257–259.

LAND, L. S., J. C. LANG, AND B. N. SMITH. 1975. Preliminary observations on the carbon isotopic composition of some reef coral tissues and symbiotic zooxanthellae. Limnol. Oceanog. 20: 283–287.

LANG, N. J. 1963A. Electron microscopy of the Volvocaceae and Astrephomenaceae. Am. J. Bot. 50: 280–300.

_____. 1963B. Electron-microscopic demonstration of plastids in *Polytoma*. J. Protozool. 10: 333–339.

_____. 1965. Electron microscopic study of heterocyst development in *Anabaena azollae* Strasburger. J. Phycol. 1: 127–134.

_____. 1968A. The fine structure of blue-green algae. Ann. Rev. Microbiol 22: 15–46.

_____. 1968B. Electron microscopic studies of extraplastidic astaxanthin in *Haematococcus*. J. Phycol. 4: 12–19.

_____, AND P. FAY. 1971. The heterocysts of blue-green algae. II. Details of ultrastructure. Proc. Roy. Soc. London B. 178: 193–203.

LANGE, W. 1970. Cyanophyta-bacteria systems: Effects of added carbon compounds or phosphate on algal growth at low nutrient combinations. J. Phycol. 6: 230–234.

LARKUM, A. W. D. 1972. Frond structure and growth in *Laminaria hyperborea*. J. Mar. Biol. Assn. U. K. 52: 405–418.

LAURET, M. 1967. Morphologie, phénologie, répartition des *Polysiphonia* marins du littoral Languedocien. I. Section *Oligosiphonia*. Nat. Monspeliensia, Sér. Bot. 18: 347–373.

_____. 1970. Morphologie, phénologie, répartition des *Polysiphonia* marins du littoral Languedocien. II. Section *Polysiphonia*. Nat. Monspeliensia, Sér. Bot. 21: 121–163.

_____. 1971. Présence de *Polysiphonia brodiaei* sur la côte Atlantique du Canada. Can. J. Bot. 645–646.

LAZAROFF, N. 1973. Photomorphogenesis and Nostocaceen development. *In* Carr, N. G., and B. A. Whitton (Ed.). The Biology of Blue-green Algae. Univ. California Press. Berkeley and Los Angeles. 676 pp.

_____, AND W. VISHNIAC. 1964. The relationship of cellular differentiation to colonial morphogenesis of the blue-green alga, *Nostoc muscorum*, A. J. Gen. Microbiol. 35: 447–457.

LEADBEATER, B. S. C. 1969. A fine structural study of *Olisthodiscus luteus* Carter. Br. Phycol. J. 4: 3–17.

_____. 1970. Preliminary observations on differences of scale morphology at various stages in the life cycle of 'Apistonema-Syracosphaera' *sensu* von Stosch. Br. Phycol. J. 5: 57–69.

_____. 1971A. Observations by means of ciné photography on the behaviour of the haptonema in plankton flagellates of the class Haptophyceae. J. Mar. Biol. Assn. U. K. 51: 207–217.

_____. 1971B. Observations on the life history of the Haptophycean alga *Pleurochrysis scherffelii* with special reference to the microanatomy of the different types of motile cells. Ann. Bot. 35: 429–39.

_____. 1972A. Fine-structural observations on some marine choanoflagellates from the coast of Norway. J. Mar. Bio. Assn. U. K. 52: 67–79.

_____. 1972B. Fine structural observations on six new species of *Chrysochromulina* from the coast of Norway with preliminary observations on scale production in *C. microcylindra* sp. nov. Sarsia 49: 65–80.

_____. 1975. A microscopical study of the marine choanoflagellate *Savillea micropora* (Norris) comb. nov., and preliminary observations on lorica development in *S. micropora* and *Stephanoeca diplocostata* Ellis. Protoplasma 83: 111–29.

_____, AND J. D. DODGE. 1967A. An electron microscope study of nuclear and cell division in a dinoflagellate. Arch. Mikrobiol. 57: 239–254.

_____, AND _____. 1967B. Fine structure of the dinoflagellate transverse flagellum. Nature 213: 421–422.

————, AND I. MANTON. 1969A. New observations on the fine structure of *Chrysochromulina strobilus* Parke and Manton with special reference to some unusual features of the haptonema and scale. Arch. Mickrobiol. 66: 105–120.

————, AND ————. 1969B. *Chrysochromulina camella* sp. nov. and *C. cymbium* sp. nov., two new relatives of *C. strobilus* Parke and Manton. Archiv. Mikrobiolog. 68: 116–132.

————, AND ————. 1971. Fine structure and light microscopy of a new species of *Chrysochromulina* (*C. acantha*). Arch. Mikrobiol. 78: 58–69.

LEBEDNIK, P. A. 1975. Biosyntematic studies of *Clathromorphum* and *Mesophyllum* with comments on the taxonomy of Corallinaceae. J. Phycol. 11 (Suppl.): 17.

————. 1977A. The Corallinaceae of northwestern North America. I. *Clathromorphum* Foslie emend. Adey. Syesis 9: 59–112.

————. 1977B. The taxonomy of the Corallinaceae with special reference to *Clathromorphum* and *Mesophyllum*. Phycologia. 16.

————, F. C. WEINMANN, AND R. E. NORRIS. 1971. Spatial and seasonal distributions of marine algal communities at Amchitka Island, Alaska. BioScience 21: 656–660.

LEE, I. K., AND M. KUROGI. 1972. Ecological observations of the members of Rhodymeniales in Hokkaido. Proc. Intern. Seaweed Symp. 7: 131–134.

LEE, J. J., M. E. MCENERY, E. M. KENNEDY, AND H. RUBIN. 1975. A nutritional analysis of a sublittoral diatom assemblage epiphytic on *Enteromorpha* from a Long Island salt marsh. J. Phycol. 11: 14–49.

LEE, K. A., AND C. L. KEMP. 1976. Chemical estimations of DNA changes during synchronous growth of *Eudorina elegans* (Chlorophyceae). J. Phycol. 12: 85–88.

LEE, K. W., AND H. C. BOLD. 1973. *Pseudocharaciopsis texensis* gen. et sp. nov., a new member of the Eustigmatophyceae. Br. Phycol. J. 8: 31–37.

————, AND ————. 1974. Phycological studies XII. *Characium* and some *Characium*-like algae. Univ. of Texas Pub., No. 7403, Austin, 127 pp.

LEE, R. E. 1971. The pit connections of some lower red algae: ultrastructure and phylogenetic significance. Br. Phycol. J. 6: 29–38.

————, AND S. A. FULTZ. 1970. Ultrastructure of the *Conchocelis* stage of the marine red alga *Porphyra leucosticta*. J. Phycol. 6: 22–28.

LEEDALE, G. F. 1967. Euglenoid Flagellates. Prentice-Hall, Inc., Englewood Cliffs, N.J. xiii + 242 pp.

————. 1970. Phylogenetic aspects of nuclear cytology in the algae. Ann. N.Y. Acad. Sci. 175: 429–453.

————. 1971. The Euglenoids. Oxford Biology Readers 5. (J. J. Head and O. E. Lowenstein, Ed.). Oxford University Press, London. pp. 1–16.

————. 1975A. Preliminary observations on nuclear cytology and ultrastructure in carbon-starved streptomycin-bleached *Euglena gracilis*. Colloques internationaux C. N. R. S. No. 240: 285–290.

————. 1975B. Envelope formation and structure in the euglenoid genus *Trachelomonas*. Br. Phycol. J. 10: 17–41.

————, AND D. E. BUETOW. 1976. Observations on cytolysome formation and other cytological phenomena in carbon-starved *Euglena gracilis*. J. Microscopie Biol. Cell. 25: 149–154.

————, B. S. C. LEADBEATER, AND A. MASSALSKI. 1970. The intracellular origin of flagellar hairs in the Chrysophyceae and Xanthophyceae. J. Cell Sci. 6: 701–719.

————, B. J. D. MEEUSE, AND E. G. PRINGSHEIM. 1965. Structure and physiology of *Euglena spirogyra*. I–VI. Arch. f. Mikrobiol. 50: 68–102; 133–165.

LEFEVRE, M. 1964. Extracellular products of algae. *In* Jackson, D. F. (Ed.) Algae and Man. Plenum Press. New York. pp. 337–367.

LEFORT, F. 1971. Sur l'appartenance à une seule et meme espece de deux Coccolithophoracées, *Cricosphaera carterae* et *Ochrosphaera verrucosa*. C. R. Acad. Sci. (Paris). 272D: 2540–2543.

LEMBI, C. A. 1975A. The fine structure of the flagellar apparatus of *Carteria*. J. Phycol. 11: 1–13.

————. 1975B. A rhizoplast in *Carteria radiosa* (Chlorophyceae). J. Phycol. 11: 219–221.

_____, AND W. R. HERNDON. 1966. Fine structure of the pseudocilia of *Tetraspora*. Can. J. Bot. 44: 710–712.

LENHOFF, H. M., L. MUSCATINE, AND L. V. DAVIS. 1968. Coelenterate biology: experimental research. Science 160: 1141–1146.

LENZENWEGER, R. 1968. Der Verlauf der Zygoten—Keimung bei *Micrasterias rotata* (Grev.) Ralfs. Arch. Protistenk. 111: 1–11.

_____. 1975. Über Konjugation und Zygotenkeimung bei *Micrasterias rotata* (Grev.) Ralfs. Beih. z. Nova Hedw. 42: 155–161.

LÉONARD, J., AND P. COMPÈRE. 1967. *Spirulina platensis* (Gom.) Geitl., algue bleue de grande valeur alimentaire par sa richesse en protéines. Bull. Jard. Bot. Nat. Belgique 37 Suppl: 1–23.

LERCHE, W. 1937. Untersuchungen über Entwicklung und Fortpflanzung in der Gattung *Dunaliella*. Arch. Protistenk. 88: 236–268.

LERSTEN, N. R., AND P. D. VOTH. 1960. Experimental control of zoid discharge and rhizoid formation in the green alga *Enteromorpha*. Bot. Gaz. 122: 33–45.

LEVINE, R. P., AND W. T. EBERSOLD. 1960. The genetics and cytology of *Chlamydomonas*. Ann. Rev. Microbiol. 14: 197–216.

_____, AND C. E. FOLSOME. 1959. The nuclear cycle of *Chlamydomonas reinhardii*. Zeit. Verebungsl. 98: 192–202.

_____, AND U. W. GOODENOUGH. 1970. The genetics of photosynthesis of the chloroplast in *Chlamydomonas reinhardtii*. Ann Rev. Genet. 4: 397–408.

LEVRING, T. 1953. The marine algae of Australia. I. Rhodophyta: Goniotrichales, Bangiales and Nemalionales. Arkiv Bot. 2: 457–566.

_____, H. A. HOPPE, O. J. SCHMID. 1969. Marine algae. A survey of research and utilization. Botanica Marina Handbooks. Vol. 1, Cramer, de Gruyter and Co., Berlin and New York. 421 pp.

LEWIN, J. C., AND C.-H. CHEN. 1971. Available iron: a limiting factor for marine phytoplankton. Limnol. Oceanog. 16: 670–675.

_____, AND R. A. LEWIN. 1960. Auxotrophy and heterotrophy in marine littoral diatoms. Can. J. Microbiol. 6: 127–134.

_____, AND _____. 1967. Culture and nutrition of some apochlorotic diatoms of the genus *Nitzschia*. J. Gen. Microbiol. 461: 361–367.

_____, _____, AND D. E. PHILPOTT. 1958. Observations on *Phaeodactylum tricornutum*. J. Gen. Microbiol. 18: 418–426.

_____, AND R. E. NORRIS. 1970. Surf-zone diatoms of the coasts of Washington and New Zealand (*Chaetoceros armatum* T. West and *Asterionella* spp.). Phycologia 9: 143–149.

_____, AND B. E. F. REIMANN. 1969. Silicon and plant growth. Ann. Rev. Plant. Phys. 19: 289–304.

LEWIN, R. A. 1950. Gamete behaviour in *Chlamydomonas*. Nature 166: 76.

_____. 1952A. Ultraviolet induced mutations in *Chlamydomonas moewusii* Gerloff. J. Gen. Microbiol. 6: 233–248.

_____. 1952B. Studies on the flagella of algae. I. General observations on *Chlamydomonas moewusii* Gerloff. Biol. Bull. 103: 74–79.

_____. 1953. The genetics of *Chlamydomonas moewusii*. J. Genetics 51: 543–560.

_____. 1954A. Sex in unicellular algae. *In* Sex in Microorganisms (D. H. Weinrich, Ed.) Am. Assn. Adv. Sci., Washington, D.C. pp. 100–133.

_____. 1954B. The utilization of acetate by wild-type and mutant *Chlamydomonas dysosmos*. J. Gen. Microbiol. 11: 459–471.

_____. 1956. Control of sexual activity in *Chlamydomonas* by light. J. Gen. Microbiol. 15: 170–185.

_____. 1957. The zygote of *Chlamydomonas moewusii*. Can. J. Bot. 35: 795–804.

_____. 1959. The isolation of algae. Revue Algol. 3: 181–197.

_____. 1972. Auxotrophy in marine littoral diatoms. Proc. Intern. Seaweed Symp. 7: 316–318.

————. 1974A. Genetic control of flagella activity in *Chlamydomonas moewusii* (Chlorophyta, Volvocales). Phycologia 13: 45–55.

————. 1974B. Biochemical taxonomy. Chapter 1 in Stewart, W. D. P. (Ed.). Algal Physiology and Biochemistry. Univ. Calif. Press, Berkeley and Los Angeles. 989 pp.

————. 1975. A marine *Synechocystis* (Cyanophyta, Chroococcales) epizoic on ascidians. Phycologia 14: 149–152.

————, AND L. CHENG. 1975. Associations of microscopic algae with didemnid ascidians. Phycologia 14: 153–160.

————, AND J. O. MEINHART. 1953. Studies on the flagella of algae III. Electron micrographs of *Chlamydomonas moewusii*. Can. J. Bot. 31: 711–717.

————, AND J. A. ROBERTSON. 1971. Influence of salinity on the form of *Asterocytis* in pure culture. J. Phycol. 7: 236–238.

————, AND N. W. WITHERS. 1975. Extraordinary pigment composition of a prokaryotic alga. Nature 256: 735–737.

L'HARDY-HALOS, M. T. 1968. Les Ceramiaceae (Rhodophyceae–Florideae) des côtes de Bretagne: 1. Le genre *Antithamnion* Näg. Rev. Algolg. n. s. 9: 152–183.

————. 1970. Recherches sur les Céramiacées (Rhodophycées–Céramiales) et leur morphogénèse. I. Structure de l'appareil végétatif et des organes reproducteurs. Rev. Gén. Bot. 77: 211–287.

————. 1971. Recherches sur les Céramiacées (Rhodophycées-Céramiales) et leur morphogenese. II. Les modalites de la croissance et les remaniements cellularies. Rev. Gén. Bot. 78: 201–256.

LICARI, G. R., AND P. E. CLOUD, JR. 1968. Reproductive structures and taxonomic affinities of some nannofossils from the Gunflint Iron formation. Proc. Nat. Acad. Sci. 59: 1053–1060.

————, ————, AND W. D. SMITH. 1969. A new Chroococcacean alga from the Proterozoic of Queensland. Proc. Nat. Acad. Sci. 62: 56–62.

LICHTLE, C., AND G. GIRAUD. 1970. Aspects ultrastructuraux particuliers au plaste du *Batrachospermum virgatum* (Sirdt)—Rhodophycees—Nemalionale. J. Phycol. 6: 281–289.

LIDDLE, L. B. 1968. Reproduction in *Zonaria farlowii*. I. Gametogenesis, sporogenesis, and embryology. J. Phycol. 4: 298–305.

————. 1972. Development of gametophyte and sporophyte populations of *Padina sanctae-crucis* Borg. in the field and laboratory. Proc. Intern. Seaweed Symp. 7: 80–82.

LIN, H., M. R. SOMMERFELD, AND J. R. SWAFFORD. 1975. Light and electron microscope observations on motile cells of *Porphyridium purpureum* (Rhodophyta). J. Phycol. 11: 452–457.

LINNAEUS, C. 1753. Species plantarum. Vols. 1, 2 *In* Stearn, W. T. 1957. Carl Linnaeus' Species Plantarum. Roy. Society. London. 1200 pp.

LIPPERT, B. E. 1967. Sexual reproduction in *Closterium moniliferum* and *C. ehrenbergii*. J. Phycol. 3: 182–198.

————. 1975. Some factors affecting conjugation in *Closterium*. Beih. z. Nova Hedw. 42: 171–177.

LIST, H. 1930. Die Entwicklungsgeschichte von *Cladophora glomerata* Kützing. Arch. Protistenk. 72: 453–481.

LITTLE, M. G. 1973. The zonation of marine supralittoral blue-green algae. Br. Phyc. J. 8: 47–50.

LITTLER, M. M. 1972. The crustose Corallinaceae. Oceanogr. Mar. Biol. Ann. Rev. 10: 311–347.

————. 1973A. The population and community structure of Hawaiian fringing-reef crustose Corallinaceae (Rhodophyta, Cryptonemiales). J. Exp. Mar. Biol. Ecol. 11: 103–120.

————. 1973B. The distribution, abundance, and communities of deepwater Hawaiian crustose Corallinaceae (Rhodophyta, Cryptonemiales). Pac. Sci. 27: 281–289.

LOEBLICH, A. R., JR., AND A. R. LOEBLICH, III. 1966. Index to the genera, subgenera, and sections of the Pyrrhophyta. Studies in Tropical Oceanography, No. 3 (Univ. of Miami). ix + 94 pp.

————, AND H. TAPPAN. 1966. Annotated index and bibliography of the Calcareous Nannoplankton. Phycologia 5: 81–216.

LOEBLICH, A. R., III. 1970. The amphiesma or dinoflagellate cell covering. North American Paleontological Convention, Chicago, 1969, Proc. G: 867–929.

————. 1974. *Pyrocystis lunula* and its relationship to *Sporodinium*. J. Protozool. 21(Suppl.): 435.

_____. 1976. Dinoflagellate evolution: speculation and evidence. J. Protozool. 23: 13–28.

_____, L. A. LOEBLICH, H. TAPPAN, AND A. R. LOEBLICH, JR. 1968. Annotated index of fossil and recent silicoflagellates and Ebridians with descriptions and illustrations of validly proposed taxa. Geol. Soc. Am. Memoir 106, 319 pp.

LOEBLICH, L. A., AND A. R. LOEBLICH, III. 1975. The organism causing New England red tides: *Gonyaulax excavata*. Proc. First Intern. Conf. on Toxic Dinoflagellate Blooms. Mass. Science and Technology Foundation. Wakefield, Mass. pp. 208–224.

LOFTHOUSE, P. F., AND B. CAPON. 1975. Ultrastructural changes accompanying mitosporogenesis in *Ectocarpus parvus*. Protoplasma 84: 83–99.

LOISEAUX, S. 1964. Sur une nouvelle espèce de *Myrionema* des environs de Roscoff et son cycle. C. R. Acad. Sci. (Paris). 258: 2383–2385.

_____. 1967A. Morphologie et cytologie des Myrionémacées. Critères taxonomiques. Rev. Gén. Bot. 74: 329–347.

_____. 1967B. Recherches sur les cycles de développement des Myrionématacées (Phéophycées). I–II. Hecatonématées et Myrionématées. Rev. Gén. Bot. 74: 529–576.

_____. 1968. Recherches sur les cycles de developpement des Myrionématacées (Pheophycées). III. Tribu des Ralfsiées. IV. Conclusions générales. Rev. Gén. Bot. 75: 295–318.

_____. 1969. Sur une espece de *Myriotrichia* obtenue en culture a partir de zoides d'*Hecatonema maculans* Sauv. Phycologia 8: 11–15.

_____. 1970A. *Streblonema anomalum* S. et G. and *Compsonema sporangiiferum* S. et G., stages in the life history of a minute *Scytosiphon*. Phycologia 9: 185–191.

_____. 1970B. Notes on several Myrionemataceae from California using culture studies. J. Phycol. 6: 248–260.

_____. 1972. Variations des cycles chez les Myrionématacées et leur signification phylogénétique. Soc. Bot. Fr., Mémoires 1972: 105–116.

LOKHORST, G. M. 1968. Kritische beschowing de over genuskriteria van *Ulothrix* Kützing en *Hormidium* Kützing (Ulotrichales). Jaarboek Kon. Ned. Bot. Ver. 1968 (1969): 49–50.

_____. 1974. Survey of taxonomic studies of the freshwater species of *Ulothrix* in the Netherlands (Abstract of a Ph. D. dissertation; Netherlands). 16 pp.

_____, AND M. VROMAN. 1972. Taxonomic study of three freshwater *Ulothrix* species. Acta Bot. Neerl. 21: 449–480.

_____, AND _____. 1974A. Taxonomic studies on the genus *Ulothrix* (Ulotrichales, Chlorophyceae). II. Acta Bot. Neerl. 23: 369–398.

_____, AND _____. 1974B. Taxonomic study on the genus *Ulothrix* (Ulotrichales, Chlorophyceae) III. Acta Bot. Neerl. 23: 561–602.

LOPPES, R., R. MATAGNE, AND P. J. STRUKERT. 1972. Complementation at the *Arg*.-7 locus in *Chlamydomonas reinhardii*. Heredity 28: 239–251.

LORCH, D. W., AND A. WEBER. 1972. Über die Chemie der Zellwand von *Pleurotaenium trabecula* var. *rectum* (Chlorophyta). Arch. Mikrobiol. 83: 129–140.

LØVLIE, A. 1968. On the use of a multicellular alga (*Ulva mutabilis* Føyn) in the study of general aspects of growth and differentiation. Nytt Mag. Zoologi 16: 39–49.

_____, AND T. BRÅTEN. 1968. On the division of cytoplasm and chloroplast in the multicellular green alga *Ulva mutabilis* Føyn. Exper. Cell Res. 51: 211–220.

_____, AND _____. 1970. On mitosis in the multicellular alga *Ulva mutabilis* Føyn. J. Cell Sci. 6: 109–129.

LUCAS, I. A. N. 1970A. Observations on the fine structure of the Cryptophyceae. I. The genus *Cryptomonas*. J. Phycol. 6: 30–38.

_____. 1970B. Observations on the ultrastructure of representatives of the genera *Hemiselmis* and *Chroomonas* (Cryptophyceae). Br. Phycol. J. 5: 29–37.

LUKAS, K. J. 1974. Two species of the chlorophyte genus *Ostreobium* from skeletons of Atlantic and Caribbean reef corals. J. Phycol. 10: 331–335.

LUND, J. W. G. 1962. Soil algae. *In* R. A. Lewin (Ed.). Physiology and Biochemistry of Algae. pp. 759–770.

LUND, S. 1966. On a sporangia-bearing microthallus of *Scytosiphon lomentaria* from nature. Phycologia 6: 67–78.

LÜNING, K. 1970. Cultivation of *Laminaria hyperborea in situ* and in continuous darkness under laboratory conditions. Helgoländ. Wiss. Meeresunters. 20: 79–88.

————. 1971. Seasonal growth of *Laminaria hyperborea* under recorded underwater light conditions near Helgoländ. *In* Proc. 4th European Marine Biology Symp. (Ed., D. J. Crisp). University Press, Cambridge. pp. 347–361.

————. 1975. Crossing experiments in *Laminaria saccharina* from Helgoländ and from the Isle of Man. Helgoländ. Wiss. Meeresunters. 27: 108–114.

————, AND M. J. DRING. 1972. Reproduction induced by blue light in gametophytes of *Laminaria saccharine*. Planta 104: 252–256.

————, AND ————. 1973. The influence of light quality on the development of the brown algae *Petalonia* and *Scytosiphon*. Br. Phycol. J. 8: 333–8.

————, AND ————. 1975. Reproduction, growth and photosynthesis of gametophytes of *Laminaria saccharina* grown in blue and red light. Mar. Biol. 29: 195–200.

————, K. SCHMITZ, AND J. WILLENBRINK. 1972. Translocation of ^{14}C-labeled assimilates in two *Laminaria* species. Proc. Intern. Seaweed Symp. 7: 420–5.

LYMAN, H., H. T. EPSTEIN, AND J. A. SCHIFF. 1959. Ultraviolet inactivation and photoreactivation of chloroplast development in *Euglena* without cell death. J. Protozool. 6: 204–265.

MACHLIS, L. 1962. The nutrition of certain species of the green alga *Oedogonium*. Am. J. Bot. 49: 171–177.

————. 1972. The coming of age of sex hormones in plants. Mycologia 64: 235–247.

————. 1973. The effects of bacteria on the growth and reproduction of *Oedogonium cardiacum*. J. Phycol. 9: 342–344.

————, G. G. C. HILL, K. E. STEINBACK, AND W. REED. 1974. Some characteristics of the sperm attractant in *Oedogonium cardiacum*. J. Phycol. 10: 199–204.

MACKIE, W., AND R. D. PRESTON. 1974. Cell wall and intercellular region polysaccharides. Chapter 2. *In* Stewart, W. D. P. (Ed.). Algal Physiology and Biochemistry. Univ. Calif. Press. Berkeley and Los Angeles. 989 pp.

MACMILLAN, C. 1902. Observations on *Pterygophora*. Minnesota Bot. Stud., ser. 2, 2: 723–741.

MAEDA, M., K. KURODA, Y. IRIKI, M. CHIHARA, K. NISIZAWA, AND T. MIWA. 1966. Chemical nature of major cell wall constituents of *Vaucheria* and *Dichotomosiphon* with special reference to their phylogenetic positions. Bot. Mag., Tokyo 79: 634–643.

MAEKAWA, F. 1960. A new attempt in phylogenetic classification of plant kingdom. J. Fac. Sci. Univ. Tokyo, Botany, 7: 543–569.

MAGNE, F. 1953. La méiose chez le *Sporochnus pedunculatus* C. A. Ag. (Sporochnale, Phéophycée). C. R. Acad. Sci. (Paris). 236: 1596–1598.

————. 1960. Le *Rhodochaete parvula* Thuret (Bangioidée) et sa reproductive sexuée. Cahiers Biol. Mar. 1: 407–420.

————. 1961. Sur le cycle cytologique de *Nemalion helminthoides* (Velley) Batters. C. R. Acad. Sci. (Paris). 252: 157–159.

————. 1964. Recherches caryologiques chez les Floridées (Rhodophycées). Cahiers Biol. Mar. 5: 461–671.

MAGUIRE, M. 1976. Mitotic and meiotic behavior of the chromosomes of the octet strain of *Chlamydomonas reinhardtii*. Genetica 46: 479–502.

MAIWALD, M. 1971. A comparative ultrastructural study of *Pyramimonas montana* Geitler and a *Pyramimonas* sp. Arch. Protistenk. 113: 334–344.

MALINOWSKI, K. C., AND J. RAMUS. 1973. Growth of the green alga *Codium fragile* in a Connecticut estuary. J. Phycol. 9: 102–110.

MANDELLI, E. F. 1968. Carotenoid pigments of the dinoflagellate *Glenodinium foliaceum* Stein. J. Phycol. 4: 347–348.

MANDRA, Y. T. 1968. Silicoflagellates from the Cretaceous, Eocene, and Miocene of California, U.S.A. Proc. Calif. Acad. Sci., 4th Ser., 36: 231–277.

MANN, K. H. 1972. Ecological energetics of the seaweed zone in a marine bay on the Atlantic coast of Canada. I. Zonation and biomass of seaweeds. Mar. Biol. 12: 1–10.

MANTON, I. 1955. Observations with the electron microscope on *Synura caroliniana* Whitford. Proc. Leeds Philosophical Soc. (Sci. Sec.) 6: 306–316.

_____. 1959. Observations on the internal structure of the spermatozoid of *Dictyota*. J. Exper. Bot. 10: 448–461.

_____. 1964A. Observations with the electron microscope on the division cycle in the flagellate *Prymnesium parvum*. Carter. J. Roy. Microsc. Soc., Ser. 3., 83: 317–325.

_____. 1964B. Further observations on the fine structure of the haptonema in *Prymnesium parvum*. Arch. Mikrobiol. 49: 315–330.

_____. 1964C. A contribution toward understanding 'the primitive Fucoid.' New Phytol. 63: 244–254.

_____. 1966A. Observations on scale production in *Pyramimonas amylifera* Conrad. J. Cell Sci. 1: 429–438.

_____. 1966B. Further observations on the fine structure of *Chrysochromulina chiton* with special reference to the pyrenoid. J. Cell Sci. 1: 187–192.

_____. 1966C. Observations on scale production in *Prymnesium parvum*. J. Cell Sci. 1: 375–380.

_____. 1967A. Further observations on the fine structure of *Chrysochromulina chiton* with special reference to the haptonema, 'peculiar' Golgi structure and scale production. J. Cell Sci. 2: 265–272.

_____. 1967B. Further observations on scale formation in *Chrysochromulina chiton*. J. Cell Sci. 2: 411–418.

_____. 1968. Observations on the microanatomy of the type species of *Pyramimonas* (*P. tetrarhyncus*) Schmarda. Proc. Linn. Soc. Lond. (Bot.) 179: 147–152.

_____. 1972A. Preliminary observations on *Chrysochromulina mactra* sp. nov. Br. Phycol. J. 7: 21–35.

_____. 1972B. Observations on the biology and micro-anatomy of *Chrysochromulina megacylindra* Leadbeater. Br. Phycol. J. 7: 235–248.

_____, AND B. CLARKE. 1956. Observations with the electron microscope on the internal structure of the spermatozoid of *Fucus*. J. Exp. Bot. 7: 416–32.

_____, _____, AND A. D. GREENWOOD. 1953. Further observations with the electron microscope on spermatozoids in the brown algae. J. Exp. Bot. 4: 319–29.

_____, K. KOWALLIK, AND H. A. VON STOSCH. 1969A. Observations on the fine structure and development of the spindle at mitosis and meiosis in a marine centric diatom (*Lithodesmium undulatum*). I. Preliminary survey of mitosis in spermatogonia. J. Microsc. 89: 295–320.

_____, _____. 1969B. Observations on the fine structure and development of the spindle at mitosis and meiosis in a marine centric diatom (*Lithodesmium undulatum*). II. The early meiotic stages in male gametogenesis. J. Cell Sci. 5: 271–298.

_____, _____, AND _____. 1970A. Observations on the fine structure and development of the spindle at mitosis and meiosis in a marine centric diatom (*Lithodesmium undulatum*). III. The later stages of meiosis I in male gametogenesis. J. Cell Sci. 6: 131–157.

_____, _____, AND _____. 1970B. Observations on the fine structure and development of the spindle at mitosis and meiosis in a marine centric diatom (*Lithodesmium undulatum*). IV. The second meiotic division and conclusion. J. Cell Sci. 7: 407–443.

_____, AND G. F. LEEDALE. 1961A. Further observations on the fine structure of *Chrysochromulina ericina* Parke et Manton. J. Mar. Biol. Assn. U. K. 41: 145–155.

_____, AND _____. 1961B. Observations on the fine structure of *Paraphysomonas vestita*, with special reference to the Golgi apparatus and the origin of scales. Phycologia. 1: 37–57.

_____, AND _____. 1969. Observations on the microanatomy of *Coccolithus pelagicus* and *Cricosphaera carterae*, with special reference to the origin and nature of coccoliths and scales. J. Mar. Biol. Assn. U. K. 49: 1–16.

_____, K. OATES, AND M. PARKE. 1963. Observations on the fine structure of the *Pyramimonas* stage of *Halosphaera* and preliminary observations on three species of *Pyramimonas*. J. Mar. Biol. Assn. U. K. 43: 225–238.

_____, AND H. A. VON STOSCH. 1966. Observations on the fine structure of the male gamete of the marine centric diatom *Lithodesmium undulatum*. J. Roy. Microsc. Soc. 85: 119–134.

_____, J. SUTHERLAND, AND B. S. C. LEADBEATER. 1975. Four new species of choanoflagellates from Arctic Canada. Proc. Roy. Soc. London, Ser. B, 189: 15–27.

MARANO, F. 1976. Étude ultrastructurale de la division chez *Dunaliella*. J. Microscopie Biol. Cell. 25: 279–282.

MARBACH, I., AND A. M. MAYER. 1970. Direction of phototaxis in *Chlamydomonas reinhardii* and its relation to cell metabolism. Phycologia 9: 255–260.

MARCHANT, H. J. 1972. Pyrenoids of *Vaucheria woroniniana* Heering. Br. Phycol. J. 7: 81–84.

_____. 1974A. Mitosis, cytokinesis and colony formation in *Pediastrum boryanum*. Ann. Bot. 38: 883–888.

_____. 1974B. Mitosis, cytokinesis and colony formation in the green alga *Sorastrum*. J. Phycol. 10: 107–120.

_____. 1977. Ultrastructure, development and cytoplasmic rotation of seta-bearing cells of *Coleochaete scutata* (Chlorophyceae). J. Phycol. 13: 28–36.

_____, AND J. D. PICKETT-HEAPS. 1970–1972. Ultrastructure and differentiation of *Hydrodictyon reticulatum*. 1970. I., Mitosis in the coenobium. Austral. J. Biol. Sci. 23: 1173–1186; II. 1971. Formation of zoids within the coenobium. Austral. J. Biol. Sci. 24: 471–486. III. 1972A. Formation of the vegetative daughter net. Austral. J. Biol. Sci. 25: 265–278. IV. 1972B. Conjugation of gametes and the development of zygospores and azygospores. Austral. J. Biol. Sci. 25: 279–291. V. 1972C. Development of polyhedra. Austral. J. Biol. Sci. 25: 1187–97. VI. 1972D. Formation of the germ net. Austral. J. Biol. Sci. 25: 1199–1213.

_____, AND _____. 1973. Mitosis and cytokinesis in *Colechaete scutata*. J. Phycol. 9: 461–471.

_____, AND _____. 1974. The effect of colchicine on colony formation in the algae *Hydrodictyon*, *Pediastrum* and *Sorastrum*. Planta 116: 291–300.

MARKHAM, J. W. 1968. Studies on the haptera of *Laminaria sinclairii* (Harvey) Farlow, Anderson et Eaton. Syesis 1: 125–131.

_____. 1972. Distribution and taxonomy of *Laminaria sinclairii* and *L. longipes* (Phaeophyceae, Laminariales). Phycologia 11: 147–157.

_____. 1973. Observations on the ecology of *Laminaria sinclairii* on three northern Oregon beaches. J. Phycol. 9: 336–341.

_____, AND P. R. NEWROTH. 1972. Observations on the ecology of *Gymnogongrus linearis* and related species. Proc. Intern. Seaweed Symp. 7: 127–130.

MARTIN, N. C., AND U. W. GOODENOUGH. 1975. Gametic differentiation in *Chlamydomonas reinhardtii* I. Production of gametes and their fine structure. J. Cell Biol. 67: 587–605.

MARTIN, T. C., AND J. T. WYATT. 1974. Extracellular investments in blue-green algae with particular emphasis on the genus *Nostoc*. J. Phycol. 10: 204–210.

MASAKI, T. 1968. Studies on the Melobesioideae of Japan. Mem. Fac. Fish., Hokkaido Univ. 16: 1–80.

MASSALSKI, A., G. F. LEEDALE. 1969. Cytology and ultrastructure of the Xanthophyceae. I. Comparative morphology of the zoospores of *Bumilleria sicula* Borzi and *Tribonema vulgare* Pascher. Br. Phycol. J. 4: 159–180.

_____, F. R. TRAINOR, AND L. E. SHUBERT. 1974. Wall ultrastructure of *Scenedesmus* culture N. 46. Arch. Mikrobiol. 96: 146–153.

MATHIESON, A. C. 1966. Morphological studies of the marine brown alga *Taonia lennebackerae* Farlow ex J. Agardh. I. Sporophytes, abnormal gametophytes and vegetative reproduction. Nova Nedw. 12: 65–79.

_____, AND R. L. BURNS. 1971. Ecological studies of economic red algae. I. Photosynthesis and respiration of *Chondrus crispus* Stackhouse and *Gigartina stellata* (Stackhouse) Batters. J. Exp. Mar. Biol. Ecol. 7: 197–206.

_____, AND _____. 1975. Ecological studies of economic red algae. V. Growth and reproduction of natural and harvested populations of *Chondrus crispus* Stackhouse in New Hampshire. J. Exp. Mar. Biol. Ecol. 17: 137–156.

_____, AND C. J. DAWES. 1974. Ecological studies of Floridean *Eucheuma* (Rhodophyta, Gigartinales). II. Photosynthesis and respiration. Bull. Mar. Sci. 23: 274–285.

_____, _____, AND H. J. HUMM. 1969. Contributions to marine algae of Newfoundland. Rhodora 71: 110–159.

_____, AND J. S. PRINCE. 1973. Ecology of *Chondrus crispus* Stackhouse. Proc. Nova Scotia Inst. Sci. 27 (Suppl.): 53–79.

_____, AND E. TVETER. 1975. Carrageenan ecology of *Chondrus crispus* Stackhouse. Aquatic Bot. 1: 25–43.

MATTONI, R. H. T. 1968. Trends in algal genetics. Chapter 11, pp. 201–211, *In* D. F. Jackson (Ed.) Algae, Man and the Environment. Syracuse University Press. Syracuse, New York. 554 pp.

MATTOX, K. R., AND H. C. BOLD. 1962. Phycological Studies. III. The taxonomy of certain Ulotrichacean algae. Univ. Texas Pub. No. 6222, Austin, Texas. 67 pp.

_____, AND K. D. STEWART. 1973. Observations on the zoospores of *Pseudendoclonium basiliense* and *Trichosarcina polymorpha* (Chlorophyceae). Can. J. Bot. 51: 1425–1430.

_____, AND _____. 1974. A comparative study of cell division in *Trichosarcina polymorpha* and *Pseudendoclonium basiliense* (Chlorophyceae). J. Phycol. 10: 447–456.

_____, _____, AND G. L. FLOYD. 1974. The cytology and classification of *Schizomeris leibeinii*. I. The vegetative thallus. Phycologia 13: 63–69.

MAY, V., AND E. J. MCBARRON. 1973. Occurrence of the blue-green alga *Anabaena circinalis* Rabenh. in New South Wales and toxicity to mice and honey bees. J. Austral. Inst. Agr. Sci. 39: 264–266.

MAYHOUB, H. 1974. Reproduction sexuée et cycle du développement de *Pseudobryopsis myura* (Ag.) Berthold (Chlorophycée, Codiale). C. R. Acad. Sci. (Paris). 278D: 867–870.

_____. 1975. Reproduction sexuée et cycle du développement de l'*Anadyomene stellata* (Wulf.) Ag. de la Mediterranee orientale. C. R. Acad. Sci. (Paris). 280D: 587–590.

MCBRIDE, D. L., AND K. COLE. 1969. Ultrastructural characteristics of the vegetative cell of *Smithora naiadum* (Rhodophyta). Phycologia 8: 177–186.

_____, AND _____. 1971. Electron microscopic observations on the differentiation and release of monospores in the marine red alga *Smithora naiadum*. Phycologia 10: 49–61.

MCBRIDE, G. E. 1968. A classroom demonstration of zoospore production in the green alga *Schizomeris leibleinii* Kuetz. J. Phycol. 4: 251–252.

_____. 1970. Cytokinesis and ultrastructure in *Fritschiella tuberosa* Iyengar. Arch. Protistenk. 112: 365–375.

_____. 1974. The seta-bearing cells of *Coleochaete scutata* (Chlorophyceae, Chaetophorales). Phycologia 13: 271–285.

_____, J. LA BOUNTY, J. ADAMS, AND M. BERUS. 1974. The totipotency and relationship of seta-bearing cells to thallus development in the green alga *Coleochaete scutata*. Laser microbeam study. Dev. Biol. 37: 90–99.

MCCANDLESS, E. L., J. S. CRAIGIE, AND J. E. HANSEN. 1975. Carrageenans of gametangial and tetrasporangial stages of *Iridaea cordata* (Gigartinaceae). Can. J. Bot. 53: 2315–2318.

_____, _____, AND J. A. WALTER. 1973. Carrageenans in the gametophytic and sporophytic stages of *Chondrus crispus*. Planta 112: 201–212.

MCCRACKEN, M. D. 1970. Differentiation in *Volvox*. Carolina Tips 33: 37–39.

_____, V. W. PROCTOR, AND A. T. HOTCHKISS. 1966. Attempted hybridization between monoecious and dioecious clones of *Chara*. Am. J. Bot. 53: 937–940.

_____, AND R. C. STARR. 1970. Induction and development of reproductive cells in the K-32 strains of *Volvox rouseletii*. Arch. Protistenk. 112: 262–282.

MCCULLY, M. E. 1966. Histological studies on the genus *Fucus*. I. Light microscopy of the mature vegetative plant. Protoplasma 62: 287–305.

_____. 1968A. Histological studies on the genus *Fucus*. II. Histology of the reproductive tissues. Protoplasma 66: 205–230.

_____. 1968B. Histological studies on the genus *Fucus*. III. Fine structure and possible functions of the epidermal cells of the vegetative thallus. J. Cell Sci. 3: 1–16.

MCDONALD, K. 1972. The ultrastructure of mitosis in the marine red alga *Membranoptera platyphylla*. J. Phycol. 8: 156–166.

_____, AND J. D. PICKETT-HEAPS. 1976. Ultrastructure and differentiation in *Cladophora glomerata*. I. Cell division. Am. J. Bot. 63: 592–601.

McElhenney, T. R., H. C. Bold, R. M. Brown, Jr., and J. P. McGovern. 1962. Algae. a cause of inhalant allergy in children. Ann. Allergy 20: 739–743.

McGovern, J. P., T. J. Hayward, and T. R. McElhenney. 1966. Airborne algae and their allergenicity. II. Clinical and laboratory multiple correlation studies with four genera. Ann. Allergy 24: 145–149.

McIntyre, A. 1967. Coccoliths as Paleoclimatic indicators of Pleistocene glaciation. Science 158: 1314–1317.

————, and A. W. H. Bé. 1967. Modern Coccolithophoridae of the Atlantic Ocean. I. Placoliths and Cyrtoliths. Deep-Sea Res. 14: 561–597.

McKay, H. H. 1933. The life-history of *Pterygophora californica* Ruprecht. Univ. Calif. Pub. Bot. 17: 111–148.

McLachlan, J. 1973. Growth Media-Marine. Chapter 2 in Stein, J. R. (Ed.). Handbook of Phycological Methods. University Press. Cambridge. xii + 448 pp.

————, and L. C.-M. Chen. 1972. Formation of adventive embryos from rhizoidal filaments in sporelings of four species of *Fucus* (Phaeophyceae). Can. J. Bot. 50: 1841–1844.

————, ————, and T. Edelstein. 1971A. The culture of four species of *Fucus* under laboratory conditions. Can. J. Bot. 49: 1463–1469.

————, ————, and ————. 1971B. The life history of *Microspongium* sp. Phycologia 10: 83–87.

————, A. G. McInnes, and M. Falk. 1965. Studies on the chitan (chitin: poly-*N*-acetylglucosamine) fibers of the diatom *Thalassiosira fluviatilis* Hustedt. I. Production and isolation of chitan fibers. Can. J. Bot. 43: 707–713.

McLaughlin, J. J. A. 1958. Euryhaline chrysomonads: nutrition and toxigenesis in *Prymnesium parvum* with notes on *Isochrysis galbana* and *Monochrysis lutheri*. J. Protozool. 5: 75–81.

————, and P. A. Zahl. 1966. Endozoic algae. *In* Symbiosis, Vol. 1 (Ed., M. S. Henry). Academic Press, New York. pp. 257–295.

McLean, R. J., 1967. Primary and secondary carotenoids of *Spongiochloris typica*. Physiol. Plant. 20: 41–47.

————. 1968. New taxonomic criteria in the classification of *Chlorococcum* species. I. Pigmentation. J. Phycol. 4: 328–332.

————, and H. B. Bosmann. 1975. Cell-cell interactions: enhancement of glycosyl transferase ectoenzyme systems during *Chlamydomonas* gametic contact. Proc. Nat. Acad. Sci. U.S.A. 72: 310–313.

————, and R. M. Brown, Jr. 1974. Cell surface differentiation of *Chlamydomonas* during gametogenesis. I. Mating and concanavalin A agglutinability. Dev. Biol. 36: 279–285.

————, C. Laurende, and R. M. Brown, Jr. 1974. The relationship of gamone to the mating reaction in *Chlamydomonas moewusii*. Proc. Nat. Acad. Sci. U.S.A. 71: 2610–2613.

————, and G. F. Pessoney. 1971. Formation and resistance of akinetes of *Zygnema*. *In* Contributions in Phycology (Ed., B. C. Parker and R. M. Brown, Jr.). Allen Press, Lawrence, Kansas. pp. 145–152.

McMurray, L., and J. W. Hastings. 1972. No desychronization among four circadian rhythms in the unicellular alga *Gonyaulax polyedra*. Science 175: 1137–1139.

————, and ————. Circadian rhythms: mechanisms of luciferase activity changes in *Gonyaulax*. Biol. Bull. 143: 196–206.

McVittie, A., and D. R. Davies. 1971. The location of the Mendelian linkage groups in *Chlamydomonas reinhardtii*. Molec. Gen. Genetics 112: 225–228.

Meeks, J. C. 1974. Chlorophylls. Chapter 5 *In* Stewart, W. D. P. (Ed.) Algal Physiology and Biochemistry. Univ. Calif. Press. Berkeley and Los Angeles. 989 pp.

Meeuse, B. J. D. 1956. Free sulfuric acid in the brown alga, *Desmarestia*. Biochim. Biophys. Acta 19: 372–374.

————. 1962. Storage products. *In* Physiology and Biochemistry of Algae (Ed., R. A. Lewin). Academic Press, New York. pp. 289–311.

Meinesz, A. 1969. Sur la reproduction sexuée de l'*Udotea petiolata* (Turra) Boerg. C. R. Acad. Sci. (Paris). 269D: 1063–1065.

_____. 1972A. Sur la croissance et la développement du *Penicillus capitatus* Lamarck forma *mediterranea* (Decaisne) P. et H. Huvé (Caulerpale, Udoteacée). C. R. Acad. Sci. (Paris). 269D: 667–669.

_____. 1972B. Sur le cycle de l'*Udotea petiolata* (Turra) Boergesen (Caulerpale, Udotéacée). C. R. Acad. Sci. (Paris). 275D: 1975–1977.

_____. 1972C. Sur le cycle de l'*Halimeda tuna* (Ellis et Solander) Lamouroux (Udotéacée, Caulerpale). C. R. Acad. Sci. (Paris). 275D: 1363–1365.

MELACK, J. M., AND P. KILHAM. 1974. Photosynthetic rates of phytoplankton in east African alkaline, saline lakes. Limnol. Oceanogr. 19: 743–755.

MELKONIAN, M. 1975. The fine structure of the zoospores of *Fritschiella tuberosa* Iyeng. (Chaetophorineae, Chlorophyceae) with special reference to the flagellar apparatus. Protoplasma 86: 391–404.

_____, AND A. WEBER. 1975. Einfluss von Kinetin auf des Wachstum von *Fritschiella tuberosa* Iyeng. (Chaetophorineae, Chlorophyceae) in axenischer Massenkultur. Z. Pflanzenphysiol. 76: 120–129.

MEREDITH, R. F., AND R. C. STARR. 1975. The genetic basis of male potency in *Volvox carteri* f. *nagariensis* (Chlorophyceae). J. Phycol. 11: 265–272.

MESLAND, D. A. M. 1976. Mating in *Chlamydomonas eugametos*; a scanning electron microscopical study. Arch. Mikrobiol. 109: 31–35.

MESSER, G., AND Y. BEN-SHAUL. 1969. Fine structure of *Peridinium westii* Lemm., a freshwater dinoflagellate. J. Protozool. 16: 272–280.

MICHANEK, G. 1971. A preliminary appraisal of world seaweed resources. Food and Agriculture Organization of the United Nations, Fisheries Circular No. 128, ii + 37 pp.

MIGNOT, J. P., L. JOYON, AND E. G. PRINGSHEIM. 1968. Compléments a l'etude cytologique des Cryptomonadines. Protistologica 4: 493–506.

MIKAMI, H. 1965. A systematic study of the Phyllophoraceae and Gigartinaceae from Japan and its vicinity. Sci. Papers, Inst. Algol. Res., Hokkaido Univ. 5(2): 181–285.

_____. 1971. *Congregatocarpus*, a new genus of the Delesseriaceae (Rhodophyta). Bot. Mag., Tokyo 84: 243–246.

MILLER, D. H., D. T. A. LAMPORT, AND M. MILLER. 1972. Hydroxyproline heterooligosaccharides in *Chlamydomonas*. Science 176: 918–920.

MILLER, M. M., AND N. J. LANG. 1968. The fine structure of akinete formation and germination in *Cylindrospermum*. Arch. Mikrobiol. 60: 303–313.

MILLIGER, L. E., K. W. STEWART, AND J. K. G. SILVEY. 1971. The passive dispersal of viable algae, protozoans, and fungi by aquatic and terrestrial Coleopters. Ann. Entomol. Soc. Am. 64: 36–45.

MILLINGTON, W. F., AND S. R. GAWLEK. 1970. Ultrastructure and initiation of wall pattern in *Pediastrum boryanum*. Am. J. Bot. 57: 552–561.

MIN-THIEN, U., AND H. B. S. WOMERSLEY. 1976. Studies on southern Australian taxa of Solieriaceae, Rhabdoniaceae and Rhodophyllidaceae. Austral. J. Bot. 24: 1–166.

MIURA, A. 1961. A new species of *Porphyra* and its *Conchocelis*-phase in nature. J. Tokyo Univ. Fish. 47: 305–311.

MIX, M. 1966. Licht- und elecktromikroskopische Untersuchungen an Desmidiaceen. XII. Zur Feinstruktur der Zellwände und Mikrofibillen einiger Desmidiaceen von *Cosmarium*- typ. Arch. Mikrobiol. 55: 116–133.

_____. 1969. Zur Feinstruktur der Zellwände in der Gattung *Closterium* (Desmidiaceae) unter besonderer Berucksichtigung des Porensystems. Arch. Mikrobiol. 68: 306–325.

_____. 1975. Die Feinstruktur der Zellwände der Conjugaten und ihre systematische Beduntung. Beih. z. Nova Hedw. 42: 179–194.

MOESTRUP, Ø. 1970. On the fine structure of the spermatozoids of *Vaucheria sescuplicaria* and on the later stages of spermatogenesis. J. Mar. Biol. Assn. U. K. 50: 513–523.

_____. 1972. Observations on the fine structure of the spermatozoids and vegetative cells of the green alga *Golenkinia*. Br. Phycol. J. 7: 169–183.

_____. 1975. Some aspects of sexual reproduction in eucaryotic algae. *In* The biology of the male gamete. (Ed., J. G. Duckett and P. A. Racey). Biol. J. Linnean Soc., Suppl. 1, 7: 23–35.

_____, AND L. R. HOFFMAN. 1973. Ultrastructure of the green alga *Dichotomosiphon tuberosus* with special reference to the occurrence of striated tubules in the chloroplast. J. Phycol. 9: 430–437.

_____, AND _____. 1975. A study of the spermatozoids of *Dichotomosiphon tuberosus* (Chlorophyceae). J. Phycol. 11: 225–235.

_____, AND H. A. THOMSEN. 1974. An ultrastructural study of the flagellate *Pyramimonas orientalis* with particular emphasis on Golgi apparatus activity and the flagellar apparatus. Protoplasma 81: 247–269.

MOEWUS, F. 1933. Untersuchungen über die Sexualität und Entwicklung von Chlorophyceen. Arch. Protistenk. 80: 469–520.

MOHR, J. L., N. J. WILIMOVSKY, AND E. Y. DAWSON. 1957. An arctic Alaskan kelp bed. Arctic 10: 45–52.

MOIKEHA, S. N., AND G. W. CHU. 1971. Dermatitis-producing alga *Lyngbya majuscula* Gomont in Hawaii II. Biological properties of the toxic factor. J. Phycol. 7: 8–13.

_____, _____, AND L. R. BERGER. 1971. Dermatitis-producing alga *Lyngbya majuscula* in Hawaii. I. Isolation and chemical characterization of the toxic factor. J. Phycol. 7: 4–8.

MOLLENHAUER, D. (Ed.). 1975. Erstes Internationales Desmidiaceen-Symposium. Sept. 1971. Beih. z. Nova Hedw. 42: 1–316.

MONAHAN, T. J., AND F. R. TRAINOR. 1970. Stimulating properties of filtrate from the green alga *Hormotila blennista*. I. Description. J. Phycol. 6: 263–269.

_____, AND _____. 1971. II. Fractionation of filtrate. J. Phycol. 7: 170–176.

MOORE, L. B. 1951. Reproduction in *Halopteris* (Sphacelariales). Ann. Bot. 15: 265–278.

MOORE, L. F., AND J. A. TRAQUAIR. 1976. Silicon, a required nutrient for *Cladophora glomerata* (L.) Kütz. Planta 128: 179–182.

MORI, M. 1975. Studies on the genus *Batrachospermum* in Japan. Jap. J. Bot. 20: 461–485.

MORNIN, L., AND D. FRANCIS. 1967. The fine structure of *Nematodinium armatum*, a naked dinoflagellate. J. Microscopie 6: 759–772.

MOSS, B. L. 1964. Wound healing and regeneration in *Fucus vesiculosus*. Proc. Intern. Seaweed Symp. 4: 117–122.

_____. 1965. Apical dominance in *Fucus vesiculosus*. New Phytol. 64: 387–92.

_____. 1966. Polarity and apical dominance in *Fucus vesiculosus*. Br. Phycol. Bull. 3: 31–35.

_____. 1967A. The apical meristem in *Fucus*. New Phytol. 66: 67–74.

_____. 1967B. The culture of fertile tissue of *Fucus vesiculosus*. Br. Phycol. Bull. 3: 209–212.

_____. 1968. The transition from vegetative to fertile tissue in *Fucus vesiculosus*. Br. Phycol. Bull. 3: 567–573.

_____. 1970. Meristems and control of growth in *Ascophyllum nodosum* (L.) LeJol. New Phytol. 69: 253–260.

_____. 1971. Meristems and morphogenesis in *Ascophyllum nodosum* ecad *Mackaii* (Cotton). Br. Phycol. J. 6: 187–193.

_____. 1974. Morphogenesis. Chapter 28 in Algal Physiology and Biochemistry (Ed., W. D. P. Stewart), Univ. of Calif. Press, Berkeley and Los Angeles. 989 pp.

MULLAHY, J. H. 1952. The morphology and cytology of *Lemanea australis* Atk. Bull. Torrey Bot. Club 78: 393–406.

MÜLLER, D. 1962. Über jahres-und lunarperiodische Erscheinungen bei einigen Braunalgen. Bot. Mar. 4: 140–155.

_____. 1967A. Generationswechsel, Kernphasenwechsel und Sexualität der Braunalge *Ectocarpus siliculosus* im Kulturversuch. Planta 75: 39–54.

_____. 1967B. Ein leicht flüchtiges Gyno-Gamon der Braunalge *Ectocarpus siliculosus*. Naturwissensch. 54: 496–497.

_____. 1968. Versuche zur Charakterisierung eines Sexual-Lockstoffes bei der Braunalge *Ectocarpus siliculosus*. I. Methoden, Isolierung und gaschromatographischer Nachweis. Planta 81: 160–168.

_____. 1972A. Life cycle of the brown alga *Ectocarpus fasciculatus* var. *refractus* (Kütz.) Ardis. (Phaeophyceae, Ectocarpales) in culture. Phycologia 11: 11–13.

_____. 1972B. Studies on reproduction in *Ectocarpus siliculosus*. Soc. Bot. Fr., Mém. 1972: 87–98.

_____. 1974. Sexual reproduction and isolation of a sex attractant in *Cutleria multifida* (Smith) Grev. (Phaeophyta). Biochem. Physiol. Pflanzen 165: 212–215.

_____. 1975. Experimental evidence against sexual fusions of spores from unilocular sporangia of *Ectocarpus siliculosus* (Phaeophyta). Br. Phycol. J. 10: 315–321.

_____, AND L. JAENICKE. 1973. Fucoserraten, the female sex attractant of *Fucus serratus* L. (Phaeophyta). FEBS Letters 30: 137–139.

_____, _____, M. DONIKE, AND T. AKINTOBI. 1971. Sex attractant in a brown alga: chemical structure. Science 171: 815–817.

MURPHY, E. B., K. A. STEIDINGER, B. S. ROBERTS, J. WILLIAMS, AND J. W. JOLLEY, JR. 1975. An explanation for the Florida east coast *Gymnodinium breve* red tide of November 1972. Limnol. Oceanogr. 20: 481–486.

MURRAY, S. N., AND P. S. DIXON. 1972. The life history of *Callophyllis firma* (Kylin) Norris in laboratory culture. Br. Phycol. J. 7: 165–168.

_____, J. L. SCOTT, AND P. S. DIXON. 1972. The life history of *Porphyropsis coccinea* var. *dawsonii* in culture. Br. Phycol. J. 7: 111–122.

MUSCATINE, L. 1967. Glycerol excretion by symbiotic algae from corals and Tridacna and its control by the host. Science 156: 516–519.

_____, J. E. BOYLE, AND D. C. SMITH. 1974. Symbiosis of the acoel flatworm *Convoluta roscoffensis* with the alga *Platymonas convolutae*. Proc. R. Soc. Lond. B. 187: 221–234.

_____, S. J. KARAKASHIAN, AND M. W. KARAKASHIAN. 1967. Soluble extra-cellular products of algae symbiotic with ciliates, a sponge and a mutant *Hydra*. Comp. Biochem. Physiol. 20: 1–12.

_____, R. R. POOL, AND R. K. TRENCH. 1975. Symbiosis of algae and invertebrates: aspects of the symbiont surface and the host-symbiont interface. Trans. Am. Micro. Soc. 94: 450–469.

MYERS, M. E. 1928. The life history of the brown alga *Egregia menziesii*. Univ. Calif. Pub. Bot. 14: 225–246.

NADAKAVUKAREN, M. J., AND D. A. McCRACKEN. 1973. *Prototheca*: an alga or a fungus? J. Phycol. 9: 113–116.

NAKAHARA, H., AND Y. NAKAMURA. 1971. The life history of *Desmarestia tabacoides* Okamura. Bot. Mag., Tokyo 84: 69–75.

_____, AND _____. 1973. Parthenogenesis, apogamy and apospory in *Alaria crassifolia* (Laminariales). Mar. Biol. 18: 327–332.

NAKAMURA, K., D. F. BRAY, J. W. COSTERTON, AND E. B. WAGENAAR. 1973. The eyespot of *Chlamydomonas eugametos*: a freeze-etch study. Can. J. Bot. 51: 817–819.

_____, _____, AND E. B. WAGENAAR. 1975. Ultrastructure of *Chlamydomonas eugametos* palmelloids induced by chloroplatinic acid treatment. J. Bact. 121: 338–343.

NAKAMURA, Y. 1965. Development of zoospores in *Ralfsia*-like thallus, with special reference to the life cycle of the Scytosiphonales. Bot. Mag., Tokyo 78: 109–110.

_____. 1972. A proposal on the classification of the Phaeophyta. *In* I. A. Abbott and M. Kurogi (Ed.). Contributions to the Systematics of Benthic Marine Algae of the North Pacific. Japanese Soc. Phycology, Kobe. pp. 147–156.

_____, AND M. TATEWAKI. 1975. The life history of some species of the Scytosiphonales. Sci. Papers, Inst. Algol. Res., Hokkaido Univ. 6(2): 57–93.

NAKAYAMA, T. O. M. 1962. Carotenoids. Chapter 24 *In* Lewin, R. A. (Ed.). Physiology and Biochemistry of Algae. Academic Press, New York. 929 pp.

NAKAZAWA, S. 1975. Physiology of *Fucus*. *In* "Advances of Phycology in Japan" (Ed. J. Tokida and H. Hirose), G. Fischer, Jena. pp. 160–170.

NALEWAJKO, C. 1966. Photosynthesis and excretion in various planktonic algae. Limnol. Oceanogr. 11: 1–10.

_____, AND D. R. S. LEAN. 1972. Growth and excretion in planktonic algae and bacteria. J. Phycol. 8: 361–366.

NATHANIELSZ, C. P., AND I. A. STAFF. 1975A. On the occurrence of intracellular blue-green algae in cortical cells of the apogeotropic roots of *Macrozamia communis*. L. Johnson. Ann. Bot. 39: 363–368.

_____, AND _____. 1975B. A mode of entry of blue-green algae into the apogeotropic roots of *Macrozamia communis*. Am. J. Bot. 62: 232–235.

NEUMAN, K. 1969A. Protonema mit Riesenkern bei der siphonalen Grünalge *Bryopsis hypnoides* und weitere cytologische Befunde. Helgoländ. Wissen. Meeresunter. 19: 45–57.

_____. 1969B. Beitrag zur Cytologie und Entwicklung der siphonalen Grünalge *Derbesia marina*. Helgoländ. Wissen. Meeresunter. 19: 355–375.

_____. 1974. Zur Entwicklungsgeschichte und Systematik der siphonalen Grünalgen *Derbesia* und *Bryopsis*. Bot. Mar. 17: 176–185.

NEUSHUL, M. 1970. A freeze-etching study of the red alga, *Porphyridium*. Am. J. Bot. 57: 1231–1239.

_____, AND A. L. DAHL. 1967. Composition and growth of subtidal parvosilvosa from California kelp forests. Helgoländ. Wissen. Meeresunter. 15: 480–488.

_____, AND _____. 1972A. Zonation in the apical cell of *Zonaria*. Am. J. Bot. 59: 393–400.

_____, AND _____. 1972B. Ultrastructural studies of brown algal nuclei. Am. J. Bot. 59: 401–410.

NEUVILLE, D., AND P. DASTE. 1972. Production de pigment bleu par la Diatomée *Navicula ostrearia* (Gaillon) Bory maintenue en culture uni-algale sur un milieu synthétique carence en azote nitrique. C. R. Acad. Sci. (Paris). 274D: 2030–2033.

_____, AND _____. 1975. Experiments regarding auxospores production in the diatom *Navicula ostrearia* (Gaillon) Bory cultured *in vitro*. C. R. Acad. Sci. (Paris). 281D: 1753–1756.

NEWROTH, P. R. 1971A. Studies on the life histories in the Phillophoraceae. I. *Phyllophora truncata* (Rhodophyceae, Gigartinales). Phycologia 10: 345–354.

_____. 1971B. Redescriptions of five species of *Phyllophora* and an artificial key to the North Atlantic Phyllophoraceae. Br. Phycol. J. 6: 225–230.

_____. 1972. Studies of life histories in the Phyllophoraceae. II. *Phyllophora pseudoceranoides* and notes on *P. crispa* and *P. heredia* (Rhodophyta, Gigartinales). Phycologia 11: 99–107.

_____, AND J. W. MARKHAM. 1972. Observations on the distribution, morphology, and life histories of some Phyllophoraceae. Proc. Intern. Seaweed Symp. 7: 120–126.

_____, AND A. R. A. TAYLOR. 1971. The nomenclature of the North Atlantic species of *Phyllophora* Greville. Phycologia 10: 93–97.

NEWTON, L. 1931. A Handbook of the British Seaweeds. The Trustees of the British Museum (Nat. Hist.). xii + 478 pp.

_____. 1951. Seaweed utilization. Sampson Low, London. 188 pp.

NICHOLS, H. W., 1964A. Culture and developmental morphology of *Compsopogon coeruleus*. Am. J. Bot. 51: 180–188.

_____. 1964B. Developmental morphology and cytology of *Boldia erythrosiphon*. Am. J. Bot. 51: 653–659.

_____. 1965. Culture and development of *Hildenbrandia rivularis* from Denmark and North America. Am. J. Bot. 52: 9–15.

_____. 1973. Growth media-freshwater. Chapter 1 In Stein, J. R. (Ed.). Handbook of Phycological Methods. Univ. Press. Cambridge. pp. 7–24.

_____, AND H. C. BOLD. 1965. *Trichosarcina polymorpha* gen. et. sp. nov. J. Phycol. 1: 34–38.

_____, AND E. K. LISSANT. 1967. Developmental studies of *Erythrocladia* Rosenvinge in culture. J. Phycol. 3: 6–18.

NICHOLSON, N. L. 1970. Field studies on the giant kelp *Nereocystis*. J. Phycol. 6: 177–182.

_____. 1976. Anatomy of the medulla of *Nereocystis*. Bot. Mar. 19: 23–31.

_____, AND W. R. BRIGGS. 1972. Translocation of photosynthate in the brown alga *Nereocystis*. Am. J. Bot. 59: 97–106.

NICOLAI, E., AND R. D. PRESTON. 1954. Cell-wall studies in the Chlorophyceae. III. Differences in structure and development in the Cladophoraceae. Proc. Roy. Soc. (London) B. 151: 244–255.

NIENHUIS, P. H. 1974. Variability in the life cycle of *Rhizoclonium riparium* (Roth) Harv. (Chlorophyceae: Cladophorales) under Dutch estuarine conditions. Hydrobiol. Bull. 8: 172–178.

_____, AND J. SIMONS. 1971. *Vaucheria* species and some other algae on a Dutch salt marsh, with ecological notes on their periodicity. Acta Bot. Neerl. 20: 107–118.

NILSHAMMAR, M., AND B. WALLES. 1974. Electron microscope studies on cell differentiation in synchronized cultures of the green alga *Scenedesmus*. Protoplasma 79: 317–332.

NITECKI, M. H. 1971. *Ischadites abbottae*, a new North American Silurian species (Dasycladales). Phycologia 10: 263–275.

NIZAMUDDIN, M. 1962. Classification and the distribution of the Fucales. Bot. Mar. 4: 191–203.

———. 1964. Phylogenetic position of Siphonocladiales. Trans. Am. Microsc. Soc. 83: 282–90.

———. 1968. Observations on the order Durvilleales J. Petrov, 1965. Bot. Mar. 11: 115–117.

———. 1970. Phytogeography of the Fucales and their seasonal growth. Bot. Mar. 13: 131–139.

NJUS, D., F. M. SULZMAN, AND J. W. HASTINGS. 1974. Membrane model for the circadian clock. Nature 248: 116–120.

NORRIS, D. R. 1969A. Thecal morphology of *Ornithocercus magnificus* (Dinoflagellata) with notes on related species. Bull. Mar. Sci. 19: 175–193.

———. 1969B. Possible phagotrophic feeding in *Ceratium lunula* Schimper. Limnol. Oceanog. 14: 448–449.

———, AND L. D. BERNER. 1970. Thecal morphology of selected species of *Dinophysis* (Dinoflagellata) from the Gulf of Mexico. Contrib. Mar. Sci. 15: 145–192.

NORRIS, R. E. 1957. Morphological studies on the Kallymeniaceae. Univ. Calif. Pub. Bot. 28: 251–317.

———. 1964. The morphology and taxonomy of South African Kallymeniaceae. Bot. Mar. 7: 90–129.

———. 1965. Neustonic marine Craspedomonadales (Choanoflagellates) from Washington and California. J. Protozool. 12: 589–602.

———. 1966. Unarmoured marine dinoflagellates. Endeavour 25: 124–128.

———. 1967A. Micro-algae in enrichment cultures from Puerto Peñasco, Sonora, Mexico. Bull. Southern Calif. Acad. Sci. 66: 233–250.

———. 1967B. Algal consortisms in marine plankton. *In* Proc. Seminar on Sea, Salt and Plants, Bhavnagar, 1965. (Ed., V. Krishnamurthy). pp. 178–189.

———. 1971. Development of the foliose thallus of *Weeksia fryeana* (Rhodophyceae). Phycologia 10: 205–213.

———, AND D. H. KIM. 1972. Development of thalli in some Gigartinaceae. *In* Contributions to the Systematics of Benthic Marine Algae of the North Pacific (Ed., I. A. Abbott and M. Kurogi). Japan. Soc. Phycol., Kobe. pp. 265–275.

———, AND B. R. PEARSON. 1975. Fine structure of *Pyramimonas parkeae*, sp. nov. (Chlorophyta, Prasinophyceae) Arch. Protistenk. 117: 192–213.

NORTH, W. J. 1961. Experimental transplantation of the giant kelp *Macrocystis pyrifera*. Proc. Intern. Seaweed Symp. 4: 248–255.

———. (Ed.). 1971. The biology of giant kelp beds (*Macrocystis*) in California. Beihefte z. Nova Hedw. 32. xiii + 600 pp.

———. 1972. Observations on populations of *Macrocystis*. *In* Contributions to the Systematics of Benthic Marine Algae of the North Pacific (Ed., I. A. Abbott and M. Kurogi). Japan. Soc. Phycol., Kobe. pp. 75–92.

NORTON, T. A., AND G. R. SOUTH. 1969. Influence of reduced salinity on the distribution of two laminarian algae. Oikos 20: 320–326.

NOVÁKOVÁ, M. 1964. *Asterococcus* Scherffel and *Sphaerellocystis* Ettl, two new genera of palmelloid green algae. Acta Univ. Carol. Biol., Praha. No. 2: 155–166.

NOVOTNY, A. M., AND M. FORMAN. 1975. The composition and development of cell walls of *Fucus* embryos. Planta 122: 67–78.

NULTSCH, W. 1956. Studien über die Phototaxis der Diatomeen. Arch. Protistenk. 101: 1–68.

———. 1974. Movements. Chapter 31 *In* Stewart, W. D. P. (Ed.) Algal Physiology and Biochemistry. Univ. Calif. Press. Berkeley and Los Angeles. 989 pp.

———, AND G. THROM. 1975. Effect of external factors on phototaxis of *Chlamydomonas reinhardtii*. Arch. Microbiol. 103: 175–79.

OCAMPO-PAUS, R. 1970. Zoosporogenesis in the Chlorococcalean alga *Radiosphaera negevensis* Ocampo-Paus and Friedmann. J. Phycol. 6: 221–222.

————, AND I. FRIEDMANN. 1966. *Radiosphaera negevensis* sp. n., a new Chlorococcalean desert alga. Am. J. Bot. 53: 663–671.

O'COLLA, P. S. 1962. Mucilages *In* Physiology and Biochemistry of Algae (Ed., R. A. Lewin). Academic Press. New York. pp. 337–356.

OGATA, E., T. MATSUI, AND H. NAKAMURA. 1972. The life cycle of *Gracilaria verrucosa* (Rhodophyceae, Gigartinales). Phycologia 11: 75–80.

O'HEOCHA, C. 1962. Phycobilins. *In* Physiology and Biochemistry of Algae (Ed., R. A. Lewin). Academic Press, New York. pp. 421–435.

————. 1971. Pigments of the red algae. Ann. Rev. Oceanog. Mar. Biol. 9: 61–82.

————, AND M. RAFTERY. 1959. Phycoerythrins and phycocyanins of cryptomonads. Nature 184: 1049–1051.

OKADA, H. AND S. HONJO. 1970. Coccolithophoridae distributed in Southwest Pacific. Pacif. Geol. 2: 11–21.

————, AND ————. 1973. The distribution of oceanic coccolithophorids in the Pacific. Deep-Sea Research 20: 355–374.

O'KELLEY, J. C. 1974. Inorganic Nutrients. Chapter 22 *In* Stewart, W. D. P. (Ed.). Algal Physiology and Biochemistry. Univ. California Press, Berkeley and Los Angeles. 989 pp.

————, AND J. K. HARDMAN. 1976. The yellow light-absorbing pigment, Py, of the *Protosiphon* photoreversible pigment system. J. Phycol. 12 (Suppl.): 30 (Abstr.).

————, AND W. R. HERNDON. 1961. Alkaline earth elements and zoospore release in *Protosiphon botryoides*. Am. J. Bot. 48: 796–802.

OLIVEIRA, L., AND T. BISALPATRA. 1973. Studies in the brown alga *Ectocarpus* in culture. I. General ultrastructure of the sporophytic vegetative cells. J. Submicro. Cytol. 5: 107–120.

OLTMANNS, F. 1922. Morphologie und Biologie der Algen. 2nd ed., Vol. II, Jena. iv + 439 pp.

ORKWIZEWSKI, K. G., AND A. R. KANEY. 1974. Genetic transformation of the blue-green bacterium, *Anacystis nidulans*. Arch. Mikrobiol. 98: 31–37.

OSAFUNE T., S. MIHARA, E. HASE, AND I. OHKURO. 1975. Electron microscope studies of the vegetative cellular life cycle of *Chlamydomonas reinhardii* in synchronous culture. III. Three-dimensional structures of mitochondria in the cells at intermediate stages of the growth phase of the cell cycle. J. Electron Microscopy 24: 247–252.

OSCHMAN, J. L. 1966. Development of the symbiosis of *Convoluta roscoffensis* (Gruff) and *Platymonas* sp. J. Phycol. 2: 105–111.

————. 1967. Structure and reproduction of the algal symbionts of *Hydra viridis*. J. Phycol. 3: 221–228.

OTT, D. W., AND R. M. BROWN, JR. 1972. Light and electron microscopical observations on mitosis in *Vaucheria litorea* Hofman ex C. Agardh. Br. Phycol. J. 7: 361–374.

————, AND ————. 1974A. Developmental cytology of the genus *Vaucheria*. I. Organization of the vegetative filament. Br. Phycol. J. 9: 111–126.

————, AND ————. 1974B. Developmental cytology of the genus *Vaucheria*. II. Sporogenesis in *V. fontinalis* (L.) Christensen. Br. Phycol. J. 9: 333–351.

————, AND ————. 1975. Developmental cytology of the genus *Vaucheria*. III. Emergence, settlement and germination of the mature zoospore of *V. fontinalis* (L.) Christensen. Brit. Phycol. J. 10: 49–56.

OTT, F. D. 1965. Synthetic media and techniques for the xenic culture of marine algae and flagellates. Virginia J. Sci. (N.S.) 16: 205–218.

————. 1967. *Rhodosorus marinus* Geitler: a new addition to the marine algal flora of the Western Hemisphere. J. Phycol. 3: 158–159.

————. 1972. A review of the synonyms and the taxonomic positions of the red algal genus *Porphyridium* Nägeli 1849. Nova Hedw. 23: 237–289.

OUTKA, D. E., AND D. C. WILLIAMS. 1971. Sequential coccolith morphogenesis in *Hymenomonas carterae*. J. Protozool. 18: 285–297.

PAASCHE, E. 1964. A tracer study of the inorganic carbon uptake during coccolith formation and photosynthesis in the coccolithophorid *Coccolithus huxleyi*. Physiol. Plantarum, Suppl. III, 82 pp.

_____. 1966. Adjustment to light and dark rates of coccolith formation. Physiol. Plantarum 19: 271–278.

_____. 1968A. Biology and physiology of coccolithophorids. Ann. Rev. Microbiol. 22: 71–86.

_____. 1968B. The effect of temperature, light intensity and photoperiod on coccolith formation. Limnol. Oceanog. 13: 178–181.

_____, S. JOHANSSON, AND L. L. EVENSEN. 1975. An effect of osmotic pressure on the valve morphology of the diatom *Skeletonema subsalsum* (A. Cleve) Bethge. Phycologia 14: 205–211.

_____, AND D. KLAVENESS. 1970. A physiological comparison of coccolith-forming and naked cells of *Coccolithus huxleyi*. Archiv Mikrobiol. 73: 143–152.

PADILLA, G. M., R. J. BRAGG, AND J. R. KENNEDY, JR. 1968. Characteristics and cellular localization of the hemolytic toxin from the euryhaline flagellate *Prymnesium parvum*. *In* Drugs from the Sea (Ed., H. D. Freudenthal). Marine Technology Soc. pp. 185–201.

PAGE, J. Z. 1973. Methods for coenocytic Algae. Chapter 6 in Stein, J. R. (Ed.) Handbook of Phycological Methods. University Press. Cambridge. pp. 105–126.

_____, AND J. M. KINGSBURY. 1968. Culture studies on the marine green alga *Halicystis parvula—Derbesia tenuissima*. II. Synchrony and periodicity in gamete formation and release. Am. J. Bot. 55: 1–11.

_____, AND B. M. SWEENEY. 1968. Culture studies on the marine green alga *Halicystis parvula—Derbesia tenuissima*. III. Control of gamete formation by an endogenous rhythm. Phycologia 4: 253–260.

PAGNI, P. S., P. L. WALNE, AND E. L. WEHRY. 1976. Fluorometric evidence for flavins in isolated eyespots of *Euglena gracilis* var. *bacillaris*. Photochem. Photobiol. 24: 373–375.

PAINE, R. T., AND R. L. VADAS. 1969. The effects of grazing by sea urchins, *Strongylocentrotus* spp., on benthic algal populations. Limnol. Oceanog. 14: 710–719.

PALIK, P. 1933. Über die Entstehung der Polyeder bei *Pediastrum boryanum* (Turpin) Meneghini. Arch. Protistenk. 79: 234–238.

PALISANO, J. R., AND P. L. WALNE. 1976. Light and electron microscopy of two permanently bleached cell lines of *Euglena gracilis* (Euglenophyceae). Hova Hedw. 27: 455–482.

PALL, M. L. 1973. Sexual induction in *Volvox carteri*. A quantitative study. J. Cell Biol. 59: 238–241.

PALMER, C. M. 1959. Algae in water supplies. U.S. Dept. Health, Ed. and Welfare. Public Health Service No. 657, 88 pp.

_____, AND C. M. TARZWELL. 1955. Algae of importance in water supplies. Public Works Magazine, June, 1955.

PALMER, J. D., AND F. E. ROUND. 1967. Persistent, vertical-migration rhythms in benthic microflora. VI. The tidal and diurnal nature of the rhythm in the diatom *Hantzshia virgata*. Biol. Bull. 132: 44–55.

PANKRATZ, H. S. AND C. C. BOWEN. 1963. Cytology of blue-green algae. I. The cells of *Symploca muscorum*. Am. J. Bot. 50: 387–399.

PAPENFUSS, G. F. 1935. Alternation of generations in *Ectocarpus siliculosus*. Bot. Gaz. 96: 421–446.

_____. 1945. Review of the *Acrochaetium-Rhodochorton* complex of the red algae. Univ. Calif. Pub. Bot. 18: 299–334.

_____. 1946. Proposed names for the phyla of algae. Bull. Torrey Bot. Club 73: 217–218.

_____. 1951A. Phaeophyta. *In* Manual of Phycology (Ed., G. M. Smith). Chronica Botanica, Waltham, Mass. pp. 119–158.

_____. 1951B. Problems in the classification of the marine algae. Svensk Bot. Tidskr. 45: 4–11.

_____. 1955. Classification of the algae. *In* A Century of Progress in the Natural Sciences, 1853–1953. Calif. Acad. Sci., San Francisco, pp. 115–224.

_____. 1957. Progress and outstanding achievements in phycology during the past fifty years. Am. J. Bot. 44: 74–81.

_____. 1960. On the genera of the Ulvales and the status of the order. J. Linn. Soc. (Bot.) 56: 303–318.

_____. 1961. The structure and reproduction of *Caloglossa leprieurii*. Phycologia 1: 8–31.

_____. 1966. A review of the present system of classification of the Florideophycidae. Phycologia 5: 247–255.

_____. 1967. Notes on algal nomenclature. V. Various Chlorophyceae and Rhodophyceae. Phykos 5: 95–105.

_____, AND Y.-M. CHIANG. 1969. Remarks on the taxonomy of *Galaxaura* (Nemaliales, Chaetangiaceae). Proc. Intern. Seaweed Symp. 6: 303–314.

PARDY, R. L. 1974. Some factors affecting the growth and distribution of the algal endosymbionts of *Hydra viridis*. Biol. Bull. 147: 105–118.

PARKE, M. 1961. Some remarks concerning the Class Chrysophyceae. Br. Phycol. Bull 2: 46–55.

_____. 1971. The production of calcareous elements by benthic algae belonging to the class Haptophyceae (Chrysophyta). Proc. II Planktonic Conf. Rome 1970 (Ed., A. Farinacci). pp. 929–937.

_____, AND I. ADAMS. 1960. The motile (*Crystallolithus hyalinus* Gaarder and Markali) and non-motile phases in the life history of *Coccolithus pelagicus* (Wallich) Schiller. J. Mar. Biol. Assn. U. K. 39: 262–264.

_____, AND P. S. DIXON. 1964. A revised check-list of British marine algae. J. Mar. Biol. Assn. U. K. 44: 499–542.

_____, J. C. GREEN, AND I. MANTON. 1971. Observations on the fine structure of zoids of the genus *Phaeocystis* [Haptophyceae]. J. Mar. Biol. Assn. U. K. 51: 927–941.

_____, AND I. MANTON. 1962. Studies on marine flagellates. VI. *Chrysochromulina pringsheimii* sp. nov. J. Mar. Biol. Assn. U. K. 42: 391–404.

_____, AND _____. 1965. Preliminary observations on the fine structure of *Prasinocladus marinus*. J. Mar. Biol. Assn. U. K. 45: 525–536.

_____, _____, AND B. CLARKE. 1955. Studies on marine flagellates II. Three new species of *Chrysochromulina*. J. Mar. Biol. Assn. U. K. 34: 579–609.

_____, _____, AND _____. 1956. Studies on marine flagellates III. Three further species of *Chrysochromulina*. J. Mar. Biol. Assn. U. K. 35: 387–414.

_____, _____, AND _____. 1958. Studies in marine flagellates IV. Morphology and microanatomy of a new species of *Chrysochromulina*. J. Mar. Biol. Assn. U. K. 37: 209–228.

_____, _____, AND _____. 1959. Studies on marine flagellates V. Morphology and microanatomy of *Chrysochromulina strobilus* sp. nov. J. Mar. Biol. Assn. U. K. 38: 169–188.

PARKER, B. C. 1961. Facultative heterotrophy in certain soil algae from the ecological viewpoint. Ecology 42: 381–386.

_____. 1964. The structure and chemical composition of cell walls in three chlorophycean algae. Phycologia 4: 63–74.

_____. 1965. Translocation in the giant kelp *Macrocystis*. I. Rates, direction, quantity of C^{14}-labeled products and fluorescein. J. Phycol. 1: 41–46.

_____. 1966. Translocation in *Macrocystis*. III. Composition of sieve tube exudate and identification of the major C^{14}-labeled products. J. Phycol. 2: 38–41.

_____. 1969. Occurrence of silica in brown and green algae. Can. J. Bot. 47: 537–540.

_____. 1970. Significance of cell wall chemistry to phylogeny in the algae. Ann. N.Y. Acad. Sci. 175: 417–428.

_____. 1971A. The internal structure of *Macrocystis*. *In* The biology of giant kelp beds (Ed., W. J. North). Beihefte z. Nova Hedw. 32. pp. 99–121.

_____. 1971B. Studies of translocation in *Macrocystis*. *In* The biology of giant kelp beds. (Ed., W. J. North). Beihefte z. Nova Hedw. 32. pp. 191–195.

_____. 1971C. On the evolution of isogamy to oogamy. *In* Contributions in Phycology (Ed., B. C. Parker and R. M. Brown, Jr.). Allen Press, Lawrence, Kansas. pp. 47–51.

_____. 1971D. Commentary on "Facultative heterotrophy in certain soil algae from the ecological viewpoint." *In* Rosowski, J. R. and B. C. Parker (Ed.). Selected Papers in Phycology. Univ. Nebraska Lincoln. 876 pp.

_____, AND H. C. BOLD. 1961. Biotic relationships between soil algae and other microorganisms. Am. J. Bot. 48: 185–197.

_____, _____, AND T. R. DEASON. 1961. Facultative heterotrophy in some chlorococcacean algae. Science 133: 761–763.

_____, AND E. Y. DAWSON. 1965. Non-calcareous marine algae from California Miocene deposits. Nova Nedw. 10:273–295.

_____, AND M. FU. 1965. The internal structure of the elk kelp (*Pelagophycus* species). Can. J. Bot. 43: 1293–1305.

PARSONS, M. J. 1975. Morphology and taxonomy of the Dasyaceae and the Lophothalieae (Rhodomelaceae) of the Rhodophyta. Austral. J. Bot. 23: 549–713.

PASCHER, A. 1916. Über die Kreuzung einzelliger, haploider Organismen: *Chlamydomonas*. Ber. Deut. Bot. Gesell. 34: 228–242.

_____. 1918. Über die Beziehung der Reduktionsteilung zur Mendelschen Spaltung. Ber. Deutsch. Bot. Gesell. 36: 163–168.

_____. 1938. Heterokontae. *In* Kryptogamen-Flora von Deutschland, Österreich und der Schweiz, Vol. II (Ed., L. Rabenhorst). Leipzig. Akad. Verlagsges pp. 481–832.

PATRICK, R., AND C. W. REIMER. 1966. The Diatoms of the United States exclusive of Alaska and Hawaii. Vol. 1. Fragilariaceae, Eunotiaceae, Achnanthaceae, Naviculaceae. Monogr. of The Acad. of Natural Sci. of Philadelphia, No. 13. xi + 688 pp.

_____, AND _____. 1975. The Diatoms of the United States exclusive of Alaska and Hawaii. Vol. 2, Part 1. Entomoneidaceae, Cymbellaceae, Gomphonemaceae, Epithemiaceae. Monogr. of The Acad. of Natural Sci. of Philadelphia, No. 13. ix + 213 pp.

PEARSE, V. B. 1974. Modification of sea anemone behavior by symbiotic Zooxanthellae: phototaxis. Biol. Bull. 147: 630–640.

_____, AND L. MUSCATINE. 1971. Role of symbiotic algae (Zooxanthellae) in coral calcification. Biol. Bull. 141: 350–363.

PEARSON, B. R., AND R. E. NORRIS. 1975. Fine structure and cell division in *Pyramimonas parkeae* Norris and Pearson (Chlorophyta, Prasinophyceae). J. Phycol. 11: 113–124.

PECK, R. E. 1953. Fossil charophytes. Bot. Review 19: 209–227.

_____, AND J. A. EYER. 1963. Pennsylvanian, Permian and Triassic Charophyta of North America. J. Paleont. 37: 835–844.

PEDERSÉN, M. 1968. *Ectocarpus fasciculatus*: marine brown alga requiring kinetin. Nature 218: 776.

_____. 1969. The demand for iodine and bromine of three marine brown algae grown in bacteria-free cultures. Physiol. Plantarum 22: 680–685.

_____. 1973. Identification of a cytokinin, 6-(3 methyl-2-butenylamino) purine in seawater and the effect of cytokinins on brown algae. Physiol. Plantarum 28: 101–105.

PENNICK, N. C., AND K. J. CLARKE. 1976. Studies of the external morphology of *Pyramimonas* 3. *Pyramimonas grossii* Parke. Arch. Protistenk. 118: 285–290.

PERCIVAL, E., AND R. H. MCDOWELL. 1967. Chemistry and Enzymology of Marine Algal Polysaccharides. Acad. Press, London and New York. xii + 219 pp.

PERROT, Y. 1968. Sur le cycle de deux formes d'*Ulothrix flacca* (Dillw.) Thuret de la région de Roscoff. C. R. Acad. Sci. (Paris) 266D: 1953–1955.

_____. 1970. Sur la spécifité et le cycle de l'*Ulothrix subflaccida*. (Wille) de la région de Roscoff. C. R. Acad. Sci. (Paris). 270D: 932–933.

_____. 1971. Sur le cycle de reproduction de l'*Ulothrix pseudoflacca* Wille de la région de Roscoff. C. R. Acad. Sci. (Paris). 273D: 858–859.

_____. 1972. Les *Ulothrix* marins de Roscoff et le problème de leur cycle de reproduction. Bull. Soc. Bot. Fr., Mém. 1972: 67–74.

PESSONEY, G. F. 1968. Field and laboratory investigation of zygnemataceous algae. Ph. D. Dissertation. The University of Texas at Austin. 179 pp.

PETERS, G. A. 1975. The *Azolla-Anabaena* relationship, III. Studies on metabolic capabilities and a further characterization of the symbiont. Arch. Microbiol. 103: 113–122.

PETERSON, J. B. 1935. Studies on the biology and taxonomy of soil algae. Dansk Bot. Arkiv., Res. Botan. 8: 1–183.

_____, AND J. B. HANSEN. 1958. On the scales of some *Synura* species. II. Biol. Medd. Dan. Vid. Selsk. 23: 1–14.

PETROV, J. E. 1965. De positione familiae Durvilleacearum et systematica classis Cyclosporophy-cearum (Phaeophyta). Novit. System. Plant. non Vascularium 1965: 70–72.

PFIESTER, L. A. 1975. Sexual reproduction of *Peridinium cinctum* f. *ovoplanum* (Dinophyceae). J. Phycol. 11: 258–265.

_____. 1976. Sexual reproduction of *Peridinium willei* (Dinophyceae). J. Phycol. 12: 234–238.

PHILLIPS, T. L., K. J. MIKLAS, AND H. N. ANDREWS. 1972. Morphology and vertical distribution of *Protosalvinia* (*Foerstia*) from the New Albany shale (Upper Devonian). *In* Advances in Paleo-zoic Botany (Ed. M. Streel, P. M. Bonamo, and M. Fairon-Demaret). Elsevier Co. New York. pp. 171–196.

PICKETT-HEAPS, J. D. 1967A. Ultrastructure and differentiation in *Chara* sp. I. Vegetative cells. Austral. J. Bio. Sci. 20: 539–550.

_____. 1967B. Ultrastructure and differentiation in *Chara sp.* II. Mitosis. Austral. J. Biol. Sci. 20: 883–894.

_____. 1968A. Ultrastructure and differentiation in *Chara sp.* III. Formation of the antheridium. Austral. J. Biol. Sci. 21: 255–274.

_____. 1968B. Ultrastructure and differentiation in *Chara* (*fibrosa*). IV. Spermatogenesis. Austral. J. Biol. Sci. 21: 655–690.

_____. 1969. The evolution of the mitotic apparatus: an attempt at comparative ultrastructural cytology in dividing plant cells. Cytobios 1: 257–80.

_____. 1970. Some ultrastructural features of *Volvox*, with particular reference to the phenomenon of inversion. Planta 90: 170–190.

_____. 1971. Reproduction by zoospores in *Oedogonium*. I. Zoosporogenesis. Protoplasma 72: 275–314.

_____. 1972A. II. Emergence of the zoospores and the motile phase. Protoplasma 74: 149–167.

_____. 1972B. III. Differentiation of the germling. Protoplasma 74: 169–193.

_____. 1972C. IV. Cell division in the germling and evidence concerning the possible evolution of the wall rings. Protoplasma 74: 195–212.

_____. 1972D. Variation in mitosis and cytokinesis in plant cells: its significance in the phylogeny and evolution of ultrastructural systems. Cytobios 5: 59–77.

_____. 1972E. Cell division in *Klebsormidium subtilissimum* (formerly *Ulothrix subtilissima*) and its possible phylogenetic significance. *Cytobios* 6: 167–184.

_____. 1972F. Cell division in *Cosmarium botrytis*. J. Phycol. 8: 343–360.

_____. 1973A. Cell division in *Tetraspora*. Ann. Bot. 37: 1017–1025.

_____. 1973B. Cell division and wall structure in *Microspora*. New Phytol. 72: 347–355.

_____. 1973C. Stereo-scanning electron microscopy of desmids. J. Microscopy 99: 109–116.

_____. 1974A. Cell division in *Bulbochaete*. I. Divisions utilizing the wall ring. J. Phycol. 9: 408–420.

_____. 1974B. Cell division in *Bulbochaete*. II. Hair cell formation. J. Phycol. 10: 148–164.

_____. 1974C. Cell division in *Stichococcus*. Br. Phycol. J. 9: 63–73.

_____. 1974D. Ultrastructural morphology and cell division in *Pedinomonas*. Cytobios 11: 41–58.

_____. 1974E. Scanning electron microscopy of some cultured desmids. Trans. Am. Micros. Soc. 93: 1–23.

_____. 1975A. Green Algae. Sinauer Associates. Sunderland, Mass. 606 pp.

_____. 1975B. Structural and phylogenetic aspects of microtubular systems in gametes and zoospores of certain green algae. Biol. J. Linnean Soc. Suppl. No. 1, 7: 37–44.

_____. 1976. Cell division in eucaryotic algae. BioScience 26: 445–450.

_____, AND L. C. FOWKE. 1969. Cell division in *Oedogonium*. I. Mitosis, cytokinesis, and cell elongation. Austral. J. Biol. Sci. 22: 857–894.

_____, AND _____. 1970. Mitosis, cytokinesis, and cell elongation in the desmid *Closterium littorale*. J. Phycol. 6: 189–215.

_____, AND _____. 1971. Conjugation in the desmid *Closterium littorale*. J. Phycol. 7: 37–50.

_____, AND H. J. MARCHANT. 1972. The phylogeny of the green algae: a new proposal. Cytobios 6: 255–264.

_____, AND K. L. McDONALD. 1975. *Cylindrocapsa*: cell division and phylogenetic affinities. New Phytol. 74: 235–241.

_____, _____, AND D. H. TIPPIT. 1975. Cell division in the pennate diatom *Diatoma vulgare*. Protoplasma 86: 205–247.

_____, AND L. A. STALHELIN. 1975. The ultrastructure of *Scenedesmus* (Chlorophyceae). II. Cell division and colony formation. J. Phycol. 11: 186–202.

PICKMERE, S. E., J. P. PARSONS, AND R. W. BAILEY. 1973. Composition of *Gigartina* carrageenan in relation to sporophyte and gametophyte stages of the life cycle. Phytochem. 12: 2441–2444.

PIENAAR, R. N. 1976. The microanatomy of *Hymenomonas lacuna* sp. nov. (Haptophyceae). J. Mar. Biol Assn. U. K. 56: 1–11.

POCOCK, M. A. 1937. *Hydrodictyon* in South Africa, with notes on the known species of *Hydrodictyon*. Trans. Roy. Soc. S. Africa 24: 263–280.

_____. 1953. Two multicellular green algae, *Volvulina* Playfair and *Astrephomeme*, a new genus. Trans. Roy. Soc. S. Africa 34: 103–127.

_____. 1956. South African parasitic Florideae and their hosts. 3. Four minute parasitic Florideae. Proc. Linn. Soc. London (Botany) 167: 10–41.

_____. 1960. *Hydrodictyon*: a comparative biological study. J. S. African Bot. 26: 167–327.

POLANSHEK, A. R. 1974. Hybridization studies of *Gigartina agardhii* and the gametophytes of *Petrocelis franciscana*. J. Phycol. 10 (Suppl.): 3 (Abstr.).

_____, AND J. A. WEST. 1975. Culture and hybridization studies on *Petrocelis* (Rhodophyta) from Alaska and California. J. Phycol. 11: 434–439.

POLDERMAN, P. J. G. 1973. *Vaucheria minuta*. A new alga from Denmark. Bot. Tidsskr. 67: 327–328.

POLLOCK, E. G. 1970. Fertilization in *Fucus*. Planta 92: 85–99.

PORTER, J., AND M. JOST. 1976. Physiological effects of the presence and absence of gas vacuoles in the blue-green alga, *Microcystis aeruginosa* Kuetz. emend. Elenkin. Arch. Microbiol. 110: 225–231.

POSTGATE, J. R. 1974. Evolution within nitrogen-fixing systems. Symposium Soc. Gen. Microbiol. 24: 279–280.

POTTHOFF, H. 1928. Zur Phylogenie und Entwicklungsgeschichte der Conjugaten. Ber. deutsch. Bot. Gesells. 46: 667–673.

POWELL, H. T. 1957A. Studies in the genus *Fucus* L. I. *Fucus distichus* L. emend. Powell. J. Mar. Biol. Assn. U. K. 36: 407–432.

_____. 1957B. Studies in the genus *Fucus* L. II. Distribution and ecology of forms of *Fucus distichus* L. emend. Powell in Britain and Ireland. J. Mar. Biol. Assn. U. K. 36: 663–693.

_____. 1963. Speciation in the genus *Fucus* L. and related genera. *In* Speciation in the Sea (Ed., J. P. Harding and N. Tebble), System. Assn. Pub., No. 5. pp. 63–77.

POWELL, J. H., AND B. D. J. MEEUSE. 1964. Laminarin in some Phaeophyta of the Pacific Coast. Econ. Bot. 18: 164–166.

PRAKASH, A., AND M. A. RASHID. 1968. Influence of humic substances on the growth of marine phytoplankton: dinoflagellates. Limnol. Oceanogr. 13: 598–606.

_____, AND J. R. TAYLOR. 1966. A "red water" bloom of *Gonyaulax acatenella* in the Strait of Georgia and its relation to paralytic shellfish toxicity. J. Fish. Res. Bd. Can. 23: 1265–1270.

PRASAD, B. N., AND P. N. SRIVASTAVA. 1963. Observations on the morphology, cytology and asexual reproduction of *Schizomeris leibleinii*. Phycologia 2: 148–156.

PRATT, R., AND J. FONG. 1940. Studies on *Chlorella vulgaris*. II. Further evidence that *Chlorella* cells form a growth-inhibiting substance. Am. J. Bot. 27: 431–436.

PRESCOTT, G. W. 1968. The Algae: A Review. Houghton Mifflin Co., Boston. 436 pp.

_____, H. T. CROASDALE, AND W. C. VINYARD. 1972. Desmidiales. I. Saccodermae, Mesotaeniaceae. N. Am. Flora Series II (6). 84 pp.

_____, _____, AND _____. 1975. A Synopsis of North American Desmids. II. Desmidiaceae: Placodermae, Section 1. Univ. Neb. Press, Lincoln. 275 pp.

PRESTON, R. D. 1968. Plants without cellulose. Sci. Am. 218: 102–108.

PRICE, I. A. 1972. Zygote development in *Caulerpa* (Chlorophyta, Caulerpales). Phycologia 11: 217–218.

PRINCE, J. S., AND J. M. KINGSBURY. 1973A. The ecology of *Chondrus crispus* at Plymouth, Massachusetts. I. Ontogeny, vegetative anatomy, reproduction, and life cycle. Am. J. Bot. 60: 956–963.

_____, AND _____. 1973B. The ecology of *Chondrus crispus* at Plymouth, Massachusetts. II. Field studies. Am. J. Bot. 60: 964–975.

_____, AND _____. 1973C. The ecology of *Chondrus crispus* at Plymouth, Massachusetts. III. Effect of elevated temperature on growth and survival. Biol. Bull. 145: 580–588.

PRINGSHEIM, E. G. 1946A. Pure cultures of algae, their preparation and maintenance. University Press, Cambridge. 119 pp.

_____. 1946B. The biphasic or soil-water culture method for growing algae and flagellata. J. Ecol. 33: 193–204.

_____. 1952. On the nutrition of *Ochromonas*. Q. J. Microscop. Sci. 93: 71–96.

_____. 1956. Contributions toward a monograph of the genus *Euglena*. Nova Acta Leopoldina 18: 1–168.

_____. 1963. Farblose Algen. Gustav Fischer Verlag. Stuttgart. 471 pp.

_____. 1966. Nutritional requirements of *Haematococcus pluvialis* and related species. J. Phycol. 2: 1–7.

_____. 1967A. Phycology in the field and in the laboratory. J. Phycol. 3: 93–95.

_____. 1967B. Zur Physiologie der farblosen Diatomee *Nitzschia putrida*. Archiv. Microbiol. 55: 60–67.

_____. 1968A. Zur Kenntnis der Cryptomonaden des Süsswasser. Nova. Hedw. 16: 367–401.

_____. 1968B. Kleine Mitteilungen über Flagellaten und Algen. XV. Zur Kenntnis der Gattung *Porphyridium*. Archiv. Mikrobiol. 61: 169–180.

_____. 1969. Die Gattungen *Chlorogonium* und *Hyalogonium*. Nova Hedw. 18: 831–867.

_____, AND O. PRINGSHEIM. 1949. The growth requirements of *Porphyridium cruentum*: with remarks on the ecology of brackish water algae. J. Ecol. 37: 57–64.

_____, AND _____. 1952. Experimental elimination of the chromatophores and eyespot in *Euglena gracilis*. New Phytol. 51: 65–76.

_____, AND W. WIESSNER. 1961. Ernährung und Stoffwechsel von *Chlamydobotrys* (Volvocales). Arch. Mikorbiol. 40: 231–246.

PRINTZ, H. 1964. Die Chaetophoralen der Binnengewässer—eine systematische Übersicht. Hydrobiol. 24: 1–376.

PROCTOR, H. N., S. L. CHAN, AND A. J. TREVOR. 1975. Production of saxitoxin by cultures of *Gonyaulax catenella*. Toxicon 13: 1–9.

PROCTOR, V. M. 1960. Dormancy and germination of *Chara* oospores. Bull. Phycol. Soc. Am. 13: 64.

_____. 1961. *Batophora* from central New Mexico. Phycologia 1: 160–163.

_____. 1966. Dispersal of desmids by waterbirds. Phycologia 5: 227–232.

_____. 1967. Storage and germination of *Chara* oospores. J. Phycol. 3: 90–92.

_____. 1970. Taxonomy of *Chara braunii*: an experimental approach. J. Phycol. 6: 317–321.

_____. 1971A. Taxonomic significance of monoecism and dioecism in the genus *Chara*. Phycologia 10: 299–307.

_____. 1971B. *Chara globularis* Thuillier (= *C. fragilis* Desvaus.): breeding patterns within a cosmopolitan complex. Limnol. Oceanogr. 16: 422–436.

_____, AND F. H. WIMAN. 1971. An experimental approach to the systematics of the monoecious-conjoined members of the genus *Chara*, series Gymnobasalia. Am. J. Bot. 58: 885–893.

PROSKAUER, J. 1950. On *Prasinocladus*. Am. J. Bot. 37: 59–66.

PROVASOLI, L. 1958. Nutrition and ecology of Protozoa and algae. Ann. Rev. Microbiol. 12: 279–308.

_____. 1963. Growing marine seaweeds. Proc. Intern. Seaweed Symp. 4: 9–17. Pergamon Press. New York.

_____. 1968. Media and prospects for the cultivation of marine algae. *In* Watanabe, A., and A. Hattori (Ed.) Cultures and collections of algae. Proc. U.S.–Japan Conf. Hakone. Sept. 1966. Jap. Soc. Pl. Physiol. 63–75.

_____, AND S. H. HUTNER, AND A. SCHATZ. 1948. Streptomycin-induced chlorophyll-less races of *Euglena*. Proc. Soc. Exp. Biol. Med. 69: 279–282.

_____, L. T. YAMASU, AND I. MANTON. 1968. Experiments on the resynthesis of symbiosis in *Convoluta roscoffensis* with different flagellate cultures. J. Mar. Biol. Assn. U. K. 48: 465–479.

PUISEUX-DAO, S. 1970. *Acetabularia* and Cell Biology. Logos Press Ltd., Great Britain. xii + 162 pp.
_____. 1975. Third Symposium on *Acetabularia*. Protoplasma 83: 167–83.

QUATRANO, R. S. 1968. Rhizoid formation in *Fucus* zygotes: dependence on protein and ribonucleic acid syntheses. Science 162: 468–470.
_____. 1972. An ultrastructural study of the determined site of rhizoid formation in *Fucus* zygotes. Exp. Cell Res. 70: 1–12.
_____. 1973. Separation of processes associated with differentiation of two-celled *Fucus* embryos. Developmental Biology 30: 209–213.
_____. 1974. Developmental biology: development in marine organisms. *In* Experimental Marine Biology (Ed., R. N. Mariscal). Academic Press, New York. pp. 303–346.

RAHAT, M., AND Z. SPIRA. 1967. Specificity of glycerol for dark growth of *Prymnesium parvum*. J. Protozol. 14: 45–48.

RAMANATHAN, K. R. 1964. Ulotrichales. Indian Council of Agricultural Research. New Delhi. 188 pp.

RAMON, E., AND I. FRIEDMANN. 1965. The gametophyte of *Padina* in the Mediterranean. Proc. Intern. Seaweed Symp. 5: 183–196.

RAMUS, J. 1969A. The developmental sequence of the marine red algae *Pseudogloiophloea* in culture. Univ. Calif. Pub. Bot. 52: 1–42.
_____. 1969B. Pit connection formation in the red alga *Pseudogloiophloea*. J. Phycol. 5: 57–63.
_____. 1969C. Dimorphic pit connections in the red alga *Pseudogloiphloea*. J. Cell Biol. 41: 340–345.
_____. 1971. Properties of septal plugs from the red alga *Griffithsia pacifica*. Phycologia 10: 99–103.
_____. 1972A. Differentiation of the green alga *Codium fragile*. Am. J. Bot. 59: 478–482.
_____. 1972B. The production of extracellular polysaccharide by the unicellular red alga *Porphyridium aerugineum*. J. Phycol. 8: 97–111.
_____, AND S. T. GROVES. 1972. Incorporation of sulfate into the capsular polysaccharide of the red alga *Porphyridum* J. Cell Biol. 54: 399–407.

RANDHAWA, M. S. 1959. Zygnemaceae. Indian Council on Agr. Research. New Delhi. 478 pp.

RAO, V. N. R., AND T. V. DESIKACHARY. 1970. McDonald-Pfitzer hypothesis and cell size in diatoms. Beihefte Nova Hedw. 31: 485–493.

RASMONT, R. 1959. L'ultrastructure des choanocytes d'eponges. Ann. Sci. Nat. (Zool.) 12: 253–262.

RAVANKO, O. 1970. Morphological, developmental and taxonomic studies in the *Ectocarpus* complex (Phaeophyceae). Nova Hedw. 20: 179–252.

RAWITSCHER-KUNKEL, E., AND L. MACHLIS. 1962. The hormonal integration of sexual reproduction in *Oedogonium*. Am. J. Bot. 49: 177–183.

RAY, S. M., AND D. V. ALDRICH. 1967. Ecological interactions of toxic dinoflagellates and molluscs in the Gulf of Mexico. *In* Animal Toxins, (Ed., F. E. Russell and P. R. Saunders). Pergamon Press, Elmsford, New York. pp. 75–83.

RAYBURN, W. R. 1974. Sexual reproduction in *Pandorina unicocca*. J. Phycol. 10: 258–265.
_____, AND R. C. STARR. 1974. Morphology and nutrition of *Pandorina unicocca* sp. nov. J. Phycol. 10: 42–49.

RAYNS, D. G. 1962. Alternation of generations in a coccolithophorid, *Cricosphaera carterae* (Braarud and Fagerl.) Braarud. J. Mar. Biol. Assn. U. K. 42: 481–484.

REES, A. J. J., G. F. LEEDALE, AND H. A. CMIECH. 1974. *Paraphysomonas faveolata* sp. nov. (Chrysophyceae), a fourth marine species with meshwork body-scales. Br. Phycol. J. 9: 273–283.

REHÁKOVÁ, H. 1969. Die Variabilität der Arten der Gattung *Oocystis* A. Braun. *In* Studies in Phycology (Ed., B. Fott), Academia. Prague. pp. 145–196.

REID, P. C. 1972. Dinoflagellate cyst distribution around the British Isles. J. Mar. Biol. Assn. U. K. 52: 939–944.

REIMANN, B. E. F., J. C. LEWIN, AND B. E. VOLCANI. 1965. Studies on the biochemistry of silica shell formation in diatoms. I. The structure of the cell wall of *Cylindrotheca fusiformis* Reimann and Lewin. J. Cell Biol. 23: 39–55.

_____, _____, AND _____. 1966. Studies on the biochemistry and fine structure of silica shell formation in diatoms. II. The structure of the cell wall of *Navicula pelliculosa* (Breb.) Hilse. J. Phycol. 2: 74–84.

REINBOLD, T. 1928. Die Meeresalgen. Deutsch. Südpolar Exped. 1901–1903. Berlin and Leipzig, 8: 179–202.

REINHARDT, P. 1972. Coccolithen. Kalkiges Plankton seit Jahrmillionen. A. Ziemsen Verlag, Wittenberg. 99 pp.

RENTSCHLER, H.-G. 1967. Photoperiodische Induktion der Monosporenbildung bei *Porphyra tenera* Kjellm. (Rhodophyta-Bangiophyceae). Planta 76: 65–74.

RETALLACK, E. T., AND K. E. VON MALTZAHN. 1968. Some observations on zoosporogenesis in the female strain of *Oedogonium cardiacum*. Can. J. Bot. 46: 769–771.

RETALLBACK, B., AND R. D. BUTLER. 1970A. The development and structure of the zoospore vesicle in *Bulbochaete hiloensis*. Arch. Mikrobiol. 72: 223–237.

_____, AND _____. 1970B. The development and structure of pyrenoids in *Bulbochaete hiloensis*. J. Cell Sci. 6: 229–241.

_____, AND _____. 1972. Reproduction in *Bulbochaete hiloensis* (Nordst.) Tiffany. I. Structure of the zoospore. Arch. Mikrobiol. 86: 265–280.

REYMOND, O. 1974. Les fibrilles de *Coelastrum cambricum* (Chlorophycees). Arch. Mikrobiol. 95: 181–186.

RHODES, R. G., AND W. R. HERNDON. 1967. Relationship of temperature to zoospore production in *Tetraspora gelatinosa*. J. Phycol. 3: 1–3.

_____, AND P. E. STOFAN. 1967. *Tetraspora*, *Chlorosaccus*, and *Phaeosphaera*, a unique example of parallel evolution in the algae. J. Phycol. 3: 87–89.

RICHARDS, J. S., AND M. R. SOMMERFELD. 1974. Gamete activity in mating strains in *Chlamydomonas eugametos*. Arch. Microbiol. 98: 69–75.

RICHARDSON, W. N. 1970. Studies on the photobiology of *Bangia fuscopurpurea*. J. Phycol. 6: 216–219.

_____. 1972. Spore classification in the genera *Bangia* and *Porphyra*. Br. Phycol. J. 7: 49–51.

_____, AND P. S. DIXON. 1968. Life history of *Bangia fuscopurpurea* (Dillw.) Lyngb. in culture. Nature 218: 496–497.

_____, AND _____. 1969. The *Conchocelis* phase of *Smithora naiadum* (Anders.) Hollenb. Br. Phycol. J. 4: 181–183.

RICKETTS, T. R. 1974. The cultural requirements of the Prasinophyceae. Nova Hedw. 25: 683–690.

RIDER, D. E., AND R. H. WAGNER. 1972. The relationship of light, temperature, and current to the seasonal distribution of *Batrachospermum* (Rhodophyta). J. Phycol. 8: 323–331.

RIETEMA, H. 1969. New type of life history in *Bryopsis* (Chlorophyceae, Caulerpales). Acta Bot. Neerl. 18: 615–619.

_____. 1970. Life-histories of *Bryopsis plumosa* from European coasts. Acta Bot. Neerl. 19: 859–866.

_____. 1971. Life-history studies in the genus *Bryopsis* (Chlorophyceae). IV. Life-histories in *Bryopsis hypnoides* Lamx. from different points along the European coasts. Acta Bot. Neerl. 20: 291–298.

_____. 1972. A morphological, developmental and cytological study on the life-history of *Bryopsis halymeniae* (Chlorophyceae). Neth. J. Sea Res. 5: 445–457.

_____. 1973. The influence of day length on the morphology of the *Halicystis parvula* stage of *Derbesia tenuissima* (De Not.) Crn. (Chlorophyceae, Caulerpales). Phycologia 12: 11–16.

RIETH, A. 1959. Periodizität beim Ausschlüpfen der Schwärmsporen von *Vaucheria sessilis*. DeCandolle. Flora 147: 35–42.

_____. 1974. Beiträge zur Kenntnis der Vaucheriaceae. XVI. *Vaucheria hercyniana*, nov. spec. und ihre Entwicklung. Arch. Protistenk. 46: 201–209.

RIETSCHEL, S. 1977. Receptaculitids are calcareous algae but no Dasyclads. *In* Fossil Algae, Recent Results and Developments. (Ed., F. Flügel). Springer-Verlag, New York. pp. 212–214.

RILEY, J. P., AND T. R. S. WILSON, 1967. The pigments of some marine phytoplankton species. J. Mar. Biol. Assn. U. K. 47: 351–362.

RINGO, D. L. 1967A. Flagellar motion and fine structure of the flagellar apparatus in *Chlamydomonas*. J. Cell Biol. 33: 543–571.

————. 1967B. The arrangement of subunits in flagellar fibers. J. Ultrastr. Res. 17: 266–277.

RIPPKA, R., A. NEILSON, R. KUNISAWA, AND G. COHEN-BAZIRE. 1971. Nitrogen fixation by unicellular blue-green algae. Arch. Mikrobiol. 76: 341–348.

————, J. WATERBURY, AND G. COHEN-BAZIRE. 1974. A *Cyanobacterium* which lacks thylakoids. Arch. Mikrobiol. 100: 419–436.

RIS, H., AND D. F. KUBAI. 1974. An unusual mitotic mechanism in the parasitic Protozoan *Syndinium* sp. J. Cell Biol. 60: 702–720.

————, AND W. PLAUT. 1962. Ultrastructure of DNA-containing areas in the chloroplast of *Chlamydomonas*. J. Cell Biol. 13: 383–391.

RIZZO, P. J. AND L. D. NOODÉN. 1972. Chromosomal proteins in the dinoflagellate alga *Gyrodinium cohnii*. Science 176: 796–797.

————, AND ————. 1974A. Isolation and partial characterization of dinoflagellate chromatin. Biochim. Biophys. Acta 349: 402–414.

————, AND ————. 1974B. Partial characterization of dinoflagellate chromosome proteins. Biochim. Biophys. Acta 349: 415–27.

ROBERTS, G., AND J. C. W. CHEN. 1975. Chromosome analysis and amitotic nuclear division in *Nitella axillaris*. Cytologia 40: 151–156.

ROBERTS, K. 1974. Crystalline glycoprotein cell walls of algae: their structure, composition and assembly. Phil. Trans. Roy. Soc. London B. 268: 129–146.

————, M. GURNEY-SMITH, AND G. J. HILLS. 1972. Structure, composition and morphogenesis of the cell wall of *Chlamydomonas reinhardii*. I. Ultrastructure and preliminary chemical analyses. J. Ultrastr. Res. 40: 599–613.

ROBERTS, M. 1967. Studies on marine algae of the British Isles. 3. The genus *Cystoseira*. Br. Phycol. Bull. 3: 345–366.

ROBINSON, D. G., AND R. K. WHITE. 1972. The fine structure of *Oocystis apiculata* W. West with particular reference to the wall. Br. Phycol. J. 7: 109–118.

————, H. SACHS, AND F. MAYER. 1976. Cytokinesis in the green alga *Eremosphaera viridis*: plasmalemma formation from open membranes. Planta 129: 75–82.

ROELEVELD, J. G., M. DUISTERHOF, AND M. VROMAN. 1974. On the year cycle of *Petalonia fascia* in The Netherlands. Neth. J. Sea Res. 8: 410–26.

ROELOFSEN, P. A. 1959. The plant cell-wall. Encyclop. Plant Anatomy 3(4), Berlin. vii + 335 pp.

ROSOWSKI, J. R., AND P. KRUGENS. 1973. Observations on the euglenoid *Colacium* with special reference to the formation and morphology of attachment material. J. Phycol. 9: 370–383.

————, AND R. L. WILLEY. 1975. *Colacium libellae* sp. nov. (Euglenophyceae), a photosynthetic inhabitant of the larval damselfly rectum. J. Phycol. 11: 310–315.

————, AND ————. 1977. Development of mucilaginous surfaces in euglenoids. I. Stalk morphology of *Colacium mucronatum*. J. Phycol. 13: 16–21.

ROSS, M. M. 1959. Morphology and physiology of germination of *Chara gymnopitys* A. Br. I. Development and morphology of the sporeling. Austral. J. Bot. 7: 1–11.

ROSS, R., AND P. A. SIMS. 1972. The fine structure of the frustule in centric diatoms: a suggested terminology. Br. Phycol. J. 7: 139–163.

ROUND, F. E. 1963. The taxonomy of the Chlorophyta. Br. Phycol. Bull. 2: 224–235.

————. 1971. The taxonomy of the Chlorophyta. II. Br. Phycol. J. 6: 235–264.

————. 1973A. The Biology of the Algae. St. Martin's Press. New York. 278 pp.

————. 1973B. On the diatom genera *Stephanopyxis* Ehr. and *Skeletonema* Grev. and their classification in a revised system of the Centrales. Bot. Mar. 16: 148–154.

RUENESS, J. 1971. *Polysiphonia hemisphaerica* Aresch. in Scandinavia. Norw. J. Bot. 18: 65–74.

————. 1973. Culture and field observations on growth and reproduction of *Ceramium strictum* Harv. from the Oslofjord, Norway. Norw. J. Bot. 20: 61–65.

————, AND M. RUENESS. 1973. Life history and nuclear phases of *Antithamnion tenuissimum*, with special references to plants bearing both tetrasporangia and spermatangia. Norw. J. Bot. 20: 205–210.

————, AND ————. 1975. Genetic control of morphogenesis in two varieties of *Antithamnion plumula* (Rhodophyceae, Ceramiales). Phycologia 14: 81–85.

RUSSELL, D. J., AND R. E. NORRIS. 1971. Ecology and taxonomy of an epizoic diatom. Pacif. Sci. 25: 357–367.

RUSSELL, G. 1963. A study in populations of *Pilayella littoralis*. Mar. Biol. Assn. U. K. 43: 469–483.

————. 1964. Systematic position of *Pilayella littoralis* and status of the order Dictyosiphonales. Br. Phycol. Bull. 2: 322–326.

————. 1966. The genus *Ectocarpus* in Britain. I. The attached forms. J. Mar. Biol. Assn. U. K. 46: 267–294.

————. 1967. The genus *Ectocarpus* in Britain. II. The free-living forms. J. Mar. Biol. Assn. U. K. 47: 233–250.

SACHS, T., AND A. M. MAYER. 1961. Studies on the relation between metabolism and phototaxis of *Chlamydomonas snowiae*. Phycologia 1: 149–159.

SAFFERMAN, R. S. 1973. Phycoviruses. Chapter 11 *In* The Biology of Blue-green Algae (N. G. Carr and B. A. Whitton, Ed.). Univ. Calif. Press, Berkeley and Los Angeles. 676 pp.

————, AND M. E. MORRIS. 1963. Algal virus isolation. Science 140: 679–680.

SAGER, R. 1955. Inheritance in the green alga *Chlamydomonas reinhardi.* Genetics 40: 476–489.

————. 1959. The architecture of the chloroplast in relation to its photosynthetic activities. Brookhaven Symposia in Biology 11: 101–117.

————. 1972. Cytoplasmic genes and organelles. Academic Press, New York. 405 pp.

————. 1974. Nuclear and cytoplasmic inheritance in green algae. Chapter 11 *In* Algal Physiology and Biochemistry. (Ed., W. D. P. Stewart) Univ. Calif. Press, Berkeley and Los Angeles. pp. 314–345.

————, AND S. GRANICK. 1954. Nutritional control of sexuality in *Chlamydomonas reinhardi*. J. Gen. Physiol. 37: 729–742.

SAITO, S. 1972. Growth of *Gonium multicoccum* in synthetic media. J. Phycol. 8: 169–175.

————, AND T. ICHIMURA. 1975. Observations of colonial multiplication in a rapidly growing alga, *Gonium multicoccum* (Volvocaceae). Bot. Mag., Tokyo 88: 145–147.

SAITO, Y. 1964. Contributions to the morphology of the genus *Laurencia* of Japan. I. Bull. Fac. Fish. Hokkaido Univ. 15: 69–74.

————. 1965. Contributions to the morphology of the genus *Laurencia* of Japan. II. Bull. Fac. Fish. Hokkaido Univ. 15: 207–212.

————. 1969A. On morphological distinctions of some species of Pacific North American *Laurencia*. Phycologia. 8: 85–90.

————. 1969B. The algal genus *Laurencia* from the Hawaiian Islands, the Philippine Islands and adjacent areas. Pac. Sci. 23: 148–160.

————, AND H. B. S. WOMERSLEY. 1974. The southern Australian species of *Laurencia* (Ceramiales: Rhodophyta). Austral. J. Bot. 22: 815–874.

SANTELICES, B. 1974. Gelidioid algae, a brief resume of the pertinent literature. Marine Agronomy U.S. Sea Grant Program, Hawaii, Tech. Report No. 1. 111 pp.

SANTOS, G. A., AND M. S. DOTY. 1971. Constituents of the green alga *Caulerpa lamourouxii*. Lloydia 34: 88–90.

SARJEANT, W. A. S. 1965. The Xanthidia. Endeavour 24: 33–39.

————. 1974. Fossil and Living Dinoflagellates. Academic Press, New York. vii + 182 pp.

SARMA, Y. S. R. K. 1963. Contributions to the Karyology of the Ulotrichales. I. *Ulothrix*. Phycologia 2: 173–183.

————, AND B. R. CHAUDHARY. 1975A. An investigation on the cytology of *Ulva fasciata*. Bot. Mar. 18: 179–181.

————, AND ————. 1975B. On a new cytological race of *Schizomeris leibleinii* Kütz. Hydrobiol. 47: 171–181.

SASNER, J. J., JR., M. IKAWA, F. THURBERG, AND M. ALAM. 1972. Physiological and chemical studies on *Gymnodinum breve* Davis toxin. Toxicon 10: 163–172.

SAUVAGEAU, C. 1915. Sur la sexualité heterogamie d'une Laminaire (*Saccorhiza bulbosa*). C. R. Acad. Sci. (Paris). 161: 769–799.

_____. 1929. Sur le développement de quelques Phéosporées. Bull. Stat. Biol. Arcachon 26: 253–420.

SAWA, T. 1965. Cytotaxonomy of the Characeae: karyotype analysis of *Nitella opaca* and *Nitella flexilis*. Am. J. Bot. 52: 962–970.

_____, AND P. W. FRAME. 1974. Comparative anatomy of Charophyta: I. Oogonia and oospores of *Tolypella* with special reference to the sterile oogonial cell. Bull. Torrey Bot. Club 101: 136–144.

SCAGEL, R. F. 1947. An Investigation on Marine Plants near Hardy Bay, B. C. Provincial Dept. Fisheries, Victoria, B.C., Canada. pp. 1–70.

_____. 1956. Introduction of a Japanese alga, *Sargassum muticum*, into the Northwest Pacific. Fish. Res. Pap. Wash. Dept. Fish. 1: 49–58.

_____. 1957. An annotated list of the marine algae of British Columbia and northern Washington (including keys to genera). Nat. Mus. Can., Bull. 150, Biol. Ser. No. 52. Queen's Printer, Ottawa. vi + 289 pp.

_____. 1966. The Phaeophyceae in perspective. Oceanogr. Mar. Biol. Ann. Rev. 4: 123–194.

_____. 1967. Guide to common seaweeds of British Columbia. British Columbia Prov. Mus., Dept. of Recreation and Conservation, Handbook No. 27, 330 pp.

_____, R. J. BANDONI, G. E. ROUSE, W. B. SCHOFIELD, J. R. STEIN, AND T. M. C. TAYLOR. 1966. An Evolutionary Survey of the Plant Kingdom. Wadsworth Pub. Co., Belmont, Calif. xi + 658 pp.

SCHANTZ, E. J. 1967. Biochemical studies in purified *Gonyaulax catenella* poison. *In* Animal Toxins. (Ed., F. E. Russell and P. R. Saunders). Pergamon Press, Elmsford, New York. pp. 91–95.

_____. 1971. The dinoflagellate toxins. *In* Microbial Toxins (Ed., S. Kadis, A. Ciegler, and S. J. Ajl). Vol. VII, Algal and Fungal Toxins, Academic Press, New York. pp. 3–26.

_____, J. M. LYNCH, G. VAYVADA, K. MATSUMOTO, AND H. RAPOPORT. 1966. The purification and characterization of the poison produced by *Gonyaulax catenella* in axenic culture. Biochem. 5: 1191–1195.

SCHERBEL, G., W. BEHN, AND C. G. ARNVED. 1974. Untersuchungen zur genetischen Funktion des farblosen Plastiden von *Polytoma mirum*. Arch. Mikrobiol. 96: 205–222.

SCHIFF, J. A. 1962. Sulfur. *In* Physiology and Biochemistry of Algae (Ed., R. A. Lewin). Academic Press, New York. pp. 239–246.

SCHLICHTING, H. E., JR. 1964. Meteorological conditions affecting the dispersal of airborne algae and Protozoa. Lloydia 27: 64–78.

_____. 1970. Airborne algae and protozoa. Carolina Tips 33: 33–34.

_____. 1971. A preliminary study of algae and protozoa in seafoam. Bot. Mar. 14: 24.

_____. 1974A. Ejection of microalgae into the air via bursting bubbles. J. Allergy and Clin. Immunology 53: 185–188.

_____. 1974B. Survival of some fresh-water algae under extreme environmental conditions. Trans. Am. Micr. Soc. 93: 610–613.

_____, AND D. F. JAMES. 1972. Algae and Medicine. Carolina Tips 35: 29–30.

SCHLÖSSER, U. 1966. Enzymatisch gesteuerte Freisetzung von Zoosporen bei *Chlamydomonas reinhardi* Dangeard in Synchronkultur. Arch. Mikrobiol. 54: 129–159.

_____. 1976. Entwicklungstadien- und sippenspezifische Zellwand-Autolysine bei der Freisetzung von Fortpflanzungs zellen un der Gattung *Chlamydomonas*. Ber. Deutsch. Bot. Ges. 89: 1–56.

_____, H. SACHS, AND D. G. ROBINSON. 1976. Isolation of protoplasts by means of a "species-specific" autolysine in *Chlamydomonas*. Protoplasma 88: 51–64.

SCHMID, R. 1976. Septal pores in *Prototaxites*, an enigmatic Devonian plant. Science 191: 287–288.

SCHMITTER, R. E. 1971. The fine structure of *Gonyaulax polyedra*, a bioluminescent dinoflagellate. J. Cell Sci. 9: 147–173.

SCHMITZ, C. J. F. 1883. Untersuchungen über die Befruchtung der Florideen. Sitzungsber. K. Preuss. Akad. Wiss. Berlin 1883(1): 215–258.

SCHMITZ, K., AND L. M. SRIVASTAVA. 1974. Fine structure and development of sieve tubes in *Laminaria groenlandica* Rosenv. Cytobiologie 10: 66–87.

————, AND ————. 1975. On the fine structure of sieve tubes and the physiology of assimilate transport in *Alaria marginata*. Can. J. Bot. 53: 861–876.

SCHNEPF, E. 1969. Leukoplastin bei *Nitzschia alba*. Österr. Bot. Z. 116: 65–69.

————, AND G. DEICHGRÄBER. 1969. Über die Feinstruktur von *Synura petersenii* under besonder Berücksichtigung der Morphogenese ihrer Kieselschuppen. Protoplasma 68: 85–106.

————, AND W. KOCH. 1966. Über die Entstehung der pulsierenden Vacuolen von *Vacuolaria viresens* aus dem Golgi-Apparat. Arch. Mikrobiol. 54: 229–236.

SCHOPF, J. W. 1970. Pre-Cambrian micro-organisms and evolutionary events prior to the origin of vascular plants. Biol. Rev. 45: 319–352.

————. 1974. Paleobiology of the Precambrian: the age of blue-green algae. Chapter 1. *In* Dobzhansky, T., M. K. Hecht and W. C. Steere (Ed.) Evolutionary Biology I. Plenum Press, New York and London. pp. 1–43.

————. 1976. Are the oldest 'fossils' fossils? Origins of Life 7: 19–36.

————, AND J. M. BLACIC. 1971. New microorganisms from the Bitter Springs formation (Late Precambrian) of the North-Central Amadeus Basin, Australia. J. of Paleontology 45: 925–960.

SCHÖTZ, F., H. BATHELT, C.-G. ARWALD, AND O. SCHINNER. 1972. Die Architektur und Organization der *Chlamydomas*—Zelle. Protoplasma 75: 229–254.

SCHRADER, H.-J. 1966. Stacheleier (Hystrichosphaeren). Mikrokosmos 55: 111–116.

SCHREIBER, E. 1925. Zur Kenntnis der Physiologie und Sexualität höherer Volvcales. Z. Botan. 17: 337–376.

SCHULTE, H. 1964. Beiträge zur Cytologie von *Vaucheria* D. C. Protoplasma 58: 227–249.

SCHULTZ, M. E., AND F. R. TRAINOR. 1968. Production of male gametes and auxospores in the centric diatoms *Cyclotella meneghiniana* and *C. cryptica*. J. Phycol. 4: 85–88.

————, AND ————. 1970. Production of male gametes and auxospores in a polymorphic clone of the centric diatom *Cyclotella*. Can. J. Bot. 48: 947–951.

SCHUSTER, F. L. 1970. The trichocysts of *Chilomonas paramecium*. J. Protozool. 17: 521–526.

SCHÜTT, F. 1896. Bacillariales (Diatomaceae). *In* Engler and Prantl, Die Natürlichen Pflanzenfamilien, Vol. 1, part 1b. W. Engelmann, Leipzig. pp. 31–150.

SCHWARZ, E. 1932. Der Formenwechsel von *Ochrosphaera neapolitana*. Arch. Protistenk 77: 434–462.

SCHWEIGER, E., H. G. WALLRAFF, AND H. G. SCHWEIGER. 1964. Endogenous circadian rhythm in cytoplasm of *Acetabularia*: influence of the nucleus. Science 146: 658–659.

SCHWEIGER, H. G. 1969. Cell biology of *Acetabulania*. Current Topics in Microbiol. and Immunol. 50: 1–36.

————, S. BERGER, K. APEL, AND M. SCHWEIGER. 1972. *Acetabularia major*, a useful tool in cell biology. Protoplasma 75: 485–86.

————, ————, K. KLOPPSTECH, K. APEL, AND M. SCHWEIGER. 1974. Some fine structural and biochemical features of *Acetabularia major*. (Chlorophyta, Dasycladaceae) grown in the laboratory. Phycologia 13: 11–20.

SCHWIMMER, D., AND M. SCHWIMMER. 1964. Algae and medicine. *In* D. F. Jackson (Ed.), Algae and Man. Plenum Press, New York. 434 pp.

SCOTT, J. L., AND P. S. DIXON. 1971. The life history of *Pikea californica* Harv. J. Phycol. 7: 295–300.

————, AND ————. 1973A. Ultrastructure of tetrasporogenesis in the marine red alga *Ptilota hypnoides*. J. Phycol. 9: 29–46.

————, AND ————. 1973B. Ultrastructure of spermatium liberation in the marine red alga *Ptilota hypnoides*. J. Phycol. 9: 85–91.

SEARLES, R. B. 1968. Morphological studies of red algae of the order Gigartinales. Univ. Calif. Pub. Bot. 43. 100 pp.

SEARS, J. R., AND R. T. WILCE. 1970. Reproduction and systematics of the marine alga *Derbesia* (Chlorophyceae) in New England. J. Phycol. 6: 381–392.

SEGI, T. 1951. Systematic study of the genus *Polysiphonia* from Japan and its vicinity. J. Fac. Fish., Univ. Mie, 1: 169–272.

SEPSENWOL, S. 1973. Leucoplast of the Cryptomonad *Chilomonas paramecium*. Exp. Cell Res. 76: 395–409.

SESSOMS, A. H., AND R. J. HUSKEY. 1973. Genetic control of development in *Volvox*: Isolation and characterization of morphogenetic mutants. Proc. Nat'l. Acad. Sci. 70: 1335–1338.

SETCHELL, W. A. 1918. Parasitism among the red algae. Proc. Am. Phil. Soc. 57: 155–172.

———, AND N. L. GARDNER. 1925. The marine algae of the Pacific coast of North America. Part III. Melanophyceae. Univ. Calif. Pub. Bot. 8: 383–898.

SHEN, E. Y. F. 1966A. Oospore germination in two species of *Chara*. Taiwania 12: 39–46.

———. 1966B. Morphogenetic and cytological investigations of *Chara contraria* and *C. zeylanica*. Ph.D. Dissertation. The University of Texas, Austin. 130 pp.

———. 1967A. Amitosis in *Chara*. Cytologia 32: 481–488.

———. 1967B. Microspectrophotometric analysis of nuclear DNA in *Chara zeylanica*. J. Cell Biol. 35: 377–384.

———. 1967C. The amount of nuclear DNA in *Chara zeylanica* measured by microspectrophotometry. Taiwania 13: 111–114.

———. 1971. Cultivation of *Chara* in defined medium. *In* Contributions in Phycology (Ed., B. C. Parker and R. M. Brown, Jr.). Allen Press, Lawrence, Kansas. pp. 153–162.

SHIHIRA, I., AND R. W. KRAUSS. 1964. *Chlorella*: Physiology and taxonomy of forty-one isolates. University of Maryland, College Park. 97 pp.

SHILO, M. 1970. The action of *Prymnesium parvum* toxin on biological membranes. Toxicon 8: 153 (Abstr.).

SHTINA, E. A. 1974. The principal directions of experimental investigations in soil algology with emphasis on the U.S.S.R. Geoderma 12: 151–156.

SHYAM, R., AND Y. S. R. K. SARMA. 1975. On certain aspects of mitotic division in *Gonium pectorale* Muller. The Nucleus 18: 129–137.

———, AND ———. 1976. Effects of colchicine on the cell division of a colonial green flagellate *Gonium pectorale* Muller. Caryologica 29: 27–33.

SIEBERT, A. E. 1973. A description of *Haplozoon axiothellae* n. sp., an endosymbiont of the polychaete *Axiothella rubrocincta*. J. Phycol. 9: 185–190.

———, AND J. A. WEST. 1974. The fine structure of the parasitic dinoflagellate *Haplozoon axiothellae*. Protoplasma 81: 17–35.

SIEGEL, B. I., AND S. M. SIEGEL. 1973. The chemical composition of algal cell walls. Critical Reviews in Microbiol. 3: 1–26.

SIEVERS, A. M. 1969. Comparative toxicity of *Gonyaulax moniliata* and *Gymnodinium breve* to annelids, crustaceans, molluscs and a fish. J. Protozool. 16: 401–404.

SILVA, P. C. 1951. The genus *Codium* in California with observations on the structure of the walls of the utricles. Univ. Calif. Pub. Bot. 25: 79–114.

———. 1962A. Comparison of algal floristic patterns in the Pacific with those in the Atlantic and Indian Oceans, with special reference to *Codium*. Proc. Ninth Pac. Sci. Cong., 1957, 4: 201–216.

———. 1962B. Classification of algae. *In* Physiology and Biochemistry of Algae (Ed., R. A. Lewin). Appendix A, Academic Press, New York. pp. 827–837.

———, AND P. CLEARY. 1954. The structure and reproduction of the red alga, *Platysiphonia*. Am. J. Bot. 41: 251–260.

———, K. R. MATTOX, AND W. H. BLACKWELL. 1972. The generic name *Hormidium* as applied to green algae. Taxon 21: 639–645.

———, AND G. F. PAPENFUSS. 1953. A systematic study of the algae of sewage oxidation ponds. State Water Pollution Control Board, Sacramento. Pub. No. 7. 35 pp.

SILVERBERG, B. A. 1975. Some structural aspects of the pyrenoid of the Ulotrichalean alga *Stichococcus*. Trans. Am. Micro. Soc. 94: 417–421.

SILVESTER, W. B., AND P. J. MCNAMARA. 1976. The infection process and ultrastructure of the *Gunnera-Nostoc* symbiosis. New Phytol. 77: 135–141.

SIMON, R. D. 1971. Cyanophycin granules from the blue-green alga *Anabaena cylindrica*: a reserve material consisting of copolymers of aspartic acid and alanine. Proc. Nat. Acad. Sci. U.S. 68: 265–267.

SIMON-BICHARD-BREAUD, J. 1971. Un appareil cinétique dans les gametocystes mâles d'un Rhodophycée: *Bonnemaisonia hamifera* Hariot. C. R. Acad. Sci. (Paris). 273D: 1272–1275.

_____. 1972. Formation de la crypte flagellaire et evolution de son contenu au cours de la gametogenèse mâle chez *Bonnemaisonia hamifera* Hariot (Rhodophycée). C. R. Acad. Sci. (Paris). 274D: 1796–1799.

SIMONETTI, G., G. GIACCONE, AND S. PIGNATTI. 1970. The seaweed *Gracilaria confervoides*, an important object for autecologic and cultivation research in the northern Adriatic Sea. Helgoländ. Wissen. Meeresunter. 20: 89–96.

SIMONSEN, R. 1970. Protoraphidaceae, eine neue Familie der Diatomeen. Beihefte z. Nova Hedw. 31: 383–394.

_____. 1972. Ideas for a more natural system of the centric diatoms. Beihefte z. Nova Hedw. 39: 37–54.

SIMPSON, T. L. 1968. The structure and function of sponge cells; new criteria for the taxonomy of Peocilosclerid sponges (Demospongiae). Peabody Mus. Nat. Hist. Yale Univ. Bull. 25. 142 pp.

SINGH, K. P. 1956. Studies in the genus *Trachelomonas*. II. Cell structure and reproduction with special reference to *T. grandis*. Am. J. Bot. 43: 274–280.

SINGH, R. N. 1942. Reproduction in *Draparnoldiopsis indica* Bharadwaja. New Phytol. 4: 262–273.

_____. 1945. Nuclear phases and alternation of generations in *Draparnaldiopsis indica* Bharadwaja. New Phytol. 44: 118–129.

_____, AND D. N. TIWARI. 1970. Frequent heterocyst germination in the blue-green alga *Gloeotrichia ghosei* Singh. J. Phycol. 6: 172–176.

SLANKIS, T., AND S. P. GIBBS. 1972. The fine structure of mitosis and cell division in the chrysophycean alga *Ochromonas danica*. J. Phycol. 8: 243–256.

SMAYDA, T. J. 1970. The suspension and sinking of phytoplankton in the sea. Oceanogr. Mar. Biol. Ann. Rev. 8: 353–414.

SMITH, D. C. 1973. Symbiosis of algae with invertebrates. Oxford Biology Readers (J. J. Head and O. E. Lowenstein, Ed.). Oxford Univ. Press, London. 16 pp.

_____, L. MUSCATINE, AND D. H. LEWIS. 1969. Carbohydrate movement from autotrophs to heterotrophs in parasitic and mutualistic symbiosis. Biol. Rev. Biol. Proc. Cambridge Phil. Soc. 44: 17–90.

SMITH, G. M. 1933. Fresh-Water Algae of the United States. McGraw-Hill Book Co., New York. 716 pp.

_____. 1944. Marine Algae of the Monterey Peninsula, California. Stanford Univ. Press, Stanford. ix + 622 pp.

_____. 1947. On the reproduction of some Pacific Coast species of *Ulva*. Am. J. Bot. 34: 80–87.

_____. 1950. The Fresh-Water Algae of the United States. 2nd ed., McGraw-Hill Book Co., New York. vii + 719 pp.

_____, (Ed.). 1951. Manual of Phycology. Chronica Botanica Co., Waltham, Mass. 375 pp.

_____. 1955. Cryptogamic Botany. Vol. I. McGraw-Hill Book Co., New York. ix + 546 pp.

SMITH, K. M., R. M. BROWN, JR., AND P. L. WALNE. 1966. Culture methods for the blue-green alga *Plectonema boryanum* and its virus with an electron microscope study of the virus-infected cells. Virology 28: 580–591.

_____, _____, AND _____. 1967. Ultrastructural and time-lapse studies of the replication cycle of the blue-green algal virus LPP-1. Virology 31: 329–337.

_____, _____, _____, AND D. A. GOLDSTEIN. 1966. Electron microscopy of the infection process of the blue-green algal virus. Virology 30: 182–192.

SMITH, R. L., AND H. C. BOLD. 1966. Phycological studies. VI. Investigations of the algal genera *Eremosphaera* and *Oocystis*. Univ. Texas Pub. 6612. Austin 121 pp.

SMITH, W. O., JR. 1974. The extracellular release of glycolic acid by a marine diatom. J. Phycol. 10: 30–33.

SNELL, W. J. 1976A. Mating in *Chlamydomonas*. A system for the study of specific cell adhesion. I. Ultrastructure and electrophoretic analysis of flagellar surface components involved in adhesion. J. Cell Biol. 68: 48–69.

————. 1976B. II. A radioactive flagella-binding assay for quantitation of adhesion. J. Cell Biol. 68: 70–79.

SÖDERSTRÖM, J. 1970. Remarks on the European species of *Nemalion*. Bot. Mar. 13: 81–86.

SOEDER, C. J. 1965. Elektronmicroscopische Untersuchungen der Protoplastteilung bei *Chlorella fusca* Shihira and Krauss. Arch. Mikrobiol. 50: 368–377.

SOLI, G. 1966. Bioluminescent cycle of photosynthetic dinoflagellates. Limnol. Oceanogr. 11: 355–363.

SOMMERFELD, M. R., AND G. F. LEEPER. 1970. Pit connections in *Bangia fuscopurpurea*. Arch. Mikrobiol. 73: 55–60.

————, AND H. W. NICHOLS. 1970. Comparative studies in the genus *Porphyridium* Naeg. J. Phycol. 6: 67–78.

————, AND ————. 1973. The life cycle of *Bangia fuscopurpurea* in culture. I. Effects of temperature and photoperiod on the morphology and reproduction of the *Bangia* phase. J. Phycol. 9: 205–210.

SOURNIA, A. 1967. Le genre *Ceratium* (Péridinien planctonique) dans le Canal de Mozambique. Contribution à une révision mondiale. Vie et Milieu, sér. A. 18: 375–440; 441–499.

————. 1973. Catalogue des espèces et taxons infraspécifiques de Dinoflagellés marins actuels publiés depuis la revision de J. Schiller. I. Dinoflagellés libres. Beihefete z. Nova Hedw. 48. X + 92 pp.

————, J. CACHON, AND M. CACHON. 1975. Catalogue des espèces et taxons infraspécifiques de Dinoflagelles marins actuels publies depuis la révision de J. Schiller. II. Dinoflagellés parasites ou symbiotiques. Arch. Protistenk. 117: 1–19.

SOUTH, G. R. 1970. Experimental culture of *Alaria* in a sub-arctic free-flowing sea water system. Helgoländ. Wissen. Meeresunter. 20: 216–228.

————, AND E. M. BURROWS. 1967. Studies on marine algae of the British Isles. 5. *Chorda filum* (L.) Stackh. Br. Phycol. Bull. 3: 379–402.

————, AND A. CARDINAL. 1970. A checklist of marine algae in eastern Canada. Can. J. Bot. 48: 2077–2095.

————, AND R. D. HILL. 1970. Studies on marine algae of Newfoundland. I. Occurrence and distribution of free-living *Ascophyllum nodosum* in Newfoundland. Can. J. Bot. 48: 1697–1701.

————, AND ————. 1971. Studies on the marine algae of Newfoundland. II. On the occurrence of *Tilopteris mertensii*. Can. J. Bot. 49: 211–213.

————, AND R. HOOPER. 1976. *Stictyosiphon soriferus* (Phaeophyta, Dictyosiphonales) from eastern North America. J. Phycol. 12: 24–29.

SOYER, M.-O. 1968. Étude cytologique ultrastructurale d'un Dinoflagellé libre, *Noctiluca miliaris* S. Trichocystes et inclusions paracristallines. Vie et Milieu 19: 305–314.

————. 1969. Rapports existant entre chromosomes et membrane nucléaire chez un Dinoflagellé parasite du genre *Blastodinium* Chatton. C. R. Acad. Sci. (Paris). 268D: 2082–2084.

————. 1971. Structure du noyau des *Blastodinium* (Dinoflagellés parasites). Division et condensation chromatique. Chromosoma 33: 70–114.

————. 1972. Ultrastructure of the nucleus of *Noctiluca* (free-living dinoflagellata) during sporulation. Chromosoma 39: 419–441.

SPARLING, S. R. 1957. The structure and reproduction of some members of the Rhodymeniaceae. Univ. Calif. Pub. Bot. 29: 319–393.

————. 1961. A report on the culture of some species of *Halosaccion*, *Rhodymenia* and *Fauchea*. Am. J. Bot. 48: 493–499.

SPRING, H., U. SHEER, W. W. FRANKE, AND M. F. TRENDELENBURG. 1975. Lampbrush-type chromosomes in the primary nucleus of the green alga *Acetabularia mediterrancea*. Chromosoma 50: 25–43.

————, M. F. TRENDELENBURG, U. SCHEER, W. W. FRANKE, AND W. HERTH. 1974. Structural and biochemical studies of the primary nucleus of two green algal species, *Acetabularia mediterranea* and *Acetabularia major*. Cytobiologie 10: 1–65.

STAEHELIN, L. A., AND J. D. PICKETT-HEAPS. 1975. The ultrastructure of *Scenedesmus* (Chlorophyceae). I. Species with the "reticulate" or "warty" type of ornamental layer. J. Phycol. 11: 163–185.

STANIER, R. Y., R. KUNISAWA, M. MANDEL, AND G. COHEN-BAZIRE. 1971. Purification and properties of unicellular blue-green algae (Order Chroococcales). Bacteriol. Rev. 35: 171–205.

STARMACH, K. 1972. *Chrysosphaera stigmatica* n. sp. (Chrysophyceae). Bull. Acad. Polon. Sci. ser. Biol. 20: 577–579.

STARR, R. C. 1953. On the morphology and reproduction of *Characium saccatum* Filarsky. Bull. Torrey Bot. Club 80: 308–313.

_____. 1954A. Heterothallism in *Cosmarium botrytis* var. *subtumidum*. Am. J. Bot. 41: 601–606.

_____. 1954B. Inheritance of mating type and a lethal factor in *Cosmarium botrytis* var. *subtumidum* Wittr. Proc. Natl. Acad. Sci. 40: 1060–1063.

_____. 1955A. Sexuality in *Gonium sociale* (Dujardin) Warming. J. Tenn. Acad. Sci. 30: 90–93.

_____. 1955B. Zygospore germination in *Cosmarium botrytis* var. *subtumidum*. Am. J. Bot. 42: 577–581.

_____. 1955C. Isolation of sexual strains of placoderm desmids. Bull. Torrey Bot. Club 82: 261–265.

_____. 1955D. A comparative study of *Chlorococcum* Meneghini and other spherical, zoospore-producing genera of the Chlorococcales. Indiana Univ. Pub. Sci. Ser. No. 20, Indiana Univ. Press, Bloomington, Ind. 111 pp.

_____. 1958A. Asexual spores in *Closterium didymotocum* Ralfs. New Phytol. 57: 187–190.

_____. 1958B. The production and inheritance of the triradiate form in *Cosmarium turpinii*. Am. J. Bot. 45: 243–248.

_____. 1962. A new species of *Volvulina* Playfair. Arch. Mikrobiol. 42: 130–137.

_____. 1963. Homothallism in *Golenkinia minutissima*. In Studies in microalgae and bacteria. Jap. Soc. Pl. Physiol., University of Tokyo Press. pp. 3–6.

_____. 1964. The culture collection of algae at Indiana University. Am. J. Bot. 51: 1013–1044.

_____. 1968. Cellular differentiation in *Volvox*. Proc. Natl. Acad. Sci. 59: 1082–1088.

_____. 1969. Structure, reproduction and differentiation in *Volvox carteri* f. *nagariensis* Iyengar, strains HK9 and 10. Arch. Protistenk. 111: 204–222.

_____. 1970. *Volvox pocockiae*, a new species with dwarf males. J. Phycol. 6: 234–239.

_____. 1971A. Sexual reproduction in *Volvox africanus* In Contributions in Phycology, (Ed., B. C. Parker and R. M. Brown, Jr.). Allen Press, Lawrence, Kansas. pp. 59–66.

_____. 1971B. The Culture Collection of Algae at Indiana University. Additions to the collection July 1966–July 1971. J. Phycol. 7: 350–352.

_____. 1972A. Control of differentiation in *Volvox*. Develop. Biology, Supplement 4: 59–100.

_____. 1972B. A working model for the control of differentiation during development of the embryo of *Volvox carteri* f. *nagariensis*. Mém. Soc. Bot. Fr. 1972: 175–182.

_____. 1973. Isolation and purification: special methods—dry soil samples. In Handbook of Phycological Methods (Ed., J. R. Stein). University Press, Cambridge. pp. 159–167.

_____. 1975. Meiosis in *Volvox carteri* f. *nagariensis*. Arch. Protistenk. 117: 187–191.

_____, AND L. JAENICKE. 1974. Purification and characterization of the hormone initiating sexual morphogenesis in *Volvox carteri* f. *nagariensis* Iyengar. Proc. Natl. Acad. Sci. 71: 1050–1054.

_____, AND W. R. RAYBURN. 1964. Sexual reproduction in *Mesotaenium kramstai*. Phycologia 4: 23–26.

STAVIS, R. L., AND R. HIRSCHBERG. 1973. Phototaxis in *Chlamydomonas reinhardtii*. J. Cell Biol. 59: 367–377.

STEELE, R. L. 1965. Induction of sexuality in two centric diatoms. BioScience 15: 298.

STEIDINGER, K. A. 1973. Phytoplankton ecology: a conceptual review based on eastern Gulf of Mexico research. CRC Critical Reviews in Microbiology 3: 49–68.

_____, M. A. BURKLEW, AND R. M. INGLE. 1973. The effects of *Gymnodinium breve* toxin on estuarine animals. In Marine Pharmacognosy (Ed., D. Martin and G. Padilla). Academic Press, New York. pp. 179–202.

_____ J. T. DAVIS, AND J. WILLIAMS. 1967. A key to the marine dinoflagellate genera of the west coast of Florida. State of Florida Bd. of Conserv. Tech. Ser. No. 52. 45 pp.

_____, AND R. M. INGLE. 1972. Observations on the 1971 summer red tide in Tampa Bay, Florida. Environ. Lett. 3: 271–277.

_____, AND E. A. JOYCE, JR. 1973. Florida red tides. Fla. Dept. Natl. Resour. Mar. Res. Lab. Educ. Ser. 17, 26 pp.

_____, AND J. WILLIAMS. 1970. Dinoflagellates. Mem. Hourglass Cruises (Florida Dept. Natural Resources) 2: 1–251.

STEIN, J. R. 1958A. A morphologic and genetic study of *Gonium pectorale*. Am. J. Bot. 45: 664–672.

_____. 1958B. A morphological study of *Astrephomene gubernaculifera* and *Volvulina steinii*. Am. J. Bot. 45: 388–397.

_____. 1963. Morphological variation of a *Tolypothrix* in culture. Br. Phycol. Bull. 2: 206–209.

_____. 1965. On cytoplasmic strands in *Gonium pectorale* (Volvocales) J. Phycol. 1: 1–5.

_____. 1966A. Growth and Mating of *Gonium pectorale* (Volvocales) in defined media. J. Phycol. 2: 23–28.

_____. 1966B. Effect of temperature on sexual populations of *Gonium pectorale* (Volvocales). Am. J. Bot. 53: 941–944.

_____. 1973. Handbook of Phycological Methods. University Press, Cambridge. 448 pp.

_____, AND C. C. AMUNDSEN. 1967. Studies on snow algae and fungi from the Front Range of Colorado. Can. J. Bot. 45: 2033–2045.

_____, AND R. C. BROOKE. 1964. Red snow from Mt. Seymour, British Columbia. Can. J. Bot. 42: 1183–1188.

STEWART, J. G. 1968. Morphological variation in *Pterocladia pyramidale*. J. Phycol. 4: 76–84.

STEWART, J. K. 1971. The biology of the green alga *Characiosiphon rivularis* Iyengar. Ph.D. Dissertation. The University of Texas at Austin.

STEWART, J. R., AND J. C. O'KELLEY. 1966. Periodic zoospore production by *Protosiphon botryoides* under alternating light-dark periods. Am. J. Bot. 53: 772–777.

STEWART K. D., G. L. FLOYD, K. R. MATTOX, AND M. E. DAVIS. 1972. Cytochemical demonstration of a single peroxisome in a filamentous green alga. J. Cell Biol. 54: 431–434.

_____, AND K. R. MATTOX. 1975. Comparative cytology, evolution and classification of the green algae with some consideration of the origin of other organisms with chlorophylls *a* and *b*. Bot. Rev. 41: 104–135.

_____, _____, AND G. L. FLOYD. 1972. Comparative cytology of the genus *Ulothrix* with special emphasis on *U. zonata*. J. Phycol. 8(suppl.): 8.

_____, _____, AND _____. 1973. Mitosis, cytokinesis, the distribution of plasmodesmata, and other cytological characteristics in the Ulotrichales, Ulvales and Chaetophorales: phylogenetic and taxonomic considerations. J. Phycol. 9: 128–141.

STEWART, V. N., H. WAHLQUIST, R. BURKET, AND C. WAHLQUIST. 1966. Observations of vitamin B_{12} distribution in Apalachee Bay, Florida. Fla. Bd. Conserv. Mar. Lab. Prof. Pap. Ser. No. 8: 34–42.

STEWART, W. D. P. (Ed.). 1974. Algal Physiology and Biochemistry. Univ. Calif. Press., Berkeley and Los Angeles. 989 pp.

_____, AND M. LEX. 1970. Nitrogenase activity in the blue-green alga *Plectonema boryanum* strain 594. Arch. Mikrobiol. 73: 250–260.

STOERMER, E. F., H. S. PANKRATZ, AND C. C. BOWEN. 1965. Fine structure of the diatom *Amphipleura pellucida*. II. Cytoplasmic fine structure and frustule formation. Am. J. Bot. 52: 1067–1078.

STOSCH, H. A. VON. 1951. Entwicklungsgeschichtliche Untersuchungen an zentrischen Diatomeen. I. Die Auxosporenbildung von *Melosira varians*. Arch. Mikrobiol. 16: 101–135.

_____. 1954. Die Oogamie von *Biddulphia mobiliensis* und die bisher bekannten Auxosporenbildungen bei den Centrales. Intern. Bot. Cong. 8, Rap. Com., Sec. 17: 58–68.

_____. 1955. Ein morphologischer Phasenwechsel bei einer Coccolithophoride. Naturwissensch. 42: 423.

_____. 1958A. Entwicklungsgeschichtliche Untersuchungen an zentrischen Diatomeen. III. Die Spermatogenese von *Melosira moniliformis* Agardh. Arch. Mikrobiol. 31: 274–282.

_____. 1958B. Kann die oogame araphidee *Rhabdonema adriaticum* als Bindeglied zwischen den beiden grossen Diatomeengruppen angesehen werden? Ber. Deutsch. Bot. Gesellsch. 71: 241–249.

_____. 1962. Über das Perizonium der Diatomeen. Vortr. Gesamtgeb. Bot. Nat. 1: 43–52.

_____. 1964A. Zum Problem der sexuellen Fortpflanzung in der Peridineengattung *Ceratium*. Helgoland. Wiss. Meeresunters. 10: 140–152.

_____. 1964B. Wirkungen von Jod und Arsenit auf Meeresalgen in Kultur. Proc. Intern. Seaweed Symp. 4: 142–150.

_____. 1965A. Sexualität bei *Ceratium cornutum* (Dinophyta). Naturwissensch. 52: 112–113.

_____. 1965B. Manipulierung der Zellgrösse von Diatomeen im Experiment. Phycolgia 5: 21–44.

_____. 1965C. The sporophyte of *Liagora farinosa* Lamour. Brit. Phycol. Bull. 2: 486–496.

_____. 1967. Haptophyceae. *In* Vegetative Fortpflanzung, Parthenogenese und Apogamie bei Algen. (Ed., W. Ruhland). Encyclopedia of Plant Physiology. 18: 646–656.

_____. 1969. Observations on *Corallina, Jania* and other red algae in culture. Proc. Intern. Seaweed Symp. 6: 389–399.

_____. 1972. La signification cytologique de la "cyclose nucléaire" dans le cycle de vie des Dinoflagellés. Soc. Bot. Fr., Mém., 1972: 201–212.

_____. 1973. Observations on vegetative reproduction and sexual life cycles of two freshwater dinoflagellates, *Gymnodinium pseudopalustre* Schiller and *Woloszynskia apiculata* sp. nov. Br. Phycol. J. 8: 105–134.

_____, AND G. DREBES. 1964. Entwickungsgeschichtliche Untersuchungen an zentrischen Diatomeen IV. Die Planktondiatomee *Stephanophyxis turris*-ihre Behandlung und Entwicklungsgeschichte. Helgoländ. Wissen. Meeresunters. 11: 209–257.

_____, AND B. E. F. REIMANN. 1970. *Subsilicea fragilarioides* gen. et spec. nov., eine Diatomee (Fragilariaceae) mit vorwiegend organischer Membran. Beihefte z. Nova Nedw. 31: 1–36.

_____, G. THEIL, AND K. V. KOWALLIK. 1973. Entwicklungsgeschichtliche Untersuchungen an zentrischen Diatomeen V. Bau und Lebenszyklus von *Chaetoceros didymum*, mit Beobachtungen über einige andere Arten der Gattung. Helgoländ. Wiss. Meeresunters. 25: 384–445.

STREHLOW, K. 1929. Über die Sexualität einiger Volvocales. Z. Bot. 21: 625–692.

STURCH, H. H. 1926. *Choreocolax polysiphoniae*, Reinsch. Ann. Bot. 40: 585–605.

SULEK, J. 1969. Taxonomische Übersicht der Gattung *Pediastrum* Meyen. *In* Fott, B. (Ed.) Studies in Phycology. Academica, Prague. pp. 197–261.

_____. 1975. Nuclear division in *Scenedesmus quadricauda* (Turp.) Bréb. Arch. Hydrobiol. Suppl. 46: 224–258.

SUNDARALINGAM, V. S. 1974. The developmental morphology of *Chara zeylanica* Willd. J. Indian Bot. Soc. 33: 272–297.

SUNDENE, O. 1959. Form variation in *Antithamnion plumula*. Nytt Mag. Botanik 7: 181–187.

_____. 1962. Reproduction and morphology in strains of *Antithamnion boreale* originating from Spitsbergen and Scandinavia. Det Norske Videnskaps–Akad. Oslo. I. Mat.–Nat. Kl. N. S. 5: 1–19.

_____. 1963. Reproduction and ecology of *Chorda tomentosa*. Nytt Mag. Bot. 10: 159–167.

_____. 1964A. The ecology of *Laminaria digitata* in Norway in view of transplant experiments. Nytt Mag. Bot. 11: 83–107.

_____. 1964B. *Antithamnion teniussimum* (Hauck) Schiffner in culture. Nytt Mag. Bot. 12: 5–10.

_____. 1966. *Haplospora globosa* Kjellm. and *Scaphospora speciosa* Kjellm. in culture. Nature 209: 937–938.

_____. 1975. Experimental studies on form variation in *Antithamnion plumula* (Rhodophyceae). Norw. J. Bot. 22: 35–42.

SUNESON, S. 1950. The cytology of bispore formation in two species of *Lithophyllum* and the significance of the bispores in the Corallinaceae. Bot. Notiser 1950: 429–450.

SUSSEX, I. M. 1967. Polar growth of *Hormosira banksii* zygotes in shake culture. Am. J. Bot. 54: 505–510.

SUTO, S. 1963. Intrageneric and intraspecific crossings of the lavers (*Porphyra*). Bull. Jap. Soc. Sci. Fish. 29: 739–748.

_____. 1972. Variation in species characters of *Porphyra* under culture conditions. *In* Contributions to the Systematics of Benthic Marine Algae of the North Pacific (Ed., I. A. Abbott and M. Kurogi). Japan. Soc. Phycology, Kobe. pp. 193–201.

SVEDELIUS, N. 1931. Nuclear phases and alternation in the Rhodophyceae. Beih. Bot. Centralbl. 48: 38–59.

————. 1933. On the development of *Asparagopsis armata* Harv. and *Bonnemaisonia asparagoides* (Woodw.) Ag. Nova Acta Reg. Soc. Sci. Upsal., Ser. 4, 9(1): 1–61.

————. 1944. *Galaxaura*, a diplobiontic Floridean genus within the order Nemalionales. Farlowia 1: 495–499.

SVENDSEN, P., AND J. M. KAIN. 1971. The taxonomic status, distribution, and morphology of *Laminaria cucullata* sensu Jorde and Klavestad. Sarsia 46: 1–22.

SWALE, E. M. F. 1969. A study of the nannoplankton flagellate *Pedinella hexacostata* Vysotskii by light and electron microscopy. Br. Phycol. J. 4: 65–86.

————. 1973. A third layer of body scales in *Pyramimonas tetrarhynchus* Schmarda Br. Phycol. J. 8: 95–99.

————, AND J. H. BELCHER. 1968. The external morphology of the type species of *Pyramimonas* (*P. tetrarhynchus* Schmarda) by electron microscopy. Proc. Linn Soc. London. 179: 77–81.

SWEENEY, B. M. 1960. The photosynthetic rhythm in single cells of *Gonyaulax polyedra*. Cold Spring Harb. Symp. Quant. Biol. 25: 145–148.

————. 1963. Biological clocks in plants. Ann. Rev. Plant Physiol. 14: 411–440.

————. 1969A. Transducing mechanisms between circadian clock and overt rhythms in *Gonyaulax*. Can. J. Bot. 47: 299–308.

————. 1969B. Rhythmic Phenomena in Plants. Academic Press, New York. 147 pp.

————. 1976A. Freeze-fracture studies of thecal membranes of *Gonyaulax polyedra*: circadian changes in the particles of one membrane face. J. Cell Biol. 68: 451–461.

————. 1976B. Red tides. Nat. Hist. 85(7): 78–83.

————, AND J. W. HASTINGS, 1957. Characteristics of the diurnal rhythm of luminescence in *Gonyaulax polyedra*. J. Cell Comp. Physiol. 49: 115–128.

————, AND ————. 1958. Rhythmic cell division in populations of *Gonyaulax polyedra*. J. Protozool. 5: 217–224.

————, AND ————. 1962. Rhythms. *In* Physiology and Biochemistry of Algae (Ed., R. A. Lewin). Academic Press, New York. pp. 687–698.

————, AND F. T. HAXO. 1961. Persistence of a photosynthetic rhythm in enucleated *Acetabularia*. Science 134: 1361–1363.

SWIFT, E., AND E. G. DURBIN. 1971. Similarities in the asexual reproduction of the oceanic dinoflagellates, *Pyrocystis fusiformis*, *Pyrocystis lunula*, and *Pyrocystis noctiluca*. J. Phycol. 7: 89–96.

————, AND C. C. REMSEN. 1970. The cell wall of *Pyrocystis* spp. (Dinococcales). J. Phycol. 6: 79–85.

————, AND W. R. TAYLOR. 1967. Bioluminescence and chloroplast movement in the dinoflagellate *Pyrocystis lunula*. J. Phycol. 3: 77–81.

SZOSTAK, J. W., J. SPARKUHL, AND M. E. GOLDSTEIN. 1973. Sexual induction in *Eudorina*: Effects of light, nutrients and conditional medium. J. Phycol. 9: 215–218.

TAI, L., AND T. SKOGSBERG. 1934. Studies on the Dinophysidae, marine armored dinoflagellates, of Monterey Bay, California. Arch. Protistenk. 82: 380–482.

TAKAHASHI, E. 1972. Studies on genera *Mallomonas* and *Synura*, and other plankton in freshwater with the electron microscope. III. On three new species of Chrysophyceae. Bot. Mag., Tokyo 85: 293–302.

————. 1975. The fine structure of the scales and flagella of the Chrysophyta. *In* "Advances of Phycology in Japan" (Ed., J. Tokida and H. Hirose). G. Fischer Verlag, Jena. pp. 67–97.

TAKATORI, S., AND K. IMAHORI. 1971. Light reactions in the control of oospore germination in *Chara delicatula*. Phycologia 10: 221–228.

TASSIGNY, M. 1971. La sexualité des Desmidiées. L'Année Biol. 10: 403–429.

TATEWAKI, M. 1966. Formation of a crustaceous sporophyte with unilocular sporangia in *Scytosiphon lomentaria*. Phycologia 6: 62–66.

————. 1972. Life history and systematics in *Monostroma*. *In* Contributions to the Systematics of Benthic Marine Algae of the North Pacific. Japanese Society of Phycology. pp. 1–15.

_____, AND K. NAGATA. 1970. Surviving protoplasts *in vitro* and their development in *Bryopsis*. J. Phycol. 6: 401–403.

TAYLOR, D. L. 1968. Chloroplasts as symbiotic organelles in the digestive gland of *Elysia viridis* [Gastropoda: Opisthobranchia]. J. Mar. Biol. Assn. U. K. 48: 1–15.

_____. 1969A. The nutritional relationship of *Anemonia sulcata* (Pennant) and its dinoflagellate symbiont. J. Cell Sci. 4: 751–762.

_____. 1969B. Some aspects of the regulation and maintenance of algal numbers in zooxanthellae-coelenterate symbioses, with a note on the nutritional relationship in *Anemonia sulcata*. J. Mar. Biol. Assn. U. K. 49: 1057–1065.

_____. 1969C. Identity of zooxanthellae isolated from some Pacific Tridacnidae. J. Phycol. 5: 336–340.

_____. 1971. Ultrastructure of the 'zooxanthella' *Endodinium chattonii* in situ. J. Mar. Biol. Assn. U. K. 51: 227–234.

_____. 1972. Ultrastructure of *Cocconeis diminuta* Pantoczek. Arch. Mikrobiol. 81: 136–145.

_____. 1973. The cellular interaction of algal-invertebrate symbiosis. Adv. Mar. Biol. 11: 1–56.

_____. 1974. Nutrition of algal-invertebrate symbiosis. I. Utilization of soluble organic nutrients by symbiont-free hosts. Proc. Roy. Soc., Ser. B, 186: 357–368.

_____, AND C. C. LEE. 1971. A new cryptomonad from Antarctica: *Cryptomonas cryophila* sp. nov. Arch. Mikrobiol. 75: 269–280.

TAYLOR, F. J. R. 1971. Scanning electron microscopy of thecae of the dinoflagellate genus *Ornithocercus*. J. Phycol. 7: 249–258.

_____. 1973. Topography of cell division in the structurally complex dinoflagellate genus *Ornithocercus*. J. Phycol. 9: 1–10.

_____. 1975. Non-helical transverse flagella in dinoflagellates (note). Phycologia 14: 45–47.

_____. 1976. Flagellate phylogeny: a study in conflicts. J. Protozool. 23: 28–40.

_____, D. J. BLACKBOURN, AND J. BLACKBOURN. 1971. The red-water ciliate *Mesodinium rubrum* and its "incomplete symbionts:" a review including new ultrastructural observations. J. Fish. Res. Bd. Canada 28: 391–407.

TAYLOR, W. RANDOLPH. 1950. Plants of Bikini and other Northern Marshall Islands. Univ. of Michigan Press, Ann Arbor. xv + 227 pp.

_____. 1960. Marine Algae of the Eastern Tropical and Subtropical Coasts of the Americas. Univ. of Michigan Press, Ann Arbor. ix + 870 pp.

_____. 1962A. Marine Algae of the Northeastern Coast of North America. Rev. edit., 2nd printing with corrections. Univ. of Michigan Press, Ann Arbor. viii + 509 pp.

_____. 1962B. Two undescribed species of *Halimeda*. Bull. Torrey Bot. Club 89: 172–177.

_____. 1962C. Observations on *Pseudobryopsis* and *Trichosolen* (Chlorophyceae-Bryopsidaceae) in America. Brittonia 14: 58–65.

_____. 1963. The genus *Turbinaria* in eastern seas. J. Linn. Soc. Lond., Bot., 58: 475–490.

_____. 1973. A new *Halimeda* (Chlorophyceae, Codiaceae) from the Philippines. Pac. Sci. 27: 34–36.

_____, AND I. A. ABBOTT. 1973. A new species of *Botryocladia* from the West Indies. Br. Phycol. J. 8: 409–412.

_____, AND C. F. RHYNE. 1970. Marine algae of Dominica. Smithsonian Contrib. Bot. No. 3. 16 pp.

TAYLOR, W. ROWLAND, H. H. SELIGER, W. G. FASTIE, AND W. D. McELROY. 1966. Biological and physical observations on a phosphorescent bay in Falmouth Harbor, Jamaica, W.I.J. Mar. Res. 24: 28–43.

THOMAS, D. L. 1971. A circumscription of the genus *Protosiphon*. *In* Contributions in Phycology (Ed., B. C. Parker and R. M. Brown, Jr.). Allen Press, Lawrence, Kansas. pp. 9–24.

_____. 1973. Electrophoretic and immunological analyses of seven chlorosarcinacean algae. J. Phycol. 9: 289–296.

_____, AND R. M. BROWN, JR. 1970A. Isoenzyme analysis and morphological variation of thirty-two isolates of *Protosiphon*. Phycologia 9: 285–292.

_____, AND _____. 1970B. New taxonomic criteria in the classification of *Chlorococcum* species. III. Isozyme analysis. J. Phycol. 6: 293–299.

THOMAS, J. P., J. C. O'KELLEY, J. K. HARDMAN, AND E. F. ALDRIDGE. 1975. Flavin as an active compound of the photoreversible pigment system of the green alga *Protosiphon botryoides* Klebs. Photochem. Photobiol. 22: 135–138.

THOMAS, W. H. 1972. Observations on snow algae in California. J. Phycol. 8: 1–9.

THOMPSON, R. H. 1975. The freshwater brown alga *Sphacelaria fluviatilis*. J. Phycol. 11 (Suppl.): 5 (Abstr.).

THOMSEN, H. A. 1975. An ultrastructural survey of the chrysophycean genus *Paraphysomonas* under natural conditions. Br. Phycol. J. 10: 113–127.

THRONDSEN, J. 1970. Marine planktonic Acanthoecaceans (Craspedophyceae) from Arctic waters. Nytt Mag. Bot. 17: 103–111.

_____. 1971. *Apedinella* gen. nov. and the fine structure of *A. spinifera* (Throndsen) comb. nov. Norw. J. Bot. 18: 47–64.

_____. 1972. Coccolithophorids from the Caribbean Sea. Norw. J. Bot. 19: 51–60.

THURSTON, E. L., AND L. O. INGRAM. 1971. Morphology and fine structure of *Fischerella ambigua*. J. Phycol. 7: 203–210.

TIFFANY, L. H. 1930. The Oedogoniaceae. The author, Columbus Ohio. pp. 1–253.

_____. 1937. Oedogoniales: North American Flora. N.Y. Bot. Gard., Vol. 11, Pt. 1. 85 pp.

_____. 1958. Algae: the Grass of Many Waters. Charles C Thomas, Springfield, Ill. 199 pp.

TILDEN, J. E. 1937. The Algae and their Life Relations. Fundamentals of Phycology. The University of Minnesota Press, Minneapolis; 2nd ed. xii + 550 pp.

TILMAN, D., AND S. S. KILHAM. 1976. Phosphate and silicate growth and uptake kinetics of the diatoms *Asterionella formosa* and *Cyclotella meneghiniana* in batch and semicontinuous culture. J. Phycol. 12: 375–383.

TINDALL, D. R., T. SAWA, AND A. T. HOTCHKISS. 1965. *Nitellopsis bulbillifera* in North America. J. Phycol. 1: 147–150.

TIPPIT, D. H., K. L. McDONALD, AND J. D. PICKETT-HEAPS. 1975. Cell division in the centric diatom *Melosira varians*. Cytobiologie 12: 52–73.

TITMAN, D. 1976. Ecological competition between algae: experimental confirmation of resource-based competition theory. Science 192: 463–465.

TOBY, A. L., AND C. L. KEMP. 1975. Mutant enrichment in the colonial alga *Eudorina elegans*. Genetica 81: 243–251.

TOMAS, R. N., AND E. R. COX. 1973A. Observations on the symbiosis of *Peridinium balticum* and its intracellular alga. I. Ultrastructure. J. Phycol. 9: 304–323.

_____, AND _____. 1973B. The symbiosis of *Peridinium balticum* (Dinophyceae). I. Ultrastructure and pigment analysis. J. Phycol. 9(Suppl.): 16.

_____, _____, AND K. A. STEIDINGER. 1973. *Peridinium balticum* (Levander) Lemmermann, an unusual dinoflagellate with a mesocaryotic and an eucaryotic nucleus. J. Phycol. 9: 91–98.

TORPEY, J., AND R. M. INGLE. 1966. The red tide. (Revision of 1955.) Fla. Board Conserv. Mar. Lab., Educ. Ser. No. 1. 27 pp.

TORREY, J. G., AND E. GALUN. 1970. Apolar embryos of *Fucus* resulting from osmotic and chemical treatment. Am. J. Bot. 57: 111–119.

TOTH, R., AND D. R. MARKEY. 1973. Synaptonemal complexes in brown algae. Nature 243: 236–237.

TOWLE, D. W., AND J. S. PEARSE. 1973. Production of the giant kelp, *Macrocystis*, by *in situ* incorporation of ^{14}C in polyethylene bags. Limnol. Oceanogr. 18: 155–158.

TÖZÜN, B. 1974. Nuclear division in the red algae *Chondria nidifica* and *Chondria tenuissima*. Br. Phycol. J. 9: 363–370.

TRAGER, W. 1970. Symbiosis. Van Nostrand Reinhold Co., New York. ix + 100 pp.

TRAINOR, F. R. 1958. Control of sexuality in *Chlamydomonas chlamydogama*. Am. J. Bot. 45: 621–626.

_____. 1959. A comparative study of sexual reproduction in four species of *Chlamydomonas*. Am. J. Bot. 46: 765–770.

_____. 1960. Mating in *Chlamydomonas chlamydogama* at various temperatures under continuous illumination. Am. J. Bot. 47: 482–484.

_____. 1961. Temperature and sexuality in *Chlamydomonas chlamydogama*. Can. J. Bot. 39: 1273–1280.

_____. 1963A. The occurrence of a *Dactylococcus*-like stage in an axenic culture of a *Scenedesmus*. Can. J. Bot. 41: 967–968.

_____. 1963B. The morphology of a *Scenedesmus* in pure and contaminated culture. Bull. Torrey Bot. Club 90: 137–138.

_____. 1963C. Culture of *Scenedesmus longus*. Bull. Torrey Bot. Club 90: 407–412.

_____. 1966. Phototaxis in *Scenedesmus*. Can. J. Bot. 44: 1427–1429.

_____. 1970. Survival of algae in a desiccated soil. Phycologia 9: 111–113.

_____. 1975. Is a reduced level of nitrogen essential for *Chlamydomonas eugametos* mating in nature? Phycologia 14: 167–170.

_____, AND H. C. BOLD. 1953. Three new unicellular Chlorophyceae from soil. Am. J. Bot. 40: 758–767.

_____, AND C. A. BURG. 1965A. Motility in *Scenedesmus dimorphus*, *Scenedesmus obliquus* and *Coelastrum microsporum*. J. Phycol. 1: 15–18.

_____, AND _____. 1965B. *Scenedesmus obliquus* sexuality. Science 148: 1094–1095.

_____, AND _____. 1965C. Detection of bristles in *Scenedesmus* species. J. Phycol. 1: 139–144.

_____, J. R. CAIN, AND L. E. SHUBERT. 1976. Morphology and nutrition of the colonial green alga *Scenedesmus*: 80 years later. Bot. Rev. 42: 5–25.

_____, AND R. L. HILTON, JR. 1964. A new species of *Hormotila* from a Connecticut soil. Phycologia 4: 99–104.

_____, AND A. MASSALSKI. 1971. Ultrastructure of *Scenedesmus* strain 614 bristles. Can. J. Bot. 49: 1273–1276.

_____, H. L. ROWLAND, J. C. LYLIS, P. A. WINTER, AND P. L. BONANOMI. 1971. Some examples of polymorphism in algae. Phycologia 10: 113–119.

TRANSEAU, E. N. 1951. The Zygnemataceae. Ohio State Univ. Press, Columbus. iv + 327 pp.

TRENCH, R. K. 1969. Chloroplasts as functional endosymbionts in the mollusc *Tridachia crispata* (Bërgh.), (Opisthobranchia, Sacoglossa). Nature 222: 1071–1072.

_____. 1971. The physiology and biochemistry of zooxanthellae symbiotic with marine coelenterates. I. The assimilation of photosynthetic products of zooxanthellae by two marine coelenterates. Proc. Roy. Soc. London B. 177: 225–235. II. Liberation of fixed ^{14}C by zooxanthellae in vitro. Proc. Roy. Soc. London B. 177: 237–250. III. The effect of homogenates of host tissue on the excretion of photosynthetic products in vitro by zooxanthellae from two marine coelenterates. Proc. Roy. Soc. London B. 177: 251–264.

_____. 1973. Further studies on the mucopolysaccharide secreted by the pedal gland of the marine slug *Tridachia crispata* (Opisthobranchia, Sacoglossa). Bull. Mar. Sci. 23: 299–312.

_____, J. E. BOYLE, AND D. C. SMITH. 1973A. The association between chloroplasts of *Codium fragile* and the mollusc *Elysia viridis*. I. Characteristics of isolated *Codium* chloroplasts. Proc. Roy. Soc. London B. 184: 51–61.

_____, _____, AND _____. 1973B. The association between chloroplasts of *Codium fragile* and the mollusc *Elysia viridis*. II. Chloroplast ultrastructure and photosynthetic carbon fixation in *E. viridis*. Proc. Roy. Soc. London B. 184: 63–81.

_____, R. W. GREENE, AND B. G. BYSTROM. 1969. Chloroplasts as functional organelles in animal tissues. J. Cell Biol. 42: 404–417.

_____, AND S. OHLHORST. 1976. The stability of chloroplasts from siphonaceous algae in symbiosis with sacoglossan molluscs. New Phytol. 76: 99–109.

_____, M. E. TRENCH, AND L. MUSCATINE. 1972. Symbiotic chloroplasts; their photosynthetic products and contribution to mucus synthesis in two marine slugs. Biol. Bull. 142: 335–349.

TRIEMER, R. E., AND R. M. BROWN, JR. 1974. Cell division in *Chlamydomonas moewusii*. J. Phycol. 10: 419–443.

_____, AND _____. 1975A. The ultrastructure of fertilization in *Chlamydomonas moewusii*. Protoplasma 84: 315–325.

_____, AND _____. 1975B. Fertilization in *Chlamydomonas reinhardi* with special reference to the structure, development and fate of the choanoid body. Protoplasma 85: 99–107.

TRIPODI, G. 1971. The fine structure of the cystocarp in the red alga *Polysiphonia sertularioides* (Grat.) J. Ag. J. Submicroscop. Cyt. 3: 71–79.

TSCHERMAK-WOESS, E. 1959. Extreme Anisogamie und ein bemerkenswerter Fall des Geschlechtsbestimmung bei einer neuen *Chlamydomonas*-Art. Planta 52: 606–622.

_____. 1962. Zur Kenntnis von *Chlamydomonas suboogama*. Planta 59: 68–76.

TSENG, C. K., AND T. J. CHANG. 1955. Studies on *Porphyra*. III. Sexual reproduction of *Porphyra*. Acta Bot. Sin. 4: 153–166.

TSUBO, Y. 1956. Observations on sexual reproduction in a *Chlamydomonas*. Bot. Mag. (Tokyo) 69: 1–6.

_____. 1961A. Sexual reproduction of *Chlamydomonas* as affected by ionic balance in the medium. Bot. Mag., Tokyo 74: 442–448.

_____. 1961B. Chemotaxis and sexual behavior in *Chlamydomonas*. J. Protozool. 8: 114–121.

TSURU, S. 1973. Preservation of marine and fresh water algae by means of freezing and freeze-drying. Cryobryology 10: 445–452.

TUPA, D. 1974. An investigation of certain Chaetophoracean algae. Beiheifte z. Nova Hedw. 46.155 pp.

TURNER, F. R. 1968. An ultrastructural study of plant spermatogenesis: spermatogenesis in *Nitella*. J. Cell Biol. 37: 370–393.

TUTTLE, R. C., AND A. R. LOEBLICH, III. 1974. Genetic recombination in the dinoflagellate *Crypthecodinium cohnii*. Science 185: 1061–1062.

TUZET, O. 1963. The phylogeny of sponges according to embryological, histological, and serological data, and their affinities with the Protozoa and the Cnidaria. *In* The Lower Metazoa. Comparative Biology and Phylogeny. (Ed., E. C. Dougherty), Univ. Calif. Press, Berkeley and Los Angeles. pp. 129–148.

TYAGI, V. V. S. 1974. Some observations on the pattern of sporulation in a blue-green alga, *Anabaena doliolum*. Ann. Bot. 38: 107–111.

_____. 1975. The heterocysts of blue-green algae (Myxophyceae). Biol. Reviews 50: 247–284.

UEDA, K., AND M. SAWADA. 1972. DNA content of akinete cells in *Cylindrospermum*. Cytologia 37: 519–523.

_____, AND S. YOSHIOKA. 1976. Cell wall development of *Micrasterias americana* especially in isotonic and hypertonic solutions. J. Cell Sci. 21: 617–631.

UMEZAKI, I. 1967. The tetrasporocyte of *Nemalion vermiculare* Sur. Rev. Algol. 9: 19–24.

_____. 1972. The life histories of some Nemaliales whose tetrasporophytes were unknown. *In* Contributions to the Systematics of Benthic Marine Algae of the North Pacific (Ed., I. A. Abbott and M. Kurogi). Jap. Soc. Phycology, Kobe. pp. 231–242.

VADAS, R. L. 1969. The ecology of *Agarum* and the kelp bed community. Dissert. Abstr. 29: 10.

_____. 1972. Ecological implications of culture studies on *Nereocystis luetkeana*. J. Phycol. 8: 196–203.

VALET, G. 1967. Sur l'origine endogène des rameaux verticillés chez certaines Dasycladales. C. R. Acad. Sci. (Paris). 265D: 1175–1178.

_____. 1968. Contribution à l'étude des Dasycladales. 1. Morphogenèse. Nova Hedw. 16: 21–82.

_____. 1969. Contribution à l'étude des Dasycladales. 2. Cytologie et reproduction. 3. Révision systématique, distribution géographique et relations phylogénétique. Nova Hedw. 17: 551–644.

VANCE, B. D. 1965. Composition and succession of cyanophycean water blooms. J. Phycol. 1: 81–86.

VANDE BERG, W. J., AND R. C. STARR. 1971. Structure, reproduction and differentiation in *Volvox gigas* and *Volvox powersii*. Arch. Protistenk. 113: 195–219.

VAN DEN DRIESSCHE, T. 1966. The role of the nucleus in the circadian rhythms of *Acetabularia mediterranea*. Biochem. Biophys. Acta 126: 456–470.

VAN DOVER, B. 1972. Studies of the ultrastructure and reproduction of several species of *Chlamydomonas* and *Carteria*. Ph. D. Dissertation. The University of Texas at Austin. 198 pp.

VAN GORKOM, H. J., AND M. DONZE. 1971. Localization of nitrogen fixation in *Anabaena*. Nature. London 234: 231–232.

VAN VALKENBURG, S. D. 1971A. Observations on the fine structure of *Dictyocha fibula* Ehrenberg. I. The skeleton. J. Phycol. 7: 113–118.

_____. 1971B. Observations on the fine structure of *Dictyocha fibula* Ehrenberg. II. The protoplast. J. Phycol. 7: 118–132.

_____, AND R. E. NORRIS. 1970. The growth and morphology of the silicoflagellate *Dictyocha fibula* Ehrenberg in culture. J. Phycol. 6: 48–54.

VASCONCELOS, L. DE, AND P. FAY. 1974. Nitrogen metabolism and ultrastructure in *Anabaena cylindrica*. I. The effect of nitrogen starvation. Arch. Mikrobiol. 96: 271–279.

VENKATESWARLU, V., AND F. E. ROUND. 1973. Observation on *Aulacodiscus amherstia* sp. nov. from the Bay of Bengal. Br. Phycol. J. 8: 163–173.

VIEN, C. 1967A. Sur l'existence de phénomènes sexuels chez un Péridinien libre, l'*Amphidinium carteri*. C. R. Acad. Sci. (Paris) 264D: 1006–1008.

_____. 1967B. Un mode particulier de multiplication végétative chez un Péridien libre, le *Prorocentrum micans* Ehrenberg. C. R. Acad. Sci. (Paris) 265D: 108–110.

_____. 1968. Sur la germination du zygote et sur un mode particulier de multiplication végétative chez le Péridinien libre *Amphidinium Carteri*. C. R. Acad. Sci. (Paris). 267D: 701–703.

VINAYAKUMAR, M., AND E. KESSLER. 1975. Physiological and Biochemical contributions to the taxonomy of the genus *Chlorella*. Arch. Mikrobiol. 103: 13–19.

WAALAND, J. R. 1973. Experimental studies on the marine algae *Iridaea* and *Gigartina*. J. Exp. Mar. Biol. Ecol. 11: 71–80.

_____, AND C. I. KEMP. 1972. Observations on the life history of *Membranoptera multiramosa* Gardner (Rhodophyceae, Ceramiales) in culture. Phycologia 11: 15–18.

WAALAND, S. D. 1975. Evidence for a species-specific cell fusion hormone in red algae. Protoplasma 86: 253–261.

_____, AND R. CLELAND. 1972. Development in the red alga, *Griffithsia pacifica* control by internal and external factors. Planta 105: 196–204.

_____, AND _____. 1974. Cell repair through cell fusion in the red alga *Griffithsia pacifica*. Protoplasma 79: 185–196.

_____, J. R. WAALAND, AND R. CLELAND. 1972. A new pattern of plant cell elongation: bipolar band growth. J. Cell Biol. 54: 184–190.

WAERN, M. 1952. Rocky-shore algae in the Öregrund Archipelago. Acta Phytogeographica Suecica 30. xvi + 298 pp.

WAGNER, G., W. HAUPT, AND A. LAUX. 1972. Reversible inhibition of chloroplast movement by cytochalazin B in the green alga *Mougeotia*. Science 176: 808–809.

WALKER, D. C., E. Y. CHI, AND M. NEUSHUL. 1975. Experimental studies of cytoplasmic organization in *Zonaria* apical cells. Am. J. Bot. 62: 901–912.

WALL, D., AND B. DALE. 1968. Modern dinoflagellate cysts and evolution of the Peridiniales. Micropaleontology 14: 265–304.

_____, AND _____. 1969. The "hystrichosphaerid" resting spore of the dinoflagellate *Pyrodinium bahamense*, Plate, 1906. J. Phycol. 5: 140–149.

_____, R. R. L. GUILLARD, AND B. DALE. 1967. Marine dinoflagellate cultures from resting spores. Phycologia 6: 83–86.

WALNE, P. L. 1966. The effects of colchicine on cellular organization in *Chlamydomonas*. I. Light microscopy and cytochemistry. Am. J. Bot. 53: 908–916.

_____. 1967. The effects of Colchicine on Cellular organization in *Chlamydomonas*. II. Ultrastructure. Am. J. Bot. 54: 564–577.

_____. 1971. Comparative ultrastructure of eyespots in selected euglenoid flagellates. *In* Contributions in Phycology (Ed., B. C. Parker and R. M. Brown, Jr.). Allen Press. Lawrence, Kansas. pp. 107–120.

_____, AND H. J. ARNOTT. 1967. The comparative ultrastructure and possible function of eyespots: *Euglena granulata* and *Chlamydomonas eugametos*. Planta 77: 325–353.

WALSBY, A. E. 1974A. The extracellular products of *Anabaena cylindrica* Lemm. I. Isolation of a macromolecular pigment-peptide complex and other components. Br. Phycol. J. 9: 371–381.

_____. 1974B. II. Fluorescent substances containing serine and threonine; and their role in extracellular pigment formation. Br. Phycol. J. 9:383–391.

_____, AND G. E. FOGG. 1975. The extracellular products of *Anabaena cylindrica* Lemm. III. Excretion and uptake of fixed nitrogen. Br. Phycol. J. 10: 339–345.

WANDERS, J. B. W., C. VAN DEN HOEK, AND E. N. SCHILLERN-VAN NES. 1972. Observations on the life history of *Elachista stellaris* (Phaeophyceae) in culture. Neth. J. Sea Res. 5: 458–491.

WANG, W. S., AND R. G. TISCHER. 1973. Study of the extracellular polysaccharides produced by a blue-green alga. *Anabaena flosaquae* A-37. Arch. Mikrobiol. 91: 77–81.

WANKA, F. 1968. Ultrastructural changes during normal and colchicine-inhibited cell division of *Chlorella*. Protoplasma 66: 105–130.

WASSMAN, R., AND J. RAMUS. 1973A. Seaweed invasion. Natural History. 82: 24–36.

_____, AND _____. 1973B. Primary-production measurements for the green seaweed *Codium fragile* in Long Island Sound. Mar. Biol. 21: 289–297.

WATABE, N., AND K. M. WILBUR. 1966. Effects of temperature on growth and calcification and coccolith form in *Coccolithus huxleyi* (Coccolithineae). Limnol. Oceanogr. 11: 567–575.

WATSON, M. W. 1975. Flagellar apparatus, eyespot and behavior of *Microthamnion kuetzingianum* (Chlorophyceae) zoospores. J. Phycol. 11: 439–448.

_____, AND H. J. ARNOTT. 1973. Ultrastructural morphology of *Microthamnion* zoospores. J. Phycol. 9: 15–29.

WAWRIK, F. 1960. Sexualität bei *Mallomonas fastigiata* var. *Kriegeri*. Arch. Protistenk. 104: 541–544.

_____. 1970A. Mixotrophie bei *Cryptomonas borealis* Skuja. Archiv. Protistenk. 112: 312–313.

_____. 1970B. Isogamie bei *Synura Petersenii* Korschikov. Arch. Protistenk. 112: 259–261.

WEARE, N. M., AND J. R. BENEMAN. 1973. Nitrogen fixation by *Anabaena cylindrica*. I. Localization of nitrogen fixation in the heterocysts. Arch. Mikrobiol. 90: 323–332.

WEHRMEYER, W. 1970A. Structure, development, and decomposition of trichocysts in *Cryptomonas* and *Hemiselmis* (Cryptophyceae). Protoplasma 70: 295–315.

_____. 1970B. Chloroplast fine structure of some photo-autotropic Cryptophyceae. Arch. Mikrobiol. 71: 367–383.

_____. 1971. Ekektronenmikroskopische Untersuchung zur Feinstruktur von *Porphyridium violaceum* Kornmann mit Bemerkungen über seine taxonomische Stellung. Arch. Mikrobiol. 75: 121–139.

WEISS, R. L., D. A. GOODENOUGH, AND U. W. GOODENOUGH. 1977. Membrane differentiations at sites specialized for cell fusion. J. Cell Biol. 72: 144–160.

WENT, J. L. VAN, A. C. VAN AELST, AND P. M. L. TAMMES. 1973. Open plasmodesmata in sieve plates of *Laminaria digitata*. Acta Bot. Neerlandica 22: 120–123.

_____, AND P. M. L. TAMMES. 1972. Experimental fluid flow through plasmodesmata of *Laminaria digitata*. Acta Bot. Neerlandica 21: 321–326.

_____, AND _____. 1973. Trumpet filaments in *Laminaria digitata* as an artifact. Acta Bot. Neerlandica 22: 112–119.

WERNER, D. 1971. The life cycle with sexual phase in the marine diatom *Coscinodiscus asteromphalus*. III. Differentiation and spermatogenesis. Arch. Mikrobiol. 80: 134–146.

WERZ, G. 1968. Differenzierung und Zellwandbildung in isoliertem Cytoplasma aus *Acetabularia*. Protoplasma 65: 349–357.

WEST, J. A. 1967. *Pilayella littoralis* f. *rupincola* from Washington: the life history in culture. J. Phycol. 3: 150–153.

_____. 1968. Morphology and reproduction of the red alga *Acrochaetium pectinatum* in culture. J. Phycol. 4: 89–99.

_____. 1969A. The life histories of *Rhodochorton purpureum* and *R. tenue* in culture. J. Phycol. 5: 12–21.

_____. 1969B. Observations on four rare marine microalgae from Hawaii. Phycologia 8: 187–192.

_____. 1970A. The life history of *Rhodochorton concrescens* in culture. Br. Phycol. J. 5: 179–186.

_____. 1970B. The conspecifity of *Heterosiphonia asymmetrica* and *H. densiuscula* and their life histories in culture. Madroño 20: 313–319.

_____. 1972A. Environmental regulation of reproduction in *Rhodochoron purpureum*. *In* Contributions to the Systematics of Benthic Marine Algae of the North Pacific (Ed., I. A. Abbott and M. Kurogi). Jap. Soc. Phycology, Kobe. pp. 213–230.

_____. 1972B. Environmental control of hair and sporangial formation in the marine red alga *Acrochaetium proskaueri* sp. nov. Proc. Intern. Seaweed Symp. 7: 377–384.

_____. 1972C. The life history of *Petrocelis franciscana*. Br. Phycol. J. 7: 299–308.

_____, AND R. E. NORRIS. 1966. Unusual phenomena in the life histories of Florideae in culture. J. Phycol. 2: 54–57.

WETHERBEE, R. 1975A. The fine structure of *Ceratium tripos*, a marine armored dinoflagellate. I. The cell covering. J. Ultrastr. Res. 50: 58–64.

_____. 1975B. The fine structure of *Ceratium tripos*, a marine armored dinoglagellate. II. Cytokinesis and development of the characteristic cell shape. J. Ultrastr. Res. 50: 65–76.

_____. 1975C. The fine structure of *Ceratium tripos*, a marine armored dinoflagellate. III. Thecal plate formation. J. Ultrastr. Res. 50: 77–87.

_____, AND M. J. WYNNE. 1973. The fine structure of the nucleus and nuclear associations of developing carposporangia in *Polysiphonia novae-angliae* (Rhodophyta). J. Phycol. 9: 402–407.

WHEELER, A. E., AND J. Z. PAGE. 1974. The ultrastructure of *Derbesia tenuissima* (De Notaris) Crouan. I. Organization of the gametophyte protoplast, gametangium, and gametangial pore. J. Phycol. 10: 336–352.

WHITE, E. B., AND A. D. BONEY. 1969. Experiments with some endophytic and endozoic *Acrochaetium* species. J. Exper. Mar. Biol. Ecol. 3: 246–274.

WHITFORD, L. A., AND G. J. SCHUMACHER. 1969. A Manual of the Fresh-water Algae in North Carolina. North Carolina Agric. Exp. Stat., Tech. Bul. No. 188. 313 pp.

WHITTLE, S. J., AND J. P. CASSELTON. 1968. Peridinin as the major xanthophyll of the Dinophyceae. Br. Phycol. Bull. 3: 602–603.

_____, AND _____. 1975A. The chloroplast pigments of the algal classes Eustigmatophyceae and Xanthophyceae. I. Eustigmatophyceae. Br. Phycol. J. 10: 179–191.

_____, AND _____. 1975B. The chloroplast pigments of the algal classes Eustigmatophyceae and Xanthophyceae. II. Xanthophyceae. Br. Phycol. J. 10: 192–204.

WICHMANN, L. 1937. Studien über die durch H-Stück-Bau der Membran ausgezeichneten Gattungen *Microspora*, *Binuclearia*, *Ulotrichopsis*, and *Tribonema*. Pflanzenforschung, No. 20. 110 pp.

WIDDOWSON, T. B. 1965. A taxonomic study of the genus *Hedophyllum* Setchell. Can. J. Bot. 43: 1409–1420.

_____. 1972A. A taxonomic revision of the genus *Alaria* Greville. Syesis 4: 11–49.

_____. 1972B. A statistical analysis of variation in the brown alga *Alaria*. Syesis 4: 125–143.

_____. 1974. The marine algae of British Columbia and northern Washington: revised list and keys. Part II. Rhodophyceae (red algae). Syesis 7: 143–186.

WIEDLING, S. 1948. Beiträge zur Kenntnis der vegetativen Vermehrung der Diatomeen. Bot. Notiser, for 1948: 322–354.

WIESE, L. 1965. On sexual agglutination and mating-type substances (gamones) in isogamous heterothallic chlamydomonads. I. Evidence of the identity of the gamones with the surface components responsible for sexual flagellar contact. J. Phycol. 1: 46–54.

_____. 1969. Algae. Chapter 4. *In* Fertilization, Vol. 2. (G. B. Metz and A. Monroy, Eds.). Academic Press, New York and London. pp. 135–138.

_____. 1974. Nature of sex specific glycoprotein agglutinins in *Chlamydomonas*. Ann. N.Y. Acad. Sci. 234: 383–395.

_____, AND P. C. HAYWARD. 1972. On sexual agglutination and mating-type substances in isogamous dioecious Chlamydomonads. III. The sensitivity of sex cell contact to various enzymes. Am. J. Bot. 59: 530–536.

_____, AND R. F. JONES. 1963. Studies on gamete copulation in heterothallic Chlamydomonads. J. Cell Comp. Physiol. 61: 265–274.

_____, AND C. B. METZ. 1969. On the trypsin sensitivity of fertilization as studied with living gametes in *Chlamydomonas*. Biol. Bull. 136: 483–493.

_____, AND D. W. SHOEMAKER. 1970. On sexual agglutination and mating-type substances (Gamones) in isogamous heterothallic Chlamydomonads. II. The effect of concanavalin A upon the mating-type reaction. Biol. Bull. 137: 88–95.

_____, AND W. WIESE. 1975. On sexual agglutination and mating type substances in isogamous dioecious chlamydomonads. IV. Unilateral inactivation of the sex contact capacity in compatible and incompatible taxa by x-mannosidase and snake venom protease. Dev. Biol. 43: 264–276.

WIK-SJÖSTEDT, A. 1970. Cytogenetic investigations in *Cladophora*. Hereditas 66: 233–262.

WILBUR, K. M., L. H. COLINVAUX, AND N. WATABE. 1969. Electron microscope study of calcification in the alga *Halimeda* (Order Siphonales). Phycologia 8: 27–35.

_____, AND N. WATABE. 1963. Experimental studies on calcification on molluscs and the alga *Coccolithus huxleyi*. Ann. N.Y. Acad. Sci. 109: 82–112.

WILCE, R. T. 1959. The marine algae of the Labrador Peninsula and N. W. Newfoundland. Natl. Mus. Canada Bull. 158, Biol. Ser. 56. iv + 103 pp.

_____. 1965. Studies in the genus *Laminaria*. III. A revision of the North Atlantic species of the Simplices section of *Laminaria*. Proc. Fifth Mar. Biol. Symp. Botan. Gothoburgensia, III: 247–256.

_____. 1966. *Pleurocladia lacustris* in Arctic America. J. Phycol. 2: 57–66.

WILDMAN, R. B., AND C. C. BOWEN. 1974. Phycobilisomes in blue-green algae. J. Bact. 117: 866–881.

WILLENBRINK, J., AND B. P. KREMER. 1973. Localization of mannitol biosynthesis in the marine brown alga *Fucus serratus*. Planta 113: 173–178.

WILLIAMS, J., AND R. M. INGLE. 1972. Ecological notes on *Gonyaulax monilata* (Dinophyceae) blooms along the west coast of Florida. Fla. Dept. Nat. Resour. Mar. Res. Lab. Leaf. Ser. Vol. 1, part 1, No. 5. 12 pp.

WILLIAMS, J. L. 1905. Studies in the Dictyotaceae. III. The periodicity of the sexual cells in *Dictyota dichotoma*. Ann. Bot. 19: 531–560.

WILSON, W. B. 1966. The suitability of sea water for survival and growth of *Gymnodinium breve* Davis; and some effects of phosphorous and nitrogen on its growth. Fla. Bd. Conserv. Mar. Lab. Prof. Pap. Ser. No. 7. 42 pp.

_____. 1967. Forms of the dinoflagellate *Gymnodinium breve* Davis in cultures. Contrib. Mar. Sci. 12: 120–134.

WILTON, J. W., AND E. G. BARHAM. 1968. A yellow-water bloom of *Gymnodinium flavum* Kofoid & Swezy. J. Exp. Mar. Biol. Ecol. 2: 167–173.

WINTER, P. A., AND P. BIEBEL. 1967. Conjugation in a heterothallic *Staurastrum*. Proc. Penn. Acad. Sci. 40:76–79.

WITMAN, G. B., K. CARLSON, J. BERLWIER, AND J. L. ROSENBAUM. 1972. *Chlamydomonas* flagella. I. Isolation and electrophoretic analysis of microtubules, matrix, membranes and mastigonemes. J. Cell Biol. 54: 507–539.

WOELKERLING, W. J. 1971. Morphology and taxonomy of the *Audouinella* complex (Rhodophyta) in southern Australia. Austral. J. Bot., Suppl. No. 1. 91 pp.

_____. 1973. The morphology and systematics of the *Audouinella* complex (Acrochaetiaceae, Rhodophyta) in northeastern United States. Rhodora 75: 529–621.

_____. 1975. Observations on *Batrachospermum* (Rhodophyta) in southeastern Wisconsin streams. Rhodora 77: 467–477.

WOLK, C. P. 1965. Heterocyst germination under defined conditions. Nature 205: 201–202.

_____. 1966. Evidence of a role of heterocysts in the sporulation of a blue-green alga. Am. J. Bot. 53: 260–262.

_____. 1967. Physiological basis of the pattern of vegetative growth in a blue-green alga. Proc. Natl. Acad. Sci. U. S. 57: 1246–1251.

_____. 1968. Role of bromine in the formation of the refractive inclusions of the vesicle cells of the Bonnemaisoniaceae (Rhodophyta). Planta 78: 371–378.

_____. 1973. Physiological chemistry of blue-green algae. Bacteriol. Reviews 37: 32–101.

WOLKEN, J. J. 1967. *Euglena*. An experimental organism for biochemical and biophysical studies. Appleton-Century Crofts, New York. 204 pp.

WOLLASTON, E. M. 1968. Morphology and taxonomy of southern Australian genera of Crouanieae Schmitz (Ceramiaceae, Rhodophyta). Austral. J. Bot. 16: 217–417.

_____. 1971. *Antithamnion* and related genera occurring on the Pacific Coast of North America. Syesis 4: 73–92.

_____. 1972A. Generic features of *Antithamnion* (Ceramiaceae, Rhodophyta) in the Pacific region. Proc. Inter. Seaweed Symp. 7: 142–145.

_____. 1972B. The genus *Platythamnion* J. Ag. (Ceramiaceae, Rhodophyta) on the Pacific coast of North America between Vancouver, British Columbia, and southern California. Syesis 5: 43–53.

WOMERSLEY, H. B. S. 1954. The species of *Macrocystis* with special reference to those on southern Australian coasts. Univ. Calif. Pub. Bot. 27: 109–132.

_____. 1964. The morphology and taxonomy of *Cystophora* and related genera (Phaeophyta). Austral. J. Bot. 12: 53–110.

_____. 1965. The Helminthocladiaceae (Rhodophyta) of southern Australia. Austral. J. Bot. 13: 451–487.

_____. 1971. The relationships of *Nizymenia* and *Stenocladia* (Gigartinales, Rhodophyta). Phycologia 10: 199–203.

WONG, J. L., R. OESTERLIN, AND H. RAPOPORT. 1971. The structure of saxitoxin. J. Am. Chem. Soc. 93: 7344–7345.

WOOD, E. J. F. 1954. Dinoflagellates in the Australian region. Austral. J. Mar. Freshw. Res. 5: 171–351.

WOOD, R. D. 1965. Monograph of the Characeae. *In* Wood, R. D. and K. Imahori. A revision of the Characeae, Vol. I. Cramer, Weinheim. 904 pp.

_____. 1968. Charophytes of North America. Bookstore, Memorial Union, Univ. Rhode Island, Kingston. 72 pp.

_____, AND K. IMAHORI. 1964. Iconograph of the Characeae. *In* Wood, R. D. and K. Imahori. A revision of the Characeae, Vol. II. Cramer, Weinheim. 6 pp., 395 plates.

WOOLERY, M. L., AND R. A. LEWIN. 1973. Influence of iodine on growth and development of the brown alga *Ectocarpus siliculosus* in axenic cultures. Phycologia 12: 131–138.

WORNARDT, W. W., JR. 1969. Diatoms, past, present, future. *In* Proc. of the First Intern. Conf. on Planktonic Microfossils. (Ed., P. Brönnimann and H. H. Renz). Leiden, Brill. pp. 690–714.

_____. 1971. Eocene, Miocene and Pliocene marine diatoms and silicoflagellates studied with the scanning electron microscope. Proc. II Planktonic Conference, Roma, 1970 (Ed., A. Farinacci). pp. 1277–1300.

WUJEK, D. E. 1969. Ultrastructure of flagellated Chrysophytes. I. *Dinobryon*. Cytologia 34: 71–79.

_____, AND J. E. CHAMBERS. 1966. Microstructure of pseudocilia of *Tetraspora gelatinosa* (Vauch.) Desv. Trans. Kansas Academy of Science 68: 563–565.

WYATT, J. T., AND J. K. G. SILVEY. 1969. Nitrogen fixation by *Gloeocapsa*. Science 165: 908–909.

WYGASCH, J. 1963. Die Blutenalge und ihre Feinbau. Mikrokosmos 52: 293–297.

WYLIE, P. A., AND H. E. SCHLICHTING. 1973. A floristic survey of corticolous subaerial algae in North Carolina. J. Elisha Mitchell Sci. Soc. 89: 179–183.

WYNNE, M. J. 1969A. Life history and systematic studies of some Pacific North America Phaeophyceae (brown algae). Univ. Calif. Pub. Bot. 50: 1–88.

_____. 1969B. *Platysiphonia decumbens* sp. nov., a new member of the Sarcomenia group (Rhodophyta) from Washington. J. Phycol. 5: 190–202.

_____. 1971. Concerning the phaeophycean genera *Analipus* and *Heterochordaria*. Phycologia 10: 169–175.

_____. 1972A. Culture studies of Pacific coast Phaeophyceae. Soc. Bot. Fr. Mémoires 1972: 129–144.

_____. 1972B. Studies on the life forms in nature and in culture of selected brown algae. *In* (Ed., I. A. Abbott and M. Kurogi) Contributions to the Systematics of Benthic Marine Algae of the North Pacific. Japanese Soc. Phycology, Kobe. pp. 133–145.

_____, D. L. McBride, and J. A. West. 1973. *Polyneuropsis stolonifera* gen. et sp. nov. (Delesseriaceae, Rhodophyta) from the Pacific coast of North America. Syesis 6: 243–253.

_____, and W. R. Taylor. 1973. The status of *Agardhiella tenera* and *Agardhiella baileyi* (Rhodophyta, Gigartinales). Hydrobiologia 43: 93–107.

Yabu, H. 1958. On the nuclear division in tetrasporangia of *Dictyopteris divaricata* (Okamura) Okamura and *Dictyota dichotoma* Lamour. Bull. Fac. Fish., Hokkaido Univ. 8: 290–296.

_____. 1969A. Observations on chromosomes in some species of *Porphyra*. Bull. Fac. Fish. Hokkaido Univ. 19: 239–243.

_____. 1969B. Mitosis in *Porphyra tenera* Kjellm. Bull. Fac. Fish., Hokkaido Univ., 20: 1–3.

_____. 1972. Observations on chromosomes in some species of *Porphyra*, III. Bull. Fac. Fish., Hokkaido Univ. 22: 261–266.

Yamagishi, T. 1963. Classification of the Zygnemataceae. Sci. Rep. Tokyo Kyoiku Daigoku. 11: 191–210.

Yarish, C. 1975. A cultural assessment of the taxonomic criteria of selected marine Chaetophoraceae (Chlorophyta). Nova Hedw. 26: 385–430.

_____. 1976. Polymorphism of selected Chaetophoraceae (Chlorophyta). Br. Phycol. J. 11: 29–38.

Yeh, P. and A. Gebar. 1970. Growth patterns and motility of *Spirogyra* sp. and *Closterium acerosum*. J. Phycol. 6: 44–48.

Zajic, J. E. 1970. Properties and products of algae. Proc. of the Symp. on the Culture of Algae Sponsored by the Division of Microbial Chemistry and Technology of the American Chemical Society, held in New York City, September 7–12, 1969. Plenum Press, New York. x + 154 pp.

Zaneveld, J. S. 1959. The utilization of marine algae in tropical south and east Asia. Eco. Bot. 13: 89–131.

Zerban, H., M. Wehner, and G. Werz. 1973. Studies on the ultrastructure of the nucleus of *Acetabularia* by the freeze-etching technique. Planta 114: 239–250.

_____, and G. Werz. 1975. Changes in frequency and total number of nuclear pores in the life cycle of *Acetabularia*. Exp. Cell Res. 93: 472–477.

Ziegler, H., and J. Ruck. 1967. Untersuchungen über die Fein-struktur des Phloems. III. Die "Trompetenzellen" von *Laminaria*-Arten. Planta 73: 62–73.

Zingmark, R. G. 1970A. Sexual reproduction in the dinoflagellate *Noctiluca miliaris* Suriray. J. Phycol. 6: 122–126.

_____. 1970B. Ultrastructural studies on two kinds of mesokaryotic dinoflagellate nuclei. Am. J. Bot. 57: 586–592.

Taxonomic Index

Boldface page numbers indicate illustrations.

Astasia, 258, 261, 266
  A. fritschii, **261**
  A. longa, 261
Asterionella socialis, 415
Asterococcus, 105, 106
  A. superbus, **108**
Asterocytis, 468, 470
Asteromphalus, 404
  A. arachna, 405
Astrephomenaceae, 71,
  100-101
Astrephomene, 101
  A. gubernaculifera, **101**
Audouinella, 481, 486
  A. floridulum, 484
  A. violacea, 486
Axilosphaera, 128, 132
  A. vegetata, **132**
Azolla, 5

Bachelotia, 269
  B. antillarum, **269**
Bacillaria paradoxa, 413
Bacillariaceae, 416
Bacillariales, 416
Bacillariophyceae, 19, 21, 270,
  357, 359, 364, 384,
  397-416, 436
Bacillariophyta, 416
Bacteriastrum, 414
  B. hyalinum, 406
Bangia, 454, 457, 460, 465,
  472, 474, 475-477
  B. atropurpurea, 477
  B. fuscopurpurea, 476, **477**
Bangiaceae, 466, 469, 471,
  473, 474-480
Bangiales, 452, 464, 471-480
Bangiophycidae, 451, 454, 455,
  460, 464, 465, 467-480
Batophora, 67, 216, 218
  B. oerstedii, **223, 224**
Batrachospermaceae, 480,
  484-487
Batrachospermum, 485,
  486-487
  B. macrosporum, 486
  B. moniliforme, 487
  B. vagum, 487
  B. virgatum, 453
Battersia, 274
Berkeleya, 414
Biddulphia, 399, 401
  B. biddulphiana, **401**
Biddulphiineae, 416
Bifurcariopsis, 349, 355
Biraphidineae, 404, 416
Blastodiniales, 424, 440
Blastodinium, 438, 439

Boldia, 474
  B. angustata, 474
  B. erythrosiphon, **473**
Boldiaceae, 466, 471, 473-474
Bonnemaisonia, 456, 458, 459,
  480, 495-497
  B. asparagoides, 495, 499
  B. hamifera, 460, 495, **496**
  B. nootkana, 497
Bonnemaisoniaceae, 458, 480,
  481, 495-497
Bonnemaisoniales, 480, 495
Bornetella, 217
  B. sphaerica, **217**
Borodinellopsis, 128, 130, 131,
  132
  B. texensis, **131**
Bossiella, 513
  B. californica, **512**
  B. californica ssp. schmittii,
  510
Bostrychia, 557, 561
Botrydiopsis, 386
  B. arhiza, **387**
Botrydium, 386, 389-390
  B. granulatum, **390**
Botryocladia, 538, 539
  B. occidentalis, 539
  B. pseudodichotoma, **539**
  B. spinulifera, 539
Botryococcus, 24
Braarudosphaera bigelowii, **25,
  383**
Bracteacoccus, 65, 116-**117**
Branchioglossum, 554
Bryopsidaceae, 191, 200-206
Bryopsidales, 190
Bryopsidiophyceae, 244
Bryopsis, 67, 166, 179, 190,
  191, 201-205, 207, 245
  B. halymeniae, **204**
  B. hypnoides, 202, **203, 204**
  B. maxima, 201, **202**
  B. plumosa, 202, **204**
Bulbochaete, 8, 159, 165-166
  B. bullardii, **166**
  B. hiloensis, **165**
Bumilleria, 389

Calcidiscus, **382**
Calciosolenia murrayi, **383**
Calliarthron, 513
  C. tuberculosum, **514**
Callithamnion, 544, 546-548
  C. byssoides, **546**
  C. pikeanum, 546
Callophyllis, 458, 462, 505,
  506, 507
  C. cristata, 507

Callophyllis (cont.)
  C. firma, 507
  C. obtusifolia, 507
  C. pinnata, 507, **508**
Caloglossa, 557
  C. leprieurii, **557**
Calosiphonia, 521
Calothrix, 3, **61**
Calyptrosphaera pirus, **383**
Carteria, 66, 71, 75, 83-84, 142
  C. crucifera, **83**
Catillochara moreyi, **27**
Caulerpa, 28, 191, **199-200**
  C. floridana, **200**
  C. prolifera, **200, 201**
  C. racemosa, **200**
  C. serrulata, 200
  C. sertularioides, **200**
Caulerpaceae, 191, 199-200
Caulerpales, 63, 69, 122,
  190-210, 245, 248
Centrales, 359, 399, 416
Centrobacillariophyceae, 416
Ceramiaceae, 456, 464, 495,
  543-552, 560
Ceramiales, 466, 495, 520,
  542-565
Ceramium, 458, 464, 544,
  550, 551
  C. pacificum, 550
  C. rubrum, 550
  C. strictum, 551
Ceratium, **32**, 420, 421,
  423-426, 432, 439, 446,
  447, 448
  C. cornutum, 424
  C. hirundinella, 439, 448
  C. horridum, 424, **425**
  C. lunula, 448
  C. tripos, **424**
  C. vultur, 448
Chaetangiaceae, 481, 492-495
Chaetoceros, **401, 409, 410**,
  414, **415**
  C. affine, **401**
  C. armatum, 415
  C. peruvianum, **401**
  C. sociale, 415
Chaetomorpha, 67, 180
  C. aerea, **182**
  C. darwinii, 183
Chaetophora, 141, 150-151
  C. incrassata, 150, **151**
Chaetophoraceae, 150-156
Chaetophorales, 69, 149-159,
  246
Chamaesiphon, 43, 49, **50**
Chamaesiphonales, 48-50
Champia, 537
  C. parvula, 537, **538**
Champiaceae, 464, 537-538

Dilophus ligulatus, 303
Dilsea, 501
Dinamoebales, 440
Dinobryon, 367-368
    D. sertularia, **367**
    D. sertularioides, **368**
    D. suecicum, **361**
Dinophyceae, 418, 439, 440
Dinophysiales, 421, 422,
    440-443, 446
Dinophysis, 423, **424**, 441, **442**
Dinothrix, 423
Dinotrichales, 423, 440
Discoaster lodoensis, **25**
Discosphaera tubifera, **383**
Dissodinium, 449
Distephanus speculum, **374**
Ditylum brightwellii, 404, 409
Draparnaldia, 151, 152-153
    D. glomerata, 151, **152**
Draparnaldiopsis, 151, 153
    D. alpina, **153**
Dudresnaya, 501
Dumontiacea, 499-502
Dunaliella, 70, 72-**73**
    D. salina, **73**
Durvillea, 343
Durvilleales, 273
Dysmorphococcus, 87
    D. globosus, **87**
    D. variabilis, **87**

Ebriales, 440
Ebriophyceae, 439, 440
Ectocarpaceae, 275, 280
Ectocarpales, 268-269, 273-282,
    286
Ectocarpus, 2, 66, **271**, 274,
    275, **276**-277, 278, 280,
    300, 302
    E. siliculosus, 275, **276**-277
Egregia, 317, 339, **341**, 342
    E. laevigata, 342
    E. menziesii, 342
Eisenia, 319, 339, 342-343
    E. arborea, **319**, **342**, 343
Elachista, 286, **287**, 288
    E. fucicola, 288
    E. stellaris, 288
Elachistaceae, 286-287
Ellipsoidion, 360, 393
    E. acuminatum, 395
Ellobiophyceae, 439-440
Emiliania huxleyi, 375
Endodinium, 435
    E. chattonii, 435
Ensiculifera loeblichii, **423**
Enteromorpha, **2**, 167, 168,
    171-172, 414
    E. intestinalis, **172**
    E. linza, 171

Eremosphaera, 133, 136-137
    E. viridis, **137**
Erythrocladia, 456, 471-472,
    474
    E. subintegra, **472**
Erythrocystis, 561
Erythropeltidaceae, 466, 469,
    471, 474
Erythrotrichia, 471-472
    E. carnea, **463**, 471-472
Eucampia, 414
Eucapsis, 44, **47**
Eucheuma, 530-534
    E. gelidium, **534**
    E. isiforme, **532**, **533**
    E. procrusteanum, 531
Eudesme, 290, **291**
Eudorina, 87, 88, 90-92, 93,
    94, 96, 99
    E. elegans, **91**
Euglena, 3, 15, 65, 255,
    258-261
    E. gracilis, 256, 257, **259**,
    261, 264, 288
    E. granulata, **260**
    E. mesnilii, **260**
    E. myxocylindracea, 257
Euglena, bleached, 257
Euglenales, 258-264
Euglenophycophyta, 17, 18,
    255-266, 268, 393
Eukaryota, 34
Eunotia, 407
Eustigmatophyceae, 270,
    357-358, 360, 387, 393-396
Euthora, 507
    E. cristata, 507
    E. fruticulosa, 507
Eutreptia, 258, 266
    E. pertyi, **259**
Eutreptiaceae, 258
Eutreptiales, 258
Exuviaella, 438, 449-450

Falkenbergia, 495-496
    F. ruflanosa, 495
Farlowia, **500**
Feldmannia, 274-275, **279**
    F. padinae, 280
Fischerella, 59, 60-61
    F. ambigua, **60**
Florideophycidae, 454-456,
    458-462, 464-465, 467, 480-
    565
Fontinalis, 50
Fosliella, 516
Fragilaria, 414
Fritschiella, 150, 153-154
    F. tuberosa, **154**
Fryeella, 538
Fucaceae, 349-354

Fucales, 267-273, 297, 343-356
Fucus, 7, 11, 16, **272**, 281, 343,
    344-346, **348**, 349, 350-352,
    353, 393, 398
    F. distichus, **347**, 350-351
    F. serratus, 347, 352
    F. vesiculosus, **344**, **350**,
    351-352
Fungi Imperfecti, 300, 481

Galaxaura, **214**, 454, 492, **494**,
    495
Gardneriella, 530
Gastroclonium, 538
    G. coulteri, 538
Gelidiaceae, 480-481, 497-499
Gelidiella, **497**, 498
Gelidium, 458, 480, **497**, **498**,
    499
    G. nudiforme, 498
    G. pusillum, 498
    G. robustum, 498
Geminella, 141, 142, 144, **145**,
    146, 149
Gephyrocapsa, 375
    G. huxleyi, 375, 378-379,
    384
Giffordia, 274-275, **278-279**,
    280, 302
Gigartina, 457-458, 462, 522,
    523, 524-527, 530
    G. agardhii, 462, **523**,
    524-525
    G. corymbifera, **523**
    G. exasperata, 523
    G. harveyana, 523
    G. papillata, 462
    G. stellata, 525, 526
    G. volans, **523**
Gigartinaceae, 521-527
Gigartinales, 456, 466, 499,
    520-536
Giraudyopsis, 374
Glenodiniales, 440
Glenodinium foliaceum, 429,
    437-438
Gloeobacter violaceus, 35
Gloeocapsa, 3, 32, 33, 34, 44,
    45, **46**
Gloeochrysis, 370
Gloeococcus, 104, 105, 106
    G. minor, 106
    G. minutissimus, **104**
Gloeocystis, 105, 106, **108**, 370
    G. ampla, 108
Gloeotrichia, 61-**62**
    G. ghosei, **42**, 43
Gloiosiphonia, 462, 499,
    503-504
    G. capillaris, **502**, **503**
    G. verticillaris, 503

Subject Index

Boldface page numbers indicate illustrations.

Fountain-type growth, **457**, 487
Fragmentation, 9, 40, 141, 146, 148, 161, 182, 184, 225, 343, 355, 362, 389
Free nuclear division, 120
Freshwater, defined, 2
Frog spittle, 1
Frustule, 398-399, 400, 403, 406, 410-412, 416
Fucoidan, 18, 268, 349, 351
Fucosan vesicles, 298, 351
Fucose, 268
Fucoserraten, 347
Fucoxanthin, 18, 267, 357, 384, 398, 417, 437, 438
Fungi, 5, 16
Furcelleran, 454
Furrowing, 144
Fusion cell, 506, **514**, 516, 530, 531, 532, **533**, 537, 538, 560

Gaffron's trace-element solution, 574
Galactan, 454, 526
Galactose, 268
Galacturonic acid, 520
Gametangial rays, 220, 222
Gametangium (a), 1, 11, 185, 192, 201, 208, 225, 229, 231, 271, 275, **300**, 390
Gamete, 1, **80**, 272, 275, **277**, 377, 406-407, **408**, 409-410, 424-426, 460 (*See also* Isogamy, heterogamy and oogamy)
Gametic union, **208**
Gametophore, 195
Gametophyte, 13, 171, 173, 175, 179, 184, 186, 188, 189, 202, 216
Gamone, 81, 82
Gas vacuoles, **38**-39
Gemma, 464
Generative auxiliary cell, 461, 466, 497
Genetic recombination, 43
Geniculum, **511**, 512-513
Genophore, **269**, 270, 298, 364
Giant nucleus, 203, 217
Girdle, 400, 410
Girdle lamellae, 270, 364, 387, 389, 395-396, 437
Gland cell, 531, **544**, 545-546
Gliding motility, 39, 413
Glossary, 579-598
Glucose amine, 18
Glucuronic acid, 520
Glycerol, 268, 435
Glycogen, 19, 37

Glycolate oxidase, 245
Glycolic acid, 386
Glycoprotein, 82, 94
Golden algae, 18, 357-374
Golgi apparatus, 67, 70, 76, 165, 262
Gonidium (a), 94, 95, 97
Gonimoblast, 461, 484, 520-521, **522**, 527, 532, 537, 543-544
Gonimoblast initial, 491, 520, 543, 560
Gonocyte, **438**, 439
Grafting, 220
Gram-negative bacteria, 35
Grana, 64
Great Salt Lake, 72
Green algae, 63-246 classification and phylogeny, 244-246
Growth, 7-8 generalized, 140, 142, 188
Gullet, 568
Guluronic acid, 268
Gynandrosporous, 163, 166
Gyrogonites, **27**

H-piece or -segment, 141, 147, 148, 385, **389**
Haematochrome, 85
Hair, 149, 156, 157, 165, 193, 459, 482
Halogen metabolism, 458-459
Haplobiontic, 12, 71, 150, 159, 167, 201, 225, 228, 238, 273
Haploid, 12, 159, 178, 218, 225, 228, 230, 238
Haplostichous, 268, 273
Haptera, 318, 325, 328, 331-333, 338, 342-343
Haptonema, **359**, 360, 374-381
Haptophytes, 374-384
Haustoria, 536
Hemiblastic, branching, **299**
Hermaphroditic, 12
Hermatypic, 436
Heterococcolith, 378, 382, 383
Heterocyst, 34, 40, **41**, **42**, 50, 55, 56, 57, 62
Heterodynamic flagella, 359, 367, 374, 381, 386, 390, 396, 417, 449, 566
Heterogamy, 11
Heteromorphic, 14, 167, 171, 273, 286, 292-293, 296, 298, 300, 316-317, 319, 462, 480, 495, 499, 523-524
Heteroplastidy, 194, 197
Heterothallism, 12, 79, 90

Heterotrichy, 149, 152, 159, 169, 268, 274, 481
Heterotrophy, 4, 15, 361, 363, 375-376, 401, 417, 436, 438, 439, 443, 446, 566
Holdfast, 7, 140, 142, 147, 154, 160, 177, 182, 196
Holoblastic branching, **299**
Holocarpic, 196
Holococcolith, 378, 382, 383
Holoplanktonic, 414
Holozoic, 417, 427, 439-441, 443-444, 448
Homodynamic flagella, 359, 374, 566
Homothallism, 12, 79
Hormogonium, 9, 39, **40**, 50, 51, 52
Hughes, Gorham and Zehnder's medium, 574 Allen's modification, 574
Humic substances, 434
Hyaline field, 403, 416
Hybridization, 352, 480, 563
Hypnospore, 10, 127, 142
Hypnozygote, 425-426
Hypocingulum, 400, **405**
Hypocone, 418, 420
Hypogynous cell, 492, 495, 502, **514**
Hypolithic, 4
Hypothallium, 283, **284**, 315
Hypothallus, 509, **511**
Hypotheca, 400, 420-421, 441, 446
Hypovalve, 400
Hystrichosphere, 426, **427**

Idioandrosporous, 163
Inducing factor, 99
Ingestion rod, 266
Intercalary band, 400, 406, 416
Intercalary cell division, 159
Intercalary meristem, 268, 509 555
Intergeniculum, 512
Internodal cells, 249, 253
Internode, 249
Intertidal, 3
Invertebrate, 433, 435, 569
Involucre, 305, 543-544, 547, **548**, 550
Iodine, 278, 453, 458
Irish moss, 28, 525
Iron, 434
Isofloridoside, 454
Isogamy, **9**, 11, 273, 274, 277, 298, 308, 362, 366-367, 389-390, 406-407, 424-425, 445

Isokontan, 65
Isomorphic, 14, 154, 167, 168, 173, 174, 182, 184, 215, 273, 283, 285, 296, 298, 302-303, 307, 462, 495, 499, 504, 514, 518, 523-524, 526, 528, 534, 538, 543, 545-546, 555, 562, 564
Isthmus, 235, 239

Karyogamy, 8, 78, 233
Kelp, 6, 7, 28, 317, 319, 321, 326, 327, 330, 331, 336-337, 340, 342-343
Keys, taxonomic, 44, 48, 50, 51, 55, 59, 61, 68, 71, 88, 103-104, 106, 111, 128, 141-142, 150, 151, 159, 168, 191, 194, 226, 231, 236, 247, 273-274, 275, 283, 286, 289-290, 308, 318, 466-467, 480-481, 499, 521, 537, 543
Kinetin, 154, 275, 278
Kinetochore, 396

Labiate process, 403, **404**, **405**
Laminaran, 18, 268, 351, 357
Lampbrush chromosomes, 222
Lateral conjugation, **227**
Laterals, 217, 218
Laver, 28
"Leaves," 250
Lens, 429
Leucoplast, 196, 210, 398, 568
Leucosin, 357, 385
Lichens, 4, 33, 117
Life cycles, **13**, 10-14, 68, 172, 174, 179, 201-202, 378, 519
   D, h + d, 14
   D^h, h + d, 14, 68, 188, 319 494
   D^i, h + d, 14, 68, 180, 182, 184, 276, 301
   H, d, 12-13, 68, 179, 200, 212
   H, h, 12-13, 68, 105, 171, 186
Lipid body, 37
List, 441
Locule, 402, **403**
Lorica, 85, 86, 261, 262, **361**, **362**, 363, 367-368, 372
Luche, 28
Luciferin-luciferase, 430, 432

Macrandrous, 163, 166

Macrogametangium, 300
Macrothallus, 288, 315-316
Major elements, 16
Malic acid, 296
Maltose, 454
Manganese, 410
Mannan, 191, 218, 454
Mannitol, 268, 322, 327-328, 385-386, 516
Mannose, 67
Mannuronic acid, 268
Manubrium, 250, 251
Marine water, 2
Mastigonemes, 77, 262, **272**, 359, 364-365
Matrices, 105, **107**, 151, 234
Medullary meristem, 317
Megacytic zone, **423**, 442-443
Megasporangium, **279**, 280, 349
Meiocyte, 425
Meiosis, 8, 12, 271, 362, 378, 393, 406-407, 424-426, 440, 444-445, 460, 462-464, 475, 492, 515, 528
   gametic, 12
   sporic, 14
   zygotic, 12
Meiosporangium, 271, 280
Meriblastic branching, 299
Meristoderm, 268, 274, 317, 320, 328
"Mermaid's wine glass," 219
Meroplanktonic, 414
Merotomy, 220
Mesochite, 346
Mesokaryotic, 417, 427, **428**, **437**, 438, **439**, 445
Mesozoic, 25
Metaboly, 255
Microfibrils, 238
Microgametangium, 300
Microplasmodesmata, 37
Microsporangium, **279**, 280, 349
Microspore, 406, **407**
Microthallus, 288, 315-316
Microtubule, 124, 125, 177, 245, 257
Microtubule-organizing center, 456
Microzoospore, 381
Miocene, 25, 26
Mississippian, 22, 27
Mitochondrion, a, 67, 70, 76, 203, 262
Mitosis, 161, 173, 180, 183, 222, 228, 236, 249, 257, 262, 391, 405, 411, 428, 432, 456
Mitosporangium, 271
Molluscans, 433

Monad, 359, 363, 365, **367**, 371, 379-380, 385-386
Monocarpogonial, 505-507
Monoecious, 12, 99, 166, 171, 192
Monopodial, 289, 543, 547, 552
Monosporangium, **271**, 272, **302**, **463**, 464, 469, 482
Monospore, 272, 456, 460, **463**, 464, 466, 469-478, 481, 493
Motility, 39
Muciferous body, 430
Mucilage duct, 321, 326-327, 340
Mucilage gland, 339-340
Mucilage-producing bodies, 256
Mucilage secretion, 238
Mucopolysaccharide, 18
Multiaxial, 191, 192
Multiaxial growth, **457**, 527, 530, 536
Multinucleate zoospore, 209
Mutants, 83, 92, 98, 99

Nannandrous, 163, 166
Nanoplankton, 372
Necridium, 40
Nemathecia, 527, 529
Nematocyst, 429-430
Neodinoxanthin, 417
Neoperidinin, 417
"Neptune's shaving brush," 197
Neritic, 375, 381, 410, 414, 434, 444
Neuston, 3, 372, 380
Nitrate, 422, 425-426, 493
Nitrogen, 82, 228, 235
Nitrogenase, 34
Nitrogen fixation, 29, 33, 34, 42, 43, 422
Nitrogen utilization, 77
Nodal cell, 249, 250, 253
Node, 249
Nonarticulated, 509, 512-513, 516
Nonprocarpial, 466, 499-501, 504-505, 514, 521
Nori, 28, 475
Nuclear cycle, 221
Nuclear cyclosis, 425
Nuclear division, 63, 64, 245
Nuclear envelope, 76, 147, 173, 222
Nucleus, giant, 203, 217
"Nuisance" seaweed, 193
Nullipore, 509